W9-BAW-471

The Meat We Eat

Twelfth Edition

The Meat We Eat

John R. Romans   *William J. Costello*

Kevin W. Jones   *C. Wendell Carlson*

South Dakota State University

and the late

P. Thomas Ziegler

The
INTERSTATE
Printers & Publishers, Inc.

Danville, Illinois

Cover design by
KEVIN W. JONES

—

Library of Congress Catalog Card No. 84-81229

2 3
4 5 6
7 8 9

ISBN 0-8134-2444-5

Preface

THE PREFACE to the 1974 edition spoke of these forces causing changes in the meat industry: "consumerism, ecology, energy, pollution, atherosclerosis, nitrosamines, isolated soy protein, leanness, muscling, sanitation, safety, and new processes and equipment to increase efficiency and save labor." Now, many of the same forces are still present, plus the effects of heightened consumer concerns with diet and health, the computer explosion, animal welfare and rights, and the continually shrinking percentage of our population who even remotely understand production agriculture, especially animal production and the processing of animal products. Furthermore, innumerable so-called "experts" are writing books on diet, health, food, and meat, in particular. The need for an honest, factual, modern, science-documented textbook on meat that can be read and clearly understood by high school and college students, producers, processors, and homemakers has never been greater. The authors sincerely believe that this text "fills the bill." The scientific base of the text has been greatly strengthened. The complex conversion of animals into human food is simplified in a clear-cut stepwise progression that leads logically from whole live animals with the physiological processes affecting tissue composition and function, through the many humanly controlled processes used to convert that living tissue into nutritious, appetizing food for humans.

Sincere appreciation is due the many friends in industry, government, and education who provided illustrations and materials for this revision. Special mention must be made of the National Live Stock and Meat Board and Ken Franklin, Ken Johnson, and John Francis; the National Pork Producers Council and Dave Meisinger; the USDA Agricultural Marketing Service and Jan Lockhard, Dan Engeljohn, and Mike May; the USDA Packers and Stockyards Administration and C. S. Smebakken; the USDA Science and Education Administration and Linda Posati and Barbara Anderson; the USDA Food Safety and Inspection Service and Karen Stuck; the U.S. Hide, Skin and Leather Association and Jerry Brieter; the IBP and Jim Lockner and Gene Anbrosen; and other colleagues in the American Meat Science Association and in the meat industry who provided helpful information. Specific mention is made of these contributors throughout the book.

Special thanks go to our colleagues here at South Dakota State University—

Dr. Jim Bailey and Dr. Ernest Hugghins for information on diseases and animal loss factors; Jerry Graslie for art work; David Coffin for photography; and LeRoy Warborg for his help, especially in the preparation of the slaughter chapters.

Classification of Our Common Meat Animals

Phyla—*Chordata* (internal skeleton and dorsal nervous system)
Subphyla—*Vertebrates* (segmented spinal column)
Class—*Mammalia* (udder secretes milk)
Subclass—*Placentates* (fetus nourished in uterus)
Order—*Ungulata* (hoofed animals)
Suborder—*Artiodactyles* (even toed)
Section—*Pecora* (true ruminants)
Family—*Bovidae* (hollow horned)
Genus—*Bos* (cattle)
Group—*Taurine* (of or like a bull)
Species—*B. taurus* (cattle)
B. indicus (humped cattle)
Genus—*Ovis* (sheep)
Group—*O. Aries* (domestic)
Genus—*Capra* (goats)
Group—*C. hirius* (domestic)
Section—*Suina* (pointed molars)
Family—*Suidae* (true swine)
Genus—*Sus*—*Sus scrofa* (wild boar)
Group—*S. domesticus* (domesticated swine)
Suborder—*Perissodactyles* (uneven toed)
Family—*Equidae* (horse family)
Genus—*Equus* (horse family)
Group—*E. caballus* (horse)
E. asinus (ass)
E. zebra (zebra)
Class—*Aves* (feathered)
Subclass—*Neornithes* (w/o teeth)
Order—*Gallinae* (fowls)
Family—*Phasianidae* (w/spurs)
Genus—*Gallus* (comb)
Species—*G. domesticus* (chicken)

Family—*Meleagridaé* (w/caruncles)
   Genus—*Meleagris* (dewbill)
      Species—*M. gallopavo* (turkey)
Family—*Anatidae* (web foot)
   Genus—*Anser*
      Species—*A. anser* (goose)
   Genus—*Anas*
      Species—*A. platyrhynchos* (duck)

Contents

Chapter 1

Introduction

APPROXIMATELY two thirds of the world's agricultural land is permanent pasture, range, and meadow; of this, at least 60 percent is unsuitable for producing crops that would be consumed directly by humans. In the United States, 44 percent of the total land area is composed of rangelands and forests which are used for grazing. This land produces cellulosic roughages in the form of grass and other vegetation that is digestible by grazing ruminant animals (cattle, sheep, goats, and deer). Cellulose is the most abundant chemical constituent in the dry substance of plants, but it cannot be digested by humans. Ruminants harbor microorganisms in the rumen portion of their four-compartment stomachs, which have the ability to utilize cellulose for energy and to synthesize essential nutrients, such as amino acids and certain vitamins. Furthermore, approximately 98 percent of the grain fed to animals in the United States consists of corn, sorghum, oats, and barley, none of which are major sources of human food in this country.[1]

It is the function of these animals (cattle, sheep, goats, and deer) to utilize grasses and grains and convert them into a more suitable and concentrated food for humans as well as materials for clothing, pharmaceuticals, and many other valuable by-products. Note Figure 1-1 depicting the central role which ruminants play in human nutrition.

Swine and poultry possess a simple stomach (one pouch like a human) and therefore cannot utilize roughage to the same degree as ruminants, but they are very efficient in converting grain into meat. These conversion abilities have made flocks and herds of prime economic importance in the development of civilization.

The importance of meat's contribution to a healthy, contemporary diet has been recognized by consumers and the industry in a U.S. Senate resolution designating the first annual National Meat Week for January 22-28, 1984. The resolution stated that "meat is a wholesome and nutritious food, one of the most valuable sources of vitamins and minerals in the human diet, and a high quality source of protein" and that "the meat industry's annual sales of $70 billion make it the largest single component of U.S. agriculture."

---

[1]*Foods from Animals.* March 1980. Quantity, Quality and Safety Council for Agricultural Science and Technology, Report No. 82.

1

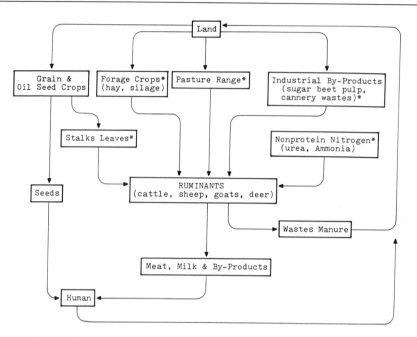

Fig. 1-1. The central role of ruminants in human nutrition. Items marked with an asterisk are converted by ruminants but are not eaten by humans. From Coun. Agr. Sci. Tech. *Ruminants as Food Producers*, Spec. Publ. No. 4, CAST, Dept. of Agron., Iowa State University, Ames, 1975, p. 4.

PEOPLE AND MEAT

A liberal meat supply has always been associated with a happy and virile people and invariably has been the main food available to settlers of new and undeveloped territories. Statistics show that per capita meat consumption varies with meat production and decreases with density of population (Tables 1-1 and 1-2). Note that the United States is the number one meat producer, producing more than twice the amount of the world's second-place producer, the USSR.

Inhabitants consume more of the food that is abundant in their area because it is easier to get and invariably cheaper. Argentina and Uruguay are cattle countries, and their people continue to be the world's heaviest consumers of beef. Australia and New Zealand are the world's most densely populated sheep countries, and their people consume approximately 30 times more lamb and mutton than do the people of the United States. The United States in turn consumes two times more pork per person than does New Zealand, because the hog population in the United States is considerably larger. A well balanced agriculture provides its people with a variety of food and consequently a better balanced diet. This type of agriculture also results in a more varied source of income.

Table 1-1. Meat Production in Specified Countries[1]

(In Thousand Metric Tons[2])

Region and Country	Beef and Veal		Pork		Lamb, Mutton, and Goat		Poultry	
	1976-80 Average	1985 Forecast	1976-80 Average	1985 Forecast	1976-80 Average	1985 Forecast	1976-80 Average	1985 Forecast
North America:								
Canada	1,058	995	655	870			484	561
Mexico	1,086	1,381	1,014	1,050			395	557
United States	11,043	10,459	6,477	6,682	149	150	5,987	7,739
Subtotal	13,187	12,835	8,146	8,602			6,866	8,857
Caribbean:								
Dominican Republic	40	49						
Central America:								
Costa Rica	74	62						
El Salvador	33	30						
Guatemala	86	65						
Honduras	53	66						
Nicaragua	74	45						
Panama	45	50						
Subtotal	365	318						
South America:								
Argentina	2,966	2,500			126	105	201	230
Brazil	2,226	2,500	870	930			916	1,520
Colombia	587	648	111	113				
Uruguay	345	349			32	41		
Venezuela	300	374	80	121			198	308
Subtotal	6,424	6,371	1,061	1,164	158	146	1,315	2,058

(Continued)

Table 1-1 (Continued)

Region and Country	Beef and Veal		Pork		Lamb, Mutton, and Goat		Poultry	
	1976-80 Average	1985 Forecast	1976-80 Average	1985 Forecast	1976-80 Average	1985 Forecast	1976-80 Average	1985 Forecast
European Community:								
Belgium-Luxembourg	291	300	658	745	3	8	124	165
Denmark	244	236	828	1,140	1	1	99	108
France	1,755	1,783	1,505	1,577	162	174	977	1,277
West Germany	1,463	1,645	2,588	2,740	27	30	343	343
Greece		85		155		123		160
Ireland	386	353	138	145	41	45	45	57
Italy	1,070	1,180	877	1,060	61	68	894	968
Netherlands	364	500	980	1,240	16	12	344	425
United Kingdom	1,049	1,040	913	960	241	295	757	870
Subtotal	6,622	7,122	8,487	9,762	552	756	3,583	4,373
Other Western Europe:								
Austria	187	218	345	384			59	67
Finland	109	120	155	173			13	19
Greece	106		128		119		114	
Portugal	86	97	140	173	23	23	130	154
Spain	411	420	819	1,230	141	140	739	830
Sweden	151	153	306	325			41	45
Switzerland	156	161	262	292			22	26
Subtotal	1,206	1,169	2,155	2,577	283	163	1,118	1,141

(Continued)

Table 1-1 (Continued)

Region and Country	Beef and Veal		Pork		Lamb, Mutton, and Goat		Poultry	
	1976-80 Average	1985 Forecast	1976-80 Average	1985 Forecast	1976-80 Average	1985 Forecast	1976-80 Average	1985 Forecast
Eastern Europe:								
Bulgaria	140	165	349	400	83	100	149	160
Czechoslovakia	421	452	808	820	7	7	181	215
East Germany	410	400	1,157	1,200	15	17	142	150
Hungary	151	145	884	1,035	6	7	309	370
Poland	827	720	1,735	1,353	23	18	375	260
Romania	298	230	851	860	56	65	362	426
Yugoslavia	335	345	740	805	60	61	246	305
Subtotal	2,582	2,457	6,524	6,473	250	275	1,764	1,886
USSR	6,827	7,300	5,009	6,100	882	850	1,832	2,800
Middle East:								
Israel	22	20					149	186
Turkey	193	230			328	390		
Subtotal	215	250						
North Africa:								
Egypt	246	320			71	88	112	175
Other Africa:								
South Africa, Republic of	580	581			164	180	296	555
South Asia:								
India	285	301			440	494		

(Continued)

Table 1-1 (Continued)

Region and Country	Beef and Veal		Pork		Lamb, Mutton, and Goat		Poultry	
	1976-80 Average	1985 Forecast	1976-80 Average	1985 Forecast	1976-80 Average	1985 Forecast	1976-80 Average	1985 Forecast
Other Asia:								
China People's Republic of (Mainland)	9	4	8,700	13,300				
Taiwan			494	657			200	355
Hong Kong			34	36			41	49
Japan	376	485	1,283	1,485			1,010	1,356
Korea, Republic of	111	158	175	330			87	177
Philippines	119	175	344	450				
Subtotal	615	822	11,030	16,258			1,338	1,937
Oceania:								
Australia	1,897	1,310	202	238	542	520	251	331
New Zealand	551	459			518	685		
Subtotal	2,448	1,769			1,060	1,205		
Total	41,642	41,664	42,614	51,174	4,337	4,697	18,624	24,299

[1]USDA, Foreign Agriculture Service. FL and P 2-84. Carcass weight.
[2]1 metric ton = 2,200 pounds.

Table 1-2. Per Capita Consumption of Red Meat,[1] Poultry,[2] and Fish[3] in Specified Countries

(In Kilograms[4])

Region and Country	Beef and Veal		Pork		Lamb, Mutton, and Goat		Poultry		Total Meat		Fish[3]
	1976-80 Average	1985 Forecast	1976-80 Average	1985 Forecast	1976-80 Average	1985 Forecast	1976-80 Average	1985 Forecast	1976-80 Average	1985 Forecast	Average 1975-77
North America:											
Canada	44.7	41.7	26.1	28.9			20.6	24.3	91.4	94.9	40.1
Mexico	14.4	18.5	13.8	14.2			5.5	7.6	20.7	40.3	10.8
United States	51.6	48.0	28.3	30.1			24.8	32.5	105.4	111.3	35.1
Caribbean:											
Dominican Republic	6.3	7.7			0.7	0.7			6.3	7.7	15.9
Central America:											
Costa Rica	14.2	14.2							14.2	14.2	9.9
El Salvador	5.8	5.6							5.8	5.6	4.8
Guatemala	8.5	7.1							8.5	7.1	1.5
Honduras	5.5	12.4							5.5	12.4	2.4
Nicaragua	12.0	12.3							12.0	12.3	9.5
Panama	21.6	23.0							21.6	23.0	21.4
South America:											
Argentina	83.2	78.4	6.8	7.3	3.2	3.0	7.1	8.2	93.5	89.6	9.0
Brazil	16.9	15.0					4.3	10.1	28.0	32.4	15.2
Colombia	21.1	23.6	4.1	4.2					25.2	27.8	7.5
Uruguay	74.2	61.0	5.8	8.0	8.1	10.2	14.4	20.4	82.3	71.2	11.0
Venezuela	23.2	26.1							43.4	54.5	22.5

(Continued)

Table 1-2 (Continued)

Region and Country	Beef and Veal 1976-80 Average	Beef and Veal 1985 Forecast	Pork 1976-80 Average	Pork 1985 Forecast	Lamb, Mutton, and Goat 1976-80 Average	Lamb, Mutton, and Goat 1985 Forecast	Poultry 1976-80 Average	Poultry 1985 Forecast	Total Meat 1976-80 Average	Total Meat 1985 Forecast	Fish[3] Average 1975-77
European Community:											
Belgium-Luxembourg	29.6	27.9	44.0	53.8	1.8	1.5	12.0	14.7	87.4	97.9	40.8
Denmark	15.4	12.3	47.8	57.6	0.6	0.6	8.4	9.8	72.2	80.3	77.4
France	31.5	30.9	31.5	35.2	3.8	4.2	15.3	17.7	82.1	88.0	48.9
West Germany	24.0	23.5	47.2	51.1	0.8	0.8	9.5	9.2	81.5	84.6	23.6
Greece	see below	22.3	see below	20.4	see below	14.1	see below	16.8	see below	73.6	
Ireland	22.4	24.7	27.9	32.8	8.6	7.2	12.6	16.7	71.5	81.4	51.3
Italy	24.5	25.8	19.9	23.8	1.3	1.5	16.2	17.5	61.9	68.6	27.3
Netherlands	19.4	17.8	32.9	36.0	0.3	0.4	7.3	12.2	59.9	66.4	29.1
United Kingdom	25.1	21.9	25.4	26.1	7.5	6.5	13.4	16.2	71.4	70.7	
Other Western Europe:											
Austria	25.2	21.8	45.6	50.5			9.4	9.9	80.2	82.2	17.2
Finland	22.8	21.4	28.4	32.2			2.7	3.9	53.9	57.5	57.5
Greece	21.2	see above	14.2	see above	13.2	see above	11.3	see above	59.9	see above	85.1
Portugal	10.9	9.9	14.6	18.1	2.3	2.2	12.9	15.3	40.7	45.5	77.8
Spain	12.2	11.6	22.4	32.8	3.7	3.7	20.0	22.3	58.3	70.4	71.6
Sweden	18.6	16.0	34.0	32.2			4.8	5.3	57.4	53.5	22.9
Switzerland	25.9	26.8	41.0	45.5			7.2	8.2	74.1	80.5	
Eastern Europe:											
Bulgaria	14.9	16.9	36.7	42.8	7.1	7.8	12.9	14.3	71.6	81.8	26.1
Czechoslovakia	27.6	25.8	52.8	51.5	0.5	0.3	12.0	12.4	92.9	90.0	17.4
East Germany	22.9	21.6	59.0	58.7	0.7	0.6	8.7	9.4	91.3	90.3	
Hungary	11.0	11.2	77.4	81.8	0.2	0.3	17.6	19.6	106.2	112.9	11.0
Poland	22.6	19.7	45.6	37.6	0.6	0.5	9.9	6.9	78.7	64.7	45.2
Romania	11.3	7.3	35.1	34.2	0.5	0.9	15.4	16.7	62.3	58.6	
Yugoslavia	14.2	13.9	32.1	33.3	2.5	2.4	10.9	12.5	59.7	62.1	6.4
USSR	26.1	28.4	18.8	23.0	3.6	3.7	7.2	11.3	55.7	66.4	63.3

(Continued)

Table 1-2 (Continued)

Region and Country	Beef and Veal		Pork		Lamb, Mutton, and Goat		Poultry		Total Meat		Fish[3]
	1976-80 Average	1985 Forecast	1976-80 Average	1985 Forecast	1976-80 Average	1985 Forecast	1976-80 Average	1985 Forecast	1976-80 Average	1985 Forecast	Average 1975-77
Middle East:											
Iraq							5.3	8.2	5.3	8.2	
Israel	15.2	18.2					34.1	44.0	49.3	62.2	24.5
Kuwait							20.6	33.3	20.6	33.3	
Saudi Arabia							14.4	30.0	14.4	30.0	
Syria							4.1	8.2	4.1	8.2	
Turkey	4.2	4.4							4.2	4.4	9.7
United Arab Emirates							9.2	11.7	9.2	11.7	
Yemen Arab Rep (Sanaa)							4.7	8.5	4.7	8.5	
North Africa:											
Egypt	6.8	10.5			1.7	2.1	3.0	6.5	11.5	19.1	
Other Africa:											
South Africa, Republic of	19.2	19.6			5.3	5.8	8.9	17.3	33.4	42.7	
South Asia:											
India	0.4	0.3			0.6	0.7			1.0	1.0	
Other Asia:											
China											
People's Republic of (Mainland)			8.4	12.8					8.4	12.8	
Taiwan	1.2	1.5	25.6	34.5			10.9	19.6	37.7	55.6	
Hong Kong	4.6	5.5	9.2	18.0			18.7	28.3	32.5	51.8	
Japan	4.4	6.3	12.3	14.7	2.1	1.3	9.0	12.3	27.8	34.6	148.6
Korea, Republic of	3.4	4.7	4.5	8.4			2.2	4.5	10.1	17.6	
Philippines	2.6	3.4	6.8	9.1					9.4	12.5	73.0
Oceania:											
Australia	59.7	41.5	13.1	15.6	19.5	20.4	16.0	21.4	108.3	98.9	32.2
New Zealand	60.8	38.3			32.0	45.3			92.8	83.6	37.3

[1]USDA, Foreign Agriculture Service. FL and P 2-84. Carcass weight. Meat *consumption* is really meat *disappearance* computed by the USDA by totaling meat production, adding imports, subtracting exports, and comparing year-ending inventories. All data are recorded as carcass weight. However, Breidenstein (National Live Stock and Meat Board, *Food and Nutrition News*, Vol. 56:3, May/June 1984) has clearly shown that *retail weight of* all red meat species averages only 82 percent of *carcass weight* due to bone and fat content, and *actual consumption* averages only 65 percent of *retail weight* due to additional fat and bone trimming, cooking loss, and plate waste.

[2]Dressed weight.

[3]National Marine Fisheries Service (Jim Roberts). Round weight is whole fish; edible weight is approximately 35 percent of round weight (latest figures available at publishing date).

[4]1 kilogram = 2.2 pounds.

Our Eating Habits

The peoples of the world as a whole are always hungry. They are hungry because of a lack of food or because of a lack of means to buy it. Under such conditions, likes and dislikes become secondary.

Those of us who dwell in a land of plenty, blessed by the elements developed through personal initiative under a system of free enterprise, have built for ourselves a rather selective standard of living. We are what we make ourselves, whether morally, spiritually, financially, or physically. These are voluntary, not decreed, acquirements.

Broadly speaking, our eating habits are governed in a large measure by spendable income, concern for our personal well-being, and doctors' advice. Statistics continue to show that U.S. consumers spend only approximately 16 percent of their total disposable income on food (12 percent at home, 4 percent away from home). Of that food dollar, 50¢ is spent on food produced by animals, and of that, 80 percent, or 40¢, is spent for meat. Table 1-3 gives a current and historical look at food consumption in the United States. Meat *consumption* is really meat *disappearance* computed by the USDA by totaling meat production, adding imports, subtracting exports, and comparing year-ending inventories. All data are recorded as retail weight. Breidenstein (National Live Stock and Meat Board, *Food and Nutrition News,* Vol. 56:3, May/June 1984) has clearly shown that *retail weight* of all red meat species averages only 82 percent of *carcass weight* due to bone and fat content, and *actual consumption* averages only 65

Table 1-3. What the Average American Eats Annually[1]

(Retail Weight in Pounds)

	1963	1968	1973	1978	1983
Civilian population in millions[2]	179.3		203.3		226.5
Meats[3]					
Beef	69.9	81.2	80.5	87.2	78.7
Veal and calf	4.1	3.0	1.5	2.4	1.7
Pork (excluding lard)	61.0	61.4	57.3	55.9	62.2
Lamb and mutton	4.4	3.3	2.4	1.4	1.5
Edible offals[4]	10.1	10.7	9.7	9.5	9.1
Total red	149.5	159.6	151.4	156.4	153.2
Fish	10.7	11.0	12.8	13.4	12.9
Poultry products					
Chickens	30.8	36.7	40.4	46.7	53.9
Turkeys	6.8	7.9	8.5	9.2	11.2
Ducks and geese	0.4	0.4	0.4	0.4	0.3
Eggs	40.3	40.2	36.8	34.5	33.1
Fats					
Lard	6.3	5.5	3.3	2.2	1.8
Margarine	9.4	10.7	11.1	11.2	10.4
Shortening	13.6	16.3	17.1	17.8	18.6
Other edible fats and oils	13.1	15.9	20.3	22.1	24.8

(Continued)

Table 1-3 (Continued)

	1963	1968	1973	1978	1983
Dairy products					
Fluid milk and cream	293.4	282.4	271.2	235.7	245.1
Butter	6.9	5.7	4.8	4.4	5.1
Cheese	13.8	15.3	18.9	21.7	24.8
Ice cream	18.0	18.4	17.3	17.4	17.9
Grains					
Wheat flour	114.1	112.8	112.8	115.2	116.0
Rice	6.6	7.9	7.0	5.7	9.8
All cereal and corn products	144.0	145.1	144.2	144.9	149.6
Sugar	116.0	122.1	131.2	136.6	142.5
Fruits					
Fresh	71.8	76.3	73.8	79.0	91.8
Canned	23.0	21.9	21.2	17.9	16.0
Frozen	3.9	3.8	3.5	3.3	3.0
Dried	2.9	2.8	2.5	2.1	2.9
Juice	18.3	22.3	29.8	31.8	33.9
Beverages					
Coffee	11.7	11.2	10.0	7.9	7.6
Tea	0.7	0.7	0.8	0.8	0.7
Cocoa	3.1	3.4	3.4	2.6	3.3
Soft drinks (gallons)	17.7	23.0	26.8	35.4	40.0
Vegetables					
Fresh	93.5	93.6	89.8	95.4	99.9
Canned	46.1	50.5	54.0	51.8	47.1
Frozen	7.2	9.6	10.6	10.7	11.1
Potatoes	86.1	78.8	70.2	73.8	77.6
Sweet potatoes	6.2	5.2	4.5	4.5	4.9
Dry beans	7.6	6.4	7.0	4.8	6.2

[1]USDA, Economic Research Service. *Food Consumption, Prices and Expenditures.* 1963-1983. Stat. Bul. No. 713. Meat *consumption* is really meat *disappearance* computed by the USDA by totaling meat production, adding imports, subtracting exports, and comparing year-ending inventories. All data are recorded as carcass weight. However, Breidenstein (National Live Stock and Meat Board, *Food and Nutrition News,* Vol. 56:3, May/June 1984) has clearly shown that *retail weight* of all red meat species averages only 82 percent of *carcass weight* due to bone and fat content, and *actual consumption* averages only 65 percent of *retail weight* due to additional fat and bone trimming, cooking loss, and plate waste.

[2]According to the census at the nearest decade.

[3]Luncheon meat consumption composed of pork, beef, and veal is included in the species data.

[4]Edible offals include pork, beef, veal, lamb, and mutton.

percent of *retail weight* due to additional fat and bone trimming, cooking loss, and plate waste.

Consumption will vary with income and comparative prices. Income also governs the consumer's selection of meat cuts. The old expression "no money, no meat" still holds. In most cases, those in the middle-income group are the heavy meat and potato consumers, particularly if their occupations require considerable physical exertion. Youths top the hamburger, hot dog, and pop-consuming group. Fortunately, they are also heavy milk and ice cream consumers.

Too few of us realize until too late that appetite is a poor governor for the operation of the intricate human engine. Opening the food throttle can result in many ailments and discomforts. Obesity and heart failure are blood relatives.

The old saw that everybody loves a fat man is a hoax. Physical well-being can mean different things to different people.

Some of us are food faddists; we make food a religion. Others follow trends such as the order of the Slim Look or the Olympian. Then we have the Society of Suffering Distaffs who, by diet tricks or pills, take off 10 pounds in 20 days and recover it in 10. They are the Lost and Found. Let's be sensible and make *well-being* and *being well* a little more congruent.

Finally, our food habits may be dictated. As babies we had no choice, no teeth, no cavities. For old age, and sometimes before, there are doctor's orders and store teeth. Our intense desire to enjoy "living it up" to the exclusion of its after-effects brings too many of us to an end which we did not want to foresee.

THE LIVESTOCK AND MEAT INDUSTRY

Historically, more than half of the U.S. farmers' cash income has been derived from animal agriculture. This holds true for the 1970s and 1980s as well, since during these decades income from livestock and livestock products has accounted for from 50 to 58 percent[2] of total farm income, depending on the relative price fluctuations between livestock and crops. As a percentage of total retail cost, the farm income portion represents 35 percent for all food, 49 percent for all meat products, 58 percent for choice beef, and 46 percent for pork.[3]

The U.S. meat industry is the nation's largest food industry, but meat packing normally ranks low among U.S. industries in net income (profit) per dollar of sales—in 1983 0.8¢ per dollar, down from 0.9¢ in 1982. Yet, meat remains the consumer's number one food cost.

Meat packers are generally defined as those companies that slaughter livestock, whether or not they also process meats and meat products. Companies that specialize in the manufacture of processed meats, including meat canners and sausage manufacturers, but which do not slaughter livestock, are defined as *meat processors*. Table 1-4 gives a breakdown showing how the meat packer's sales dollar is spent for a recent year, and back through a decade. Obviously livestock and other raw materials are the major cost items. Wages, salaries, and employees' benefits compose the major operating expenses. Table 1-5 compares costs for beef and pork packers. Note that a considerably higher gross margin was realized by pork packers, but the net earnings were not greatly different between the beef and pork packers. Historically, pork packers have done more processing of products prior to sale, thus more costs in supplies and containers, etc. Now more processing is being done in beef plants, rather than the selling of hanging carcasses, as was prevalent in the past. Table 1-6 shows that meat processing can boost net income by adding more value to the product being sold.

---

[2]USDA and South Dakota Department of Agriculture Statistical Reporting Services.
[3]*Agricultural Outlook*. Dec. 1982. Economic Research Service, USDA.

Table 1-4. Sales, Raw Materials Costs, Expenses, and Income of the Meat Packing Industry, 1973, 1978, and 1983[1]

Item	Million Dollars		
	1973	1978	1983
Total sales	33,225	43,625	49,975
Cost of livestock and other raw materials	26,935	34,425	39,160
Gross margin	6,290	9,200	10,815
Operating expenses			
Wages and salaries	2,540	3,465	3,807
Employee benefits			
Retirement expense	111	207	166
Payroll taxes	151	242	312
Insurance and hospitalization	136	264	431
Vacation, holiday, sick leave	168	252	248
All other benefits	40	54	64
Total benefits	606	1,019	1,221
Interest	105	139	174
Depreciation	200	307	358
Rents	86	129	158
Taxes[2]	65	93	66
Supplies and containers	845	1,290	1,627
All other expenses	1,240	2,190	2,697
Total operating expenses	5,687	8,632	10,108
Income before taxes	603	568	707
Income taxes	263	235	321
Net income	340	333	386
	Percent of Total Sales		
Total sales	100.0%	100.0%	100.0%
Cost of livestock and other raw materials	81.1	78.9	78.4
Gross margin	18.9	21.1	21.6
Operating expenses			
Wages and salaries	7.6	7.9	7.6
Employee benefits			
Retirement expense	0.3	0.5	0.3
Payroll taxes	0.5	0.6	0.6
Insurance and hospitalization	0.4	0.6	0.9
Vacation, holiday, and sick leave	0.5	0.6	0.5
All other benefits	0.1	0.1	0.1
Total benefits	1.8	2.4	2.4
Interest	0.3	0.3	0.3
Depreciation	0.6	0.7	0.8
Rents	0.3	0.3	0.3
Taxes[2]	0.2	0.2	0.1
Supplies and containers	2.6	3.0	3.3
All other expenses	3.7	5.0	5.4
Total operating expenses	17.1	19.8	20.2
Income before taxes	1.8	1.3	1.4
Income taxes	0.8	0.5	0.6
Net income	1.0	0.8	0.8

[1]American Meat Institute. *Annual Financial Review*. 1984.

[2]Other than social security and income taxes.

Table 1-5. Percentage Breakdown of Meat Packers' Sales Dollar by Species Slaughtered, 1982-83[1]

Item	Cattle Packers[2]		Hog Packers[2]	
	1982	1983	1982	1983
Total sales	100.0%	100.0%	100.0%	100.0%
Cost of livestock	86.5	86.3	68.3	69.3
Gross margin	13.5	13.7	31.7	30.7
Operating expenses				
Wages and salaries	5.1	5.0	11.0	11.0
Employee benefits				
Retirement expense	0.1	0.1	0.7	0.6
Payroll taxes	0.4	0.4	0.8	0.8
Insurance and hospitalization	0.4	0.5	1.3	1.3
Vacation, holiday, sick leave	0.2	0.2	0.9	0.8
All other benefits	Sm.	0.1	0.1	0.3
Total benefits	0.9	0.9	4.1	3.8
Interest	0.3	0.2	0.5	0.6
Depreciation	0.5	0.5	1.0	1.1
Rents	0.3	0.2	0.3	0.4
Taxes	0.1	0.1	0.1	0.2
Supplies and containers	1.8	1.9	5.3	5.4
All other	2.6	3.3	7.7	6.8
Total operating expenses	12.0	12.7	29.7	29.3
Income before taxes	1.5	1.0	2.0	1.4
Income taxes	0.7	0.4	0.9	1.0
Net income	0.8	0.6	1.1	0.4

[1]American Meat Institute. *Annual Financial Review*. 1984.

[2]Refers to AMI Annual Financial Survey participants whose slaughter of identified species represented at least 75 percent of their total live-weight slaughter.

PACKERS AND STOCKYARDS ACT[4]

The purpose of the Packers and Stockyards Act, as passed in 1921, and, as continually updated, is to provide for uniform rates and practices and to prevent unfair, unjustly discriminatory, or deceptive practices. This is particularly helpful to farmers who are unable to personally supervise the sale of their livestock. Federal supervision under the act extends to trade practices, commissions, feed and yardage rate charges, weighing and scale testing, and to other services rendered at stockyards and packing plants. Carcass weight and grade sales are covered under the act, as are all live sales. Since payment is made on the basis of carcass weight and grade after the animals are slaughtered, such factors as place and time of slaughter, carcass trim, grading and prices, condemnation terms, identification, accounting, and promptness of payment must be specified clearly prior to the sale. A final written report, accurately explaining all these factors, must accompany payment to the seller. Carcass weights for payment are the hot

[4]USDA, Packers and Stockyards Administration, Washington, D.C., Sept. 1982.

Table 1-6. Percentage Breakdown of the Sales Dollar by Company Classifications, 1983[1]

Item	Meat Packing Companies				Meat Processing Companies
	National	Regional	Local	Total	
Total sales	100.0%	100.0%	100.0%	100.0%	100.0%
Cost of livestock and raw materials	79.0	77.4	80.7	78.7	61.3
Gross margin	21.0	22.6	19.3	21.3	38.7
Operating expenses					
Wages and salaries	7.4	7.8	7.8	7.5	10.1
Employee benefits					
Retirement expense	0.4	0.3	0.2	0.4	0.7
Payroll taxes	0.6	0.7	0.6	0.6	0.8
Insurance and hospitalization	0.8	0.9	0.6	0.9	0.9
Vacation, holiday, sick leave	0.5	0.5	0.3	0.5	0.7
All other benefits	0.2	0.1	0.1	0.2	0.2
Total benefits	2.4	2.5	1.8	2.4	3.3
Interest	0.4	0.3	0.4	0.4	0.5
Depreciation	0.7	0.7	0.8	0.7	1.4
Rents	0.3	0.3	0.2	0.3	0.5
Taxes[2]	0.1	0.1	0.2	0.1	0.3
Supplies and containers	3.5	3.2	2.1	3.4	6.3
All other expenses	5.2	5.7	4.6	5.3	12.2
Total operating expenses	19.9	20.8	17.9	20.1	34.6
Income before taxes	1.1	1.7	1.4	1.3	4.1
Income taxes	0.6	0.8	0.5	0.6	1.8
Net income	0.5	0.9	0.9	0.6	2.3

[1]American Meat Institute. *Annual Financial Review.* 1984.

[2]Other than social security and income taxes.

weights prior to shrouding. Carcasses must be final graded before the close of the second business day following slaughter. Payment must be made for the full purchase price with all methods of sale by the close of the next business day following transfer of possession, or in the case of grade and yield purchases, following determination of the purchase price.

Other regulations under this act include prohibition against certain dual ownership relationships between custom feedlots and packing plants and bonding requirements for meat packers to prevent livestock producers from taking heavy losses if a meat packer should declare bankruptcy after livestock is delivered but before payment is made. The entire act is being reviewed to determine if certain regulations should be consolidated, abolished because they are out of date, or revised or alleviated to put less burden on the livestock industry.

THE PRESENT TREND

The trend in consumer preference for leaner meats continues. Unfortunately, the consumers are now bombarded from many sides with various admonitions concerning the composition of their diets, especially regarding the fat content. In the early 1950s, it was reported that the cholesterol level in the human circulatory system may be raised by the ingestion of certain fats. Cholesterol is found in the human blood stream and, when triggered by a factor yet unknown, may be deposited on the inside walls of arteries and veins, thereby hindering the free movement of blood to and from the heart.

Yet, after 25 years of research costing several billion dollars, scientists are still divided on the diet-heart question. Some insist that diet manipulations to control levels of cholesterol and saturated fat intake are essential to control coronary heart disease. Animal fats are generally more highly saturated than vegetable fats; thus, these scientists condemn animal fat in the diet. Other scientists, by now the larger of the two opposing groups, cite evidence that factors other than diet, such as lack of exercise, high blood pressure, cigarette smoking, obesity, and heredity, have an equal or higher relationship to heart disease. Furthermore, cholesterol is manufactured by the human body itself and is essential in normal body functions. Fat does serve as a rich source of energy; so therefore, many of us rather sedentary American folks should not over-consume fat, for if we do, we will become obese. Other roles of fat in the human body are to serve as carriers of vitamins A, D, E, and K to protect body tissues and vital organs and to regulate body temperature. Thus fat is essential and by no means a nutrient to be avoided but rather to be consumed at proper levels.

The consumer demand for leaner meat has made certain demands on the producer, packer, and retailer. The producer must select and breed for quality meat animals that are more heavily muscled and that will reach a market weight at an earlier age without the extended feeding period. Progeny testing is essential in order to find the strains that efficiently produce more edible meat of acceptable quality and less waste fat. The packer must compensate these efforts by paying a premium for animals of superior meat type and quality and by discounting those that produce overdone carcasses and cuts and that require costly fat trimming. Packers are doing more further processing of carcasses and cuts, utilizing the latest techniques that have been developed to produce products uniform in size and quality. The retailer should in turn give the matters of trim and improved display conditions further consideration.

The consumer has recognized that fat in excess of 0.2 inch over the outside of a chop, steak, or roast is simply waste. However, this same consumer desires a tender, juicy, and flavorful meat cut. Marbling (fat within the muscle) is becoming less important to overall palatability as we continue to market younger animals. In fact, recent research has shown little relationship between marbling and tenderness in such animals. However, a certain minimum level of marbling is necessary to give meat its characteristic aroma, flavor, and juiciness. This minimum level is an amount of fat that would by no means cause a health hazard in the diet of any normal consumer.

There should be no doubt in our minds that eating in moderation of those foods that furnish us a balanced diet is the answer to most of our overweight ills and that does not mean the exclusion of animal products.

Although the authors confess to being moderately heavy meat eaters, it is not to be concluded from what is to follow that meat is something magic. A mixed diet that includes sufficient grains and cereals, a liberal amount of vegetables, fresh fruit, milk, a savory serving of one of over a hundred possible tasty meat dishes, and as little of those complicated sweets that end up a meal as possible is the sensible one. But even a sensible meal can be made destructive by over-eating. Over-work and over-distension of the digestive tract can be likened to an over-inflated tire run at high speed—it may result in a blowout. A quotation from *Exchange* is worth repeating at this point: "Some businessmen make more of a feature of their eating than of their business. Their business is merely an interval between meals. These men who hate to let business interfere with eating usually come to the day when their eating puts them out of business."

A Natural Food

The term *natural* takes on an added meaning to most consumers in the present day, since much has been made of *organic* or *natural* foods in the popular press. Yet, there is no official definition for *organic* or *natural* foods. The heading to this section does not imply any meaning other than the following: Lean muscle is composed of approximately 20 percent protein, 9 percent fat, 70 percent moisture, and 1 percent ash. An edible portion of meat with a fat covering of about 0.2 inch would be composed of 17 percent protein, 20 percent fat, 62 percent moisture, and 1 percent ash. The lean and fat tissues are very similar to human body tissue, and because of this, they are highly digestible and can be easily and rapidly assimilated by the human digestive tract. Milk and eggs are other foods of economic importance that have similar qualities. A chemical analysis may show that a food is rich in certain food elements, but these elements may not be very digestible, or the body may not be able to absorb and utilize them as efficiently as it would the same elements in another food. This is known as the biological value of a food nutrient.

The biological value of a food nutrient is determined under controlled conditions by accounting for the total nutrient intake in terms of what is retained in the body tissues versus that which is excreted in the feces and urine.

Meat proteins have a high biological value. Some products made from soybeans, which are now widely used as meat extenders, lack the complete array of essential amino acids in the proper proportions alone, and thus blend well with meat proteins to raise the biological value of the final product. The concentrated nature of meat and its ability to be readily and almost wholly absorbed from the intestines makes it a highly desirable, if not essential, food for humans.

A Virile and Protective Food

Experience is a great teacher. It has taught the soldier, the laborer, and the trainer of athletes that meat has something besides "fill-in value." Before scientists revealed its rather broad vitamin content and the high biological value of its proteins and fats, its merits were expounded by the expression "It sticks to the ribs." The 10 amino acids which are considered essential for human life are all found in meat. Its proteins have, by individual analyses for the amino acids, been found to be biologically complete.

The human requirements for energy secured through the medium of carbohydrates and fat can be supplied in large measure by the fat in meat, since fat has 2.25 times the energy value of carbohydrates.

Nutrient density is a measure of the concentration of a nutrient per calorie of food consumed. Selecting foods that are dense sources of nutrients assists one in obtaining needed daily vitamins, minerals, and protein while minimizing calorie intake. Lean meat is a dense source of protein.

With the exception of calcium, meat contains all the necessary minerals for human body metabolism. Add to this list of nutritive elements the daily discoveries, through extensive research, that meat is also rich in many of the vitamins so necessary to a normal, healthy body, and the completeness of meat as a food is rather evident. The meat diet of the hardy Eskimo attests to this fact.

The height and weight of humans are governed in large part by the available food supply. The human life span rises with a balanced diet and improved medical knowledge and facilities. The people of some nations consume rather low amounts of red meat and relatively high amounts of seafood. The importance of food overshadows every material need of our people. It is responsible for the world's number one business and is its prime political sedative.

A Palatable Food

Brushing aside for the moment all that has been said, and forgetting high-sounding names, over-zealous scientists, and doctors' admonitions, let us revert to plain hungry mortals seated around the festive board. Instinctively we look for the platter of meat, which to most of us is not only the king but the whole royal family of appetite appeal. It transcends all other foods in aroma, causing a watering of the mouth and a conscious glow in the most bulbous organ of the gastro-intestinal tract. It is a psychological stimulus that causes a flow of saliva and gastric juice, preparing the food chamber for the royal guest. And it does not beguile us; it satisfies. It accomplishes this by supplying what it advertises to our nostrils before we consume it. As we crunch its juicy fibers between permanent or removable ivories, we receive our first pleasant realization of a previous longing sensation. As we swallow the tasty mass, we begin to radiate satisfaction in our eyes, in our speech, and in our actions. We become more amiable, more clear-minded, and more reasonable—certainly a most honorable tribute to any food product.

Chapter 2

Preparations for Slaughtering—
Safety and Sanitation

AS INDICATED in the 1962 edition of this text, the practice of raising, slaughtering, and processing the farm pork supply has been more or less traditional in rural America. Speculation was made at that time regarding the possible continuance or increase of this tradition and the inclusion of other types of meat animals. Since 1962, vast changes have occurred in the agricultural and general economy and in consumer consciousness regarding the wholesomeness of the food supply that probably fewer people will have the opportunity to slaughter their own meat animals and process the carcasses "on the farm."

The Wholesome Meat Act of 1967 (to be discussed in detail in the following chapter) made all state inspection systems at least equal to the federal inspection system. This, in reality, eliminated many small locker operators from business who in the past processed farm-slaughtered meats or oftentimes went to the farms to slaughter livestock and subsequently brought resulting carcasses into their locker plants for processing. Although federal law now permits this practice if the product is strictly utilized for the producer and the producer's family and not for sale, absolutely no product for sale can be processed in exactly the same facility unless the facility and all equipment are completely sterilized before processing the product for sale. Furthermore, the two types of product *cannot* be mixed. As a practical matter, it is very difficult for an operator to maintain two completely separate operations.

Thus, before proceeding with any slaughter activity on their own, persons should contact local and state authorities to be assured that they are not violating existing laws designed to protect the health and welfare of all U.S. citizens.

The passage of the Wholesome Meat Act strengthened the entire meat industry in the United States, but in so doing probably eliminated some locker operators who catered to the farm-slaughtered custom business. Thus, opportunities for new people to engage in farm slaughter are becoming more limited, although interest has increased due to recent price inflations.

Nevertheless, whatever a particular situation may be regarding meat animal slaughter, for example, on the farm or in an approved locker or meat packing plant, two big S's are of prime importance—*Safety* and *Sanitation*.

SAFETY

A publication entitled *Meat Industry Safety Guidelines* produced by the National Safety Council[1] should be read by anyone who is responsible for meat plant safety. Each person working in a meat plant or with meat processing in any form or aspect should be responsible for safety, but of course, within any organization, large or small, certain key people must assume major responsibility for implementing a safety policy.

According to surveys conducted by the National Safety Council, meat packing has historically ranked high in the number of disabling injuries, but near the average of all industries in terms of severity of accidents. An injury is termed *disabling* if the injured person cannot return to the same work the next scheduled shift. Mining ranked higher in frequency, and the communications and automobile industries were the lowest in frequency. Unfortunately, the statistics for meat packing do not routinely differentiate between slaughter and processing nor between species slaughtered. Such a differentiation would be helpful in terms of applying appropriate regulatory activities to the individual phases of the industry, since slaughtering is more hazardous than wrapping meat, etc. Other references appropriate for securing further detail about safety programs are: *Health and Safety Guide for Meat Packing, Poultry Dressing and Sausage Manufacturing Plants,*[2] *OSHA Handbook for Small Businesses,*[3] and *What to Do About OSHA.*[4]

Without giving a detailed discussion of an entire safety program, which is contained in the National Safety Council publication, we will instead lay out a basic framework for the prevention of accidents. Since knife accidents are the most prominent cause of disabling injuries in the meat industry, a more complete discussion of knife use and safety is included.

Prevention of accidents and injuries in general is achieved through attention to two broad areas:

- Control of conditions—the working environment.
- Control of the actions of people—behavior.

Many accidents and injuries are the result of a combination of environmental and personal causes.

---

[1]425 N. Michigan Avenue, Chicago, IL 60611.

[2]U.S. Dept. of Health and Human Services, Public Health Service, Center for Disease Control, National Institute for Occupational Safety and Health (NIOSH), 1600 Clifton Road N.E., Atlanta, GA 30333.

[3]U.S. Dept. of Labor, Occupational Safety and Health Administration (OSHA), Room N-3641, 3rd and Constitution Avenue N.W., Washington, DC 20210.

[4]Chamber of Commerce of the United States, 1615 H Street N.W., Washington, DC 20062.

Knife Safety

According to the National Safety Council, knife safety can best be obtained by applying the following four control measures:

Utilization of Protective Equipment

Protective equipment consists of plated or stainless steel metal gloves for thumb and two fingers or for the whole hand. A new fibrous glove has stainless wire woven in and is lighter and cheaper than the metal mesh gloves. Other protective equipment includes properly engineered arm guards made of metal, plastic, or neoprene and abdominal protectors made of stainless steel, aluminum mesh, neoprene, or suede leather with steel studs.

Proper Selection, Sharpening, and Guarding of Knives

This means choosing the right knife for each job and proper sharpening and utilization of knife guards to prevent the user's hand from sliding over the handle onto the blade.

Safe Handling of Knives When Not in Use

This dictates that when not in use, the knife be carried in a knife scabbard worn around the worker's waist rather than being placed on the table or in table slots, which has been proven to be unsafe. The scabbard must be composed of a material that can be routinely cleaned and sterilized (Figure 2-1).

Training in Knife Safety

This training falls upon the responsible person or persons to organize the work, engineer the facilities, and maintain a consistency of environment conducive to knife safety. One old but nevertheless reliable rule is "If you drop your knife, *do not* grab for it!"

Knife Selection

As noted above, proper knife selection is basic to knife safety. In order to be able to select the proper knife for a given task, one should know the parts of a knife. The two main parts are the handle and the blade. The handle may be made of rosewood, beechwood, hard rubber, plastic, or various other composition materials. Important considerations in selecting a knife handle are size, safety grip, balance, ease of sanitation, and resistance to strong detergents and the action of dishwashers. Beechwood is soft and not as satisfactory as rose-

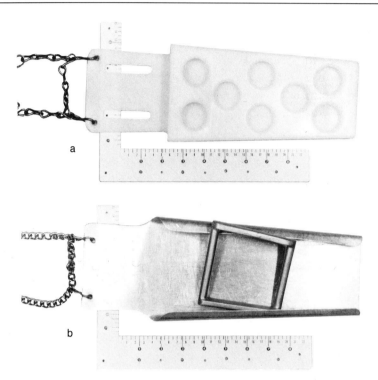

Fig. 2-1. (a) Plastic knife scabbard and (b) metal knife scabbard. Note that both scabbards are held to the worker's body by metal chains for ease of cleaning and sanitizing each entire unit.

wood, since the latter is more attractive and will not readily absorb grease. Hard rubber and plastic handles will not shrink, warp, splinter, or crack but may lose some luster after repeated washings as will most wooden handles. The shape and size of the handle is a personal matter; for example, persons with large hands favor a thick handle. *Ergonomics* is the technique of designing tools so that the work of the user is kept at a minimum. An example of this is a knife design that is hoped will eliminate the wrist strain syndrome (tenosynovitis) in persons who work steadily on boning lines. The uniqueness of this knife is the angle of the blade/handle attachment (Figure 2-2).

In all knives, the blade is fastened to the handle by the portion of the blade called the "tang." The tang may extend to the butt end of the handle or only part way. The handle is fastened to the tang, usually with three rivets or steel pins. The pins must fit tightly and remain so in order to prevent meat particles and moisture from accumulating, which could cause microbial growth and the swelling and cracking of wooden handles. Plastic handles are usually molded directly around the tang to prevent this.

Knife blades vary in length, width, thickness, stiffness, curve, and design (Figure 2-3). The blade of a knife should be a good bit longer than the meat portion to be cut. Thus boning knives and skinning knives are generally made in 5-

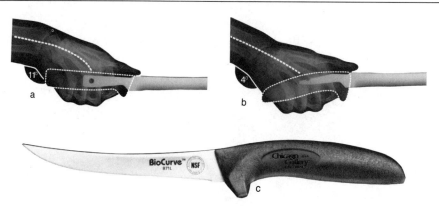

Fig. 2-2. (a) X-ray photo showing how gripping a straight-handled knife requires the wrist to bend at an 11° angle. This "ulnar deviation" is the cause of stress and tension in the wrist and forearm. The range of wrist motion is also reduced. (b) X-ray photo showing how the handle attached at an angle to the blade produces only a 4° bend in the wrist. More energy can be passed through the unlocked (natural) wrist, thus reducing tension and stress. (c) Six-inch BioCurve™ curved boner. (Photo courtesy Chicago Cutlery, Minneapolis)

and 6-inch blade lengths, while steak knives, breaking knives, slicing knives, and butcher knives come in 8-, 10-, and 12-inch blade lengths. Personal preferences exist among butchers and meat cutters. Some prefer straight boners, others prefer curved; some prefer flexible, others prefer stiff; some prefer narrow, others prefer wide. A wide, thick blade will wear longer, but it will drag against the meat. A thin, flexible, narrow blade is usually preferred by boners, because it turns easily in tight places, can be sharpened more easily than a thick blade, and moves through meat with less resistance.

The steel in almost all knife blades is approximately 80 percent iron and 1 percent carbon. The remainder is composed of 14 to 17 percent chromium in stainless steel blades, with other metals, such as manganese, silicon, molybdenum, nickel, tungsten, and vanadium, entering into the various alloys which compose particular knife blades. The more carbon in the steel, the tougher the steel, which makes it more difficult to work into a blade. The measure of hardness of steel is known as the "Rockwell number." An American metallurgist named Rockwell developed a machine with a diamond-pointed cone which is pressed to a standard depth into the metal, registering the force required on a dial. Rockwell 57 is somewhat standard, while Rockwell 67-68 definitely indicates a metal too tough for sharpening.

An interested person once visited the research lab of one of our largest steel producers and asked three highly skilled research metallurgists to recommend a steel for knife blades. He received three different answers. Even the experts cannot agree on the best alloy for knife blades. Theoretically, the perfect blade would never rust, never break, and never have to be resharpened. There really isn't such a blade. However, modern knife manufacturers have formulated special steels which come close to the perfect blade described. There are many rep-

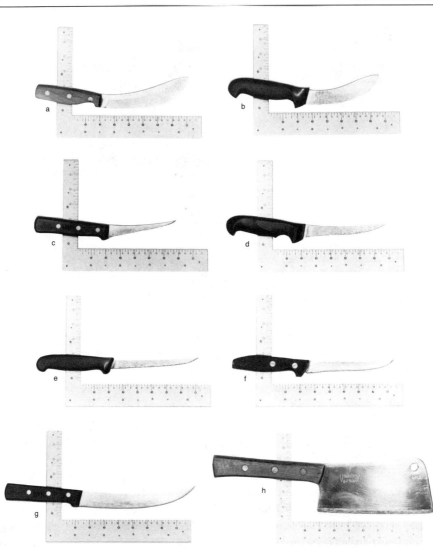

Fig. 2-3. (a) Six-inch skinning knife; rosewood handle. (b) Five-inch skinning knife; molded plastic handle. (c) Five-inch curved boner; medium flex, rosewood riveted handle. (d) Six-inch curved boner; medium flex, molded plastic handle. (e) Six-inch straight boner; flexible, narrow, molded plastic handle. (f) Six-inch straight boner; stiff, wide, hard rubber riveted handle. (g) Eight-inch breaker; stiff, wide, rosewood riveted handle. (h) Eight-inch stainless steel cleaver; rosewood riveted handle. This tool is used less frequently in modern meat processing.

utable knife manufacturers from which dealers for the meat cutter and butcher may purchase knives.

The Sharpening of Knives

A dull knife is inefficient and ineffective except for cutting oneself. Those of us who marvel at the speed and dexterity of persons working in the slaughter rooms of large or small packing houses must remember that they work with sharp tools and must be experts to hold their jobs.

The conventional steps of grinding, honing, and steeling are described here, followed by a description of a new diamond-imbedded hand tool that can replace all three of these steps.

Grinding

Manufacturers of knives may or may not market them sharpened for immediate use. Grinding is done to get extra thinness to the cutting edge. Some knives may need no further grinding on a coarse stone because of type of grind (V, hollow, concave, or diamond-sharpened) already on the blades (Figure 2-4). A sand stone, an emery stone, or a diamond sharpener may be used for grinding. The stones should be water- or oil-cooled to avoid heating the knife and to float the metal particles away.

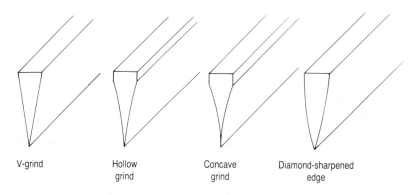

V-grind   Hollow   Concave   Diamond-sharpened
               grind   grind   edge

Fig. 2-4. Types of grinds normally found on knife blades.

The blade usually need not be ground back more than ¼ inch from the edge, forming what is termed a *bevel*. This bevel should be the same on both sides of a skinning knife so the operator may use it with either hand in siding (removing the hide from the side of a carcass). If the bevel is only on one side of the knife, the bevel side must be next to the hide when skinning.

It is advisable to grind the knife by running the stone with the edge of the knife, rather than at a right angle against the edge (Figure 2-5). This method is safer, and one is less likely to scar the blade any further back than the actual bevel itself, and since the smooth finish of a knife is less likely to rust than the ground surface, this is rather important in maintaining the neat appearance of the blade. A sand stone of medium fine grade is preferable to an emery stone for grinding knives. The use of an electric emery belt grinder can be disastrous to the inexperienced, who may inadvertently grind a good portion of the blade away before realizing what is happening.

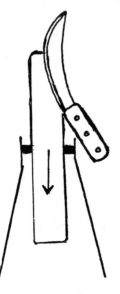

Fig. 2-5. Grinding.

Honing

Honing is done to put a fine, smooth edge on a blade. All knives must be honed initially and periodically thereafter as needed.

Honing may be done on a flat, fine carborundum stone on which water, liquid soap, or oil is used to maintain a clean, abrasive surface (Figure 2-6) or on a wooden-handled hone resembling a steel, except that it is a carborundum or ceramic stone (Figure 2-7).

The flat stone should be set in a block of wood or else placed on a damp cloth to keep it from sliding (Figure 2-6). The handle of the knife is grasped with one hand, and the heel of the knife blade is placed on the end of the stone nearest to the operator. The blade of the knife should be tilted up enough to make

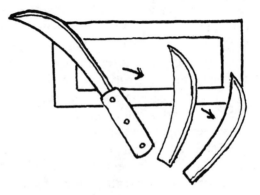

Fig. 2-6. Honing on a flat stone.

the bevel lie flat with the stone. The angle of tilt can vary from almost flat to no more than 20°, exactly the same as the bevel on the knife. The finger tips of the other hand are placed on the flat side of the blade near the back edge to exert pressure on the blade. With a sweeping motion toward the end of the stone, the knife is drawn completely across and inward against the cutting edge of the blade.

The knife is then turned over in the palm of the hand by a twist of the thumb and index finger and drawn across the stone toward the operator. Honing against the edge of the knife avoids the formation of a wire edge. To finish the sharpening process, the knife should be tested for sharpness and smoothness of edge by running the edge of the blade lightly over the flat of the thumbnail. If the knife slides easily, it lacks the proper sharpness. A sharp edge will pull on the nail, and a rough or wire edge will rasp the nail. Another method is to move the ball of the thumb lightly over the edge, but this is not recommended for amateurs.

The technique for the use of the round, hand-held hone is similar to that used in steeling a knife blade, a description of which follows. A word of caution:

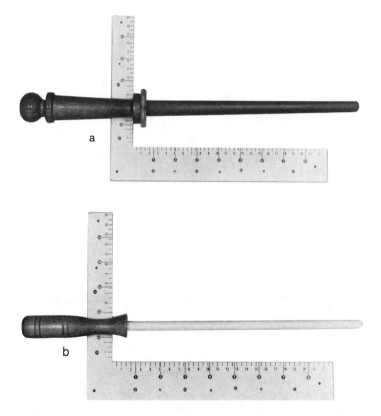

Fig. 2-7. Round, hand-held hones. (a) A carborundum hone that will break if dropped. (b) A ceramic hone known as a "fiddlestick" that is tougher and thus resistant to damage by dropping.

"When using the hand-held carborundum hone, *don't drop it*; it will easily break into several pieces, rendering it useless." The ceramic stone is considerably tougher and more resistant to breakage.

Steeling

Steeling is done to line up the tiny microscopic teeth on the blade and to straighten a wire edge that may have developed during honing.

There are steels of various types on the market which are adapted to certain uses. The carborundum and ribbed steels are primarily for kitchen use where knives need not be razor sharp. The mirror smooth steel for a razor sharp edge is the one best suited to slaughter house and retail meat dealer needs. Steels range in length from 8 to 14 inches, the 10- and 12-inch lengths being the most popular (Figure 2-8). With the proper wrist and elbow action, about 7 or 8 inches of the steel is all that is used except when a steak knife is being sharpened.

The steel is held firmly in the palm of the hand with the thumb in line with the fingers rather than using a "surrounding grip" (Figure 2-9) in a position al-

Fig. 2-8. (a) Ten-inch steel; ⅜-inch diameter, smooth polish, molded handle with guard. Light weight and easily carried on waist chain opposite scabbard. (b) Same style and size steel as (a) except that it has a stylish rosewood handle and a larger diameter steel tapering to ⅜ inch, thus it is heavier and more suitable for home and demonstration carving and cutting.

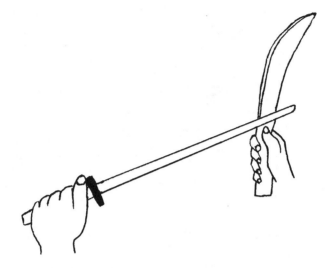

Fig. 2-9. Steeling a knife. Proper way to grip steel. Note how thumb is in line with fingers and thus protected.

most diagonal to the body but with a slight upward tilt. This is important as it permits the free movement of the knife across the steel without drawing it too close to the hand holding the steel. The heel of the blade is placed against either the near side or the far side of the tip of the steel at an angle ranging from almost flat to no more than 20°, exactly the same as the bevel, and the blade is brought down across the steel toward the other hand with a quick swinging motion of the wrist and forearm. The entire blade should pass lightly over the steel. Then the knife is brought in position on the opposite side of the steel, and the same motion is repeated. Once the single stroke, just explained, is mastered, the double stroke will come easily. This consists of placing the knife on the steel for the backward stroke in addition to the downward stroke. Less than a dozen strokes of the knife should be sufficient to realign the edge on a knife that is not very dull.

Sharpening any tool requires the removal of metal, and diamond, the hardest known substance, can work against even the hardest metals with little effect on its own structure. A patented process by which diamond particles are permanently attached to a strong metal-backing surface to keep the diamonds in precise alignment for sharpening has produced a handy new tool that can replace grinder, hone, and steel. The diamond sharpener (Figure 2-10) is strictly a hand tool, but with it one can sharpen (cut metal) as well as or better than with a conventional hand or mechanical knife sharpener.

Sharpening steps using the diamond sharpener:[5]

• Lay the knife nearly flat on the sharpener. The preferred angle is approx-

---

[5]EZE-LAP Diamond Products, Box 2229, Westminster, CA 92683.

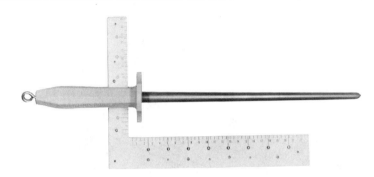

Fig. 2-10. Diamond sharpener.

imately 7°, although 85 percent of all new knives have angles varying between 10° and 40°.

- Alternating between the sides, rub in circles with moderate to heavy pressure to establish the edge (Figure 2-11). Figure 2-12 gives micro-photo comparisons of knife edges sharpened by this method at different angles.
- To finish, use the same procedure, but with light to very light pressure. Taking a few final light strokes into the cutting edge at the same flat angle creates micro-sawteeth aligned in one direction. These micro-sawteeth hook into the material being cut, just as a hand saw's teeth hook into wood.

The natural wobble in hand and arm motion will give the perfect angle. To prove this fact to yourself, use a black, non-permanent felt tip marking pen to mark each side of the cutting edge. Perform the sharpening as instructed above, then observe the brightness where the ink has been removed. It is obvious that since the ink is ground away with the metal all along the area behind and on the cutting edge, the technique really works.

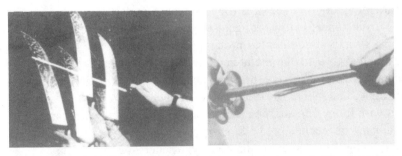

Fig. 2-11. (Left) The rotary motion and angularity to be used while sharpening with the diamond sharpener. The angle should be no steeper than 5° to 7° between the sharpener and blade. (Right) Alternately, the sharpener may be used as a file by holding the knife still and rubbing the sharpener on the knife. (Photo courtesy EZE-LAP Diamond Products, Westminster, California)

This circular sharpening technique eliminates one of the biggest problems in edge maintenance, the thickness developed immediately behind the cutting edge, normally called "the shoulder" of the knife, created by using the age-old technique of steeling.

MICROPHOTO COMPARISONS OF SHARPENED EDGES (Magnification 200x)

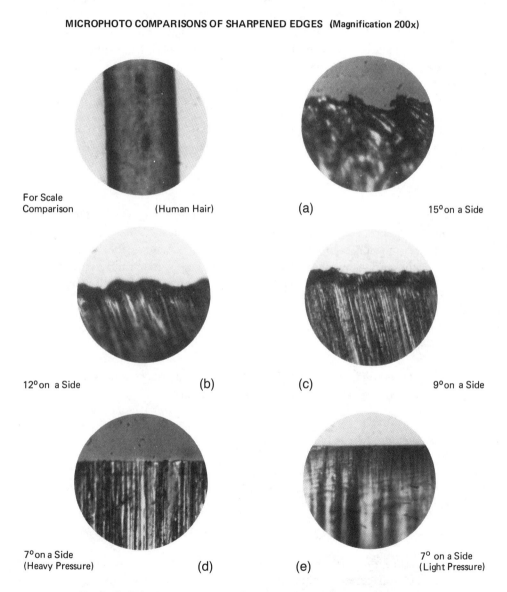

Fig. 2-12. This shows the results of sharpening at several angles. A human hair is included for size-scale comparison. Photo (a) shows metal particle pullout, which will occur with any abrasive during sharpening a large angle. Photos (b), (c), and (d) show how the edge improves as the sharpening angle is progressively reduced down to approximately 7°. The edges shown in (d) and (e) were made with a single EZE-LAP Sharpener, with a lightening of pressure to achieve the razor edge. (Photo courtesy EZE-LAP Diamond Products, Westminster, California)

It is generally known that standard steels do not remove metal, but only re-align the cutting edge. The steel actually may dull because after each stroke the fine microscopic cutting edge is bent from side to side. The cutting edge breaks off just like a piece of fine wire bent back and forth between the fingers. The result is an edge that must be reshaped by grinding. The fine sawtooth or wire edge may be eliminated by laying the sharpened blade perfectly flat against another blade, then rubbing the blades together in small circles. There should be absolutely no angle between the knife blades. The wobble from the circular motion creates a perfect angle to eliminate any wire edge or sawtooth. A steel really is never required.

Other Aspects of Safety (from the National Safety Council)

Buildings and Equipment

Buildings and equipment considerations involve electrical hazards—fired and unfired pressure vessels; railings, stairways, floor surfaces, aisles; guarding of power transmission equipment—lifts and hoists, chains, cables; many other items of equipment—water hoses, steak-tenderizing machines, frozen meat slicers, meat grinders, stuffers, patty-forming machines; and the occurrence of fumes and gases.

Materials Handling

Materials handling involves lifting, use of hand trucks and power trucks, operation of elevators, and maintenance of storage areas.

Housekeeping

Housekeeping is not an occasional push broom effort but an orderly arrangement of operations, tools, equipment, storage, facilities, and supplies.

Maintenance

Maintenance operations include personnel protection, use of ladders and portable tools, refrigeration maintenance, and welding precautions.

Fire Protection

Principles of fire protection involve the evaluation of factors causing fires and the correct use of fire extinguishers.

Cleaning Compounds

Use of cleaning compounds is an exceedingly complex area involving only the utilization of chemicals authorized by the USDA Food Safety and Inspection Service (FSIS) and then being absolutely sure of proper uses and amounts to avoid alkali hazards, acid hazards, and germicide hazards.

First-Aid and Medical Procedures

First-aid and medical procedures include knowledge of general rules, emergency procedures, and vision and hearing conservation.

For a complete discussion of the above items, refer to the National Safety Council's *Meat Industry Safety Guidelines,* mentioned at the beginning of this chapter.

SANITATION

Sanitation regulations are spelled out in detail in two publications by the USDA, FSIS:[6] *The Meat and Poultry Inspection Manual* and *Meat and Poultry Inspection Regulations.* A brief discussion of these regulations appears in Chapter 3. Yet, if one has only a fundamental knowledge of the rich nutritional composition of meat on which certain microorganisms can thrive as do humans, and of the microorganisms themselves, which do thrive on meat under certain conditions, one recognizes that a sanitation program and sanitation rules and habits become very necessary and meaningful. Thus one does not generate a sanitation program merely to satisfy a state or federal meat inspector, but rather to improve the efficiency, service, and, yes, profitability of the meat operation. Ultimately, of course, consumer satisfaction is the goal of a meat operation, and foremost among consumer concerns is wholesomeness of the products that are consumed.

Later, in Chapter 18 in which meat preservation is considered, a somewhat detailed discussion of meat microbiology is included. However, our purpose here is to draw the reader's attention to the importance of sanitation before we get into the details of slaughter and processing, because rightfully, the consideration of sanitation must be given *first.*

Back in 1957, in "A Perspective of the Sanitation Problems in the Meat Industry,"[7] C. F. Niven, Jr., wrote of the importance of the meat "sanitarian" in the production of safe and wholesome food, and the difficult position such a person is in because of the need to prove to top management that sanitation efforts pay rather than cost. Unless a disastrous food-spoilage episode had been experienced, the need for the many precautions against hazards was not readily appar-

---

[6]USDA, Food Safety and Inspection Service (FSIS). *Meat and Poultry Inspection Program,* Washington, D.C.

[7]American Meat Institute Foundation. Circ. No. 39, Nov. 1957.

ent. Now, all top management, supervisors, and workers must be concerned about sanitation, since it requires an all-out effort by all persons involved to produce wholesome, top-quality products.

More recently, Union Carbide Films Packaging Division[8] and Dr. Joe Rakowsky, formerly of Central Soya, Chemurgy Division,[9] published small brochures describing the need for sanitation in plain, easy-to-understand terms. Union Carbide says to keep it *clean, cold,* and *covered* to maintain product quality. Dr. Rakowsky indicates that the three greatest offenders causing food poisoning or disease are *filth, fingers,* and *flies.* Fingers are necessary for a countless variety of purposes and thus must be cleaned between nonfood and food uses. Flies serve no useful purpose and must be completely eliminated from animal slaughter and meat processing areas by screens, air curtains, and approved insecticides.

To summarize sanitation needs, we can expand on the following:

- *Clean* refers to equipment, buildings, people, wrapping materials, ingredients, and *anything* that comes in contact with the meat food. What looks *clean* may very well not be, since bacteria, yeasts, and molds (microorganisms) can be seen only through a microscope. It would take 25,000 of them laid end to end to make 1 inch.

  A sanitation program must be well organized and efficient. First the contaminants must be identified in order to know what detergents will be required to remove them, and the proper equipment must be available for use. The first step in equipment and floor cleaning is to physically remove the chunks and small particles of meat and offal. No liquid is needed, but rather a dry squeegee, brush or broom, and shovel. It is extremely wasteful to use large amounts of water to move this material which would eventually clog drains and cause septic tanks, cesspools, and various waste-treatment facilities to fail. Since the ultimate pH of meat is 5.6 (acid) a basic (alkaline) cleanser is used first and routinely to neutralize and remove the remaining tiny meat residues. Water temperature should be 140° F or slightly higher to cut the fat residues. Periodically, once or twice a week on a daily cleanup schedule, the basic detergent should be followed immediatley with an acidic cleanser to remove mineral deposits which may build up. After being cleaned, equipment and facilities should be sanitized with hot water (180° F) or an approved chlorine or iodine rinse. Finally, metal equipment should be lightly sprayed with an edible mineral oil to prevent oxidation of the metal (rusting).

  For workers to be *clean* means that they must wash with approved germicidal soaps and warm water *many* times during the day, especially after breaks. In some plants, workers completely change clothes when

---

[8]*The 3 C's of Plant Sanitation.* Union Carbide Films Packaging Division, Chicago.
[9]*Sanitation Simplified.* Central Soya, Chemurgy Division, Chicago.

going into the plant to work with meat. They may even wear masks like those that surgical doctors and nurses wear in the hospital operating room. So it is logical that employees should not use tobacco, pick their noses, or scratch their heads while working with meat food. Furthermore, the *water* supply must be potable; that is, pure and non-contaminated. That is a basic requirement, and is discussed first in the FSIS regulations. Thus, when we speak of being *clean* we have eliminated *filth* sometimes caused by *fingers* and *flies*.

- *Cold* means *without question* holding meat at a temperature below 40° F, but better yet, down very near to the point where meat begins to freeze, at 28° F. Most microorganisms that cause food-borne intoxication and infection grow best at temperatures between 40° and 140° F, so it is important that meat be kept at either below 40° F or above 140° F, and when it is cooking, it should pass through this range from 40° to 140° F in a time space no longer than four hours. The old saying "life begins at 40" really refers to food microorganisms. The closer to 28° F you hold the meat during storage, transport, or display, the longer the shelf life; that is, the maintenance of top quality. Thus, federal regulations require that processing rooms and equipment used in rooms where temperatures may exceed 50° F must be completely cleaned and sanitized every five hours. This is an example of a very useful regulation that processors can and must abide by that benefits their operation by minimizing contamination, thus maintaining a high level of quality control.

- *Covered* means excluding airborne microorganisms and vermin by the use of screens or other suitable barriers between processing areas. It means proper covering of meats, carcasses, wholesale cuts, and primal cuts during transport. It means the proper wrapping of retail cuts for display or for preservation by freezing. It also means workers not coughing, sneezing, or spitting tobacco juices in the presence of meat food.

Thus, sanitation is basic and important, and just good common sense can go a long way toward establishing and maintaining satisfactory sanitation in any meat-processing situation.

Chapter 3

Meat Inspection and
Animal Loss Factors

THE WHOLESOMENESS of each American's meat and poultry supply is protected by the USDA. Federal meat inspection dates back to June 30, 1906, with the passage of the Meat Inspection Act. Previous to that time, a limited form of federal inspection had been started in 1891, but this was only a voluntary inspection of cattle and hogs intended for export. The Meat Inspection Act of 1906 has been continually evaluated and improved over the years, but the most sweeping change was the passage of the Wholesome Meat Act on December 15, 1967. The major thrust of this 1967 law, which still ties to the original 1906 act, was to make all of the various state inspection systems at least equal to the federal inspection system. Previous to this time, federal meat inspection applied *only* to meat and meat products in interstate or foreign commerce. Many states had their own inspection systems for meat and meat products moving within their own state borders, but others did not. Thus, consumers in the various states may have had varying levels of protection.

The 1967 law gave the states three years, until December 15, 1970, to inaugurate an inspection system equal to the federal system. If any state could not or would not develop its own system, the federal inspection system took over in that state after December 1970. The following 23 states, Arkansas, California, Colorado, Connecticut, Idaho, Kentucky, Maine, Massachusetts, Michigan, Minnesota, Missouri, Montana, Nebraska, Nevada, New Hampshire, New Jersey, New York, North Dakota, Oregon, Pennsylvania, Rhode Island, Tennessee, and Washington, do *not* have state inspection within their borders, but rather *all* plants, both *intra*state and *inter*state, are federally inspected. The USDA also does all the poultry inspection in four other states: Georgia, South Dakota, Utah, and West Virginia. The federal government assists those states that continue to operate their own inspection systems (equal to federal) by assuming 50 percent of the cost of these state systems. An alternative, which looks desirable to some state legislatures and governors interested in finding new ways to balance their states' budgets, is to forego state inspection, turn it over to the federal system, and thus save all of the state cost.

ADMINISTRATION OF THE ACT

Meat inspection was first administered by the Bureau of Animal Industry, which became a part of the USDA's Agricultural Research Service. Later, meat inspection came under the USDA's Consumer and Marketing Service (C & MS). On April 2, 1972, the Secretary of Agriculture established the Animal and Plant Health Inspection Service (APHIS) and assigned to this new agency the meat and poultry inspection functions heretofore carried out by the C & MS.

On March 14, 1977, the new Secretary of Agriculture established the Food Safety and Quality Service (FSQS) and assigned to it the responsibility for food inspection and grading. In addition to meat and poultry inspection, the new agency was responsible for the inspection and grading of dairy products, eggs, and fresh and processed fruits and vegetables. Yet another name change took place with a new administration in June 1981, when the meat grading and purchasing functions were transferred back to the Agricultural Marketing Service. The name *FSQS* was dropped, and meat, poultry, and egg inspection became the responsibility of the USDA's Food Safety and Inspection Service (FSIS).

Meat inspection is *not* synonymous with meat grading. *Meat inspection* guarantees meat wholesomeness, safety, and accurate labeling. All meat that is sold *must* by law be inspected. Meat inspection is paid for by our taxes. Every citizen, regardless of his or her meat-eating habits, pays for meat inspection through taxes. Conversely, *meat grading* designates expected quality and yield of carcasses and cuts and is optional. Grading is paid for by meat packers and processors and ultimately by the consumer, in the price of meat. (See Chapter 11.) The only way in which grading relates to the wholesomeness and safety of meat is in the regulation that meat must pass inspection before it can be graded. In some instances, the same properly trained and licensed USDA employee may perform both inspection and grading functions.

THE COST OF MEAT INSPECTION

Meat is a highly perishable food that takes approximately 40¢ of our food dollar, thus about 6 percent of our take-home pay.

Unlike any other food, meat is very much like the human body. Animals are subject to many of the same diseases as humans. For our protection, we must have rigid standards of inspection of the health of each animal and its products and the sanitary conditions under which it is processed. Who should pay for this inspection? Some say the meat packer (processor); others say Uncle Sam. If it is the packer, the added cost will be reflected in the price of the product and we pay by proxy.

Back in 1947, Congress included a provision in the USDA appropriation bill to pass along the cost of federal inspection to packers and processors. This shift of the cost of meat inspection from the government to the packer was in effect until the passage of the Kem bill (S2256), when the cost reverted again to the

USDA, effective July 1, 1948. Since that time, the meat packing industry has been reimbursing the government for overtime or special services only, at a current base time rate of $21 per hour. Politically or otherwise, a one-year trial was sufficient. Probably the main reason why shifting the cost of inspection to the packer did not succeed was the fact that those who did not have federal inspection were able to undersell those who had to bear the cost of inspection. The cost of meat inspection for each consumer remains amazingly low—approximately 13¢ per month.

FOOD SAFETY AND INSPECTION SERVICE ORGANIZATION AND RESPONSIBILITIES[1]

The Food Safety and Inspection Service (FSIS) of the USDA assures that meat and poultry products moving in interstate and foreign commerce for use as human food are safe, wholesome, and unadulterated and are accurately marked, labeled, and packaged. The laws also protect meat producers by insuring that no one will be able to gain an unfair economic advantage by putting unwholesome or misbranded products on the market.

Meat inspection is composed of several facets:

- Ante-mortem inspection (inspection of live animals prior to slaughter)
- Post-mortem inspection (inspection of the carcasses and associated tissues following slaughter)
- Reinspection during processing
- Sanitation
- Facilities and equipment
- Labels and standards
- Compliance
- Pathology and epidemiology
- Residue monitoring and evaluation
- Federal-state relations
- Foreign programs

Each of these facets is explained more fully in the following material.

Ante-mortem Inspection

Ante-mortem inspection procedures must be accomplished in a manner that will detect and remove any animal that will produce unwholesome meat. Ani-

---

[1]USDA, FSIS. *Meat and Poultry Inspection, 1983.* Report of the Secretary of Agriculture to the U.S. Congress, March 1, 1984.

mals which are down, disabled, diseased, or dead (4 d's) are unquestionably unfit for human consumption. But other animals may be only "suspect," that is, suspected of being sick. The meat packer must provide a paved, drained, covered suspect pen for such animals and must provide some means of restraining an animal; for example, a chute and nose tongs. The animal-holding area, where routine inspection is performed, must have a minimum light intensity of 10 footcandles, while the suspect area must have a minimum of 20 foot-candles. These light intensities are measured 3 feet above the floor. The packer must also provide a rectal thermometer, tags, hog rings, and a ringer for tagging.

The inspector observes the animals at rest in the pen and from each side, as animals are being slowly moved around. With poultry, an inspection is made on the day of slaughter to the extent necessary to detect disease and/or other harmful conditions.

Suspect animals are restrained, identified, temperatured, and held for a detailed examination. Normal rectal temperatures (° F) are 100 horses, 101 cattle, 102.3 sheep, 102.5 swine, 103.8 goats, 104.9 turkeys, and 107 chickens. Animals to be condemned are either tagged "U.S. Condemned" and killed immediately by an establishment employee or released to approved authorities for treatment under official supervision.

Animal identity is set up whereby the inspector can check that all animals being slaughtered have had ante-mortem inspection on the day of slaughter.

Animals showing signs of disease must be separated from the other animals. A company employee shall be available for moving, sorting, restraining, and identifying the animals. After an animal has had a detailed examination, it may be released without restriction or held after slaughter for further examination and disposition by the post-mortem inspector.

Dead meat animals and dead poultry must be immediately tagged "U.S. Condemned" and tanked or effectively denatured under the supervision of an inspector.

Denaturing involves completely marking by pouring a denaturant (dye), such as crude carbolic acid or finely powdered charcoal, over the animal or its parts.

Post-mortem Inspection

Post-mortem inspection procedures must be accomplished in a manner that will detect and remove any unwholesome carcass, part, or organ from human food channels. Again, lighting on the kill floor is especially important so that inspectors can make the most accurate appraisals of disease and contamination.

All areas of head inspection and head washing for cattle, calves, hogs, and sheep must be illuminated to a minimum of 50 foot-candles. For poultry, a minimum of 50 foot-candles at the lowest inspection point and final post-mortem areas is required.

Head Inspection

Head inspection is very important for cattle, calves, and hogs. In cattle, the head must first be examined on all surfaces for pathological conditions and contamination; then inspectors must properly incise and examine the mandibular, atlantal (if present), suprapharyngeal, and parotid lymph nodes (two each) for lesions of tuberculosis (page 84), actinomycosis (lumpy jaw), neoplasms, etc. (Table 3-2). The external and internal muscles of mastication must be inspected by incising them in such a manner as to split the muscles in a plane parallel to the lower jawbone for tapeworm cysts (cysticercosis). The tongue is observed and palpated along its entire length for bruises, hair sores, etc.

With calves, visual inspection is made to determine if heads are free of hair, hide, horns, and contamination. Incisions for node inspection may be confined to the suprapharyngeal lymph nodes, unless there is reason to believe that other nodes should be examined.

In swine, both mandibular lymph nodes are sliced and examined for abnormalities, such as tuberculosis lesions (page 84).

Viscera Inspection

CATTLE

- *Lungs*. Incision and inspection of the tissue of the right and left bronchial and anterior, middle, and posterior mediastinal lymph nodes, palpation of the parietal or curved surface, and observation of the ventral surface for tumor, abscess, or pneumonic conditions must be completed.
- *Heart*. The inner and outer surfaces of the heart must be examined, and the muscles of the left ventricle and interventricular septum must be incised and examined.
- *Liver*. Inspection includes surface observation and palpation for abscesses and other abnormalities, opening of the bile duct for liver flukes, and incision of the portal lymph nodes.
- *Spleen*. The spleen, along with the mesenteric lymph nodes and abdominal viscera, should be observed, and the ruminoreticular junction should be palpated. Mesenteric lymph nodes should be incised if necessary.

CALVES

Lungs, heart, and liver must be observed and palpated, and the viscera, including paunch and intestines, must be carefully observed.

SWINE

Lung inspection includes observation of the parietal and ventral surfaces and palpation of the parietal surface, bronchial, and mediastinal lymph nodes. Both sides of the liver must be observed, and the parietal surfaces and portal lymph nodes should be palpated and incisions made, if necessary. The heart, spleen, and mesenteric lymph nodes should be incised, if necessary.

SHEEP AND GOATS

Observation and palpation of the lungs and related lymph nodes, heart, spleen, and liver must be made. The main bile duct must be opened by a meat plant worker and examined for parasites. The viscera must be carefully observed.

Rail Inspection of the Carcass

CATTLE

All surfaces must be observed for pathology and cleanliness. The superficial inguinal (super-mammary), internal iliac, lumbar and renal lymph nodes, exposed kidney and pillars (supporting tissues), and flat portion of the diaphragm are to be palpated and observed. Incisions are to be made if necessary.

SWINE

All parts of the carcass must be observed. Remnants of the liver and lungs, bruises, wounds, and other abnormalities shall be removed by a meat plant worker before rail inspection can be completed. The kidneys should be palpated and observed for evidence of pathology, particularly kidney worms.

CALVES

There should be a visual inspection and observation of the carcass. The exposed kidneys and iliac nodes should be palpated, when necessary. The back of "hide-on" calves must be palpated to detect grubs and dirt.

SHEEP AND GOATS

There should be an observation of the internal and external surfaces of the entire carcass. The prefemoral, superficial inguinal, popliteal, iliac lymph nodes, and diaphragm, kidneys, spleen (if present), and prescapular lymph nodes should be palpated.

POULTRY

The inspector must observe all external and internal surfaces of the cavity of each carcass, including a careful examination of the air sacs, kidneys, and sex organs and must observe and palpate the legs, heart, liver, and spleen. The spleen is to be crushed on adult birds. The inspector shall signal the trimmer of action to be taken with regard to removing defective parts and recording condemnations.

Disposition of Carcasses and Parts

Contaminated or diseased heads shall be condemned or retained. When heads show pathological conditions (page 41), corresponding carcasses as well as the heads must be tagged for examination by a veterinarian. The veterinarian may permit the food inspector to dispose of viscera with localized lesions (abscessed livers, contaminated viscera, etc.). Carcasses which are contaminated or bruised or have other abnormal conditions shall be retained for trimming or examination by a veterinarian before being passed for final washing.

With poultry, questionable birds must be retained for final post-mortem inspection by a veterinarian.

Final Inspection Procedures and Dispositions

All organs, body cavities, and surfaces of the carcasses must be checked and examined. The lymph nodes must be exposed and incised, if necessary (TB, malignancies, etc.).

With poultry, all carcasses and viscera retained for final inspection must be thoroughly examined by a veterinarian. Sufficient checks must be made on all condemned as well as edible product to assure that proper dispositions are being made by the post-mortem line inspectors.

Disposition of carcasses and parts must be made in a professional manner on the basis of scientific training and reason and according to meat inspection law and regulations. If necessary, samples shall be sent to the laboratory for final pathological diagnosis to aid with dispositions.

Chilling and Moisture Control (Poultry)

Since poultry is often chilled with ice, certain restrictions apply. Immediately after slaughter, evisceration, and washing, carcasses must be chilled by an approved method that will preclude adulteration. This can be accomplished by using potable ice. An internal temperature of 40° F or less must be achieved. Moisture pickup must not exceed 8 percent.

Control of Restricted Products, Animal Food Products, Condemned Carcasses, and Inedible Materials

• All *restricted products,* those meat products or ingredients which are undergoing tests for wholesomeness, must be under direct control or under lock or seal at all times until they are rendered acceptable for human consumption.

• *Animal food products,* those meat products suitable only for animals, must be under the inspector's control until they are packed and identified or denatured. Like products cannot be saved for human consumption and animal food simultaneously.

• *Condemned carcasses* or parts must be under the inspector's control until they are tanked (rendered safe through severe heat treatment) or properly denatured; that is, completely marked by a denaturant, such as crude carbolic acid or finely powdered charcoal, being poured over the product.

• *Inedible material* must be handled in a prompt, efficient manner by being placed in properly marked containers, under the inspector's supervision or control, until it is tanked or properly denatured, or until it is identified as food other than for humans.

Table 3-1 shows the number of animals and carcasses federally inspected and condemned and the percent condemned during fiscal year 1983. Table 3-2

Table 3-1. Animals and Carcasses Federally Inspected and Condemned, for Fiscal Year 1983[1]

Species	Inspected	Condemned	Condemned as a Percent of Those Inspected
Cattle	33,405,565	122,895	.37
Calves	2,682,907	36,132	1.35
Swine	78,803,598	189,145	.24
Goats	81,225	515	.63
Sheep	6,194,959	31,347	.50
Equine	138,295	722	.52
Total meat	121,306,549	380,756	.31
Total poultry	4,533,669,000	42,821,638	.94

[1]MP Form 9300 Data. USDA, FSIS. *Meat and Poultry Inspection Program.* Statistical Summary for 1983. FSIS-14. Issued April 1984.

presents the causes of condemnation and the number of carcasses condemned for each cause during fiscal year 1983. Table 3-3 provides the number of retained carcasses passed for food after removal of affected parts and the causes of retention during fiscal year 1983. Table 3-4 specifies the number of livers condemned and the causes of condemnation during fiscal year 1983. Table 3-5 tells the number of poultry federally inspected during 1981, 1982, and 1983. Table 3-6 indicates the number and percent of poultry condemned on post-mortem federal inspection and the causes of condemnation during fiscal year 1983.

Table 3-2. Animals/Carcasses Condemned, for Fiscal Year 1983[1]

| Cause of Condemnation | Number of Carcasses Condemned | | | | | |
	Cattle	Calves	Sheep and Lambs	Goats	Swine	Equine
Degenerative and dropsical conditions:						
Emaciation	3,860	1,585	3,522	64	538	51
Miscellaneous	3,337	219	75	1	267	6
Infectious diseases:						
Actinomycosis, actinobacillosis (lumpy jaw)	806	2	4	—	2	—
Caseous lymphadenitis (cheesy inflammation of lymph glands)	—	—	7,080	148	—	—
Coccidioidal granuloma (fever, inflammation of lungs especially)	21	—	2	—	—	—
Swine erysipelas (skin inflammation)	—	—	—	—	2,641	—
Tetanus (lock jaw)	33	65	—	—	—	—
Tuberculosis nonreactor	33	—	—	—	2,752	—
Tuberculosis reactor	28	—	—	—	—	—
Miscellaneous	168	10	3	—	74	4
Inflammatory diseases:						
Arthritis (inflammation of joints)	1,566	1,288	1,077	5	19,479	5
Eosinophilic myositis (muscle leukocyte)	4,044	10	44	—	6	—
Mastitis (udder)	706	—	4	3	54	—
Metritis (uterus)	1,462	—	20	1	592	3
Nephritis, pyelitis (kidney)	3,298	167	283	—	1,667	6
Pericarditis (heart)	4,477	55	64	1	1,080	1
Peritonitis (cavity of abdomen)	4,372	1,176	114	10	8,892	16
Pneumonia (lungs)	11,061	3,720	4,215	19	13,071	81
Uremia (kidney)	789	28	1,160	—	797	4
Miscellaneous	1,402	667	60	1	626	8
Neoplasms (tumors, cancers):						
Carcinoma (general)	4,379	12	32	—	672	47
Epithelioma (skin)	18,389	3	—	—	5	6
Malignant lymphoma (lymph glands)	10,755	103	23	4	1,066	11
Sarcoma (connective tissue, cartilage, bone)	279	1	7	6	438	8
Miscellaneous	656	18	15	1	1,293	113
Parasitic conditions:						
Cysticercosis (tapeworms)	86	—	400	—	5	—
Myiasis (maggots)	21	2	—	—	—	—
Miscellaneous	181	1	2,595	—	423	—

(Continued)

Table 3-2 (Continued)

Cause of Condemnation	Number of Carcasses Condemned					
	Cattle	Calves	Sheep and Lambs	Goats	Swine	Equine
Septic conditions (decay):						
Abscess, pyemia (pus forming, blood poisoning)	9,408	458	1,331	15	26,561	32
Septicemia (blood poisoning)	8,438	3,416	332	9	7,338	50
Toxemia (toxic substances in blood)	4,018	489	266	1	2,941	16
Other:						
Central nervous system disorders	133	27	21	—	1,205	4
Contamination	—	—	490	10	6,307	—
Deads	9,279	15,905	6,214	189	65,474	94
Icterus (jaundice, yellowing of skin)	621	3,603	1,280	5	8,327	4
Injuries	4,677	827	322	12	3,567	33
Moribundity (to be dying)	9,198	1,121	208	5	7,022	18
Pigmentary conditions	214	13	9	5	633	65
Pyrexia (fever)	264	27	3	—	136	2
Residue	97	1,025	—	—	89	—
Sexual odor	—	—	—	—	105	—
Miscellaneous general	326	87	69	—	2,996	1
Other reportable diseases	13	2	1	—	4	45
Total	122,895	36,132	31,347	515	189,145	722

[1]MP Form 9300 Data. USDA, FSIS. *Meat and Poultry Inspection Program.* Statistical Summary for 1983. FSIS-14. Issued April 1984.

Reinspection During Processing

Inspection and control of processed products must assure that only sound, wholesome products are distributed into human food channels. This includes acceptable procedures for destroying trichinae in products containing pork muscle. These procedures include using only wholesome ingredients and approved quantities of acceptable chemicals, providing adequate protection during processing and storing, and controlling restricted products.

Facilities—Sufficient Lighting in All Areas

MEAT

All inspection areas (boning tables, grinders, bacon presses, slicers, choppers, etc.) must have 50 foot-candles of lighting. All other areas must contain at least 20 foot-candles, except dry storage, where lighting sufficient for purpose is acceptable, and 10 foot-candles at front shank level of carcass in coolers. All lights must have protective coverings in processing rooms and areas where product is exposed.

Table 3-3. Retained Carcasses Passed for Food After Removal of Affected Parts, for Fiscal Year 1983[1]

Cause of Retention	Number of Carcasses					
	Cattle	Calves	Sheep and Lambs	Goats	Swine	Equine
Degenerative and dropsical conditions:						
Miscellaneous	4,047	7	37	3	45	63
Infectious diseases:						
Actinomycosis, actinobacillosis (lumpy jaw)	181,637	556	64	—	313	—
Caseous lymphadenitis (cheesy inflammation of lymph glands)	—	—	46,375	974	—	—
Coccidioidal granuloma (fever, inflammation of lungs, especially)	2,145	16	—	—	2	—
Tuberculosis nonreactor	—	—	—	—	303,047	—
Miscellaneous	1,622	311	44	—	3,544	4
Inflammatory diseases:						
Arthritis (inflammation of joints)	152,916	12,320	34,015	400	765,866	479
Eosinophilic myositis (muscle leukocyte)	3,106	2	4	—	338	—
Mastitis (udder)	16,720	—	34	4	7,287	—
Metritis (uterus)	3,989	—	9	—	819	1
Nephritis, pyelitis (kidney)	19,875	8,130	1,715	30	30,043	276
Pericarditis (heart)	16,533	439	1,152	23	91,239	44
Peritonitis (cavity of abdomen)	10,922	660	255	10	61,613	7
Pneumonia (lungs)	122,951	61,695	55,495	2,167	392,198	5,827
Miscellaneous	115,550	8,358	27,600	772	102,887	27
Neoplasms (tumors, cancers):						
Epithelioma (skin)	75,837	393	—	—	35	—
Miscellaneous	5,512	237	26	8	10,324	214
Parasitic conditions:						
Cysticercosis (tapeworms)	—	—	30,574	58	—	—
Myiasis (maggots)	1,313	51	—	—	32	—
Miscellaneous	25,538	501	6,424	603	68,505	3,123
Septic conditions (decay):						
Abscess, pyemia (pus forming, blood poisoning)	378,363	14,603	23,796	497	1,967,304	4,694
Other:						
Contamination	854,001	93,230	236,567	3,589	2,524,391	3,901
Injuries	563,503	42,214	32,518	1,313	488,824	4,404
Pigmentary conditions	6,898	193	22	16	3,897	1,747
Residue	904	177	—	—	389	1
Skin conditions	4,566	30,365	4,449	13	124,301	4
Miscellaneous general	11,649	3,584	1,218	72	32,134	179
Other reportable diseases	—	—	—	—	—	—
Total	2,580,097	278,042	502,393	10,552	6,979,377	25,043

[1]MP Form 9300 Data. USDA, FSIS. *Meat and Poultry Inspection Program.* Statistical Summary for 1983. FSIS-14. Issued April 1984.

Table 3-4. Livers Condemned, for Fiscal Year 1983[1]

| Cause of Condemnation | Number/Pounds of Livers | | | | | |
	Cattle	Calves	Sheep and Lambs	Goats	Swine	Equine
	.... (number)		.... (pounds)			(number)
Abscess	3,216,957	19,527	—	—	—	1,506
Carotinosis (excessive orange/yellow pigment)	53,554	6,757	—	—	—	297
Cirrhosis (liver)	34,763	2,846	—	—	—	1,258
Degenerative condition	62,502	8,172	—	—	—	942
Distoma (worms)	1,539,439	13,090	—	—	—	785
Melanosis (excessive black pigment)	10,062	1,128	—	—	—	162
Other parasitic conditions	162,575	2,799	—	—	—	24,977
Sawdust	100,957	1,533	—	—	—	38
Telangiectasis (abnormal dilation of capillaries)	682,040	4,666	—	—	—	27
Miscellaneous[2]	591,532	16,837	2,532,145	25,012	35,833,954	2,804
Total	6,454,381	77,355	2,532,145	25,012	35,833,954	32,796

[1]MP Form 9300 Data. USDA, FSIS. *Meat and Poultry Inspection Program*. Statistical Summary for 1983. FSIS-14. Issued April 1984.

[2]Individual causes not recorded for pounds of sheep and lamb, goat, and swine livers condemned (see discussion in text).

Table 3-5. Number of Poultry Federally Inspected, 1981-83[1]

Class	1981	1982	1983
	.... (thousands)		
Young chickens	4,058,280	4,079,196	4,155,861
Mature chickens	205,374	196,111	190,417
Fryer-roaster turkeys	9,353	6,309	4,339
Young turkeys	153,233	153,602	160,024
Old turkeys	1,381	1,245	1,265
Ducks	17,924	19,404	20,644
Other	1,446	984	1,119
Total	4,446,991	4,456,851	4,533,669

[1]USDA, FSIS. *Meat and Poultry Inspection, 1983*. Report of the Secretary of Agriculture to the U.S. Congress, March 1, 1984.

Table 3-6. Number and Percent of Birds Condemned on Post-mortem Inspection by Class and Cause, for Fiscal Year 1983[1]

Cause of Condemnation	Number and Percent of Carcasses Condemned							
	Young Chickens	Mature Chickens	Fryer-Roasters (Turkeys)	Young Turkeys	Old Turkeys	Ducks	Other Poultry	Total
Tuberculosis	—	5,703	—	26	—	—	3	5,732
		0.00		0.00	0.00	0.00	0.00	0.00
Leukosis	3,989,491	293,673	380	2,786	37	—	28	4,286,395
	0.10	0.15	0.01	0.00	0.00		0.00	0.09
Septicemia	14,626,360	2,101,944	16,580	980,584	15,506	133,337	3,857	17,878,168
	0.35	1.10	0.38	0.61	1.23	0.65	0.34	0.39
Airsacculitis	8,494,922	24,855	802	233,796	3,310	49,238	998	8,807,921
	0.20	0.01	0.02	0.15	0.26	0.24	0.09	0.19
Synovitis	300,251	11,114	260	150,557	426	19,559	7	482,174
	0.01	0.01	0.01	0.09	0.03	0.09	0.09	0.01
Tumors	1,429,907	1,914,107	26	2,326	452	3,251	133	3,350,202
	0.03	1.01	0.00	0.00	0.04	0.02	0.01	0.07
Bruises	708,046	346,707	600	15,350	1,285	14,316	185	1,086,669
	0.02	0.18	0.01	0.01	0.10	0.07	0.02	0.02
Cadaver	1,380,119	246,105	3,938	46,782	536	17,748	400	1,695,628
	0.03	0.13	0.09	0.03	0.04	0.09	0.04	0.04
Contamination	2,231,768	487,862	4,558	50,677	502	15,204	698	2,791,269
	0.05	0.26	0.11	0.03	0.04	0.07	0.06	0.06
Overscald	504,118	37,022	371	9,623	90	1,889	173	553,286
	0.01	0.01	0.01	0.01	0.01	0.01	0.02	0.01
Other	1,298,636	479,185	491	35,787	467	68,971	657	1,884,194
	0.03	0.25	0.01	0.02	0.04	0.33	0.06	0.04
Total	34,963,618	5,948,277	28,006	1,528,474	22,611	323,513	7,139	42,821,638
	0.84	3.12	0.65	0.96	1.79	1.57	0.64	0.94

[1]MP Form 513 Data. USDA, FSIS. *Meat and Poultry Inspection Program.* Statistical Summary for 1983. FSIS-14. Issued April 1984. Includes *only* post-mortem inspection.

POULTRY

There must be 30 foot-candles in operating areas, 50 foot-candles at inspection stations, and 10 foot-candles in storage areas and coolers.

*Management Controls—Products Received
from Acceptable Source*

MEAT

Carcasses, cuts, and manufacturing meats must bear legible marks of inspection where slaughtered and/or last processed.

POULTRY

Carcasses must be properly labeled with inspection legends.

NONMEAT

Supplies must be properly identified and labeled as required by the federal Food and Drug Administration.

Use and Handling—Storage of Raw Meat

MEAT

Raw meat, emulsions, and the finished, perishable product must be stored at a room temperature of 50° F or lower, accessible to inspection, and handled in a manner to avoid contamination.

POULTRY

Raw poultry meat, emulsions, and the finished perishable products must be held at an internal temperature of 40° F or less and handled in a manner to avoid contamination.

Formulation Control and Identification

All ingredients, emulsions, mixtures, liquids, etc., must be identified through all phases of processing. All formulas and formulating procedures must be readily available for review by operating personnel and inspectors. The quantity of meat and nonmeat materials is controlled to produce a product in compliance with published standards and label declarations. Controls must be such that the inspector can evaluate adequacy of the formulation.

Processing Controls for Curing
and Pumping and Smoking

CURING AND PUMPING

Restricted ingredients (nitrates, nitrites, phosphates, ascorbates, corn syrup) are to be used according to specific standards. (See Chapters 18 and 20.) All formulas for pickle and curing solutions and all pumping and curing procedures must be readily available for review by plant management and inspector. The product must be uniformly pumped and cured.

SMOKING

Uniform procedures are used to shrink product into compliance with applicable regulations. (See Chapters 18 and 20.) All pork products, except bacon, must be heated to a temperature not lower than 137° F, and the method and control used must be known to insure such results. Bacon must be heated to 128° F, since it is fried prior to eating. When poultry rolls are heat-processed in any manner, cured and smoked poultry rolls must reach an internal temperature of 155° F prior to being removed from the cooking media, and all other poultry rolls must reach an internal temperature of at least 160° F prior to being removed. The additional heat required to assure poultry safety is based on its higher moisture content.

Trichinae Control

Pork products that are not customarily cooked in the home or elsewhere before being consumed must be subject to an approved treatment for destruction of trichinae. The treatment consists of heating, controlled freezing, or curing, according to section 318.10 of the FSIS regulations:

1. Heating to a temperature not lower than 137° F.
2. a. Freezing at temperature indicated. (Note: freezing will not destroy trichinae in bear meat.)

Temperature (° F)	6" Thick or Less (days)	More Than 6" but Less Than 27" Thick (days)
5	20	30
−10	10	20
−20	6	12

   b. Internal product temperature at center of meat pieces.

Temperature (° F)	Hours
0	106
−5	82
−10	63
−15	48
−20	35
−25	22
−30	8
−35	½

3. Curing. Six methods for curing sausages and three methods for curing hams and shoulders are spelled out in *Meat and Poultry Regulations* (USDA, FSIS).

Generally, the combined effects of time, temperature, amount of salt, and diameter of product interact to destroy any trichinae that may be present. Small products and products containing salt that are heated to high temperatures take less processing time to become trichina free. A discussion covering the incidence of *Trichinella spiralis* follows later in this chapter.

Knowledge of Management Controls

The inspector must be knowledgeable of procedures and controls used by management in the manufacture and formulation of all finished products.

Security of Brands, Certificates, and Seals

All brands and devices used for marking articles with the inspection legend, self-locking seals, official certificates, or other accountable items shall be kept under adequate security, such as lock or seal, and an up-to-date inventory shall be maintained of such security items (does not include printed labels).

Table 3-7 lists the pounds of meat and meat products that were condemned and destroyed on reinspection in fiscal year 1983 and the reasons for condemnation. These amounts represent less than 0.04 percent of the total amount that was reinspected. Table 3-8 lists the pounds of materials other than meat that were rejected in fiscal year 1983 and the reasons for rejection.

Sanitation

A general common-sense discussion of sanitation was given in Chapter 2. This section is more specific in regard to FSIS sanitation requirements.

Operational sanitation must permit production of wholesome products and must also permit product handling and processing without undue exposure to contaminants. Facilities and equipment must be properly cleaned at regular intervals. All personnel must practice good personal hygiene, and management must provide the necessary equipment and materials to encourage such hygiene. Particular emphasis should be placed on product and product zones.

Floors, Walls, Ceilings

Floors must be free of an accumulation of fats, blood, and other foreign material. Walls must be free of dirt, mold, blood, scaling paint, and other contaminants. Ceilings and overhead must be free of dust, scaling paint, scaling plaster, mold, rust, condensation, leaks, etc.

Table 3-7. Pounds of Meat and Meat Products Condemned and Destroyed on Reinspection, for Fiscal Year 1983[1]

Cause	Beef	Veal	Mutton and Lamb	Goat Meat	Pork	Total Less Horse Meat	Horse Meat	Grand Total
Tainted, sour, or putrid	5,928,436	111,818	100,984	4,009	6,708,930	12,854,177	46,587	12,900,764
Rancid	749,663	12,767	4,751	515	663,745	1,431,441	2,982	1,434,423
Molds or foreign odors	247,416	2,318	1,340	649	586,801	838,524	127	838,651
Unsound canned goods	836,309	58	23,016	—	357,711	1,217,094	—	1,217,094
Unclean or contaminated	3,359,433	34,248	26,204	2,721	3,440,117	6,862,723	3,344	6,866,067
Misc. path. condition	203,778	6,423	117	—	64,274	274,592	673	275,265
Misc. parasitic condition	6,589	125	—	—	171	6,885	—	6,885
Total	11,331,624	167,757	156,412	7,894	11,821,749	23,485,436	53,713	23,539,149

[1]MP Form 407 Data. USDA, FSIS. *Meat and Poultry Inspection Program.* Statistical Summary for 1983. FSIS-14. Issued April 1984.

Table 3-8. Pounds of Materials Other Than Meat Rejected for Use, for Fiscal Year 1983[1]

Materials	Noncompliance with Federal Regulations	Contamination	Odor, Color, or Taste	Sour and/or Moldy	Unsound Canned Goods	Miscellaneous	Total
Spices and seasonings	1,552	11,629	500	665	—	810	14,656
Flours and grains	350	85,358	200	1,260	—	6,330	93,798
Dairy and egg products	14,950	122,338	8,747	13,929	—	2,680	154,097
Fruits and vegetables	—	5,108		55,975	972	2,961	73,763
Soaps, oils, and cleaners	1,080	6,065	560	—	—	—	7,145
Casings	50	700		—	—	—	1,310
Curing agents	200	—	2,232	—	—	—	2,432
Miscellaneous	2,931	41,469	—	1,933	—	18,388	64,721
Total	21,113	272,667	12,239	73,762	972	31,169	411,922

[1]MP Form 407-4 Data. USDA, FSIS. *Meat and Poultry Inspection Program.* Statistical Summary for 1983. FSIS-14. Issued April 1984.

Equipment

All equipment must be in good condition and free from contaminants, that is, rust, dust, dried blood, scrap meat, grease, etc. The following equipment must be sanitized after use on each carcass.

CATTLE

Brisket saw, weasand rods, front shank tie-down chains, and dehorning equipment.

CALVES

Brisket saw or cleaver.

SWINE

Knife or other tools used to partly sever the head and a brisket-splitting device and a saw or cleaver, if carcass splitting occurs before viscera inspection is completed.

Personnel, Clothing, and Personal Equipment

- Establishment employees must wear clean and washable or disposable outer clothing. Street clothing should be properly covered. The wearing of sleeveless garments that would permit the exposure of the underarm should not be permitted. Suitable head coverings applicable to both sexes must be worn to adequately cover the hair. FSIS employees must appear neat and clean and demonstrate good working and sanitary practices expected of food inspection employees.
- All personnel must be required to wash their hands as often as necessary to prevent product contamination, and always after returning from lavatory rooms. The use of tobacco in any form—spitting or smoking—is not permitted in rooms where edible product is handled. Any practice which may be considered unsanitary should be prohibited.
- Personnel equipment (knives, scabbards, steels, tool boxes, gloves, etc.) must be kept in a sanitary condition at all times.

Employee Welfare Facilities; for Example,
Lunchroom, Locker, and Lavatory Facilities

Dressing rooms equipped with lockers or suitable alternate devices, lavatory rooms (showers in meat slaughter plant), urinals, and other than hand-operated wash basins with soap and towels are required. Lavatory rooms and lavatory-room vestibules, which shall be separated from adjoining dressing rooms, should have solid, self-closing doors. These areas must be free of odor, properly maintained, and kept clean at all times.

Coolers, Rails, Hooks, Drains, and Equipment

- Equipment in coolers must be free of corrosion, rust, dust, dry blood, scrap meat, and accumulation of fat. Also, overhead pipes, beams, and light fixtures, as well as ceilings and walls, must be free of contaminants. Walls and ceilings should be free of mold and condensation.
- Rails must be clean, free of flaking paint, excessive oils and grease, rust, etc. Hooks must be clean and in good repair.
- Equipment must be clean, in good repair, and free of debris.
- All drains and gutters should be properly installed with approved traps and vents.

Inedible and Condemned Rooms

MEAT

The area and equipment of inedible or condemned products handling should be adequate for the quantity of product. It must be separate from the edible products department and be properly maintained. An acceptable area for truck sanitizing should be available.

POULTRY

Refuse facilities should be entirely separate from other rooms in the establishment. They must be properly constructed and vented, drained as required, and kept in good repair. Acceptable water connections for cleanup must be provided.

Offal Rooms and Coolers—Facilities and Equipment

Coolers must be free of condensation, and floors, walls, and ceilings must be free of the accumulation of dry blood, fat, scrap meat, mold, dirt and dust, and nuisances. Chutes, tables, pans, etc., must be constructed of rust-resistant materials and maintained in a clean and acceptable manner.

Product Handling

- The product must be handled in a clean and acceptable manner. Cooked, ready-to-eat products and raw meat, emulsions, and finished perishable products must be stored at a room temperature of 50° F or lower (poultry, 40° F) and be accessible to inspection.

- Finished frozen products must be maintained in a frozen state under non-fluctuating temperature, reasonably free of overhead frost, and accessible to inspection.
- Inventory of nonmeat material, where applicable, must be approved by the federal Food and Drug Administration and properly identified. Product must be received only at a specified area until it is reviewed by the inspector. Nonmeat approval stickers must be applied as applicable.
- A suitable compartment or refrigerated area for holding return product pending disposition should be equipped for sealing in order to maintain security.
- Well arranged and adequate facilities for handling inedible and condemned material must be provided. Layout must be such that will allow positive control of condemned materials.
- Unpackaged custom products must be held separately from inspected products (separate rail, racks, etc.).
- A thorough cleanup and the sanitizing of equipment are required after slaughter or processing of custom-exempt product prior to resuming handling of inspected product.
- Viscera separation and product handling must be conducted in a sanitary manner. Community baths are forbidden for all products. Paunches must be emptied without contaminating outer surfaces; however, in some instances, paunches can be sold full for immediate use as animal feed. Accumulation of offal is not permitted. Pork hearts must be opened completely and all blood clots removed. Pork stomachs, chitterlings, and/or ruffle fat must be clean and free of ingesta or any other contaminants.

Carcass Cleanliness and Prevention of Contamination
(Sanitary Dressing Procedures)

HEAD HANDLING

- *Cattle.* Head and corresponding carcass shall be identified by duplicating numbered tags or by other acceptable means and removed in a manner to avoid soilage with rumen and contents. Horns and all pieces of hide shall be removed before washing the outer surfaces of the head.
- *Sheep and Goats.* Heads must be flushed and washed in a cabinet if they are being saved for edible purposes.

CARCASS PREPARATION

- *Swine.* The carcass must be free of hair and scurf after passing through the scalding and dehairing equipment, and the hind feet should be clean of hair and scurf before being gambrelled. If hogs are dipped in rosin, the nostrils and mouth must be closed with rubber bands or other acceptable means prior to dipping. No shaving is permitted after the head is dropped.

- *Calves.* Calves may be showered before stunning to aid in washing the hide, and must be washed clean before any incisions (except stick wounds) are made.

CARCASS SKINNING

- *Cattle.* The area of the skinning bed must be acceptably clean before the carcass is lowered. The head skin should be manipulated so that the neck is protected. The front and hind feet are removed before any other incision is made. The carcass must be removed from the skinning bed in a manner to prevent contamination. Lactating udders must be removed in a way to prevent soilage of carcass with udder contents. The supramammary lymph nodes must be left attached to the carcass until inspection is complete. The dropping of bung should be made part of rumping operations. The rectum must be tied, and the bladder must be tied or removed to prevent contamination.

- *Calves.* The establishment has the responsibility for skinning and handling calf carcasses in a sanitary manner. When skinning operations start, the entire carcass should be skinned.

☞   NOTE: In cases where the establishment handles "hide-on" car-   ☜
casses, the operation must be conducted in a sanitary manner. Hair-to-carcass contact is not permitted. Calf carcasses skinned after chilling must be examined closely to detect injection lesions, foreign bodies, parasites, bruises, or other pathology. All abnormal tissue must be removed.

- *Sheep and Goats.* All operations in removing the pelt shall be done in a manner to prevent contamination of the carcass.

CARCASS EVISCERATION

Carcass evisceration shall be done without contaminating carcasses or organs. The rectum must be tied to prevent soilage. Viscera shall be presented in an orderly manner to facilitate inspection.

CARCASS WASHING

All carcasses must be thoroughly and properly washed.

Slaughter, Scalding, and Picking

POULTRY

- *Procedures.* A continuous intake of water must be sufficient to maintain acceptably clean scalding water and provide a minimum overflow of 1 quart of water per bird per minute. A complete removal of hair and feathers with a final wash of potable water must be made.
- *Product Washing.* The product must be effectively washed inside and out to remove excess blood, loose tissue particles, or any foreign material. Contamination of any tissue, other than the external skin surface, must be removed by trimming. All product must be clean before being chilled.
- Carcasses must be protected against possible contamination from floor cleanup or any fixed objects.

Potable Water

When water is used in areas where edible products are slaughtered, eviscerated, dressed, processed, handled, or stored, it must be potable (drinkable).

To determine potability, the plant must have a local authority certification on an analysis of samples taken from within the facility. The certification must be on an annual basis if the supply is from a municipal source, and if the supply is from a private well, cistern, spring, etc., the certification must be on a semiannual basis. Additional testing and certification are required when there is reason to believe that water is being contaminated—cross connection of potable and nonpotable lines, back siphonage, surface drainage or ineffective drainage, floods, etc.

The usual test to determine potability is the total coliform count, a bacteriological test which, when positive, indicates that the water has been contaminated with human or animal feces.

VACUUM BREAKERS

Vacuum breakers of an acceptable type must be provided on waterlines connected to various equipment, where necessary, to prevent contamination of waterlines by back siphonage.

ICE MUST BE FROM AN ACCEPTABLE SOURCE

Ice must be made from potable water, certified by the appropriate local or state health agency, and handled and stored in a manner to avoid contamination. Block ice should be washed immediately before crushing.

NONPOTABLE WATER

The use of nonpotable water is permitted only where it cannot come in contact with edible products or potable water. Adequate identification of nonpotable lines is required.

Sewage and Waste Disposal Control

Sewage and waste disposal systems must effectively remove sewage and waste materials—manure, paunch contents, trash, garbage, and paper. Such systems must also prevent undue accumulation or development of odors and must not serve as harbors for rodents or insects. Systems must be approved by local or state health authorities for official plants. If there is no local or state agency with jurisdiction, or if the system is hooked directly into municipal lines, documentation must be provided.

ONSITE HANDLING

Conditions for onsite handling must be acceptable. Sanitary problems created by the accumulation of objectionable materials and by the harboring of rodents or other nuisances are unacceptable.

Pest Control

The plant's pest control program must be capable of preventing or eliminating product contamination. Plant management must make reasonable efforts to prevent entry of rodents, insects, or animals into areas where products are handled, processed, or stored—including effective closures to outside openings (doors, screens, windows)—by the use of exterminating procedures, sprays, baits, etc. Only approved insecticides and rodenticides (*List of Chemical Compounds Authorized for Use Under USDA Inspection Programs*) may be used and must be applied in an approved manner.

Facilities and Equipment

A division of FSIS develops standards for facilities and equipment which will assure that products produced in a plant will be sanitary and wholesome. The division is responsible for approving drawings and specifications of meat and poultry facilities and equipment prior to their use in federally inspected plants. During the fiscal year 1983, the Facilities, Equipment, and Sanitation Division of Technical Services reviewed 3,153 blueprints of plants and 2,371 drawings of equipment.

Labels and Standards

The Standards and Labeling Division reviews and approves all labels for federally inspected meat and poultry products.

Label reviewers make sure that the label is truthful and not misleading and that the product contains appropriate ingredients. The division also develops formal product standards which specify the meat or poultry content and ingredients of processed products. During the fiscal year 1983, the Standards and Labeling Division of Technical Services approved only 100,271 labels of the 115,430 total labels submitted.

Additives

Ingredients aimed at improving physical qualities, such as flavor, color, and shelf life of a product, must be approved by USDA before being used in inspected meat and poultry products. USDA sees that additives used:

- Are approved by the federal Food and Drug Administration and are limited to specified amounts.
- Meet a specific, justifiable need in the product.
- Do not promote deception as to product freshness, quality, weight, or size. Paprika, for example, is not permitted in fresh meat, since its red color can make raw meat look leaner and fresher than it is.
- Are truthfully and properly listed on the product label.

Labels Help Consumers Know What They're Paying For

- Labels on all inspected products must be approved by USDA.
- Labels must contain accurate product name; list of ingredients, in order from greatest to smallest amount; name and place of business of packer, manufacturer, or person for whom product is prepared; net weight; mark of federal inspection.
- Labels may include nutritional information. Although this information is not yet required for all foods, it is helpful to consumers who want to know how much of the Recommended Daily Allowance (RDA) they can expect from a serving. More and more consumer products are displaying this information on the label.

To be labeled with a particular name, a federally inspected meat or poultry product must be approved as meeting specific product requirements. Standards are set by USDA so that consumers will get what they expect when they shop. For example:

- "Beef with gravy" must contain at least 50 percent beef (cooked basis), while the minimum meat content for "gravy with beef" is 35 percent beef (cooked basis).

- "Ham salad" must contain at least 35 percent ham (cooked basis).
- "Hot dogs" and "bologna" are limited to 30 percent fat.
- "Chicken soup" must have at least 2 percent chicken meat (cooked, deboned basis).
- "Turkey pot pie" must contain at least 14 percent turkey meat (cooked, deboned basis).

Consumers Participate in Setting Standards

When new or revised standards and labeling rules are being considered, USDA makes this information available to news outlets. Consumers seeing such items should let USDA know their views.

Compliance

The Compliance Division monitors businesses engaged in interstate food marketing and distribution. The division investigates violations of the inspection laws; controls violative products through detentions, civil seizures, and voluntary recalls; and assures that appropriate criminal, administrative, and civil sanctions are carried out.

This compliance program operates between the processor and the consumer as a "second line of defense" against unfit products in marketing channels. The aim is to prevent fraudulent or illegal practices once the product has left the processing plant by having compliance officers check for uninspected meat or poultry, counterfeit inspection stamps, inaccurate labels, and contamination or spoilage of products.

Approximately 14,000 meat and poultry product handlers are periodically reviewed by compliance officers. Adjustable risk categories determine the frequency of scheduled reviews, and additional reviews are conducted randomly. Total reviews for fiscal year 1983 numbered approximately 40,000. Table 3-9 summarizes related enforcement actions.

Pathology and Epidemiology

The Pathology and Epidemiology Division develops the pathology, epidemiology, and serology programs that support meat and poultry inspection. The division provides laboratory and investigative services, studies infectious agents associated with food, and develops serological tests for infectious and toxic agents found in meat and poultry products. The division operates the Meatborne Hazard Control Center, which investigates reports of potential health hazards.

This division works with local, state, and federal public health agencies to

Table 3-9. Enforcement Actions by USDA FSIS Compliance Division,
During Fiscal Year 1983[1]

Action	Number	Pounds
Detention of suspect product	748	7,047,473
Monitoring of product recalls	14	794,350
Court seizures initiated by compliance	4	107,944
Irregularities reported to inspection supervisors	1,130	
Cases prepared by compliance	674	
Cases referred to inspector general	13	
Cases requiring consultation with general counsel	56	
Letters of warning issued	1,012	

[1]USDA, FSIS. *Meat and Poultry Inspection, 1983.* Report of the Secretary of Agriculture to the U.S. Congress, March 1, 1984.

control food poisoning outbreaks by speeding identification of products responsible for human health hazards and has found the major cause to be improper handling of products at institution, restaurant, or home level during preparation for serving. Some samples of improper handling are:

- Inadequate cooking.
- Storage at warm, "median" temperatures which allows bacteria and other organisms to multiply rapidly (between 40° and 140° F).
- Failure to keep raw and cooked products separate during preparation.
- Contamination by human carriers of bacteria.
- Poor sanitation practices.

The division has found that perfringens, salmonella, staphylococcus, and trichina poisonings result from improper handling. (See pages 557-560.)

Residue Monitoring and Evaluation

The Residue Evaluation and Planning Division develops and coordinates the FSIS role in controlling unsafe drug and chemical residues that may occur in meat and poultry. The division develops residue monitoring and surveillance programs for both the domestic and the import inspection programs. It samples animals, flocks, and herds suspected of illegal residues and alerts drug and pesticide enforcement agencies when violations are found. It also has primary responsibility for the Residue Avoidance Program, a cooperative educational effort involving producer organizations and the USDA Extension Service.

Table 3-10 summarizes laboratory analyses of meat and poultry samples by FSIS during fiscal year 1983. Of the samples, approximately 113,400 were taken from processed products such as hams, sausages, cured meats, and similar items.

Table 3-10. Analyses of Meat and Poultry Samples by FSIS, During Fiscal Year 1983[1]

Category of Samples and Analyses	Number of Samples
Food chemistry	96,817
Food microbiology and species	18,543
Chemical residues	35,806
Antibiotic residues	21,108[2]
Pathology	3,598
Pathology (nonresidue)	13,060
Food additives and nonfoods	13,352
Serology	9,945
Total	212,229

[1]USDA, FSIS. *Meat and Poultry Inspection, 1983.* Report of the Secretary of Agriculture to the U.S. Congress, March 1, 1984.
[2]Includes 9,433 STOP (Swab tests on premises) tests.

Federal-State Relations

Inspection is marked by continuing and expanding federal and state cooperation. The federal government shares the cost of state inspection by picking up 50 percent of the costs and offers technical, laboratory, and training aid. In order to continue operating state inspection programs for intrastate plants, and in order to continue receiving federal funding assistance, states must maintain inspection requirements at least equal to those of the federal program.

The federal-state relations staff provides this technical support and direction to state governments to assure that state inspection programs enforce requirements at least equal to those of the federal inspection laws. Currently, state-inspected plants may sell their products only within the state. However, a change in the law to allow interstate shipment of state-inspected meat is pending. It is logical, since the state inspection program must in fact be equal to the federal program. FSIS staff also gives technical assistance to plants operating under the Talmadge-Aiken Act. These plants have federal inspection conducted by state inspectors. During 1983, 1,799 intrastate plants were reviewed by federal field supervisors in accordance with requirements of the federal inspection laws.

Foreign Programs

Imported meat and poultry products must meet the same standards as those produced in the United States. One measure of a country's inspection effectiveness is its body of laws, regulations, procedures, administration, and operations which must first meet U.S. standards; USDA then approves the overall program. Individual plants within that country then apply to their own government for certification to export to the United States. Each certified plant is subject to continuous inspection by inspectors of the foreign country's government. This evaluation is supplemented by the work of 20 FSIS veterinary medical officers with considerable experience in the domestic meat-inspection system. They conduct periodic onsite reviews of certified foreign plants to assure that the same standards of inspection are enforced as those enforced in federally inspected U.S. plants. Ten of these officers are stationed in countries that are major exporters to the United States, including two in Australia and one each in Canada, Costa Rica, Denmark, Mexico, the Netherlands, New Zealand, Uruguay, and West Germany. The other reviewers are stationed in Washington, D.C., but travel to other countries when necessary.

The frequency of onsite review is determined by plant size, nature and complexity of operations, and anticipated volume of exports to the United States. Plants that export large volumes, or those that are of special concern, are reviewed at least four times annually; other certified plants are reviewed at least once a year.

Imported meat and poultry products are reinspected as they arrive in this country. They must bear prominent marking as to their country of origin. A meat-inspection certificate issued by the responsible official of the exporting country must accompany each shipment of meat offered for entry into the United States. The certificate identifies the product by origin, destination, shipping marks, and amounts. It certifies that the meat comes from veterinary ante-mortem and post-mortem inspection; that it is wholesome, not adulterated or misbranded; and that it is otherwise in compliance with U.S. requirements.

To assure that the certifications made by foreign officials are correct, USDA inspectors, at the ports of entry and at destination points of inspection, inspect each lot of imported meat and poultry products.

A description of each lot arriving at U.S. ports is entered into the Automated Import Information System (AIIS) computer. This system centralizes inspection and shipping information from all ports and allows FSIS to set the inspection requirements based on the compliance history of each establishment. Information stored in the system includes:

- Amount of product offered from each establishment and the amount refused entry.
- Results of samples tested for pesticides, hormones, heavy metals, antibiotics, and other drug analyses, which often show up first in fat and organs. A regulation became effective March 14, 1983, which requires countries exporting meat to the United States to include random tests on internal organs and fat of slaughtered livestock in their residue-testing programs.
- Results of samples tested for excess water, fat, percent of meat, fillers (non-fat dry milk, soy, and other flours), net weight, and species verification.
- Results of inspections for contamination, processing defects (bone, skin, and glands), off-condition, and pathological defects.

- Results of samples analyzed for maximum internal temperature or sub-jected to incubation (to assure product stability and prevent the introduction of foreign animal diseases).

To assure that representative samples are selected, statistical sampling plans are applied to each lot of product to be inspected. The sampling plans and criteria for acceptance or rejection of imports are the same as those used for U.S. federally inspected meat.

Although the sampling plans are generated by the AIIS to guide the inspection of imported lots, an inspector may hold product and require additional samples or inspection procedures where it is considered necessary. As a further check, imported meat which is subsequently used in domestic processed products receives additional examination in U.S. plants.

Analytical and Scientific Support

All the activities just discussed need a solid scientific base. The FSIS science program furnishes the analytical support and scientific guidance to the meat and poultry inspection program. Science support services are designed to assure that meat and poultry products are safe from disease, microorganisms that cause food poisoning, harmful chemicals, and toxins. Laboratory analysis enables the FSIS to detect unsanitary preparation and economic adulteration (the substitution of cheaper or less desirable ingredients for those required).

The science group in the FSIS cooperates with other federal agencies (notably the Food and Drug Administration, the Environmental Protection Agency, and the Center for Disease Control) and with state and local health authorities in carrying out its responsibilities. It develops and maintains close ties with national and international scientific communities in order to keep abreast of scientific and technological advances and to open new avenues for the exchange of scientific information.

Its chemistry division develops and improves practical, analytical procedures for detecting adulterants and chemical residues in meat and poultry products. This division performs highly complex chemical analyses, coordinates an accredited laboratory program, and conducts onsite technical reviews of chemistry field-service laboratories to assure the quality and integrity of analytical results. In addition, the division participates with the FDA in evaluating new animal drug applications.

The microbiology division provides analytical services to federal, state, and local agencies and advises other FSIS divisions of the significance of laboratory results. The division develops economical and efficient analytical screening methods for use in laboratories, in plants, and on the farm. This division also carries out special investigations on the safety and quality of products and processes.

EXEMPTIONS FROM MEAT INSPECTION

The requirements of the Wholesome Meat Act and the regulations for inspection do not apply to the following.

1. The slaughtering by any individuals of livestock of their own raising, and
   the preparation by them, and transportation in commerce of the carcasses, parts thereof, meat and meat food products of such livestock exclusively for use by them and members of their households and their
   nonpaying guests and employees.

2. The custom slaughter by any person of cattle, sheep, swine, or goats delivered by the owners thereof for such slaughter, and the preparation by
   such slaughterer, and transportation in commerce of the carcasses, parts
   thereof, meat and meat food products of such livestock, exclusively for
   use, in the households of such owners, by them and members of their
   households and their nonpaying guests and employees, nor to the custom
   preparation by any person of carcasses, parts thereof, meat or meat food
   products derived from the slaughter by any individuals of cattle, sheep,
   swine, or goats of their own raising or from game animals, delivered by
   the owners thereof for such custom preparation, and transportation in
   commerce of such custom-prepared articles, exclusively for use in the
   households of such owners, by them and members of their households
   and their nonpaying guests and employees. Such meat processors are
   only subject to occasional inspections, perhaps biweekly or monthly, depending on the availability of inspectors.

   However, if custom operations are conducted in an official establishment, all of the provisions of the act shall apply to such establishment,
   including the following.

   a. If the custom operator prepares or handles any products for sale, they
      are kept separate and apart from the custom prepared products at all
      times while the latter are in the operator's custody.

   b. The custom-prepared products are plainly marked "Not for Sale" im-
      mediately after being prepared and are kept so identified until deliv-
      ered to the owner.

   c. If exempted custom slaughtering or other preparation of products is
      conducted in an official establishment, all facilities and equipment in
      the official establishment used for such custom operations shall be
      thoroughly cleaned and sanitized before they are used for preparing
      any products for sale.

   d. The exempted custom-prepared products shall be prepared and han-
      dled in accordance with the provisions of the act and shall not be
      adulterated.

3. Operations of types traditionally and usually conducted at retail stores
   and retail-type establishments in any state or organized territory, for sale

in normal retail quantities, or service of such articles to consumers at such establishments. Operations of types traditionally and usually conducted at retail stores and restaurants are the following.

a. Cutting up, slicing, and trimming carcasses, halves, quarters, or wholesale cuts into retail cuts such as steaks, chops, and roasts and freezing such cuts.
b. Grinding and freezing products made from meat.
c. Curing, cooking, smoking, or other preparation of products, except slaughtering, rendering, or refining of livestock fat or the retort processing of canned products.
d. Breaking bulk shipments of products.
e. Wrapping or rewrapping products.
f. Any quantity or product purchased by a consumer from a particular retail supplier shall be deemed to be a normal retail quantity if the quantity so purchased does not in the aggregate exceed one-half carcass. The following amounts of product will be accepted as representing one-half carcass of the species identified.

	One-Half Carcass (lb.)
Cattle	300
Calves	37.5
Sheep	27.5
Swine	100
Goats	25

g. A retail store is any place of business where the sales of a product are made to consumers only; at least 75 percent in terms of dollar value of total sales of a product represents sales to household consumers and the total dollar value of sales of product to consumers other than household consumers does not exceed $30,200 for meat and $23,100 for poultry per year. These figures are adjusted in accordance with Consumer Price Index changes when the amount of adjustment equals or exceeds $500. USDA is currently drafting a proposal that would remove all dollar limitations on such sales.

NUMBER OF MEAT AND POULTRY PLANTS[2]

Federally Inspected Plants

Table 3-11 presents the number of meat and poultry slaughtering and/or

---

[2]USDA, FSIS. *Meat and Poultry Inspection, 1983.* Report of the Secretary of Agriculture to the U.S. Congress, March 1, 1984.

processing plants that operated under inspection in each state or U.S. territory as of September 30, 1983. Only federally inspected plants may sell their products in interstate or foreign commerce. Talmadge-Aiken plants are federally inspected, but staffed by state employees.

Table 3-11. Meat and Poultry Plants Operating Under Federal Inspection in Each State or Territory as of September 30, 1983[1]

State or Territory	Meat Plants	Poultry Plants	Meat/Poultry Plants	Total
Alabama	17	25	17	59
American Samoa	1	—	—	1
Arizona	20	—	12	32
Arkansas	69	37	48	154
California	376	64	338	778
Colorado	89	6	53	148
Connecticut	69	8	46	123
Delaware	5	5	2	12
District of Columbia	11	4	6	21
Florida	51	8	31	90
Georgia	31	43	35	109
Guam	1	—	3	4
Hawaii	1	—	1	2
Idaho	45	—	37	82
Illinois	206	14	100	320
Indiana	47	16	25	88
Iowa	58	6	33	97
Kansas	38	1	24	63
Kentucky	116	6	62	184
Louisiana	19	6	15	40
Maine	14	2	19	35
Mariana Islands	1	—	3	4
Maryland	27	13	16	56
Massachusetts	108	19	75	202
Michigan	298	6	98	402
Minnesota	53	21	110	184
Mississippi	5	18	12	35
Missouri	174	27	105	306
Montana	26	—	43	69
Nebraska	87	8	57	152
Nevada	6	3	17	26
New Hampshire	11	3	17	31
New Jersey	145	13	107	265
New Mexico	10	—	15	25
New York	380	35	277	692
North Carolina	36	22	23	81
North Dakota	25	—	14	39
Ohio	84	11	54	149
Oklahoma	28	3	18	49
Oregon	82	4	36	122
Pennsylvania	458	53	199	710
Puerto Rico	69	2	31	102

(Continued)

Table 3-11 (Continued)

State or Territory	Meat Plants	Poultry Plants	Meat/Poultry Plants	Total
Rhode Island	34	7	16	57
South Carolina	22	10	12	44
South Dakota	15	3	6	24
Tennessee	120	15	74	209
Texas	142	16	135	293
Utah	12	5	17	34
Vermont	4	—	7	11
Virginia	28	16	26	70
Virgin Islands	2	—	3	5
Washington	82	10	63	155
West Virginia	7	2	6	15
Wisconsin	50	10	41	101
Wyoming	—	—	1	1
Subtotal	3,915	606	2,641	7,162
Talmadge-Aiken plants	167	9	111	287
Total	4,082	615	2,752	7,449
Plants Summarized by Type of Operation				
Slaughtering	331	187	2	520
Processing	2,527	279	2,313	5,119
Slaughtering and processing	1,057	140	326	1,523
Subtotal	3,915	606	2,641	7,162
Talmadge-Aiken plants	167	9	111	287
Total	4,082	615	2,752	7,449

[1]USDA, FSIS. *Meat and Poultry Inspection, 1983*. Report of the Secretary of Agriculture to the U.S. Congress, March 1, 1984.

State Program Data

Table 3-12 summarizes the number of states at the end of fiscal year 1983 with intrastate inspection programs for meat (27) and poultry (23). "M" after the name of the state indicates that the state conducted a meat inspection program; "M & P" indicates that the state conducted meat and poultry inspection programs.

In order to continue operating state inspection programs for intrastate plants, and in order to continue receiving federal funding assistance, states must maintain inspection requirements at least equal to those of the federal program. During 1983, 1,799 intrastate plants were reviewed by field supervisors in accordance with requirements of the federal inspection laws.

Table 3-13 presents the number of meat and poultry plants that were inspected under Talmadge-Aiken agreements as of September 30, 1983. The department is responsible for inspection in such plants; however, federal inspection is carried out by state employees.

Table 3-12. State Inspection Program[1]

State	Plants					
	Official			Exempt		
	Meat	Poultry	Combination	Meat	Poultry	Total
Alabama M&P	106	8	0	50	0	164
Alaska M&P	7	0	7	3	0	17
Arizona M&P	62	7	1	45	0	115
Delaware M&P	9	0	0	3	0	12
Florida M&P	274	10	1	64	0	349
Georgia M[2]	161	0	0	56	—	217
Hawaii M&P	68	5	0	1	0	74
Illinois M&P	453	54	0	33	13	553
Indiana M&P	127	16	48	46	11	248
Iowa M&P	198	8	0	191	21	418
Kansas M&P	187	8	8	39	4	246
Louisiana M&P	113	6	40	77	2	238
Maryland M&P	54	8	0	24	7	93
Mississippi M&P	92	3	0	22	3	120
New Mexico M&P	35	1	0	32	1	69
North Carolina M&P	223	18	0	103	0	344
Ohio M&P	378	45	0	140	22	585
Oklahoma M&P	100	9	5	127	0	241
South Carolina M&P	74	11	28	0	0	113
South Dakota M[2]	50	0	0	93	—	143
Texas M&P	504	9	0	162	1	676
Utah M[2]	37	0	0	77	—	114
Vermont M&P	22	0	0	25	7	54
Virginia M&P	26	3	0	174	2	205
West Virginia M[2]	50	0	0	56	—	106
Wisconsin M&P	218	12	105	164	6	505
Wyoming M&P[3]	32	0	0	41	0	73
Total	3,660	241	243	1,848	100	6,092
California[4]	—	—	—	434	16	450
Minnesota[4]	—	—	—	425	11	436

[1]USDA, FSIS. *Meat and Poultry Inspection, 1983*. Report of the Secretary of Agriculture to the U.S. Congress, March 1, 1984.

[2]Poultry program under federal jurisdiction.

[3]Does not accept federal funds for inspection program.

[4]Official plants are under federal jurisdiction. Custom-exempt facilities are reviewed under state jurisdiction.

Table 3-13. Meat and Poultry Plants Inspected Under Talmadge-Aiken Agreements as of September 30, 1983[1]

State	Meat Plants	Poultry Plants	Combination Plants	Total
Alabama	6	—	6	12
Alaska	—	—	—	—
Delaware	7	—	1	8
Florida	—	—	—	—
Georgia	36	1	24	61
Hawaii	2	—	—	2
Illinois	22	2	6	30
Indiana	1	—	3	4
Maryland	8	—	13	21
Mississippi	9	—	5	14
New Mexico	1	—	4	5
North Carolina	37	2	8	47
Ohio	—	—	—	—
Oklahoma	5	—	14	19
Texas	6	—	5	11
Utah[2]	2	—	6	8
Vermont	—	—	—	—
Virginia	25	4	16	45
Wyoming	—	—	—	—
Total	167	9	111	287

[1]USDA, FSIS. *Meat and Poultry Inspection, 1983*. Report of the Secretary of Agriculture to the U.S. Congress, March 1, 1984.

[2]Utah relinquished its Talmadge-Aiken agreement January 10, 1983. These plants are now under full federal inspection.

In addition to its mandatory inspection activities, the FSIS provides voluntary, reimbursable inspection services in establishments. They may include plants that request bison and reindeer inspection, identification service, approved warehouse inspection, animal food inspection, and food inspection.

VIOLATION OF WHOLESOME MEAT ACT

If any of the following conditions exist, the plant must be designated as endangering public health (EPH), and corrective action must be taken immediately.

- Use of nonpotable water in edible products departments.
- Improper sanitation that results in bacterial growth and development in or on product, foreign matter entering product, or failure to control vermin and insects.
- Presence of carcasses or parts showing sufficient evidence to identify a systemic diseased condition or containing evidence of bearing a disease transmissible to humans.
- Use of unsound meat/poultry in processing meat/poultry food products.

- Presence of harmful chemicals and preservatives in excess of permitted tolerances.
- Failure to properly treat or destroy trichinae.

Inspectors should take immediate action to correct deficiencies in all phases of the operation within their purview.

When it has been determined that a plant is endangering public health (EPH), it must be surveyed for corrective action after five working days. When a plant is deficient in one or more of the basic requirements but is not in an EPH category, it must be resurveyed no later than before the end of the succeeding quarter.

Each violation involving intent to defraud or any distribution or attempted distribution of an article that is adulterated shall make the person or persons representing a firm or corporation subject to imprisonment for not more than three years or to a fine of not more than $10,000.

THE MEAT INSPECTOR

FSIS personnel are divided into several classifications. There are professional inspectors, administrators, laboratory scientists, etc., trained in veterinary science, chemistry, microbiology, meat science, muscle biology, and related disciplines who have passed required civil service examinations which qualifies them for the GS 11-12 rank and nonprofessional or lay inspectors who are required to pass a civil service examination and are designated as GS 5-9.

In an effort to streamline the operation and costs of government regulatory agencies, the FSIS, the USDA division responsible for inspection, and the Agricultural Marketing Service (AMS), the USDA division responsible for grading, have signed an agreement to cross-utilize employees between the two agencies. This means that persons can be properly trained and licensed in both inspecting and grading, thus eliminating double staffing in some plants.

THE MARKS OF FEDERAL AND STATE INSPECTION

Each establishment under federal or state inspection is granted an official number which appears on the inspection stamp and identifies the product wherever it is found (Figures 3-1 and 3-2).

Fig. 3-1. Examples of two state inspection stamps.

This is the federal stamp put on meat carcasses. It is only stamped on the major cuts of the carcass, so it may not appear on the roast or steak you buy.

You will find this mark on every pre-packaged processed meat product—soups to spreads—that has been federally inspected.

This is the mark used on federally inspected fresh or frozen poultry or processed poultry products.

Fig. 3-2. Examples of federal inspection stamps.

HUMANE SLAUGHTER

The federal Humane Slaughter Act went into effect on July 1, 1960, and was updated effective October 11, 1979. Under the act, all official establishments slaughtering livestock under provision of the federal Meat Inspection Act must employ humane methods of slaughtering and handling livestock. Failure may result in temporary suspension of inspection, meaning operations must cease until the violation is corrected. Many states have enacted laws similar to the federal law. Ritual slaughter methods, such as Kosher killing, are exempt from the act.

The federal act recognizes three methods of immobilization: mechanical, chemical, and electrical. To comply with the law, use of any one of the three methods must produce complete unconsciousness with a minimum of excitement and discomfort.

The mechanical stunners consist of the penetrating type and the concussion type. In the former, the captive bolt enters the skull, while in the concussion type, the force is delivered through a mushroom head on the end of the bolt. Blank cartridges with different powder loads for different-sized animals are triggered off by contact to propel the bolt into the head. The least costly is the captive bolt pistol which resembles a pistol. The stunning can be administered to the forehead or in back of the poll.

The chemical method employs carbon dioxide gas, which, because it is heavier than air, can be held in a pit with a minimum loss of gas. This method is used for hogs and calves which ride in individual compartments down an incline into the chamber which holds a concentration of 65 to 75 percent carbon dioxide. The time of exposure to render unconsciousness depends upon the size of the animal and the production rate. As the animals emerge, they may be stuck in the prone position or shackled and stuck on the rail.

In one electrical method in a large plant, hogs move along by conveyor to a squeeze box which halts them until the electric probe can be applied to the head for from one to four seconds, depending upon the weight of the hog (Figure 4-1). A small, hand-held electric stunner can be used in smaller hog plants and for sheep (Figures 4-2 and 6-2). A five- to seven-second interval between stunning and sticking is advisable, since a longer time interval may result in blood spattering; that is, capillary rupture in the muscle causes unsightly blood spots in the meat. The prone bleed is favored to reduce the time lapse (Figure 4-6).

The advantages of any of the above methods, aside from the humane factor, are the elimination of excitement, fewer internal bruises, safer and better working conditions, more economical operation, and an improved product.

For on-the-farm slaughter, some persons still use the rifle. The bullet is aimed at a point in the center of the forehead about 1 inch higher than the eye or right behind the ear, if the shooting occurs from the side. Extreme caution must be exercised when live firearms are used. Furthermore, heads become contaminated and unfit for human consumption due to the lead sponge iron or any brittle slug entering the head.

THE KOSHER STAMP

The Humane Slaughter Act of 1960 does not apply to *ritual* slaughter methods such as Kosher killing where the live animal is suspended (unstunned) and bled by an incision across the throat (cut throat) made by a specially trained rabbi or shohet. The knife has a razor sharp, 14-inch blade. The throat of the animal must be washed free of any grit or foreign material so it will not nick the knife. The shohet makes no inspection of the carcass prior to placing the Kosher stamp in script or block letters on the carcass. The stamp carries no implication of the health or grade of the animal; only that it is proper, according to the law, and clean.

The law referred to is found in Leviticus 17:14, which states, "You shall not eat the blood of any creature, for the life of every creature is its blood; whoever eats it shall be cut off."

Fig. 3-3. Kosher stamp.

The explanation given by those of the Jewish faith as to why they consume forequarter meat primarily is because it contains less blood than the hindquarter and is more easily veined.

ANIMAL DISEASES AND OTHER LOSS FACTORS

Anyone doing a limited amount of slaughtering, whether it be for home use or for subsequent sale, should be able to recognize unhealthy or unthrifty animals and should know something about the effect of the more common ailments on the wholesomeness, quality, and value of carcasses. Whenever there is any doubt concerning the health of an animal, a veterinarian should be consulted. The diseases, parasites, and other loss factors covered here are those that directly affect the amount, quality, and wholesomeness of meat. Many other diseases and conditions affect animal health, growth, and thriftiness but are too numerous to be discussed here.

Pregnancy

Animals should not be slaughtered in the advanced stages of pregnancy when it is obvious that they are in the preparatory stage of parturition. This preparatory stage is characterized by dilation of the cervix and rhythmic contractions of the longitudinal and circular muscles of the uterus. During parturition, increased levels of the female hormones oxytocin and estrogen work to increase the irritability of the uterine musculature allowing it to contract. The physiological condition of the female is thus disturbed, which may affect her whole musculature. Furthermore, her temperature most likely will have risen above normal. Following parturition and passage of the placenta, a female can be slaughtered, if desired, when her body temperature returns to normal (page 40).

Accidental Death or Injury

A *healthy* animal that is killed or injured through accident, suffocates from bloat, or dies from a heart puncture caused by a nail or wire is fit for food, providing some knowledgeable person is there to cut the throat and bleed the animal. This should be done immediately in the case of death, and within minutes, or one or two hours, in the case of an injury, depending on the nature and severity of the injury. By *no* means is a dead or diseased animal suitable for human food. Once the formerly healthy animal is bled, the slaughtering procedure should be completed and the carcass chilled as soon as possible, preferably within the hour.

Swine Porcine Stress Syndrome (PSS)

This condition was rather widespread in the mid-1960s and early 1970s, but its occurrence has diminished in recent years as pork producers have selected against it. Affected pigs are unable to withstand the stress of management procedures that involve handling and crowding, transportation, or sudden environmental change. When these pigs are subjected to such stressful situations, they show a reaction that sometimes results in death.

The following sequence of events is typical, if a pig suffering from PSS is subjected to stress. The pig may show signs of trembling or muscle tremors and be difficult to move. These initial symptoms are followed by irregular blotching of the skin of white pigs, labored breathing, and increased body temperature. The terminal stage of the syndrome is total collapse and a shock-like death of the animal. The condition is genetically controlled, so producers do well to select against the trait.

The PSS condition is usually associated with low-quality or pale, soft, exudative (PSE) pork. Dark, firm, dry (DFD) pork may also result from the PSS condition. The occurrence of either muscle condition depends on the rate of muscle glycogen (starch) breakdown. Muscle has only about 0.5 to 1.0 percent glycogen. When this glycogen breaks down in muscle, it forms acid. The acid condition in muscle alters the muscle protein structure and color in such a way that a very acid condition (low pH) produces pale, soft, watery muscle, while an alkaline condition (high pH) produces dark, firm, dry muscle. When animals (swine or cattle) are stressed, some, not all, will begin to break down glycogen and form acid in the muscle. If they are slaughtered when the muscle is very acid, the meat will be pale, soft, and exudative (watery). If, on the other hand, they are not slaughtered during the time the muscle is acid, but sometime later, after the lactic acid has been removed from the muscle by the living animal's metabolism but before more glycogen is restored in muscle, the muscle will be dark, firm, and dry. For further discussion, see Chapter 21, pages 711, 714-715.

Proper Livestock Shipping—Bruise and Death Loss Prevention

The growers of our nation's meat supply face innumerable problems in raising the animals for market. The fight against birth losses, disease, predatory animals, the elements—heat, cold, and drought—and nutrient and vitamin deficiencies is a continuous one. After producers have made these investments in feed, buildings, medications, equipment, and months of care, a shocking number of livestock never make it to market alive, or if they do, they are severely discounted for bruises and/or broken limbs, etc. A bruise results from the hemorrhaging of a blood vessel under the hide. Bruised meat cannot be used for human food. The livestock industry loses approximately $50 million annually from bruises on cattle, hogs, and sheep.

The National Livestock Loss Prevention Board, organized in 1934, was reorganized in 1952 under the name "Livestock Conservation Institute"[3] for the purpose of studying the causes of livestock losses on the farm and on the way to market. Most bruises occur during loading and unloading. In cattle, the major sites of bruising are in the valuable sirloin/hip area (31 percent of all cattle bruises), ribs (13 percent), and shoulders (36 percent). In swine, 66 percent of all bruises occur in the ham, followed by shoulders (10 percent) and loin (7 percent). In sheep, the major site is the legs (27 percent), followed by the loin (17 percent). Horns cause a great deal of the bruises in cattle. Also, gates being swung into the sides of the animals, the animals being forced to turn short corners (out of or into the side of a semi), and their being forced through too narrow an opening result in bruises. Ham damage in hogs results from human kicking and slick floors that cause a hog to split or spread its legs violently out to the sides. A canvas slapper should be used for moving hogs. The most common cause of bruises in sheep is their being grabbed by the wool or by the hind leg. When possible, a Judas goat or pet sheep should be used to lead them.

Loading facilities are very important. The recommended slope on loading chutes is 20°, although a 25° slope may be a reasonable compromise for maximum. Ramps should have solid sides to block out distractions from outside the chute that may spook the animals. A flat landing at the top of the chute is helpful for ease in getting animals to walk onto the truck.

Death loss in shipping is a special concern in hogs, although extra care in cattle shipments could result in reduced shrinkage. Shrink has run 5 and 8 percent in fed cattle hauled 50 and 750 miles respectively. Thus, most shrink comes in the early stages of a trip. All species shrink more in hot weather, and range livestock not accustomed to people shrink more than other livestock. Hog transit deaths are much higher in the summer months, running from 48 to 60 per 100,000 hogs versus 33 to 38 per 100,000 during the other seasons. Attention should be paid to the Livestock Weather Safety Index, which is widely disseminated over the media, when livestock shipments are being planned. Hot weather and high humidity are deadly, especially for hogs.

Sand should be used in the bottom of trucks and should be wet down in hot weather and covered with straw in cold weather.

The Livestock Conservation Institute has provided space recommendations for shipping livestock.

Hogs need more room in a truck during hot weather. Hogs weighing 200 pounds need a minimum of 3.5 square feet per animal. A 230-pound hog needs up to 4.4 square feet when the humidity is high and the temperature is over 75° F.

---

[3]239 Livestock Exchange Building, South St. Paul, MN 55075.

Table 3-14 provides a rule-of-thumb guide per running foot of truck floor (based on a 92-inch inside truck width) for varying weights of hogs when temperatures are below 75° F. When the Livestock Weather Safety Index is in the Alert range, 10 percent fewer hogs should be loaded, and when it is in the Danger zone, 20 percent fewer hogs should be loaded.

Table 3-14. Minimum Hog Space Requirements

Average Weight	Number Hogs per Running Foot of Truck Floor (92-in. Truck Width)
(lb.)	
100	3.3
150	2.6
200	2.2
250	1.8
300	1.6
350	1.4
400	1.2

Examples (250-lb. hogs):

44-ft. trailer—44 × 1.8 = 79 hogs when temperature is mild. If the Livestock Weather Safety Index is in the Alert range, load 71. If the Weather Safety Index is in the Danger zone, ship only at night and load 63 hogs.

44-ft. double-deck trailer—88 × 1.8 = 158 during mild weather, 142 head if the Weather Safety Index is in the Alert range and 126 head if in the Danger zone.

44-ft. possum-belly (two 10-ft. front decks, three 25-ft. middle decks, and two 9-ft. rear decks, 113 ft. of floor space)—113 × 1.8 = 203 head in mild weather, 183 head when the Index is Alert and 163 head if the index says Danger.

Overloading of trucks is a major cause of bruises in loads of horned cattle or cattle which have had their horns tipped. Two or three too many horned 1,000-pound steers on a semitrailer can double the bruising. Tables 3-15 and 3-16 provide rule-of-thumb guides for the number of cattle or calves which can be loaded per running foot of truck floor in a truck with a standard 92-inch width. The figures apply to cows, range animals, and feedlot animals with horns or tipped horns. For feedlot steers or heifers without horns or with horn growths due to improper dehorning (scurs), the load can be increased by 5 percent.

Table 3-17 provides a rule-of-thumb guide for loading sheep.

Swine Abscesses

In the early 1960s, the swine industry, including members of the Livestock Conservation Institute (LCI), became concerned about the incidence of losses of pork carcasses and carcass parts at packing plants due to swine abscesses. These are largely jowl abscesses, although some packers have found shoulder abscesses extending into the muscle as far as the needle was inserted when an earlier vaccination was given. These losses have been increasing steadily since the late 1940s. A national survey conducted by LCI indicated an annual product loss at the processing level of $12 million. The LCI Swine Abscess Committee coordinated the findings of four state research groups and one USDA group and prepared and distributed a leaflet containing suggested control measures. The

Table 3-15. Truck Space Requirements for Cattle (Cows, range animals, or feedlot animals with horns or tipped horns; for feedlot steers and heifers without horns, increase by 5 percent)

Average Weight	Number Cattle per Running Foot of Truck Floor (92-in. Truck Width)
(lb.)	
600	.9
800	.7
1,000	.6
1,200	.5
1,400	.4

Examples (1,000-lb. cattle):

44-ft. single-deck trailer—44 × 0.6 = 26 head horned, 27 head polled.

44-ft. possum-belly (four compartments, 10-ft. front compartment; two middle double decks, 25 ft. each; 9-ft. rear compartment, total of 69 ft. of floor space)—69 × 0.6 = 41 head of horned cattle and 43 head of polled cattle.

Table 3-16. Truck Space Requirements for Calves (Applies to all animals in 200- to 450-pound weight range)

Average Weight	Number Calves per Running Foot of Truck Floor (92-in. Truck Width)
(lb.)	
200	2.2
250	1.8
300	1.6
350	1.4
400	1.2
450	1.1

Examples (450-lb. calves):

44-ft. single-deck trailer—44 × 1.1 = 48 head.
44-ft. double-deck trailer—88 × 1.1 = 97 head.

Table 3-17. Truck Space Requirements for Sheep
(Use for slaughter sheep; load 5 percent
fewer if sheep have heavy or wet fleeces)

Average Weight	Number Sheep per Running Foot of Truck Floor (92-in. Truck Width)
(lb.)	
60	3.6
80	3.0
100	2.7
120	2.4

Example (120-lb. sheep):

44-ft. triple-deck trailer—44 × 3 × 2.4 = 317 shorn sheep,
302 wooly sheep.

bacterial group E *Streptococcus* (GES) was found to be the principal cause of
jowl abscesses. More recently, *Streptococcus equisimilis* (group C), one of the
"strep" commonly associated with domestic animals, has been the major organism isolated from swine abscesses. These abscesses were 1 to 20 mm in diameter, mostly in intermuscular tissue, although a few were in the muscle itself and
had thick, fibrous walls, with yellowish, cheesy, odorless pus.

Abscesses induced in swine by GES occur primarily in the lymph nodes of
the head and neck following infection via the oral or nasal routes. Involvement
of lymph nodes in other areas of the body is unusual but may occur if GES enters the bloodstream. After ingestion, the organisms are quickly carried to the
regional lymph nodes. Organisms have been found in mandibular and cervical
lymph nodes as early as two hours after oral exposure. Microscopic abscesses
begin forming in lymph nodes as early as 48 hours after exposure. These abscesses can be seen with the naked eye on cut surfaces of dissected lymph nodes
as early as seven days after exposure and can usually be felt or seen externally
within two weeks. Susceptibility of swine to abscesses is influenced by age. The
disease occurs primarily in post-weaning swine. Swine can become infected with
GES by nasal contact with infected swine, via drinking water, and through
feces. The presence of abscesses which drain to the outside skin surface are also
a source of infection. GES persists in pasture and swine pen soil for several
weeks, especially at cold temperatures. Swine can continue to transmit the disease as long as 2½ years after initial infection, but usually the carrier state lasts
10 months or less. Swine become immune to GES following recovery from the
disease, and their serum will provide temporary passive immunity when injected
into susceptible swine. Infected sows transmit passive immunity to their pigs via
colostrum.

A commercially produced vaccine (Jowl-Vac), prepared from a living non-
abscess-inducing strain of GES, is effective for protection of pigs against ab-

scesses, according to experimental evidence. Manufacture has been discontinued, apparently due to lack of demand.

To cut down on costly abscesses resulting from various injections, the injections should be made just behind the pig's jawbone in the higher portion of the jowl (Figure 3-4).

Injections at the site indicated seem to cause less tissue reactions, and if the needle is contaminated, any tissue reaction at this particular area causes less tissue trim than if injected higher or further back.

Sanitary procedures must be observed while the animals are being vaccinated. Animals contaminated with fecal material should be cleaned before being injected.

To prevent contamination of the bacterin in the vaccine bottle, the needle that was used for injecting a pig should not be inserted in the bottle.

☞   CAUTION: Directions for withdrawal periods, as given on the label,   ☜
        should be carefully observed.

After reaching a peak in the mid-1960s, the incidence of condemnations due to swine abscesses declined steadily; largely because of the application of knowledge gained from research.

Fig. 3-4. Injections at the site indicated seem to cause less tissue reactions, and if the needle is contaminated, any tissue reaction at this particular area causes less tissue trim than if injected higher or further back. (Source: LCI 83-8-14, *Vaccination Abscesses*)

Cattle Liver Abscesses

Liver abscesses in feedlot cattle have been attributed to an abrupt increase in the intake of high-energy feeds.[4] A high-grain ration causes the rumen pH to become more acidic. Some cattle suffer from rumenitis (inflamation of the rumen wall) because of the acidic condition. The rumen wall becomes weaker and somewhat porous, allowing rumen bacteria *(Spherophorus necrophorus)* to pass out into the blood stream.[5] These organisms eventually cause abscesses in the liver, which, in severe cases, can spread to other organs, sometimes resulting in adhesions. Losses in some cases can go as high as $60 per animal in organ and carcass damage, to say nothing of the cost in poor gains and efficiency.

Swine Arthritis[6]

Arthritis is commonly recognized as being a major factor in swine lameness. The disease is brought about by infection of the joint and the surrounding tissues by bacteria or mycoplasmas (nonmotile microorganisms without cell walls that are intermediate in some respects between viruses and bacteria). USDA meat inspection records indicate that trimming parts of pork carcasses and discarding whole carcasses due to arthritis are leading causes of loss at slaughter. Greater loss probably occurs on the farm because of slower and less efficient gains and reduced performance by adult breeding stock and lactating sows. Death loss occurs in some instances but is not generally considered to be a major factor.

One type of arthritis comes about as a result of erysipelas (diamond skin disease). This type can be prevented by vaccination. Another type that affects baby pigs is streptococcal arthritis that can be prevented with penicillin if treated early enough. Two species of mycoplasmas have been shown to cause arthritis in swine. Stressful conditions and practices should be avoided to guard against these types. Diseases, such as pneumonia and diarrhea, should be controlled, since weakened pigs are more susceptible to mycoplasma arthritis. Furthermore, breeding stock with good leg conformation and action that come from arthritis-free herds should be selected. General good management and sanitation are important in avoiding arthritis.

In the acute stages of arthritis, joints show swelling, and pigs are lame and doing poorly. The infection can spread throughout the body, not often resulting in death but often causing such carcasses to be condemned by FSIS personnel upon slaughter. When the whole carcass is not infected, only the joint filled with excess fluid and the muscle surrounding that joint must be removed and condemned. If the arthritic condition is localized and arthritis passes into the chronic stage before the animal is slaughtered, the affected joints become enlarged,

---

[4]H. Brown *et al.* 1973. *Journal of Animal Science,* 37:1085.
[5]H. A. Smith. 1944. *American Journal of Veterinary Research,* 5:234.
[6]*Pork Industry Handbook.* PIH 36.

stiffened, hard, and nonfunctional, because connective tissue and bone have been deposited there. Such animals do not do well but when slaughtered, they have less loss of product than those slaughtered in the acute stage.

Brucellosis

Brucellosis, caused by the bacteria *Brucella abortus,* inflicts a serious economic drain on the livestock industry. The cost to the producer resulting from cattle and swine abortions and delayed breeding is sizable, and many times it goes unrecognized. Packers are affected because brucellosis reduces the supply of slaughter animals and increases their operating costs. Furthermore, brucellosis can create a health hazard for livestock producers, veterinarians, and packing house workers. Since infected animals are normally culled from the herd and slaughtered, packing house workers are among those persons more apt to become exposed to the disease; however, the disease poses no threat to the wholesomeness of the meat processed from infected females. According to the Public Health Service's Center for Disease Control, in 1978, nearly half the reported cases of brucellosis involved meat processors, veterinarians, livestock owners, and others who are in direct contact with cattle or swine. Even though brucellosis is considered an occupational disease, an additional 9 percent of the cases reported resulted from eating unpasteurized or foreign dairy products. In 1982, there were 154 cases of human brucellosis reported nationally.

Since the symptoms of human brucellosis—fatigue, fever, chills, weight loss, and body aches—are very similar to those of other common human infections, the disease is often overlooked or misdiagnosed. Chronic forms may cause more adverse effects, such as emotional disturbances, arthritis, and attacks on the central and optic nervous systems.

At the present time, there is no human vaccine against brucellosis—called undulant fever in humans—although relapses of human brucellosis are common. Prompt diagnosis and treatment can shorten the course of the disease and prevent complications.

Large economic losses for livestock producers, as well as the probability of contracting brucellosis, can be reduced by wearing protective gloves and scrubbing well afterwards, when assisting in calving; avoiding touching the eyes, nose, and mouth after touching newborn calves or raw milk; burning or burying aborted fetuses and contaminated placental tissue; disinfecting all contaminated areas, such as calving pens or areas where an abortion occurred; and testing all cattle suspected of brucellosis as well as any additions to a herd.

Tuberculosis

Various types of tubercle bacilli exist, and they most commonly infect a single species of animal. These bacilli are capable, however, of causing disease in a wide range of hosts—including humans and domestic animals. We are concerned primarily with three types, human, bovine, and avian.

The human type tubercle bacilli are usually associated with the familiar lung disease. They can also cause disease in other parts of the body and in other animals, including cattle, hogs, dogs, cats, canaries, and monkeys.

The bovine bacilli are usually associated with tuberculosis in cattle. Many other species, including humans, may also become afflicted—if they come in contact with infected cattle or drink unpasteurized milk from infected cattle.

The avian bacilli infect chickens primarily, but they may cause the disease in varying degrees in a wide variety of birds and mammals.

Hogs are susceptible to infection by all three types of tubercle bacilli. Odds are against hogs becoming infected with human or bovine types in the United States; however, this is due primarily to an extensive campaign to eradicate bovine tuberculosis. Laws requiring the cooking of all garbage before it can be fed to hogs have played a large part.

Avian tuberculosis, on the other hand, still causes condemnation of pork at slaughter houses. (See Table 3-2.)

In February 1972, the following additional meat inspection regulations went into effect pertaining to tuberculosis:

Carcasses Passed Without Restriction

- To be passed without restriction, a cattle carcass must be found to be free of TB lesions during post-mortem inspection and not identified as a tuberculin test reactor.
- A swine carcass may be passed without restriction as long as the lesions are localized and limited to one primary site. Primary sites are cervical, mesenteric, and mediastinal lymph nodes.

Carcasses Passed for Cooking

Passed for cooking requires holding the product at 170° F for 30 minutes. The loss to the packer in lowered quality of this cooked meat is approximately two thirds of its uncooked market value. The cooking requirement is more severe than that required for milk pasteurization (161° F for 15 seconds or 143° F for 30 minutes). Thus researchers are working to bring these regulations into synchronization by proving that the lower pasteurization temperatures are safe for meat also.

- A cattle carcass with a localized lesion of tuberculosis in any location must be passed for cooking.
- A swine carcass with a localized lesion of tuberculosis in any two primary sites must be passed for cooking. Therefore, a lesion in the head and intestinal tract would result in that carcass being passed for cooking.

• Tuberculin test reactors must be passed for cooking, even if they are free of tuberculosis lesions or when such lesions are localized.

(See Table 3-18 for a summation of the carcasses passed for food after cooking in fiscal year 1983.)

Table 3-18. Retained/Restricted Carcasses Passed for Food After Cooking or Refrigeration, in Fiscal Year 1983[1]

	Number of Carcasses					
Cause of Retention	Cattle	Calves	Sheep and Lambs	Goats	Swine	Equine
Infectious diseases:						
Caseous lymphadenitis (cheesy inflammation of lymph glands)	—	—	31	1	—	—
Tuberculosis nonreactor	84	—	—	—	22,094	—
Tuberculosis reactor	611	—	—	—	—	—
Inflammatory diseases:						
Eosinophilic myositis (muscle leukocyte)	156	—	—	—	43	—
Parasitic conditions:						
Cysticercosis (tapeworms)	6,226	1	—	—	—	—
Miscellaneous	362	—	1	—	477	—
Other:						
Sexual odor	—	—	—	—	217	—
Total	7,439	1	32	1	22,831	—

[1]MP Form 9300 Data. USDA, FSIS. *Meat and Poultry Inspection Program.* Statistical Summary for 1983. FSIS-14. Issued April 1984.

Condemnations

Causes for condemnation as described by the pre-1972 regulations still are in effect when the lesions of tuberculosis are generalized; that is, distributed so widely that it could only happen by entry of the tubercle bacilli into the systemic circulation. Furthermore, lesions of TB in the hepatic lymph node which drains the liver indicate that the condition is generalized and that the carcass must be condemned.

Cattle Grubs

Cattle grubs are the immature forms of two species of warble flies, usually known as heel flies. The common cattle grub, *Hypoderma lineatum,* attacks cattle in all parts of the United States, but the northern cattle grub, *Hypoderma bovis,* is not usually found south of a line extending across northern New Mexico and northern Oklahoma. North of that line, untreated cattle are infested with both species, but the common cattle grub appears in the back two weeks to two months earlier than the northern cattle grub. Similarly, the adults of the common cattle grub are active a month or so earlier in any locality than those of the northern cattle grub.

Cattle growers are most apt to be aware of these parasites at two points in the life cycle of the grubs—first, when heel flies are chasing cattle in the spring and early summer and, again, the following winter when grubs appear in the animals' backs after nine or more months inside their bodies.

The adult insects (heel flies) lay their eggs on the heels, legs, and other body parts of cattle. The eggs hatch into larvae (grubs) in three or four days.

Soon after hatching, the young grubs burrow into the skin and slowly work their way through the animal's body until they reach the gullet (common cattle grub) or spinal canal (northern cattle grub). The grubs remain in the gullet or spinal canal several months before starting another migration, this time to the muscles in the animal's back.

When the grubs reach the animal's back, they settle just beneath the hide and cut breathing holes through it. At this time, the cattle grower may notice swellings, often called warbles or wolves, forming beneath the hide. The grubs remain in the animal's back for about six weeks. During this period, they gradually enlarge their breathing holes.

When full grown, the spiny grubs work their way out through the breathing holes and drop to the ground, where they change to pupae. Three to 10 weeks later, the time depending upon the temperature, the adult heel flies emerge from the pupal cases and are ready for mating and egg laying. The entire life cycle takes about a year, 8 to 11 months of which are spent as grubs in the bodies of cattle.

The losses begin when heel flies lay their eggs on the cattle. The heel flies cause no pain to cattle, but they frighten the animals and make them difficult to manage. When attacked, cattle run about wildly with their tails in the air and are often injured in this wild stampeding. Cattle find some relief from heel flies by standing for hours in deep shade or water, but then while there, they are unable to eat.

The hide is damaged by the breathing holes cut by the grubs in the back, and two or more years are required for these holes to heal. Hides with five or more grub holes are downgraded, and a hide with a large number of holes is virtually worthless.

The grubs produce abscesses on the inner surface of the skin, and these abscesses cause large, discolored swellings to form on the adjacent loin muscle and its covering. When the cattle are slaughtered, the yellowish-green, jelly-like substance must be trimmed from the carcass. The economic loss results both from the meat that is trimmed off and from the downgrading of the carcass and its sale at a lower price per pound.

Total economic loss is difficult to estimate because of variation in losses from place to place and year to year, but it has been estimated nationally to be approximately $192,000 per year.[7]

The most commonly used and most practical control measure for cattle grubs is the application of a systemic insecticide. It is called systemic because it is distributed inside the body of the animal. The circulatory system carries the insecticide to the site where the grubs occur.

Several chemicals have been successfully and extensively used for grub control by the pour-on technique. Some of these materials can also be used to spray or to dip cattle. Also, commercial mineral mixes and feed additives that contain a systemic insecticide are available if the livestock owner prefers to use this method. Since the exact formulations often change, specific recommendations for the use of these systemic insecticides should be obtained each year from the local county agent, the veterinarian, the state university, or the product manufacturer.

Internal Parasites (Worms)

Many kinds of internal parasites (worms) cause damage and financial loss to the meat and livestock industry. The loss to the producer takes the form of retarded growth, failure to convert feed efficiently, and increased susceptibility to other diseases. Losses occur to the packer as a result of organ and carcass condemnations.

In swine, of the many worms that utilize the pig as a host, the two that cause the most organ and carcass damage are the large roundworm (*Ascaris suum*) and the kidney worm (*Stephanurus dentatus*).[8] Large roundworms occur in weanling pigs on up to and including mature sows and boars. Its migrating larvae damage the liver and lungs and create conditions favorable for the development of bacterial and viral pneumonias, diarrhea, and occlusion of the intestine. It is largely this worm that causes liver condemnations in slaughter animals (Table 3-4). Sanitary procedures aid in the control of all worms by removing the infective stage, but this lasts only a short time until the hogs are reinfected through manure, dirt clinging to sows, and even through the sow's first milk, colostrum. Roundworms can be controlled by employing a life-cycle deworming program utilizing a broad spectrum dewormer, one that will eliminate several worm species in addition to the large roundworms.

More of a problem in the southern United States, kidney worms generally appear in older pigs and breeding swine. These worms damage the kidneys, liver, tissues surrounding the kidneys, and ureters. Adhesions may also form between the liver and other organs. No treatment has been cleared for use against them. Thus considerable damage can result in the organs and carcass.

In cattle, the internal parasite *Cysticercus bovis* (tapeworm) is a potential

---

[7]*Great Plains Beef Cattle Handbook.* GPE 3351.

[8]*Pork Industry Handbook.* PIH 44.

concern for humans.[9] Thus meat inspectors incise the external and internal muscles of mastication, looking carefully for white tapeworm cysts which could be ingested by humans eating contaminated and "rare" beef. The condition is known scientifically as *cysticercosis* (Table 3-2), but the common designation is "beef measles," and the human form of the tapeworm is *Taenia saginata*. The cycle is completed between humans and cattle when birds pick up tapeworm eggs while drinking from open human sewage disposal plants, from effluent, or otherwise contaminated water and transmit them by defecating on feedstuffs later eaten by fattening cattle. The tapeworm eggs pass through the birds completely unadulterated. After the human tapeworm egg is swallowed by a beef animal, it passes into the animal's intestine where a tiny larva hatches. The larva penetrates the intestinal wall and travels to all parts of the animal's body. Many larvae lodge in muscle tissue where they develop into ¼-inch white cysts which contain the head of a new tapeworm. If these cysts were not discovered during post-slaughter inspection, the life cycle of the tapeworm would be completed if infested beef were eaten "rare" by any unwitting human. "Well done" cooking destroys the cysts as does carcass freezing (Table 3-18). In heavily infected carcasses, condemnation is the rule (Table 3-2). See doneness chart in center section.

In infected humans, the tapeworm containing cyst or cysticercus turns inside out and attaches itself to the wall of the small intestine. After the bladder or cyst is digested, the tapeworm begins producing segments which contain eggs. When egg production is completed within a segment, it may contain as many as 100,000 eggs. At intervals, the oldest segments detach themselves and pass out of an infested human through fecal material. Eggs, about ⅟₆₀₀ of an inch in size, can live for a year or longer under favorable environmental conditions. The human tapeworm has been known to reach a length of 70 feet, and, contrary to popular opinion, infected persons are usually obese because they are always hungry. In the United States, the beef measles problem is mostly centered along the Mexican border—in California, Arizona, New Mexico, and Texas.

Perhaps cattle have been infected prior to crossing the U.S.–Mexico border for finish and ultimate slaughter in this country. However, unsanitary toilet habits by feedlot personnel, campers, hikers, and picnickers all add to the tapeworm spread and threat. Humans are the sole host. More stringent efforts to eliminate or screen out birds around feedlots are recommended for tapeworm control.

Avermectins,[10] a group of cyclic lactones produced by fermentation of *Streptomyces avermitilis* are effective against a wide range of helminth and anthropod parasites. This is a systemic control agent that can be administered by subcutaneous injection, orally, or by a slow release implant. Ivermectin, the commercial preparation of avermectin developed by Merck & Co., has been experimentally shown to control stomach worms, intestinal worms, lung worms,

[9]L. W. Dewhirst. 1974. University of Arizona Agricultural Experiment Station, Tucson.

[10]LCI 83-9-8. Research reports on Ivermectin.

lice, scabies, mange, ticks, and grubs in cattle; stomach and intestinal worms in sheep; round and red stomach worms, intestinal thread worms, lung worms, mange mites, and lice in pigs; and many internal parasites of horses. Be sure to check with your veterinarian and/or county agent before using this compound, since it is not cleared for all the uses mentioned above.

Trichina[11, 12]

The microscopic parasite *Trichinella spiralis,* or trichina, is not a widespread parasite in this country and can be controlled. The encapsulated larvae are found in the muscles of rats, dogs, cats, swine, game (especially bear), and humans. Epidemics of human trichinosis have been traced back to raw horse meat in France and Italy. Humans acquire the parasite by eating the improperly prepared meat of an infected animal. Such animals receive their infection by eating infected rats or uncooked or partially cooked viscera or flesh of infected animals, generally through the medium of garbage.

The encysted larvae are liberated in the stomach of the host and pass on to the small intestine where they reach sexual maturity in a few days. After mating, the female penetrates the lining of the intestine and gives birth to young larvae which are carried by the blood stream to the striated muscles where they attain maturity after several weeks. Encapsulation then takes place and, if the cycle is not repeated, the larvae in the cysts eventually die. About 250 of these minute parasites measure an inch.

In humans, the severity of the clinical disease is dependent upon the level of infection. A light infection of 1 to 10 larvae would probably not prompt a visit to the clinic. A moderate infection of 50 to 500 larvae would cause concern, while a severe infection of 1,000 or more larvae would be considered life-threatening. About 2 percent of humans who contract it die. Trichinosis can mimic a wide variety of diseases. Most mild cases are misdiagnosed as influenza or other viral fevers. The disease is characterized by fever, gastrointestinal symptoms, myositis (inflammation of the muscles), swollen eyelids, and an abnormal increase in the number of leukocytes in the blood (eosinophilia).

Due to the widespread improvement of sanitary conditions in our large hog-producing areas and to state and federal legislation requiring all garbage to be cooked before being fed to swine (Swine Health Protection Act, Public Law 96-468, 1980, amended 1984), the likelihood of any trichina infection is very small. The law prohibits the feeding of garbage to swine except when it is properly heat-treated at a licensed treatment facility. Garbage must be heated throughout at boiling (212° F or 100° C at sea level) for 30 minutes while being agitated. The law does not prevent the feeding of uncooked waste from ordinary household operations when it is fed directly to swine on the same premises

[11]Livestock Conservation Institute, 239 Livestock Exchange Building, St. Paul, MN 55075.
[12]USDA Animal and Plant Health Inspection Service, Federal Building, Hyattsville, MD 20782.

where this household is located. Also, the federal law does not apply to garbage consisting of any of the following: rendered products, bakery waste, candy waste, eggs, domestic dairy products (including milk), fish from the Atlantic Ocean within 200 miles of the continental United States or Canada, or fish from inland waters of the United States or Canada which do not flow into the Pacific Ocean.

At present, the following 16 states, with more than 50 percent of the national swine population, prohibit the feeding of garbage in any form: Alabama, Delaware, Georgia, Idaho, Illinois, Iowa, Louisiana, Maryland, Mississippi, Nebraska, New York, South Carolina, South Dakota, Tennessee, Virginia, and Wisconsin. Nevertheless, the incidence of trichinosis in swine is greater in the United States than in many other countries (over the past 20 years averaging 0.11 percent of all swine) and remains a problem for complete consumer acceptance of pork in this country.

From 1975 to 1981, there were 1,066 human trichinosis cases reported in the United States. In 86.1 percent of the cases reported, pork was incriminated, with 79.1 percent coming directly from pork products and 7 percent from pork-contaminated ground beef. The remaining 13.9 percent was due to the ingestion of meat from wild animals. Trichinella will not be present in beef muscle, since the animal is a herbivore; however, if the processing of beef follows the processing of pork, with the same equipment being used, and if strict cleanup is not observed in the plant before the beef is processed, pork scraps become mixed with the beef, thus increasing the chance for cross-infection.

In 70.3 percent of all cases the meat was known to have been eaten raw. For the remainder, the meat was believed to have been inadequately cooked.

The National Pork Producers Council has announced January 1, 1987, as the target date for attaining a trichina-free pork status in the United States and has adopted the following steps to attain this objective:

- Strict enforcement of garbage-cooking laws.
- Identification and traceback of all U.S. swine from packing plant back to owner.
- In-depth producer education.
- Research on all phases of trichina control.

Part of the research will involve methods of detection. A trichinoscope (microscope), which can detect one trichina/gram, the level which can induce trichinosis in humans, has been used in Germany since 1866 to determine the presence of trichina in pork muscle. A digestion system developed in the United States by Dr. William J. Zimmerman of Iowa State University can detect one trichina/45 to 50 grams of diaphragm. More recently, breakthroughs in antibody methodology have allowed the development of a rather quick (25 minutes) and sensitive (many times more sensitive than the trichinoscope) blood test, called the enzyme-linked immunosorbent assay (ELISA), which may provide an avenue for the elimination of trichinae from U.S. pork.

Other research efforts will involve low-level irradiation treatment of pork carcasses using spent fuel rods to provide gamma radiation from cesium 137 or cobalt 60 to render the pork trichina free. Also efforts are being made to breed pigs which are resistant to trichina.

Recall that heating to 137° F, freezing, and salting are accepted by FSIS (pages 51-52) as trichina control procedures.

In addition to spreading trichinosis, garbage can serve as a means of transmission of numerous infectious or communicable foreign and domestic diseases of swine including, but not limited to, African swine fever (ASF), hog cholera, foot-and-mouth disease, swine vesicular disease, and vesicular exanthema of swine. All these diseases can be spread through infected meat scraps in improperly treated garbage that is fed to swine or through material that has been associated with such meat scraps. U.S. officials are conducting an intensified program to inspect meat and related products entering the United States, especially from countries in the Western Hemisphere with ASF. Complete surveillance is impossible considering the tremendous volume of international traffic, especially between the Caribbean Islands and the United States. A single contaminated meat product in garbage that reaches susceptible hogs could cause an outbreak. Thus, control of garbage is tremendously important to the U.S. swine industry.

Chapter 4

Hog Slaughter

TABLE 4-1 lists the states in the order of commercial pork slaughter volume in 1983. There were 87,584,259 hogs slaughtered commercially in the United States and approximately 500,000 slaughtered on the farm, making a total of 88,084,259[1] hogs slaughtered in the United States in 1983. This number of hogs produced 15.2 billion pounds of carcass, averaging 173 pounds per hog. Nearly 97 percent of the commercial hog slaughter was done under federal inspection.

Table 4-1. Commercial Hog Slaughter, Number of Head (in Thousands), Ranked by States, 1983[1,2]

Rank	State	Number	Rank	State	Number
1	Iowa	21,427.6	24	Washington	338.1
2	Illinois	7,795.5	25	Arkansas	286.3
3	Minnesota	5,793.5	26	New Jersey	284.4
4	Michigan	5,284.8	27	Delaware and	
5	Nebraska	4,626.6		Maryland	260.7
6	Virginia	3,925.4	28	Arizona	196.3
7	Missouri	3,774.0	29	Utah	194.6
8	Ohio	3,631.8	30	Oklahoma	185.7
9	Indiana	3,285.7	31	Colorado	175.7
10	South Dakota	3,009.3	32	Oregon	168.5
11	Kentucky	2,531.1	33	Florida	124.6
12	Pennsylvania	2,448.1	34	Idaho	81.7
13	Wisconsin	2,436.2	34	Louisiana	81.7
14	Mississippi	2,404.0	35	New England[3]	78.2
15	North Carolina	2,387.5	36	North Dakota	70.2
16	Georgia	2,163.2	37	New York	65.8
17	Tennessee	2,018.6	38	Hawaii	44.0
18	California	1,824.0	39	West Virginia	29.3
19	Kansas	1,417.2	40	New Mexico	9.3
20	Texas	1,280.4	41	Wyoming	9.1
21	South Carolina	552.7	42	Nevada	3.3
22	Alabama	525.2	Total		87,584.0
23	Montana	354.1			

[1]*Annual Livestock Slaughter*. March 1984. Crop Reporting Board, SRS, USDA.

[2]Includes slaughter in federally inspected and in other slaughter plants, but excludes animals slaughtered on farms.

[3]New England includes Connecticut, Maine, Massachusetts, New Hampshire, Rhode Island, and Vermont.

[1]*Annual Livestock Slaughter*. March 1984. Crop Reporting Board, SRS, USDA.

Packers purchased 79 percent of these hogs directly, 11 percent through terminal markets, and 10 percent through auction markets. Of all these purchases, 14 percent were made on a carcass grade and weight basis.[2]

Before proceeding to slaughter hogs, be sure that you have proper and safe equipment; that your equipment is ready to use (sharp and clean); and that the animal you are about to slaughter is healthy. That is, you should study and master the first three chapters before going ahead with this chapter.

FAST (DRIFT) PRIOR TO SLAUGHTER

Hogs should be held off feed for 16 to 24 hours prior to slaughter to assure completeness of bleeding and to ease evisceration. The intestinal mucosa (mucous lining of the intestine) normally prevents bacteria of the gut from penetrating the blood stream and thus infecting the meat. However, if the intestines are distended with feed when the pig encounters the stress and excitement of slaughter, the mucosa is put under exceptional pressure and may "give way" to the organisms. Also, when the intestinal tract contains a heavy load of ingesta, it is simply harder to handle during evisceration and is more likely to burst or be accidentally cut, thus contaminating the carcass.

WEIGHING AND RECORDING[3]

The live hog or a group (draft) of hogs should be weighed on a balanced scale prior to slaughter in order that yield (dressing percent) can be accurately accounted for. The sex, breed, and ID number should be recorded along with the live slaughter weight. Dressing percent is calculated as follows: Carcass weight/live weight $\times$ 100 = dressing percent (yield). Routinely, carcass weight is "hot," or "warm," meaning it was taken prior to carcass chilling. Chilling causes carcasses to lose 1 to 2 percent due to moisture evaporation, thus the dressing percent would be lower if chilled carcass weight were used. Furthermore, animals weigh less immediately prior to slaughter than at purchase, since animals "shrink" in transit and rarely gain back to their purchase weight while standing in yards awaiting slaughter, even if they have feed and water before them. Thus a dressing percentage calculated on slaughter weight would be higher than if it were calculated on purchase weight. Realistically, purchase weight should be used in order to compute financial returns, but slaughter weight should be recorded as well to be able to determine what shrink the animals have undergone between purchase and slaughter. Hogs that are scalded and have their hair

---

[2]*Annual Financial Review*. Sept. 1984. American Meat Institute, P. O. Box 3556, Washington, DC 20007.

[3]Dan Engeljohn and John Romans. 1979. *Pork Slaughter in Small Plants: Dehairing and Skinning Methods*. Slide set E 70 prepared by the College of Agriculture Instructional Resources, University of Illinois at Urbana–Champaign.

removed, in general, dress approximately 70 percent. Heavier, fatter hogs will dress higher in the 70s. Excessively filled light weight hogs will dress in the high 60s. Thus fat, fill, and weight are three important factors that affect dressing percentages. Hogs that are skinned and have their feet removed will dress approximately 10 percent less than scalded hogs.

IMMOBILIZATION

Hogs must be stunned with a federally acceptable device (mechanical, chemical, or electrical) so that they are rendered unconscious prior to being hoisted and stuck. Whichever method is used, the time interlude between stunning and sticking should be minimized to prevent blood splashes in the muscle. These are caused by capillary rupture resulting from increased blood pressure in the constricted capillaries while the heart continues to pump. After stunning, the quicker the pressure is released through sticking the less chance there is of capillary rupture. Sticking the hog while it is in a prone position, rather than in a suspended position, cuts down this time interval and thus often prevents blood splashes.

Note the use of a commercial electrical stunner in Figure 4-1, a scissors-type electrical stunner in Figure 4-2, and the commercial use of CO_2 in Figures 4-3 and 4-4.

Fig. 4-1. A commercial electric stunner. Hogs are driven into a squeeze whose moving sides propel the subject to the stunner. (Courtesy Cincinnati Butchers' Supply Co., Cincinnati)

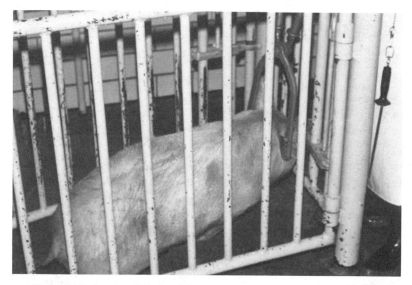

Fig. 4-2. An inexpensive electric stunner suitable for use in smaller slaughter plants. (University of Illinois photo)

Fig. 4-3. Hogs passing through a lane toward the CO_2 chamber. Hesitant hogs are urged forward by the light application of an electric prod. (Courtesy G. A. Hormel & Co.)

Fig. 4-4. Hogs emerging from the CO_2 chamber in an immobile and unconscious form after being in the chamber 45 seconds. The attendant is positioning the hogs on the conveyor so the heads are over the bleeding trough. (Courtesy G. A. Hormel & Co.)

STICKING (EXSANGUINATION)

A 6-inch sticking knife, sharpened on both sides of the tip, is large enough for the ordinary hog. A straight, rigid, 6-inch boning knife will work as a sticking knife. For large hogs (400 to 600 pounds), the 7-inch blade is desirable. If the hog is suspended, you should steady it by placing the flat of your hand on the hog's shoulder (never by grasping a leg). Locate the cranial (front) tip of the sternum bone. Insert the knife straight in initially, then with the point of the knife directed toward the tail (this is very important), give an upward thrust toward the anus, dipping the point until it strikes the backbone, thereby severing the carotid arteries and jugular veins, and then withdraw the knife. The stick hole should be no wider than the knife blade to avoid contamination in the scalding tank. Exercise care in keeping the knife midway between the shoulders to avoid a shoulder stick. No twisting or cross cutting of the knife is necessary. If you insert the knife too far cranially to (in front of) the sternum bone, the animal will not bleed freely and will die slowly.

When the hog is stuck properly (Figure 4-5), blood will gush out of the opening in a steady stream (Figure 4-6) due to the severing of the carotid arteries (bright red blood) and the jugular veins (dark blood). A proper stick means a fast bleed.

Figures 4-7 and 4-8 show prone sticking in a commercial plant.

Blood should not be allowed to run into the regular sewer, since it has a high biological oxygen debt (BOD), meaning it would hinder complete sewage degradation by microbes in a sewage system. Rather, blood should be saved

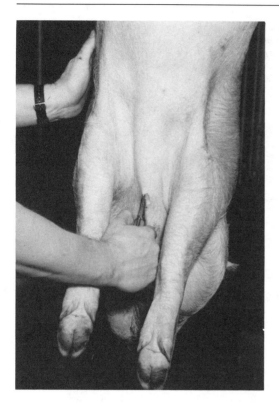

Fig. 4-5. Properly sticking a suspended hog.

Fig. 4-6. Blood gushing out of the
small opening is caught in a large, funneled vat.

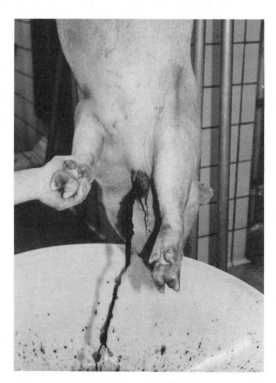

through a special drain or removed from the animals into vats or barrels (Figure 4-6), to be processed and used for animal feed or fertilizer. Only if it is kept completely sterile by removal from the animal through a tube or syringe can it be used for human food. (See Chapter 10.) Blood volume in animals may approxi-

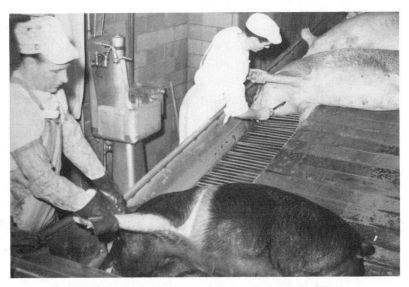

Fig. 4-7. All hogs on the sticking conveyor face in the same direction, backs toward the overhead belt conveyor and head over the bleeding trough. The sticker is about to stick a hog as it is in this position. (Courtesy G. A. Hormel & Co.)

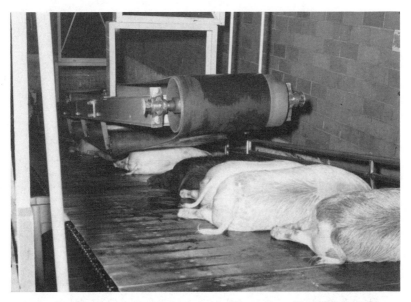

Fig. 4-8. Bled hogs emerging from the double conveyor. The overhead belt conveyor with movable weight rollers that press the belt against the hog in place during the entire bleeding process. (Courtesy G. A. Hormel & Co.)

mate 6 to 8 percent of live weight with considerable individual variation. A good stick results in removal of approximately 50 percent of the total blood in an animal.

SCALDING

Sufficient water of the right temperature (143° F ± 2; 61° C ± 1) and removal of the hog from the water when the hair slips easily will assure a good, quick job. A safe scalding temperature is 135° to 160° F, the lower temperature requiring more time. Water up to 180° F can be used but is not recommended since you must withdraw the hog as soon as the hair slips easily. Initial heating causes the protein in the hair follicle to denature, thus loosening the hair. Overscalding causes the skin to contract around the base of the bristles, holding them tight, and is referred to as "setting the hair." Even further overscalding begins to cook the skin, and it may deterioriate, allowing the contaminated scalding water to enter the meat. Severely "cooked" hogs must be condemned. A rule of thumb (or finger) for determining the proper scalding temperature, if a thermometer is not readily available, is to carefully place your finger in the water and begin slowly counting—one-thousand-one, one-thousand-two, one-thousand-three—*and then* if you *must* remove your finger at about one-thousand-four or one-thousand-five, that indicates the temperature is approximately 143° F. This 4- to 5-second rule works for most people; however, some are more or less heat tolerant so it is not completely foolproof, but it does give a good safety range to avoid overscalding a hog.

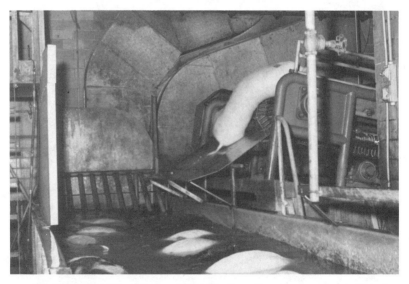

Fig. 4-9. Hogs sliding into a scalding tub. From there they are elevated into the dehairing machine. (Courtesy G. A. Hormel & Co.)

Once the hog is in the water, keep it under water, and continually move it and turn it to get uniform scalding. In large plants, hogs enter the scalding tub and are moved through the tub by a conveyer set at the proper speed to allow the proper scalding time. Thus, if the water temperature is right, everything is "automatic" (Figure 4-9). These large plants may run the temperature at 139° F, which would allow 4½ minutes before overscalding, during the normal season. During the hard-hair season (September, October, November), the water temperature should be 139° to 140° F and the immersion period 4 to 4½ minutes, while in the easy-hair season (February, March), a temperature of 136° F for 4 minutes is preferable. However, in small plants without the automation, hair condition must be checked periodically during the scalding period. Check the ham, flank, belly, and head regions. Once you determine that the hair is loose, get the hog out of the water at once. At 145° F, scalding time may be 2 to 3 minutes, but red and black hogs may take longer. Also the time of the year has an influence, as noted above.

If you are scalding the hog in a barrel on a farm, insert a hook in the side of the mouth (for light hogs) or between the lower jaw bones, and scald the rear half of the hog. After this half is scraped and the hind feet are shaved clean, open the tendons, and insert the gambrel. Using the gambrel for manipulating the hog, immerse the front half, and scald and scrape it, then hoist the carcass, shave the remainder of the hair from the carcass, and rinse it.

HAIR AND SCURF REMOVAL

Various machines, sometimes called "polishers," are manufactured to remove hair from scalded hogs. The large daily slaughter of hogs by packers has made it necessary to devise mechanical equipment to handle large numbers of hogs per hour. Today, a single dehairing machine will handle from 150 to 500 or more hogs per hour, depending upon its length. Large plants are equipped with twin machines that will handle up to 1,000 hogs per hour. These machines are constructed of heavy V-shaped bars, a heavy steel frame, and two shafts to which belt scrapers with metal tips are attached. The lower shaft runs from 55 to 60 rpm (revolutions per minute), and the upper shaft runs around 100 rpm. Both shafts run in the same direction. Hot water, 140° F, is sprayed on the hogs as they pass through the dehairer toward the discharge end.

Medium and small plants may have smaller dehairing machines that handle one or two hogs at a time. These machines have belt scrapers attached to rotating bars which cause the hog to rotate, while the scrapers knock and pull the hair out (Figure 4-10). The action of the flippers and the turning of the hog causes the hair to be removed, provided the hog was properly scalded. After about 15 to 30 seconds in the machine, most of the hair should be removed. A machine is now available for small plants that combines the scalding and dehairing (polishing) operations (Figure 4-11).

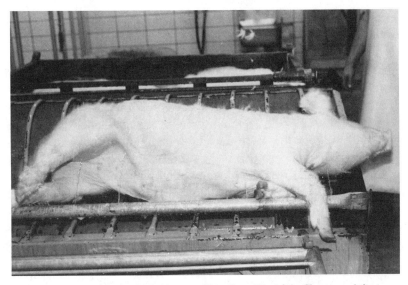

Fig. 4-10. In this small dehairing machine, the action of the flippers and the turning of the hog causes the hair to be removed.

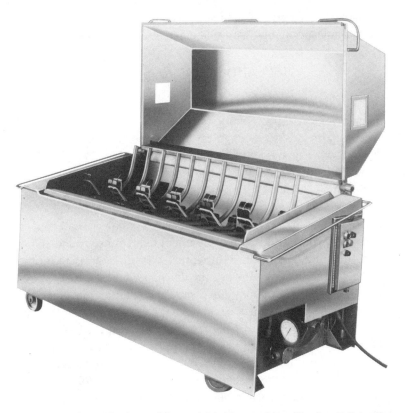

Fig. 4-11. A combination scalding and dehairing machine. (Courtesy E-Zuber Engineering & Sales, Inc., Minneapolis)

After dehairing, and before the hog cools down, remove the toe nails with a toe nail puller. If possible, engage both hooks of the puller at the top of the nail (Figure 4-12), and pull the nail. Once the animal cools off, the nails are harder to pull. In the case of tough, hard-to-remove toe nails, a pruning shear may be used to clip the toes off.

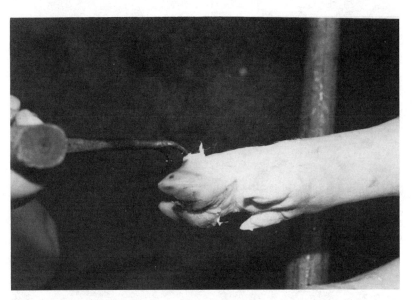

Fig. 4-12. Toe nails are removed by using a toe nail puller.

Once the nails are removed from the front and hind feet, remove the skin and hair between the toes on all four feet.

The two tendons in the hind foot must be exposed so that the trolley hook or gambrel can be inserted to hang the hog on the rail. On the back side of the hind foot make two cuts. One cut should be off to the left next to the dewclaw, and the other cut should be off to the right of the center of the foot next to the other dewclaw (Figure 4-13). Make the cut as deep as possible. By cutting to the left and to the right of the center, you will avoid cutting the tendons. If the cut is properly executed, you will be able to place your finger between the bone and the two tendons.

A bell-type scraper that is fairly sharp is an important factor in the effective removal of scurf. The round working surface permits rapid dehairing, and plenty of pressure can be applied to remove the bristles and dirt (Figure 4-14). All evidence of hair, including eyebrows and hair on the lips and inner ear, must be removed before head removal and evisceration can take place.

If the head is free of hair and debris and if the animal has been stunned in an approved manner so that no part of the head is contaminated (a bullet or penetrating captive bolt would contaminate the head), the head can be utilized for human consumption. A hog's inner ear contains wax and dirt, which would con-

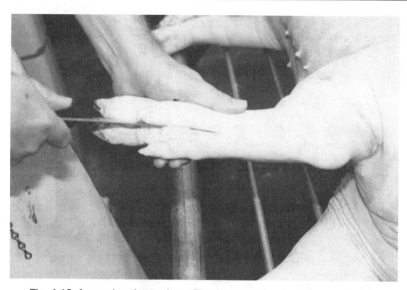

Fig. 4-13. Loosening the tendons. The small and large tendons on the back of the hocks are used for supporting the carcass on the gambrel.

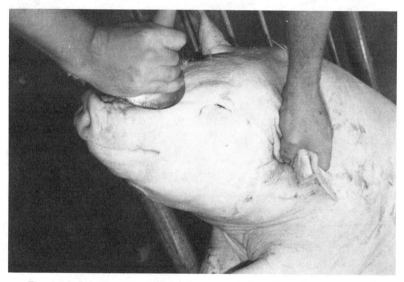

Fig. 4-14. A bell scraper with its sharp rounded working edge removes bristles and dirt.

taminate any meat product made from the hog's head, so it must be removed. To remove the inner ear, hold the ear with one hand, and insert knife with the other hand. Cut in a circle around the inside of the ear, as though you were coring an apple (Figure 4-15). Once you have cut around the inner ear, pull this section with one hand and cut the inner section off at the base with your knife in order to clean the ear of contaminates. Next, remove the hair and skin surrounding the

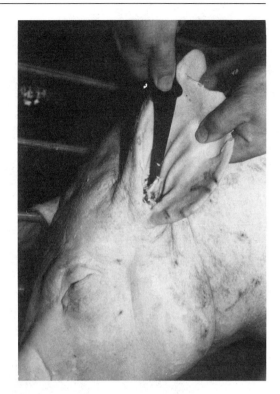

Fig. 4-15. The inner ear, contaminated with wax and dirt, must be removed if the head is to be saved for human consumption.

eye, including the eyelid. At a point directly above the eye, make a cut as deep
as you can, until you hit the skull. Next, pull the skin surrounding the cut above
the eye, and continue cutting towards the eye (Figure 4-16). You may cut directly over the eye, or you may cut the eye out. Continue cutting with one hand
and pulling with the other until the skin surrounding the eye is removed. Try to
remove only the skin, not the lean tissue. To remove the lips, first place a knife
directly above the corner of the mouth, and make a cut as deep as possible until
hitting the jawbone. By placing your finger in the cut you just made, pull the lips
towards you and continue cutting lips (Figure 4-17). You should be able to remove both the upper and the lower lips at the same time. Again, remove only
the skin, and leave the lean tissue intact if possible.

The rest of the carcass must be scraped in order to remove the remaining
hair. Use your skinning knife as if it were a razor blade. Rub your hand back
and forth on the skin surface to detect areas you have missed. Be sure to scrape
under the legs. If the hair is hard to remove or if a bruise or callus is evident,
remove that area by cutting it out with your knife. When the hog is completely
cleaned, it is ready to be hoisted up and landed onto the rail. Place a gambrel in
the tendons that were previously opened. The hog should now be scraped once
more, starting at the ham end and working down to remove clinging water and
loose hair. Small, fine, and tough hairs that remain can be removed by singeing
them with a propane torch. The burning action of the torch causes the hair that

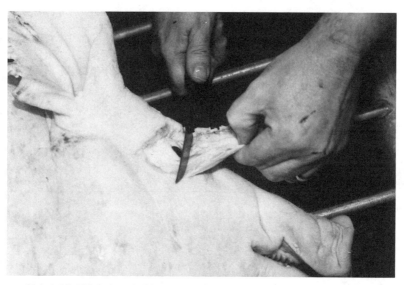

Fig. 4-16. The hair and skin surrounding the eye, including the eyelid, must be removed.

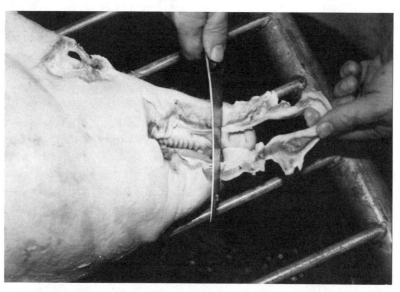

Fig. 4-17. Lips are removed because they are very difficult to clean completely.

was missed to char and become evident. Be sure to singe the hind feet area well, as the hair is hard to see there. Also, the head region is difficult to clean, therefore be sure and lift the front legs and ears while singeing. Do not hold the flame in one area too long, as you will burn the skin.

Once you have singed the remaining hair on the entire carcass, wash and scrape the charred hair off. Scrape as if you were shaving the animal. You can feel the brittle ends of the hair if you miss a spot. Be sure to lift the legs up and clean around them and the head before finishing the scraping process. Once the hog is free of singed hair and brittle ends of hair, wash the hog thoroughly from the hind feet to the head.

DEPILATING HOG CARCASSES

At one time the practice of dipping each carcass after it was dehaired into a hot solution (250° to 300° F) of rosin and cottonseed oil for a period of six to eight seconds to remove hair, stubble, and roots was quite popular. When the rosin coating plasticized after cooling, it was stripped by pull-rolling it down the carcass, taking with it the remaining hair, stubble, and roots. However, in recent years, shortages of trees, wood, and rosin have caused many packers to discontinue its use and turn instead to mechanical brushes and torches to completely clean dehaired pork carcasses.

PULLING HOG SKINS IN A SMALL PLANT[4]

The following method of slaughtering a hog fits in well where there is a scarcity of labor. It is a one-person job and requires neither hot water nor scalding equipment. Conventional hand skinning involves considerable knife skill. Because the fat is soft and easily cut, the knife must be kept tight against the skin at all times. Deep gashes made by uncontrolled strokes of the knife do considerable damage to hams and bacons that are to be cured. Furthermore, the soft, thin skin is easily cut and, if cut, is worthless as garment material and must be sold at low prices for manufacture into animal protein feeds.

An easier method of removing the pigskin from the carcass by pulling results in a saleable pigskin and, at the same time, produces a nice, smooth carcass free of gouges. The light area shown in Figure 4-18 is the area which is to be skinned. Most persons are inclined to skin too far back into each side with a knife, thus needlessly damaging the hide and scoring the carcass.

The hog is humanely stunned and bled in the conventional manner. After bleeding is complete, the hog is placed on the skinning rack or cradle. Depending upon what use is to be made of the pig's feet, they may be skinned out (Figure 4-19), discarding only the hooves and dewclaws and saving the feet for human food—or they may be completely cut off without skinning (Figure 4-20) and discarded into the inedible barrel to be processed into animal food. When removing the hind foot, be sure to cut below the hock joint itself at the flat joint to avoid destroying the tendon anchor used to hang the carcass. The value of pigs' feet is not standard in all areas, so it may not pay to spend the time and effort to

---

[4]Method perfected by Bill Condradt, Plains Processing, Shiocton, Wisc., and Wieland Kayser, Kayser's Butchering—Lena Maid Meats, Lena, Ill. Slide set E 71 prepared by the College of Agriculture Instructional Resources, University of Illinois at Urbana–Champaign.

clean them properly for human food. Feet with the skin on will average approximately 1 pound each. At any rate, the feet must be removed before the hide is removed in order to prevent contamination of the skinned carcass with particles of dirt and/or manure dropped from the feet.

Fig. 4-18. When the skin is pulled, the lighter area is the *only* area that should be hand skinned.

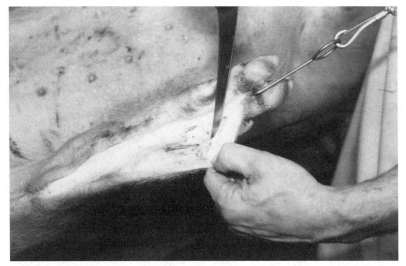

Fig. 4-19. Feet may be skinned out more easily by holding them taut with a shroud pin attached to a scabbard chain.

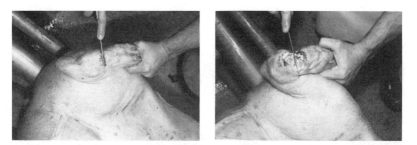

Fig. 4-20. Removing whole unskinned foot. (Left) Applying down pressure to locate the flat joint. (Right) Retaining tendon anchor for carcass hanging.

Once the feet are removed, an opening is made in the hide down the belly from the tip of the jaw to the tail. An opening is made down the rear of each hind leg, meeting in the center several inches ventral to the anus. Start skinning the inside of the ham. Follow up forward just past the teat line toward the shoulder, skinning part of the hollow dimple behind the front leg. Begin skinning under the jowl, where the original stick opening was made. Skin down on the jowl and head as far as possible before turning the hog to complete head skinning in the cradle later. Skin down the shoulder to the point of the shoulder. Then start on the other side. Skin out as much of the head as possible. Skin down on the outside of the shoulder to the point and past the dimple behind the leg. Be careful here *not* to cut through the skin. Skin toward the other ham just to clear the teat line and then skin the inside of the ham. An electric or air powered skinning knife is useful; however, one should avoid doing too much skinning—just clear the teat line and leave the skin on the rear of the hams.

To fully prepare the carcass for skin pulling, you must completely skin out the head. This is a critical area, and care should be taken not to cut through the skin. Turn the pig in the cradle to gain access to the top of the head and neck area. Skin past the eyes and ears. Around the ears the skin is very thin and close to the flesh. Skin past the hump of fat on the crown of the skull and around the ear area just far enough to get past the lean connecting tissues. Skin only until you get through the lean and you are back into the fat again. Figure 4-21 shows the head properly skinned. As an alternate method for skinning the head, insert a stainless steel meat hook into the lower jaw bone, raising the head part way by hoist in order to gain easier access to skin out the back of the skull.

Figure 4-22 shows the rear view of a hog properly prepared for hide pulling. To position the hog for pulling its skin, insert a gambrel in the rear legs, and raise the hog. If two hoists are available, one may be used to keep tension on the carcass while the second pulls the skin off. A ring should be securely fastened in the floor directly under the point of the jaw. Insert a strong meat hook into the ring in the floor and into the lower jaw bone of the hog. Raise the first hoist to take up the slack, but do not exert too much tension as you might pull the head right off. An alternate method is one in which only one hoist is used per hog, so that the carcass can land on a rail, thus freeing that hoist for pulling the skin. A

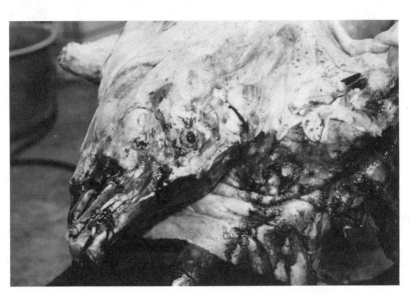

Fig. 4-21. The head has been properly skinned prior to pulling.

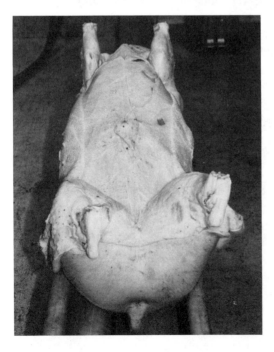

Fig. 4-22. A hog properly "pre-skinned," ready for hide pulling.

chain is cinched from the S-hook to the floor ring and tightened by hand, or with a cinching device, such as a trucker's load tightener (Figure 4-23). Gather and roll together as much of the loose head and shank skin as possible, and place a short piece of chain around it, or use the hoist chain itself (Figure 4-24). Raise the hoist slowly at first to put pressure onto the pigskin, and check that the skin is starting to pull off properly (Figure 4-25). If it is pulling off into the fat, stop and carefully use your skinning knife to release the skin. Proceed to pull the rest of the skin up and off the carcass (Figure 4-26).

Figure 4-27 shows the smooth carcass free of gouges remaining after employing this method of pulling the skin. Figure 4-28 compares a carcass after pulling the skin this way with one that has been conventionally hand skinned by an experienced butcher.

All animals do not pull the same, just as all hogs do not scald the same. Even with patience and the desire to change over to a new system after doing it for years another way, this method does frustrate a person at first. The first couple of days are hard on the nervous system. One fault most persons may have is that of overskinning by hand.

The skin represents from 6 to 8 percent of the weight of the live hog. For other than home consumption, skinning has not previously been economical because of the weight loss due to skin removal, a poor market for skins, and also because it was a slower process, compared to dehairing in large commercial ma-

Fig. 4-23. A trucker's load tightener can be used to maintain tension on the carcass as the skin is being pulled.

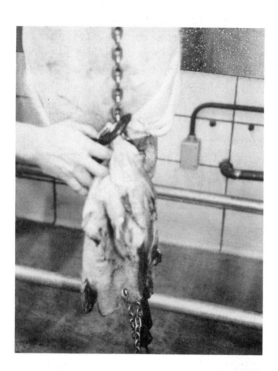

Fig. 4-24. Loose head and shank skin is gathered within the hoist chain for pulling.

Fig. 4-25. The hoist is raised slowly and somewhat tentatively at first to check the tension and to make sure the skin is starting to pull off properly.

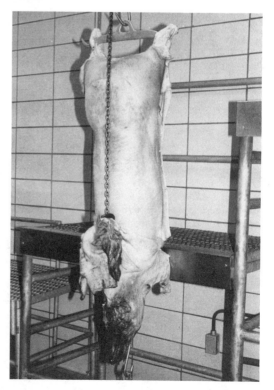

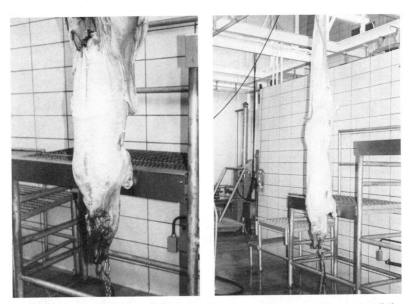

Fig. 4-26. Once the skin is properly started, a steady pull is maintained until the skin is completely pulled from the carcass.

Fig. 4-27. A smooth, gouge-free carcass following skin pulling. The tiny dark spots on the ham are blood from fat surface capillaries that have ruptured when the skin pulled off. These spots have no effect on the meat quality since they are only on the exterior fat surface which is always trimmed off.

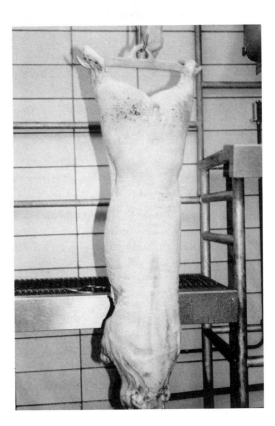

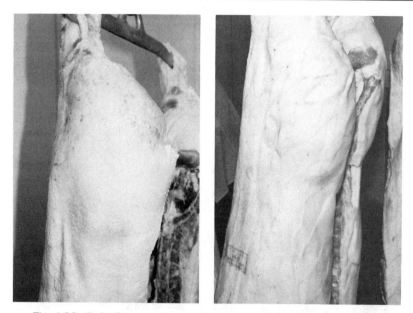

Fig. 4-28. (Left) Closeup of a carcass from which the skin was pulled by the method described under "Pulling Hog Skins in a Small Plant." (Right) Closeup of a carcass from which the skin was removed by an experienced butcher. The advantage in carcass smoothness is clearly evident.

chines. For home use, pork skin or rind is not necessary either in the curing or in the subsequent keeping of the cured meat. Curing tests conducted at The Pennsylvania State University have shown that skinned pork cuts take the cure faster and keep as well as unskinned cuts. Other advantages are that the task can be done by one person, no hot water is required, and no rinding (removal of skin) of pork fat for rendering is necessary.

Steps must be taken to prepare pigskins for storage until they are picked up by the purchaser. Immediately after the skin is pulled from the carcass, trim the head off right behind the ears and trim out the tail and anus at the other end. Salt the skins with a liberal application of ice cream salt or coarse sack or stock salt after placing skin, with the flesh side up, on a piece of plywood. Salt and fold so as to hold salt on hide, and place in a 50-gallon drum. When the drum is full, sprinkle a little salt on top. Salted pigskins do not smell or cause a fly problem. Pigskins should be salted as soon as they are available. They should never be held over unsalted.

In recent years, as new equipment has been developed, a number of commercial hog skinning operations have begun in the United States. Among them is the Jimmy Dean Meat Co., which skins hogs for the production of its whole-hog pork sausage (Figures 4-29 to 4-32). The practice of skinning hogs has increased, since a majority of hams and bellies are now merchandised skinless. Furthermore, the hide is more valuable when removed intact before it is cured, as portions are used for human burn treatments and clothing.

Fig. 4-29. Power clippers are used to remove hind feet. The beginning of the skinning operation is the folding over of skin on the ham region. (Courtesy *The National Provisioner* and the Jimmy Dean Meat Co.)

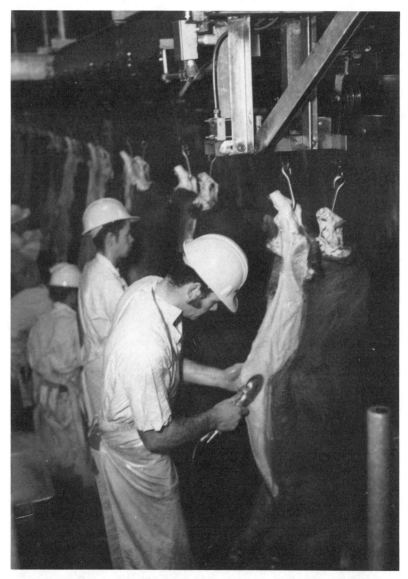

Fig. 4-30. Employee continues by folding skin down to the chest area, using a power knife. Skinning continues down across the chest. The front feet are removed, and the head and shoulders are completely skinned out. (Courtesy *The National Provisioner* and the Jimmy Dean Meat Co.)

Fig. 4-31. Carcass at right of photo shows the loose skin around the head, which is gathered up in a chain. Note the carcass anchors between jaw bone and platform. Next carcass is in the process of being skinned, and the process has been completed on the remaining carcasses. (Courtesy *The National Provisioner* and the Jimmy Dean Meat Co.)

Fig. 4-32. Skin take-off, at this instant, is at the point of bung area. Note the large machines used for this commercial operation. (Courtesy *The National Provisioner* and the Jimmy Dean Meat Co.)

HEAD REMOVAL

Remove the head at the atlas joint. Enter just behind the ears at the natural seam, and follow the seam down both sides of the face, leaving the jowls completely on the carcass. With practice, removing the head is quite simple; however, for the beginner, it tends to be frustrating and complicated. A sharp knife is especially important for the process. First, make a cut directly above the ears (Figure 4-33). Grasp one ear with one hand and push the head down. This helps separate the opening. Next, position your knife straight up and down, and cut into the seam, outlining the jowl (Figure 4-34). Cut as close to the jaw bone as possible, leaving as much of the jowl on the carcass as you can. Do this to both sides of the head.

Now that you have freed the head of the jowl and tissues in front of the back bone, you are ready to actually cut the joint that holds the head. This is called the atlas joint (the first cervical vertebra-skull connection) and is shaped somewhat like a U. Your knife must follow this path ····ʝʊ̈ʟ···· to sever the joint. To be sure that you are located near this joint, stick your knife directly into the center of the joint. If the blade goes in as far as the handle, you are in the right place. If it does not, you are probably one joint too high. You can sever the joint either by inserting your knife inside the center of the joint and cutting both directions from the inside out or by locating the opening and cutting upward and then downward from the outside in (Figure 4-35).

Once you are through the atlas joint, cut through the esophagus and trachea, allowing the head to drop further. To avoid cutting through the mandibular lymph nodes needed for inspection, sever the windpipe and esophagus directly

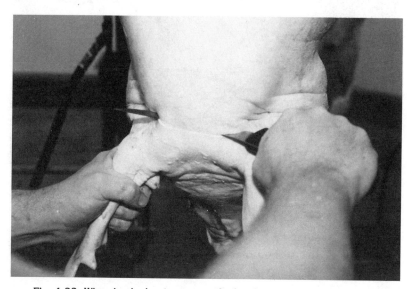

Fig. 4-33. When beginning to remove the head, grasp one ear and pull down while making an inward cut directly above the ears to the backbone.

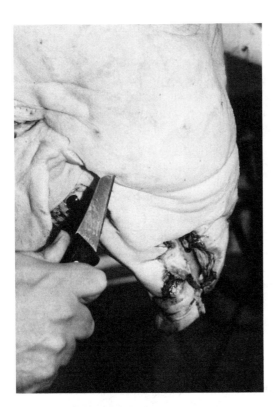

Fig. 4-34. Cutting vertically, follow the natural seam down both sides of the face, separating the head from the jowl and carcass.

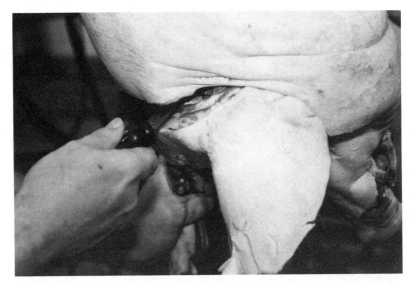

Fig. 4-35. Severing the atlas joint.

above (caudal) the pharynx, leaving the pharynx with the head. This facilitates eviscerating later. Leave head attached by skin, or if it is removed before inspection is complete, keep proper identification. The salivary glands are rather large and are important to identify as they are located directly above the mandibular lymph nodes, which the inspector must examine (Figure 4-36). Head inspection is important to determine if the pig had tuberculosis. The center of the lymph node is sandy if tuberculosis is present and clear if the animal is healthy.

The mandibular lymph nodes are sliced to check for the grainy TB condition. Clean, clear glands indicate a healthy pig. After the head has been inspected and passed by the inspector, you must ''drop the tongue.'' With the hog head resting on a rack, place the fingers of one hand into the windpipe and pull toward you. With your knife, begin cutting around the mouth cavity. Be sure to cut between the salivary gland and the jaw bone so that the salivary gland can be removed. Once all the connective tissue and muscle are freed, the tongue will drop, or you can pull it partly out. Since the salivary glands and pharynx are present on the end of the tongue, they must be removed. Cut directly across the end of the tongue. After the head is cleaned and washed, the inspection stamp must be applied to the forehead. Head weights from forty-two 218-pound Illinois hogs averaged almost 11 pounds each, or 5 percent of the live weight.

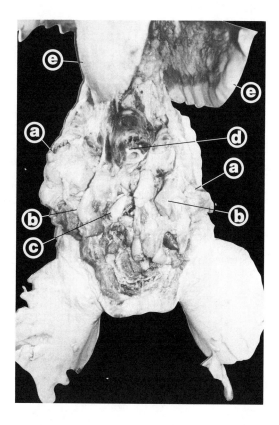

Fig. 4-36. Pig head hanging by skin attachments at the jowl. (a) *Mandibular* lymph nodes (sliced). (b) Salivary glands (right gland sliced approximately in half). (c) One side of *atlas* joint. (d) Epiglottis (Adam's apple). (e) Jowls.

EVISCERATION

The first step in evisceration of a barrow is to remove the pizzle, or penis. Make a superficial cut between both hams (Figure 4-37). Continue cutting down towards the pizzle, cutting a little deeper. As you cut you should notice a white cord. This cord is part of the pizzle. When you get down to the end of the pizzle (the preputial pouch), cut the skin around it (Figure 4-38). Now it is necessary to cut a little deeper in the opening you made so that the pizzle cord will be exposed. Remain outside the abdominal wall but beneath the pizzle. You should cut deeply enough so that you can completely free the pizzle (Figure 4-39). Care must be taken not to cut too deeply when loosening the pizzle, as the intestines are located directly dorsal to the pizzle. When the whole pizzle is free, except at the point of origin which is next to the aitch bone, pull the pizzle back towards the anus and cut away the lean and connective tissue which adheres to the penis. Do not cut the pizzle off. Just loosen the connections around it until you can tell when you are at its point of attachment. If you cut the pizzle off, it is possible that the bladder will leak urine and contaminate the carcass.

A gilt, of course, has no pizzle to remove, thus these steps are omitted, if you are working on a gilt.

Make a small opening through the abdominal wall. Care must be taken not

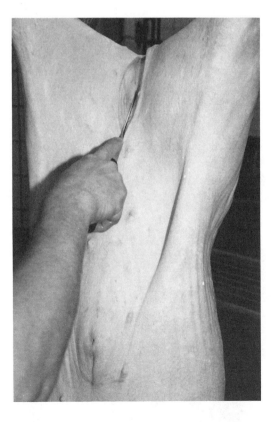

Fig. 4-37. A superficial cut is made between the hams.

Fig. 4-38. The skin around the preputial pouch is cut.

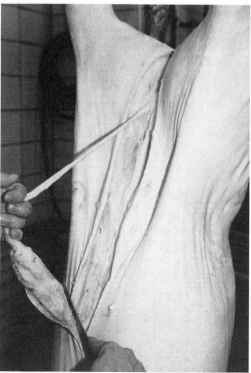

Fig. 4-39. The pizzle is completely freed.

to puncture any part of the intestine. The aitch bone is located directly between the two hams. Separate the hams at the midline indicated by a silvery membrane. In males, this is exactly at the center of the pizzle root. You can feel the exact center of the aitch bone by placing your finger on it (Figure 4-40). The aitch bone can be split at its midpoint where it is joined by cartilage. In Figure 4-41, the knife is positioned in the center of the aitch bone. Press carefully and firmly with a sharp knife. The aitch bone will split easily in a young pig if you are exactly on the center cartilaginous seam. If you cannot get it split, check your location. Do not be overzealous in applying force on your knife, because it may slip and sever the large intestine, which lies immediately behind the aitch bone. The aitch bone is just a few inches long.

Bunging, or loosening the anus and the reproductive tract, is the next step. It may be easier for a beginner to bung from the backside of the animal where the detail of the anus can be easily seen; however, experienced butchers bung from the abdominal side of the carcass. Efficiency in work habits makes this a necessity so that one learns to feel the critical portions of the anus and to cut around them without seeing exactly where the knife is. Holding your knife vertically, cut to the backbone on both sides of the anus (Figure 4-42). The anus is attached to the backbone; therefore, it is necessary to cut behind the bung in order to avoid cutting the intestinal wall (Figure 4-43).

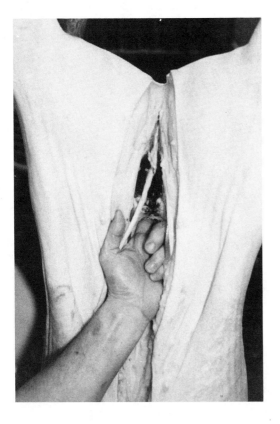

Fig. 4-40. After opening the abdominal wall one can feel the ridge where the aitch (pelvic) bone is joined in cartilage.

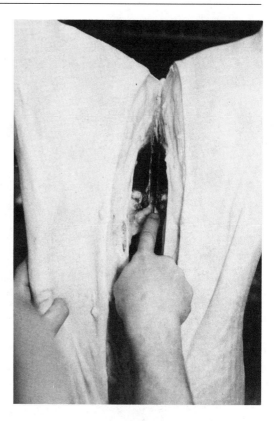

Fig. 4-41. Knife positioned in the very center of the aitch bone where the bone splits easily through the cartilaginous joint.

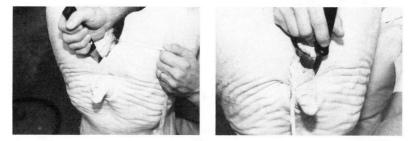

Fig. 4-42. In bunging, a cut is made on first one side and then on the other side of the anus.

You should be aware of the location of the knife tip so that you do not puncture the bung (large intestine). When all the connective tissues of the bung are severed, you should be able to pull the bung up and out to the ventral side, where you can examine carefully for punctures which should *not* be present. Although the sphincter muscle prevents fecal matter from escaping, you may tie off the bung for extra safety. String should always be available on the slaughter floor. Figure 4-44 illustrates the properly loosened bung, including penis (pizzle) and preputial pouch.

Splitting the sternum is the next step. First, open the skin from the stick

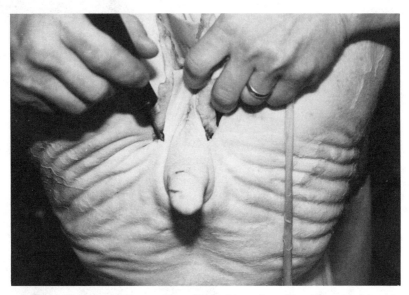

Fig. 4-43. Separate the anus from the backbone by cutting, being careful not to puncture the intestinal wall.

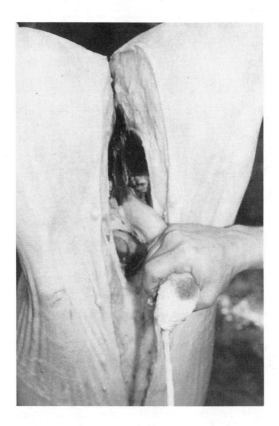

Fig. 4-44. A properly loosened bung, including penis and preputial pouch.

hole to the bottom of the jowl (Figure 4-45). Be sure only one opening is made. Extend the stick hole (the original opening). The sternum bone extends from the stick hole to approximately the first teat. Cut through the skin and fat to the bone (Figure 4-46). You may split the sternum by using a hand saw held with the blade facing up (Figure 4-47). You must saw on the center of the sternum bone, and be careful not to puncture the intestines with the saw at the top end of the sternum. Alternately the sternum may be split with a knife; however, you must cut off-center to get through the cartilaginous connections of the ribs, thus making unequal spareribs on the two sides, with one side having all the sternum bone.

Now you are ready to open the ventral side. It is very important to protect the intestines from the edge of the knife. Have your thumb and index finger at the heel of the knife blade. When cutting, your fingers will separate the knife edge from the intestines. Insert your knife, handle first, into the upper opening where you had split the hams, and with the blade protruding outward, cut through the abdominal wall for the entire ventral length of the carcass. Keep your fingers next to the wall of the abdomen and in front of the intestines. Put pressure on your knife, and cut *in one smooth, continuous motion* down through the opened sternum (Figure 4-48). Once you start, you should not stop. If you stop, the intestines will begin to fall out of the cavity, and you will have difficulty finishing opening the ventral side without puncturing the intestines. Once

Fig. 4-45. Before splitting the breast bone, make an opening by extending the stick opening. Never make a second opening, as it will expose the meat unnecessarily.

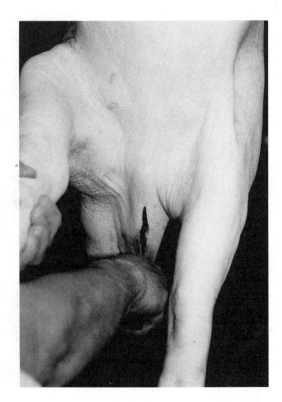

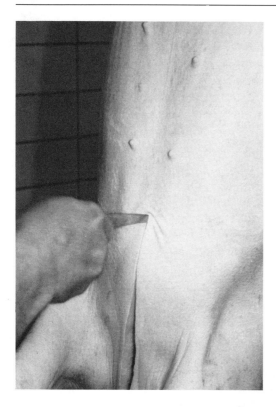

Fig. 4-46. Cut through the skin and fat to the bone, and continue to a point near the first teat.

Fig. 4-47. You may split the sternum by using a hand saw held with the blade facing up.

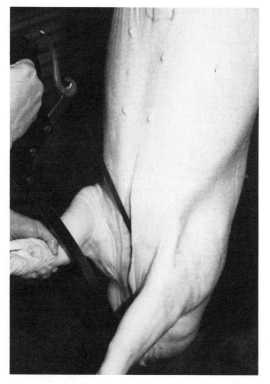

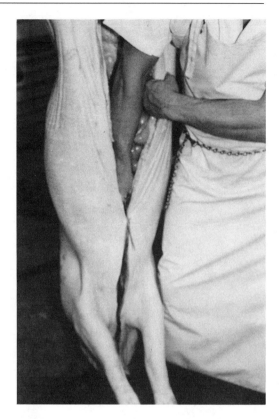

Fig. 4-48. In opening the ventral side, keep your fingers next to the abdominal wall and in front of the intestines. Cut in one smooth, continuous motion.

you have completed the ventral side opening, the intestines will fall out only partially, not completely. Grasp the bung and pull firmly. You will be able to pull the bung a little; however, you will notice two white cords that are connected to the kidneys. These cords are the ureters, which are connections between the kidneys and the bladder. Cut the ureters, allowing the bung to be easily pulled (Figure 4-49). A silvery membrane which is part of the diaphragm separating the abdominal cavity from the thoracic cavity must be cut from the top to the bottom. The membrane goes with the intestines, and the diaphragm muscle should be left in the carcass to be saved for human food (Figure 4-50). Once both sides of the silvery membrane have been cut, you must sever the membranes and connective tissues that are connected to the backbone. Make a cut just above the liver all the way to the backbone. Also, the kidneys and kidney fat are still present in the carcass at this point. As soon as you make this cut, all the internal organs, except the kidneys, will fall out. Therefore, you must be prepared to catch and hold all the intestines and organs. A firm grasp of the stomach and part of the intestines will assure you of not dropping the remaining viscera (Figure 4-51). The pluck, or contents of the thoracic cavity, will not generally come out easily. Therefore, you must cut to the left, inside the thoracic cavity, cutting the pericardium (heart sack) and then push down. By pushing firmly down on the windpipe, the remaining organs will come free. Place the entire tract on a

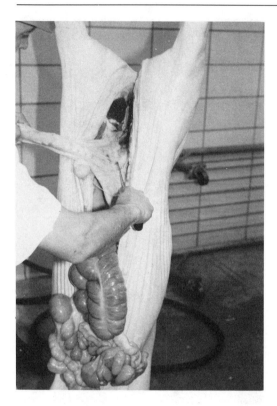

Fig. 4-49. One must sever the ureters in order to successfully pull the bung out of the body cavity.

Fig. 4-50. Cutting the diaphragm to release viscera. (a) Central diaphragm muscle. (b) Silvery diaphragm membrane (the portion to the left has been cut). (c) Liver. Care must be taken to avoid cutting the esophagus, which extends through the diaphragm. Here the knife is posed several inches above the esophagus.

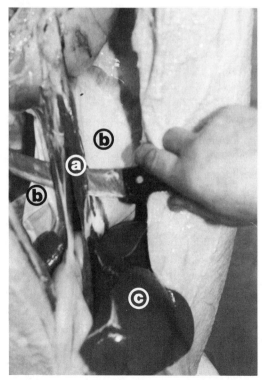

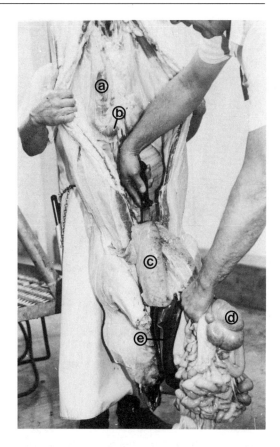

Fig. 4-51. Completing evisceration. A firm grasp on the stomach and intestines is required to prevent the viscera from falling to the floor while the remaining support tissues are cut. (a) Tenderloin *(psoas)* showing through under the leaf fat. (b) Kidney. (c) Lungs. (d) Blind gut *(caecum).* (e) Liver.

table or tray (not on the floor) for inspection. The liver must be removed from the rest of the viscera after it passes inspection. The pale green bag is the gall bladder which is removed and discarded. The gall bladder contains bile salts. To remove it, first sever the white bile duct as far away from the bladder as possible by making only a superficial cut across the bile duct. Next, pull the bile duct and the gall bladder off of the liver. The duct and gall bladder pull off quite easily, if you free the duct and pull firmly and steadily on it (Figure 4-52). It is important that you do not puncture or spill the gall bladder contents. Once the gall bladder is free, discard it in the inedible barrel, and wash the liver.

SPLITTING, INSPECTION, WEIGHING, AND REFRIGERATION

The carcass must be split in half in order for it to adequately chill and for ease of cutting and fabricating cuts. A power circular saw may be used by experienced, qualified workers, as long as it is equipped with an automatic brake in the blade, or a safer, small Wellsaw may be used (Figure 4-53). The safest procedure, if you are a student, is to use the hand saw (Figure 4-54). Begin at the tail bone and work down. Lay the saw on the sacrum so that it will be at a 45° angle.

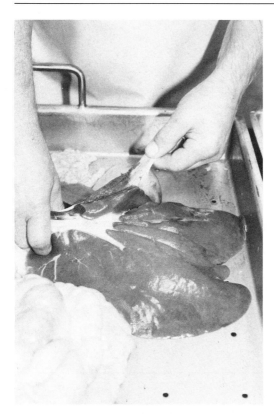

Fig. 4-52. The pork liver is three-lobed. The gall bladder is removed by loosening the duct away from the bladder and pulling steadily on it.

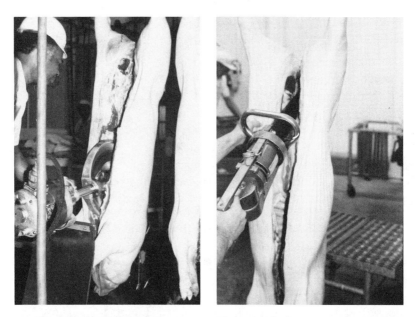

Fig. 4-53. (Left) Halving the carcass with an electric disc saw, such as packers use. (Right) A small Wellsaw is often used in small plants to split hog carcasses. (Left, University of Illinois photo)

Fig. 4-54. The safest procedure for students is to use the hand saw for splitting.

Keep this angle as you saw through the backbone. The carcass should be split precisely in half. Watch for the spinal cord as you are sawing, since this will show that you are in the center as you split each vertebra. You will want to be especially careful when completing the splitting of the shoulder. Sometimes the skin and fat are left slightly attached between the sides to prevent the two separated sides from slipping off the gambrel ends. After the carcass has been split, remove the glands and blood clots that hang in the neck and jowl regions.

Next, remove the leaf fat and kidneys. The leaf fat is a high-quality fat lining the abdominal wall, commonly called kidney fat in beef and sheep. The leaf fat is easily removed. First, pull the end of the fat from the belly at a point just above the diaphragm muscle (Figure 4-55). Continue pulling up. You can also pull the kidneys and kidney fat out with the leaf fat. At the top near the ham, be careful not to damage the femoral artery, which may be used later to cure the hams (Figure 4-56). Facing the ham will be the next step. This process decreases the dressing percentage for the carcass; however, it increases the attractiveness and lowers the fat trim on the carcass. Begin cutting a strip of skin and fat directly above the last teat. Continue pulling with one hand and cutting with your knife (Figure 4-57). The skin and fat may be removed back to the tail bone. Finally, wash the carcass, starting at the top, and spray downward towards the jowl. Once both sides are clean, ask the inspector to examine them for final inspec-

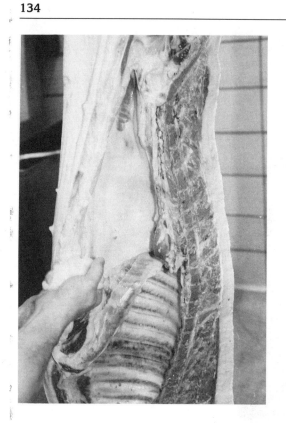

Fig. 4-55. Leaf (kidney) fat is re-moved by starting where it joins the di-aphragm muscle and pulling up and away.

Fig. 4-56. Using care in removing the leaf fat so that the *femoral* artery (b) is not damaged. (a) Aitch bone. (c) Dor-sal *aorta*.

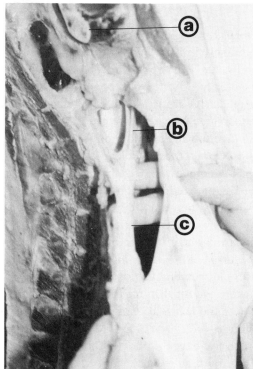

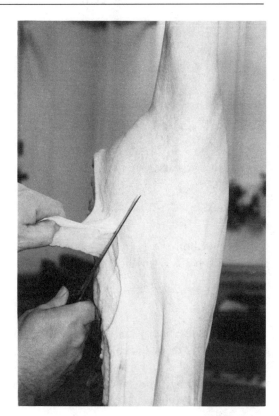

Fig. 4-57. Hams are faced; that is, the fat collar is removed to improve appearance.

tion. The inspector will be looking for hair and other contaminants, as well as abscesses and other abnormal growths. After the inspector has examined the carcass, weigh it, subtracting the weight of the trolley. At the direction of the inspector, apply the inspection stamp to each wholesale cut. Make sure the surface to be stamped is rather dry to prevent running of the ink. This may be done by preblotting the moisture with a sanitary paper towel. The stamp insures that the carcass has been inspected for wholesomeness. Notice the shape and the information contained on the stamp, including the establishment number unique to the plant where the pig was slaughtered (Figure 4-58). All the wholesale cuts are stamped except the loin, spareribs, and Boston butt, and the carcass is marked with an ID number. The loin and spareribs should be stamped on the inside with a smaller-sized stamp, where it will not be trimmed off when the carcass is cut. Place the carcass in the 32°-34° F (0°-1° C) cooler for a 24-hour chill period, unless accelerated (hot) processing is scheduled, in which case the carcass may be tempered (held at intermediate temperatures; that is, 60° F) for several hours or boned immediately. (See Chapter 21.) Where large numbers of warm carcasses are handled, the chill room is generally precooled to a temperature several degrees below freezing to compensate for the heat from the carcasses which raises the cooler temperature considerably.

The cooler shrink on a 24-hour chill will average between 1 and 2 percent,

Fig. 4-58. The inspection stamp (in this case the state stamp) is applied to each wholesale cut.

depending upon the humidity of the cooler. The inside temperature of the ham should reach 38° F for thorough chilling.

WILTSHIRE SIDE

A Wiltshire is an English style side of pork that must conform to rather rigid specifications in order to satisfy the English market. It consists of the entire side, or half, of a hog carcass, minus the head, feet, aitch bone, backbone, tenderloin, and skirt. The sparerib and neck rib are left in the carcass.

The ideal side weighs approximately 60 pounds; minimum 40 pounds; maximum 80 pounds. The ideal length from the fore part of the first rib to the fore end of the aitch bone is 29 inches; minimum 26 inches; maximum 32 inches. Wiltshires are cured and then packed, either in bales or in boxes, for export. They are smoked after they reach their destinations.

Chapter 5

Cattle Slaughter

TABLE 5-1 lists the states in the order of commercial cattle slaughter volume in 1983. There were 36,648,868 cattle slaughtered commercially in the United States and 325,000 slaughtered on the farm, making a total of 36,973,868[1] cattle slaughtered in the United States in 1983. This number of cattle produced 23.5 billion pounds of carcass, averaging 636 pounds per animal. About 95 percent of the commercial cattle slaughter was done under federal inspection. Packers pur-

Table 5-1. Commercial Cattle Slaughter, Number of Head (in Thousands), Ranked by States, 1983[1,2]

Rank	State	Number	Rank	State	Number
1	Texas	6,109.8	24	Utah	258.4
2	Nebraska	5,071.1	25	Alabama	253.7
3	Kansas	4,709.0	26	Georgia	216.5
4	Iowa	3,296.3	27	New Mexico	167.5
5	Colorado	2,154.8	28	Kentucky	164.4
6	California	1,684.8	29	North Dakota	159.0
7	Wisconsin	1,246.2	30	New Jersey	152.1
8	Minnesota	1,164.8	31	North Carolina	146.3
9	Illinois	1,156.4	32	South Carolina	123.8
10	Pennsylvania	948.9	33	Oregon	110.9
11	Washington	931.6	34	Virginia	100.6
12	Missouri	765.2	35	Montana	99.1
13	Idaho	743.5	36	New England[3]	94.8
14	South Dakota	682.4	37	Arkansas	90.6
15	Ohio	630.1	38	Louisiana	67.1
16	Oklahoma	540.6	39	Delaware and	
17	Michigan	536.5		Maryland	65.0
18	Florida	385.3	40	Hawaii	56.0
19	Arizona	361.3	41	West Virginia	38.4
20	Tennessee	324.5	42	Wyoming	13.3
21	Indiana	299.3	43	Nevada	6.2
22	Mississippi	264.6	Total		36,648.9
23	New York	259.0			

[1]*Annual Livestock Slaughter*. March 1984. Crop Reporting Board, SRS, USDA.

[2]Includes slaughter in federally inspected and in other slaughter plants, but excludes animals slaughtered on farms.

[3]New England includes Connecticut, Maine, Massachusetts, New Hampshire, Rhode Island, and Vermont.

[1]*Annual Livestock Slaughter*. March 1984. Crop Reporting Board, SRS, USDA.

chased 78 percent of these cattle directly, 7 percent through terminal markets, and 15 percent through auction markets. Of all these purchases, 33 percent were made on a carcass grade and weight basis.[2] In 1962, direct purchases of cattle by packers accounted for only 43 percent of total purchases, while more than half were purchased in terminal markets. Since that time, the number of cattle purchases in terminal markets has declined, while the number of direct purchases has risen. Carcass grade and weight purchases are a form of direct purchase based on actual carcass grade and weight. Direct sources for steers and heifers are higher (approximately 88 percent) than the total average, while auction sources are more important (approximately 51 percent) for cows and bulls.

Before proceeding to slaughter cattle, be sure that you have proper and safe equipment; that your equipment is ready to use (sharp and clean); and that the animal you are about to slaughter is healthy. That is, you should study and master the first three chapters prior to attempting the study and activities described in this chapter.

FAST (DRIFT) PRIOR TO SLAUGHTER

Cattle will lose from 3 to 4 percent of their weight if kept off their feed for 24 hours. This is referred to as "shrink," or "drift." It has become a general practice for buyers to demand this shrink, either as a mathematical deduction from full-fed weight or as an actual off-feed practice. If reputation means anything at all to feeders, they will never salt or fill their cattle before sale time. A cattle buyer is seldom fooled and then only once by the same person. The carcass yield will betray the perpetrator.

Thus, cattle should be kept off feed at least 24 hours previous to slaughter. Results of a test conducted at the Pennsylvania Agricultural Experiment Station to determine the effects of fasting beef animals for periods ranging from 24 hours to 48 hours on the yield and appearance of the carcasses were definitely in favor of the longer fast period. The fasted animals bled out more thoroughly and were easier to dress, and the carcasses were brighter in appearance than those from cattle allowed feed up to the time of slaughter. (Note previous discussion on hogs.) Undue rough handling or excitement causes the blood to be forced to the outermost capillaries, from which it will be unable to drain as thoroughly as it would under normal heart action. This results in a fiery carcass (pink tinge to the fat), which has lower keeping qualities due to the retained blood.

WEIGHING AND RECORDING

The live beef animal should be weighed on a balanced scale prior to slaughter in order that yield (dressing percent) can be accurately accounted for. *Car-*

---

[2]*Annual Financial Review*. Sept. 1984. American Meat Institute, P. O. Box 3556, Washington, DC 20007.

cass yield is often confused with *USDA Yield Grade*, but the two terms signify distinctly different aspects of animal value. USDA Yield Grade refers to the percent of the carcass that is available for sale as retail product, while carcass yield refers only to the weight of the entire carcass (muscle, fat, and bone), expressed as a percent of the animal's live weight. The other term, *carcass yield*, is more clearly designated as dressing percent. Hot dressing percent (hot carcass weight/live weight × 100) is a function largely of fill and fat; thus, fatter cattle will usually dress higher. However, muscle is more dense than fat, so a lean, heavily muscled, animal may dress higher than a fat, less-muscled animal. In any case, the animal with more fill will dress lower. All cattle will average around 60 percent. Heifers run slightly lower in dressed yield, due mainly to a greater amount of internal fat surrounding the intestines, which does *not* stay with the carcass. Buyers discount prices offered for open heifers as insurance against pregnancy. Pregnancy lowers dressing percentage considerably, in some cases, as much as 10 percent.

STUNNING

One of a number of compression guns can be used on a properly restrained animal. They are made with long or short handles and are of the penetrating or non-penetrating type. The non-penetrating type has a mushroom head. These guns operate on the forehead or behind the poll. Several different types of stunners and the locations of stunning are shown in Figures 5-1, 5-2, and 5-3.

STICKING (EXSANGUINATION)

Standard Method

After the animal is hoisted, make an incision through the hide only, between the point of the brisket and the jaw (Figure 5-4). Insert the knife in front of the brisket at a 45° angle (Figure 5-5), and sever the carotid arteries and jugular vein (Figure 5-6).

Kosher Method

Hoist the animal without stunning it, and cut across the throat in one continuous stroke with a 16-inch knife (chalaf). The killing is done by the rabbi or shohet. If properly done, the animal loses consciousness within three seconds.

ELECTRICAL STIMULATION

Electrical stimulation refers to the passage of an electrical current through a carcass at some point during the slaughtering process. The electrical stimulation of pre-rigor muscle causes the carcass to undergo a rapid series of muscle contractions and relaxations. This rapid series of contractions and relaxations accel-

Fig. 5-1. A power-actuated stunner. It uses .22 caliber rim fire power loads in five graded strengths for effective stunning of all weights of cattle, calves, hogs, and sheep. This packing house tool meets all legislative and humanitarian requirements and delivers effective stunning blows to pates (crown) or backs of heads. Automatic penetrator rod retraction, lever-type trigger, sleeve-style bolt, and lightweight and compact design are among the features of the stunner, made for easy portability and comfortable, one-hand operation. (Courtesy Remington Arms Co., Bridgeport, Connecticut)

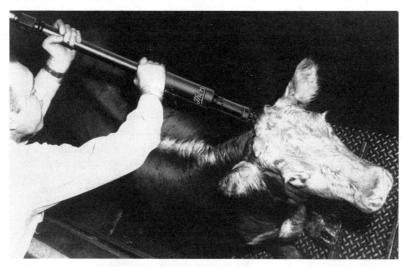

Fig. 5-2. A compression stunner for cattle. In this case, the stunner (penetrating type) is aimed at the medulla oblongata. (Courtesy Thor Power Tool Co., Aurora, Illinois)

Fig. 5-3. × marks the spot to stun the forehead, which is the point at which imaginary lines from each eye to the opposite horn root, or poll, cross.

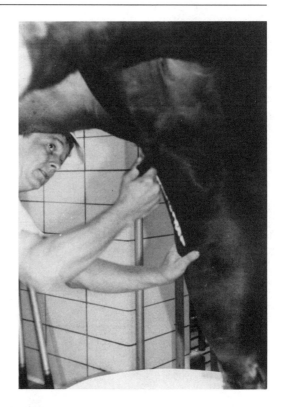

Fig. 5-4. Make an incision through the hide only, between the jaw and the point of the brisket.

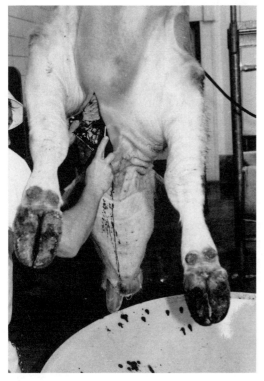

Fig. 5-5. Insert the knife in front of the brisket at a 45° angle.

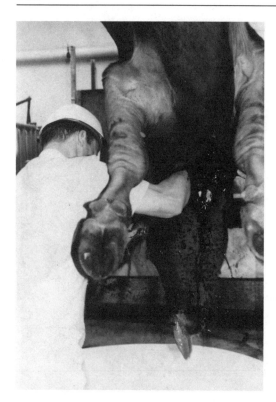

Fig. 5-6. Sever the carotid arteries and jugular vein. Blood is being caught in a large, funneled vat.

erates the rigor process (stiffening of the carcass) by speeding up normal bio-chemical and physical events which occur in muscle after death, that is, depletion of energy stores and decline in muscle pH. For a more detailed discussion of the effects of electrical stimulation on post-mortem muscle changes and meat quality, the reader is referred to pages 716-718.

Since 1978, electrical stimulation has become widely practiced in the beef packing industry. However, it was first recognized for its tenderizing effect on meat in 1951, by Harsham and Deatherage.[3] There was limited commercial interest in the process (in the United States) until scientists from the Texas Agricultural Experiment Station determined that the use of stimulation had a beneficial effect on quality grade designation for beef carcasses. (See Chapters 11 and 21.) Since our marketing system for beef carcasses is highly dependent on the quality grade a carcass receives, this finding provided the economic incentive for the development of commercial electrical stimulation equipment. Early in 1978, the LeFiell Company developed the first commercial stimulator, and rapid adoption of this technology occurred within the industry.

Two methods of electrical stimulation are recognized: high-voltage stimulation (greater than 500 volts) and low-voltage stimulation (less than 75 volts). The high-voltage stimulation equipment is designed for larger packers with continu-

---

[3]A. Harsham and F. E. Deatherage. 1951. *Tenderization of Meat.* U.S. Patent 2,544,581.

ous chain operations that can handle up to 350 head per hour. These systems are considerably more expensive than low-voltage systems and have complex safety mechanisms built in to prevent accidental electrocution of workers. The low-voltage electrical stimulation systems are much cheaper, because the complex safety equipment is not needed. However, low-voltage systems are presently considerably slower and have not been well adapted to high-speed, continuous chain operations. Low-voltage stimulation operates on a slightly different principle than high-voltage stimulation. Low-voltage stimulation must be completed no later than 15 minutes after stunning, because this method relies on the still functioning nervous system which triggers muscle contractions, while high voltage directly stimulates the muscle itself.

The benefit of improved tenderness is no longer recognized as the primary benefit of electrical stimulation. In fact, improvements in tenderness are generally so little that they are often not detected by the human palate. There are, however, a number of other advantages for electrically stimulating beef carcasses. These advantages include dissipation of heat ring (see pages 716 and 717); improved lean color, firmness, texture, and marbling score (see Chapters 11 and 21); improved bleeding of the carcass; and greater ease of hide pulling.

Figure 5-7 illustrates a steer being stimulated with a low-voltage stimulator.

Fig. 5-7. Low-voltage electrical stimulation of beef. (Left) Steer in relaxed state with no electrical current applied. (Right) Steer in state of contraction with 75 volts (1.5 amps) of electricity applied. Electricity is generally "pulsed" through carcasses at 0.5- to 1.5-second intervals. The electricity travels from the two probes (one inserted in each hock) to the nose clamp (ground). Note upward contracted state of forelegs, spreading of hind legs, erect tail, and ears in the photo to the right. The steer's forelegs were secured with a rope to prevent them from violently swinging about.

Note that stimulation is done immediately after exsanguination in this case. Violent muscle contractions cause the forelegs to point upward, the tail and ears to become erect, and the hind legs to spread apart. Also note how excess blood is being "pumped" by the heart due to rigorous contractions of the heart muscle.

RODDING THE WEASAND

This separation of the esophagus from the trachea must be made to allow the abdominal cavity organs and contents and the esophagus to be pulled out of the body cavity separately from the thoracic cavity organs without breakage during evisceration. In some cases, this "rodding of the weasand" is done immediately following bleeding. In a completely on-the-rail slaughter system, animals are never lowered to a skinning cradle. Also, the butcher may feel it is more convenient to do it at this point. A metal rod, with a handle on one end and formed into several spiral loops on the other end, is threaded onto the esophagus just behind the Adam's apple and forced toward the rumen, thus separating the esophagus and trachea (Figure 5-8). Alternately, this step can take place on the skinning cradle after the brisket has been split (see Figure 5-19). A knot is made in the esophagus, or a string or rubber band is put around it to prevent spillage of rumen contents.

Fig. 5-8. Rodding the weasand. (Top) Metal rod with spiral loops is threaded into esophagus and (bottom) forced toward the rumen, thus separating the esophagus and trachea.

HEADING

Open the hide from one horn to one nostril. Continue the opening from the stick down through the center of the jaw and skin out one side of the face (Figure 5-9). Skin out the front of the face (Figure 5-10). Turn the head and skin the opposite side (Figure 5-11). Grasp the jaw in one hand, bend the head back on its poll, and remove the head by cutting through the Adam's apple and the atlas joint (Figure 5-12). Head weights on forty-three 1,037-pound Illinois steers averaged 29 pounds and ranged from 1.5 percent to 4.5 percent of live weight.

Fig 5-9. Heading. Skinning out one side of face. (Courtesy University of Illinois)

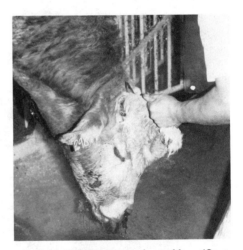

Fig 5-10. Skinning out front of face. (Courtesy University of Illinois)

Fig 5-11. Skinning out other side of face. (Courtesy University of Illinois)

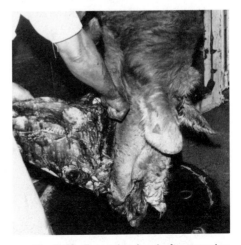

Fig 5-12. Removing head after severing atlas joint. (Courtesy University of Illinois)

SHANKING OR LEGGING

Place the skinning rack under the withers, and lower the animal onto the rack. The feet must be removed prior to opening up the hide to prevent contamination of the carcass with manure and dirt dropped from the hooves. Open the hide on the rear of the foreshank and the rear of the hindshank, continuing the cut to the midline to be made on the belly from the neck to the bung. On the foreshank sever the tendon to release tension by cutting across the shank. Skin out the shanks in order to see the joint clearly, and remove them at the flat, smooth joint below the knee and hock. This joint is at the enlargement, about 1 inch below the knee joint, just where it tapers down to the cannon bone (Figure 5-13). You have found the correct joint if you expose two flat bones in each joint face. If you expose three rounded bones in each joint, you are too high into the knee joint itself. On the hock, it is about 1 inch from where the taper takes

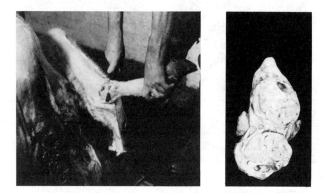

Fig. 5-13. (Left) Removing the foreshank. (Right) Close-up of flat joint, proper location of foreshank removal. Note two flat bones in each joint face.

Fig. 5-14. Removing the hindshank.

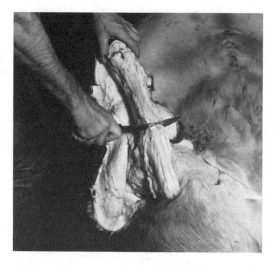

place. A decided groove is evident when the knife rests at the proper spot. It is *especially* important to remove the hindshank at the proper location below the hock joint itself in order to preserve the tendon anchors for carcass hanging.

Cut around to either side, and then grasp the shank near the foot, and give a sharp thrust downward and outward from the stifle joint (Figure 5-14). Occasionally this flat joint will have ossified in aged animals, in which case it will be necessary to use a saw.

SIDING

Open the hide down the middle from the stick opening to the bung. Open the hide in an M configuration over the brisket by cutting the hide diagonally from the point of the brisket to the inside base of the foreshank on each side.

Fig. 5-15. Siding can be started at the fore end or rear end. This beef is held on its back by means of a skinning rack.

Siding can be started at the fore end or rear end (Figure 5-15). To side, grasp the hide firmly with an upward pull, and with long, smooth strokes of the skinning knife, remove the hide down over the sides. Air or electrically powered rotary skinning knives are popular in many plants, since they make the job of skinning much easier. The bevel of either knife must be flat to the hide to avoid making cuts or scores (Figure 5-16). This is one of the most difficult tasks in the skinning operation and requires considerable experience before satisfactory progress can be made. Note the proper hide pattern in Figure 5-17. Attempt to avoid scoring or cutting the hide, as this lowers its value for leather.

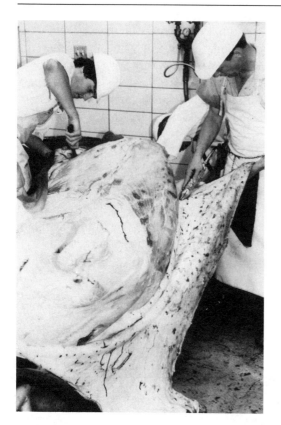

Fig. 5-16. In siding, with either a conventional hand knife or a power rotary skinner, hold the hide tightly, and keep the bevel of your knife flat to the hide in order to avoid making cuts or scores in the hide.

Fig. 5-17. The pattern of the hide. The dotted line indicates where the opening is made in the hide. (See also Figure 10-3.) (Courtesy USDA)

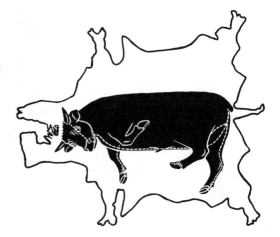

OPENING

After the hide has been sided down to the rail on the skinning rack on each side, the brisket is opened. Cut through the fat and muscle of the brisket at its center with a knife, and then saw through the sternum (Figure 5-18).

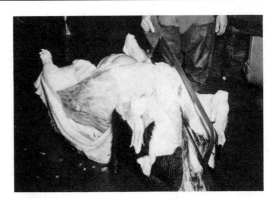

Fig. 5-18. Splitting the brisket.

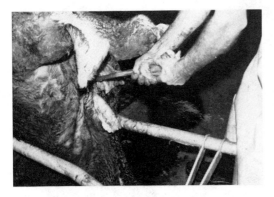

Fig. 5-19. Rodding the weasand. Esophagus (weasand) has been separated from the trachea and is here tied in a knot to prevent rumen spillage on carcass. (Courtesy University of Illinois)

Once the chest cavity is opened, rod the weasand; that is, separate the esophagus from the trachea (Figure 5-19), unless this was accomplished earlier, while the animal was suspended.

A beef tree (a large, spreading gambrel) or beef trolleys, designed to operate in conjunction with a power lift and spreader, may be inserted in the hocks so that the carcass can be raised to a convenient height for splitting the aitch bone, bunging, tailing, and removing the hide from the round and rump. In the case of a male, the pizzle (penis) must be removed first. Cut it loose from the belly wall and back to its origin at the pelvic junction. Do not cut it off while pressure is on the urinary bladder, or the carcass will become contaminated with urine. It is possible to split the aitch bone on very young cattle with your knife, but to do it successfully, you must be *exactly* on center. Also, you run the risk of breaking your knife blade and/or puncturing an intestine. Thus the saw is preferred for its speed, safety, and accuracy. With the pizzle out of the way, proceed first with your knife to separate the rounds. Each top round muscle is covered with a tough membrane, and where the two join over the high point of the pelvis, they form a decidedly heavy, white-appearing membrane. Follow the membrane with your knife, and avoid cutting into the muscle. When you reach the pelvis, use the saw and carefully saw through the center of the aitch bone (Figure 5-20).

In bunging a male, pull the pizzle taut posteriorly, and loosen the anus by cutting completely around it (Figure 5-21). Be careful while severing all support-

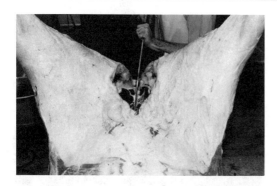

Fig. 5-20. Splitting the aitch bone with a saw, while the carcass is raised to a convenient height.

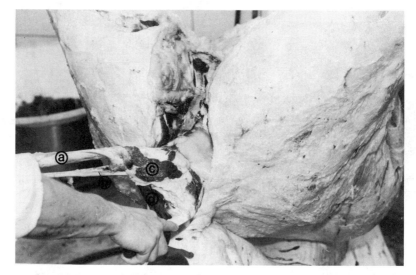

Fig. 5-21. Bunging a male by holding tension on the pizzle while cutting completely around the anus (bung gut). Care is taken while severing all supporting connections that the intestine itself is not punctured. (a) Penis. (b) Retractor penis muscle. (c) Thin layer of *gracilis* muscle from inside round, adhering to bung. (d) Anus.

ing connections that the intestine itself is not punctured. When bunging a female, you must grasp the anus itself, in lieu of the penis to hold it steady for bunging. Once loosened, the bung is tied with a cord and pushed through into the abdominal cavity, where it can be reached from the belly side.

To remove the tail, skin around its base and split the hide the entire length of the tail. Sever the tail two joints from the body, and skin entirely around its base. By placing a paper towel over the skinned base stub, the tail can be easily pulled out the remainder of the way (Figure 5-22).

Some butchers prefer to start splitting the carcass at this half-hanging stage. The carcass is more stable in this position and thus promotes more accurate splitting of the sacral vertebrae section of the backbone.

Before eviscerating, you should completely remove the hide from the carcass. Removal of the hide from the back is called backing. This operation consists of running the knife around the back between the hide and the carcass and letting the hide drop of its own weight (Figure 5-23). The hide may also be removed by cutting in from either side as a completion of the siding operation.

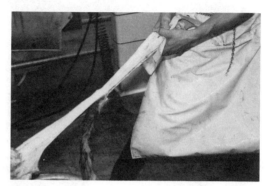

Fig. 5-22. Removing the tail.

Fig. 5-23. Backing, that is, letting the hide drop of its own weight after lightly cutting the connective tissue attachment between the hide and carcass.

EVISCERATING

Insert the handle of the knife in the abdominal cavity with the blade leaning upward and your fist protecting the intestines and rumen. Open the belly cavity, using one continuous motion (Figure 5-24).

Loosen the fat and membrane that hold the bung gut and bladder to backbone, and cut the ureters that connect the kidneys and the bladder (Fig. 5-25). The kidneys and kidney fat remain in a beef carcass. The reason is largely custom, although the kidney fat does protect the valuable tenderloin muscle from drying out and darkening during aging. Loosen the liver with the hands, and then sever it from the backbone with a knife. Place on an inspection tray for observation by inspector. Following inspection, remove the gall bladder by cutting across the top of the bile duct at the center of the liver and peeling it, rather than cutting it out.

Pull the previously loosened esophagus up through the diaphragm to allow the abdominal viscera to fall freely into the inspection cart.

The membrane separating the abdominal from the thoracic cavity is called

Fig. 5-24. Opening the belly cavity in one continuous motion by inserting the knife handle into the opening at the cod area, pointing the blade upward, and using the fist and handle to protect the intestines and rumen.

Fig. 5-25. A few well placed cuts of the supporting membranes and ureters connecting the bladder and kidneys will allow the paunch and intestines to drop into the gut cart.

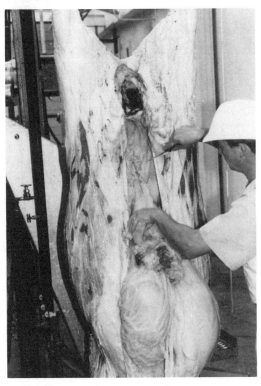

the diaphragm and consists of the diaphragm muscle and the membrane joining the muscle. Cut out only the membrane, as the diaphragm muscle is good, edible meat and is commonly known as the *skirt* or hanging tenderloin. The organs that lie in the thoracic cavity are called the *pluck* and consist of the heart, lungs, and windpipe (Figure 5-26).

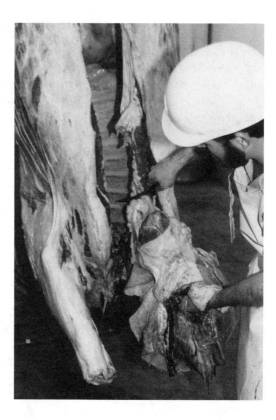

Fig. 5-26. Removing the pluck, consisting of the heart, lungs, and windpipe. The diaphragm muscle remains in the carcass.

HALVING (SPLITTING)

Splitting the beef into sides by sawing through the exact center of the backbone may be accomplished by using a hand or an electric beef-splitting saw. Figure 5-33 shows an electric saw in use in a commercial plant.

Using a hand saw, stand on the belly side of the carcass, and saw through the caudal vertebrae to the sacrum and through the sacrum and the lumbar vertebrae. Special care should be taken to split each superior spinous process of each vertebra, since this has an important bearing on the weight of wholesale and retail loin, rib, and chuck cuts. Since the feather bones in the dorsal region (rib area) of the backbone are quite long and narrow, it is desirable to first saw through the main body of the backbone. This is done by pointing the saw toward the tail and sawing at an angle to the backbone. Sawing through each individual dorsal vertebra at a 45° angle assures the operator of a 50-50 split of the feather bones, since the spine (if the saw is in the center) steers the saw blade and holds it steady. After the splitting of the forequarters is completed, the vein is removed from the inside of the neck.

HIDE YIELD

The weight of the hide varies with the breed of cattle. Herefords carry the heaviest hides, followed by the Angus, Holsteins, and Brown Swiss. Shorthorns and the dairy breeds carry the lightest weight hides. Hides from the average run of cattle slaughtered by the large packing concerns average 7 percent of the live weight of the animals. Slaughter records of the purebred steers killed and dressed in the meats laboratory of The Pennsylvania State University and verified at the University of Illinois show that Hereford hides average 8.5 percent, Angus 7.5 percent, and Shorthorn 6.5 percent of the live weight of the animals.

Hides should be trimmed (ears, lips, and fat off), spread out, hair side down, in a cool place and given a liberal application of salt. It requires from 15 to 18 pounds of salt to cure a 50- to 70-pound hide. For best results, the hide cellar should have a uniform temperature of 50° to 60° F and lower for long-time storage. Hides will shrink from 15 to 25 percent in curing and from 40 to 50 percent if left to dry.

WASHING AND SHROUDING (CLOTHING)

All blood should be washed off both the inside and the outside of the carcass, and the shoulders should be pumped by working the shanks up and down. Cold or lukewarm water is used to wash the carcass. Shrouds made from unbleached duck cloth are immersed in warm water, stretched tightly, and pinned over the outside of the sides of warm beef before they are moved into the cooler. The shrouds absorb the blood, smooth the external fat covering, and cause the fat to appear white and dense. The shrouds, or cloths, are removed after the carcass has chilled. A practice that has been utilized is to wet the clothing in a 14° to 18° salimeter strength salt brine at a temperature of 115° to 125° F. It is claimed that brined cloth has greater adhesiveness and that it helps the cloth to absorb blood and more of the bruise discoloration which may be on a carcass. Thus, the main purpose of the shroud is to improve the appearance of the carcass for the potential buyer. With the advent of increased fabrication of beef carcasses into wholesale and retail cuts in packing plants, some packers have discontinued the use of the shroud in an effort to save unnecessary costs. Cooler shrinkage averages from 1 to 3 percent in the first 48 hours or from 4 to 6 percent if aged for two weeks. The amount of cooler shrinkage is dependent upon cooler humidity and the finish of the carcass.

COMMERCIAL BEEF SLAUGHTERING OPERATIONS

It is not unusual to find operations in the United States killing some 2,000 cattle in an eight-hour day, or 250 per hour. To accomplish such an endeavor, considerable use is made of mechanized equipment as can be noted in Figures 5-27 through 5-33.

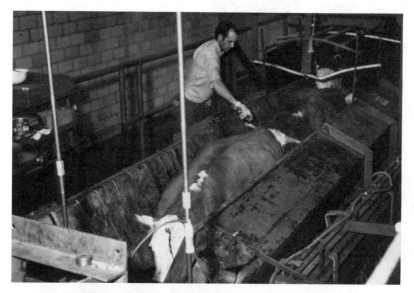

Fig. 5-27. Easily and quickly, employee places stunning instrument on head of restrained animal in commercial operation. (Courtesy *The National Provisioner*)

Fig. 5-28. Making full use of power tools, employees sever front feet and horns, if any, from bled carcasses. (Courtesy *The National Provisioner*)

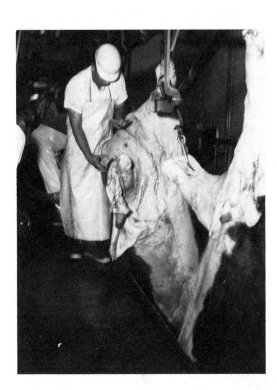

Fig. 5-29. At second transfer station, carcasses are shown being taken up as employee performs butting operation on next carcass. (Courtesy *The National Provisioner*)

Fig. 5-30. Final step in hide preparation for strip-off includes freeing of small of the back as shown. (Courtesy *The National Provisioner*)

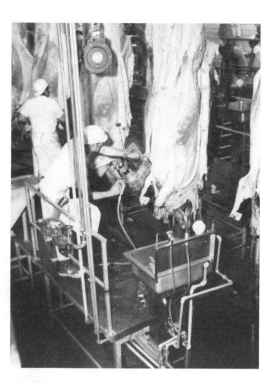

Fig. 5-31. With power tool, employee at hide removal station opens brisket. (Courtesy *The National Provisioner*)

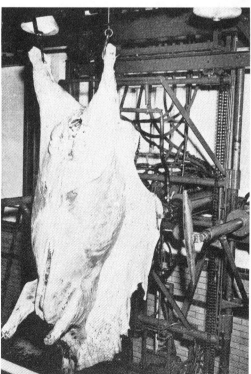

Fig. 5-32. Can-Pak hide puller.

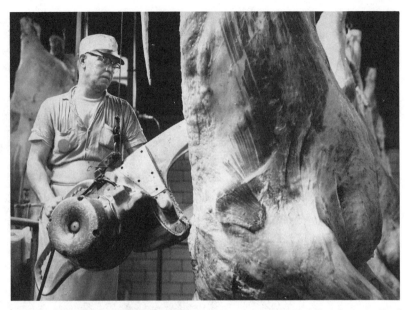

Fig. 5-33. At this station, a cantilever platform moves the butcher downward and forward as he splits the carcass. Note this commercial operator is splitting from the back side because this method fits into the company's completely mechanized operation. (Courtesy Monfort of Colorado, Inc.)

Canada Packers, Ltd., was the designer of a system of on-the-rail dressing of beef that has eliminated the operator having to stoop while working. There are 14 stations in the operation. Driving, knocking, and shackling the beef are performed by one person at station one. Sticking and scalping (heading) are performed at station two. One worker legs, butts, inserts trolley, and removes the shackle at the third station, while a fourth worker legs, butts, and inserts trolley in the second leg at the fourth station. At the fifth station, one worker removes the front foot, performs work on the brisket, dehorns, and removes the head. One worker rims, clears shanks, and works on the chuck and neck at the sixth station. At station seven, work is performed by a single worker on the rump; the bung is dropped, and the tail is pulled. The rosette (*cutaneous trunchi* muscle on the surface of the outside chuck) is cleared, using an air-operated skinning tool, flanking is performed, and the hide is pulled by the machine operated by one worker at station eight. The brisket is sawed, and the carcass is eviscerated at station nine, and completion of the hide pull is made at station ten. In the pull-off, the hide drops directly into a chute. The splitter and scriber at station eleven operates on an elevating bench, using a foot switch. Trimming is done at station twelve, weighing at station thirteen, and shrouding at station fourteen.

Several newer hide-pulling machines are marketed now, some of which pull the hide down. One advantage of a down pull is that hide-borne contaminants, such as loose hair, dirt, manure, etc., either fall to the floor or drop progressively away from the carcass to the lower portions of the hide.

CARE OF THE BEEF CARCASS

A thorough chilling during the first 24 hours is essential, otherwise the carcass may sour. This occurs first at the hip joint—a deep-seated joint from which the heat is slow to escape.

A desirable temperature for chilling warm carcasses is 33° F. Since a group of warm carcasses will raise the temperature of a chill room considerably, it is good practice to lower the temperature of the room to 5° below freezing (27° F) before the carcasses are moved in. Temperatures more severe than this can cause cold shortening, an intense shortening of muscle fibers, causing a toughening effect (see pages 711 and 715).

Chapter 6

Sheep and Lamb Slaughter

TABLE 6-1 lists the states in the order of commercial sheep and lamb slaughter volume in 1983. There were 6,619,383 sheep and lambs slaughtered commercially in the United States and 173,000 slaughtered on the farm, making a total of 6,792,383[1] sheep and lambs slaughtered in the United States in 1983. This

Table 6-1. Commercial Sheep and Lamb Slaughter, Number of Head (in Thousands), Ranked by States, 1983 [1, 2]

Rank	State	Number	Rank	State	Number
1	Colorado	1,527.7	24	Kansas	6.0
2	California	1,464.4	25	Oklahoma	5.2
3	Texas	751.5	26	Louisiana	4.4
4	Iowa	551.0	27	Montana	4.2
5	South Dakota	508.0	28	Nebraska	3.2
6	Minnesota	476.3	29	Nevada	3.0
7	Michigan	332.9	30	Wyoming	2.5
8	Illinois	169.0	31	Arizona	2.0
9	Washington	162.2	32	Arkansas	1.7
10	Pennsylvania	150.8	33	Tennessee	1.6
11	New Mexico	141.0	34	West Virginia	1.4
12	New York	52.8	35	North Carolina	1.2
12	New Jersey	52.8	36	North Dakota	1.0
13	Delaware and Maryland	52.1	37	Florida	.7
14	Utah	31.1	38	Georgia	.6
15	Missouri	30.9	39	South Carolina	.2
16	Ohio	30.1	39	Alabama	.2
17	New England[3]	27.3		Hawaii	—
18	Oregon	17.9		Mississippi	—
19	Indiana	13.0			
20	Wisconsin	11.6	Total[4]		6,619.6
21	Kentucky	11.1			
22	Idaho	8.9			
23	Virginia	6.1			

[1]*Annual Livestock Slaughter*. March 1984. Crop Reporting Board, SRS, USDA.

[2]Includes slaughter in federally inspected and in other slaughter plants but excludes animals slaughtered on farms.

[3]New England includes Connecticut, Maine, Massachusetts, New Hampshire, Rhode Island, and Vermont.

[4]States with no data printed are still included in U.S. total; not printed to avoid disclosing individual operations.

[1]*Annual Livestock Slaughter*. March 1984. Crop Reporting Board, SRS, USDA.

number of sheep and lambs produced 380.4 million pounds of carcass, averaging 56 pounds per head. Of the commercial slaughter of sheep and lambs, 97 percent was done under federal inspection. Packers purchased 81 percent directly, 5 percent through terminal markets, and 14 percent through auction markets. Of all these purchases, 29 percent were made on a carcass grade and weight basis.[2]

Somewhat in contrast to the decentralization occurring in hog and cattle marketing and slaughtering, sheep and lamb slaughtering is concentrating in fewer and larger, more specialized plants. Many large areas of the country are no longer served by either adequate live markets or slaughtering facilities.

Before proceeding to slaughter sheep and lambs, be sure that you have proper and safe equipment; that your equipment is ready to use (sharp and clean); and that the animal you are about to slaughter is healthy. That is, you should study and master the first three chapters before going ahead with this chapter.

FAST (DRIFT) PRIOR TO SLAUGHTER

A 24-hour fast previous to slaughter is probably of greater importance with sheep than with other forms of livestock. It not only facilitates the eviscerating process but also adds materially to the bright appearance of the carcass. See previous discussions on hogs and cattle. The removal of the pelt is made somewhat easier by a limited fasting period; however, water must be present during the fasting to avoid dehydration, which results in tissue shrink and difficulty in pelt removal.

HANDLING

Never lift a sheep by grasping the fleece, as this causes a surface bruise on the carcass. Instead, place one hand under the jaw and the other at the dock, and lead the lamb. Grasp a sheep by the leg when catching it.

YIELD (DRESSING PERCENT)

Dressing percent is calculated as follows: Carcass weight/live weight × 100 = D. P. Most sheep and lambs will average in the 50 percent range. Normally, dressing percent is calculated on hot carcass weight compared to purchase weight of the sheep. Heavy-muscled and fatter lambs can go up to 54 to 55 percent, depending on weighing conditions. Lambs that are full will dress less, as will thin ewes; lightweight, unfinished lambs; and sheep that are carrying fleece.

---

[2]*Annual Financial Review*. Sept. 1984. American Meat Institute, P.O. Box 3556, Washington, DC 20007.

TOOLS AND EQUIPMENT

A 5-inch, curved boning knife or a thin, well ground skinning knife is best adapted for the job of pelting a sheep.

A trough-like skinning rack on legs about 18 inches high for holding a sheep on its back is very handy. The trough is 6 inches wide at the bottom with sloping sides 6 inches high. These racks are also used for veal and are called lamb or veal cradles. In the absence of a rack, use a table or a platform.

STUNNING AND STICKING (EXSANGUINATION)

A sharp blow on top of the poll will stun sheep that do not have horns. The use of the captive bolt powered by cartridges or air with the mushroom head (Figure 6-1) or the use of an electric stunner (Figure 6-2) is recommended.

Hoist or place the stunned sheep on a table, or on a sheep and veal rack. Grasp the jaw or ear with one hand, insert the knife behind the jaw, blade-edge outward (Figure 6-3), and draw the knife out through the pelt (Figure 6-4), severing the jugular veins and carotid arteries. Blood yield on lambs will average about 3 to 5 percent of live weight.

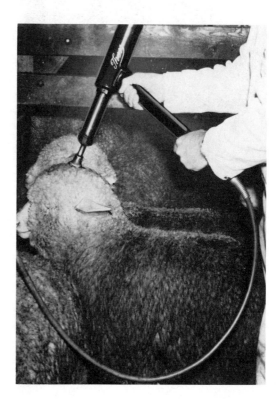

Fig. 6-1. A compression stunner. This particular model uses compressed air—no cartridges. Its high velocity blow is effective in painlessly stunning cattle, hogs, and sheep. It is furnished with either a penetrating or a non-penetrating head. The above illustrates the mushroom, or non-penetrating, type head used for veal and lamb. (Courtesy Thor Power Tool Co., Aurora, Illinois)

Fig. 6-2. An inexpensive electric stunner. (Courtesy University of Illinois)

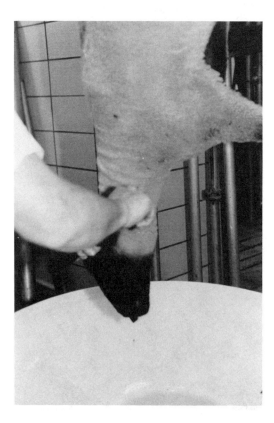

Fig. 6-3. When sticking a lamb, grasp the jaw or ear with one hand, and insert the knife behind the jaw, blade edged outward.

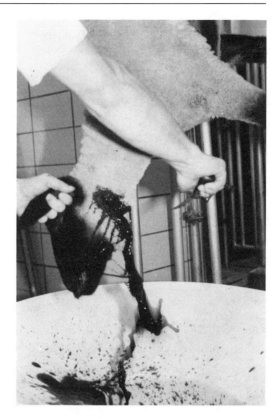

Fig. 6-4. Draw the knife out through the pelt, severing the jugular veins and carotid arteries.

PELTING

With the sheep lying on its back in the rack, grasp a foreleg, or secure foreleg to scabbard chain with a bent shroud pin, and open the pelt down the front of the leg from the break joint to the breast (Figure 6-5). Do the same on the other foreleg, having the two cuts meet in a point in front of the breast. Skin out the forelegs at this time. The front foot is removed at the "break joint" in lambs (Figure 6-6). The break joint is recognized by a swelling in the cannon bone at its lower extremity just above the hoof (wrist or spool joint). In yearling mutton, and mutton older than 15 months, the break joint is ossified, and the front foot is removed at the spool joint. A comparison of a spool and break joint is shown in Figure 6-7.

Grasp a hind leg and open it down the back of the leg from the hoof to the bung. The knife should be held fairly flat to the carcass and leg in making the opening in order to avoid cutting the tendon and the fell or exposing the muscle. Remove the foot at the lowest possible joint (Figure 6-8) in order to leave an intact tendon anchor for hanging the carcass. Loosen the tendon over the back of the hock, and then proceed to skin out the opposite hind leg.

Grasp the cut edge of the pelt and pull, at the same time using the fist of your other hand to fist the pelt loose (Figure 6-9). Loosen it around the flank,

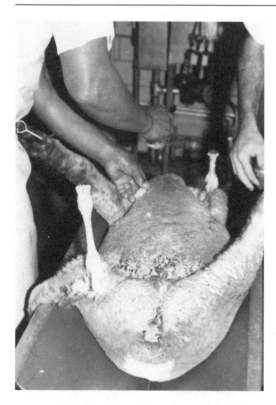

Fig. 6-5. Technique used to skin out forelegs and hind shanks. Butchers hold legs taut with shroud pin hook while skinning. Note break joint on foreleg, spool joint on hind leg. (Courtesy University of Illinois)

Fig. 6-6. Skinning out a foreleg. (A) Location of spool or mutton joint. (B) Location of lamb or break joint. (Courtesy University of Illinois)

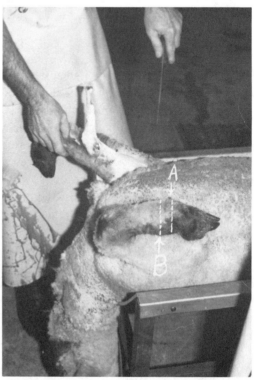

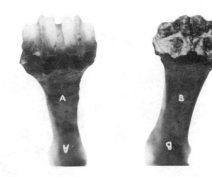

Fig. 6-7. (A) A spool or mutton joint. (B) A lamb or break joint.

Fig. 6-8. Disjointing the foot on the hind leg. The dotted line across marks the break joint. If the foot were removed at this joint, the anchorage of the tendons (A) would be weakened, and the legs would have to be tied together at the hocks instead of the tendons.

cod, or udder area and over the brisket. The pelt may then be fisted over the belly by turning and pushing the fist against the pelt, *not* against the carcass.

A strong cord may be used to tie the tendons of both hind legs together, or an S-hook may be inserted through the tendons of both legs to engage in an overhead trolley or rack. In order to prevent back and groin injuries to yourself, use a mechanical hoist to hoist the lamb to the overhead trolly or rack. If a mechanical hoist is not available, get help. The job should be done by two persons.

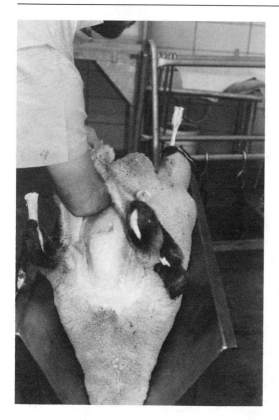

Fig. 6-9. Fisting the pelt off the belly while the sheep remains on the cradle.

Open the pelt down the center of the belly, and fist it loose around the side and up the leg, being careful to avoid breaking the fell—a thin, colorless, connective tissue membrane that separates the pelt from the carcass and that should always be left intact on the carcass (Figure 6-10). It is safer to fist up the leg than to pull the pelt down the leg. Unless the skin is started exactly right, pulling the pelt down the leg may tear the protective fell and expose the muscle. Sever the bung by cutting across it where it is attached to the pelt, and pull and fist the pelt from the tail. Then, fist the pelt over the shoulder and pull off the back and neck. It is necessary to maintain clean hands when fisting to avoid contaminating the lamb carcass with wool and dirt from the pelt.

Sever the head at the atlas joint. Skin out the head and remove the tongue, if there is a market for edible sheep heads and tongues. In many areas there is no market for these items, so they are simply removed and discarded to be made into animal food and fertilizer. Head weights from forty-two 118-pound Illinois lambs averaged almost 6 pounds and ranged from 4.5 to 5.5 percent of live weight.

The trachea (windpipe) and esophagus must be separated (Figure 6-11). This allows the abdominal cavity organs and contents to be removed separately from the thoracic cavity organs later during evisceration. The loosened esophagus (properly tied off or knotted) slips easily out of the thoracic cavity while still se-

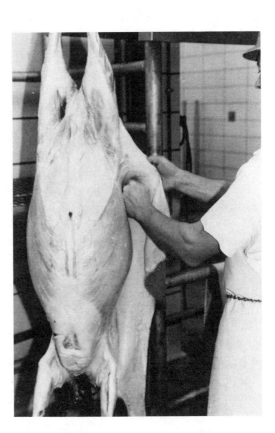

Fig. 6-10. Fisting a lamb on the rail, being careful not to break the fell.

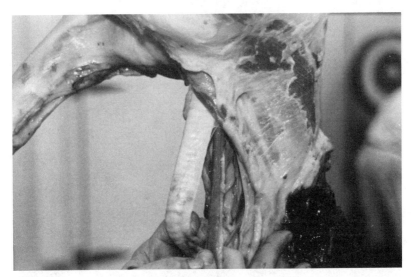

Fig. 6-11. The trachea (white, ribbed vessel) and esophagus (dark, smooth vessel) must be separated for ease of evisceration.

curely fastened to the rumen. You must split the breast prior to evisceration and remove both the abdominal and the thoracic cavity contents at the same time, if this prior separation was not completed. Either step must be taken to avoid breaking the digestive tract and contaminating the carcass during evisceration.

EVISCERATING

Since the carcass is not split in lamb slaughter, the pelvic bone is not split, because the small size allows complete chilling of the intact carcass. Some butchers prefer to split the breast during slaughter and prior to evisceration. The safest method is to use the hand saw as is done in hog slaughter. The split breast allows for complete evisceration of abdominal and thoracic cavities at one time.

Some butchers leave the breast intact, in which case they must have previously loosened the esophagus from the trachea in order to eviscerate the abdominal cavity separately.

Taking a position to the rear of the carcass, cut around the bung, and loosen it (Figure 6-12). On a male carcass, loosen the pizzle (penis) and remove it from the surface of the abdomen back to its anchor at the pelvic junction (aitch bone). Holding it taut to keep it out of the way, make an opening at the cod (Figure 6-13). Cut the pizzle off as near as possible to its root deep in the crotch. Insert the first and second fingers to guard the point of the knife, or insert the handle into the abdominal cavity as in pork and beef, and continue the opening to the breast (Figure 6-14). Grasp the loose bung. Use the knife to sever the ureters that lead to the kidneys. These are strong and will tear out the kidneys and the kidney fat if they are not cut (Figure 6-15). The stomach and intestines are now easily pulled out. The liver is removed from the abdominal tract, and the bile

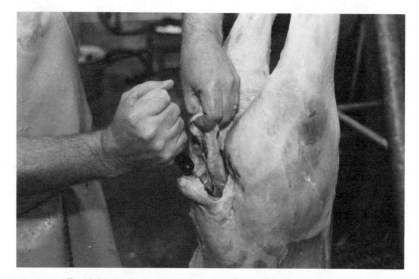

Fig. 6-12. Bung the sheep by carefully cutting around the anus.

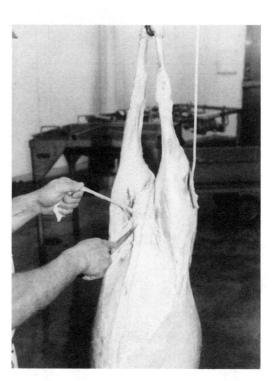

Fig. 6-13. Hold the loosened pizzle tight to keep it out of the way while making a small opening at the cod. Then cut the pizzle off as near as possible to its root, deep in the crotch.

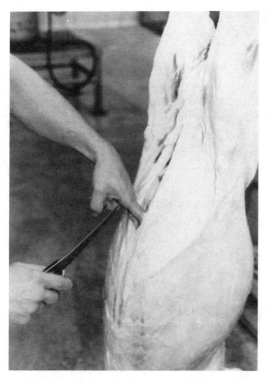

Fig. 6-14. Insert first and second fingers to guard the point of your knife as you open the belly.

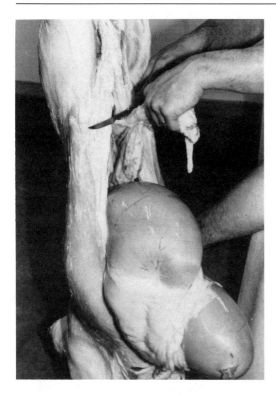

Fig. 6-15. Cutting the ureters that lead from the kidneys to the bladder, and other supporting tissues, in order to release the intestinal tract.

duct is removed from the liver. Split the chest with a knife or saw, and remove the pluck, if this has not already been done. Packers do not remove the spleen or melts.

Wash the inside and outside of the carcass, and especially wash the blood out of the neck and chest cavity (Figure 6-16). Trim all scrag ends off the neck, and double the foreshank up against the arm, using a tendon from the foreshank (Figure 6-17) to hold it. This operation plumps the shoulder and keeps the foreshank out of the way in a crowded commercial lamb cooler.

PELTS

The pelt is rubbed with fine salt on the flesh side to preserve it until the time it is sold. In packing houses, a depilatory paste made of lime and sodium sulfide is spread on the skin side of pelts, and the next day the wool can be pulled. The skins are sold to tanners. Wool is the chief by-product of sheep slaughter and ranges from 2 to 8 pounds of pulled wool per head.

HOTHOUSE LAMBS

Hothouse lambs are rated by epicureans as being the most delectable of the lamb age groups. They are dropped during the months of October, November, December, and January and are marketed between the ages of 6 to 10 weeks.

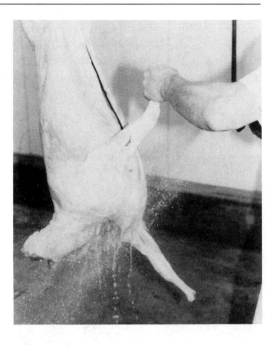

Fig. 6-16. Washing the neck and breast.

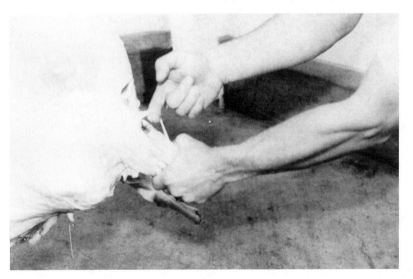

Fig. 6-17. Using the loosened tendon to hold the doubled up foreshank.

The name "hothouse" is rather ambiguous, but its name indicates that these lambs have been housed in barns or sheds where they are protected from the cold weather.

Hothouse lambs may be defined as "lambs that are dropped out of the regular lambing season and marketed at live weights ranging from 25 to 60 pounds." This makes the hog-dressed weight, which is about 70 percent of the live weight, range between 18 and 42 pounds. "Hog dressed" means dressed with head and pelt on but with feet and viscera removed (Figure 6-18). The object of this method of dressing is to hold down shrinkage and thus aid in maintaining the pink color of baby lambs. The full-dressed weight is 48 to 55 percent of the live weight. New York City prefers lambs weighing 30 to 40 pounds hog dressed (head and pelt on).

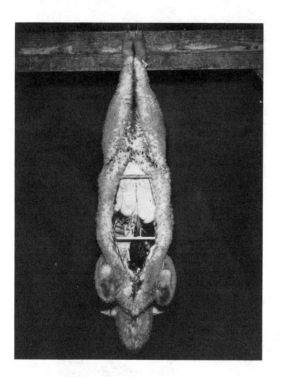

Fig. 6-18. Hothouse lamb (hog dressed). This lamb graded "extra fancy" on the New York City market. It was sired by a purebred Southdown ram and was out of a Hampshire-Dorset-Merino cross-bred ewe. It had an abundance of kidney fat, which is one of the most important factors in judging the finish and determining the grade. Hothouse lambs are dressed "pluck in." The pluck, in this case, consists of the liver, heart, lungs, gullet, and windpipe. This lamb weighed 39 pounds alive and 28 pounds, hog dressed.

COMMERCIAL LAMB SLAUGHTER

The majority of sheep and lamb slaughter takes place in large, highly mechanized slaughter plants, each capable of handling 200,000 to 700,000 head per year. With automated equipment, including a continuously or intermittently moving rail, the rate of slaughter generally ranges from 250 to 350 animals per hour. A glimpse of such a mechanized operation is provided in Figures 6-19 through 6-22.

AGING LAMB AND MUTTON

Lamb and mutton rank with beef in keeping qualities, dependent upon the temperature of the refrigerator and the amount of fat covering the carcass. The higher grades can be aged for several weeks, and choice yearling carcasses are best if aged three or four weeks. However, most packers move lamb out of their coolers and into the marketing sequence as soon as possible after killing, that is, 24 hours or less after slaughter.

Fig. 6-19. Lambs are suspended by all four limbs to position for initial legging. (Courtesy John Morrell & Co.)

Fig. 6-20. Pelt is pulled by hand as lambs move at approximately 200 per hour. (Courtesy John Morrell & Co.)

Fig. 6-21. Federal inspector (in white) examines each carcass following evisceration. (Courtesy John Morrell & Co.)

Fig. 6-22. Partial view of some 2,000 to 3,000 lamb carcasses in cooler containing the day's kill. They are crowded but adequately spaced for chilling. (Courtesy John Morrell & Co.)

Chapter 7

Veal and Calf Slaughter

TABLE 7-1 lists the states in the order of commercial veal and calf slaughter volume in 1983. There were 3,076,733 calves and veal slaughtered commercially in the United States and 85,000 slaughtered on the farm, making a total of 3,161,733[1] calves and veal slaughtered in the United States in 1983. This number of calves and veal produced 407.9 million pounds of carcass, averaging 129

Table 7-1. Commercial Veal and Calf Slaughter, Number of Head (in Thousands), Ranked by States, 1983[1,2]

Rank	State	Number	Rank	State	Number
1	New York	635.8	23	Oklahoma	5.3
2	Wisconsin	444.0	24	Virginia	5.0
3	Pennsylvania	302.9	25	Minnesota	4.9
4	Texas	258.1	26	Alabama	3.8
5	California	238.1	27	Arkansas	3.5
6	Illinois	192.3	28	Arizona	2.7
7	New England[3]	174.8	29	Idaho	1.7
8	Florida	113.7	30	Kentucky	1.0
9	Indiana	109.2	31	West Virginia	.8
10	New Jersey	107.8	32	North Dakota	.4
11	Louisiana	105.6	33	Kansas	.3
12	Michigan	74.5	34	Nebraska	.2
13	Washington	66.0	34	South Dakota	.2
14	Delaware and Maryland	46.1	35	Colorado	.1
15	South Carolina	41.8	35	Utah	.1
16	Tennessee	39.0	35	Wyoming	.1
17	Ohio	29.0		Montana	—
18	Mississippi	24.9		Nevada	—
19	Georgia	19.2		New Mexico	—
20	Oregon	8.9		Iowa	—
21	Missouri	8.2		Hawaii	—
22	North Carolina	6.5	Total[4]		3,076.7

[1]*Annual Livestock Slaughter.* March 1984. Crop Reporting Board, SRS, USDA.

[2]Includes slaughter in federally inspected and in other slaughter plants, but excludes animals slaughtered on farms.

[3]New England includes Connecticut, Maine, Massachusetts, New Hampshire, Rhode Island, and Vermont.

[4]States with no data printed are still included in U.S. total; not printed to avoid disclosing individual operations.

---

[1]*Annual Livestock Slaughter.* March 1984. Crop Reporting Board, SRS, USDA.

pounds per head. Of the commercial slaughter of calves and veal, 91 percent was done under federal inspection. Packers purchased 40 percent directly, 7 percent through terminal markets, and 53 percent through auction markets. Of all these purchases, 22 percent were made on a carcass grade and weight basis.[2]

Considerable confusion is evident in circles outside the livestock industry as to the distinction that exists between a vealer and a calf. The USDA defines a vealer as an immature bovine animal, usually not over 3 months of age, that has subsisted largely on milk or milk replacers, thus making the color of its lean light grayish pink. Such veal has the characteristic trimness of middle associated with limited paunch development. A calf is defined as an immature bovine animal between 3 and 8 months of age, which, for a considerable period of time, has subsisted in part or entirely on feeds other than milk and has thus developed a heavier middle. Grayish red is the typical color of calf carcass lean.

Since it is rather difficult to determine the age of a vealer or calf, no set age can be given as a definite dividing line. Weight and conformation are used more as a basis for determining their classification, with weight being the determining price factor among vealers and calves of equal conformation, finish, and quality.

VEALERS

There is no sex classification made for vealers, since they are not old enough for sex conditions to have had any influence on their physical characteristics. They are sold on the market for slaughter purposes only. The greatest supply of vealers comes from dairy farms during the spring and fall months. The large market centers for veal are New York City, Buffalo, Chicago, Detroit, Milwaukee, and South St. Paul.

Immature Veal

The practice on many dairy farms which do not have purebred stock is to allow the calf to suckle the dam for several days to remove the colostrum milk. The calf is then sold to a dealer for slaughter, or it may be sold to a farmer or dealer who keeps some cows for the purpose of vealing calves. The carcasses of these immature vealers are usually designated as *bob veal*. To discourage the sale of immature veal, most states have legislation regulating the legal age at which veal can be slaughtered. "Bob" or immature veal, although not unwholesome, is an uneconomical buy because of (1) the high moisture content, (2) the large proportion of bone to lean, and (3) the low quality.

The skins from immature or "bob veal" are called "deacon skins" and generally weigh under 9 pounds.

---

[2]*Annual Financial Review*. Sept. 1984. American Meat Institute, P. O. Box 3556, Washington, DC 20007.

The skins from stillborn calves are called "slunk skins" and have short, fine hair. Cattle dealers have them tanned and made into jackets and vests.

The term *vealing calves* refers to the feeding of young calves, either by hand or by letting them suckle strange cows, and supplementing the milk ration with a grain gruel. The most recent calf feeding practice is to use milk replacements. When such feeding has produced calves of the desired weight, they are sent to slaughter.

SLAUGHTER CALVES

Sex conditions have caused some changes in the physical characteristics of calves over 3 months of age that are not evident in vealers, and hence the market classifies calves as to sex. Size and weight are important in the selection of calves, either for slaughter purposes or for further feeding.

Before proceeding to slaughter a veal or calf, be sure you have proper and safe equipment, that your equipment is ready to use (sharp and clean), and that the animal you are about to slaughter is healthy. That is, you should study and master the first three chapters, before going ahead with this chapter.

METHODS OF DRESSING VEAL AND CALVES

Veal is dressed with either "skin off" or "skin on" (hog dressed). In the past, the "hog style" carcass was popular because it prevents the outer surface of the carcass from becoming dark and dry, and it is still being used in some parts of the United States. However, a problem arises in the cooler with the loss of hair, which can become affixed to other meat and thus contaminate it. So more and more plants are removing the skins on the kill floor and wrapping the carcass in a pliofilm wrap or bag to prevent dehydration and darkening and to preserve the fresh appearance and bloom. Calf carcasses are dressed generally in the same manner as beef, with skin off and split into sides.

HANDLING, STUNNING, AND STICKING

Vealers and calves should be kept off feed for 18 hours before slaughter. They should be handled with care to avoid bruises and undue excitement.

Any of the mechanical stunners can be used (Figure 7-1), and electric stunning also works well on calves.

Two methods of sticking are common. The one is to "Kosher stick," or cut the throat just back of the jaw. The other is to stick in front of the brisket, as in the sticking of beef (Figures 7-2 and 7-3). Since calves struggle for a longer period after sticking than other classes of livestock, it is well to hoist them before sticking. This keeps them clean and makes it easier to skin out the head and foreshanks.

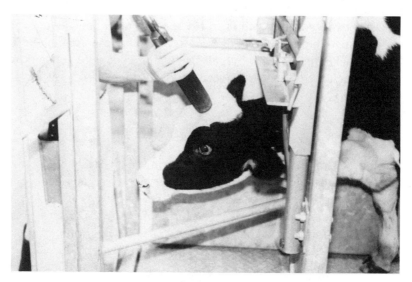

Fig. 7-1. Stunning a calf with a light load in a captive bolt stunner.

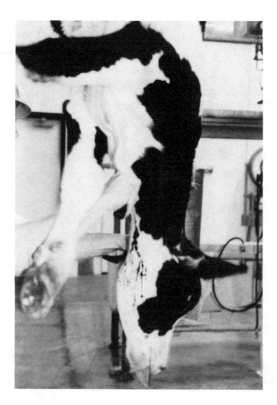

Fig. 7-2. In sticking, first open the hide from the jaw to the brisket.

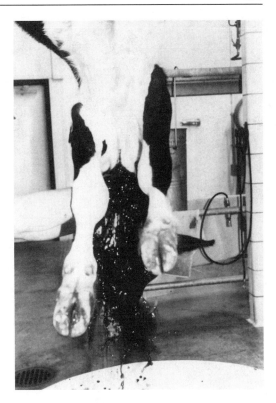

Fig. 7-3. Enter the opening in front of the brisket, and sever the jugular vein and carotid arteries. Note that blood is being caught in a funneled vat rather than being allowed to run down the floor drain.

RODDING THE WEASAND

The esophagus (weasand) must be separated from the trachea in order for the abdominal cavity organs and contents to be removed during evisceration. This is called "rodding," because a rod is used (Figure 7-4). Once the rod has encircled the weasand, it is pushed until the end reaches the rumen (Figure 7-5) so that the trachea (windpipe) is completely separated from the esophagus along its entire length. The weasand is then tied off to prevent spillage of rumen contents.

DRESSING

Skin Off

The calf or veal is placed in a cradle, and the method of opening the skin is the same as in beef (Figure 7-6). Since a calf skin is thinner and softer and more readily scored or cut than is a beef hide, the better plan is to fist (Figure 7-7) or pull off the sides and back of the carcass (Figure 7-8).

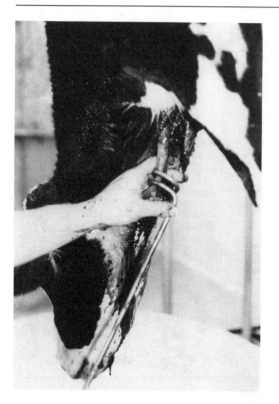

Fig. 7-4. Rodding the weasand. The loops on the rod have been threaded onto the esophagus (weasand).

Fig. 7-5. As the rod is pushed upward toward the stomach, it separates the esophagus and trachea.

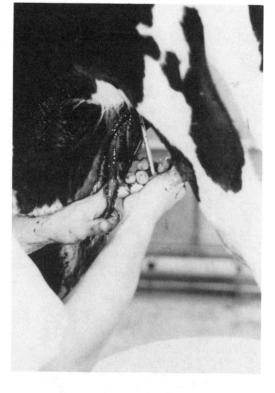

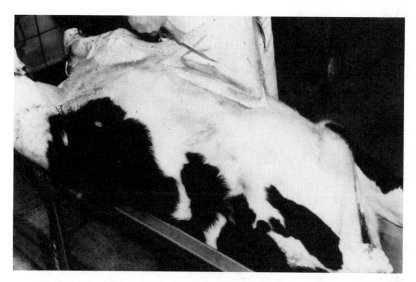

Fig. 7-6. The calf has been placed in a cradle, the feet removed, hide opened down the centerline, and siding begun.

Fig. 7-7. The calf hide may be fisted free of the carcass, since it is thin and soft. This method may not be possible on older calves, but when used, it prevents nicks and gouges in hide and carcass.

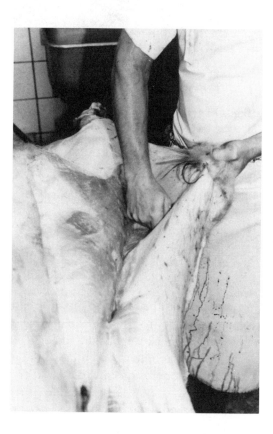

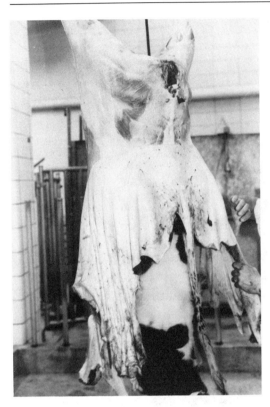

Fig. 7-8. After landing on the rail, the hide may be pulled off the back.

Skin On (Hog Dressed)

The skin is opened from the hoof to the knee on the foreshank and to the hock on the hindshank. Skin out the foreshanks and hindshanks, and remove them at the break joint. Skin out the head, and remove it at the atlas joint. Split the skin and carcass over the median line of the belly from the back end of the brisket to the cod or udder.

EVISCERATION

Cut around the bung, and let it drop into the abdominal cavity (Figure 7-9). Remove the entrails from the abdominal cavity (Figure 7-10), but leave the liver in the carcass. The gall bladder must be removed from the liver.

Cut the diaphragm, and remove the pluck (Figure 7-11). Care must be exercised in this operation in order to keep from mutilating the thymus gland or sweetbreads. The sweetbreads and liver are considered part of a veal carcass and are weighed with the carcass. They are removed from calf carcasses.

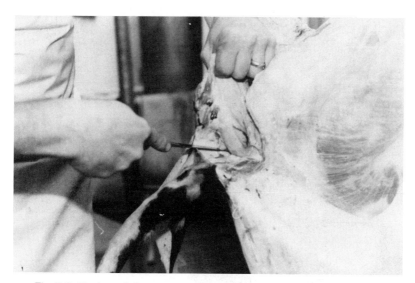

Fig. 7-9. The bung is loosened by carefully cutting around both sides and finally cutting the attachment to the backbone.

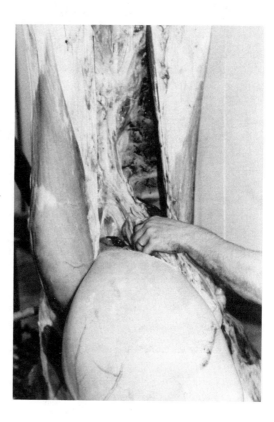

7-10. The abdominal cavity contents, including the rumen in the foreground, are removed by making a few well placed cuts through the supporting tissues.

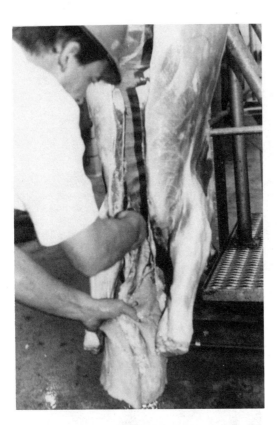

7-11. The diaphragm membrane is cut, leaving the muscle intact. This allows the pluck, composed of the heart, sweatbreads (thymus gland), trachea, and lungs to be removed.

Fig. 7-12. A general view of a well illuminated, compact veal and calf dressing department. Note hearts and livers on rack in right foreground. (Courtesy *Meat Processing*)

HALVING

Because of their small size, veal are not split, but calves are split for ease of handling.

COMMERCIAL VEAL AND CALF SLAUGHTER

A modern veal and calf processing plant, utilizing a continuous, mechanically powered rail system, is depicted in Figures 7-12 and 7-13.

Fig. 7-13. Two views of an eviscerator at work. (Courtesy *Meat Processing*)

Chapter 8

Poultry Processing

SINCE 1925 there has been a spectacular rise in the production of commercial broilers in the United States. Broiler production started to expand in the Delmarva section of Delaware, Maryland, and Virginia and increased in the United States from 34 million (4 percent of the total chicken meat supply) in 1934 to 4.13 billion in 1983, accounting for over 94 percent of all chicken meat consumption. Combined, broilers, fowl, and turkey accounted for over 31 percent of all meat consumption in 1983.[1]

A highly commercialized operation for killing, dressing, eviscerating, cutting up, packaging, and transporting chickens has been developed that provides a constant supply of ready-to-cook poultry in supermarkets across the United States. So rapid and constant are the present supply channels that less than 10 percent of this product is sold frozen, the major portion being sold fresh chilled. Turkeys are somewhat seasonal, with 60 percent being produced during the last half of the year and 80 percent being frozen for future consumption, while the corresponding figures for ducks are 70 and 40 percent respectively. Almost 100 percent of the geese are processed from October to January, and most of them are marketed frozen.

STANDARDS AND GRADES[2]

The difference between standards of quality and grades is sometimes misunderstood. The standards of quality enumerate the various factors that determine the grade. These factors, such as fat covering, fleshing, exposed flesh, discolorations, etc., when evaluated collectively, determine the grade of the bird.

The U.S. Consumer Grades for Poultry are by far the most important, since they are the grades used at the retail level. The U.S. Consumer Grades are: U.S. Grade A, U.S. Grade B, and U.S. Grade C.

The U.S. Procurement Grades are designed primarily for institutional use. These grades are: U.S. Procurement Grade I and U.S. Procurement Grade II.

---

[1]*Poultry and Egg Outlook and Situation.* Aug. 1984. USDA.

[2]*Poultry Grading Manual.* Agricultural Handbook No. 31, USDA.

The procurement grades are based on relative quality and tolerances within the lot.

Modern breeding, feeding, and management practices have resulted in improved poultry. These factors, together with more efficient marketing and procurement practices, have made it unnecessary to have wholesale poultry grades and live poultry grades. Individual quality standards for live poultry are used on a very limited basis and mostly in an academic setting.

CLASSES OF POULTRY[3]

"Kind" refers to the different species of poultry, such as chickens, turkeys, ducks, geese, guineas, and pigeons. The kinds of poultry are divided into "classes" by groups which are essentially of the same physical characteristics such as fryers or hens. These physical characteristics are associated with age and sex.

The kinds and classes of live, dressed, and ready-to-cook poultry listed here are in general use in all segments of the poultry industry.

The following provisions of the official U.S. standards apply to individual carcasses of ready-to-cook poultry in determining the kind of poultry and its class.

Chickens

The following are the various classes of chickens:

Rock Cornish Game or Cornish Game Hen

A Rock Cornish game hen or Cornish game hen is a young, immature chicken (usually five to six weeks of age) weighing not more than 2 pounds (ready-to-cook weight), a Cornish chicken, or a Cornish chicken cross.

Rock Cornish Fryer, Roaster, or Hen

Rock Cornish fryer, roaster, or hen pertains to the progeny of a cross between a Cornish and a Plymouth Rock chicken. The *fryer, roaster,* or *hen* term should apply only if the age and characteristics are appropriate according to the following paragraphs.

Broiler or Fryer

A broiler or fryer is a young chicken (usually under 13 weeks of age) of either sex with tender meat; soft, pliable, smooth-textured skin; and flexible

---

[3]*Ibid.*

breastbone cartilage. Most broilers are now marketed at between 6½ and 7½ weeks of age, for a 2½- to 3½-pound ready-to-cook weight.

Roaster

A roaster is a young chicken (usually three to five months of age) of either sex, with tender meat; soft, pliable, smooth-textured skin; and breastbone cartilage that may be somewhat less flexible than that of a broiler or fryer.

Capon

A capon is a surgically unsexed male chicken (usually under eight months of age), with tender meat and soft, pliable, smooth-textured skin.

Hen Fowl or Baking Chicken

A hen fowl or baking chicken is a mature female chicken (usually more than 10 months of age), with meat less tender than that of a roaster and with a non-flexible breastbone tip.

Cock or Rooster

A cock or rooster is a mature male chicken, with coarse skin, toughened and darkened meat, and a hardened breastbone tip.

Turkeys

The following are the various classes of turkeys:

Fryer-Roaster Turkey

A fryer-roaster turkey is a young, immature turkey (usually under 16 weeks of age) of either sex, with tender meat; soft, pliable, smooth-textured skin; and flexible breastbone cartilage.

Young Turkey

A young turkey is a turkey (usually under eight months of age), with tender meat, soft, pliable, smooth-textured skin; and breastbone cartilage that is somewhat less flexible than a fryer-roaster turkey. Sex designation is optional.[4]

---

[4]For labeling purposes, the designation of sex within the class name is optional, and the two classes of young turkeys may be grouped and designated as "young turkeys."

Yearling Turkey

A yearling turkey is a fully matured turkey (usually under 15 months of age), with reasonably tender meat and reasonably smooth-textured skin. Sex designation is optional.[5]

Mature Turkey or Old Turkey (Hen or Tom)

A mature or old turkey is a turkey (usually in excess of 15 months of age) of either sex, with coarse skin and toughened flesh.

Ducks

The following are the various classes of ducks:

Broiler Duckling or Fryer Duckling

A broiler duckling or fryer duckling is a young duck (usually under 8 weeks of age) of either sex, with tender meat, a soft bill, and a soft windpipe.

Roaster Duckling

A roaster duckling is a young duck (usually under 16 weeks of age) of either sex, with tender meat, a bill that is not completely hardened, and a windpipe that is easily dented.

Mature Duck or Old Duck

A mature or old duck is a duck (usually over 6 months of age) of either sex, with toughened flesh, hardened bill, and hardened windpipe.

Geese

The following are the various classes of geese:

Young Goose

A young goose may be of either sex, with tender meat and a windpipe that is easily dented.

---

[5]*Ibid.*

Mature Goose or Old Goose

A mature or old goose may be of either sex, with toughened flesh and hardened windpipe.

Guineas

The following are the various classes of guineas:

Young Guinea

A young guinea may be of either sex, with tender meat and flexible breastbone cartilage.

Mature Guinea or Old Guinea

A mature or old guinea may be of either sex, with toughened flesh and hardened breastbone.

Pigeons

The following are the various classes of pigeons:

Squab

A squab is a young, immature pigeon of either sex, with extra tender meat.

Pigeon

A pigeon is a mature pigeon of either sex, with coarse skin and toughened flesh.

THE GRADING SERVICE

Poultry grading services of the USDA are permissive in that individuals, firms, or governmental agencies that desire to utilize them may request them of their own volition. Services are performed on the basis of regulations promulgated by the Secretary of Agriculture. These regulations have been developed in cooperation with the industry, including all affected or related groups, such as health and marketing officials, producers, processors, and consumers.

Grading generally involves the sorting of products according to quality and size, but it also includes the determination of the class and condition of products. For poultry, grading may be for determining class, quality, quantity, or condition, or for any combination of these factors.

Grading for quality can be accomplished by examining each carcass in the lot, or by examining a representative sample of the lot of poultry to be graded. Only ready-to-cook poultry that is first inspected for wholesomeness and then graded on an individual-bird basis may be individually marked with an official grade mark.

Ready-to-cook poultry must have been officially inspected for condition and wholesomeness before it is graded for quality, whether the grading is done in an approved plant or elsewhere. The U.S. Consumer Grades are applicable only to poultry which has been graded on an individual carcass basis by a grader or by a limited licensed grader working under the supervision of a grader. The U.S. Procurement Grades are generally applied when the poultry has been graded on the basis of an examination of a prescribed sample of the lot.

STANDARDS OF QUALITY FOR DRESSED AND READY-TO-COOK POULTRY[6]

A Quality

Conformation

The carcass or part is free of deformities that detract from its appearance or that affect the normal distribution of flesh. Slight deformities such as slightly curved or dented breastbones and slightly curved backs may be present.

Fleshing

The carcass or part has a well developed covering of flesh. The breast is moderately long and deep and has sufficient flesh to give it a rounded appearance, with the flesh carrying well up to the crest of the breastbone along its entire length. The leg is well fleshed and moderately long, thick and wide at the knee and hip joint area, showing a plump appearance, with thigh and wings moderately fleshed.

Fat Covering

The carcass or part, considering the kind, class, and part, has a well developed layer of fat in the skin. The fat is well distributed so that there is a noticeable amount of fat in the skin in the areas between the heavy feather tracts.

---

[6]*Poultry Grading Manual.* Agricultural Handbook No. 31, USDA.

Defeathering

The carcass or part has a clean appearance, especially on the breast. The carcass or part is free of pinfeathers, diminutive feathers, and hair which is visible to the inspector or grader.

Exposed Flesh

Parts are free of exposed flesh resulting from cuts, tears, and missing skin (other than slight trimming on the edge). The carcass is free of these defects on the breast and legs. Elsewhere the carcass may have exposed flesh due to slight cuts, tears, and areas of missing skin, providing the aggregate of the areas of flesh exposed does not exceed the diameter as specified in Table 8-1.

Table 8-1. Maximum Exposed Flesh Area for A Quality

Carcass Weight		Maximum Aggregate Diameter Permitted	
Minimum	Maximum	Breast & Legs	Elsewhere
None	1 lb. 8 oz.	None	¾ in.
Over 1 lb. 8 oz.	6 lb.	None	1½ in.
Over 6 lb.	16 lb.	None	2 in.
Over 16 lb.	None	None	3 in.

Disjointed and Broken Bones and Missing Parts

Parts are free of broken bones. The carcass is free of broken bones and has no more than one disjointed bone. The wing tips may be removed at the joint, and, in the case of ducks and geese, the parts of the wing beyond the second joint may be removed, if removed at the joint, and both wings are so treated. The tail may be removed at the base. Cartilage separated from the breastbone is not considered as a disjointed or broken bone.

Discolorations of the Skin and Flesh

The carcass or part is practically free of such defects. Discoloration due to bruising shall be free of clots (discernible clumps of red or dark cells). Evidence of incomplete bleeding, such as more than an occasional slightly reddened feather follicle, is not permitted. Flesh bruises and discolorations of the skin such as ''blue back'' are not permitted on the breast or legs of the carcass or on these individual parts, and only lightly shaded discolorations are permitted elsewhere. The total areas affected by flesh bruises, skin bruises, and discolorations such as ''blue back'' singly or in any combination shall not exceed one half of the total aggregate area of permitted discoloration. The aggregate area of all discolorations for a carcass or a part therefrom shall not exceed the diameter as specified in Table 8-2.

Table 8-2. Maximum Discolorations for A Quality

Carcass Weight		Maximum Aggregate Diameter Permitted		
Minimum	Maximum	Breast & Legs	Elsewhere	Part
None	1 lb. 8 oz.	½ in.	1 in.	¼ in.
Over 1 lb. 8 oz.	6 lb.	1 in.	2 in.	¼ in.
Over 6 lb.	16 lb.	1½ in.	2½ in.	½ in.
Over 16 lb.	None	2 in.	3 in.	½ in.

Freezing Defects

With respect to consumer-packaged poultry, the carcass, part, or specified poultry food product is practically free from defects which result from handling or occur during freezing or storage. The following defects are permitted if they, alone or in combination, detract only very slightly from the appearance of the carcass, part, or specified poultry food product.

- Slight darkening over the back and drumsticks, provided the frozen bird or part has a generally bright appearance.
- Occasional pockmarks due to drying of the inner layer of skin (derma); however, none may exceed the area of a circle ⅛ inch in diameter for poultry weighing 6 pounds or less, and ¼ inch in diameter for poultry weighing over 6 pounds.
- Occasional small areas showing a thin layer of clear or pinkish-colored ice.

B Quality

Conformation

The carcass or part may have moderate deformities, such as a dented, curved, or crooked breast; or crooked back; or misshapen legs or wings, which do not materially affect the distribution of flesh or the appearance of the carcass or part.

Fleshing

The carcass or part has a moderate covering of flesh considering the kind, class, and part of the bird. The breast has a substantial covering of flesh with the flesh carrying up to the crest of the breastbone sufficiently to prevent a thin appearance. The leg, thigh, and wing all have a sufficient amount of flesh to prevent a thin appearance.

Fat Covering

The carcass or part has sufficient fat in the skin to prevent a distinct appearance of the flesh through the skin, especially on the breast and legs.

Defeathering

The carcass or part may have a few nonprotruding pinfeathers or vestigal feathers which are scattered sufficiently so as not to appear numerous. Not more than an occasional protruding pinfeather or diminutive feather shall be in evidence under a careful examination.

Exposed Flesh

Parts may have exposed flesh resulting from cuts, tears, and missing skin, provided that not more than a moderate amount of the flesh normally covered by skin is exposed. The carcass may have exposed flesh resulting from cuts, tears, and missing skin, provided that the aggregate of the areas of flesh exposed does not exceed the diameter as specified in Table 8-3.

Table 8-3. Maximum Exposed Flesh Area for B Quality

Carcass Weight		Maximum Aggregate Diameter Permitted	
Minimum	Maximum	Breast & Legs	Elsewhere
None	1 lb. 8 oz.	¾ in.	1½ in.
Over 1 lb. 8 oz.	6 lb.	1½ in.	3 in.
Over 6 lb.	16 lb.	2 in.	4 in.
Over 16 lb.	None	3 in.	5 in.

Notwithstanding the foregoing, a carcass meeting the requirements of A Quality for fleshing may be trimmed to remove skin and flesh defects, provided that no more than one third of the flesh is exposed on any part and that the meat yield of any part is not appreciably affected.

Disjointed and Broken Bones and Missing Parts

Parts may be disjointed but are free of broken bones. The carcass may have two disjointed bones or one disjointed bone and nonprotruding broken bone. Parts of the wing beyond the second joint may be removed at a joint. The tail may be removed at the base. The back may be trimmed in an area not wider than the base of the tail to halfway to the hip joints.

Discolorations of the Skin and Flesh

The carcass or part is free of serious defects. Discoloration due to bruising shall be free of clots (discernible clumps of red or dark cells). Evidence of incomplete bleeding shall be no more than very slight. Moderate areas of discoloration due to bruises in the skin or flesh and moderately shaded discoloration of the skin such as "blue back" are permitted, but the total areas affected by such discolorations singly or in any combination may not exceed one half of the total aggregate area of permitted discoloration. The aggregate area of all discolorations for a carcass or a part therefrom shall not exceed the diameter as specified in Table 8-4.

Table 8-4. Maximum Discolorations for B Quality

Carcass Weight		Maximum Aggregate Diameter Permitted		
Minimum	Maximum	Breast & Legs	Elsewhere	Part
None	1 lb. 8 oz.	1 in.	2 in.	½ in.
Over 1 lb. 8 oz.	6 lb.	2 in.	3 in.	1 in.
Over 6 lb.	16 lb.	2½ in.	4 in.	1½ in.
Over 16 lb.	None	3 in.	5 in.	1½ in.

Freezing Defects

With respect to consumer-packaged poultry, the carcass, part, or specified poultry food product may have moderate defects which result from handling or which occur during freezing or storage. The skin and flesh shall have a sound appearance but may lack brightness. The carcass or part may have a few pockmarks due to drying of the inner layer of skin (derma). However, no single area of overlapping pockmarks may exceed that of a circle ½ inch in diameter. Moderate areas showing layers of clear or pinkish- or reddish-colored ice are permitted.

C Quality

A part that does not meet the requirements for A or B Quality may be of C Quality if the flesh is substantially intact.

A carcass that does not meet the requirements for A or B Quality may be of C Quality. Both wings may be removed or neatly trimmed. Trimming of the breast and legs is permitted, but not to the extent that the normal meat yield is materially affected. The backs may be trimmed in an area not wider than the base of the tail and to the area between the hip joints.

U.S. PROCUREMENT GRADES FOR
READY-TO-COOK POULTRY[7]

The U.S. Procurement Grades for ready-to-cook poultry are applicable to carcasses of ready-to-cook poultry when they are graded as a lot by a grader on the basis of an examination of each carcass in the lot or each carcass in a representative sample thereof.

U.S. Procurement Grade I

Any lot of ready-to-cook poultry composed of one or more carcasses of the same kind and class may be designated and identified as U.S. Procurement Grade I in the following situations:

1. When 90 percent or more of the carcasses in such lot meet the requirements of A Quality, with the following exceptions:
   a. Fat covering and conformation may be as described for B Quality.
   b. Trimming of skin and flesh to remove defects is permitted to the extent that not more than one fourth of the flesh is exposed on any part, and the meat yield of any part is not appreciably affected.
   c. Discoloration of skin and flesh may be as described for B Quality.
   d. One or both drumsticks may be removed if severed at the joint.
   e. The back may be trimmed in an area not wider than the base of the tail and extending to the area between the hip joints.
   f. The wings or parts of wings may be removed if severed at a joint, and the tail may be removed at the base.
2. When the balance of the carcasses meet the same requirements, except that they may have only a moderate covering of flesh.

U.S. Procurement Grade II

Any lot of ready-to-cook poultry of the same kind and class which fails to meet the requirements of U.S. Procurement Grade I may be designated and identified as U.S. Procurement Grade II, provided that (1) trimming of flesh from any part does not exceed 10 percent of the meat and (2) portions of a carcass weighing not less than one half of the whole carcass may be included if the portion approximates in percentage the meat-to-bone yield of the whole carcass.

THE INSPECTION SERVICES

Inspection refers to the condition of poultry and its healthfulness and fitness

---

[7]*Ibid.*

for food. It is not concerned with the quality or grade of poultry. The inspection mark on poultry and poultry products means that they have been examined during processing by a veterinarian or by a qualified lay person under the supervision of a veterinarian. Plants which apply for inspection service and are approved are known as official plants or establishments.

A complete discussion of meat and poultry inspection and an illustration of a poultry inspection stamp were provided in Chapter 3.

DRESSING POULTRY

Selection of Birds for Dressing

The value of poultry and turkey for meat varies considerably with the strain and breed. Some are thin-meated, some are deep-sided and rangy, while others are thick-meated and particularly full-breasted. The thick-meated, full-breasted, well finished bird will be a top grade bird on any market, provided it has been properly dressed. To avoid having too many birds of the lower grades, thin chickens should be retained on a finishing ration. Turkeys should be fed until the skin no longer appears blue. "Produce what the market demands, not what is convenient at the time" is good advice to follow.

HANDLING PREVIOUS TO DRESSING

Birds, like animals, should receive careful handling to avoid bruises, abrasions, and broken limbs. Temperature extremes should be avoided in the crates or holding pens and they should be well ventilated. The birds should have free access to water. Water is a heat regulator and helps the birds to eliminate waste products. Birds denied water for too long a time lose weight and dress poorly. Feed should be withheld for 8 to 10 hours before the chickens are killed, since a bird full of feed will not bleed as well, will be harder to eviscerate, and the full crop is unsightly and wasteful in an undrawn bird. A marked loss in dressing yield has been shown to occur after 12 hours of fasting.[8]

The fowl should be caught by the leg below the thigh and not allowed to strike its breast on a hard surface. One wing should be held when a bird is picked up by the shank to prevent its struggling. Over-heated, over-excited birds will bleed poorly, producing a carcass of higher blood content and lower keeping quality.

TOOLS AND EQUIPMENT

The old method of chopping off the heads of the fowl with a hatchet or cleaver and tossing the headless birds into a barrel, one on top of the other as

---

[8]*Factors Affecting Poultry Meat Yields*. Bul 630, S.D.S.U. AES, NCM-46 Regional Technical Committee.

rapidly as the beheader could swing the hatchet, was considered the acme of speed. But this practice disappeared into limbo as science streamlined and mechanized the commercial dressing practices of animals and fowl. In fact, the mechanical processing of fowl was patterned very closely after the methods employed by the meat packing industry in the slaughter of hogs. The commercial poultry processing plant has overhead tracks (much lighter in structure and lower than those used for animals) which take the form of a belt chain to which oval link chains are suspended to hold the shackles for suspending the bird. These belt chains move at a controlled speed in the same manner as those employed for hogs.

Other equipment may consist of such items as an automatic stunner, automatic killer, scalder, picker, hock picker, automatic hock cutter, outside bird washer, eviscerating trough, oil sac cutter, opening cut machine, automatic drawing machine (Figure 8-1), automatic gizzard processing system, automatic lung remover, automatic neck breaker, combination washer, automatic head cutter and neck skin cutter, automatic continuous chill system, continuous giblet chiller, automatic giblet wrapper, automatic sizing system, automatic cutup machines (Figure 8-2), and numerous other items of equipment which allow poultry processors to process some 3,000 birds per hour with a single system of machines.

Fig. 8-1. An automatic drawing machine which is a completely self-contained unit mechanically driven from an overhead conveyor, thus eliminating synchronizing problems. Specially designed drawing spoons are washed and sanitized automatically after each drawing operation. Two machines replace four to six operators on eviscerating lines of a 6,000-bird-per-hour operation. (Courtesy Gainesville Machine Co., Inc., Gainesville, Georgia)

Fig. 8-2. The new USDA-approved Gainesville Automatic Poultry Cut-up System consists of two machines, one to cut up the front half of the bird (left) and one to cut up the back half of the bird (right).

The complete systems have been field proven in many plant operations throughout the country since July 1970.

The two machines can make either an eight-piece cut (two wings and two pieces of breast on the front half machine and two drumsticks and two thighs on the back half machine) or a nine-piece cut (two wings and three pieces of breast on the front half machine and two drumsticks and two thighs on the back half machine). The back half machine can be further adjusted to make a strip cut varying from ⅜ inch to as wide as necessary.

This system, which gives a more uniform cut than hand operations, works as fast as the operator can feed birds into it. Some plants are cutting up to 44 birds per minute.

Improved sanitation is achieved, because each machine has a built-in continuous water rinse. End of day clean-up can be handled in 20 to 30 minutes by one operator.

The Gainesville Automatic Cut-up System reduces labor costs by requiring fewer workers, who may be unskilled.

Increased safety reduces downtime due to absenteeism. (Courtesy Gainesville Machine Co., Inc., Gainesville, Georgia)

Where only a few birds are dressed, the equipment may consist of a scalding tub, a shackle for holding the bird, and probably a bleeding cup. The knives necessary in each case are a sticking knife (a 3-inch blade for chickens and a 4-inch blade for turkeys), pinning knife (a paring knife will do), several sizes of boning knives (3½-inch to 6-inch blades) for eviscerating, and a linoleum knife for splitting the back (or a power meat saw, if available). A thermometer for testing water temperature and bone shears for severing heads, necks, and shanks are desirable. Although the blades of the knives used for sticking and braining are narrow (¼ inch wide), the handles should be of standard grip.

BLEEDING PRACTICES

- Severing the neck (chopping-block method or barrel method).
- Dislocating the neck (simple and sanitary but not recommended except for home use). It is accomplished by placing the thumb on the top of the neck in back of the comb and the fingers under the lower jaw and giving a quick downward pull and a backward jerk of the head by compressing the third and fourth fingers in the opposite direction of the thumb.
- Cutting the throat. This is the approved and most widely practiced method employed today. The bird is suspended by the legs with shackles made of heavy wire, which not only hold the feet in the V-shaped vise but also spread the legs. For turkeys and some larger chickens, the use of an electric stunning knife or automated stunning device is recommended for aid in feather removal and improved bleeding. Stunning should immediately precede cutting the throat, which is cut as close to the head as possible but so as to sever the jugular vein. Alternatively, a sash cord with a 2-inch square wooden block attached is a simple homemade shackle, but it holds the legs together and hampers feather removal, unless one leg is freed. The feet of the fowl should be level with the eyes of the worker for convenience of operation.
- Cutting the veins. Grasp the head, and hold it firmly in the left hand (if you are right handed), pressing the thumb and forefinger on both sides of the junction of the upper and lower beaks. This forces the mouth open so

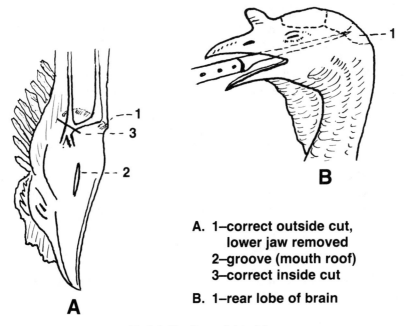

A. 1—correct outside cut, lower jaw removed
   2—groove (mouth roof)
   3—correct inside cut

B. 1—rear lobe of brain

Fig 8-3. Bleeding and debraining.

the point of the sticking knife can be inserted, sharp edge downward, to the base of the skull. Press the point of the knife into the flesh, lift the handle upward, and cut downward and to the right, severing the veins. If a good bleed does not result, try again until there is free bleeding.

DRY-PICKING AND DEBRAINING

If the fowl is to be dry-picked, hold the head of the bird in the same position, insert the knife blade (blade edge up) into the cleft in the roof of the mouth and force it through to the rear lobe of the brain *(medulla oblongata)*. The chicken gives a peculiar squawk if properly debrained, whereas the turkey relaxes its wings, and the main tail feathers spread out fan shaped. The puncturing of the brain relaxes the feather muscles, causing the feathers to become loose and easily plucked. This condition lasts for two to three minutes at which time the muscles again begin to tighten up due to rigor mortis, making it necessary to have rapid and orderly plucking. "Roughing" the bird before the muscles reset is the process of removing the major part of the plumage in the order in which the parts of the bird bleed out. It consists of twisting out the tail and main wing feathers and plucking the breast, neck, back, thighs, and legs. This is followed by the more tedious task of pinning. The object is to pluck and handle the bird so that the outside layer of skin is free from tears, abrasions, or bruise spots and maintains its natural bloom. Dry-picking has been largely replaced by semi-scalding, except for Kosher-type processing.

SCALDING PRACTICES

Hard, or Hot, Scalding

Hard, or hot, scalding is one of the earliest methods used for the quick removal of the feathers. It is still the common practice employed in home dressing birds of all kinds. The speed at which feathers are loosened is dependent upon the temperature of the water and the period of immersion. The hotter the water, the quicker the feathers are loosened. But the temperature of the water to use also is dependent upon the age and nature of the fowl to be scalded. Young birds with tender skins should not be scalded in water over 150° F, whereas mature birds scald well at 155° to 160° F. Mature fowl can be scalded in water around 185° F, but the immersion period must be short to avoid cooking the skin. When high scalding temperatures are used, it is well to immerse the scalded bird in cold water as soon as the feathers are loosened. This stops further scalding action.

The hot scald works well on birds having a large number of pinfeathers. The practice gives good results if the operator is careful not to overscald. The latter causes the skin to tear and discolor and gives the bird a cooked appearance, producing a carcass that lacks bloom and turns brown rapidly (or bright red upon freezing).

Fat birds will hold their natural color longer, because the melted fat forms a film over the skin, excludes the air, and retards dessication. Where a deep yellow color is desired on fat birds of the yellow-skinned variety, one practice is to dip the dressed fowl into boiling water and then douse it immediately into cold water. The hot water melts the fat and draws it, along with the yellow pigment, to the surface of the skin. The cold water causes the fat to harden and the color to set in the fat. Hot scalding is not practiced on birds destined for the commercial market.

Semi-scalding (Slack Scalding)

The semi-scalding method was developed in the late 1920s and is now universally used in all the large poultry packing plants. It lends itself well to mechanization and has the advantage of lower labor costs. The appearance of the birds is improved and they do not turn red or brown but retain a natural bloom. This, incidentally, throws more birds into the higher grades, resulting in greater financial return.

The temperature of the water should be 125° to 126° F for young, tender-skinned birds, 127° to 128° F for roasters and young turkeys, and 130° to 132° F for aged birds. These temperatures do not loosen the feathers as much as or as rapidly as the hot scald but neither do they cook or cause the outer skin to peel. Because the skin is not weakened in strength, this type of scalding makes it possible to use the mechanical picker. These picking machines consist of a drum upon whose circumference are mounted innumerable fingers made of rubber. As the drum revolves, the semi-scalded birds are held against its outer surface, and the feathers are rubbed off.

The period of immersion varies from 20 to 30 seconds for broilers and up to 60 seconds for older and larger birds. The birds should be suspended from shackles, if plucked by hand, and the plucking cannot be done by rubbing as in the hot-scald method. The feathers must be pulled out as in dry-picking. The scalding vats should be equipped with thermostatically controlled steam jets in order to keep the desired water temperature. A detergent added to the water aids heat penetration.

WAX PICKING

Wax picking works very well and is usually employed in combination with the semi-scalding method for ducks and geese.

After the birds are roughed on the picking machine, they are dried by passing them through a drying machine. The wax will stick to the dry feathers and stubs more tightly than it will to wet feathers.

The dried birds are then dipped, by hand or automatically, into a preparation of melted wax (patented) at a temperature of 125° to 130° F for a period of 30 to 60 seconds. When birds are moving along a processing rail, the bucket containing the heated wax moves up and envelops the bird that is suspended by the head and feet. This specially prepared wax has a melting point of around 120° F and is hardened by passing the bird under a cold water spray or through a cool air blast. Frequently the birds are wax-dipped a second time. The hardened wax is then pulled from the bird with the feathers, pinfeathers, hair, and scale encased in it, producing an attractively dressed product. This wax is renovated for reuse by boiling out the water (if water spray was used for cooling) and straining off the feathers.

CHILLING

Subjecting the birds to a temperature of 32° to 36° F is highly essential for the immediate removal of body heat. Freezing warm birds before the animal heat has been removed is to be avoided, since "cold shortening" may occur, causing the meat to be less tender as the actin and myosin filaments of the muscle fibers slide and lock together. The most common procedure is to place the birds in tanks of ice slush containing phosphate. Phosphates aid in plumping the birds, chilling them, absorbing water, and reducing cooking losses. In-line chillers are in vogue today.

In former years, when birds were thoroughly chilled, their heads were wrapped with paper, their wings were folded against their bodies, and they were packed in paper-lined boxes or barrels for shipment. Birds processed to this point were designated as blood-and-feather-dressed, or New York–dressed. Very limited numbers are so handled today.

EVISCERATING

To complete the processing procedure, the birds should be eviscerated. The different operations in their proper order are as follows:

- Remove the head. Using the bone shears, cut through the back of the skin on the neck, peel the skin down, and sever the neck close to the shoulders.
- Remove the crop and windpipe by hooking the short gullet (between crop and gullet) with the index finger and peeling the crop loose from the skin by working it forward, cutting at the lower end of the gullet.
- Use the index finger to loosen the lungs from the chest wall by inserting it between the ribs and the lungs.
- Remove the feet at the hock joint.
- Make an incision from the rear end of the keel bone to the rectum, and cut around the rectum.

- Draw the intestinal tract, including the heart, lungs, and liver through this opening. Chilled or partially chilled birds are easier to draw than warm ones.
- Remove the bile sac from the liver, separating it from the intestines.
- Cut away the gizzard from the intestines, and split along the edge of the fleshy part of the gizzard sufficiently deep to cut the muscle but not the inner lining. Proper pressure of both thumbs pulling the halves apart should permit peeling without breaking the lining and spilling the contents of the gizzard. Gizzards are easier to peel if they have been partially chilled in ice water.

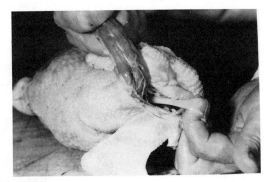

Fig. 8-4. Remove the crop.

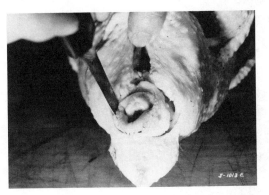

Fig. 8-5. Cut around the vent with the opening made to the rear of the keel bone.

Fig. 8-6. Remove the entrails.

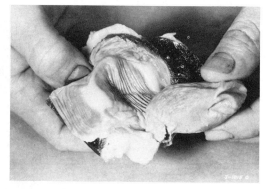

Fig 8-7. Make the opening cut on the fleshy side of the gizzard.

Fig. 8-8. Peel the gizzard.

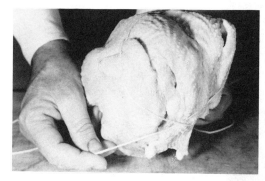

Fig. 8-9. The first step in trussing.

Fig. 8-10. Complete the trussing.

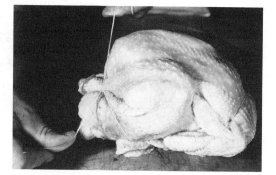

The bird should be shaped to give it a plump, compact appearance. The wings are compressed against the sides of the carcass, and the legs are brought together at the vent. A length of cord is drawn over the fore part of the breast and over the wings and crossed over the back, brought over the ends of the drumsticks, and tied tightly at the back of the rump. This style of trussing is employed on a bird for roasting. Wire trusses can be used.

Broilers may receive a different treatment. Use the linoleum knife to cut along either side of the backbone, beginning at the rear and cutting forward. This leaves the backbone and neck in one piece. Remove the neck with a bone shears. The two halves of the fowl are laid open sufficiently to remove the entrails. Split through the breast with a cleaver, or preferably a power saw, to halve the bird. The halves can be quartered, if desired.

CHICKEN PARTS

Broilers are being sold as parts in ever-increasing proportions. Some consumers prefer white meat, so an all-breast pack of broiler stock best meets their need. Others prefer dark meat, so all-thighs, drumsticks, or leg parts are most popular for them. Parts are most adaptable for specialty dishes and frequently for barbecues.

Most of the parts are made by cutting at certain joints, as can be seen by examining the skeleton and relating to the parts shown. (See Figures 8-11 through 8-17.)

Grade standards are dependent in part on the carcass grade given before cutting into parts, but the parts also must conform to the following specifications:

Breasts and Wishbones

Breasts are separated from the backs at the shoulder and by a cut from that point along the junction of the vertebral and sternal ribs. Ribs may or may not be removed and the breasts may be cut along the breastbone into halves after the sternum or calcified portion has been snapped out and removed. The breasts may be labeled "with ribs" if the ribs were not removed. A third part, the wishbone, may be produced by severing it from the breast halfway between the front point of the breast bone and the end of the wishbone (clavicle) to a point where the wishbone joins the shoulder. Neck skin is not included.

Legs, or Drumsticks and Thighs

Legs include the thigh *(femur)* and drumstick *(tibia)* as removed from the hip joint with the *metatarsus* removed at the hock joint. Back skin or pelvic bones are not included, but the pelvic meat may be included. Drumsticks may be

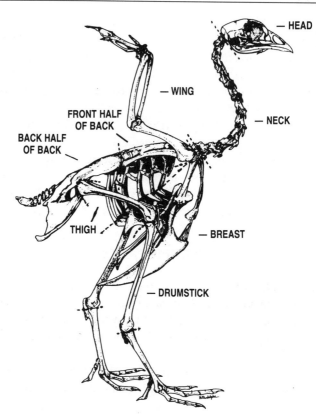

Fig. 8-11. Skeleton of the fowl.

separated from the thigh by a cut through the knee joint (*femorotibial* and *patellar* joint).

Wings

Wings include the entire wing (*humerus, radius, ulna,* and *metacarpus*) with all muscle and skin tissue intact. The wing tip (*phalanges*) may be removed.

Backs

Backs include the pelvic bones and all the vertebrae posterior to the shoulder joint. They may be cut into two parts as shown in Figure 8-17.

- Quality backs would meet all the applicable provisions and include the meat on the *ilium* (oyster), pelvic meat and skin, and vertebral ribs and *scapula,* with meat and skin.
- Quality backs may have one of the meat portions noted above removed.
- Quality backs shall include the meat and skin from the pelvic bones only (not oyster) with the remaining skin essentially intact.

Fig. 8-12. Open the back by splitting along either side of the backbone. The side can be quartered, if desired.

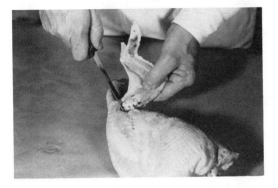

Fig. 8-13. Cut up chicken. First—Remove the wing.

Fig. 8-14. Second—Remove the leg.

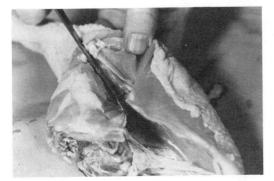

Fig. 8-15. Third—Remove the tail piece (not shown). Fourth—Separate the rib and neck piece from the breast.

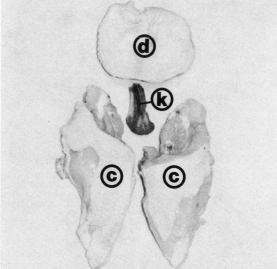

c. Half breast
d. Wishbone breast
k. Sternum—discarded

Fig. 8-16. Fifth—Divide the breast into three sections.

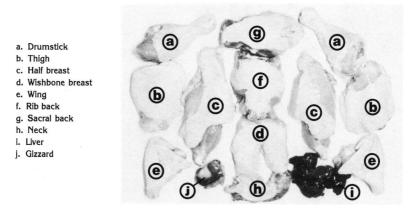

a. Drumstick
b. Thigh
c. Half breast
d. Wishbone breast
e. Wing
f. Rib back
g. Sacral back
h. Neck
i. Liver
j. Gizzard

Fig. 8-17. The cut-up chicken.

Necks

The neck should not have the skin removed.

The cuts, in order of their monetary value, are breasts, legs (or thighs and drumsticks), wings, backs, necks, and gizzards. The livers and hearts may bring a good price.

FABRICATED OR FURTHER-PROCESSED TURKEY

The utilization of large tom turkeys for roasting is not practical in many American homes. Their chief outlet is to the hotel and restaurant trade. To widen the market for large toms and keep their price more in line with hen turkeys, which better meet the weight requirements of the average family, investigators have devised several variations of turkey processing, which are presently in use in industry.

The binding of pieces or chunks of meat together to form rolls or loaves has received considerable attention by the poultry industry, especially in the utilization of turkey meat. It is estimated that over 60 percent of all turkey meat produced is used in the production of these convenience items.

Usually, the breast meat is used for turkey rolls or steaks, and the thigh meat may be formed, cured, and smoked to simulate ham. The drumsticks, wings, and necks may be mechanically deboned and the resultant product used for turkey burgers, turkey patties, turkey frankfurters, or turkey sausages. Cornell University workers[9] have summarized current information on roll manufacture and have indicated that the addition of dried egg albumen, wheat gluten (vital), or gliadin, all protein materials, aided significantly toward binding meat chunks together into a simulated roast to be easily sliced and served by the consumer.

Half and Quarter Turkey

Halving a bird and roasting one half with the dressing underneath has been well received.

Quartering a large bird by making the division midway between the wing and leg is a practice that works out well for small families. The idea that turkey is something that must be roasted has somewhat retarded the sale of "cut-up" turkey.

Dressing Percent

The overnight fasting shrink may vary from 2 percent in chickens to 7 percent in heavy turkeys.

The dressing loss, whether blood-and-feather-dressed or full-dressed, depends upon the weight and condition of the bird. Chickens under 5 pounds will lose an average of 11 percent blood-and-feather-dressed as against 27 percent full-dressed. Chickens over 5 pounds will average 9 and 25 percent loss respectively. Male turkeys weighing between 13 and 17 pounds will lose an average of 10 percent blood-and-feather-dressed and those over 20 pounds will lose about 8 percent. The same birds full-dressed will average a loss of 20 and 18 percent respectively. Blood-and-feather-dressed female turkeys weighing under 10 pounds average 10 percent loss, those between 10 and 12 pounds average 9 percent, and those weighing between 12 and 15 pounds average 7.5 percent loss. These same

[9]Vadehra and Baker. *Food Technology*. 1970. 24:766.

birds will show a full-dressed loss of approximately 19 to 21 percent, or slightly more than the toms.

DUCKS AND GEESE

Waterfowl are very tight-feathered, which makes them difficult to scald. One good method is to steam them. This is accomplished by churning them in water that is near the boiling point or by wrapping a burlap sack around the birds and then immersing them in hot water. Water temperatures as low as 160° F can be used, but the scalding time is considerably longer.

Properly bled and debrained waterfowl may be readily dry-picked, but the fine down feathers must be removed by scalding or waxing. Waxing limits the value of the down feathers and is often not employed until after most of the down has been otherwise removed. (See page 205.) How to produce geese with minimal amounts of pinfeathers is a major unsolved problem.

Long Island had been a large production center for green ducks (8 to 12 weeks of age). This production has now largely moved to the Midwest. Long Island Style ducklings have become a legend among connoisseurs of food.

Except for limited processing and sale as fresh items around holidays, most waterfowl are frozen.

GUINEA FOWL AND PHEASANTS

Most guineas are dressed in the same manner as chickens—dry-picked or semi-scalded. The young birds are marketed at dressed weights of 2 pounds or under.

Pheasants produced commercially are probably one of the meatiest birds for their size of any bird used for human consumption. They are marketed full-dressed and generally semi-scalded before being picked, eviscerated, and frozen. They weigh from 1 to 2 pounds ready-to-cook, and the major portion of the meat is on the breast. The bones are quite small in proportion to the amount of edible meat.

Game birds may be skinned, as described in the next chapter.

STORAGE

Almost all chicken is sold fresh, primarily to avoid the bone darkening that occurs when it is being frozen. It is usually packed in ice for shipment and short-time holding. After being vacuum packed, most other poultry is frozen, first in a hard chill ($-20°$ F) with air blast or brine and then completed, and the birds are held at a constant temperature ($0° \pm 5°$ F).

Chapter 9

Game Processing

DEER AND OTHER BIG GAME

THE CONSERVATION of game by legislation, which restricts the period when such game may be legally shot, is making venison more abundant and, therefore, a more noticeable addition to the family larder. Its scarcity, however, makes it all the more obvious that better care should be taken of the carcass to prevent the flagrant waste that is sometimes evident in its preservation and utilization. State and provincial regulations regarding the hunting of deer and other big game animals vary considerably throughout the United States and Canada. It is the hunter's obligation to know and abide by these regulations. Every effort should be made to make a quick and efficient kill for optimal meat quality and to minimize the suffering of the animal.

Precautions at Time of Kill[1]

This may sound silly, but make sure your deer is dead! Many hunters have leaned their guns against a tree and prepared to field dress the deer only to see it spring up and bound off. A good rule of thumb is to wait 15 minutes before approaching a downed deer or 30 minutes before trailing a wounded deer. A wounded deer will generally travel less than 1 mile before lying down *if* it feels that it is not being followed. Given ample time (the 30-minute wait), the deer will often die where it has bedded down, or at least can be approached more easily for a second shot as it becomes less wary. However, when a wounded deer is trailed immediately, it will generally keep running (sometimes 5 miles or more), which significantly reduces the hunter's chance of finding it. The good hunter will avoid wounding game animals by knowing his or her capabilities in placing a shot for a quick, clean kill.

Approach the deer with caution from its tail end. Have your gun ready for another shot, and nudge the deer with your foot. It may have been wounded and simply dropped from exhaustion. If it's still alive, you will know it. Keep out of

---

[1]Excerpts from *Field Dressing Your Deer*. Illinois Dept. of Conservation.

reach of the legs until you are certain the deer is dead. Adhere to the state laws regarding tagging and reporting the kill.

Bleeding

It is not necessary to bleed your deer unless it has been shot in the head, and often not then. A shotgun slug in the neck or body cavity does a good job of bleeding it for you. If you want to bleed your deer, and if it happens to be a cornfed trophy buck, don't ruin the head by slashing the neck. Make a small cut at the base of the neck. Insert the blade of the hunting knife (4½- to 5½-inch blade) several inches in front of the point of the breast with the point of the blade aimed toward the tail. Plunge it up to the hilt, press the blade downward to the backbone, and withdraw it with a slicing motion. Elevate the rear portion of the deer to permit the blood to drain by gravity.

Field Dressing and Handling

Many hunters think that you should cut the glands from a deer's legs. You can if you want to but it's not necessary. Deer have glands not only on their legs but between their toes and at the corners of their eyes. All glands are inactive after the animal's death. Carry a small sharp knife to use for this purpose. Don't handicap yourself with a big knife, unless you plan on bagging a deer in hand-to-hoof combat. Those Bowies with the 10- and 12-inch blades may look good swinging from your belt, and they may be just the tools for Green Berets or commandoes, but when it comes to field dressing a deer, they are nothing except awkward. A clasp knife or a sheath knife with a sharp 4- or 6-inch blade is all you need.

A 6- to 8-foot section of good clothesline is handy to have with you. With it you can tie the deer's leg to a tree to give yourself working space. If you do not have a rope, you can spread the deer's hind legs by inserting a small branch about 3 feet long between them. With the animal on its back, block it on either side with logs or rocks to keep it in place.

If you have shot a buck, cut the genitals from the body. Carefully cut the hide in the abdominal area. Minimize the opening when field dressing the deer, since in dragging or carrying the deer, it is exposed to weeds, soil, and insects, which contaminate it. Open the body cavity slightly. Tilt the animal sideways and let the blood drain from the body cavity. Do not split the aitch bone now, but cut the bung loose as you would with a lamb (page 170), by cutting around the bung and pulling it back through the abdominal opening. Split the aitch (pelvis) bone when you get the deer home where you can work under more sanitary conditions. Having loosened the rectum, remove all of the intestinal tract, using extreme caution to minimize internal fecal contamination of the carcass. Remove the pluck (heart, lungs, gullet, and windpipe) and liver by cutting the gullet and windpipe as far forward in the chest cavity as possible. Carry a plastic bag with you to place the liver and heart in, as they will make a very nutritious and palatable camp meal.

In parts of the country, warm temperatures and insects can be a problem. Flies often become a problem in southern deer ranges and western antelope ranges, due to the earlier seasons. To remedy fly and other insect problems, you can wrap the carcass in cheese cloth to prevent the entry of flies into the opened body cavity. Some hunters will rub ground black pepper on the inside of the carcass to further distract flies from laying eggs. In warm weather, meat spoilage is a problem that can ruin an otherwise successful hunt. Meat spoilage can be prevented by getting the carcass to a locker operator with a refrigerated cooler within a day of the kill. If this is not possible, and borderline spoilage temperatures exist (above 40° F), hang the carcass in a well shaded spot, and prop the body cavity open with a clean stick to facilitate chilling.

Tagging and Transporting

Many hunters prefer to tag their whitetail as soon as they find it. Others prefer to tag it after it is dressed. The important thing is to be sure that it is tagged before you load it into your car. There is no easy way to get your deer out of the timber into your car, but one method seems to require the least effort. Tie the front legs of your deer to its head and drag it. Remember, normally, you must check in your deer at the check station specified for your county on the same day it was killed. Check your local laws.

Cornfed whitetails are fine eating, but you must take good care of the carcass. Keep it as cool as possible. The field-dressed deer is often transported on the outside of the hunter's car to give mute evidence of the hunter's good fortune and to serve as refrigeration in transit, but it must not rest over the hood, where it will absorb engine heat. At the journey's end, after being checked in, the deer should be skinned, unless it is to be hung and aged in the cold for a week. Leaving the skin on the carcass during aging holds down shrinkage and avoids discoloration. The aging temperature should be 32° to 38° F.

Disposition of the Carcass

If you intend to process the carcass yourself, you must be especially aware that the temperature of the carcass should not exceed 40° F while it is aging.

If you intend to deposit the carcass with a locker operator for processing, be sure to make previous arrangements. There are strict federal and state regulations governing the processing of game in meat processing establishments as listed below.

- Wild game carcasses shall be dressed prior to entering the processing or refrigerated areas of the licensed establishment.

- Wild game carcasses stored in the refrigerated areas of the licensed establishment shall be contained and handled in a manner that will assure complete separation of wild game from domestic meat and meat products. This may be accomplished by, but not limited to, the following: (1) the use of separate coolers, (2) the enclosing of game in metal cages, and (3) the complete enclosing of game carcasses with plastic or shrouds.
- A written request shall be made by the establishment to the responsible governmental agency for a listing of the days and time of day wild game carcasses may be processed.
- All equipment used which comes in contact with wild game shall be thoroughly cleaned and sanitized prior to its use on domestic animal or poultry carcasses.

Skinning

Open the skin over the rear of the hock and down the back of the leg to the rectum. Skin around the hock, and remove the leg at the break joint on the lower part of the hock. Make an opening between the tendon and the hock, and insert a hog gambrel. Raise the carcass until the haunches are at shoulder height, and proceed to pull the pelt from the rounds (Figure 9-1). Use the fist to remove the pelt from the sides, and continue to pull it down the back. Remove the forelegs at the smooth joint (just below the knee joint). Very little knife work is necessary, since the pelt can be pulled and fisted from the carcass. If the head is to be mounted, the skin on the neck (cape) should be opened on the topside of the neck and behind the shoulder (Figure 9-2). A properly caped deer (Figure 9-3) should have sufficient shoulder skin for the taxidermist to work with. It is far better to leave too much skin with the cape than too little.

The head is removed at the atlas joint after the neck has been caped out such that cape and head are removed in one piece. After the pelt and head have been removed, split the underside of the neck, and remove the gullet and windpipe and the remainder of the pluck, if this has not been done previously. Use a stiff brush and plenty of clean water to wash the hair and soil from the inside of the carcass. Place the carcass under refrigeration. Figure 9-4 shows a full-dressed carcass.

Pelts

Care of Hide and Head

Rub the skin side with a liberal amount of fine salt, and apply plenty of salt to the head. Let the salt be absorbed for 24 to 48 hours, then fold the pelt, hair side out, and tie securely with strong cord. Tag it according to law, and ship it to a taxidermist for mounting and tanning. Use clean table salt to avoid mineral

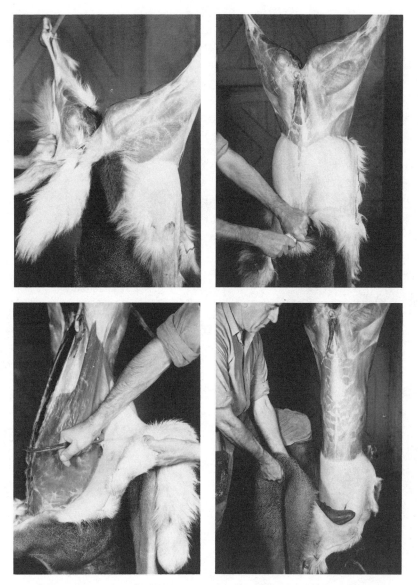

Fig. 9-1. Skinning a deer. (Top left) Pulling the skin from the haunch. (Top right) Pulling the pelt from the loin. (Bottom left) Fisting over the side. (Bottom right) Using body weight to pull pelt over back and shoulder. (Courtesy The Pennsylvania State University)

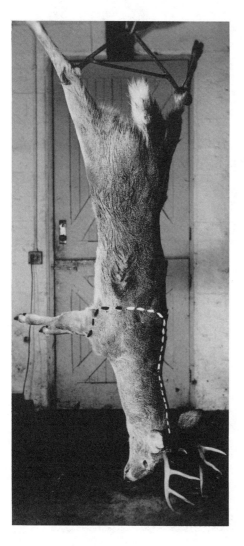

Fig. 9-2. The dotted white and black line indicates the opening to be made when skinning out the cape for a head that is to be mounted. (Courtesy The Pennsylvania State University)

Fig. 9-3 Don't ruin your trophy deer by cutting the cape too short. Note how the prominence of the shoulder adds to the beauty and balance of the finished mount. (Photo by Kevin W. Jones)

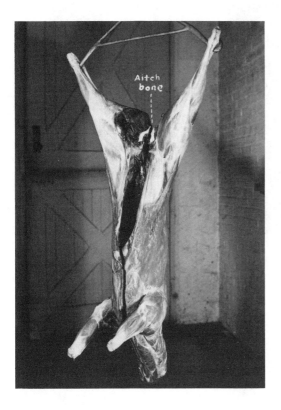

Fig. 9-4. The full-dressed carcass showing the aitch bone. (Courtesy The Pennsylvania State University)

stains, particularly if the pelt is to be made into buckskin. Save time and money by discarding badly torn or scored pelts.

Cutting the Carcass

Split the carcass through the center of the backbone, dividing it into two sides; however, if the neck is to be used as a pot roast or neck slices, remove it before splitting the carcass. Place the side of venison on the table, inside down, and remove the hind leg by cutting in front of and close to the hip bone. Move forward to the shoulder, and remove it by cutting between the fourth and fifth ribs. The back with breast attached must have the breast removed. Cut across the ribs about 3 inches from the backbone on the blade end to the loin end. The ribs are separated from the loin by cutting directly behind the last rib. The leg is placed on the table, aitch bone on top. Cut parallel to the aitch bone, and remove the rump. Remove the flank. Figure 9-5 illustrates where these cuts are made.

Fig. 9-5. One method of cutting a venison carcass: (1) rear shank; (2) heel; (3) round steak; (4) rump and hip loin; (5) loin chop or roast; (6) rib chop or roast; (7) top of shoulder or chuck; (8) neck pot roast; (8A) neck slices; (9) arm roast; (10) foreshank; (11) breast; (12) flank. The letters in white indicate the line where wholesale cuts are made. The above carcass weighed 92 pounds and was in excellent finish. (Courtesy The Pennsylvania State University)

Venison rib chops, round steaks, and rolled shoulder roasts are illustrated in Figure 9-6. The shanks, breast, and flank are usually boned and ground into deerburgers or incorporated with pork for sausage. The neck and heel portions of the carcass are also often used for sausage or ground venison.

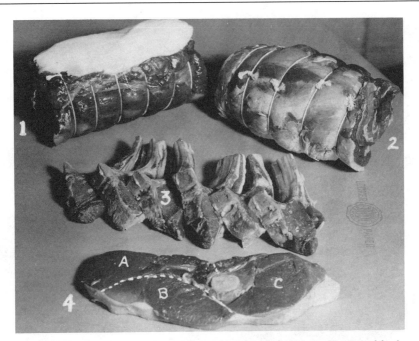

Fig. 9-6. Some cuts of venison. (1) Rolled shoulder of buck with slice of fresh pork fat for self-basting; (2) rolled shoulder of fat doe (it has sufficient fat); (3) rib chop; (4) venison round steak indicating (A) inside round, (B) outside round, and (C) round tip. (Courtesy The Pennsylvania State University)

Boneless Cuts

An alternate method of cutting is to make all cuts boneless, including the rib and loin chops. Butterfly chops from these tender cuts are highly desirable. Furthermore, by making boneless cuts, you eliminate the need for a saw, which is a desirable factor in home cutting. The reader may refer to Chapter 15 on lamb cutting and apply many of those principles to cutting a venison carcass. Figures 9-7 through 9-12 depict a venison carcass and the boneless sub-primal cuts obtained from it.

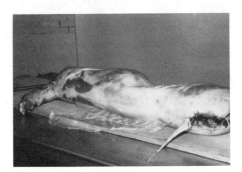

Fig. 9-7. A 96-pound field-dressed doe yielded this 73.5-pound carcass, with a loin eye of 4.35 square inches. The thin exterior muscle was removed prior to the *longissimus* muscle.

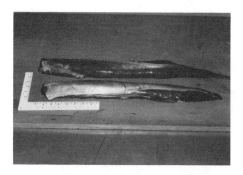

Fig. 9-9. The tenderloins weighed 0.6 pound.

Fig. 9-11. The round (sirloin) tip roasts totaled 3.8 pounds. These are ideal for roasts and quite satisfactory for steaks.

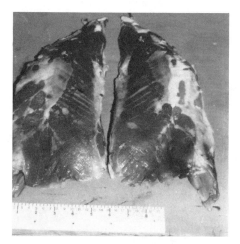

Fig. 9-8. The doe's boneless loin muscles weighed 5.1 pounds. The thin, connective tissue cover must be removed for palatable chops.

Fig. 9-10. The top round roasts totaled 4.3 pounds.

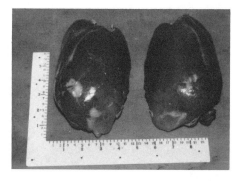

Fig. 9-12. The bottom round roasts weighed 4.1 pounds. These roasts can be ground with the shoulder muscles and the remaining usable trim from the less desirable thin rough cuts (flank, plate, shank) to be mixed with pork trim for fresh sausage or processed into summer sausage. Or if your family likes roast venison, the bottom makes a delicious pot roast.

A University of Illinois study involved seven deer (three does, four bucks) with the following results:

Average for Seven Deer

	Pounds	
Field-dressed weight	112	
Estimated age (years)		1.6
Carcass weight to cut (Figure 9-7)	87	
Dressing percent		78
Loin-eye area at 12th rib (sq. in.)		4.5

		% of Carcass
Lean trim (excluding mutilated)	27.4	31.5
Boneless loin and rib (Figure 9-8)	7.0	8.0
Tenderloin (Figure 9-9)	1.0	1.1
Boneless top round (Figure 9-10)	5.3	6.1
Boneless round tip (Figure 9-11)	4.6	5.3
Boneless bottom round (Figure 9-12)	4.4	5.2
Edible meat	49.7	57.2
Mutilated (not usable)	9.8	11.2
Bone	22.7	26.1
Fat	3.6	4.1
Nonedible	36.1	41.4
Total (edible and nonedible)	85.8	98.6

The loin-rib muscle *(longissimus)* and tenderloin were made into 1¼-inch chops and the top round into 1¼-inch steaks. Panel members tasted chops and steaks with and without the thinly adhering fat cover. No preference was shown in regard to fat cover or animal age, but doe meat was slightly preferred over buck meat.

Mutilated or Bloodshot Areas

If a large area of the carcass is affected by the shot, portions of it may be salvaged by washing it free from hair and soaking it in a weak brine made by dissolving ½ pound of salt in 1 gallon of water. The salt will draw out most of the blood overnight, and the best meat will be suitable for grinding or stewing. Badly mutilated meat can be used for dog food, if all bone chips and slugs or shot are removed.

Sausages

A variety of tasty sausages and processed products can dramatically improve the palatability of the trimmings that would otherwise be used for ground venison. In cases where older bucks are taken, particularly those which have not had access to grain diets, the entire carcass can be better utilized in sausage. This will help reduce the extreme "gaminess" of the meat that is often present in such carcasses. One taste of sausage made from the less-tender cuts of venison ground with fat pork trimmings will assure anyone that no part of the venison carcass should be wasted. Some creative ideas on how to utilize this venison are given in the following recipes. If you are intent on home-processing your deer and making sausage, you will need some equipment and supplies. Your local locker operator might provide some assistance in locating a supplier.

Fresh Venison Sausage

10 lb. lean venison	12 g ground sage (1 tsp.)
10 lb. 50% fat pork trim	10 g mace (2 tsp.)
175 g salt (9 tbs.)	10 g coriander (2 tsp.)
20 g ground black pepper (4⅛ tsp.)	10 g hot mustard (2 tsp.)

Grind lean venison and fat pork through a ⅜-inch coarse grinder plate. Mix uniformly with salt and spices. Regrind sausage through a ⅛-inch fine grinder plate, and stuff into sheep or hog casings. This sausage can also be wrapped in freezer paper in bulk form and subsequently made into patties before cooking, if desired.

Unlike the other sausages described here, fresh venison sausage is not cured or smoked and is thus the easiest to make. This sausage will only keep for three to five days (in a refrigerator) and thus should be sharp frozen until ready for consumption.

NOTE: For conversions
1.0 pound = 454 grams
1.0 ounce = 28.4 grams

Polish Venison Sausage

10 lb. lean venison	17 g ground ground coriander (3½ tsp.)
10 lb. 50% fat pork trim	10 g garlic powder (2 tsp.)
1 lb. cold water (1 pt.)	15 g paprika (3 tsp.)
200 g salt (¾ cup)	11 g ground nutmeg (2¼ tsp.)
40 g sugar (8¼ tsp.)	11 g ground ginger (2¼ tsp.)
22 g certified cure (4½ tsp.)*	6 g ground caraway (1¼ tsp.)
30 g ground black pepper (6¼ tsp.)	

Grind venison through a ⅜-inch coarse grinder plate, and regrind through a ⅛-inch fine plate. Blend venison with salt, water, and cure in a mixer for two to

---

*Certified cure is a nitrite cure which assures that you will have the correct nitrite level of 150 ppm (parts per million) when used according to directions, usually 1 ounce per 25 pounds of meat. *It is very important that this ingredient is weighed out accurately and that you do not use more than the amount called for.* Certified cure cannot be purchased through a grocery store or a drug store, but rather through a local locker operator, who deals with a spice company.

three minutes. Next, grind pork trim through the coarse and the fine grinder plates, add to mixer, and mix an additional four minutes while adding the remaining ingredients. Stuff into sheep or hog casings, and link casings by twisting in opposite directions every 6 inches (Figure 20-3). Hang the linked polish sausage on smokehouse sticks, and cook at 130° F for one hour. After this drying period, smoke product for two to three hours, while gradually increasing temperature to 155° F. Increase temperature of smokehouse to 165° F, and cook until internal product temperature reaches 152° F.

Venison Summer Sausage

10 lb. lean venison
10 lb. 50% fat pork trim
2 lb. cold water (1 qt.)
225 g salt (¾ cup)
100 g dextrose or 200 g corn syrup
   (5 tbs. or 1 cup)
22 g certified cure (4½ tsp.)

30 g ground black pepper (6¼ tsp.)
16 g ground coriander (3¼ tsp.)
4 g ground nutmeg (¾ tsp.)
4 g ground red pepper (¾ tsp.)
4 g whole mustard seed (¾ tsp.)
0.5 g ground ginger (⅛ tsp.)
Starter culture (if available)

Grind venison through a ⅜-inch coarse grinder plate, and regrind through a ⅛-inch fine plate. Mix ground venison with salt, water, and cure for three minutes. Add pork trim that has been ground through a ⅛-inch fine grinder plate, and add remaining nonmeat ingredients while mixing. The summer sausage should be mixed an additional four minutes after all ingredients have been added. Stuff product into 3-inch fibrous casings, and clip or tie the ends of the casings. Care should be taken to avoid air pockets within the stuffed casings. If a commercial started culture has been used, the product can go straight into the smokehouse and be fermented for 12 to 18 hours at 80° to 100° F. If no starter culture has been used, hold the product in a 38° F cooler for three days before following the same fermentation schedule. For more information on fermentation processes, refer to the section in Chapter 20 on dry and semi-dry sausages. The product should be smoked during the fermentation process. After the fermentation process, gradually increase the smokehouse temperature to 165° F, and cook the product to an internal temperature of 152° F. Immediately after cooking, rinse with cold water to remove the greasy film on the product surface and to shrink casing to prevent formation of wrinkles, and place in a 34° to 38° F cooler. If a ''tangier'' summer sausage is desired, increase the level of dextrose in the formulation, and ferment for a longer period of time.

Venison Jerky

3 lb. lean venison round muscle
250 g water (1 cup)
100 g soy sauce (¾ cup)
20 g seasoned salt or barbecue salt
   (4⅛ tsp.)

10 g Accent or MSG (2 tsp.)
6 g garlic powder (1¼ tsp.)
6 g black pepper (1¼ tsp.)
6 g Worcestershire sauce (1¼ tsp.)
2 g liquid smoke (optional) (½ tsp.)

The round muscle should be in one piece and free of any fat. Semi-freeze the meat, and slice it with the grain into ⅛-inch slices, and cut into strips of the desired size. Mix remaining ingredients together to form the marinade. Pour the sauce over the meat. Marinate for 24 hours, stirring occasionally. Lay venison strips on oven racks in smokehouse, and smoke for six to eight hours at 110° to 120° F, or until the desired degree of chewiness is achieved. The jerky can also be dried in an oven at the lowest setting, with the door ajar.

Venison Smoky Sausage Sticks

20 lb. lean venison
250 g salt (¾ cup + 1 tbs.)
22 g certified cure (4½ tsp.)
35 g ground black pepper (7¼ tsp.)

10 g ground red pepper (2 tsp.)
4 g ground coriander (¾ tsp.)
4 g garlic powder (¾ tsp.)

Grind venison through a ⅜-inch coarse grinder plate, and regrind through a ⅛-inch fine plate. Mix ground venison with salt and cure until meat is very tacky (three to five minutes). Add the remaining spices, and mix for an additional four to five minutes. Stuff the sausage into the smallest diameter casing available. Sheep casings will work; however, collagen casings that are ½-inch diameter work best. It is important that an edible casing be used. There is no need to link this sausage, since it will be cut into 6-inch "sticks" when finished. Place the stuffed sausage in a cooler at 38° F for three days. Smoke the sausages on smokehouse sticks for six hours at 80° to 100° F, using wet or damp sawdust. Raise the smokehouse temperature to 120° F for an additional two hours, and finish the product at 150° F for two more hours. Cut into 6-inch sticks when it is cooled.

For more information on home sausage production, the reader is referred to Chapter 20.

GAME BIRDS

Pheasant, quail, grouse, partridge, and wild turkeys account for the majority of game birds taken by hunters. Turtle doves, although very small, are also popular among hunters in many regions of the country. With the exception of the wild turkey, game birds are generally cleaned by removing the skin with the feathers, since it is much easier and faster than plucking.

Pheasants

Pheasants are probably the most popular of the upland game birds and, for their size, are one of the meatiest birds used for human consumption. These birds commonly will have a 1- to 2-pound carcass, with the majority of the meat on the breast. State hunting regulations vary considerably with hunting season, bag limit, and possession limit. It is illegal to shoot a hen pheasant in most states. For this reason, many states require the head or feet to be left on the carcass when it is being transported. It is the obligation of the hunter to know and abide by the regulations of the state in which he or she is hunting.

There are two popular methods of cleaning pheasants. After the bird has been skinned, follow the same procedures outlined for dressing poultry (pages 206-207). An alternative method of processing pheasants is faster and neater; however, this procedure may not be legal if heads are required to be left on the carcass when being transported across state lines. This alternative dressing procedure is as follows:

- Skin the bird by making a small lateral incision on the underside of the breast and pulling the skin and feathers off the carcass.
- From the top (dorsal) side, cut down both sides of the back, starting at the cranial end and cutting through to the last rib.
- Separate the carcass by pulling the breast apart from the neck, back, and legs. The intestinal tract, heart, lungs, and liver will remain attached to the back portion.
- Remove the feet and lower legs at the joint below the drumsticks.
- Cut each leg off from the back portion by cutting immediately adjacent to the back and through the ball and socket joint.

Dressing pheasants in this manner will produce three pieces (two leg-thigh pieces and one breast) from each bird (Figure 9-13). The back and neck are discarded, since they contain very little meat (Figure 9-14). This method is preferred by many hunters, because they don't have to handle the entrails.

Fig. 9-13. The *three-piece* dressing method for pheasants yields one breast-wing portion and two leg-thigh pieces. This method is preferred by many experienced pheasant hunters because it is faster and cleaner than the conventional method.

Fig. 9-14. The pheasant's neck, back, and viscera remain in one piece and are discarded. There is very little salvageable meat on a pheasant's neck or back.

Quail, Grouse, Partridge, Dove

These smaller game birds, like pheasants, have the majority of their carcass weight in the breast. Most hunters prefer removing the skin with the feathers rather than performing the more tedious task of dry-picking. These birds are generally dressed by removing the entrails from the whole carcass. In the case of doves (and sometimes quail) only the breast portion is kept for human consumption, due to the small size of these birds and the limited value of the remaining portions.

Waterfowl

Ducks and geese are among the most difficult of birds to clean. If the skin is removed, the meat will dry out during roasting. For this reason, ducks and geese are usually scalded or dry-picked and dipped in wax to remove the feathers. Refer to the poultry processing chapter for details.

RABBITS

Wild rabbits, taken as part of the hunter's bag, can be dressed in the same manner as domestic rabbits. Wild rabbits will exhibit a darker meat that is often tougher than the meat of domesticated rabbits.

The domestic rabbit is one of the most efficient meat producers of all animals. Rabbits have very high reproductive rates, averaging 48 to 64 offspring per year. Young rabbits can reach 4- to 5-pound market weights at eight weeks of age. Rabbits are also highly efficient converters of feed into meat, requiring approximately 2½ pounds to produce 1 pound of gain. Rabbits can be fed high-forage, low-grain diets, which are not competitive with human food needs, and still maintain high levels of production efficiency. Rabbit meat is generally higher in protein content and lower in fat content than meat from other species.

Consumption of rabbit meat in the United States remains very low, despite the high efficiency and nutritional quality of rabbit meat. This is likely due to the perception of the rabbit as a pet animal rather than a meat animal. At the present time, there is only one USDA inspected rabbit slaughter plant in the United States.

Classes of Ready-to-Cook Domestic Rabbit[2]

Fryer or Young Rabbit

A fryer or young rabbit is a young domestic rabbit carcass weighing not less than 1½ pounds and rarely more than 3½ pounds, processed from a rabbit usually less than 12 weeks of age. The flesh of a fryer or young rabbit is tender and fine grained and is a bright pearly pink color.

Roaster or Mature Rabbit

A roaster or mature rabbit is a mature, or old, domestic rabbit carcass of any weight, but usually over 4 pounds, processed from a rabbit usually eight months of age or older. The flesh of a roaster or mature rabbit is more firm and coarse grained, and the muscle fiber is slightly darker in color and less tender, and the fat may be more creamy in color than that of a fryer or young rabbit.

Quality Grades[3]

A carcass found to be unsound, unwholesome, or unfit for food shall not be included in any of the quality designations.

A Quality

- Is short, thick, well rounded, and full-fleshed.
- Has a broad back; broad hips; broad, deep-fleshed shoulders; and firm muscle texture.
- Has a fair quantity of interior fat in the crotch and over the inner walls of the carcass and a moderate amount of interior fat around the kidneys.
- Is free of evidence of incomplete bleeding, such as more than an occasional slight coagulation in a vein. Is free from any evidence of reddening of the flesh due to fluid in the connective tissues.
- Is free from all foreign material (including, but not being limited to, hair, dirt, and bone particles) and from crushed bones caused by removing the head or the feet.
- Is free from broken bones, flesh bruises, defects, and deformities. Ends of leg bones may be broken due to removal of the feet.

---

[2]Poultry Division, Agricultural Marketing Service, USDA.
[3]*Ibid.*

B Quality

- Is short, fairly well rounded, and fairly well fleshed.
- Has a fairly broad back, fairly broad hips, fairly broad and deep-fleshed shoulders, and fairly firm muscle texture.
- Has at least a small amount of interior fat in the crotch and over the inner walls of the carcass, with a small amount of interior fat around the kidneys.
- Is free of evidence of incomplete bleeding, such as more than an occasional slight coagulation in a vein. Is free from any evidence of reddening of the flesh due to fluid in the connective tissues.
- Is free from broken bones and practically free from bruises, defects, and deformities. Ends of leg bones may be broken due to removal of the feet.

C Quality

A carcass that does not meet the requirements of A or B Quality may be of C Quality and such carcass:

- May be long, rangy, and fairly well fleshed.
- May have thin, narrow back and hips, and soft, flabby muscle texture.
- May show very little evidence of exterior fat.
- May show very slight evidence of reddening of the flesh due to blood in the connective tissues.
- Is free from all foreign material (including but not being limited to, hair, dirt, and bone particles) and from crushed bones caused by removing the head or feet.
- May have moderate bruises of the flesh, moderate defects, and moderate deformities. May have not more than one broken bone in addition to broken ends of leg bones due to removal of the feet and may have a small portion of the carcass removed because of serious bruises. Discoloration due to bruising in the flesh shall be free of clots (discernible clumps of dark or red cells).

Inspection

Rabbit inspection is identical in scope and completeness to poultry inspection, which is discussed in detail in Chapter 3. The mark of inspection is identical for poultry and rabbit and is illustrated in Chapter 3.

Handling

Never lift a rabbit by the ears or legs. Grasp a fold of skin over the rabbit's shoulders, support the rump with your free hand, and hold the back of the rabbit against your body.

Dressing Rabbits

The method of slaughtering rabbits consists of the following:

- Give a sharp blow on the top of the rabbit's head to stun it.
- Make an incision at the rear of the hock between the bone and the tendon.
- Suspend the carcass by hanging it on a hook through the hock.
- Sever the head at the atlas joint.
- Remove the free rear leg at the hock joint.
- Remove the tail and forelegs (knee joint).
- Cut the skin on the rear of the loose leg to the base of the tail and up the rear of the suspended leg. (Figure 9-15.)
- Pull the edges of the cut skin away from the flesh and down over the carcass. Make no other cuts in the skin.
- Eviscerate by opening the median line of the belly, leaving the heart, liver, and kidneys in the carcass.
- Remove the suspended rear leg, and rinse the carcass in cold water to remove any hair or blood.
- Joint the carcass by removing the forelegs and hind legs, cutting the loin in one piece, and separating the shoulders.

Fig. 9-15. The procedure followed in skinning a rabbit (Courtesy U.S. Fish and Wildlife Service)

Pelts

Rabbit pelts have a fur skin value. The small ones should be stretched on a thin board or wire stretcher 24 inches long and 4 inches wide at the narrow end and 7 inches wide at the base. The skins of 10- and 12-pound rabbits need a board 30 inches long, 4 inches wide at the narrow end, and 9 inches wide at the base. Stretch the warm skin on the board with the fore part over the narrow end, and smooth out the wrinkles. Have both front legs on one side of the shaping board. Remove any surplus fat, and make sure that the skin dries flat. Do not dry in the sun or artificial heat, and do not use salt. When dry, the skins can be stored in a tight box, but each layer should be sprinkled with naphtha flakes or moth balls to ward off moths.

A communicable disease known as tularemia is prevalent in wild rabbits. While dressing rabbits having the disease, humans can become infected through abrasions in their skin. It is transmitted from one rabbit to another by the rabbit louse or tick. The disease has not been observed in domestic rabbits, and the disease organism is destroyed by cooking.

Chapter 10

Packing House By-products

HUMANS have utilized animals as a food source throughout history. Early men and women also made use of animal skins for clothing and shelter; bones and horns for tools; tendons and intestines for weapons, tools, and bindings; teeth, claws, feathers, or hair for ornaments; stomachs, bladders, and skins for containers. Nothing was wasted following a successful hunt for meat. Modern society learned well from those ancestors, for today's meat industry still utilizes the nonmeat portions of livestock. Mr. Tony Javurek, who guided two tours daily at John Morrell's Sioux Falls packing plant for more than 20 years, expressed the meat industry philosophy about by-products during each tour. Tony always said, "We use all parts of the pig except the squeal and the curl in its tail."

Meat slaughter by-products (offal) include all parts of the animal that are not included in the carcass. Cutting and processing of the carcass results in additional nonmuscle by-products, namely, fat, bone, and other connective tissues.

By-products play a dual role as both assets and liabilities in the late twentieth century meat industry. This chapter will demonstrate that modern living is enhanced by many nonmeat products, which originate from the meat industry. However, environmental concerns, energy costs, and increased volume have combined to make by-product handling a major economic and management problem in the meat industry. A hog kill operation with a 1,000-head-per-hour capacity must be able to process approximately 72,000 pounds of by-product material per hour. The by-product varies from edible muscle and organ meats to hair and manure, therefore requiring a myriad of physical movement and sanitation and processing technologies.

Meat industry by-product handling practices have changed since 1896, when *The National Provisioner* reported that a man in South Omaha made $10,000 yearly by skimming grease from the sewers which carried refuse from the big packing houses into the Missouri River. That same year, the Butchers Board of Trade of San Francisco was considering a sanitary improvement in the way of a sewer which would empty far out into the bay. Offal from the slaughter houses was being dropped into the bay near shore and was drifting to and fro with the tide, causing unpleasant odors.[1]

---

[1]*The National Provisioner.* Meat for the Multitudes II. Vol. 185 (1):60. July 4, 1981.

Today, economics, modern technology, and industry's concern for the environment result in maximum salvage and utilization of all by-product materials. Much of the air and water effluent (flowing out) from meat operations is as clean or cleaner than the water and air entering the plant.

Processed by-products have been a significant source of income to the meat processing industry. Present-day beef slaughter operations cover the slaughter costs, profits, and part of the purchase cost of the livestock with sales of by-product material. (See Chapter 12.) The meat industry continues to seek new uses and markets for the many nonmeat items that result from the disassembly lines.

Modern conditions make it almost impossible to cut production and distribution expenses for the majority of commodities; hence, one of the most important opportunities for gaining competitive advantage, or even for enabling an industry or individual business to maintain its position in this new competition, is to reduce its manufacturing expense by creating new credits for products previously unmarketable. From the viewpoint of individual business, this manufacture of by-products has turned waste into such a source of revenue that, in many cases, the by-products have proved more profitable per pound than the main product.

The remainder of this chapter will attempt to inform the reader of some of the purposes that meat by-products serve.

EDIBLE BY-PRODUCTS

Edible by-products, sometimes referred to as "variety meats," include livers; hearts; tongues; brains; oxtails; kidneys; sweetbreads (thymus and/or pancreas glands); tripe (stomach); chitterlings and natural casings (intestines); lamb and calf fries, often called mountain oysters (testicles); and, in some societies, blood. Meat trimmings derived from heads are used in processed meats and are characterized as "edible offal" or "edible by-product items." Edible fats removed during slaughter include pork leaf fat (kidney fat) and the caul fat shrouding the rumen and stomach. All edible by-products must be wholesome and must be segregated, chilled, and processed as specified by the FSIS (Food Safety and Inspection Service) (Chapter 3). Several edible offal items are illustrated in Figure 10-1. The yields of offal items for beef and pork are listed in Table 10-1.

VARIETY MEATS

Variety meats are economical sources of valuable nutrients (Table 10-2). Liver is the most nutritious of all meat items. The *nutrient density* of liver exceeds that of muscle meats, which are high. One 3½-ounce (100-gram) serving of beef liver provides 22 to 23 grams of high-quality protein, while contributing fewer than 150 calories to the diet. An excellent source of readily digested heme iron, liver also provides B vitamins, particularly B_{12}, as well as vitamin A to

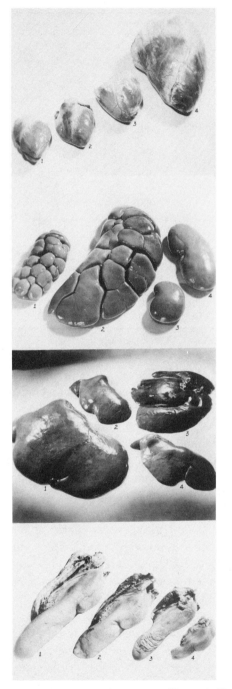

Hearts

1. Lamb
   2. Pork
      3. Veal
         4. Beef

Kidneys

1. Veal
   2. Beef
      3. Lamb
         4. Pork

Livers

1. Beef
   2. Lamb
      3. Pork
         4. Veal

Tongues

1. Beef
   2. Veal
      3. Pork
         4. Lamb

Fig. 10-1. Selected variety meats. (Courtesy National Live Stock and Meat Board)

Table 10-1. By-product Yields

	1,000-Pound Beef		230-Pound Hog	
	(lb.)	*(%)*	*(lb.)*	*(%)*
Blood	40.0	4.0	7.0	3.04
Edible kill fat	35.0	3.5	8.0	3.5
Feet			5.0	2.17
Head and cheek meat	4.5	0.45	1.38	0.6
Heart	4.0	0.40	0.5	0.20
Hide or hair	70.0	7.0	2.0	0.87
Intestine			4.2	1.83
Kidney	1.2	0.12	0.37	0.16
Liver	11.0	1.10	2.5	1.10
Lungs	4.5	0.45	1.0	0.43
Spleen	1.5	0.15		
Sweetbreads	0.3	0.03		
Tail	1.5	0.15	0.25	0.1
Tongue	3.5	0.35	0.75	0.3
Tripe (stomach)	20.0	2.0	1.5	0.65
Weasand (esophagus)	0.4	0.04	0.13	0.05
Manure	45.0	4.5	4.0	1.74
Inedible raw material	130.0	13.0	23.0	10.0
Rendered fat, edible	25.0	2.5	28.0	12.17
Rendered fat, inedible	40.0	4.0	6.0	2.61
Cracklings	30.0	3.0	5.0	2.17

Source: Packers Engineering and Equipment Co., Inc., catalog.

consumers who enjoy its unique flavor. (Heme is the O_2 carrying component of hemoglobin and myoglobin and is a source of iron more readily absorbed by the human digestive system than most dietary iron forms.) Variety meats are relatively high in protein, with the exception of brain. More extensive use of meat animal by-products for human food has been proposed as one method to solve world nutrition problems. In addition to enhancing human nutrition, new developments in meat by-product utilization would increase the overall efficiency of livestock production.

Although many consumers in the United States purchase and use variety meats in their menus, larger amounts of some by-products are utilized in processed meats, principally sausage. (See Chapter 20.) It should be noted that FSIS regulations require that all processed meats containing variety meats indicate in the label ingredient list each specific variety meat that is included in the formulation.

People in other societies of the world, particularly those in Europe and Japan, consume much more variety meats per capita than the average person in the United States. Therefore, the American meat industry has developed overseas markets for those nutritious products of animal agriculture. Table 10-3 indicates that edible offal items (variety meats) produced a positive international trade balance of more than $255 million for the United States in 1983.

A complete discussion of the uses of variety meats could fill the pages of a

Table 10-2. Proximate Protein Calorie and Fat Content of 100 Grams of Cooked Variety Meats[1]

	Protein	Fat	Calories
	.... (grams)		
Beef			
Brain	11.5	9.0	130
Heart	28.9	6.0	177
Kidney	24.7	3.6	138
Liver	22.9	4.3	137
Lung	20.3	3.7	120
Pancreas	27.1	17.2	271
Spleen	25.1	4.2	145
Thymus (sweetbreads)	20.5	24.9	312
Tongue	22.2	21.5	288
Veal			
Brain	10.5	7.4	111
Heart	26.3	4.5	153
Kidney	26.3	5.9	165
Liver	21.5	7.6	160
Lung	18.8	2.6	104
Pancreas	29.1	14.6	256
Spleen	23.9	2.6	125
Thymus (sweetbreads)	18.4	2.9	105
Tongue	26.2	8.3	187
Pork			
Brain	12.2	8.7	130
Heart	23.6	4.8	144
Kidney	25.4	4.7	151
Liver	21.6	4.7	135
Lung	16.6	3.1	99
Pancreas (sweetbreads)	28.5	10.8	219
Spleen	28.2	3.2	149
Tongue	24.1	18.6	270
Lamb			
Brain	12.7	9.2	137
Heart	21.7	5.2	140
Kidney	23.1	3.4	129
Liver	23.7	10.9	199
Lung	20.9	3.0	116
Pancreas	23.3	9.6	186
Spleen	27.3	3.8	150
Tongue	21.5	20.5	277

[1]From the Meat Board's *Meat Book*. 1977.

text, so only a very brief description of some preparation techniques will be included here.

Liver

Liver is most often sliced, relatively thin, after the outer connective tissue membrane is removed. The slices may be cooked by a variety of methods, such as frying, broiling, sautéing, or braising. Liver may be ground or chopped and added as an ingredient to loaves, sausages, spreads, or other dishes.

Table 10-3. Value of United States Imports and Exports of Meat By-products, 1983

	Imports	Exports
	.... (million dollars)	
Variety meats	4.4	260.6
Tallow, greases, and lard	3.8	587.2
Casings	52.7	19.0
Hides and skins	66.1	777.3
Wool and mohair	175.4	53.3
Total	302.4	1,697.4

Source: *Meat Facts*. 1984. American Meat Institute.

Heart

Heart is generally less tender than liver and requires moist heat and long-term cookery. Hearts which have been slashed open for inspection may be sewn and the cavities filled with a dressing and then roasted like a turkey. The meat has an excellent flavor, and the dressing takes on the "hearty" flavor of the meat, resulting in memorable eating experiences.

Tongue

After braising, both heart and tongue (Figure 10-2), thinly sliced, are excellent cold sandwich meats. Garnishes, mustard, horseradish, or other dressings make those cold sandwiches truly unique snacks or lunch fare. Removal of the tough outer membrane of the tongue is facilitated by blanching (short exposure to boiling water) prior to long-term, moist-heat cookery.

Brains, Sweetbreads, and Fries

Brains, sweetbreads, and fries are often sliced thinly and dipped in batter or flour (breaded) before being deep fried. Precookery aids in removing the outer membranes of sweetbreads and brains (Figure 10-2) as well as "setting" the soft, delicate products for easier slicing. These variety meats may be mixed with scrambled eggs by breaking them up into small pieces.

Kidneys

Kidneys are sometimes left in lamb and veal loins to become part of the loin or "kidney chops." Kidneys may be included as one of the ingredients of meat casseroles, stews, and pies. Lamb and veal kidneys may be broiled or skewered, because they are more tender than beef kidneys.

Fig. 10-2. Beef tongues and brains are individually bagged, placed in shipping boxes, and frozen after a post-mortem chill period. (Courtesy IBP Corp., Dakota City, Nebraska)

Oxtail

Oxtail is an unsurpassed source of rich, meaty flavor and texture for soups and stews. The tail sections should be browned (not essential), then allowed to simmer until the meat is tender and separated from the bone. The bone portions can be removed prior to serving. The stock may be added to other soup ingredients.

Tripe

The first *(rumen)* and second *(reticulum)* stomachs of cattle are washed, soaked in lime water, and scraped to remove the inside walls. They are then cooked (and sometimes pickled in vinegar) and sold as tripe. Tripe may be cooked and served with sauces and dressings or may be utilized in meat dishes, as indicated for kidneys.

Chitterlings

Pork intestines are thoroughly cleaned and cooked with sauces to be served as "chitterlings."

Intestines, as well as other portions of the digestive tract, are used as casings for sausage products. (See Chapter 20.) Many specialty sausages are derived largely from edible offal items. Jelled products, such as headcheese, souse, and scrapple, may rely on high collagen content meat sources, such as pork skin trimmings and head trims for unique identifying characteristics. "Haggis" is a Scottish food made from the hearts, lungs, and livers of sheep or calves, highly seasoned, mixed with oatmeal, and boiled in a sheep stomach.

ANIMAL FATS

Not only have animal fats served the human race as a food energy source, but also they have served in many other ways. Melted fat, or grease, rubbed into the hides made the tepee (*tipi* in Sioux) leather supple and waterproof. Bear grease, or other grease, provided a sheen and a unique odor to the hair of both the male and the female of many generations of our ancestors. Today, the lanolin in skin preparations and cosmetics of all kinds is a component of the lipid extracted from wool. Animal and fish oils provided the fuel for the oil lamps that preceded the kerosene, gasoline, propane, and electric lamps of modern years. At one time during the current century, the value of the fat derived from hog slaughter and cutting was equal to the meat value. Of course, hogs were fatter and meat was cheaper in those days than they are now.

Rendering

The meat industry has extracted (and still does) the fat from almost all byproducts that have no other use or market potential. The extraction technique used is called "rendering." Rendering usually involves the heating or cooking of the raw material to liquefy the fat and to break down membranes or other structures which may hold the fat. The melted fat can then be drawn off, leaving the solid or semi-solid nonfat materials. In animal tissue, the nonfat material is mainly protein and mineral. Rendering serves two purposes: it separates fat and protein, and it cooks the animal tissue. The separated fat and protein have greater value than the fresh raw material. Cooking significantly increases the storage stability of the fat and protein products by killing the microorganisms which were present in the raw material and by facilitating removal of most of the moisture.

The rendering process has evolved along with all other aspects of the meat industry. Early rendering was accomplished by direct application of heat to containers holding the animal tissue. The next step in the evolution of the rendering process was the injection of live steam into the product. The steam increased the rate of temperature increase and speeded the process, but it added moisture and resulted in overheating some portions of the product. The tank water which accumulates in wet rendering must be evaporated and reduced to the consistency of molasses. That product, referred to as "stick" or "stick liquor," is mixed with the tankage (precipitated solids) and dried.

Confinement of the steam in a jacket around the vessel, which contained the constantly stirred product, was a further improvement. By closing the vessel and drawing a vacuum on the inside of the steam-jacketed chamber, the boiling point of the moisture was reduced, and the moisture vapor was evacuated. The rendering process was therefore further shortened, and the temperatures required were reduced. Low-temperature rendering produces higher quality fats and protein products. This method, dry rendering, is used extensively in the manufacture of lard and in the extraction of edible and inedible fats and oils. Additional advantages over wet rendering (live-steam injection into the product) are that it eliminates the expense and problems of handling tank water and reduces odor problems.

Many rendering operations today utilize large, steam-jacketed tanks, which can be charged with thousands of pounds of fresh raw materials, closed, and vacuumized. The product is constantly agitated during the cooking process. At the end of the cycle, the liquid fat is removed, and the solid protein material is mechanically pressed to squeeze out as much fat as possible. The pressed solid is ground and sold as high-protein inedible meat meal. The areas of the plant where the rendering takes place are referred to as "tank houses" or "tank rooms." Product destined to be rendered is to be "tanked."

Although rendering has historically been a batch process, a continuous rendering system is now in use. The particle size of the edible raw material is reduced, and the product is preheated to allow it to be pumped through a heat exchanger, which increases the product temperature in a short time interval. The liquefied material is separated into solids, water, and fat by centrifugation. Not only is the fat produced from this system of exceptional quality, but also the minimal heat denaturation of the protein results in a raw material which may be utilized in some edible meat products. The protein material is called "partially defatted beef fatty tissue (PDBFT)" or, in the case of pork, "PDPFT."

Solvent extraction systems for inedible fatty materials have been used successfully in the meat and rendering industries.

LARD

Trends in Lard Production and Consumption

Lard production and consumption followed a steady decline (Table 10-4) until the late 1970s, when per capita consumption stabilized at slightly more than 2 pounds per year. A dramatic change has come about in composition of our market hogs away from the "lard bucket" types toward the lean, muscular, fast-growing market animals which are bred to produce lean, edible products. The fourth column in Table 10-4 shows lard production per slaughtered hog, which demonstrates the progress that has been made. The decrease in production, coupled with increased competition from fats of vegetable origin, has resulted in a decrease in consumption since 1950.

Table 10-4. Trend in U.S. Lard Production and Consumption[1]

Year	Total Hog Slaughter	Total Lard Production	Lard Production per Hog[2]	Consumption[3] Total	Consumption[3] per Person	Civilian Population July 1
	(1,000 hd.)	(mil. lb.)	(lb.)	(mil. lb.)	(lb.)	(mil.)
1950	79,263	2,631	33.2	1,891	12.6	150.2
1955	81,051	2,660	32.6	1,639	10.1	162.3
1960	84,150	2,562	30.4	1,358	7.6	178.1
1965	76,458	2,045	26.7	1,225	6.4	191.5
1970	87,052	1,913	22.0	939	4.7	201.9
1971	95,648	1,960	20.5	880	4.3	204.9
1972	85,865	1,559	18.2	787	3.8	207.5
1973	77,890	1,254	16.1	705	3.4	209.6
1974	83,083	1,366	16.4	681	3.2	211.6
1975	69,880	1,012	14.5	632	3.0	213.8
1976	74,959	1,060	14.1	585	2.7	215.9
1977	78,442	1,037	13.2	502	2.3	218.9
1978	78,417	1,006	12.8	497	2.3	220.5
1979	90,179	1,155	12.8	569	2.5	233.0
1980	97,174	1,217	12.5	588	2.6	225.6
1981	92,475	1,167	12.6	573	2.5	227.7
1982	82,844	919	11.1	586	2.5	231.1
1983	88,084	978	11.1	424	1.8	232.3

[1]*Livestock and Meat Statistics*. Livestock Division, Agricultural Marketing Service, USDA, Stat. Bul. No. 522; *Fats and Oils Outlook and Situation*. USDA, FOS 310; *Meat Facts*. 1984. AMI; and *Livestock and Poultry Outlook and Situation*. USDA.

[2]Computed by dividing total lard production by total hog slaughter.

[3]As lard direct only and not including lard used in manufactured foodstuff.

Official Definitions[2]

Lard, Leaf Lard

(a) Lard is the fat rendered from clean and sound edible tissues from swine. The tissues may be fresh, frozen, cooked, or prepared by other USDA approved processes which will not result in the adulteration or misbranding of the lard. The tissues shall be reasonably free from blood and shall not include stomachs, livers, spleens, kidneys, and brains or settlings and skimmings. "Leaf lard" is lard prepared from fresh leaf (abdominal) fat.

(b) Lard (when properly labeled) may be hardened by the use of lard stearin or hydrogenated lard or both and may contain refined lard and deodorized lard, but the labels of such lard shall state such facts as applicable.

(c) Products labeled "lard" or "leaf lard" must have the following identity and quality characteristics to insure good color, odor, and taste of finished product:

(1) Color .... White when solid. Maximum 3.0 red units in a 5¼-inch cell on the Lovibond scale.

[2]*Meat and Poultry Inspection Regulations*. USDA.

(2) Odor and taste . . . .   Characteristic and free from foreign odors and
                                        flavors.
(3) Free fatty acid . . . .   Maximum 0.5 percent (as oleic) or 1.0 acid val-
                                        ue, as milligrams KOH per gram of sample.
(4) Peroxide value . . . .   Maximum 5.0 (as milliequivalents of peroxide
                                        per kilogram fat).
(5) Moisture and volatile
    matter . . . .   Maximum 0.2 percent.
(6) Insoluble impurities . . . . .   By appearance of liquid, fat, or maximum 0.05
                                        percent.

(d) Product found upon inspection not to have the characteristics specified in paragraph (c) of this section but found to be otherwise sound and in compliance with paragraph (a) of this section may be further processed for the purpose of achieving such characteristics.

Rendered Animal Fat or Mixture Thereof

"Rendered animal fat," or any mixture of fats containing edible rendered animal fat, shall contain no added water, except that "puff pastry shortening" may contain not more than 10 percent water.

Packing House Lard

Pork packers have three grades of pork fat: (1) the killing fats (intestinal), (2) the leaf fats, and (3) the cutting fats. Usually these are rendered within 24 hours after the hogs have been slaughtered.

The three most important sources of lard obtained from a hog carcass are (1) leaf fat, (2) fat trimmings, and (3) fat backs and plates.

A market-weight hog grading U.S. No. 1 yields about 7 pounds of lard per 100 pounds live weight, whereas the same weight hog grading U.S. No. 3 yields about 14 pounds.

Kettle-rendered Lard

Steam-jacketed kettles with mechanical agitators are used in the kettle-rendered lard method. Leaf fat rendered by this process is known as "open-kettle–rendered leaf lard" and is the highest grade of commercial lard outside of the neutral lards or the new processed lards. Trimming fats go into kettle-rendered lard.

Steam-rendered Lard

The steam-rendered lard process consists of bringing live steam into direct contact with the fat in a closed vertical tank or cylinder under a pressure of 30 to 50 pounds. Mostly killing and trimming fats are used. If it is bleached with fuller's earth and refined, it is known as "refined lard."

Dry-processed Rendered Lard

In dry-processed rendered lard, fats are cooked in horizontal steam-jacketed tanks under a vacuum. The three kinds of fat may be rendered separately, or all kinds of pork fat may be converted into lard under this method.

Neutral Lard

Neutral lard consists generally of leaf or back fats that are rendered in a water-jacketed kettle by slowly melting them at 126° F. Neutral lard is white in color, bland in flavor, and finds wide use in the manufacture of butter substitutes.

Lard Substitutes

Lard substitutes are made of a combination of (1) lard and other animal fats (lard compound); (2) vegetable oils with animal fats; and (3) hydrogenated vegetable oils, the most prominent of which are cottonseed, soybean, peanut, coconut, sunflower, and safflower.

Lard Oil and Stearin

Fat may be stored at high temperatures, usually 90° to 100° F to permit the liquid (lard oil) to separate from the solid (stearin). Stearin is the white solid material composed of glycerin and stearic acid left after the pressing operation forces out the lard oil. Lard oil consists mainly of olein and is made from prime steam lard. It is used in the manufacture of margarine, as a burning oil, and as a lubricant for thread-cutting machines.

Modern Lard

One of the accomplishments of the packing industry during the period of World War II was the improvement made in lard to meet the competition of vegetable shortenings. This new type of lard is no longer rendered pork fat as such. It has been given new treatments and new names.

The processing necessary to produce this new lard has added to its cost and, like all improvements, must be paid for by the consumer. Because the average consumer objected to the blue color (which is the natural color of pure lard), the manufacturers had to decolorize it; because of the natural odor, they had to deodorize it; so that it would not become too soft at room temperature, they had to add hydrogenated lard flakes and raise its melting point; in order that it would keep on the shelf as well as in the refrigerator, processors had to give it added stability by adding an antioxidant; and, finally, they had to place it in a container that would preserve these added qualities.

Lard as a Shortening

Pure lard is a natural fat—nothing has been added and nothing removed. This is an important factor in its digestibility and nutritive value. All fats are highly digestible, but their ease of digestion is reported to depend on their melting point. In order to convert liquid vegetable oils to lard-like consistency and give them increased stability, manufacturers use the hydrogenation process to raise the melting point. Lard is practically liquid at body temperature; hydrogenated fats are not.

Fats contain varying amounts of linoleic and arachidonic fatty acids, which are essential to the human body. Since the body cannot synthesize these fatty acids, they must be supplied in the diet. In hydrogenation, linoleic acid, the nutritionally valuable substance, is converted to a more saturated fatty acid to make fat more resistant to the development of rancidity. Most lard that consumers buy is not hydrogenated and contains an adequate amount of the essential fatty acids.

Lard is composed of a mixture of liquid and solid fats, which gives it a wider plastic range under lower temperatures than hydrogenated fats. This makes it possible to use lard right out of the refrigerator.

Repeated tests have indicated that lard has greater shortening value than hydrogenated fats. By shortness is meant the force necessary to break a standard cracker or pie crust. The lower the breaking strength, the greater the shortening value. This shortening ability results in a flakier and lighter crust.

Antioxidants for Animal Fats

FSIS, USDA, has to approve any antioxidant used in animal fats sold in interstate commerce. To date, these approved antioxidants, which must be tasteless, odorless, and non-toxic and must stabilize the fat by retarding rancidity as claimed, are: (1) BHA (butylated hydroxyanisole), (2) BHT (butylated hydroxytoluene), (3) glycine, (4) propyl gallate, (5) resin guaiac, and (6) tocopherols.

An illustration of the use of (4) is the combination of propyl gallate, lecithin, corn oil, and citric acid marketed by Griffith Laboratories of Chicago, under the trade name of G-4. It is sold in both regular and concentrated form, the recommended amount being 2 ounces of the regular to 100 pounds of lard or fat and 6 to 8 ounces of the concentrate to 1,000 pounds of fat. It is also available with salt (5¾ ounces of antioxidant to 100 pounds of salt). This is recommended by the company in seasoning sausage, fried pork skins, potato chips, nut meats, popcorn, peanut butter, etc.

A.M.I.F.–72, developed by H. R. Kraybill and staff of the American Meat Institute, has BHA as its main ingredient and, when used alone, is not unusually effective in increasing the stability of lard as measured by the Active Oxygen Method, but it does have the unusual property of "carrying through" and protecting the foods made with lard from becoming rancid. When 70 parts of propylene glycol, 6 parts of propyl gallate, and 4 parts of citric acid are added to 20 parts of BHA, it has very definite antioxidant qualities. It is commercially available under the trade name Tenox II, made and sold by Tennessee Eastman Corporation, Kingsport, Tennessee. The recommendations are for the use of 1 pint of Tenox for each 2,000 pounds of lard. Add the Tenox II to the melted lard, and stir thoroughly to insure complete distribution.

Rendering the Pork Fat

Lard rendered at home may be a combination of leaf and trimming fat, which is cooked either with or without the rind (skin). Killing fat (intestinal fat) is usually rendered separately for soap-making purposes. Removal of the skin (rinding) is suggested. Lard yield based on rinded fat should average 80 to 85 percent as compared to 75 to 78 percent for unrinded fat, when extracted by the ordinary hand-press method.

The fat should be cut into pieces of uniform size, not over an inch square, or run through a chopper. The smaller the pieces, the more rapid and thorough the rendering.

The longer the fat is held before it is rendered, the greater will be the free fatty acid content of the lard. This lowers the keeping quality as well as the smoke point of the lard. The smoke point (250° to 425° F, depending upon its free fatty acid content) is the temperature at which lard begins to give off smoke in cooking.

Overcooking lard increases the free fatty acid content, lowers its keeping quality, develops a more pronounced flavor and odor, and darkens the color. Particles of meat adhering to the fat will also cause a discoloration. The temperature of the lard, when rendering, should not exceed 240° F. The longer lard is hot, the more free fatty acid is formed; therefore, quick cooling, to inhibit this action, is a recommended practice. Lard should not be rendered in copper or rusty iron kettles or run through brass valves or fittings, because some of the copper and rust are dissolved and combined with the fats, forming oxidative salts that will lower the stability of the lard. Stainless steel is considered an ideal receptacle for rendering. Rust-free iron kettles are in general use, and aluminum is very satisfactory. If rendering is done with an open fire, it will be necessary to stir the fat frequently and watch the fire closely to avoid scorching the lard. Continual agitation by stirring gives the most rapid and thorough rendering. The lard is ready to be "drawn off" when the cracklings have become amber in color and no more moisture rises from the lard.

Lard rendering can be done on the stove or in the oven when smaller volumes are home-rendered. Oven rendering has some advantages, especially if the temperature controls on the oven are accurate. Whenever lard or grease is heated, either during rendering or during cooking, precautions to prevent spilling and burns must be observed. Severe burns result when hot grease spills and clings to exposed skin surfaces or clothing. Removing lard-filled containers from the oven is a particularly hazardous operation.

Pressing and Cooling

The lard press should be sweet and clean. A slight coating of last year's rancid lard on the press or old, ill-smelling containers will lower the quality and hasten the deterioration of the new batch. Lard containers should be thoroughly scrubbed and dried. Several thicknesses of cheese cloth are sufficient to strain out the sediment.

If lard is allowed to cool without being stirred, the lard oil in it tends to separate from the stearin and causes a grainy texture, a characteristic of country lard. To get a smooth lard, set it away in a cool place, and when it becomes creamy, stir it well with a paddle and let it harden. Pork packers plasticize lard by cooling it rapidly on a chill roll or in a Votator chilling machine. Rapid chilling and agitating cause the formation of small crystals which produce a firm, smooth lard. The lard storage temperature should be 40° F.

Stabilizing Home-rendered Lard

To improve the keeping quality of home-rendered lard, the USDA's Eastern Regional Research Laboratory has recommended the addition of 2 to 3 pounds of hydrogenated vegetable shortening to every 50 pounds of lard at rendering time, just before settling and separating the cracklings. The shortening should be stirred into the hot lard to get a thorough mix. The vegetable oils used in making vegetable shortening contain vitamin E (tocopherol), which is an antioxidant. The addition of the antioxidant to home-rendered lard in this manner provides a cheap and easy method to follow without tangling with vexatious chemical terms.

Cracklings

The protein-rich solid material which remains after the lard has been pressed from rendered pork is called cracklings. Some people add salt to cracklings and enjoy them as snack items. They are much like the pork "skin puffs" that are sometimes seen in the supermarket and on snack counters.

TALLOWS AND GREASES

Classification and Grading

Table 10-5 lists the standards for tallow and grease, as recognized by the industry and marketers of those products. The criteria for the classes and grades of animal fat products are based upon certain chemical and physical characteristics. The source of the raw material and the processing procedure may influence those characteristics.

Table 10-5. Trading Standards for Tallow and Grease

	Minimum Titer[1] (° C)	Maximum FFA[2] (%)	MIU[3] (%)	FAC Color[4] (Score)	Lovibond Color[5] (Score)
Tallow					
Edible	—	1.0	0.73	1	—
Fancy	41.5	4.0	1.0	7	10
Bleachable fancy	41.5	4.0	1.0	—	100
Prime	40.5	6.0	1.0	11B	125
Special	40.5	10.0	1.0	11C	180
No. 1	40.5	15.0	2.0	33	400
No. 2	40.0	35.0	2.0	No color	—
Grease					
Lard	—	0.5	0.18	1	—
Rendered pork fat	—	1.0	0.80	—	10
Choice White	37.0	4.0	1.0	11B	100
A. White	37.0	6.0	1.0	15	125
B. White	36.0	10.0	1.0	11C	180
Yellow	36.0	15.0	2.0	37	400
House	37.5	20.0	2.0	39	—
Brown	37.5	50.0	2.0	No color	—

[1]Crystallization temperature: (+) 40° C = tallow, (−) 40° C = grease.
[2]Free fatty acid percentage.
[3]Moisture, impurities, and unsaponifiables.
[4]Fat Analysis Committee—standards are matched (white).
[5]Standards matched against product.
Source: Packers Engineering and Equipment Co., Inc., catalog.

The major classifications are *edible* and *inedible,* which refers to human consumption. Edible fats may be produced only from edible carcass parts maintained under approved USDA conditions. The beef and pork fats utilized in margarine manufacture are examples of edible fats, as is lard. Nearly all other fats are inedible.

- Minimum titer refers to the temperature developed as the liquid fat cools to a solid. A higher titer is more desirable. Fats with titers equal to or over 104° F (40° C) are *tallows* and those with titers under 104° F (40° C) are *greases*.
- FFA (free fatty acid content). A lower percentage of free fatty acid is more desirable.

- MIU (moisture, insolubles, and unsaponification). A lower percentage of MIU, indicating less "contamination" with useless material, is preferred.
- FAC (Fat Analysis Committee) color. Color designation varies in descending desirability from the lightest and whitest to the darkest as follows: Fancy, Bleachable Fancy, Choice White, Choice White A, Choice White B, Yellow, House (intermediate between Yellow and Brown), and Brown.
- Lovibond color. The Lovibond color system uses a set of glass standards to match the color of the product. The fats are graded lower as color intensity increases.

Edible Fats

Edible tallows and lard are used in oleomargarine (margarine), shortenings, and cooking fats. The original name *oleomargarine* is derived from "oleo" which designates beef fat in the industry terminology. Edible lamb tallow may be utilized in oleomargarine and other shortening manufacture, while inedible lamb fats are utilized for the manufacture of feeds, soaps, and lubricants. Although margarine is mainly of vegetable origin, significant amounts of edible animal fats are merchandised and consumed in the form of margarine (Table 10-6).

Table 10-6. Margarine: Fats and Oils Used in Manufacture, U.S., 1970-1983

Year	Selected Reported Fats and Oils Consumed in Margarine Manufacture				
	Soybean	Cottonseed	Corn	Animal Fats[1]	Total[2]
	.... *(million pounds)*				
1970	1,410	68	185	99	1,792
1971	1,385	63	186	169	1,831
1972	1,461	65	194	138	1,885
1973	1,491	63	213	80	1,889
1974	1,457	58	188	167	1,905
1975	1,568	46	188	52	1,917
1976	1,671	51	218	44	2,091
1977	1,585	44	243	80	2,026
1978	1,593	42	211	74	1,997
1979	1,643	25	222	86	1,985
1980	1,653	25	223	104	2,016
1981	1,685	25	214	78	2,012
1982	1,718	21	220	29	1,996
1983[3]	1,549	34	212	41	1,836

[1]Lard and beef fats.

[2]Includes small quantities of peanut, coconut, palm, and sunflower oils.

[3]Preliminary.

Sources: *Fats and Oils Outlook and Situation.* 1982. USDA; and *Oil Crops Outlook and Situation.* 1984. USDA.

The meat inspection regulations concerning margarine production are rather extensive, due to the wide variety of ingredients and the extensive blending and processing required.

The shortening products, used in pastries and many other foods and used as a cooking media for frying, represent a larger market for edible fats (Table 10-7). Whereas only 4 to 5 percent of the margarine produced in 1980 to 1982 was animal fat, the shortenings produced in those same years were 6 to 9 percent lard and 16 percent beef fat. One fourth to one fifth of the shortening tonnage in 1980 to 1982 represented a total of 1 billion pounds per year.

Table 10-7. Shortening: Fats and Oils Used in Manufacture, U.S., 1971-1982

Calendar Year	Selected Reported Fats and Oils Consumed in Shortening Manufacture						
	Soybean	Cottonseed	Coconut	Palm	Lard	Beef Fats	Total[1]
				(million pounds)			
1970	2,182	276	45	NA	430	546	3,599
1971	2,047	168	57	140	520	575	3,537
1972	2,043	168	77	205	441	610	4,091
1973	2,268	199	86	184	341	536	3,696
1974	2,177	194	61	270	317	637	3,725
1975	2,025	154	106	604	166	602	3,728
1976	2,322	128	128	532	156	622	3,938
1977	2,279	160	78	371	185	748	3,855
1978	2,480	189	75	266	220	808	4,059
1979	2,680	169	93	222	316	713	4,213
1980	2,660	189	103	188	378	673	4,200
1981	2,767	136	125	217	315	724	4,304
1982[2]	2,948	158	140	190	251	679	4,391

[1]Includes small quantities of corn oil, peanut oil, safflower oil, and sunflower oil.
[2]Preliminary.
Source: *Agricultural Statistics*. 1983. USDA .

Table 10-4 demonstrates the effect of reduction in the amount of lard produced per hog in reducing the total lard production in the industry. Conversely, edible tallow production (Table 10-8) has increased significantly (approximately 800 million pounds in 1977 to approximately 1,300 million pounds in 1983), while numbers of cattle slaughtered declined from 42 million head in 1977 to 37 million head in 1983. Since fat content (per carcass) has been decreasing, there must be another explanation for the 40 percent increase in tallow production. The industry transition to centralized breakdown of beef into primals and sub-primals (boxed beef) resulted in increased salvage and edible processing of beef-cutting fats, which formerly were discarded totally or converted to inedible material by local retail cutting operations and renderers. Export marketing and prices, both world and domestic, are influenced to a greater extent by fluctuations in seed oil supply than in animal fat supply. That fact helps explain the failure of export sales tonnages to reflect production levels of animal fat. Generally, the world had an oversupply of fats and oils, keeping all prices low in the early 1980s.

Table 10-8. Production and Utilization of Edible Lard and Tallow

	Lard			Tallow		
Year	Production	Export	Domestic Use	Production	Export	Domestic Use
	.... (million pounds)					
1977	966	132	861	795	18	768
1978	1,072	97	966	921	51	862
1979	1,200	94	1,126	982	70	915
1980	1,159	144	1,023	1,122	137	995
1981	1,057	114	955	1,101	94	996
1982	962	95	851	1,206	95	1,110
1983	950	55	920	1,300	100	1,225

Sources: *Fats and Oils Outlook and Situation.* 1982. USDA; and *Oil Crops Outlook and Situation.* 1984. USDA.

Inedible Fats

Meat slaughter and processing plants that have rendering facilities must have two separate rendering units. The two units must be separated physically to prevent any intertransfer of raw material, product, or contamination from the inedible-rendering area to the edible-rendering area. Equipment used to handle, contain, or process inedibles must be clearly marked "Inedible," and any edible product contacting any so marked pans, trucks, or conveyors immediately changes its classification to inedible. In meat and poultry operations, all soft tissue, some bones, sweepings, scrapings, and skimmings that are not classified as edible or do not have other uses, are cooked and processed into inedible fat and meal by rendering. In most cases, with the exception of the beef hide, dead or condemned animal products are processed by inedible rendering. Table 10-9 lists the production statistics and some information on the usage of inedible tallow and grease in the United States. In 1982, the price of inedible tallow averaged 13¢ per pound. Therefore, the 6 billion pounds of inedible fat produced in 1982 represented nearly $800 million return to the industry.

Use in Animal Feeds

A major domestic use of tallows and grease is in animal feeds (Table 10-9). These fats are usually stabilized with an approved antioxidant to prevent rancidity from developing which would make the feed unpalatable. Fat is the richest food nutrient in terms of energy, and as such has been used successfully in cattle, poultry, swine, and pet feeds.

In addition to the energy value, fats reduce the dust, improve the color and texture, enhance the palatability, increase pelleting efficiency, and reduce machinery wear in the production of animal feeds.

During the early 1980s, the pet food industry produced approximately 9 bil-

Table 10-9. Inedible Tallow and Grease: Supply and Disposition, United States

	1977	1978	1979	1980	1981	1982
	.... (million pounds)					
Supply						
Stocks Jan. 1	355	344	347	390	413	451
Production	6,106	5,815	5,836	5,916	6,124	6,026
Imports	4	3	NA	NA	NA	NA
Total	6,465	6,162	6,183	6,306	6,537	6,477
Exports	2,885	2,698	2,795	3,254	3,134	3,035
Disposition						
Factory use	3,180	3,220	3,117	2,979	2,985	2,898
Soap	737	695	663	666	520	545
Feed	1,330	1,390	1,239	1,246	1,323	1,418

Source: *Agricultural Statistics*. 1983. USDA.

lion pounds of product per year at a retail value of $4.5 billion. Comparable figures were $350 million in 1958 and $1.6 billion in the mid-1970s. Animal fats are used as energy sources, and meat meal is used extensively in some products for protein and mineral sources as well as for palatability enhancement. The pet and animal food industry utilizes some fresh by-products (uncooked) for canned and fresh frozen pet and special animal foods (zoo foods, mink food, racing and guard dog food, and fox food).

Fatty Acids

Fatty acids obtained from animal fats today are being used in ever increasing quantities in the manufacture of scores of familiar products and some not so familiar.[3] The list includes such things as abrasives, shaving cream, asphalt tile, lubricants, candles, caulking compounds, cement additives, cleaners, cosmetics, deodorants, paints, polishes, perfumes, soaps, detergents, plastics, printing inks, synthetic rubber, and water-repellent compounds. The importance of fatty acids to allied U.S. industries was noted recently when a looming beef shortage caused a large rubber manufacturing company to become concerned about a possible shortage of beef tallow.

Oils

Lard oil, made from "A" white grease, is used for making a high-grade lubricant used on delicate running machine parts. The oil from "B" white grease is sometimes called "extra neat's-foot oil" and is used in giving viscosity to mineral oils. The oils made from the brown grease are used in compounding cutting

---

[3]Werner R. Boehme. *Render*. Feb. 1974. P. 20.

oils, heavy lubricating oils, special leather oils, illuminating oils, and stearic acid and are combined with paraffin in candle making.

Pure neat's-foot stock is made from the shin bones and feet of cattle. This stock is grained and pressed to secure the pure neat's-foot oil, which is used in the leather and textile industries.

Soap

Prior to 1965, the greatest utilization of tallow and grease had been in soap making.

The two main classes of soap are boiled soap and cold or semiboiled soap. The bulk of our soap supply is boiled soap, which appears on the market as hard soap (soda base).

Soft soaps (potash base) are made by the semiboiled method and find wide medical use (green soap). They are more expensive than the soda soaps, but they leave fabrics and polished woodwork surfaces in a less harsh condition.

The glycerin remains in cold or semiboiled soaps, but is a separate and important by-product in the making of boiled soap. It is in the form of a syrupy red liquid which is freed from the fat and settles to the bottom of the kettle during the soap-making operation. Glycerin is purified by distillation, and one of its uses is in the medical profession as a vehicle for certain medicines to be applied externally. It also finds wide use in the manufacture of nitroglycerin and other high explosives. One hundred pounds of animal fat will yield sufficient glycerin to make 24½ pounds of nitroglycerin.

Raw materials for the manufacture of soap are animal fats (lard oil and tallow) and vegetable oils (coconut, olive, corn, palm, palm kernel, cottonseed, soybean, peanut, linseed, rapeseed, and sesame).

Other greases used for making cheap grades of soap for special purposes are extracted from garbage, wool scourings, bone boiling, glue making, and hide trimmings. The garbage grease is changed into fatty acids and distilled, resulting in a very white fat of good odor that is used by practically all large soap manufacturers.

Fats, composed of fatty acids combined with glycerin, will unite with sodium and potassium hydroxide to form soap, glycerin, and water. A pure soap, regardless of the fat used, is a neutral soap in which there is an exact balance between the fat and the alkali used.

Boiled soaps appear on the market as milled soaps, framed soaps, or in the form of chips, flakes, powders, etc. Toilet soaps are examples of scented milled soaps; chips, flakes, and powders are the unpressed milled soaps; and bar laundry soaps are examples of framed soaps.

Floating soaps contain many minute air bubbles that make them lighter than water but with a 15 to 20 percent higher moisture content than milled soaps. Cleanser is a mixture of soap, alkaline salts, and a mineral abrasive.

Washing powder is a mixture of pulverized soaps containing a large amount of filler, usually soda ash.

Dry-cleaning soaps are partially saponified lime soaps dispersed in organic solvents, insoluble in water but soluble in gasoline, benzine, naphtha, etc.

Soft soaps contain about 50 percent water.

Efficiency of soaps for certain purposes is increased by the addition of fillers, either alkaline salts, such as soda, ash, sal soda, and trisodium phosphate, or inert fillers, such as talc, starch, and clay. The alkaline salts act as water softeners and detergents.

From 1947 until 1964, soap production declined some 2 billion pounds, due primarily to the increased use of detergent powders and liquids. Domestic use of animal fats in soap has declined in recent years, due to the replacement by plant lipid sources as well as the development of other cleaning agents.

Scientists from the USDA Agricultural Research Service developed and evaluated completely biodegradable soap-based formulations that were 60 percent tallow. They reported that the tallow-based soaps worked in hard water and in cool or lukewarm water because they contained a lime-soap dispersing agent and a nonphosphate builder. The supply of tallow, which the scientists noted was abundant, is constantly replenished and should be sufficient, even if the proposed products should completely replace today's detergents. These soaps would not be dependent upon the world's nonrenewable petroleum supplies.

Objections to conventional detergents do not apply to the tallow-based products. They will not cause eutrophication of streams, because they contain no phosphate or other chemicals that would fertilize the stream bed and promote algae growth. The new soaps proved to be no more toxic to fish and no more irritating to the eyes and skin of guinea pigs, mice, and rabbits than the phosphate-based detergents to which they were compared.

Although the research was done some years ago, the tallow-based products are not produced in the United States. France and Japan are currently producing and using this type of product. Since Japan, in particular, is an excellent market for U.S. export by-products, these tallow-based products may provide increased markets for U.S. tallow.

Meat Meals

The dry, defatted, high-protein material which results from rendering may vary, depending on the raw materials used and on the processing technique employed. Some of the products are listed here.

Tankage, Digester Tankage, and Wet-rendered Tankage

Meat animal soft-tissue by-products and dead animal–rendered products from direct steam-pressure (wet-rendering) systems. High in protein (55 to 60 percent) but reduced availability of and low in certain essential amino acids. A good source of calcium and phosphorus. Most of the residue from packing plants is made into feeds. Product from rendering plants (dead animals) is more likely to be used as fertilizer or for other nonfeed purposes.

Tankage, with Bone

Similar to tankage, with greater amount of bone, resulting in increased calcium and phosphorus levels, with corresponding decreased protein level.

Meat Scrap(s), and Meat Meal

Raw materials are similar to tankage, but rendering is completed in steam-jacketed tanks (dry rendering). Protein quality is improved, and no dried blood is added to meat meal, as is often true for tankage.

Meat and Bone Scrap or Meal

Additional bone increases calcium and phosphorus and reduces protein. May be less valuable as a feed ingredient.

Feather Meal

Feathers cooked at high pH (wet-rendered), ground to form a high-protein (80 percent) meal. Although very digestible, the protein lacks certain essential amino acids.

Poultry By-product Meal

Much like meat scrap in composition, appearance, and value.

Blood Meal

Dried ground blood, high in protein (80 percent), especially the amino acid lysine. Unpalatable as a feed ingredient with reduced digestibility. Flash dried (atomized into hot vacuum chamber), blood is a better quality feed source.

Fish Meal

Much like meat scraps but may vary, depending on the type of fish processed. High content (60 percent) of good-quality protein.

Bone Meal, Steamed Bone Meal, and Special Bone Meal

Bones are ground and wet-rendered to remove the fat and moisture, and the largely mineral (7 to 15 percent protein) remainder is ground. Composition may vary due to difference in raw materials or processing techniques.

The products listed above are utilized as organic fertilizers, but more extensively as animal feeds (Table 10-10). Some adhesives utilize animal proteins, especially blood meal or dried blood, as base materials. Bone meal is utilized in the production of china, instrument keys, steel alloys, glass, water-filtering agents, and enamels.

Table 10-10. Consumption of High-Protein Animal Feeds at 44 Percent Protein Equivalents

	1965/66	1970/71	1975/76	1980/81	1982/83
	.... (thousand tons) ....
Dairy	45	38	37	31	23
Poultry	2,255	2,272	2,175	2,458	1,975
Hog	960	1,068	913	953	694
Other livestock	514	459	398	334	209
Total	3,774	3,836	3,522	3,777	2,902

Source: *Fats and Oils Outlook and Situation*. ERS, USDA. FOS-310:14.
Note: Figures have been rounded.

HIDES AND PELTS

Classification of Beef Hides

Hides, pelts, and skins are classified according to (1) the species and class from which they come, (2) their weight, (3) the nature and extent of branding, and (4) the type of packer producing them.

The general term *hide* refers to beef "skin" weighing more than 30 pounds. Those weighing less than 30 pounds are classified as skins and are subdivided into calfskins (the lightest) and kipskins (intermediate between calfskins and hides). Hides are designated as steer or cow and may further be classified by weight into extra-light (ex-light), 30 to 48 pounds; light, 48 to 58 pounds; and heavy, 58 pounds and up.

Brands and their locations affect the value of hides, since those branded on the side and butt (rear quarter) cause damaging scar tissue in this most valuable area when they are processed into leather. The term *Colorado* or *Texas* is applied to those hides branded on the side, the Texas hides generally being plump and close grained. Butt-branded hides are often indicated as such in the market. Unbranded hides are known as "natives" (Table 10-11).

Hides are further classified into *major packer, small packer,* and *country* hides. The principal differences exist in the manner of take off and subsequent handling. Major packer hides are uniform in pattern, have a minimum of cuts and scores, are free from dung, and have been cured, sorted, and stored under

Table 10-11. Beef Hide Classifications Based on Sex, Weight, and Brand Location

Selection	Cured Wt.	Green Wt.
	.... *(lb.)*	
Cows		
Light native cows	30¾-53	36½-63
Heavy native cows	54 and up	64 and up
Branded cows	30¾ and up	36½ and up
Steers		
Extra-light native steers	30¾-48	36½-57
Light native steers	49-58	57-69
Heavy native steers	59 and up	70 and up
Light and extra-light butts	30¾-58	36½-69
Heavy butts	59 and up	70 and up
Light and extra-light Colorados	30¾-58	36½-69
Heavy Colorados	59 and up	70 and up
Light and extra-light Texas	30¾-58	36½-69
Heavy Texas	59 and up	70 and up
Bulls		
Native and branded bulls	59 and up	70 and up
Calves		
Light calves	9½ and down	11 and down
Heavy calves	9½-15	11-18
Kips		
Kips	15-25	18-28
Overweight		
Overweight	25-30	28-36

Source: *Hides and Skins*. 1979. Edited by John Minnoch and S. R. Minnoch.

Note: All green weights above are net (tared for mud and manure), as are the cured weights (tared for mud, manure, and salt).

standard conditions. Small packer and country hides generally are not uniform in take off (pattern), contain more cuts and scores, and because of various systems of handling by different individuals, they are sometimes undersalted, showing hair slips and maggot infestation. Hides are graded 1 (no deep scores or cuts), 2 (up to four cuts), or 3 (five or more cuts with some loss of hair), Grade 1 being the highest quality hide.

Hides that have been removed from dead animals (those from rendering plants, those that have been range salvaged, etc.) are called "renderer" or "murrain" hides.

Handling Hides and Pelts

The ultimate destination of most pelts is the tannery. The condition of the pelt and the directness of marketing affect the price received by the farmer. Packers have a decided advantage in that (1) they deal directly with the tannery buyer, (2) their hides have a good take-off with a minimum of cuts and scores, and (3) they have large quantities of the different classes and grades with which to attract large buyers.

Since the hide is the most valuable by-product of cattle, the livestock producers should consider the hide as well as the flesh in their livestock management programs. Lice, ticks, grubs, fleas, mange, scabies, pox, ringworm, and warts, most of which can be controlled by dipping, sorely lower the value of hides for leather; so do the branding iron, thorn and wire gashes, and horn and nail scores.

A seller who deliberately allows large, hard clumps of dung to adhere to the hide; who fails to remove the dewclaws, sinews, lean, fat, lips, ears, and tail bone; who uses dirty salt or insufficient salt and allows the hide alternately to freeze and thaw; who is careless in removing the hide, making deep scores or cuts; and who binds up the hide with baling wire (leaving a rust stain) is responsible for making country hide prices what they are.

A Voluntary Hide Trim Standard

A hide trim that is designed to encourage economy in shipping hides and improve the quality of leather products was developed at the University of Cincinnati and became effective April 4, 1965. (See Figure 10-3.) It is designated:

> Commercial Standard CS268-65
> Hide Trim Pattern for Domestic Cattlehides
> (Office of Commodity Standards)
> (U.S. Department of Commerce)

Its use has been accepted and is currently being widely used by the industry.
The trim is as follows:

Head

Ears, ear butts, snouts and lips, fat, and muscle tissue should be removed. The pate should be removed from the pate (crown) side of the head by cutting through the eye hole. The narrow side of the head should be trimmed through the eye in a similar manner. All ragged edges should be removed.

Kosher Heads

These should be removed by cutting across at the top of the Kosher cut. (Headless Kosher hides are to be put into their respective weight classifications

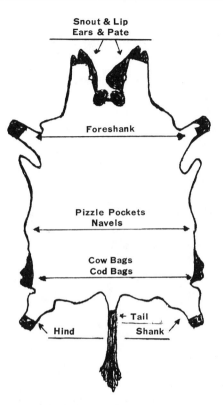

Snout & Lip
Ears & Pate

Foreshank

Pizzle Pockets
Navels

Cow Bags
Cod Bags

← Tail

Hind   Shank

Modern Hide Trim Standard

Fig. 10-3. The dark portions must be trimmed off in order to conform to the trim standard. See also Figure 5-17.

by lowering the testing weights 3 pounds in the case of cow hides and ex-light steer hides and 5 pounds in the case of light and heavy steer hides.)

Shanks

Foreshanks should be trimmed straight across through the knee at a point which will eliminate the cup. Hindshanks should be trimmed just above the dew-claws.

Cow Bags, Teats, Cow Navels, and Cod Bags

Cow bags, teats, cow navels, and cod bags should be removed straight with the belly line.

Cow Bags

Cow hides that have the bags trimmed out due to government regulations are not considered as being off-pattern if properly shaped in other respects and are deliverable as No. 1 hides if conforming to other trade standards of that designation.

Pizzle Pocket

The pizzle pocket should be split through the center, but left on the hide for identification.

Tails

Maximum tail length should be no more than 6 inches cured, measured from the root. (Removal of the switch is necessary because it is difficult to cure. The presence of a poorly cured appendage will cause the hide proper to spoil.)

Cheek Brands

When a hide has a cheek brand and no other, that portion of the head should be trimmed off and the hide placed in a native selection.

Wire in Hides

Green hides should be examined carefully, and all wire hog-rings should be removed before the hides are salted. Their presence can be damaging to hide processing and tanning machinery and to leather.

CURING PRACTICES

Hide Curing

Deterioration of all parts of the animal begins at the time of slaughter. Hides and skins are more susceptible to deterioration than the carcass, because they are highly contaminated with bacteria, mud, blood, and fecal material and are often wet when removed from the carcass. As with all other products of meat animal slaughter, control of the hide deterioration rate is imperative in order to maximize the value of the final product. Hide processing, therefore, serves two purposes. It prevents or reduces the rate of deterioration, and it transforms the raw hide or skin into a usable product. The steps in hide processing are (1) preserving the hide temporarily by curing; (2) removing the hair, fat, and other tissue; (3) preparing the dehaired hide, chemically, for the tanning process; and (4) tanning the hide into leather.

Curing hides and skins is accomplished in much the same manner as curing the meat. Salt is used to extract some of the moisture from the hide and is absorbed by the hide to achieve salt levels toxic to many bacteria. A green hide will be almost two-thirds water (60 to 65 percent), whereas after curing, the water content should be no more than 45 to 48 percent, and the salt level should be at least 14.5 percent. The hide should be approximately 85 percent salt-saturated to prevent spoilage. Curing is more effective if the hides have been washed to cool them and to remove dirt and manure before curing is initiated. Some processors *flesh* the hide prior to curing. A properly cured hide can be stored and shipped without deterioration.

Salt Pack Curing

In salt pack curing, largely a process of the past, hides are spread hair side down in the hide cellar or hide room and 1 pound of rock salt per pound of hide is sprinkled evenly over the flesh side. Another hide is spread over the salted hide and given a similar application of salt. This progresses until the day's kill is salted. The hides are allowed to cure for approximately 30 days, after which they are freed of any undissolved salt. They are sorted as to weight; graded on the basis of cuts, scores, grub holes, and slips; and folded and tied individually. The undissolved salt is washed and reused on a fresh pack with the addition of some fresh salt. During the curing process, 100 pounds of green hide can lose up to 35 pounds of water and gain up to 6 pounds of salt.

Brining Hides

The more rapid methods of curing hides have resulted from dissolving the salt to produce a saturated solution (brine) into which the hides are introduced. The pack-curing procedure required that the salt first extract water from the hide, then the salt was subsequently dissolved by the extracted water, and the resulting salt solution penetrated the hide. Salt penetration is speeded by agitating the hides immersed in a saturated curing solution. Brine curing can be done in large vats, raceways (large oval vats with a center island, much like a race track), revolving drums, or hide processors. Depending upon conditions, hides may be satisfactorily brine cured overnight in agitated vats and raceways and, routinely, within 24 hours. Most of the salt-cured beef hides produced in the United States are raceway-cured hides (Figure 10-4).

Hide Processor Curing

The curing of hides in wooden tanner's drums has been replaced largely by "hide processors." Hide processors are large, inclined axis, stationary concrete-mixer or washing-machine type units with specially designed spiral blades and wash screens. The capacities are from 250 hides to 400 hides or more. Hides direct from the hide pullers or from the fleshing machines are dumped into the mixer for an initial chilling/washing cycle. After the wash water has been removed from the drum, the hides are dry cured by fresh salt which is added at the rate of 20 to 24 percent of the total hide weight. The drums revolve slowly for about two hours, after which they revolve about five minutes per hour for four or five hours. The excess brine extracted from the hides is removed, and the curing is complete in six to seven hours. The use of brine instead of dry salt improves the appearance of the cured hides but does not influence the final leather quality. Brine curing, when compared to dry salt curing in the hide processor, increases the time required to completely cure the hides. Cured hides are tied and palletized (Figure 10-5).

Further Preparation of the Hides

To produce the desired qualities in the various leathers requires many different manipulations, which necessarily make tanning a rather complicated process. The several steps in preparation prior to tanning hides for leather are, briefly, as follows:

Trimming

Ears, shanks, and tails are removed. The trimmings from calfskins are sold to be manufactured into gelatin. (May be done before the hides are cured.)

Splitting (Halving)

Beef hides are split lengthwise into two sides.

Soaking

Dried skins are soaked in fresh water (changed every 24 hours) from two to five days at a temperature of 52° F.

Washing

The soaked hides must be washed thoroughly to remove dung, dirt, and blood. (May be done before the hides are cured.)

Fleshing

Adhering fat, flesh, and membranes are removed by a fleshing machine designed for the purpose. The person doing home tanning should use a sharp, flexible-bladed knife.

Dehairing

Hides consist mainly of the proteins keratin (hair and epidermis) and col-

Fig. 10-4. A raceway hide-curing system. Hides are being unloaded from the raceway (top of picture) and hooked to the conveyor to be transferred to the grading area. (Courtesy IBP Corp., Dakota City, Nebraska)

Fig. 10-5. Hide grading and palletizing area. Note hides in transit to processing area on overhead conveyors. Fleshing machine operators are on the balcony. Hides are spread on tables in background for trimming, grading, and folding. Grades are sorted on different pallets. (Courtesy IBP Corp., Dakota City, Nebraska)

lagen (body of the hide). Although there may be some overlapping effect, acids principally degrade collagen and alkalies degrade keratin. Hides and wooled pelts are therefore dehaired or dewooled by the use of the alkali lime.

Hides immersed in a milk-lime solution, made by dissolving 8 pounds of hydrated lime in 4 gallons of water, can be dehaired in three to four days. The addition of sodium or calcium hydrosulfide materially hastens the loosening of hair and wool and is used in commercial tanneries. Sheep pelts are dewooled by making a lime paste and covering the skin side with a layer ¼ inch thick. The pelt so treated is folded along the line of the backbone (limed sides together) and placed in a warm room until the wool slips.

Scudding

The dehaired skins are placed, grain side up, over a rounded wooden beam, and with a specially designed two-handled knife, the beamster (the person who uses the knife) pushes the knife over the skin, forcing out any material remaining in the hair follicles.

Deliming

The hide is soaked in a sulfuric acid bath (½ pound of sulfuric acid and 9 gallons of water for 50 to 60 pounds of hide) and left to soak for one hour, with the position being changed frequently. The object is to remove the lime.

Bating

This is the process of digesting the degradation products and the elastin fibers in the skin. Prior to 1910, practically all bating was accomplished by the use of manures. When hen or pigeon manure was used, the process was called "bating," but the use of dog manure was called "puering." The active principle in these dung bates was found to be protein-digesting enzymes which caused the skin to become soft and pliable. The enzymes of the pancreatic gland and those produced from wood have displaced the objectionable dung bates. Dessicated pancreas mixed with a dry deliming agent (ammonia salt) can be purchased under the trade name of Oropon (Rohm and Haas Company). Hides and skins are bated in revolving drums at a temperature of about 90° F for 1 to 1½ hours, after which they are washed in clear water.

Pickling

If the bated skins are to be preserved for storage and future tanning, they are placed in a pickle bath (7 pounds of salt, 5 gallons of water, and ¼ pound of concentrated sulfuric acid) for several hours.

The steps listed above were seldom accomplished by a meat packer 20 years ago. In those days, the hide-processing steps beyond salt curing were accomplished by the tanner or by an intermediary operation called a "hide processor." Although most packers today do ship cured hides, a few may be taking hides through the pickling procedure and some through the first chrome-tanning stage to produce what the industry refers to as a "wet blue" hide.

The advantages of further processing to the industry include not only reduced shipping weights and costs due to trimming and moisture loss at the site of production but also the salvage of trimmings and fleshings through the packing plant rendering operation. Pollutants resulting from dehairing, bating, and, particularly, the chrome process are produced in a more remote area and within an operation well versed in dealing with pollution. With more processing being done at the packer level, the added value at a high-volume position in the process should have a positive economic impact on the whole industry.

Commercial Tanning

Tanners produce finished leather, whether they start with fresh, cured, pickled, or wet blue hides. Some tanners may also fabricate consumer products from the leather. Several steps are necessary to produce finished leather, and they are listed here.

Tanning

The common tanning processes may be grouped into vegetable, chrome, and miscellaneous.

VEGETABLE TANNING

Tannin is a chemical substance found in certain classes of plant life that is capable of combining with animal protein and converting it into leather. The tannin is extracted by leaching the wood or bark in water. Commercial tanbark is the shredded, spent bark from which tanneries have leached the tannin. In order of rank, the species of plants that furnish the major quantity of tannin for the leather industry are quebracho, American chestnut, mangrove, myrobalans, wattle bark, valonia, spruce, oak, hemlock, gambier, and sumac.

Hides require from 30 to 45 days to tan. They are immersed in tan liquors of a given strength and moved daily through a series of rocker vats and a similar number of layer vats. Harness leather receives the shorter tan, sole leather the longer tanning period. Especially thick hides may be split into several layers to hasten tannin penetration.

CHROME TANNING

Sodium dichromate ($Na_2Cr_2O_7$), a red crystalline compound, is the tanning material used in chrome tanning. When converted into chromic sulfate, it combines with hide protein to produce a leather with a greater resistance to heat and abrasion than vegetable-tanned leather. Chrome tanning is rapid, requiring only a few days. Unlike vegetable tanning, which imparts various shades of tan to the leather, chrome tanning imparts shades from green to blue. Chrome tanning is sometimes combined with aluminum salts to produce a white leather that has some of the desirable characteristics of chrome leather. Vegetable-tanned leathers are often chrome retanned to secure a leather that combines the good qualities of both leathers.

Blue stock, which is properly prepared through chrome tanning, can produce almost any type of leather. While the process has demonstrated versatility, tanners should be cautioned about its high level of pollution. Adequate effluent control requires the manufacturer's ability to interpret and apply analytical data. Knowledge of the function of chemicals can reduce total effluent, but big gains are the result of reuse of wash waters and process liquors containing chemicals, according to Thomas C. Blair, Rohm and Haas Company.

MISCELLANEOUS TANNAGES

A number of "syntans" (synthetic tannins) are on the market. They produce a white leather that lacks the durable qualities of the vegetable- and chrome-tanned leathers, but which may be combined with the latter processes to produce serviceable leather suitable for special uses. Some common syntans are Leukonal, Tanigan, Tanak, Tanasol, Mertanol, and Arkotan. Other methods of tanning that serve special needs are Calgon, alum, aldehyde, oil, quinone, and tungsten tanning.

Whatever the tanning process, there are subsequent treatments required to finish the leather for almost any purpose.

Setting Out

Setting out is the mechanical extraction of excess water from the hides as they come from the tanning vats. The same thing is accomplished by hand by means of a "hand slicker" (metal scraper).

Splitting

Thick hides that are not to be used for sole leather, but made into upholstery leather, are split by a machine into two or three layers. The wool sides of pickled sheep skins are split into grain layers or skivers, which are used for making hat bands, bookbinding, etc. The flesh layers are used for making chamois leather.

Shaving

Light skins need not be split and are run through shaving machines to even the thickness and make them smooth and clean.

"Shoddy" leather is made by grinding waste leather to a pulp and pressing it into solid sheets, either with or without the addition of a binding material.

Fatliquoring

In order to prevent the cohesion of the leather fibers during the drying process, it is necessary to rub or "stuff" the damp leather with a fat or oil. This keeps the leather from drying out hard and stiff. Neat's-foot oil and glycerin are recommended for home tanning.

Staking

Flexing the tanned hide or skin over a rounded metal blade set in the end of a block of wood (3 feet high) fastened to the floor is known as "hand staking." This is an important operation, because it stretches and flexes the leather fibers and makes them pliable.

Dyeing, Drying, Buffing, Glazing,
Plating, and Finishing

Other operations consist of dyeing, drying, buffing (sandpapering to produce a nap), glazing, plating, and finishing. Patent leather, also called "japanned" or "enameled" leather, is a chrome-tanned leather that has received three separate coatings of a linseed oil varnish. Chamois skin is the flesh side of an oil-tanned sheep skin. Cod, whale, seal, or shark oil is used.

CALF SKINS

Skins are the pelts of small animals, wild or domestic, such as calves, sheep, goats, muskrats, foxes, minks, etc.

Calf and kip skins produce leather that some experts refer to as the "elite of the industry."[4] To insure the attainment of leather quality, these fragile hides must be processed carefully. Curing must be accomplished quickly and adequately, but excessive salting should be avoided. Trimming a calf hide is very important, due to its small size and the importance of a proper pattern (see Figure 10-3). Storage intervals of calf and kip skins should be minimized to reduce

---

[4]*Hides and Skins.* 1979. Edited by John Minnoch and S. R. Minnoch, distributed by National Hide Association, printed by Eakin Press, P.O. Box 178, Burnet, TX 78611.

the probability of deterioration. The skins of near-term unborn calves are termed *slunk* skins and command high prices due to scarcity and leather quality.

PIGSKINS

As indicated in Chapter 4, most U.S. pork processors have used in the immediate past and are presently using a dehairing technique, which leaves the skin on pork carcasses until they are processed into wholesale or retail cuts. With such a system, the only fresh skin available of any consequence is that resulting from the fatback and the hams, and this skin is largely used in gelatin production. Most whole pigskins for tanning are imported.

Pigskin has been replaced by cattle hides as a covering for footballs. Its most common use is in leather for gloves, wallets, handbags, brief cases, toiletry cases, tobacco pouches, book bindings, and leggings.

Wolverine World Wide Incorporated[5] developed a machine to remove scalded skin from the belly and backfat portions of the carcass. Single-side units, capable of removing a 3- to 3½-square-foot piece of skin extending from the ham cut-off line to the shoulder cut-off line, are used as well as double-side units, which remove the skin from the middle section of the whole carcass in one piece. The pieces are processed and tanned, and much of the leather is merchandised in the form of sueded footwear.

The pigskin leather is tough and produces scuff-resistant footwear. The hog bristle (hair) is unique in that it grows through the skin from the follicle in the subcutaneous fat layer. The holes, or pores, through which the hairs pass, result in a naturally "air-conditioned" type of leather.[6] Wolverine uses significant amounts of pigskins, particularly when sueded leathers are in style.

Scalded skins can be pulled, using skinning equipment, but the leather is 10 percent thinner and has less tensile strength than leather from unscalded skins. The heat damage incurred during scalding makes many skins unsuitable for use as leather; however, with the introduction of the mechanical pig skinner (Chapter 4), several processors are gearing up to hog-skinning operations. Curing procedures for pork skins utilize the hide-processor technique in some plants. In some cases, skins may be transferred fresh to commercial processors or directly to a tanning company, such as Wolverine. Defleshers (fleshing machines) smaller than the ones used for beef hides are available. Many are constructed of stainless steel, because the pork fleshings are edible and can be converted to edible lard, if processed through FSIS approved equipment and procedures.

A 1978 research trial at Purdue University,[7] using values current at that time, demonstrated a value advantage of about $1.50 per hog for skinning com-

---

[5]*Ibid.*

[6]*Ibid.*

[7]M. D. Judge, C. P. Salm, and M. R. Okos. 1978. *Hog Skinning Versus Scalding.* Proceedings of Meat Industry Research Conference:155.

pared to scalding. Their data assumed a hide-market value of $5 per hide. Although scalding was more labor efficient during the slaughter phase, labor was required to remove the skins at later stages of processing.

Specially selected and treated hog skins, because of their similarity to human skin, are used in the treatment of humans suffering from massive burns and injuries that have removed large areas of skin and in the healing of persistent skin ulcers. Hog skins are cut into strips or patches, shaved to remove the hair, split to 0.008 to 0.020 inch in thickness, and then cleansed, sanitized, and packaged.

The skins are applied directly to the injured areas to decrease pain, inhibit infection, and prevent loss of body fluids. Closely adhering "porcine" dressings help prepare the patient for permanent skin grafting by promoting the development of granulation tissue, so essential before skin grafting can begin. They also ease the flexing of joints and the stretching of scar tissue early in the treatment procedure, a vital factor in the patient's return to full physical capability.[8]

Skins for porcine dressings are sent from two source plants to Genetic Laboratories, Inc., of Minneapolis. Upon arrival there, the skins are shaved, split, and packaged in a variety of forms in surgically clean conditions and chill-stored or frozen to await shipment to hospitals and burn-treatment centers throughout the nation.

In explaining the rapidity of this process, the director of the Skin Bank at Genetic Laboratories said that within 24 hours from the time a hide is removed from a hog, the porcine skin made from the hide can be used as a dressing on a burn victim. It has been estimated that these porcine skin dressings save at least one life per day in this country, and it is known that the dressings alleviate much pain and shorten the length of hospitalization for other severe burn victims.

SHEEP PELTS

Sheep pelts are classified by the length of wool they carry as follows: full wool, 1½ inches or more; fall shorn, 1 through 1½ inches; Shearling No. 1, ½ through 1 inch; Shearling No. 2, ¼ through ½ inch; Shearling No. 3, ⅛ through ¼ inch; and Shearling No. 4, ⅛ inch.

Sheep pelts are further classified as to their origin as follows:

> River—Missouri River area
> Southwest—Texas, Oklahoma, Arizona area
> Northern—Minnesota and the Dakotas area

Normally, sheep pelts are cured with the wool left on and are then sent to a "pulling" operation which removes the wool. Depending on the wool length, it is used in a wide variety of ways. The resulting skin is processed into leather,

---

[8]*Hog Is Man's Best Friend.* 1978. National Live Stock and Meat Board.

using techniques similar to those described for beef hides. The leather produced from sheep skins is used for suedes, shoe uppers, linings, garments, and accessories. Some sheepskins are processed and tanned with the wool left on to provide coat and boot linings, as well as rugs, decorative pelts, and seat covers. A great deal of sheep skin is used for bookbinding, hat sweat bands, shoe linings, gloves, and chamois skins. Because they are larger and wear better, goat skins are more valuable than sheep skins.

Sheep skins of long-wooled breeds from which the wool has been pulled supply the best wearing leather; skins from the Merino breed supply the poorest. Pulled wools constitute about one seventh of all the wool produced in the United States. Approximately 3.75 pounds of grease wool is required to make 1 pound of woolen cloth. The degreasing of wool removes a wool oil that may be 15 percent of the weight of the wool, which, when treated, produces some valuable products, including lanolin, which are used as bases for ointments and cosmetics, leather dressings, and fiber lubricants. The potassium carbonate removed in the wash water in wool cleaning represents a significant part of the by-product value of degreasing.

Approximately 20 years ago, a new use for shearlings became economically important. Tests in professional institutions convinced the Agricultural Research Service scientists of the USDA that glutaraldehyde-tanned shearling bed pads were far superior to any other product.

The specially tanned shearlings, sheepskins with the wool evenly clipped, were found to be effective nursing aids for preventing and healing bedsores. The painful sores develop in patients who cannot move and must lie for long hours in one position.

Shearlings have long been recognized as ideal bed pads, but before the development of glutaraldehyde tanning,[9] the pads were tanned by conventional methods and shrank and hardened after a few launderings. The glutaraldehyde-tanned shearlings, however, can be run through a washer and dryer as many as 54 times and still retain their original shape and resiliency.

Both patients and staff in hospitals, nursing homes, and outpatient clinics highly praised the new washable bed pads. The shearlings were serviceable for as long as 28 months and proved more effective than bed pads made from synthetic fibers.

When in use, the wool of the shearling is placed in direct contact with the patient's skin. This allows for free circulation of air and absorption of perspiration. It also minimizes skin abrasion, thereby aiding in preventing and in healing the bedsores. The pads are resilient and distribute the weight of the patients evenly. The wool is highly flame-retardant—an important safety factor for the bedridden or disabled patient.

ARS chemist William F. Happich and his associates at the Eastern Marketing and Nutrition Laboratory, who developed the glutaraldehyde tanning proc-

---

[9]USDA. *Agricultural Research*. Dec. 1965. P. 8.

ess for shearlings, conducted the tests in the Philadelphia area in close cooperation with nursing administrators.

HOME TANNING

Although available equipment and number and kind of hides or skins to be processed will determine technique to some degree, a few suggestions for the home tanner are offered. One procedure for preparation of fresh sheep skins for tanning would be to start by soaking the pelt for 10 to 15 minutes in a tub while working out the blood and manure. It may take more than one cycle to complete the task. A young sheep-pelt processor suggested that the pelt then be placed in a washing machine and run through the complete wash cycle, using a mild dish soap. When the pelt is removed from the washer, the small particles of dirt, hair, and loose tissue can be wiped, carded, or combed off the skin surface. The hide is salted and cured for an interval up to 30 days.

Several simple methods of tanning skins have been devised which make it possible for a novice to make a serviceable product.

Salt Alum[10] Tanning

Soak the salted sheep skin in water until it is soft, and then place it on a table or beam where you can trim it and flesh it. Washing can be done either in a tub or on the table. Any good soap powder will do. The water should be warm (125° F). Rub and rinse repeatedly until it is clean, and then extract the water either by hand or by using an ordinary washing machine wringer. Place the pelt on the table, fleece side down, and give it a thorough rubbing with a mixture of one part of powdered alum to two parts of common salt (4 ounces of alum and 8 ounces of salt will tan the average-sized sheep pelt). Leave the pelt in this position overnight, but the next morning, hang it over a rail, skin side up. The following morning, sponge the skin to remove the unabsorbed salt, and rub 1 to 2 ounces of neat's-foot oil or glycerin into the damp, soft skin. As the skin dries, during the next two days, it must be staked several times to keep it from becoming hard. When the skin is dry, buff it, using a coarse-grade sandpaper fastened over a block of wood. Card the wool, using an ordinary wool card, and the pelt is ready for use. It makes a comfortable cover for the hard, cold seats on farm implements.

Alum Tanning

This method is suitable for fur skins, since the aluminum sulfate does not color the hair or skin. Dissolve 1 pound of aluminum sulfate in 1 gallon of water,

---

[10]Alum is (a) potassium aluminum sulfate–$KAl(SO_4)_2 \cdot 12 H_2O$ or (b) ammonium aluminum sulfate–$NH_4 Al(SO_4)_2 \cdot 12 H_2O$ or (c) aluminum sulfate–$Al_2(SO_4)_3$.

and dissolve 4 ounces of crystallized sodium carbonate (soda ash) and ½ pound of salt in ½ gallon of water and pour this slowly into the alum solution, while stirring it vigorously. Place the prepared skin (soaked, fleshed, and washed) in the solution for two to four days, depending upon its thickness, and then rinse it and put it through a wringer. Now, rub the damp skin with glycerin, and, as it dries, stake it and buff it. Before it is rubbed with oil or glycerin, retanning the skin with 1 pound of Leukanol (a syntan made by Rohm and Haas Company), dissolved in 1 gallon of water, makes the skin tougher and softer without discoloring it.

Vegetable Alum Tanning

Dissolve ½ pound of aluminum sulfate and ½ pound of salt in a small quantity of water. Dissolve 2 ounces of gambier[11] or Terra Japonica[11] in a little boiling water. Mix the two solutions, and add sufficient water to make 1 gallon. Use sufficient flour with the 1 gallon of tanning liquor to make a moderately thin paste. Take a properly prepared pelt (soaked, trimmed, fleshed, and washed) and apply three coatings about ¼ inch thick at two-day intervals to the skin side of the pelt, removing each previous coating before applying the next. When the pelt is practically dry, rinse it in warm water containing some borax, and then rinse in fresh water. Squeeze out the water and slick the skin with a dull knife. Apply a coating of glycerin and hang up to dry. Stake the pelt several times while drying and then buff with coarse sandpaper.

This method produces a yellow skin of good tensile strength.

Salt Acid Tanning

Make up a solution of 3 ounces (fluid) commercial sulfuric acid and 2 pounds of common salt per gallon of soft water in a wooden or another nonmetallic container and in sufficient volume to immerse the prepared hides (soaked, fleshed, and washed). Small, thin (calf) hides should remain in the tanning solution 12 to 24 hours. Cow or steer hides may require two to seven days, depending upon the size and thickness of the hides. The hides should be stirred or moved in the solution every few hours to assure even tanning. Wash hides in cold water to remove excess acid and salt, and stretch the hides to drain. Sheep pelts should be spread wool down and the tanning solution daubed on with a rag or a handful of wool. The application should be repeated several times to assure that the hides are thoroughly treated with the solution. After 12 hours, wipe off any excess tanning solution, and wash the surface with a damp cloth. When the hides are nearly dry, work them (staking), and apply a coat of neat's-foot oil. Al-

---

[11]Name for a yellowish, dry, resinous astringent substance obtained from a Malayan woody vine.

low to dry, then moisten thoroughly and work them while they are drying. This technique is said to produce a strong white leather.

Hides can be tanned with the hair left on in most of the procedures described or can be dehaired prior to tanning by immersing them in a solution of lime water made by dissolving 2 pounds of lime in 5 gallons of water. The hides should remain in the solution until the hair slips, which should occur in three to five days. Scrape the hair off with a fleshing knife, and wash all the lime water from the hides. The dehaired hides are ready for tanning.

LEATHER, PAST AND FUTURE

At one time, the greatest use for leather in the United States was to manufacture harnesses for the many horses that provided power for work and transportation. With the advent of mechanized power and transportation the need for harness leather decreased rapidly. Leather was used extensively to transfer the horsepower from engine pulleys to machine pulleys by leather belts. But again technology found new and better methods. There remained, however, a brisk demand for leather for shoe manufacture, until gradually, synthetic composition soles began to take over, and by 1960, only 35 percent of the shoes manufactured in the United States had leather soles. Technology was able to produce the composition soles cheaper; however, it has been estimated that 75 to 80 percent of the domestic leather use in 1983 was for the production of shoes. The resilient, pliable wearing qualities of leather are utilized in the shoe uppers, if not in the soles. The hides from more than 100,000 cattle are required every year for leather accessories in the sports field.

U.S. hide producers are exploring new frontiers in domestic hide consumption and at the same time eyeing the growing demand for hides for export. To do this, greater use of research and technology is being made. The United States is a prime source of raw materials for countries that manufacture leather (Japan, Republic of Korea, and others), as evidenced by the export of 20.5 million cured whole hides out of the estimated 37.5 million produced in 1983 (Table 10-12). An additional 1 million wet blue hides were exported in 1983 by some estimates. Table 10-12 demonstrates that increasingly greater numbers of U.S. hides are being exported, while less are being domestically processed each year.

The use of leather in the United States includes 15 to 20 percent of production for garments and gloves and 5 to 10 percent miscellaneous products in addition to 75 to 80 percent for footwear. Leather garments are popular and more practical than in the past, as a result of leather's resistance to the effects of drycleaning solvents.

Hides and skins exported to other countries return to the United States as leather, or as consumer products made of leather. Tanning and leather fabrication are labor-intensive activities. Countries with large, low-cost labor forces can utilize raw materials obtained from the United States to profitably produce prod-

Table 10-12. Hides and Skins: U.S. Production, Domestic
Consumption, and Export for Selected Years

| Year | Production | | U.S. Consumption | Export | U.S. Leather Production[1] |
	Cattle	Calves	Cattle	Cattle	
.... (1,000 pieces)					
1978	39,552	4,170	16,303		20,089
1979	33,678	2,823	11,327		18,051
1980	33,807	2,588	15,880	19,000	17,636
1981	34,953	2,798	16,939	19,500	18,959
1982	35,826	3,020	12,608[2]	22,800	16,500
1983[2]	37,500			20,500	

[1]Production of cattle, calf, kip, goat, sheep, lamb, and horse leathers.
[2]Estimates.
Source: Report of fourth Annual Meeting of U.S. Hide, Skin and Leather Association. 1983.

Table 10-13. Leather: U.S. Production, Imports for Consumption,
Exports of Domestic Merchandise, and Apparent
Consumption, 1978-82

Year	Production	Imports	Exports	Apparent Consumption	Ratio (Percent) of Imports to Consumption
.... (million dollars)					
1978	1,456	222	190	1,488	15
1979	1,803	284	243	1,844	15
1980	1,823	217	259	1,781	12
1981	1,761	354	266	1,849	19
1982	1,461	318	275	1,504	21

Source: Compiled from official statistics of the U.S. Department of Commerce and reported by the
U.S. Hide, Skin and Leather Association. 1983.

ucts sold in the United States at prices lower than U.S. manufacturers can produce the product (Table 10-13).

Although technology continues to produce materials to replace leather in
current uses, new uses or the improvement of traditional leather products also
results from technological advances. New domestic uses for hides are being developed, as evidenced by the previous discussion concerning the medical uses of
pigskins and lamb shearlings.

Is there a leather future? Indications are that the single most valuable by-
product will remain so.

PHARMACEUTICAL BY-PRODUCTS

The medical arts have used animal products in the healing process for centuries. In fact, some animal products have held "magical" healing powers for

certain societies throughout history. Similar conditions exist today in that minute portions of certain animal extractives used each day can literally be the difference between life and death for many humans.

Scattered through various parts of the animal body are a number of internally secreting, ductless endocrine glands. The substance secreted by each exercises some specific control over the conduct, character, and development of the body. Their functions are so interrelated that under- or over-secretion of any one of several of the glands will cause abnormalities. Most of the magical products which are derived from animal tissues saved by the meat industry and which are extracted, purified, and prepared for consumers by the pharmaceutical industry are *hormones*. Enzymes and other types of chemicals are also derived from animal slaughter by-products.

The authors sincerely appreciate the cooperation of the National Live Stock and Meat Board in the preparation of *The Meat We Eat*. Much of the following information about the pharmaceutical by-products has been quoted directly from two National Live Stock and Meat Board publications: *Hog Is Man's Best Friend* and *The Good Things We Get from Cattle Besides Beef*.

Adrenals

The adrenals are also called the suprarenal glands and are two in number. They are bean-shaped in the sheep, measure approximately 1 inch by ½ inch and may be some 2 inches from the kidney. In the ox, they are located near the center of the animal (medial), anterior (towards the front) of the kidney, and are triangular or heart-shaped. In the pig, they are long and narrow and are located on the medial border of the kidney. They are located astride the kidneys in humans and are larger than most endocrine glands. They are reddish brown in color and are somewhat bean-shaped. The cortex (outer portion) produces steroid secretions essential to life maintenance; a lack of these secretions causes Addison's disease. The medulla (inner portion) of the gland produces epinephrine, which constricts the blood vessels and increases heart action. The valuable drug epinephrine is secured from the adrenals and is used in surgical operations to arrest hemorrhage and stimulate heart action. It requires the adrenals of 13,000 head of cattle to produce 1 pound of epinephrine. Each adrenal gland weighs approximately ½ ounce (14 grams).

Hog adrenal glands until recently were an important source of many different hormones which physicians used to treat illnesses or chemical imbalances in the human body. Now many adrenal compounds are being made synthetically.

Corticosteroids, from the cortex or outer shell of the hog's adrenal glands, influence and regulate the human body's utilization of minerals and nutrients. They regulate water, nitrogen, potassium, and sodium (salt) balance in the body. They are used in treating shock, deficiencies of the adrenal glands, and Addison's disease—a debilitating ailment brought on by malfunction of the human adrenal cortex which causes an underweight condition, dark-skin pigmentation, loss of strength, and loss of body sodium.

Cortisone, one of the corticosteroids, influences fat, sugar, carbohydrate, and water metabolism (utilization of food and other materials within the body), improves muscle tone, and reduces pain caused by calcium deposits in humans. It has many therapeutic uses, such as treatment for shock, arthritis, and asthma.

Adrenal medulla, the inner core of the adrenal glands, produces epinephrine and norepinephrine hormones, which are also produced synthetically.

Epinephrine, also called adrenalin, is used to stimulate body processes in the utilization of food. This drug is used to relieve some of the symptoms of hay fever, asthma, and some forms of allergies affecting the mucous membrane of the nasal passages. It may be used to shrink blood vessels during certain types of surgery, control bronchial asthma spasms, and reduce inner-eye pressure during glaucoma treatment. It is also used to restore heart rhythm in cardiac arrest and is used by dentists to prolong the effects of local anesthetics.

Norepinephrine helps shrink blood vessels in humans, reducing the flow of blood through the body and slowing the pace of rapid heartbeats.

Blood

Blood fibrin extract from hog blood is used to make amino acids used in parenteral (infused as intravenous) solutions for nourishing certain types of surgical patients.

Fetal pig plasma is important in the manufacture of vaccines and in tissue culture media. The reason is that it contains no antibodies.

Plasmin, a hog blood enzyme which has the unique ability to digest fibrin in blood clots, is used to treat patients who have suffered heart attacks.

Thrombin helps create significant blood coagulation. It is valuable in the treatment of wounds, particularly in cases in which the injury is in an inaccessible part of the body, such as the brain, bones, or gastrointestinal tract (as in the case of peptic ulcers). Thrombin is also used in skin grafting to help keep the graft in place and to "cement" gaps where tissues have been surgically removed.

Fibrinolysin is combined with desoxyribonuclease from the pancreas to aid in the removal of dead tissue that results from certain vaginal infections. It is a valuable cleansing agent for infected wounds or clotted blood and can speed up the healing of skin damaged by ulcers or burns. (Blood is also used in cancer research, protein hair conditioners, fertilizers, and animal feed.)

Brain

Hog brains are a potential source of cholesterol, the raw material from which vitamin D_3 is made. Vitamin D_3 is the sunshine vitamin so necessary in building bones and teeth and is used to treat rickets in children and premature infants.

The hypothalamus produces hormone-releasing hormones, relatively small molecules that cause the release of various hormones from the pituitary gland. Isolating these hormone-releasing hormones from animal hypothalami and determining their effects is a recent scientific breakthrough in hormone research.

Thromboplastin, made from the brains of cattle, is used as a blood coagulant in surgery.

Gall Bladder

Chenodeoxycholic acid, from hog bile acid, is given to humans to dissolve gallstones, thus eliminating the hazards and complications of major surgery.

Ox bile extract from liver bile or a component (dehydrocholic acid) is used in the treatment of indigestion, constipation, and bile tract disorders resulting from disease or surgery.

The gall bladder of the average beef contains about 4 ounces of bile. The galls should be slashed and the bile emptied into a barrel in the freezer where it is allowed to freeze. Each day's production is put on top of the previous day's output so the barrel consists of layers of frozen bile. The bile is shipped in the frozen state. Four beef galls yield 1 pound of bile.

Cortisone from bile has been found to relieve pain by reducing inflammation in joints of sufferers of arthritis. The gall of 100 cattle (25 pounds) is needed to produce sufficient cortisone to treat the average patient for one week. Cortisone is also secured in small quantities from sheep and calves. The animal source of cortisone has now been replaced by a synthetic product. A unique market has developed in China for gallstones, where they are thought to have mystical values. In the early 1980s, the price for gallstones in the United States was $155 per ounce. One large beef packer, slaughtering more than 100 cattle per hour, accumulated less than 1 pound of gallstones in two years.

Heart

Hog heart valves, from young pigs to full-sized market hogs, are specially preserved and treated and surgically implanted in humans to replace heart valves that have been weakened or injured by rheumatic fever or through birth defects.

Hog valves are superior to mechanical valves in several ways. For one, a vast majority of patients with mechanical valves usually require constant infusion of anticoagulant drugs to prevent the valve from sticking, and in many patients, the use of anticoagulants over long periods of time causes many types of undesirable side effects.

But the hog valve is a naturally formed and functioning organ from a living animal which has been turned into an innate object that continues to function normally in the human body, generally with fewer anticoagulants required. If a problem develops with a hog valve, malfunction is not fatal because early warning symptoms alert patient and physician in time for surgery. Also, anticoagulants are generally not prescribed for children because they tend to retard physical growth. Hog valves, implanted in children with a minimum of anticoagulants, usually correct congenital heart defects without disturbing physical development.

Since the first operation in 1971, many thousands of heart valves have been implanted in men, women, and children ranging in ages from less than 1 year to more than 70 years. Since hog valves don't grow while implanted in the human heart, three surgeries may be indicated for youngsters, as they grow and develop, using progressively larger valve replacements each time.

Bovine pericardial tissue (the membrane enclosing and attaching the heart within the chest cavity) is processed and utilized to "patch" the patient's pericardial tissue while closing following bypass or other heart surgery. Adhesions of the pericardium to the sternum, which often follows cardiac surgery, are prevented by the use of the treated bovine tissue.

Intestines

Heparin, classed as one of the "essential" pharmaceuticals and obtained almost exclusively from the inner (mucosa) lining of the hog's small intestine and from lungs, is a natural anticoagulant used to thin the blood and dissolve, prevent, or retard clotting during surgery, especially organ transplants. Heparin is also used as a gangrene preventative in cases of frostbite and as a burn treatment.

Enterogastrone, a hormone taken from the hog duodenum (beginning of the small intestine), is used to regulate gastric secretions in the stomach. It is also being used experimentally to speed the emptying time of the stomach.

Secretin hormone, also from the duodenum, stimulates pancreas glands to produce pancreatic juices. It is injected in humans to test for disease of the pancreas.

Much of the "cat gut" used for surgical suture is derived from sheep and other meat animal intestines.

Liver

In 1926, researchers reported that patients with pernicious anemia showed marked improvement when lightly cooked animal liver—a source of vitamin B_{12}—was included in their diets. After that, pernicious anemia patients were treated by including cooked liver in their diets. Funded by a grant from the National Live Stock and Meat Board in 1924, Dr. G. H. Wipple of the University of Rochester researched the importance of liver as a treatment for pernicious anemia. The practice of prescribing the eating of raw liver was started by Dr. George Minot of Harvard in 1926. Today, patients may be given vitamin B_{12} by injection which bypasses the need for the intrinsic factor, deficient in pernicious anemia patients. Desiccated liver, containing added nutrients, is used as a nutritional supplement.

Liver extract was sometimes combined with folic acid and injected into the blood stream to treat various types of anemia, including pernicious anemia. Liver injections were also used to treat sprue, a long-term condition associated with diarrhea, weakness, emaciation, and anemia.

Ovaries

Hog ovaries are a source of progesterone and estrogens used to treat various reproduction problems in humans. Sow ovaries are the major source of relaxin, a hormone often used during childbirth. It requires the slaughter of 145 female hogs to produce 1 pound of fresh ovaries from which corpus luteum and ovarian extracts are prepared.

Pancreas

The pancreas is more commonly known as the pork sweetbread but should not be confused with the commercial veal sweetbreads (thymus gland). The pancreas has both internal and external secretions, the latter passing into the small intestines to effect the digestion of starch, protein, and fat. The internal secretion (insulin) regulates sugar metabolism. Failure of the pancreas to regulate this sugar metabolism results in the affliction known as diabetes.

Insulin, first isolated by Doctors Banting and Best, is secured from specialized groups of cells in the pancreas known as the islets of Langerhans and is used extensively in treating diabetes. Other extracts made from the pancreas, such as pancreatin, are used as a remedy for intestinal disorders.

Dr. Sanger and associates were able to establish the primary structure of insulin, that is, they determined the number and sequence of amino acids that make up the protein insulin. Thus it was theoretically possible to synthesize insulin from a "test-tube," and such a task has been accomplished. However, the procedure is detailed and costly and at present will not contribute significantly to the world supply of insulin for treating diabetics.

A great need for insulin has developed throughout the world, and the supply has never been too great. It takes between 12 and 26 beef pancreas glands to make sufficient insulin to treat a diabetic for a year. To make 1 pound of insulin crystals yielding 11.5 million units of insulin requires 13,000 to 20,000 cattle.

The following specifications for the collection and subsequent handling of pancreas glands for insulin manufacture are those of Eli Lilly & Co., Indianapolis.

Grade A

GLANDS

- Pancreas glands are to be taken from healthy animals. The beef and calf pancreas gland is located in the ruffle fat and lies attached to the liver near the gut. The pork pancreas is located between the small and large intestine imbedded in the ruffle fat.
- Calf pancreas glands are those taken from calves not over six months old.

COLLECTION, TRIMMING, AND DELIVERY TO FREEZER

- Glands are to be plucked from the viscera as soon as possible. Long exposure to water spray should be avoided.
- Glands should be collected in small buckets, perforated to permit drainage, or in small buckets surrounded with cracked ice. The glands should never be allowed to stand in water or directly in contact with ice. Water may dissolve the insulin in the glands.
- Glands are to be clean and trimmed closely, with all surface fat and connective tissue removed. Particular care should be taken in removing fat from pork pancreas; this is best done by stripping with the fingers. It is important with all pancreas, particularly beef and calf, that all of the tail of the gland be saved as *this portion is richest in insulin.*
- Glands are to be trimmed promptly and delivered *directly* to the sharp freezer. Not more than one hour should elapse between removal of glands from viscera and placing them on trays in the freezer.
- It is important in all of the above operations that glands be handled in batches in order of removal from viscera so that each gland will be frozen in minimum time.

FREEZING

- Glands should be promptly spread on clean, prechilled trays in the freezing chamber at the lowest available temperature, 0° F or less if available, but not higher than 15° F in any case.
- Glands should be *individually* frozen so that no two glands touch.
- Within 48 hours after freezing hard, glands should be removed from trays and either stored in temporary *covered* containers or packed in shipping cases.

Hog pancreas glands are also a source of insulin hormone. Today the pancreas glands from approximately 60,000 hogs produce 1 pound of pure insulin, enough to treat 750 to 1,000 diabetics for one year. A year's production of 85 million market hogs could be the source of 1,400 pounds of insulin, more than enough to supply 1.25 million people who take daily injections of insulin just to stay alive.

Hog insulin is especially important because its chemical structure most nearly resembles that of humans. This is significant because approximately 5 percent of all diabetics are allergic to other forms of insulin and can tolerate only insulin from hogs.

There are more than 6 million identified diabetics in the United States today, plus an estimated additional 4 million undetected diabetics who are not being treated. Diabetes was a killer disease before it was discovered that animal insulin could be used in humans.

Chymotrypsin is an enzyme used to cleanse wounds and to remove dead tissue where ulcers and infections occur. It can be used in treating serious injury or following surgery when localized inflammation and swelling result due to excess fluids. Chymotrypsin is generally used for the removal of devitalized tissue in eye surgery.

Glucagon is given to raise the blood sugar level and to treat insulin overdoses in diabetics, or when a low blood sugar episode is caused by alcoholism. It has a specialized use in the treatment of some psychiatric disorders.

LPH (lipotropic hormone) is used as a digestive aid and is important in the digestion and absorption of fats and oils.

Pancreatin is a mixture of pancreatic enzymes used to treat faulty digestion in humans. Because of its high-fat digestive capability, pancreatin is also used in the treatment of cystic fibrosis, a disease afflicting approximately 4 million people in this country.

Trypsin is a digestive aid that helps break down food by aiding in the hydrolysis of protein in the upper part of the small intestine.

Trypsin and its sister enzyme chymotrypsin are prescribed to remove dead and diseased tissue from wounds and to speed healing after surgery or injury.

Parathyroids

Parathyroids consist of four small glands the size of a grain of wheat, which are located close to the thyroid gland. Their secretions regulate the calcium content of the blood stream and maintain the tone of the nervous system. The complete removal of the parathyroids causes death within a few weeks. To secure 1 pound of parathyroid extract requires the slaughter of approximately 3,600 animals. Each parathyroid gland weighs 1 gram or less.

Parathyroid hormone is used to compensate for the human body's inability to naturally produce this hormone. Without it, parathyroid deficiency can result in convulsions, painful muscular spasms, loss of calcium from the bones, abnormal tooth development, and cataracts.

Pineal

The pineal gland is about one third the size of the pituitary, is reddish in color, and is located in a brain cavity behind and just above the pituitary. Its secretion regulates child growth—hastening or retarding puberty and maturity.

The hog's pineal gland secretes the hormone melatonin, which is being studied as a possible treatment for schizophrenia. Melatonin presently is being tested as a possible stimulant for mental and physical development of some types of mental retardation in children. It also affects color of the skin and the formation of freckles.

Pituitary

Located at the base of the brain and well protected in a separate bone cavity, the pituitary gland is about the size of a pea and is grayish yellow in color. It is made up of an anterior and a posterior lobe which have distinct functions. The *anterior lobe* is known to produce (1) the growth-promoting hormone, (2) the thyroid-stimulating hormone, (3) the mammary-stimulating hormone or prolactin, (4) the gonad-stimulating hormone, and (5) the adrenal-cortex–stimulating hormone (ACTH). The *posterior lobe* exerts principles that (1) control blood pressure and pulse rate, (2) regulate the contractile organs of the body, and (3) govern energy metabolism.

Pituitary glands in hogs produce a great number of hormones used to control human growth and metabolism problems and to regulate activity of the body's other endocrine glands.

ACTH (adrenocorticotropic hormone) is used to treat rheumatism and arthritis, eye inflammation, certain skin disorders, and multiple myeloma, a terminal form of leukemia. Interestingly, 1,800 hogs are needed to produce 1 pound of pituitary glands, which contains only ¾ of a gram (about ¹⁄₄₀ of an ounce) of ACTH.

Corticotropin (ACTH) is a valuable diagnostic tool. Its most important medical use is to assess the operation of the adrenal glands. It can also be used in the treatment of psoriasis, the control of severe allergic reactions (rhinitis and bronchial asthma), eye inflammation due to allergies, certain respiratory diseases, anemia, infectious mononucleosis, and leukemia. It takes the pituitary glands from 10,000 cattle to produce 1 pound of this valuable pharmaceutical.

ADH (antidiuretic hormone, or vasopressin) helps regulate the body's water losses by the kidneys through urine production. Failure of the body to produce ADH results in excessive water loss in urine—a disease called "diabetes insipidus." ADH also influences blood pressure by constricting the muscles in blood vessels.

Vasopressin is used in testing for renal functions. It is also employed to stimulate proper movement of material through the intestinal tract following operations and to dispel "gas shadows" when abdominal X-rays are being made.

Prolactin stimulates milk secretion in the mammary glands, and it may play a role in the future treatment of breast cancer.

Oxytocin hormone is used to treat obstetrical complications, induce labor, increase uterine muscle contractions, cause milk release by mammary glands, and lower the body's blood pressure to control uterine bleeding at childbirth. Also used as a wound-closer by professional boxers.

TSH (thyroid-stimulating hormone, or thyrotropic hormone) is used in conjunction with isotopes, etc., to locate small particles of thyroid cancer which may have spread to other parts of the body. Thyrotropin (TSH) is a hormone that stimulates the thyroid gland. It is used as a diagnostic tool to determine if a patient is suffering from hypothyroidism caused by an anterior pituitary failure or by complete failure of the thyroid gland. (In the event of anterior pituitary failure, the drug will stimulate proper functioning of the gland.)

Growth hormone (GH) has shown great potential for increasing animal production; however, most of it is now being produced by genetic engineering techniques.

Skin

Gelatin from hog skin collagen is used for coating pills and making capsules. Gelatin is taken orally, theoretically to improve fingernail strength. See discussion of pork skins as burn bandages earlier in this chapter. A porcine collagen product has been developed to stimulate clotting during surgery. The product is applied directly on the surface of the bleeding tissue.

Spleen

Splenin fluid affects capillary permeability and blood clotting time and speeds up recovery from inflammatory conditions (redness and swelling).

Stomach

Linings of the hog's stomach contain proteins and enzymes used in many commercially produced digestive aids and antacids.

The pyloric lining of the hog stomach is rich in "intrinsic factor," which must be present before the human body can utilize vitamin B_{12} from food or vitamin preparations to relieve or prevent pernicious anemia.

The pink mucous lining of the hog stomach is the richest natural source of pepsin which is used in the treatment of achylia gastrica, the failure of the stomach to produce gastric juices (acid and pepsin). Achylia gastrica is often present in cases of pernicious anemia and stomach cancer accompanied by achlorhydria, or lack of hydrochloric acid.

Mucin is used to treat peptic and duodenal ulcers. It also lubricates food movement through the digestive tract and is considered a valuable adjunct to many digestive products.

Rennet (rennin), a mild enzyme, is used to help infants digest milk and is used in cheese making.

Thymus

In veal, the thymus gland has a commercial food value. It is cream in color and is located in the neck near the chest cavity and has two lobes, the second lying within the chest cavity. The thymus functions primarily in youth by inhibiting the activity of the sex glands and is considered to be a source of factors affecting the ability of the body to resist infections or react to the presence of foreign bodies. It atrophies after the age of puberty.

Thyroid

In the sheep, the thyroid gland is dark with a long, ellipsoidal outline measuring about 2 inches by ½ inch. It is located on the first five or six rings of the trachea. In the ox, the gland has two lateral lobes connected by an isthmus and is located just below the larynx. In a calf, it is about 3 inches long and dark in color. In swine, the thyroid is triangular shaped, is about 2 inches across, may be located some distance from the larynx, has no isthmus, and somewhat adjoins the esophagus. It is smaller in cattle than in humans, and its secretion is an iodine-containing compound termed *thyroxin*. In the young, a deficiency of thyroid tissue causes a condition known as "cretinism," resulting in physical deformity and defective mentality, or idiocy; in the adult, it causes a condition known as "myxedema," defined as "severe thyroid deficiency" *(hypothyroidism)*, characterized by dry skin and hair and loss of physical and mental vigor.

A deficiency of iodine in the diet or water supply may cause a simple goiter. Goiter in humans and animals can be treated by supplying the necessary iodine. Over-secretion of the thyroid increases basal metabolism causing the afflicted to become nervous and thin. The action of thyroid secretions is interrelated with other glands.

The dessicated thyroid is used extensively in keeping hypothyroid patients from the slow-moving, slow-talking, inactive existence they would otherwise lead. It is one of the few glandular substances that is effective when taken orally. It requires 40 beef thyroids to make a pound (14 to 21 grams per gland). Hog thyroids are equally valuable. Thyroids are handled very similarly to pancreas.

Calcitonin is given to lower calcium and phosphate levels in the blood and to regulate the heartbeat. It is also used in the treatment of Paget's disease, a painful malady of the bone.

Thyroglobulin, obtained exclusively from hogs, is given as a supplement to persons with under-active thyroids.

Nervous System

Cholesterol comes from the spinal cord. It is essential in the synthesis of male sex hormones which are used when natural development of male characteristics does not occur. These hormones are also used to treat menopausal syndromes and to prevent swelling of breasts and milk production when a mother does not nurse a newborn baby.

☞   NOTE: in the case of many pharmaceuticals, it has become less ex-   ☜
pensive to synthesize the product than to refine it from animal sources. However, in some cases (as with male sex hormones) another animal by-product is used in the process. In many areas, synthesis has been only partial, and animal sources remain extremely important. Such is the case with the protein drugs (insulin, parathyroid hormones, and pituitary hormones) which are so complex that chemists shudder at the problems of attempting a synthesis.

Bone Cartilage

Plastic surgeons may use the cartilage from the breastbone of young cattle to replace flat bones, such as facial bones. The specially processed xiphoid or xiphisternal cartilage permits bone damage repair.

OTHER BY-PRODUCTS

Glue and Gelatin

Both glue and gelatin are colloidal proteins. They are chemically and physically similar and differ mainly in that gelatin is made from clean, sweet materials prepared under sanitary conditions to make it edible.

The raw materials used to produce gelatin and glue are high in collagen. They are connective tissue; skin or hide trimmings; sinews, horn piths, lips, ear tubes, pizzles, and cartilage; beef and calf bones; mammary glands; heads of cattle, calves, and sheep; and knuckles and feet. Pigskins are a good source of gelatin.

Glue and Gelatin Stocks

The three main types of glue are hide glue, bone glue, and blood albumin glue. The latter is water resistant and is used widely in the manufacture of plywood.

The oldest and widest use for glue is in the furniture and veneer industry. Glue has so many varied uses that it has been said that glue holds the world together. It is used in sizing paper; in the manufacture of wool, silk, and other fabrics; in sizing straw hats; in sizing walls that are to be painted; in sizing barrels or casks that are to contain liquids; on the heads of matches to make an air-tight cap over the phosphorus; in the manufacture of sand and emery paper to hold the abrasive on the paper; in the manufacture of dolls, toys, and ornaments; in the making of picture frames, mirror frames, rosettes, billiard balls, composition cork, imitation hard rubber, printing rolls, mother-of-pearl, gummed tape, paper boxes, kalsomine, automobile bodies, caskets, leather goods, and bookbinding; and many other products.

The two types of gelatin according to their source are hide gelatin and bone gelatin.

Gelatin finds wide use in the manufacture of ice cream; in the making of certain pharmaceutical preparations and capsules for medicine; in the coating of pills; in the making of mayonnaise dressings and emulsion flavors; in the clarifying of wine, beer, and vinegar; in the making of court plaster; in photography; in electroplating; as a bacteria culture medium; and for various other uses.

A large percentage of gelatin comes from the bones of veal. The heads of veal calves find favor with some people of foreign extraction who cook them and use the head meat and broth with noodles or as gelled meat.

Blood

According to USDA, FSIS, no blood which comes in contact with the surface of the body of an animal or is otherwise contaminated can be used for food purposes. Only blood from inspected animals may be used for meat food products. The defibrination of blood intended for food purposes shall not be performed with the hands.

Research has shown ways to prepare blood proteins resulting in their utilization as emulsifiers in human foods, but the procedures are not utilized in U.S. industry. In the Soviet Union, 40 percent of all slaughter blood is recovered for human food.

Blood contains around 17 percent ammonia, of which 14 percent is nitrogen. If the blood is allowed to coagulate, the jelly-like fibrin is cooked and dried, and the residue is pressed into cakes. The cakes are finely ground and disposed of as blood meal or mixed with low-grade tankage for stock feed. When mixed with potassium or phosphoric acid, blood makes a very rich fertilizer. The serum is clarified and dried and sold as blood albumin.

One hundred pounds of beef blood treated with an anticoagulant and centrifuged will yield about 40 pounds of solid (cells) material and 60 pounds of plasma, or 16 pounds of dried solids, 3.4 pounds of dried serum, and 3.5 pounds of wet fibrin.

Blood albumin is used in certain malt extracts and in fixing pigment colors in cloth, in finishing leather, in clarifying liquors, and in manufacturing glue. Dried blood or blood meal is a very concentrated stock food containing about 87 percent protein.

Inferior grades of blood are used in feeds and fertilizers and in the manufacture of buttons and imitation tortoise shell articles. Blood and its products are finding added medicinal and therapeutic uses.

Hair

A substitute for camel's-hair brushes is made from the delicate hairs on the inside of the ears of cattle. Hog bristles for making brushes were formerly imported from China but are now being produced in the United States in increasing amounts. It requires considerable hand labor to collect the proper length hair which is found over the shoulder and back of the hog. The fine hair of the bulk of our domestic hogs is not suitable for brush making and is processed and curled for upholstering purposes.

The hog hair is cooked in vats in (1) plain water, (2) water with acetic acid added (100 to 1), or (3) water with detergent (76 percent solid caustic soda). The cooking loosens the cuticle and scurf and requires from four to six hours. The hair passes from the cookers into high-speed cylinders studded with $3\frac{1}{2}$-inch steel pins into which hot water and steam are fed. These washers and pickers remove the cuticle and scurf, and the picked hair passes by conveyor to the feed apron of a drying machine where blasts of hot air dry and fluff it. A suction tube draws the dried and fluffed hair into a winnowing machine which removes the dust and fine hair. A cutaway intake in the suction tube permits the toenails to drop through. The finished hair is baled in burlap bags. The yield of finished hair on the basis of a 10 percent moisture content is 35 percent in summer and 75 percent in winter.

Bones, Hooves, and Horns

Shin bones of cattle, with knuckles removed, are cooked to remove the meat and neat's-foot oil and are washed and air-dried. They are then sawed into flat slabs from which crochet needles, bone teething rings, pipestems, dice, chess pieces, electrical bushings, washers, collar buttons, flat buttons, knife handles, and many other articles are made. Some other uses for bone are in the case hardening of steel; in the manufacture of bone black used as a bleach for oils, fats, waxes, sugar, or pharmaceutical preparations; as a stock food (ground bone meal, steamed bone meal); and as a fertilizer.

Cattle horns can be split into thin strips, pressed in heated molds of various patterns, and colored to make imitation tortoise shell. Horns are used for making napkin rings, goblets, tobacco boxes, knife and umbrella handles, and many other articles.

White hooves are used for making imitation ivory products. Black hooves find use in the manufacture of potassium cyanide for extracting gold.

HORSE MEAT

The human consumption of horse meat in the United States is very small, but horse meat is used in the manufacture of pet foods. The Horse Meat Act was approved July 24, 1919, making federal inspection necessary for horses slaughtered for interstate or foreign shipment, if the meat is to be sold for human consumption. The slaughter establishment must be separate from those slaughtering other animals or where the meat products of other animals are handled. The inspection stamp on horse meat is applied with green ink rather than the purple found on other meat species. The act provides that such meat must be conspicuously labeled, marked, branded, or tagged "horse meat" or "horse meat product." Horse meat is also regulated by the federal Food, Drug and Cosmetic Act, and the use of horse meat in every state is regulated by the U.S. Wholesome Meat Act. In 1983, 15 inspected equine slaughter plants were listed by the USDA.

The average annual export of horse meat for the period 1930-1940 was 3 million pounds as against 41½ million pounds in 1946. The U.S. horse population declined from 21½ million in 1919 to slightly over 3 million head in 1959 where it stabilized and has perhaps increased. Accurate horse census figures are practically non-existent, but the horse population is not adequate to support a major horse meat industry.

The horse meat business is not the highly profitable business it was in the early 1940s, because the cost of the horse has risen to a point where the margins are about the same as in the meat packing industry. However, in recent years, horse slaughter has increased again, possibly due to the unsettled beef market (Table 10-14). The meat is not the only source of revenue. Horse hides are used for leather; tallow for soap; hair, mane, and tail for hair goods; the glands for pharmaceuticals; and the edible offal as food for fish or carnivorous animals. Dead horses are converted into oils, glue, and fertilizer. Horse meat is important as human food in most European countries, the USSR, and Japan.

**Table 10-14. Horses Slaughtered Under Federal Inspection,
1943-1982[1, 2]**

Year	Number	Year	Number	Year	Number
1943	39,935	1959	88,100	1977	317,000
1945	59,674	1964	38,447	1978	323,000
1947	276,290	1972	72,419	1979	304,000
1949	307,794	1973	88,493	1980	274,000
1951	340,287	1974	201,447	1981	219,000
1953	270,533	1975	230,764	1982	149,000
1955	196,106	1976	293,000		

[1]*Federal Meat and Poultry Inspection Statistical Summary for 1975*. USDA, FSQS (1943-1975).
[2]*Agricultural Statistics*. 1983. USDA (1976-1981).

Chapter 11

Federal Meat Grading and
Its Interpretations

FEDERAL MEAT GRADING

THE FEDERAL Meat Grading Service was established by the Sixty-eighth Congress of the United States on February 10, 1925; however, tentative standards were formulated for grades of dressed beef in 1916. They provided the basis for uniformly reporting the dressed beef markets according to grades, which became a voluntary grading service early in 1917. The grade specifications were improved from time to time as experience gained through their use indicated what changes were necessary. They were published first in mimeographed form in June 1923. After slight changes, they were included in Department Bulletin No. 1246, *Market Classes and Grades of Dressed Beef,* which was published in August 1924. Grade standards for veal and calf carcasses and vealers and slaughter calves were introduced in 1928. The official standards for grades of lamb and mutton carcasses were made effective on February 16, 1931. Tentative standards for grades of pork carcasses and fresh pork cuts were issued by the USDA in 1931. These tentative standards were slightly revised in 1933.

Federal meat grading was administered by the Livestock Division of the Agricultural Marketing Service (AMS) of the USDA until 1977, when it was placed under the USDA's new Food Safety and Quality Service (FSQS). In 1981, the FSQS was abolished, and the grading service was returned to the Livestock Division of the AMS.

The Purpose of the Act

Producer and Processor

The act authorized the establishment of a grading service for perishable farm products for the purpose of making available to individuals, organizations, and establishments, an agency that would certify the class, quality, and condition of the products examined to conform with uniform standards.

Retailer and Consumer

The federal grade stamp on meat provides consumers with a reliable guide to quality and quantity. Each quality grade name is associated with a specific degree of quality, thus enabling consumers to utilize meat most efficiently by preparing it in the manner for which it is best suited. Federally graded meat is widely found in retail stores. Most retailers sell only the grade or grades that are requested or deemed to meet the needs of their customers. The quality grade name appearing on the grade stamp can be used to serve the consumer in two important ways: as a guide to quality and as a guide to preparation. Meat of each grade will provide a satisfactory dish if it is appropriately cooked.

A consumer must learn, either by study or from experience, what government grade names mean and represent, and the same is true of packer brands that represent these grades. Quality in meat is quite variable and difficult for the average person to recognize in the retail cut. Because of this fact, the consumer has come to depend on items bearing brand names. If the particular brand meets a consumer's approval, a new customer has been added—until some better graded product is tried and accepted. The manufacturer of a food product may be unknown to the public, but the brand name of the product, if good, is on every tongue. There are so many brand names, however, that it would appear that the terminology used by the U.S. grading service would be a pleasant relief.

Yield grades are more meaningful to the retailer since they indicate expected yield of edible meat from a carcass and its various wholesale cuts. If a consumer chooses to buy a side or quarter of beef, the yield grade is extremely important, since it will give an expected yield of edible meat from the hanging carcass.

Inspection Requirements

Products, to be eligible for grading service, must be prepared under federal inspection or other official inspection acceptable to the administration.

Types of Service

The Federal Meat Grading and Certification Branch engages in two main types of activities: (1) the grading and identification for grade of beef, veal, pork, lamb, and mutton for sale on a grade basis through regular commercial channels and (2) the examination and certification, for conformance with specifications for grade and other factors, of meats offered for delivery to federal, state, county, and municipal institutions which purchase meat on the basis of contract awards. This latter service covers all kinds of meats, meat products, and by-products. Vendors and others requesting certification service for compliance of meat and meat food products with approved specifications must apply to the Meat Grading and Certification Branch, Livestock Division, AMS, USDA.

BEEF CARCASS GRADING

Development of the Standards[1]

- Standards were first formulated in 1916.
- Standards were first published in 1923.
- Standards were slightly changed in 1924.
- Voluntary beef grading and stamping service was begun in May 1927.
- The official standards were amended in July 1939 to provide a single standard for the grading and labeling of steer, heifer, and cow beef according to similar inherent quality characteristics. The amendment also changed certain grade terms for steer, heifer, and cow beef from *Medium, Common,* and *Low Cutter* to *Commercial, Utility,* and *Canner,* respectively.
- An amendment in November 1941 made similar changes in the grade terms for bull and stag beef and established the following grade terminology for all beef: *Prime, Choice, Good, Commericial, Utility, Cutter,* and *Canner.*
- Compulsory grading by the Office of Price Administration was in effect from 1941 to 1946 during World War II.
- An amendment in October 1949 eliminated all references to color of fat.
- In December 1950, the official standards for grades of steer, heifer, and cow beef were amended by combining the Prime and Choice grades and designating them as Prime, renaming the Good grade as Choice, and dividing the Commercial grade into two grades by designating the beef produced from young animals included in the top half of the grade as Good while retaining the Commercial grade designation for the remainder of the beef in that grade.
- In June 1956, the official standards for grades of steer, heifer, and cow beef were amended by dividing the Commercial grade into two grades strictly on the basis of maturity, with beef produced from young animals being designated as Standard, while Commercial was retained as the grade name for beef produced from mature animals.
- In June 1965, the official standards for grades of steer, heifer, and cow beef were revised to place less emphasis on changes in maturity in the Prime, Choice, Good, and Standard grades.

    The rate of increase in required marbling to offset increasing maturity was changed, and the minimum marbling permitted was reduced for more mature carcasses by as much as 1 to 1.5 degrees in Prime, 1 degree in Choice, and 0.75 degree in Good and Standard. Consideration of the 2 degrees of marbling in excess of that described as abundant was eliminated.

---

[1]*Official United States Standards for Grades of Carcass Beef.* Title 7, Chapter XXVIII, Part 2853, Sections 2853.102-2853.107, of the Code of Federal Regulations, reprinted with amendments effective Oct. 6, 1980.

The manner of evaluating conformation was clarified by providing that carcasses may meet the conformation requirements for a grade either through a specified development of muscling or through a specified development of muscling and fat combined.

The requirement was established that all carcasses be ribbed prior to grading.

Established standards for yield grades of carcasses and certain wholesale cuts of all classes of beef.

- In July 1973, the official standards were revised to provide separate quality grades for beef from young (A maturity) bulls in a class designated as *bullock*. The quality grade standards for bullock were the same as those for steer, heifer, and young cow beef. *Bull* was retained as the class designation for beef from more mature bulls, but the quality grades for bull and stag beef were eliminated, leaving only the yield grade standards to apply. *Stag* beef was redesignated as *Bullock* or *Bull,* dependent on its evidences of maturity.

- In April 1975 (not implemented until February 1976 due to court actions), consideration of maturity within A maturity was eliminated. The minimum marbling requirement was increased 0.5 degree for the youngest Good but reduced 0.5 degree at the A/B maturity line for Good and reduced 1 degree at the A/B line for Prime, Choice, and Standard. Conformation was eliminated as a quality grade factor. All carcasses graded were required to be identified for both quality and yield grades. Maximum maturity permitted in Good and Standard was reduced to the same as that permitted in the Prime and Choice.

- In October 1980, several changes were made. Removal of the yield grade stamp from an officially graded beef carcass is allowed, provided the fat thickness does not exceed ¾ inch. Carcasses that have had the characteristics of the rib eye or thickness of fat over the rib eye altered are not eligible for grading. Grading can be accomplished only when the beef is in carcass form, in the plant where the animal was slaughtered, and not before at least a 10-minute period has elapsed following ribbing.

- On November 8, 1984, the USDA Agricultural Marketing Service proposed the elimination of kidney, pelvic, and heart fat as one of the four factors now considered in determining the yield grade of beef carcasses. Yield grades would be based on the three other factors now considered—external fat thickness, hot-carcass weight, and rib-eye area. Quality grade standards would not be affected. With KPH fat no longer a factor in yield grading, the reverse roll would be used to indicate that 1 percent or more of KPH fat had been left in. (See Figure 11-51.) This proposal would only go into effect after the USDA has received comments and testimony supporting it conclusively. Thus the material presented in this text on beef carcass grading is based on current standards in effect at this writing.

Classes of Beef Carcasses

The first step in beef carcass grading should be to determine the class.

Class determination of beef carcasses is based on evidences of maturity and apparent sex condition at the time of slaughter. The classes of beef carcasses are steers, bullocks, bulls, heifers, and cows. Carcasses from males—steers, bullocks, and bulls—are distinguished from carcasses from females—heifers and cows—as follows:

Steer, bullock, and bull carcasses have a "pizzle muscle" (attachment of the penis) and related "pizzle eye" adjacent to the posterior end of the aitch bone (Figures 11-1 and 11-3).

Steer, bullock, and bull carcasses have, if present, rather rough, irregular fat in the region of the cod. In heifer and cow carcasses, the fat in this region—if present—is much smoother (Figures 11-1 and 11-2).

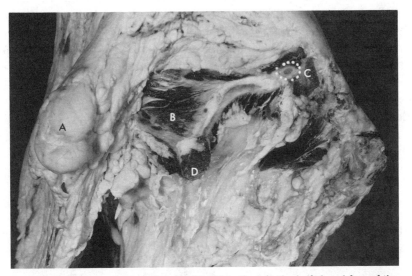

Fig. 11-1. Steer carcass. The cod fat (A) to the left, the half-closed face of the *gracilis* muscle (B), and the pizzle ring (C) at the right of the aitch bone (D) identify this as a male.

In steer, bullock, and bull carcasses, the area of lean exposed immediately ventral to the aitch bone is much smaller than in heifer and cow carcasses (Figures 11-1, 11-2, and 11-3).

Steer, bullock, and bull carcasses are distinguished by the following:

In steer carcasses, the "pizzle muscle" is relatively small, light red in color, and fine in texture, and the related "pizzle eye" is relatively small.

In bullock and bull carcasses, the "pizzle muscle" is relatively large, dark red in color, and coarse in texture, and the related "pizzle eye" is relatively large.

Bullock and bull carcasses usually have a noticeable crest, caused by a

shortening and bunching of the neck muscles. Male hormones cause this secondary sex characteristic to become evident as a bull matures.

Bullock and bull carcasses also usually have a noticeably more highly developed round muscle *(gluteus medius)* adjacent to the hipbone, commonly referred
to as the "jump muscle." However, in carcasses with a considerable amount of
external fat, the development of this muscle may be obscured.

Although the development of the secondary sex characteristics is given primary consideration in distinguishing steer carcasses from bullock or bull carcasses, this differentiation is also facilitated by consideration of the color and
texture of the lean. In bullock and bull carcasses, the lean is frequently at least
dark red in color with a dull, "muddy" appearance—and in some cases it may
have an iridescent sheen. Also, it is frequently coarse and has an "open" texture.

The distinction between bullock and bull carcasses is based solely on their
evidences of skeletal maturity. Carcasses with the maximum maturity permitted
in the bullock class must still qualify for A maturity and have slightly red and
slightly soft chine bones, and the cartilages on the ends of the thoracic vertebrae
have some evidences of ossification; the sacral vertebrae are completely fused;
the cartilages on the ends of the lumbar vertebrae are nearly completely ossified;
and the rib bones are slightly wide and slightly flat. Bull carcasses have evidences of more advanced maturity.

Heifer and cow carcasses are distinguished by the following:

Heifer carcasses have a relatively small pelvic cavity and a slightly curved
aitch bone. In cow carcasses, the pelvic cavity is relatively large, and the aitch
bone is nearly straight.

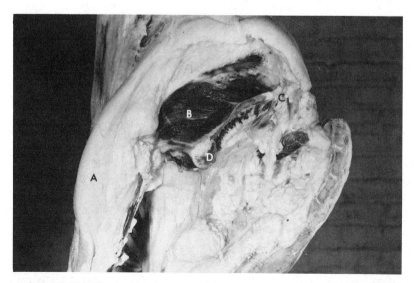

Fig. 11-2. Heifer carcass. Note the presence of the udder (A), the exposed face
of the *gracilis* muscle (B), and the lack of a pizzle ring (C) at the right of the aitch
bone (D).

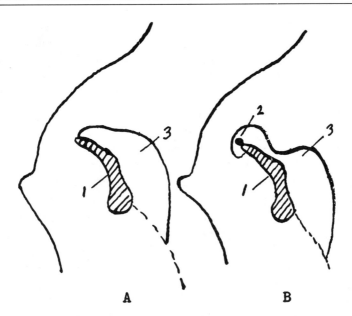

Fig. 11-3. A—The contour of the lean area of the *gracilis* muscle on heifer carcasses. B—The same area on steer carcasses. (1) Aitch bone, (2) pizzle eye, (3) lean area of *gracilis*. This shows the method of identifying steer from heifer rounds when the rump and shank have been removed.

In heifer carcasses, the udder usually will be present. In cow carcasses the udder usually will have been removed. However, neither of these is a requirement.

Two Considerations—Quality and Yield

The grade of a steer, heifer, cow, or bullock carcass is based on separate evaluations of two general considerations: The indicated percent of trimmed, boneless major retail cuts to be derived from the carcass, referred to as the *yield grade*, and the palatability-indicating characteristics of the lean, referred to as the *quality grade*. When graded by a federal meat grader, the grade of a steer, heifer, cow, or bullock carcass must consist of both the quality grade and the yield grade. The grade of a bull carcass consists of the yield grade only.

The carcass beef grade standards are written so that the quality and yield grade standards are contained in separate sections. Eight quality grade designations—Prime, Choice, Good, Standard, Commercial, Utility, Cutter, and Canner—are applicable to steer and heifer carcasses. Except for Prime, the same designations apply to cow carcasses. The quality grade designations for bullock carcasses are Prime, Choice, Good, Standard, and Utility. There are five yield grades applicable to all classes of beef, denoted by numbers 1 through 5, with Yield Grade 1 representing the highest degree of major retail cut yield.

When officially graded, bullock and bull beef will be further identified for their sex condition; steer, heifer, and cow beef will not be so identified. The designated grades of bullock beef are not necessarily comparable in quality or yield with a similarly designated grade of beef from steers, heifers, or cows. Neither is the yield of a designated yield grade of bull beef necessarily comparable with a similarly designated yield grade of steer, heifer, cow, or bullock beef.

Ribbing the Carcass

See Figure 14-3 for illustration and further discussion.

To determine the quality grade or yield grade of a carcass, it must be split down the back into two sides, and one side must be partially separated into a hindquarter and forequarter by cutting it with a saw and knife insofar as practicable, as follows: A saw cut perpendicular to both the long axis and the split surface of the vertebral column is made across the twelfth *thoracic* vertebra at a point which leaves not more than one half of this vertebra on the hindquarter. The knife cut across the rib-eye muscle starts—or terminates—opposite the above-described saw cut. From that point it extends across the rib-eye muscle perpendicular to the outside skin surface of the carcass at an angle toward the hindquarter, which is slightly greater (more nearly horizontal) than the angle made by the thirteenth rib with the vertebral column of the hindquarter posterior to that point. As a result of this cut, the outer end of the cut surface of the rib-eye muscle is closer to the twelfth rib than is the end next to the chine bone. Thus the cut surface of the rib-eye muscle is perpendicular to its long axis. Beyond the rib eye, the knife cut shall continue between the twelfth and thirteenth ribs to a point which will adequately expose the distribution of fat and lean in this area. The knife cut may be made prior to or following the saw cut but must be smooth and even, such as would result from a single stroke of a very sharp knife.

Other methods of ribbing may prevent an accurate evaluation of quality grade and yield grade determining characteristics. Therefore, carcasses ribbed by other methods will be eligible for grading only if an accurate grade determination can be made by the official grader under the standards.

Beveling of the fat over the rib eye, application of pressure, or any other influences which alter the area of the rib eye or the thickness of fat over the rib eye may prevent an accurate yield grade determination. Therefore, carcasses subjected to such influences may not be eligible for a yield grade determination. Also carcasses with more than minor amounts of lean removed from the major sections of the round, loin, rib, or chuck will not be eligible for a yield grade determination.

Conformation and Quality

Conformation, the manner of formation of the carcass, refers to the thickness of muscling and to an overall degree of thickness and fullness of the carcass. Conformation has been eliminated as a beef quality grade factor. Justification for its elimination is well documented, since we have known for many years that conformation in itself has nothing to do with meat palatability. It was kept in the grading standards in the past because it was vital to the buying, selling, and distribution of carcasses. A certain conformation grade gave the purchasers mind's-eye views of their purchases, and they would expect to receive exactly what they ordered. The USDA proposed in 1962 to remove conformation from the standards, but was unsuccessful, due to meat industry disagreement. Now, more than three fourths of our beef is distributed in a box. (See Chapters 12 and 14.) When loins, ribs, chucks, and rounds arrive "knife-" or "saw-ready," sealed in a bag inside a box, the critical factors are the palatability indicating traits—marbling, maturity, color, texture, and firmness—which remain in the quality standards and the yield factors—trimness and muscling which are included in the USDA Yield Grade standards.

Nevertheless, the following paragraph remains in order that we do not lose sight entirely of conformation, which some segments of our modern meat industry still value quite highly. And something can be said in behalf of the consumer who prefers a shapely, symmetrical top loin or rib steak, rather than one which is long and narrow.

Conformation is evaluated by averaging the conformation of the various parts of the carcass, considering not only the proportion that each part is of the carcass weight but also the general value of each part as compared with the other parts. Thus, although the chuck and round are nearly the same percentage of the carcass weight, the round is considered the more valuable cut. Therefore, in evaluating the overall conformation of a carcass, the development of the round is given more consideration than the development of the chuck. Similarly, since the loin is both a greater percentage of the carcass weight and also generally a more valuable cut than the rib, its conformation receives more consideration than the conformation of the rib. Superior conformation implies a high proportion of meat to bone and a high proportion of the weight of the carcass in the more valuable parts. It is reflected in carcasses which are very thickly muscled, very full and thick in relation to their length, and which have a very plump, full, and well rounded appearance. Inferior conformation implies a low proportion of meat to bone and a low proportion of the weight of the carcass in the more valuable parts. It is reflected in carcasses which are very thinly muscled; very narrow and thin in relation to their length; and which have a very angular, thin, sunken appearance.

Quality—Maturity

After class determination, the next step in quality grading should be the determination of *maturity*.

For steer, heifer, and cow beef, *quality* of the lean is evaluated by considering its *marbling* and firmness as observed in a cut surface in relation to the apparent *maturity* of the animal from which the carcass was produced. The *maturity* of the carcass is determined by evaluating the size, shape, and ossification of the bones and cartilages—especially the split chine bones—and the color and texture of the lean flesh. In the split chine bones, ossification changes occur at an earlier stage of maturity in the posterior portion of the vertebral column (*sacral* vertebrae) and at progressively later stages of maturity in the *lumbar* and *thoracic* vertebrae. The ossification changes that occur in the cartilages on the ends of the split *thoracic* vertebrae are especially useful in evaluating maturity, and these vertebrae are referred to frequently in the carcass beef standards. Unless otherwise specified in the standards, whenever the ossification of cartilages on the *thoracic* vertebrae is referred to, this shall be construed to refer to the cartilages attached to the *thoracic* vertebrae at the posterior end of the forequarter (Table 11-1). The size and shape of the rib bones also are important considerations in evaluating differences in maturity. In the very youngest carcasses considered as beef (A − maturity), the cartilages on the ends of the chine bones show no ossification, cartilage is evident on all of the vertebrae of the spinal column, and the *sacral* vertebrae show distinct separation. In addition, the split vertebrae usually are soft and porous and very red in color. In such carcasses, the rib bones have only a slight tendency toward flatness. In progressively more mature carcasses, ossification changes become evident first in the bones and cartilages of the *sacral* vertebrae, then in the *lumbar* vertebrae, and still later in the *thoracic* vertebrae. In beef, which is very advanced in maturity (E + maturity), all the split vertebrae will be devoid of red color and very hard and flinty, and the cartilages on the ends of all the vertebrae will be entirely ossified. Likewise, with advancing maturity, the rib bones will become progressively wider and flatter until in beef, from very mature animals, the ribs will be very wide and flat.

In steer, heifer, and cow carcasses, the range of maturity permitted within each of the grades varies considerably. The Prime, Choice, Good, and Standard grades are restricted to beef from young cattle; the Commercial grade is restricted to beef from cattle too mature for Good or Standard; and the Utility, Cutter, and Canner grades include beef from animals of all ages. By definition, bullock carcasses are restricted to those whose evidences of maturity do not exceed A maturity.

Table 11-1 should provide a useful summary of necessary information for determining maturity groupings. The entire visible skeleton including the rib cage should be observed and studied and the rib-eye muscle observed as well before coming to a composite final evaluation of maturity. The extremes at either end of the maturity span are A minus (A−) at the youngest end (referred to as A^0 when considering width of the A maturity area in the grading chart, Figure 11-5, in terms of percentages) and E plus (E +) or E^{100} at the extreme oldest end of the maturity span. Either end of the span is relatively easy to identify, but to determine the exact location within the span is more difficult. Two critical points exist in the span: (1) *the A^{100}/B^0 junction* after which marbling level must in-

Table 11-1. Maturity Descriptions

Approximate Chronological Age	Maturity Group	Vertebral Ossification			Ribs	Rib-Eye Muscle
		Sacral	Lumbar	Thoracic		
9-10 months	A – (A⁰)	*Distinct separation*	*Cartilage evident on all vertebrae*	*Cartilages evident on all vertebrae; soft, porous, and very red chine bones*	Slight tendency toward flatness	Light grayish in color; very fine in texture
30 months	A + /B – (A¹⁰⁰/B⁰)	Completely fused	*Nearly completely ossified*	*Cartilages have some evidence of ossification; slightly red and slightly soft chine bones*	Slightly wide; slightly flat	Light red in color; fine in texture

(Continued)

Table 11-1 (Continued)

Approximate Chronological Age	Maturity Group	Vertebral Ossification			Ribs	Rib-Eye Muscle
		Sacral	Lumbar	Thoracic		
42-48 months	B + (B¹⁰⁰)	Completely fused	Completely ossified	*Cartilages partially ossified*; chine bones tinged with red	Slightly wide; slightly flat	Tends to be fine in texture
42-48 months	C − (C⁰)	Completely fused	Completely ossified	*Cartilages moderately (20%-30%) ossified*; chine bones tinged with red	Slightly wide; slightly flat	Tends to be fine in texture

(Continued)

Table 11-1 (Continued)

Approximate Chronological Age	Maturity Group	Vertebral Ossification			Ribs	Rib-Eye Muscle
		Sacral	Lumbar	Thoracic		
60-70 months	$C+/D-(C^{100}/D^{0})$	Completely fused	Completely ossified	*Cartilages show considerable ossification but outlines are plainly visible;* moderately hard, rather white chine bones	Moderately wide and flat	Moderately dark in color; slightly coarse in texture
Older	$D+/E-(D^{100}/E^{0})$	Completely fused	Completely fused	*Cartilages barely visible;* hard, white chine bones	Wide and flat	Dark red and coarse textured

(Continued)

Table 11-1 (Continued)

Approximate Chronological Age	Maturity Group	Vertebral Ossification			Ribs	Rib-Eye Muscle
		Sacral	Lumbar	Thoracic		
Oldest	$E + (E^{100})$	Completely fused	Completely fused	Completely ossified	Very wide and flat	Very dark red and coarse textured

THORACIC VERTEBRAE

crease equally with maturity to maintain a given grade and (2) *the B^{100}/C^0 junction* after which cattle are no longer eligible for the Prime, Choice, Good, or Standard grade.

If any one skeletal area is more important in determining maturity grouping, it would be the *thoracic* area; thus, diagrams of that area are included in Table 11-1. Nevertheless, a composite evaluation must be made, including an appraisal of conformation, since angularity implies maturity. Descriptions are listed for points of merger between the maturity groupings, so by comparing such points, you can locate a carcass within the maturity span and estimate the percentage of distance between the points of merger. Also included in Table 11-1 is an approximation of chronological age of cattle which will produce carcasses of each given maturity.

In determining a subjective trait such as carcass maturity, it takes a good deal of judgment and at least some experience. By using a percentage system, you can evaluate the separate factors as listed in the outline and then put them together for a composite evaluation of maturity. When you are undecided, skeletal development takes precedence over muscle firmness, texture, and color.

Quality—Color and Texture

In steer, heifer, and cow beef, the *color* and *texture* of the lean flesh also undergo progressive changes with advancing *maturity*. In the very youngest carcasses considered as *beef,* the lean flesh will be very fine in texture and light grayish red in color. In progressively more mature carcasses, the texture of the lean will become progressively coarser and the color of the lean will become progressively darker red. In very mature beef, the lean flesh will be very coarse in texture and very dark red in color. Since color of lean also is affected by variations in quality, references to color of lean in the standards for a given degree of maturity vary slightly with different levels of quality. In determining the maturity of a carcass in which the skeletal evidences of maturity are different from those indicated by the color and texture of the lean, slightly more emphasis is placed on the characteristics of the bones and cartilages than on the characteristics of the lean. In no case can the overall maturity of the carcass be considered more than one full maturity group different from that indicated by its bones and cartilages.

Bullock beef carcasses having darker colors of lean than specified in the standards for the quality level for which they would otherwise qualify are evaluated on the basis of skeletal characteristics only, and the final grade will be determined in accordance with the procedures specified in the standards for grading dark-cutting beef.

Dark-cutting Beef

References to color of lean in the standards for steer, heifer, and cow beef involve only colors associated with changes in maturity. They are not intended to apply to colors of lean associated with so-called *dark-cutting beef*. Dark-cutting beef results from a reduced glycogen (starch) content of the lean at the time of slaughter. As a result, this condition does not have the same significance in grading as do the darker shades of red associated with advancing maturity. The dark color of the lean associated with dark-cutting beef is present in varying degrees from that which is barely evident to so-called *black cutters* in which the lean is actually nearly black in color and usually has a gummy texture. There is little or no evidence which indicates that the dark-cutting condition has any adverse effect on palatability, although the condition does favor microbial growth due to the elevated pH. It is considered in grading because of its effect on acceptability and value. A more complete discussion of this phenomenon is contained in Chapter 21. Depending on the degree to which this characteristic is developed, the final grade of carcasses which otherwise would qualify for the Prime, Choice, or Good grade may be reduced as much as *one full grade*. In beef otherwise eligible for the Standard or Commercial grade, the final grade may be reduced as much as *one half of a grade*. In the Utility, Cutter, and Canner grades, this condition is not considered.

Quality—Marbling (Intramuscular Fat)

The second major factor used to determine quality grade is marbling. A certain level of marbling is necessary to assure optimum palatability, especially in terms of juiciness and flavor. The degrees of marbling, in order of descending quantity, are abundant (Ab), moderately abundant (MA), slightly abundant (SA), moderate (Md), modest (Mt), small (Sm), slight (Sl), traces (Tr), and practically devoid (PD). Illustrations of the lower limits of eight of the nine degrees of marbling considered in grading beef appear in Figure 11-4.

Colored reproductions of rib eyes containing SA, Md, Mt, Sm, Sl, and Tr amounts of marbling appear in the center of this textbook.

Except for the youngest maturity group (A), within any specified grade, the requirements for marbling and firmness increase progressively with evidences of advancing maturity. However, firmness is seldom a limiting grade factor. To make it easier to equalize advancing maturity with higher levels of marbling, the standards recognize five different maturity groups (see previous section) and nine degrees of marbling. Marbling can be scored in percentages equally as convenient as maturity. Note the depth of the marbling segments on the beef-grading chart (Figure 11-5), and compare the marbling photos with the rib eye in question. If you determine, for instance, that the marbling level is higher than the minimum small pictured, but less than the minimum modest, you know you are somewhere between Sm^0 and Sm^{100}, perhaps 50 percent of the way to modest; thus, the marbling level is Sm^{50}, or typical small. Percentages should be designated in no smaller units than 10, and as such will make combination with maturity for final grade very straightforward.

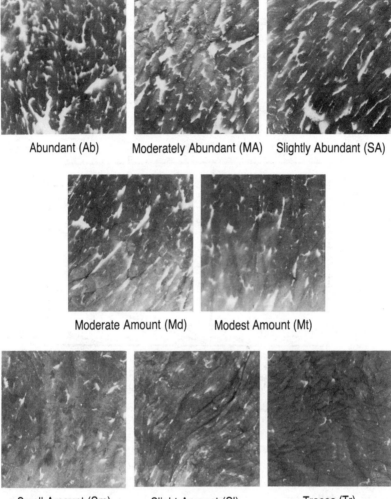

Fig. 11-4. Degrees of marbling. This series of black and white photos was used as the USDA standard until 1981, when the National Live Stock and Meat Board published a set of colored pictures which now serves as the USDA standards. See color section in the center of this text for colored reproductions of these standards. (Courtesy USDA)

Relationship Between Marbling, Maturity, and Quality

The relationship between marbling, maturity, and quality is shown in Figure 11-5. From this figure it can be seen, for instance, that the minimum marbling requirement for Choice does not increase from a minimum small amount for the A maturity carcasses but does increase to a maximum small amount for carcasses having the maximum maturity permitted in Choice. Likewise, in the Commercial grade, the minimum marbling requirement varies from a minimum small amount in beef from animals with the minimum maturity permitted to a maximum moderate amount in beef from very mature animals. No consideration is given to marbling beyond that considered maximum abundant. The marbling and other lean flesh characteristics specified for the various grades are based on their appearance in the rib-eye muscle of properly chilled carcasses that are ribbed between the twelfth and thirteenth ribs.

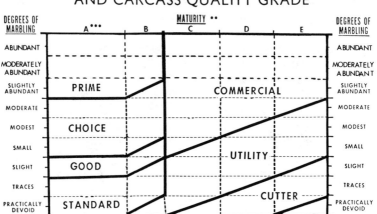

*Assumes that firmness of lean is comparably developed with the degree of marbling and that the carcass is not a "dark cutter."
**Maturity increases from left to right (A through E).
***The A maturity portion of the Figure is the only portion applicable to bullock carcasses.

Fig. 11-5. Relationship between marbling, maturity, and quality.

In certain instances (*e.g.*, meat-grading contests and carcass-evaluation demonstrations), it is necessary to determine the grade to the nearest third. Combining maturity and marbling using percentages is straightforward, provided a few rules are followed according to which grades and maturity levels are under consideration:

In A maturity, marbling is the complete determinant of quality grade. The *Prime* and *Choice* grades are each *three* marbling levels deep (see Figure 11-5) or 300 percent for the full grade in terms of marbling. Thus,

$$Sm^{90} = 90/300 = Choice^{30} = Low\ Choice$$
$$SA^{90} = 90/300 = Prime^{30} = Low\ Prime$$
$$Mt^{80} = 180/300 = Choice^{60} = Av\ Choice$$
$$MA^{80} = 180/300 = Prime^{60} = Av\ Prime$$
$$Md^{70} = 270/300 = Choice^{90} = High\ Choice$$
$$Ab^{70} = 270/300 = Prime^{90} = High\ Prime$$

The *Good* grade is *one* marbling level deep, thus percentage slight marbling = percentage grade, *i.e.*,

$$Sl^{50} = Good^{50} = Av\ Good$$

The *Standard* grade is *two* marbling levels deep or 200 percent for the full grade in terms of marbling. Thus:

$$PD^{60} = 60/200 = Standard^{30} = Low\ Standard$$
$$Tr^{50} = 150/200 = Standard^{75} = High\ Standard$$

In *B, C, D, and E maturity,* as maturity increases, an equal increase in marbling is required in order to maintain a given grade (see Figure 11-5). Logical steps are: (1) set up a minimum marbling requirement for the grade called for by the maturity and marbling combination in question; (2) subtract the minimum marbling percent from the actual marbling percent, giving excess marbling above the minimum requirement; and (3) adjust excess marbling percentage by dividing by the number of marbling levels associated with the grade in question. In the good grade (one marbling degree deep), no adjustment is needed as this equals position in grade, but in the standard grade (two marbling levels deep) and all other grades (three marbling levels deep), the excess percent must be adjusted by dividing as was done for A maturity carcasses. *Caution:* Use only 10 percent increments.

Examples:

Maturity	Marbling	Grade	Minimum Marbling for Grade	Marbling Excess	Adjustment for Marbling Depth	Quality Grade
B^{10}	Sm^{10}	Ch	Sm^{10}	0	$0/300 =$	$Ch^0 = $ Low Choice
B^{10}	Sm^0	G	Sl^{10}	90	none	$G^{90} = $ High Good
B^{50}	Mt^{70}	Ch	Sm^{50}	120	$120/300 =$	$Ch^{40} = $ Av Choice
D^{20}	Tr^{20}	Ut	Tr^{20}	0	$0/300 =$	$Ut^0 = $ Low Utility
D^{20}	Sl^{40}	Ut	Tr^{20}	120	$120/300 =$	$Ut^{40} = $ Av Utility
E^{50}	Md^{50}	Comm	Md^{50}	0	$0/300 =$	$Comm^0 = $ Low Commercial
E^{40}	SA^{60}	Comm	Md^{40}	120	$120/300 =$	$Comm^{40} = $ Av Commercial
B^{20}	PD^{60}	Std	PD^{20}	40	$40/200 =$	$Std^{20} = $ Low Standard
B^{30}	Sl^{50}	G	Sl^{30}	20	none	$G^{20} = $ Low Good
C^{50}	Sl^{60}	Ut	PD^{50}	210	$210/300 =$	$Ut^{70} = $ High Utility
D^{30}	Md^{20}	Comm	Mt^{30}	90	$90/300 =$	$Comm^{30} = $ Low Commercial

Quality Specifications for Bullocks, Bulls, and Stags

Bullock, by definition, includes only carcasses within the A maturity range. Specifications for bullock beef are thus identical to those for steer and heifer beef in the Prime, Choice, Good, Standard, and Utility grades where spelled out for A maturity.

Bulls and stags (all exceeding A maturity) *are not* quality graded but may be yield graded.

Specifications for Official U.S. Standards for Grades of Carcass Beef (Quality—Steer, Heifer, Cow)[2]

Prime (Cow beef not eligible for Prime grade)

As the name implies, beef of this grade is highly acceptable and palatable. Prime grade beef is produced from young and well fed beef-type cattle. The youth of the cattle and the careful intensive feeding which they have had combine to produce very high-quality cuts of beef. Such cuts have liberal quantities of fat interspersed within the lean (marbling). These characteristics contribute to the juiciness and flavor of the meat. Rib roasts and loin steaks of this grade are consistently tender, flavorful, and juicy, and cuts from the round and chuck should also be highly satisfactory:

Maturity groups: A and B
Marbling minimum requirements:
    A maturity minimum marbling is level A^0 to A^{100}
    B maturity minimum marbling increases equally as B^0 increases to B^{100}

Grade	Minimum Marbling	FOR A^{0-100}	Maturity B^{0-100}
Low Prime (P^{0-33})	slightly abundant	SA^0	SA^{0-100}
Av Prime (P^{34-66})	moderately abundant	MA^0	MA^{0-100}
High Prime (P^{67-100})	abundant	Ab^0	Ab^{0-100}

Examples for quality grade computation are given on the previous page.

Choice

This grade is preferred by most consumers because it is of high quality but usually has less fat than beef of the Prime grade. More of this grade of beef is stamped than of any other grade. Choice beef is usually available the year-round

---

[2]*Ibid.*

in substantial quantity. Roasts and steaks from the loin and rib are tender and juicy, and other cuts, such as those from the round or chuck which are more suitable for braising and pot roasting, should be tender with a well developed flavor.

Maturity groups: A and B
Marbling minimum requirements:
    A maturity minimum marbling is level A^0 to A^{100}
    B maturity minimum marbling increases equally as B^0 increases to B^{100}

Grade	Minimum Marbling	FOR	Maturity A^{0-100}	B^{0-100}
Low Choice (Ch^{0-33})	small		Sm^0	Sm^{0-100}
Av Choice (Ch^{34-66})	modest		Mt^0	Mt^{0-100}
High Choice (Ch^{67-100})	moderate		Md^0	Md^{0-100}

    Examples for quality grade computation are given on page 309.

Good

    This grade pleases thrifty consumers who seek beef with little fat but with an acceptable degree of quality. Although cuts of this grade lack the juiciness associated with a higher degree of fatness, their relative tenderness and high proportion of lean to fat make them the preference of some people.

Maturity groups: A and B
Marbling minimum requirements:
    A maturity minimum marbling is level A^0 to A^{100}
    B maturity minimum marbling increases equally as B^0 increases to B^{100}

Grade	Minimum Marbling	FOR	Maturity A^{0-100}	B^{0-100}
Low Good (G^{0-33})	slight		Sl^0	Sl^{0-100}
Av Good (G^{34-66})	slight, small		Sl^{34}	$Sl^{34}-Sm^{34}$
High Good (G^{67-100})	slight, small		Sl^{67}	$Sl^{67}-Sm^{67}$

    Examples for quality grade computation are given on page 309.
Due to the narrowness of this grade, it is often divided in halves rather than thirds for grading contests, the grade is divided at Sl^{50} in A^{0-100} maturity and at $Sl^{50}-Sm^{50}$ in B^{0-100} maturity.

Standard

Standard grade beef has a very thin covering of fat and appeals to consumers whose primary concern is a high proportion of lean. However, less than 1 percent of the beef graded is standard. When properly prepared, such beef is usually relatively tender. It is mild in flavor and lacks the juiciness usually found in beef with more marbling. Nevertheless, consumers rarely see this grade at retail markets, for most of it is included with the some 40 percent ungraded beef which is further processed for use in the hotel and restaurant trade. (See Chapters 12 and 17.)

Maturity groups: A and B
Marbling minimum requirements:
    A maturity minimum marbling is level A^0 to A^{100}
    B maturity minimum marbling increases equally as B^0 increases to B^{100}

Grade	Minimum Marbling	FOR	Maturity A^{0-100}	B^{0-100}
Low Standard (St^{0-33})	practically devoid		PD^0	PD^{0-100}
Av Standard (St^{34-66})	practically devoid, traces		PD^{66}	PD^{66}-Tr^{66}
High Standard (St^{67-100})	traces, slight		Tr^{33}	Tr^{33}-Sl^{33}

Examples for quality grade computation are given on page 309.

Commercial

Beef that is graded Commercial is produced from older cattle and usually lacks the tenderness of the higher grades. Cuts from this grade, if carefully prepared, can be made into satisfactory and economical meat dishes. Most cuts require long, slow cooking with moist heat to make them tender and to develop the rich, full beef flavor characteristic of mature beef. This grade is also rarely seen at retail markets, since ribs and loins are tenderized for hotel and restaurant use as less expensive steaks and chucks; rounds and rough cuts are processed into sausages, ground beef, and related products. (See Chapters 17, 20, and 21.)

Maturity groups: C, D, and E

Marbling minimum requirements increase equally as maturity increases in all three maturity groups.

Grade	Minimum Marbling	FOR C^{0-100}	Maturity D^{0-100}	E^{0-100}
Low Commercial (Co^{0-33})	small, modest, moderate	Sm^{0-100}	Mt^{0-100}	Md^{0-100}
Av Commercial (Co^{34-66})	modest, moderate, slightly abundant	Mt^{0-100}	Md^{0-100}	SA^{0-100}
High Commercial (Co^{67-100})	moderate, slightly abundant, moderately abundant	Md^{0-100}	SA^{0-100}	MA^{0-100}

Examples for quality grade computation are given on page 309.

Utility

Beef of this grade is produced mostly from cattle somewhat advanced in age and is usually lacking in natural tenderness and juiciness. The cuts of this grade rarely appear in the retail markets and carry very little fat but provide a palatable, economical source of lean meat for pot roasting, stewing, and boiling or for ground-meat dishes. For satisfactory results, long, slow cooking by moist heat is essential. Ribs and loins may be tenderized for hotel and restaurant (fast-food) use as less expensive steaks and chucks; rounds and rough cuts are processed into sausages, ground beef, and related products. (See Chapters 17, 20, and 21.)

Maturity groups: A, B, C, D, and E

Marbling minimum requirements increase equally as maturity increases in all five maturity groups.

Grade	Minimum Marbling	FOR A^{0-100}	B^{0-100}	Maturity C^{0-100}	D^{0-100}	E^{0-100}
Low Utility (Ut^{0-33})	devoid, practically devoid, traces, slight	Dev^0	Dev^{0-100}	PD^{0-100}	Tr^{0-100}	Sl^{0-100}
Av Utility (Ut^{34-66})	practically devoid, traces slight, small	Dev^0	Std not Ut	Tr^{0-100}	Sl^{0-100}	Sm^{0-100}
High Utility (Ut^{67-100})	devoid, slight, small, modest	Dev^0	Std not Ut	Sl^{0-100}	Sm^{0-100}	Mt^{0-100}

Examples for quality grade computation are given on page 309.

Cutter

Three maturity groups are recognized. In all groups, the rib-eye muscle is devoid of marbling, is soft and watery, and is dark red in color.

Since this grade of beef is not presented to the buying public in our retail markets, but instead appears in frankfurters, bologna, and hamburger, it will not be discussed beyond this point.

Canner

This grade includes only those carcasses that are inferior to the minimum requirements specified for the Cutter grade and are utilized in the same manner.

Live and Carcass Illustrations of Beef Quality Grades

Live and carcass pictures for beef, lamb, and pork grades (Figures 11-6 through 11-17, 11-31 through 11-48, and 11-54 through 11-62) were generously provided by the University of Illinois Animal Science and Vo Ag Services departments, South Dakota State University, and the National Live Stock and Meat Board. The carcass pictures appear in color in the *Meat Evaluation Handbook,* published by the Meat Board, and were originally secured by Meat Board staff and American Meat Science Association members who worked tirelessly to produce a most useful handbook. The handbook is highly recommended for the serious student of meat judging and grading.

The beef carcass quality grades are based on marbling and maturity (see previous discussion). In live cattle, the trait of maturity (age) is somewhat easily recognized after some training and experience. Marbling, however, is very difficult to estimate on a live animal without the use of supplemental information, such as length of time on feed, composition of ration, and genetic background. Thus many cattle are bought on "reputation," that is, buyers learn to know the way various feeders feed a certain kind of cattle in order to produce carcasses of a given grade.

Some characteristics can be observed on live cattle, which relate to overall fatness and therefore indirectly to marbling, but there are some "fat" cattle that don't marble and some "lean" cattle that do marble. Therefore, the pictures depicting especially the live grades of Prime, Choice, and Good must be recognized as typical examples only; and furthermore, the final carcass grade is not known until the carcass is ribbed and the marbling evaluated. Certainly breeds and types other than those shown for each grade qualify daily in the United States for the various grades, especially in view of the removal of conformation from the grading standard.

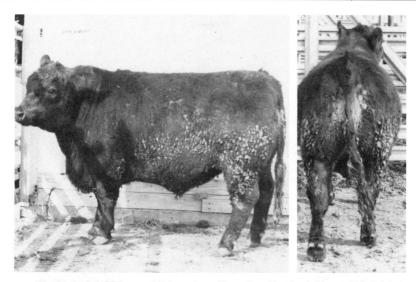

Fig. 11-6. A 1,090-pound Prime steer. From the side, the fullness of his brisket and the fore and rear flanks is evidence of his having been fed. Fullness here represents fat. From the rear, his width through his middle and fullness of twist again indicate feed, fat, and the potential to marble.

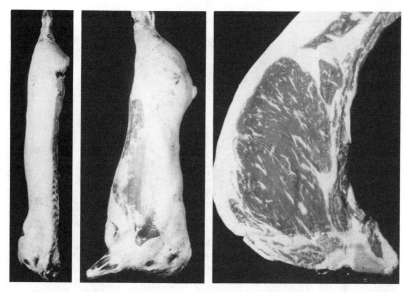

Fig. 11-7. A U.S. Prime beef carcass. This typical A maturity carcass has a light red color of lean; red, porous chine bones; large cartilaginous "buttons" on the *thoracic* vertebrae; and the *sacral* vertebrae are nearly completely fused, all traits which cannot be seen in these photos. The moderately abundant marbling and firm rib-eye muscle combine with A maturity to indicate Prime quality, an Average Prime carcass. The carcass is wide and thick in relation to its length and is thickly muscled throughout—note the plumpness in the round and the thickness in the loin, rib, and chuck.

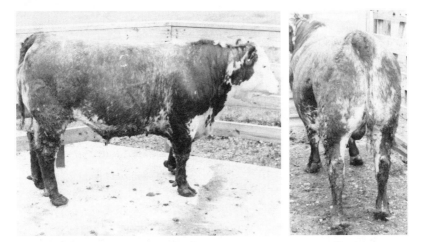

Fig. 11-8. A 1,150-pound Choice steer. From the side, you can see that his brisket and fore flank are moderately full, indicating some fatness. From the rear, he appears to be trim in the twist and lower round but shows evidence of some fatness over the edge of his loin.

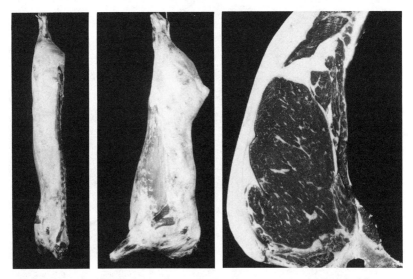

Fig. 11-9. A U.S. Choice beef carcass. The rib-eye muscle of this typical A maturity carcass displays Average Choice quality. It has a weak moderate (moderate [30]) amount of marbling and slightly firm lean. This quality, irrespective of the average muscling evidenced by a moderately plump round and a moderately thick loin, rib, and chuck, results in an Average Choice grade carcass.

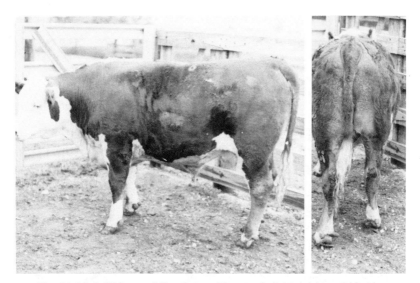

Fig. 11-10. A 980-pound Good steer. The steer's light weight and his trimness about the brisket and flanks as well as behind the shoulders indicate a short time on feed. This steer is narrow behind, lacking muscling and bulge to the round.

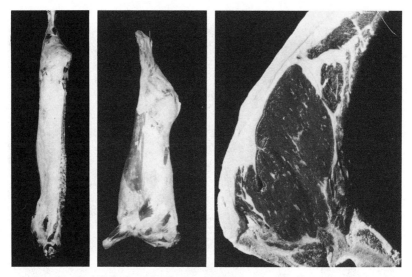

Fig. 11-11. A U.S. Good beef carcass. This very young (A− maturity) carcass has distinct separation of the *sacral* vertebrae; red, porous chine bones; and a very light red color of lean, none of which can be seen in these illustrations. The slight amount of marbling in the moderately soft rib-eye muscle qualified it for Average Good. This carcass has slightly plump rounds and slightly thick and full loins, ribs, and chucks.

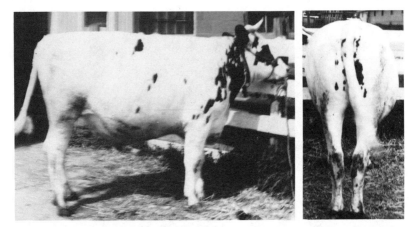

Fig. 11-12. A 1,125-pound Standard steer. Young steers of dairy breeding predominate here, although young cattle of any breed which are somewhat underfinished or because of genetics do not marble readily qualify for the Standard grade. On the other hand, many dairy steers move up to the Good and Choice grades, since conformation no longer holds them back. The steer is narrow behind and shows little evidence of external finish.

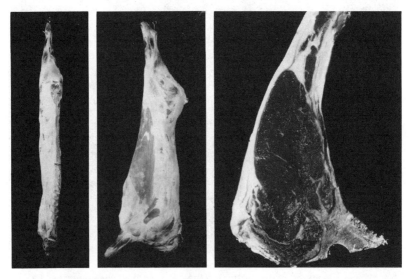

Fig. 11-13. This carcass is in the younger maturity group (A). The rib-eye muscle, which is moderately soft and has traces of marbling, is typical of Average Standard grade. This carcass is slightly thinly fleshed throughout.

Fig. 11-14. A 1,120-pound Commercial cow. The advanced age of this cow is apparent from her angular conformation. The fullness in her brisket and behind her shoulders indicates a well finished cow.

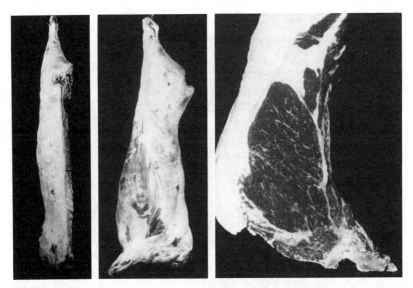

Fig. 11-15. A U.S. Commercial beef carcass. This very hard-boned carcass with no cartilages on the ends of the chine bones because they are completely ossified and flinty is E+ maturity. This is also indicated by the dark red, coarse-textured rib-eye muscle. The moderately abundant amount of marbling indicates Average Commercial grade quality.

Fig. 11-16. An 840-pound Utility cow. From the side, her extreme angularity from front to rear is obvious. Note her protruding ribs indicating practically a complete void of cover. From the rear, concave rounds and sunken sirloin indicate a lack of muscling. Normally such cows are dry before going to market. This cow was slaughtered to obtain an active udder as part of a research project, thus her obviously preponderant udder.

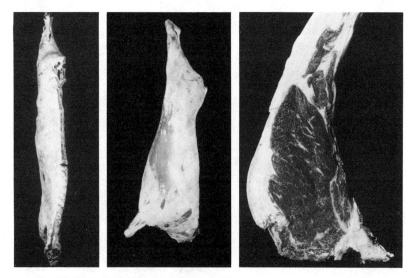

Fig. 11-17. A U.S. Utility beef carcass. The modest amount of marbling in the dark red, coarse-textured rib-eye muscle of this E+ maturity carcass qualified it for Average Utility. This carcass has slightly concave rounds, slightly thin sunken loins, and flat, thinly fleshed ribs.

Yield Grades[3]

Yield grades indicate the quantity of meat, that is, the amount of retail, consumer-ready, ready-to-cook, or edible meat that a carcass contains. Whichever term you choose, the final amount available for eating is extremely important economically. Do not confuse yield grade with dressing percent, often called *yield*. Yield grade refers to the amount of edible product from the carcass. Dressing percent *(yield)* refers to the amount of carcass from a live animal.

The *lower* the number of the yield grade, the *higher* the percent of the carcass in total retail product including all roasts, steaks, and lean trim (ground beef). However, yield grades were originally determined by measuring only the boneless, closely trimmed roasts and steaks from the round, loin, rib, and chuck (Table 11-2).

Table 11-2. Beef Yield Grade Boneless Primal Roast and Steak Yield Equivalents

Yield Grade	Percent of Carcass in Closely Trimmed Roasts and Steaks from the Round, Loin, Rib, and Chuck
1	52.6-54.6
2	50.3-52.3
3	48.0-50.0
4	45.7-47.7
5	43.5-45.4

Determination of a Yield Grade

The determination is made on (1) the amount of external fat; (2) the amount of kidney, pelvic, and heart (KPH) fat; (3) the area of the rib-eye muscle (REA); and (4) the hot carcass weight (HCW).

- The amount of *external fat* on a carcass is evaluated in terms of thickness of this fat over the rib-eye muscle measured perpendicular to the outside surface at a point three fourths of the length of the rib eye from its chine bone end (Figure 11-18). This measurement may be adjusted, as necessary, to reflect unusual amounts of fat on other parts of the carcass. In determining the amount of this adjustment, if any, particular attention is given to the amount of fat in such areas as the brisket, plate, flank, cod or udder, inside round, rump, and hips in relation to the actual thickness of fat over the rib eye. Thus, in a carcass which is fatter over other areas than is indicated by the fat measurement over the rib eye, the measurement is adjusted upward. Conversely, in a carcass which has less fat over the other areas than is indicated by the fat measurement over the rib eye,

---

[3]*Ibid.*

the measurement is adjusted downward. In many carcasses, no such adjustment is necessary; however, an adjustment in the thickness of fat measurement of 0.1 inch or 0.2 inch is not uncommon. In some carcasses, a greater adjustment may be necessary. As the amount of external fat increases, the percent of retail cuts decreases. Each 0.1-inch change in adjusted fat thickness over the rib eye changes the yield grade by 25 percent of a yield grade.

• The amount of *kidney, pelvic, and heart fat* considered in determining the yield grade includes the kidney knob (kidney and surrounding fat), the lumbar and pelvic fat in the loin and round, and the heart fat in the chuck and brisket area, which are removed in making closely trimmed retail cuts (note Chapter 14). The amount of these fats is evaluated subjectively and expressed as a percent of the carcass weight. As the amount of kidney, pelvic, and heart fat increases, the percent of retail cuts decreases. A change of 1 percent of the carcass weight in these fats changes the yield grade by 20 percent of a yield grade.

• The *area of the rib eye* is determined where this muscle is exposed by ribbing (Figure 11-18). This area usually is estimated subjectively; however, it may be measured. Area of rib-eye measurements may be made by means of a grid calibrated in tenths of a square inch or by a compensating planimeter measurement of an acetate tracing. An increase in the area of rib eye increases the percent of retail cuts. A change of 1 square inch in area of rib eye changes the yield grade by approximately 30 percent of a yield grade.

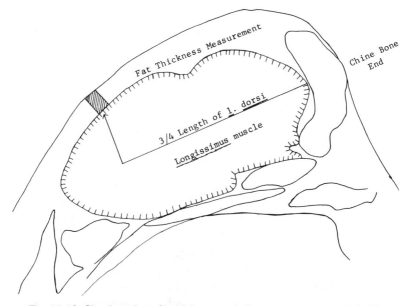

Fig. 11-18. Showing where fat thickness and rib-eye area are measured. (Courtesy American Meat Science Association) See pages 298 and 450 for details on ribbing the carcass.

- *Hot carcass weight* (or chilled carcass weight × 102 percent) is used in determining the yield grade. As carcass weight increases, the percent of retail cuts decreases. A change of 100 pounds in hot carcass weight changes the yield grade by approximately 38 percent of a yield grade.

The standards include a mathematical equation for determining yield grade group. This group is expressed as a whole number; any fractional part of a designation is always dropped. For example, if the computation results in a designation of 3.9, the final yield grade is 3—it is not rounded to 4. *Equation:* 2.50 + (2.50 × adjusted fat thickness, inches) + (0.20 × percent kidney, pelvic, and heart fat) + (0.0038 × hot carcass weight, pounds) − (0.32 × rib-eye area, square inches).

Examples of calculation of yield grades:

- Let us suppose that we have a 610-pound (hot weight) Prime grade carcass that has 0.7 of an inch of fat thickness over the rib eye (other external fat covering normal), has 4.5 percent of its carcass weight in kidney, pelvic, and heart fat, and has 10.4 square inches of rib eye. This is an old-fashioned steer but one still seen too many times. Use the equation: 2.50 + (2.50 × 0.7) + (0.20 × 4.5) + (0.0038 × 610) − (0.32 × 10.4) = 2.50 + 1.75 + 0.9 + 2.32 − 3.33 = 4.14. Result—Yield Grade 4.
- A modern steer having a 700-pound Low-Choice carcass with 0.5-inch fat thickness, 2.5 percent KPH fat, and 13 square inches in REA calculates: 2.50 + (2.50 × 0.5) + (0.2 × 2.5) + (0.0038 × 700) − (0.32 × 13.0) = 2.50 + 1.25 + 0.5 + 2.66 − 4.16 = 2.75. Result—Yield Grade 2.

The federal meat grader rather accurately determines the yield grade rating for a beef carcass by a visual appraisal, aided by a pocket scale graduated in tenths of an inch and the preliminary yield grades associated with fat thickness. The grader is provided with a card that contains the following:

- Determine a *preliminary yield grade* (PYG) from a schedule. (See Table 11-3.) The fat thickness is a single measurement of the fat over the rib eye three fourths of the length of the rib eye from its chine bone end. This measurement may be adjusted, either upward or downward as necessary, to reflect unusual amounts of fat on other parts of the carcass or cut.
- Determine the final yield grade (1 to 5) by adjusting the PYG, as necessary, for variations in kidney, pelvic, and heart (KPH) fat from 3.5 percent and for variations in area of rib eye (REA) from the weight area of rib-eye schedule. (See Table 11-4.) Such adjustments are rounded to the nearest tenth of a grade.
- Rate of adjustment for REA in relation to weight: For each square inch more than area indicated in the weight area of rib-eye schedule, subtract 0.3 of a grade from the PYG. For each square inch less than area indicated in the weight area of rib-eye schedule, add 0.3 of a grade to the PYG.

Table 11-3. Preliminary Beef Yield Grade by Fat Thickness

Thickness of Fat over Rib Eye	Preliminary Yield Grade
(in.)	
0.0	2.0
0.2	2.5
0.4	3.0
0.6	3.5
0.8	4.0
1.0	4.5
1.2	5.0
1.4	5.5
1.6	6.0

- Rate of adjustment for percent of KPH fat: For each percent of KPH fat more than 3.5 percent, add 0.2 of a grade to the preliminary yield grade. For each percent of KPH fat less than 3.5 percent, subtract 0.2 of a grade from the PYG.

Table 11-4 is useful in learning how to estimate rib-eye area (REA) as it gives an expected REA according to carcass weight. Heavily muscled carcasses may have REAs equal to or exceeding 2 square inches per 100 pounds of carcass weight.

Table 11-4. Beef Carcass Weight—Area of Rib-Eye Schedule

Hot Carcass Weight	Area of Rib Eye	Hot Carcass Weight	Area of Rib Eye
(lb.)	(sq. in.)	(lb.)	(sq. in.)
		675	11.9
350	8.0	700	12.2
375	8.3	725	12.5
400	8.6	750	12.8
425	8.9	775	13.1
450	9.2	800	13.4
475	9.5	825	13.7
500	9.8	850	14.0
525	10.1	875	14.3
550	10.4	900	14.6
575	10.7	925	14.9
600	11.0	950	15.2
625	11.3	975	15.5
650	11.6	1,000	15.8

Guide to rib-eye area for other weights of carcasses between the 25-pound graduations:

		600-608 lb.	11.0 sq. in.
600	11.0	609-616 lb.	11.1 sq. in.
625	11.3	617-624 lb.	11.2 sq. in.
		625-633 lb.	11.3 sq. in.

In actual practice, USDA and company graders, other industry personnel, and teachers and students preparing for intercollegiate grading competition use a simplified system such as the one described below:

- Start with a PYG based on actual fat thickness, which may be adjusted up or down for abnormal fat deposits on other parts of the carcass.
- Adjust PYG for HCW, REA, and KPH a constant 0.1 unit of variation from a base as follows:

	Base	*Amount to Equal 0.1 YG*
HCW	600 lb.	25 lb.
REA	11.00 sq. in.	0.33 sq. in.
KPH fat	3.5%	0.5%

Round off all estimates and calculations to 0.1. Example (using carcasses graded by equation on page 323):

	Carcass A	*Carcass B*
Fat thickness, in.	0.7	0.5
HCW, lb.	610	700
REA, sq. in.	10.4	13.0
KPH fat, %	4.5	2.5
PYG (from Table 11-3)	3.75 ⟶ 3.8	3.25 ⟶ 3.3
Adjusted for		
HCW	0.0	+ 0.4
REA	+ 0.2	− 0.7
KPH	+ 0.2	− 0.2
Yield grade	4.2	2.8

The equation is the most accurate, while the grader's method is off a few hundredths due to rounding. The method used by graders works most efficiently and quickly, since the grader knows the requirements *verbatim* and sometimes must apply them *and* a quality grade to beef carcasses passing on a moving chain that often runs in excess of 350 carcasses per hour.

Specifications for Official U.S. Standards for Grades of Carcass Beef (Yield)

The following descriptions provide a guide to the characteristics of carcasses in each yield grade to aid in determining yield grades subjectively.

Yield Grade 1

Carcasses have a thin layer of external fat over the ribs, loins, rumps, and clods and slight deposits of fat in the flanks and cods or udders; a very thin layer of fat over the outside of the rounds and over the tops of the shoulders and necks.

Muscles are usually visible through the fat in many areas of the carcass.

(Description of two different weight carcasses near the borderline of Yield Grades 1 and 2. These descriptions facilitate the subjective determination of the yield grade without making detailed measurements and computations. The yield grade for most beef carcasses can be determined accurately on the basis of a visual appraisal by a trained and experienced grader.)

- A 500-pound carcass might have 0.3 inch of fat over the rib eye, 11.5 square inches of rib eye, and 2.5 percent of its weight in kidney, pelvic, and heart fat.
- An 800-pound carcass might have 0.4 inch of fat over the rib eye, 16 square inches of rib eye, and 2.5 percent of its weight in kidney, pelvic, and heart fat.

Yield Grade 2

Carcasses are completely covered with fat, but the lean is plainly visible through the fat over the outside of the rounds, the tops of the shoulders, and the necks.

There is a slightly thin layer of fat over the loins, ribs, and inside rounds, with a slightly thick layer of fat over the rumps, hips, and clods.

Small deposits of fat occur in the flanks and cods or udders.

- A 500-pound carcass (near the borderline of Yield Grades 2 and 3) might have 0.5 inch of fat over the rib eye, and 10.5 square inches of rib eye, and 3.5 percent of its weight in kidney, pelvic, and heart fat.
- An 800-pound carcass might have 0.6 inch of fat over the rib eye, 15 square inches of rib eye, and 3.5 percent of its weight in kidney, pelvic, and heart fat.

Yield Grade 3

Carcasses are usually completely covered with fat, and the lean usually is visible through the fat only on the necks and lower part of the outside of the rounds.

There is a slightly thick layer of fat over the loins, ribs, and inside rounds, with a moderately thick layer over the rumps, hips, and clods.

Slightly large deposits of fat occur in the flanks and cods or udders.

- A 500-pound carcass (near the borderline of Yield Grades 3 and 4) might have 0.7 inch of fat over the rib eye, 9.5 square inches of rib eye, and 4 percent of its weight in kidney, pelvic, and heart fat.
- An 800-pound carcass might have 0.8 inch of fat over the rib eye, 14 square inches of rib eye, and 4.5 percent of its weight in kidney, pelvic, and heart fat.

Yield Grade 4

Carcasses in this group are usually completely covered with fat. The only muscles visible are those on the shanks and over the outside of the plates and flanks.

There is a moderately thick layer of fat over the loins, ribs, and inside rounds and a thick covering of fat over the rumps, hips, and clods.

Large deposits of fat occur in the flanks and cods or udders.

- A 500-pound carcass (near the borderline of Yield Grades 4 and 5) might have 1 inch of fat over the rib eye, 9 square inches of rib eye, and 4.5 percent of its carcass weight in kidney, pelvic, and heart fat.
- An 800-pound carcass might have 1.1 inches of fat over the rib eye, 13.5 square inches of rib eye, and 5 percent of its weight in kidney, pelvic, and heart fat.

Yield Grade 5

A carcass in this group has more fat on all the various parts, a smaller area of rib eye, and more kidney, pelvic, and heart fat than a carcass in Yield Grade 4.

Yields and Value Differences Between Yield Grades

Value differences between yield grades naturally fluctuate from week to week and month to month due to overall meat price fluctuations and variations in cutting and trimming methods. Further discussion on yield of cuts and value differences between carcasses is found in Chapters 12 and 14.

In 1983, 56.5 percent of the total federally inspected beef slaughter, 71.2 percent of the steers and heifer slaughter, and 1.4 percent of the cow slaughter were graded by the USDA. Of the steers and heifers, 93 percent graded Choice, 4 percent Good, and 3 percent Prime. Of the cows, 83 percent graded Utility, 16 percent Commercial, and 1 percent Cutter. Of these same carcasses, 53.4 percent of the steers and heifers yield graded YG 3, 37.4 percent YG 2, 5.6 percent YG 4, 3.0 percent YG 1, 0.6 percent YG 5, while 63.4 percent of the cows graded YG 2, 28.4 percent YG 3, 6.8 percent YG 1, 1.3 percent YG 4, and 0.1 percent YG 5. A total of 1,396,000 pounds (carcass weight) of bulls and bullocks were graded by the USDA in 1983.[4]

Live and Carcass Illustrations of Beef Yield Grades

Beef carcass yield grades are based on four factors (see previous discussion): (1) carcass weight; (2) loin-eye area; (3) external fat thickness; and (4) percent kidney, pelvic, and heart fat. Since these final yield grade factors are more objective (that is, may be actually measured) than are those for the quality grades, live appraisal for yield grade is somewhat more straightforward than live appraisal for quality grades.

For instance, we look at certain key areas in the live animal for indications of trimness and muscling. Those animals which are deep and full in the twist, flanks, and brisket and those that are wide and flat over the top of the back will be *fat,* because those places fill up with fat as an animal becomes overfat. Muscling indicators in the live animal are width through the lower round in the area of the stifle joint, width through the chest, leg placement well spaced on all four corners, evidence of forearm muscling, and a rounding, muscular appearance over the loin and back. Muscles are not square, but rounded. Squareness over the back indicates fatness, not muscling.

Muscling and fatness are the two big yield grade factors, and they can be reasonably appraised. Carcass weight is, of course, a function of live weight and can easily be considered when evaluating the live animal. Kidney, pelvic, and heart fat is somewhat difficult to estimate. (Note on page 294 that in November 1984 it was proposed by the USDA that KPH percent no longer be used for YG.) Cattle with wasty middles may have excessive internal fats, but this is easily confused with *fill* (feed in the digestive tract, primarily in the rumen). Thus, an estimate of kidney, pelvic, and heart fat based on overall fatness is probably the most useful. Dairy animals are known for carrying heavier than normal amounts of kidney fat.

In actual evaluations, movement of the animal is extremely helpful in evaluating muscling and trimness. Obviously, these pictures (Figures 11-19 through 11-28) don't provide movement, and thus serve only as a general guide to evaluation. When evaluating, the student should always observe animals in motion.

---

[4]*National Summary of Meats Graded.* Dec. 1984. Livestock Division, Meat Grading and Certification Branch, USDA.

Fig. 11-19. A 1,020-pound Yield Grade 1 steer. From the side, trimness in his flank and brisket is apparent. Muscling in the shoulder and bulge to the lower round can be observed from this view. From the rear, note the leg placement squarely on the corners. He lacks depth of twist, indicating trimness, but is wide and thick through the lower round, indicating muscling.

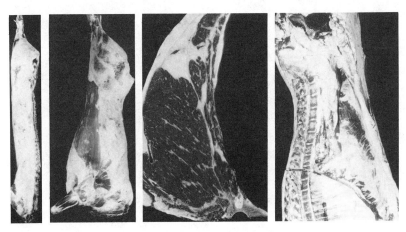

Fig. 11-20. Yield Grade 1 beef carcass.

Carcass weight . . . . 645 pounds
External fat thickness. . . . . 0.2 inch
Rib-eye area. . . . . 13.9 square inches
Kidney, pelvic, and heart fat 2.5%
Yield Grade. 1.5
Quality grade . . . . High Choice

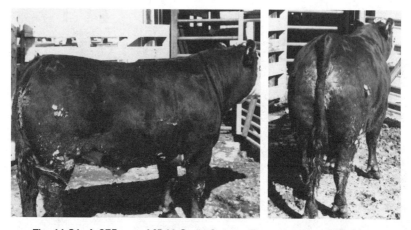

Fig. 11-21. A 975-pound Yield Grade 2 steer. From the side, this steer shows considerable forearm muscling and is moderately trim in the rear flank. His stance from behind is somewhat closed, although he does show width through the center of his rounds and a muscular turn to his loin.

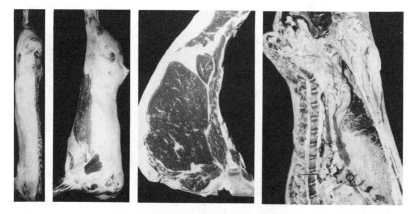

Fig. 11-22. Yield Grade 2 beef carcass.

Carcass weight . . . . 605 pounds
External fat thickness. . . . . 0.4 inch
Rib-eye area. . . . . 12.3 square inches
Kidney, pelvic, and heart fat 3.0%
Yield Grade. 2.5
Quality grade. . . . . Average Choice

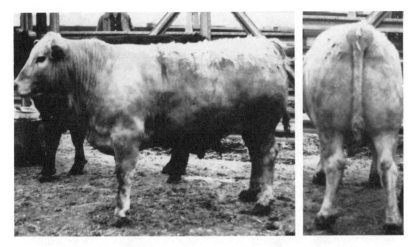

Fig. 11-23. A 1,120-pound Yield Grade 3 steer. From the side, this steer appears upstanding but does show some flesh in his brisket. Forearm and rear quarter muscling are evident. From the rear, evidences of excess fat appear around the tail head, and his turn of loin indicates possible excess finish. Muscling from this view appears adequate but not exceptional.

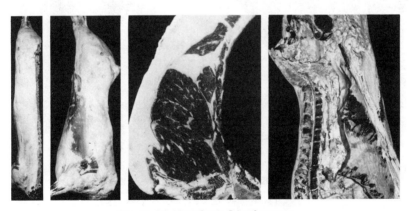

Fig. 11-24. Yield Grade 3 beef carcass.

Carcass weight .... 700 pounds
External fat thickness..................... 0.6 inch
Rib-eye area............................. 11.8 square inches
Kidney, pelvic, and heart fat 3.5%
Yield Grade.............................. 3.6
Quality grade........................... Average Choice

Fig. 11-25. A 1,060-pound Yield Grade 4 steer. From the side, excessive development of the brisket is apparent. The steer appears very heavy in the front quarter and lacks muscle in the valuable hind quarter. From the rear, his broad, flat-topped back indicates excessive finish.

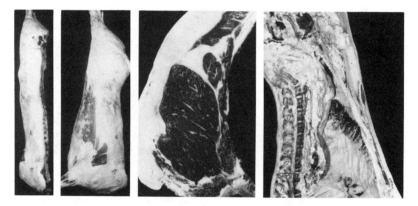

Fig. 11-26. Yield Grade 4 beef carcass.

Carcass weight . . . . 665 pounds
External fat thickness. . . . . 0.9 inch
Rib-eye area . . . . 10.5 square inches
Kidney, pelvic, and heart fat 3.5%
Yield Grade. 4.6
Quality grade . . . . Average Choice

Fig. 11-27. A 1,200-pound Yield Grade 5 steer. This big steer from the side appears to have adequate muscling, but excessive fat is especially evident over the tail head and sirloin. From the rear, he is much wider in his middle and up over his back than he is through his rounds, indicating excessive fat. His rear stance is closed, giving evidence of poor muscling.

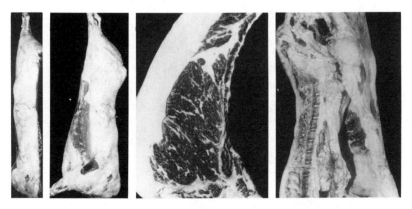

Fig. 11-28. Yield Grade 5 beef carcass.

Carcass weight .... 750 pounds
External fat thickness..................... 1.1 inchs
Rib-eye area............................. 10.9 square inches
Kidney, pelvic, and heart fat 5.0%
Yield Grade............................. 5.6
Quality grade........................... High Choice

Oftentimes, judges like to get their hands on an animal to feel the fatness, but such an opportunity is seldom available during the cattle-marketing sequence. Thus top cattle buyers perfect the "eye ball" method of evaluation.

VEAL AND CALF CARCASS GRADING

Per capita consumption of veal and calf has fallen from 4.6 pounds in 1962 to 1.7 pounds in 1983 (Table 1-3), while commercial veal and calf slaughter has fallen from a high of 13,270,000 in 1954 to the present level of 3,076,700 (Table 7-1).

The high demand for beef has caused the onset of the practice of castrating and feeding male dairy calves for subsequent slaughter as beef steers. Most of the more thickly muscled calves are selected for further feeding, leaving calves with less well developed conformation for the veal and calf market.

Grading standards have been continually updated to reflect industry changes. The most recent change, in October 1980, allows for grading only in carcass form after the hide is removed and only in the establishment where hide removal occurs. In 1971, changes were made which (1) increased the emphasis placed on the color of lean in classing veal; (2) reduced the conformation requirements one full grade in both veal and calf; (3) reduced the quality requirements by varying degrees from a maximum of one grade for veal to no change for the oldest calves; and (4) eliminated the Cull grade for both classes, leaving Prime, Choice, Good, Standard, and Utility as the carcass grades for veal and calf.

Changes in standards for grades of veal and calf carcasses previously made effective March 10, 1951, coincided with revisions made in the live grades and (1) combined the former Choice and Prime grades under the name *Prime;* (2) renamed the Good grade *Choice;* (3) established a new grade called *Good,* which included meat from the top half of the former Commercial grade; (4) continued the remainder of the Commercial grade as *Commercial;* and (5) left the Utility and Cull grades unchanged.

A further revision was made in October 1956, changing the grade name *Commercial* to *Standard* and making certain changes in the phrasing of the standards designed to facilitate their interpretations.

Differentiation Between Veal, Calf, and Beef Carcasses[5]

Differentiation between veal, calf, and beef carcasses is made primarily on the basis of the color of the lean, although such factors as texture of the lean; character of the fat; color, shape, size, and ossification of the bones and cartilages; and general contour of the carcass are also given consideration. Typical

---

[5]*Official United States Standards for Grades of Veal and Calf Carcasses.* Title 7, Chapter I, Part 53, Sections 53.107-53.111, of the Code of Federal Regulations, reprinted with amendments effective Oct. 6, 1980.

veal carcasses have a grayish pink color of lean that is very smooth and velvety in texture, and they also have a slightly soft, pliable character of fat and marrow and very red rib bones. By contrast, typical calf carcasses have a grayish red color of lean, a flakier type of fat, and somewhat wider rib bones with less pronounced evidences of red color.

Classes of Veal and Calf Carcasses

Class determination is based on the apparent sex condition of the animal at the time of slaughter. Hence, there are three classes of veal and calf carcasses—steers, heifers, and bulls. While recognition may sometimes be given to these different classes on the market, especially calf carcasses from bulls that are approaching beef in maturity, the characteristics of such carcasses are not sufficiently different from those of steers and heifers to warrant the development of separate standards for them. Therefore, the grade standards which follow are equally applicable to all classes of veal and calf carcasses.

Application of Standards

Veal and calf carcasses are graded on a composite evaluation of *conformation* and *quality. Conformation,* or the manner of formation of the carcass, refers to its thickness or fullness.

Quality of lean—in all veal carcasses, all unribbed calf carcasses, and ribbed calf carcasses in which their degree of marbling is not a consideration—usually can be evaluated with a high degree of accuracy by giving equal consideration to the following factors, as available: The amount of feathering (fat intermingled within the lean between the ribs) and the quantity of fat streakings within and upon the inside flank muscles. (In making these evaluations, the grader considers the amounts of feathering and flank fat streakings in relation to color [veal] and maturity [calf].) In addition, however, consideration also may be given to other factors if, in the opinion of the grader, this will result in a more accurate quality assessment. Examples of such other factors include firmness of the lean, the distribution of feathering, the amount of fat covering over the diaphragm or skirt, and the amount and character of the external and kidney and pelvic fat.

When ribbed calf carcasses are graded, the quality evaluation of the lean is based entirely on its characteristics as exposed in the cut surface.

Figure 11-29 illustrates how the factors are combined to arrive at the quality grades shown in Figure 11-30.

The *final grade* of a carcass is based on a composite evaluation of its conformation and quality. Conformation and quality often are not developed to the same degree in a carcass, and it is obvious that each grade will include various combinations of development of these two characteristics. The principles governing the compensations of variations in the development of quality and conformation are as follows: In each of the grades, a superior development of quality is permitted to compensate, without limit, for a deficient development of conformation. In this instance, the rate of compensation in all grades is on an equal basis—a given degree of superior quality compensates for the same degree of deficient conformation. The reverse type of compensation—a superior development of conformation for an inferior development of quality—is not permitted in the Prime and Choice grades. In all other grades, this type of compensation is permitted, but only to the extent of one third of a grade of deficient quality. The rate of this type of compensation is also on an equal basis—a given degree of superior conformation compensates for the same degree of deficient quality.

In 1983, 9.8 percent of federally inspected veal and calf slaughter was quality graded by the USDA. Of this 9.8 percent, 75 percent graded Choice, 10 percent Good, 5 percent Prime, and 2 percent Standard.[6]

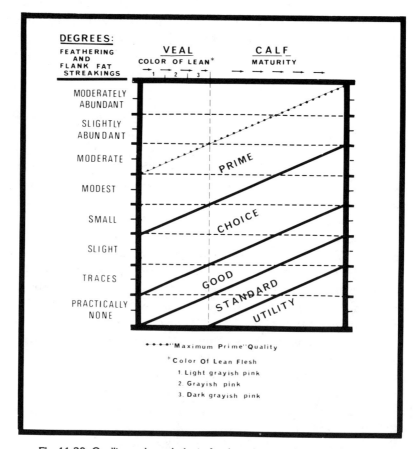

Fig. 11-29. Quality grade equivalent of various degrees of feathering and flank fat streakings in relation to color of lean (veal) or maturity (calf).

[6]*National Summary of Meats Graded.* Dec. 1984. Livestock Division, Meat Grading and Certification Branch, USDA.

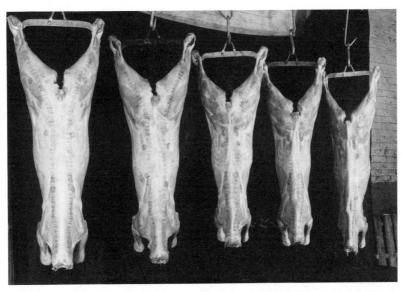

Fig. 11-30. Five grades of veal carcasses (left to right): Prime, Choice, Good, Standard, and Utility.

LAMB, YEARLING MUTTON, AND MUTTON CARCASS GRADING

Development of the Standards[7]

- In February 1931, grades for lamb and mutton carcasses were made effective.
- In October 1940, the standards were amended so as to change the grade designations Medium and Common to *Commercial* and *Utility*, respectively.
- In April 1951, the official standards were again amended when Prime and Choice grades were combined and designated as *Prime*. The Good grade was renamed *Choice*, which also became the highest grade for carcasses of mutton older than yearlings.

  The top two thirds of the Commercial grade was designated as *Good*. The lower one third of the Commercial grade was combined with the top two thirds of the Utility grade and designated as *Utility*, thereby eliminating the Commercial grade name. The lower one third of the Utility grade was combined with the Cull grade and designated as *Cull*.
- In February 1957, the standards for grades of lamb carcasses were amended by reducing the quality requirements for Prime and Choice grade carcasses from more mature lambs.

---

[7]*Official United States Standards for Grades of Lamb, Yearling Mutton and Mutton Carcasses.* Title 7, Chapter I, Part 54, Sections 54.121-54.127, of the Code of Federal Regulations, reprinted with amendments effective Oct. 17, 1982.

The quality requirements for the Good grade were increased slightly, particularly for carcasses from very young lambs.

Carcasses with quality indications equivalent to the lower limit of the upper third of the Good grade were permitted to be graded *Choice*, provided they had a development of conformation equivalent to the mid-point of the Choice grade or better.

Practically all references to quantity of external and kidney and pelvic fats were also eliminated.

- In March 1960, both the conformation and the quality requirements for the Prime and Choice grades were reduced about one-half grade. In addition, a minimum degree of external fat covering was prescribed for the Prime and Choice grades.

  The emphasis placed on internal factors considered in evaluating quality was decreased by reducing the emphasis on feathering between the ribs, eliminating consideration of overflow fat, and increasing the emphasis on firmness of fat and lean.

- On March 1, 1969, yield grades were adopted for lamb, yearling mutton, and mutton carcasses.

- In October 1980, the standards were changed to require grading only in carcass form and only in the establishment where slaughtered or initially chilled.

- In October 1982, the standards were revised to allow a carcass with one or two break joints to be classed as lamb. Feathering was dropped as a quality grade criterion.

Differentiation Between Lamb, Yearling Mutton, and Mutton Carcasses

Differentiation between lamb, yearling mutton, and mutton carcasses is made on the basis of differences that occur in the development of their muscular and skeletal systems (Table 11-5).

Typical lamb carcasses must have at least one front shank break joint.

Yearling mutton carcasses may have either break joints or spool joints on their front shanks. (See Figures 6-6 and 6-7.)

Typical mutton carcasses always have spool joints on their front shanks.

Two Considerations—Quality and Yield

The grade of a lamb, a yearling mutton, or a mutton carcass is based on separate evaluations of two general considerations: (1) The indicated percent of trimmed, boneless major retail cuts to be derived from the carcass, referred to as the *yield grade* and (2) the palatability-indicating characteristics of the lean and

Table 11-5. Lamb, Yearling Mutton, and Mutton Maturity Indicators

	Young Lambs	Mature Lambs	Yearling Mutton	Mutton
	(A maturity)	(B maturity)		
Rib bones	Moderately narrow, slightly flat	Slightly wide, moderately flat	Moderately wide Tend to be flat	Wide Flat
Break joints	Moderately red, moist, and porous	Slightly red, dry, and hard	May or may not be present	Spool joints
Color of inside flank mucles	Slightly dark pink	Light red	Slightly dark red	Dark red
Texture of lean	Fine	Moderately fine	Coarse	Coarse

conformation, referred to as the *quality grade*. When graded by a federal meat grader, the grade of a lamb, yearling mutton, or mutton carcass may consist of the quality grade, the yield grade, or a combination of both the quality grade and the yield grade.

The terms *quality grade* and *quality* are used throughout the standards. The term *quality* is used to refer only to the palatability-indicating characteristics of the lean. As such, it is one of the factors considered in determining the quality grade. Although the term *quality grade* is used to refer to an overall evaluation of a carcass based on both its quality and its conformation, this is not intended to imply that variations in conformation are either directly or indirectly related to differences in palatability, since research has shown no relationship.

Conformation is the manner of formation of the carcass with particular reference to the relative development of the muscular and skeletal systems, although it is also influenced, to some extent, by the quantity and distribution of external finish.

Quality of the lean flesh is best evaluated from consideration of its texture, firmness, and marbling, as observed in a cut surface, in relation to the apparent maturity of the animal from which the carcass was produced. However, in grading carcasses, direct observation of these characteristics is not always possible. Therefore, the quality of the lean is evaluated indirectly by giving equal consideration to both the streaking of fat within and upon the inside flank muscles and the firmness of the fat and lean in relation to the apparent evidence of maturity.

Apparent *sex condition* of the animal at the time of slaughter is not normally considered in ovine carcass grading. However, carcasses which have thick, heavy necks and shoulders, typical of uncastrated males, are discounted in grade in accord with the extent to which these characteristics are developed. Such discounts may vary from less than one-half grade in carcasses from young lambs, in which such characteristics are barely noticeable, to as much as two full grades in carcasses from mature rams, in which such characteristics are very pronounced.

Lamb Carcass Quality Grades

Prime

Carcasses possessing the minimum qualifications for the Prime grade are moderately wide and thick in relation to their length and have moderately wide and thick backs, moderately plump and full legs, and moderately thick and full shoulders.

Requirements for firmness of lean and fat vary with changes in maturity.

- Young lambs—small quantity of fat streaking on inside flank muscle.
- Mature lambs—modest amount of fat streaking on inside flank muscle.

The lean flesh and the exterior finish should be not less than tending to be moderately firm. The minimum external fat requirements for Prime are at least a very thin covering of external fat over the top of the shoulders and the outsides of the center parts of the legs, and the back must have at least a thin covering of fat; that is, the muscles of the back may be plainly visible through the fat.

A development of quality which is superior to that specified as minimum for Prime may compensate, on an equal basis, for a development of conformation which is inferior to that specified as minimum for Prime. For example, a carcass which has evidence of quality equivalent to the midpoint of the Prime grade may have conformation equivalent to the midpoint of the Choice grade and remain eligible for Prime. However, in no instance may a carcass be graded Prime which has a conformation inferior to that specified as minimum for the Choice grade.

Choice

To minimize repetition, substitute the word *slightly* for *moderately* in the description of the Prime grade. A carcass which has conformation equivalent to at least the midpoint of the Choice grade may have quality equivalent to the minimum for the upper third of the Good grade and remain eligible for Choice. Superior quality may compensate on an equal basis for a conformation which is inferior to that specified as minimum for Choice. For example, top Choice quality in a carcass with top Good conformation makes it eligible for the Choice grade. However, in no instance may a carcass be graded Choice which has a conformation inferior to that specified for the Good grade.

Good

Lamb carcasses possessing minimum conformation qualifications for the Good grade are slightly thin muscled throughout; are moderately narrow in relation to their length; and have slightly thin, tapering legs and slightly narrow, thin backs and shoulders. The young lambs are practically devoid of fat streaking on the inside flank muscles. Their lean flesh and exterior finish must be not less than slightly soft.

The more mature lambs have traces of fat streaking on the inside flank muscles. Also, their lean flesh and external finish must be not less than slightly soft.

A carcass which has conformation equivalent to the midpoint of the Good grade and quality equivalent to the minimum for the upper one third of the Utility grade is eligible for Good. Also, a quality which is superior to that specified as the minimum for the Good grade may compensate for a conformation which is inferior to that specified as minimum for Good on the basis of one-half grade of superior quality for one-third grade of deficient conformation. However, in no instance may a carcass be graded Good which has a conformation inferior to that specified as minimum for the upper one third of the Utility grade.

Utility

The Utility grade includes those lamb carcasses whose characteristics are inferior to those specified as minimum for the Good grade.

Yearling and Mutton Carcasses

The grades are Prime, Choice, Good, and Utility for yearlings and Choice, Good, Utility, and Cull for mutton carcasses.

In 1985, 92.5 percent of federally inspected lamb and mutton slaughter was quality graded by the USDA. Of this 92.5 percent, 11 percent was graded Prime, 88 percent Choice, and 1 percent Good.[8]

Live and Carcass Illustrations of Lamb Quality Grades

Lamb carcass quality grading is based on conformation, maturity, and evidences of quality as seen in flank streaking and flank fullness and firmness (see previous discussion and Figures 11-31 through 11-38). Conformation can be readily determined in live lambs, and maturity can be checked by mouthing. However, the quality traits must be estimated in the live animal by evaluation of overall finish or fatness. Sight alone is often deceiving, especially when evaluating wooled lambs, so most buyers will "wade through" the group making spot checks of fatness over the rump, ribs, and back on a number of the lambs.

---

[8]*National Summary of Meats Graded.* Dec. 1985. Livestock Division, Meat Grading and Certification Branch, USDA.

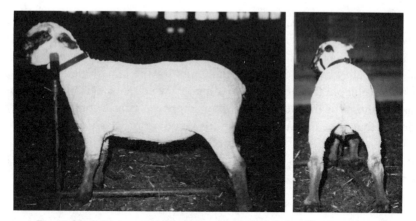

Fig. 11-31. A Prime wether. From the side, this black-faced lamb appears moderately thickly fleshed. The shoulders and hips are moderately smooth. From the rear, he is moderately wide over the back, loin, and rump. The twist is moderately deep and full, and the legs are moderately large and plump.

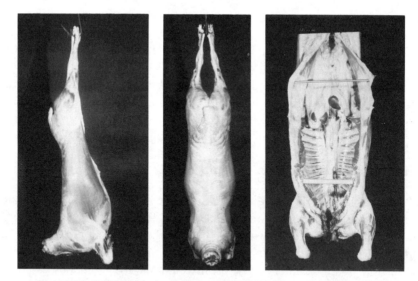

Fig. 11-32. A U.S. Prime lamb carcass. This lamb carcass has typical A maturity as indicated by the moderately narrow, slightly flat rib bones and the slightly dark pink color of the inside flank muscles. Its plump, full legs; wide, thick back; and thick, full shoulders qualify its conformation for Average Prime. The modest streakings of fat in the inside flank muscles, and the moderately full and firm flanks combine to indicate Average Prime quality. When the conformation and quality grades are combined, this carcass qualifies for Average Prime.

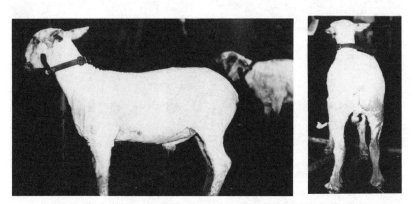

Fig. 11-33. A Choice crossbred wether. The one individual shown here is only that *one* of many breeds and types that annually qualify for this grade. The shoulders and hips show some prominence, indicating that he may be carrying a thin fat covering. From the rear, he is not as wide over the back, loin, and rump as the Prime lamb nor is his leg as fully developed.

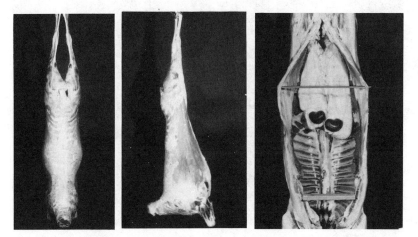

Fig. 11-34. A U.S. Choice lamb carcass. This typical A maturity lamb carcass has Average Choice conformation as evidenced by slightly plump and full legs, a slightly wide and thick back, and slightly thick and full shoulders. The slight streakings of fat in the inside flank muscles, coupled with slightly full and firm flanks, combine to meet the quality requirements for Average Choice. With Average Choice conformation and quality, the final grade of this carcass is Average Choice.

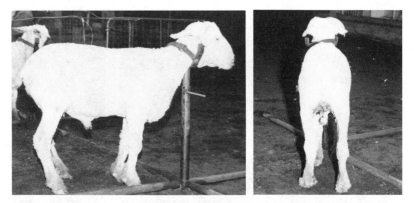

Fig. 11-35. A Good wether. From the side, this western lamb is moderately rangy and thinly fleshed. His hips and shoulders are moderately prominent. From the rear, he is slightly narrow over the back, loin, and rump. The twist is slightly shallow, and the legs are slightly small and thin.

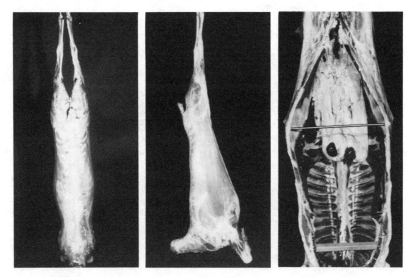

Fig. 11-36. A U.S. Good lamb carcass. This is a more mature lamb carcass (B− maturity) as indicated by the slightly wide, moderately flat rib bones and light red color of the inside flank muscles. It has Average Good conformation in that the legs tend to be slightly thin and tapering, and the back and shoulders tend to be slightly narrow and thin. Likewise, Average Good quality is indicated by the strong traces of streaking of fat in the inside flank muscles, and the slightly full and firm flanks. Average Good conformation and Average Good quality combine to qualify this carcass for Average Good.

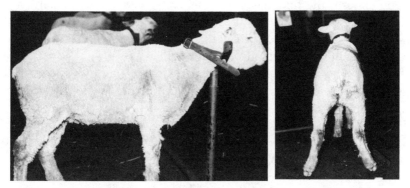

Fig. 11-37. A Utility ewe lamb. This is a very rangy and angular lamb that is very thinly fleshed. She is very narrow over the back, loin, and rump and very shallow in the twist. The legs are very small and present a slightly concave appearance.

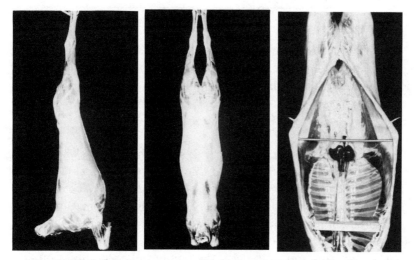

Fig. 11-38. A Utility lamb carcass. This lamb carcass has slightly wide, moderately flat rib bones and light red color of the inside flank muscles, indicating B− maturity. It has Low Utility quality since there are no fat streakings in the inside flank muscles, and the flanks are soft and watery. This carcass is slightly narrow and thin, qualifying it for Average Good conformation. Therefore, when the Average Good conformation and the Low Utility are combined, this carcass qualifies for Average Utility.

Yield Grades[9]

Yield grades indicate the quantity of meat, that is, the amount of retail, consumer-ready, ready-to-cook, or edible meat that a carcass contains. Whichever term you choose, the final amount available for eating is extremely important. Do not confuse yield grade with dressing percent, often called *yield*. Yield grade refers to the amount of edible product from the carcass. Dressing percent *(yield)* refers to the amount of carcass from a live animal.

The *lower* the number of the yield grade, the *higher* the percent of the carcass in boneless, closely trimmed retail cuts from the leg, loin, rack, and shoulder (Table 11-6).

Table 11-6. Lamb Yield Grade Retail Yield Equivalents

Yield Grade	Percent of Warm Carcass in Closely Trimmed Primal Cuts
1	47.6-49.4
2	45.8-47.2
3	44.0-45.4
4	42.2-43.6
5	40.4-41.8

The yield grade of an ovine carcass is determined by considering three characteristics: The amount of external fat, the amount of kidney and pelvic fat, and the conformation grade of the legs.

External Fat

The amount of *external fat* for carcasses with a normal distribution of this fat is evaluated in terms of its actual thickness over the center of the rib-eye muscle and is measured perpendicular to the outside surface between the twelfth and thirteenth ribs. On intact carcasses, fat thickness is measured by probing. This measurement may be adjusted, as necessary, to reflect unusual amounts of fat on other parts of the carcass. In determining the amount of this adjustment, if any, particular attention is given to the amount of external fat on those parts where fat is deposited at a faster-than-average rate, particularly the rump, outside of the shoulders, breast, flank, and cod or udder. Thus, in a carcass which is fatter over these other parts than is normally associated with the actual fat

---

[9]*Official United States Standards for Grades of Lamb, Yearling Mutton and Mutton Carcasses.* Title 7, Chapter I, Part 54, Sections 54.121-54.127, of the Code of Federal Regulations, reprinted with amendments effective Oct. 17, 1982.

thickness over the rib eye, the measurement is adjusted upward. Conversely, in a carcass which has less fat over these other parts than is normally associated with the actual fat thickness over the rib eye, the measurement is adjusted downward. In many carcasses, no such adjustment is necessary; however, an adjustment in the thickness of fat measurement of 0.05 or 0.10 inch is not uncommon. In some carcasses, a greater adjustment may be necessary. As a guide in making these adjustments, the standards for each yield grade include an additional related measurement—body wall thickness, which is measured 5 inches laterally from the middle of the backbone between the twelfth and thirteenth ribs. As the amount of external fat increases, the percent of retail cuts decreases—each 0.05-inch change in adjusted fat thickness over the rib eye changes the yield grade by one third of a grade.

Kidney and Pelvic Fat

The amount of kidney and pelvic fat considered in determining the yield grade includes the kidney knob (kidney and surrounding fat) and the *lumbar* and pelvic fat in the loin and leg, which are removed in making closely trimmed retail cuts. The amount of these fats is evaluated subjectively and expressed as a percent of the carcass weight. As the amount of kidney and pelvic fat increases, the percent of retail cuts decreases—a change of 1 percent of the carcass weight in kidney and pelvic fat changes the yield grade by one fourth of a grade.

Leg Conformation

The conformation grade of the legs is evaluated as described in the quality standards. The evaluation is made in terms of thirds of grades and coded using 15 for High Prime and 1 for Low Cull. An increase in the conformation grade of the legs increases the percent of retail cuts—a change of one third of a grade changes the yield grade by 5 percent of a grade.

The Yield Grade Equation

The yield grade of an ovine carcass or side is determined on the basis of the following equation: yield grade = 1.66 − (0.05 × leg conformation grade code) + (0.25 × percent kidney and pelvic fat) + (6.66 × adjusted fat thickness over the rib eye in inches). The application of this equation usually results in a fractional grade. However, in normal grading operations any fractional part of a yield grade is dropped. For example, if the computation results in a *yield grade* of 3.9, the final yield grade is 3—it is not rounded to 4.

Example Calculation of Yield Grades

- Overfat, lightly muscled lamb

  Low Choice leg conformation, 5 percent kidney and pelvic (KP) fat, and fat thickness of 0.35 inch.
  Use the equation: $1.66 - (0.05 \times 10) + (0.25 \times 5) + (6.66 \times 0.35) =$
  $$1.66 - 0.5 + 1.25 + 2.33 = 4.74 = \text{Yield Grade 4}$$

- Trim, heavily muscled lamb

  Average Prime leg conformation, 2.5 percent KP fat, and fat thickness of 0.14 inch.
  $1.66 - (0.05 \times 12) + (0.25 \times 2.5) + (6.66 \times 0.14) =$
  $1.66 - 0.6 + 0.625 + 0.93 = 2.61 = \text{Yield Grade 2}$

The federal meat grader rather accurately determines the yield grade rating for a lamb carcass by visual appraisal, aided by a pocket scale graduated in tenths of an inch, which is sometimes used to probe fat thickness, since lamb carcasses are not normally ribbed. The grader has knowledge of preliminary yield grade relationships to fat thickness and the adjustment factors for leg conformation and percent of kidney and pelvic fat. Stepwise, the grader:

1. Determines a preliminary yield grade (PYG), based on the schedule in Table 11-7, to reflect the external fatness of the carcass.
2. Determines the final yield grade (1 to 5) by adjusting the preliminary yield grade, as necessary, for variations in kidney and pelvic fat from 3.5 percent and for variations in leg conformation grade from Average Choice.
   a. Rate of adjustment for percent of kidney and pelvic fat:
      (1) For each percent of kidney and pelvic fat more than 3.5 percent, add 0.25 of a grade to the preliminary yield grade.
      (2) For each percent of kidney and pelvic fat less than 3.5 percent, subtract 0.25 of a grade from the preliminary yield grade.
   b. Rate of adjustment for leg conformation grade:
      (1) For each one third of a grade that the conformation of the legs exceeds Average Choice, subtract 0.05 of a grade from the preliminary yield grade.
      (2) For each one third of a grade that the conformation of the legs is less than Average Choice, add 0.05 of a grade to the preliminary yield grade.

Examples that use the same carcasses are worked through the above equations.

	Overfat, Lightly Muscled Carcass		Trim, Heavily Muscled Carcass	
Fat thickness		0.35 in.		0.14 in.
PYG (from Table 11-7)		4.33		2.9
KP fat percent		5.0%		2.5
Standard		3.5%		3.5
Excess		1.5%	Under	1.0%
× 0.25 for each percent		+ 0.37		− 0.25
PYG adjusted for KP fat		4.70		2.65
Leg conformation		Low Choice (10)		Average Prime (14)
Under standard		1/3	Over	3/3 = 1
× 0.05 for each 1/3		+ 0.05		− 0.15
Yield grade		4.75 = 4		2.54 = 2

The equations are the most accurate, while the grader's method is off a few hundredths due to rounding. The method used by graders works most efficiently and quickly, since the grader knows the YG requirements *verbatim* and must grade at a reasonably high rate of speed.

Table 11-7. Preliminary Lamb Yield Grade by Fat Thickness

Fat Thickness over Rib Eye[1]	Preliminary Yield Grade
(in.)	
0.05	2.33
0.10	2.67
0.15	3.00
0.20	3.33
0.25	3.67
0.30	4.00
0.35	4.33
0.40	4.67
0.45	5.00
0.50	5.33
0.55	5.67
0.60	6.00

[1]This fat thickness measurement over the rib-eye muscle should be adjusted, as necessary, to reflect unusual amounts of fat on other parts of the carcass.

Yield Grade Descriptions

The following descriptions provide a guide to the characteristics of carcasses in each yield grade to aid in determining yield grades subjectively.

Yield Grade 1

A carcass in Yield Grade 1 usually has only a thin layer of external fat over the back and loin and slight deposits of fat in the flanks and cod or udder. There is usually a very thin layer of fat over the tops of the shoulders and the outside of the legs. Muscles are usually plainly visible on most areas of the carcass.

A carcass of this yield grade which is near the borderline of Yield Grade 1 and Yield Grade 2 might have 0.1 inch of fat over the rib eye, 1.5 percent of its weight in kidney and pelvic fat, and an Average Prime leg conformation grade. Such a carcass with normal fat distribution may also have a body wall thickness of 0.4 inch.

Yield Grade 2

A carcass in Yield Grade 2 usually has a slightly thin layer of fat over the back and loin and the muscles of the back are not visible. The top of the shoulders and the outside of the legs have a thin covering of fat, and the muscles are slightly visible. There are usually small deposits of fat in the flanks and cod or udder.

A carcass of this yield grade which is near the borderline of Yield Grade 2 and Yield Grade 3 might have 0.2 inch of fat over the rib eye, 2.5 percent of its weight in kidney and pelvic fat, and a Low Prime leg conformation grade. Such a carcass with normal fat distribution may also have a body wall thickness of 0.7 inch.

Yield Grade 3

A carcass in Yield Grade 3 usually has a slightly thick covering of fat over the back. The tops of the shoulders are completely covered with fat, although the muscles are still barely visible. The legs are nearly completely covered, although the muscles on the outside of the lower legs are visible. There usually are slightly large deposits of fat in the flanks and cod or udder.

A carcass of this yield grade, which is near the borderline of Yield Grade 3 and Yield Grade 4, might have 0.3 inch of fat over the rib eye, 3.5 percent of its weight in kidney and pelvic fat, and a High Choice leg conformation grade. Such a carcass with normal fat distribution may also have a body wall thickness of 0.9 inch.

Yield Grade 4

A carcass in Yield Grade 4 usually is completely covered with fat. There usually is a moderately thick covering of fat over the back and a slightly thick covering over the shoulder and legs. There usually are large deposits of fat in the flanks and cod or udder.

A carcass in this yield grade, which is near the borderline of Yield Grade 4 and Yield Grade 5, might have 0.4 inch of fat over the rib eye, 4.5 percent of its weight in kidney and pelvic fat, and an Average Choice leg conformation grade. Such a carcass with normal fat distribution would also have a body wall thickness of 1.2 inches.

Yield Grade 5

A carcass in Yield Grade 5 usually has more external and kidney and pelvic fat and a lower conformation grade of leg than a carcass in Yield Grade 4.

Live and Carcass Illustrations of Lamb Yield Grades

Lamb carcass yield grading is based on leg conformation, external fat thickness, and kidney and pelvic fat percent (see preceding discussion and Figures 11-39 through 11-48). Fat thickness is by far the most important factor, each 1/10 change representing 2/3 of one yield grade. Thus the appraisal of fatness must be made as accurately as possible, meaning a "hands-on" approach. A light touch over the back, ribs, and rump with closed fingers can reveal a tremendous amount of yield grading information.

Yields and Value Differences Between Yield Grades

Value differences between yield grades naturally fluctuate from week to week and month to month due to overall meat price fluctuations and variations in cutting and trimming methods. Further discussion on yield of cuts and value difference between carcasses is found in Chapters 12 and 15.

Applying the Grade Stamp

The federal meat grader weighs all the factors that have been discussed, decides upon the grade, and makes an identifying mark with a stamp designating the grade: xxx for Prime, xx for Choice, etc. After the grading task is ended, the grader or an assistant applies the ribbon grade stamp over the entire length of the carcass with a roller (Figure 11-49) and at such other points that all principal retail cuts will bear the grade label. That is why federally graded beef and lamb are often referred to as "rolled." The grade designations are in full grades. The Cutter and Canner beef grades are not rolled. If the buyer wishes to know whether the rolled beef of the grade being bought is in the top, middle, or low bracket of the grade, he or she can demand a grading certificate which will so indicate.

Do not confuse grading with inspection or the grading stamp with the inspection stamp used to designate that the process has been completed. Every animal that is slaughtered for meat sale *must* by law be inspected. The federal inspection stamp is round and appears on each wholesale cut (Figure 11-50). Grading is *not* required by law.

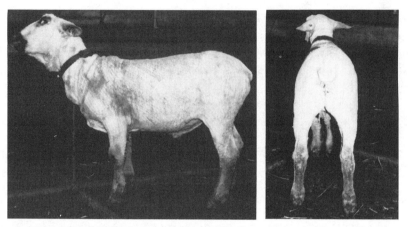

Fig. 11-39. A Yield Grade 1 wether. Only a very small percentage of market lambs qualify for this grade, since it requires a high degree of trimness and muscling. This crossbred lamb is restricted just behind the shoulders, indicating that this area has not filled with fat. Also he is especially trim in the breast and flank. From the rear, leanness is apparent by the lack of fullness in the twist. Excellent muscling is indicated by his being wider and thicker through the center of his leg than he is over his back.

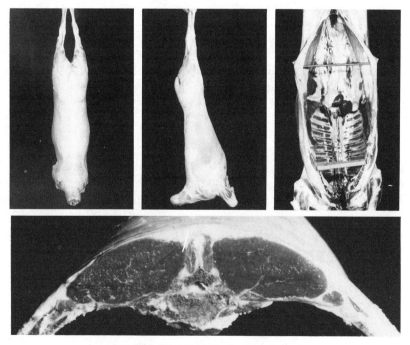

Fig. 11-40. A Yield Grade 1 carcass.

External fat thickness .... 0.05 inch
Kidney and pelvic fat 1.5%
Leg conformation grade .... Low Prime
Yield Grade 1.7

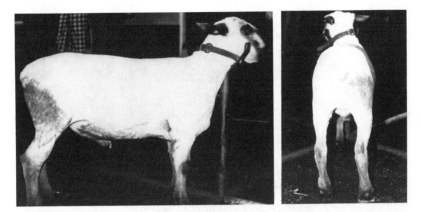

Fig. 11-41. A Yield Grade 2 wether. From the side, this wether is very trim in his breast and flank, but is uniformly smooth behind the shoulders, indicating that this area is somewhat filled with fat. This well muscled lamb is wider through the center of his legs than up over his top, but is carrying more finish in his twist than the Yield Grade 1 lamb.

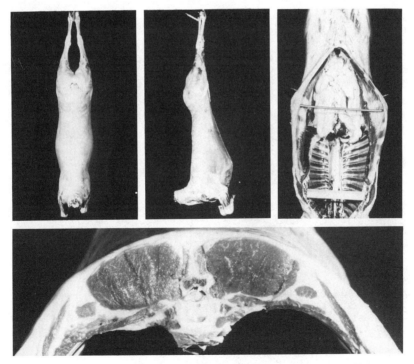

Fig. 11-42. A Yield Grade 2 lamb carcass.

External fat thickness . . . . 0.10 inch
Kidney and pelvic fat 3.0%
Leg conformation grade . . . . Average Prime
Yield Grade 2.4

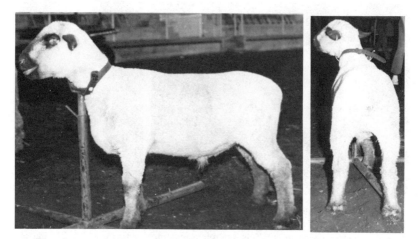

Fig. 11-43. A Yield Grade 3 wether. From the side, increased deposits of fat can be observed in the areas previously discussed, and from the rear view as well.

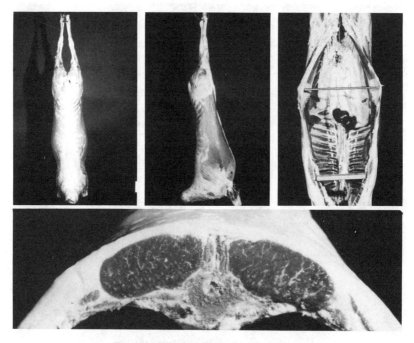

Fig. 11-44. A Yield Grade 3 lamb carcass.

External fat thickness . . . . 0.25 inch
Kidney and pelvic fat 3.5%
Leg conformation grade . . . . High Choice
Yield Grade 3.6

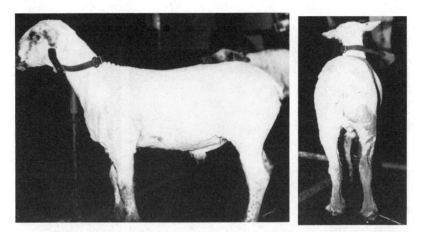

Fig. 11-45. A Yield Grade 4 wether. Although this wether is long and appears trim in the breast region, the overall smooth appearance from shoulder to sirloin and rump gives evidence of his overall fatness. From the rear, he is narrow and light muscled in his legs, and his twist is filled with fat.

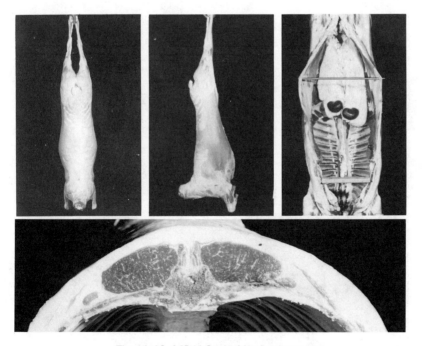

Fig. 11-46. A Yield Grade 4 lamb carcass.

External fat thickness . . . . 0.35 inch
Kidney and pelvic fat 4.5%
Leg conformation grade . . . . Low Choice
Yield Grade 4.6

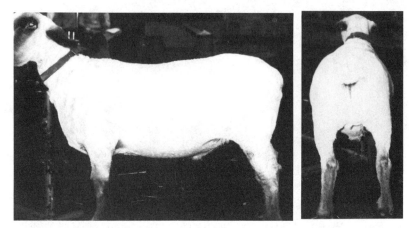

Fig. 11-47. A Yield Grade 5 wether. The heavy middle on this wether is an indication of excessive kidney fat. His fullness of flank and breast portrays his overall fatness. Although the lamb stands wide behind and has plump, heavy legs, the depth of twist and squareness of dock indicate fatness.

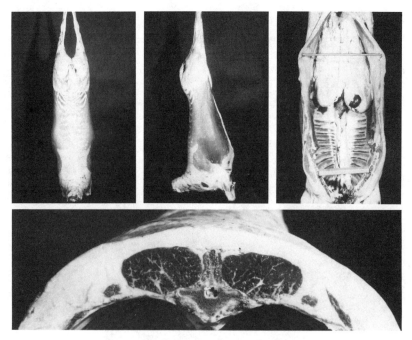

Fig. 11-48. A Yield Grade 5 lamb carcass.

External fat thickness . . . . 0.45 inch
Kidney and pelvic fat 5.5%
Leg conformation grade . . . . Average Choice
Yield Grade 5.5

Fig. 11-49. Applying the grade stamp. Note the round inspection stamp adjacent to the rolled grade stamp. (Courtesy Livestock Division, Agricultural Marketing Service, USDA) (See also Figures 11-50 and 11-51.)

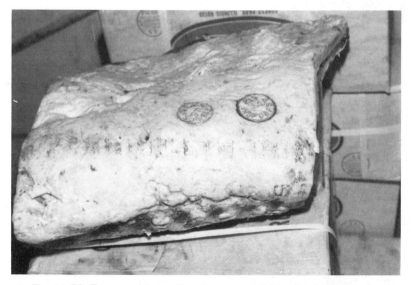

Fig. 11-50. Two round inspection stamps and the grade stamp showing the combined quality and yield roll. Two inspection stamps indicate that the animal was slaughtered in one plant, cut into a wholesale rib in a second plant, and shipped to a third plant or store (see also Figure 11-51).

Fig. 11-51. (Left) Closeup of a hindquarter showing the combination yield, quality roll, and inspection stamp. "E P" are the initials of the grader who graded the carcass. Note normal presence of kidney knob. (Right) Closeup of hindquarter from which kidney knob has been removed, making this carcass a Yield Grade 2. To indicate that trimming has been done, the grade is rolled on upside down. Approval of a USDA proposal would eliminate KPH as a YG factor and use the reverse roll illustrated here to indicate KPH was present (see page 294).

PORK CARCASS GRADING

Development of the Standards[10]

Tentative standards for grades of pork carcasses and fresh pork cuts were issued by the USDA in 1931. These tentative standards were slightly revised in 1933.

New standards for grades of barrow and gilt carcasses were proposed by the USDA in 1949. These standards represented the first application of objective measurements as guides to grades for pork carcasses. Slight revisions were made in the proposed standards prior to adoption, as the Official United States Standards for Grades of Barrow and Gilt Carcasses, effective September 12, 1952.

The official standards were amended in July 1955, by changing the grade designations Choice No. 1, Choice No. 2, and Choice No. 3 to U.S. No. 1, U.S. No. 2, and U.S. No. 3, respectively. In addition, the backfat thickness requirements were reduced for each grade, and the descriptive specifications were reworded slightly to reflect the reduced fat thickness requirements and to allow more uniform interpretation of the standards.

On April 1, 1968, the official standards were again revised to reflect the im-

---

[10]*Marketing Bulletin No. 49.* Agricultural Marketing Service, USDA.

provements made since 1955 in pork carcasses. The minimum backfat thickness requirement for the U.S. No. 1 grade was eliminated and a new U.S. No. 1 grade was established to properly identify the superior pork carcasses then being produced. The former No. 1, No. 2, and No. 3 grades were renamed No. 2, No. 3, and No. 4 respectively. The former Medium and Cull grades were combined and renamed U.S. Utility. Also, the maximum allowable adjustment for variations from normal fat distribution and muscling was changed from one half to one full grade to more adequately reflect the effect of these factors on yields of cuts.

In January 1985, the standards were further revised. The new grades are based on last rib backfat thickness and muscling. Only three levels of muscling (thick, average, and thin) are recognized. The width of the grades remains 3 percent of four lean cuts, but the expected yield from each grade is 7.4 percent higher than previously. The lines between Grades 1 and 2, 2 and 3, and 3 and 4 are set at 1, 1.25, and 1.5 inches of last rib backfat thickness, respectively.

Pork Carcass Classes

The five classes of pork carcasses, comparable to the same five classes of slaughter hogs, are barrow, gilt, sow, stag, and boar.

"Because of the relationships between sex and/or sex condition in pork and the acceptability of the prepared meats to the consumer, separate standards have been developed for (1) barrow and gilt carcasses and (2) sow carcasses. There are no official standards for grades of stag and boar carcasses.

"The determination of sex condition is based on the following:

- Barrow carcasses are identified by a small 'pizzle eye' [Figure 11-1 illustrates a beef pizzle eye] and the typical pocket in the split edge of the belly where the preputial sheath was removed.
- Gilt carcasses are recognized by the smooth split edge of the belly, the absence of the 'pizzle eye,' and the lack of development of mammary tissue.
- Sow carcasses exhibit the smooth split edge of the belly characteristic of females. They differ from gilts in that mammary tissue has developed in connection with advanced pregnancy or lactation.
- Stag carcasses have the pocket in the split edge of the belly typical of males, and the 'pizzle eye' is larger and more prominent than in barrows. In addition, other distinguishing characteristics that often may be noted are rather heavy shoulders, thick skin over the shoulders, large bones and joints, and a dark red color of lean.
- Boar carcasses have the same distinguishing characteristics as stag carcasses but to a more pronounced degree."[11]

---

[11]*Meat Evaluation Handbook*. National Live Stock and Meat Board.

Grades for Barrow and Gilt Carcasses

Differences in barrow and gilt carcasses due to sex condition are minor, and the grade standards are equally applicable for grading both classes.

Grades for barrow and gilt carcasses are based on two general conditions: quality-indicating characteristics of the lean and expected combined yields of the four lean cuts (ham, loin, picnic shoulder, and Boston shoulder).

Quality

With respect to quality, two general levels are considered: one for carcasses with characteristics which indicate that the lean in the four lean cuts will have an acceptable quality and one for carcasses with characteristics which indicate that the lean will have an unacceptable quality.

The quality of the lean is best evaluated by a direct observation of its characteristics in a cut surface and when a cut surface of major muscles is available, this shall be used as the basis for the quality determination. The standards describe the characteristics of the loin-eye muscle at the tenth rib. However, when this surface is not available, other exposed major muscle surfaces can be used for the quality determination based on the normal development of the characteristics in relation to those described for the loin-eye muscle at the tenth rib.

When a major muscle cut surface is not available, the quality of the lean shall be evaluated indirectly based on quality-indicating characteristics that are evident in carcasses. These include firmness of the fat and lean, amount of feathering between the ribs, and color of the lean. The standards describe a development of each of these factors that is normally associated with the lower limit of acceptable lean quality. The degree of external fatness, as such, is not considered in evaluating the quality of the lean.

Carcasses which have characteristics indicating that the lean in the four lean cuts will not have an acceptable quality or which have bellies too thin to be suitable for bacon production are graded U.S. Utility. Also graded U.S. Utility—regardless of their development of other quality-indicating characteristics—are carcasses which are soft and oily. Belly thickness is determined by an overall evaluation of its thickness with primary consideration being given to the thickness along the navel edge and thickness of the belly pocket at the flank end of the belly.

A numerical system for designating pork quality, as observed in the cut surfaces of major muscles, was published by the University of Wisconsin in 1963 in *Special Bulletin 9, Pork Quality Standards*. Although not used as such in federal descriptions, these standards are based on five levels of color, marbling, and firmness and served as the forerunner of the National Pork Producers Council (NPPC)[12] standards published in *Procedures to Evaluate Market Hog Performance* (2nd edition, 1983) which are widely recognized in the industry today.

---

[12]P.O. Box 10383, Des Moines, IA 50306.

The NPPC standards now are composed of three levels of acceptance, with the intermediate level in all quality traits of color, marbling, and firmness being preferred. The criteria are listed in Table 11-8 and are illustrated in the section near the center of this textbook.

Table 11-8. National Pork Producers Council Pork Quality Standards

	Score	Description	Level of Acceptance
Color	1	Pale (white or gray)	Unacceptable
	2	Light pink, reddish pink	Preferred
	3	Dark red	Unacceptable
Marbling	1	Traces, slight amounts	Acceptable
	2	Small, moderate amounts	Preferred
	3	Abundant amounts	Excessive, unacceptable
Firmness	1	Soft and watery	Unacceptable
	2	Intermediate	Preferred
	3	Firm	Acceptable

Yield

Four grades—U.S. No. 1, U.S. No. 2, U.S. No. 3, and U.S. No. 4—are provided for carcasses which have indications of an acceptable lean quality and acceptable belly thickness. These grades are based entirely on the expected carcass yields of the four lean cuts, and no consideration is given to a development of quality superior to that described as minimum for these grades. The expected yields of the four lean cuts for each of these four grades are shown in Table 11-9.

Table 11-9. Expected Yields of the Four Lean Cuts Based on Chilled Pork Carcass Weight, by Grade[1]

Grade	Yield
U.S. No. 1	60.4% and over
U.S. No. 2	57.4% to 60.3%
U.S. No. 3	54.4% to 57.3%
U.S. No. 4	Less than 54.4%

[1]These yields will be approximately 1 percent lower if based on hot carcass weight.

The yields shown are based on cutting and trimming methods used by the USDA in developing the standards (these methods generally specify 0.2-inch fat trim, and copies may be obtained from the Livestock Division, Agricultural Marketing Service, USDA, Washington, DC 20250). Other cutting and trimming methods may result in different yields. For example, if more fat is left on the four lean cuts than prescribed in the USDA methods, the yield for each grade will be higher than indicated. However, such a method of trimming, if applied uniformly, should result in similar differences in yields between grades.

Carcasses vary in their yields of the four lean cuts because of variations in their degree of fatness and in their degree of muscling (thickness of muscling in relation to skeletal size).

Objective Measure

In these standards, the actual thickness of backfat is a single measurement including the skin made opposite the last rib and perpendicular to the skin surface (Figure 11-52). For carcasses that have been skinned, 0.1 inch is added to the measurement to compensate for the loss of the skin, if the skin has been smoothly and evenly removed. The yield of four cuts from skinned carcasses will be higher than indicated in Table 11-9.

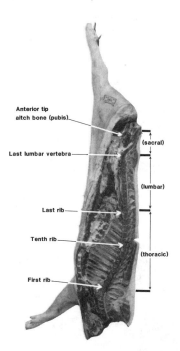

Fig. 11-52. Backfat is measured opposite the last rib. Length is measured from the front edge of the first rib where it joins the backbone to the front tip of the aitch bone on a split, non-ribbed carcass. Length is no longer required for a USDA grade but nevertheless is of interest to pork producers. This carcass has been ribbed between the tenth and eleventh ribs to expose the loin eye and the fat cover over it. Although not required by the USDA for carcass grading, exposing the loin eye and fat cover in this manner provides a very useful indicator of carcass quality and cutability that is widely used.

Degree of Muscling

The degree of muscling specified for each of the four grades decreases progressively from the U.S. No. 1 grade through the U.S. No. 4 grade. This reflects

DEGREES OF MUSCLING

Thick ⟵-------- Average--------⟶ ⟵------ Thin------⟶

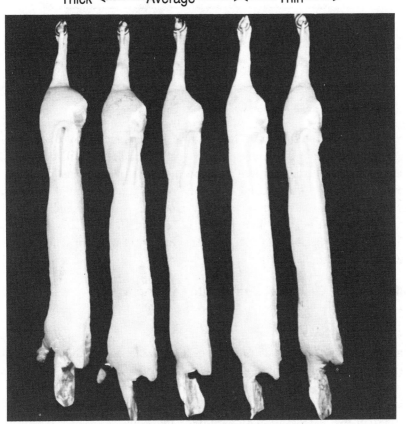

Fig. 11-53. Pre-1985 degrees of muscling. The current thick (superior) muscling includes only those carcasses previously classed above as very thick. Current
average muscling includes the previous thick and moderately thick degrees above,
and the current thin (inferior) muscling includes the previous slightly thin and thin degrees. (Courtesy Livestock Division, Agricultural Marketing Service, USDA)

the fact that among carcasses of the same weight, fatter carcasses normally have
a lesser degree of muscling. For purposes of these standards, three degrees of
muscling are recognized: thick (superior), average, and thin (inferior). These are
intended to cover the entire range of muscling present among pork carcasses
currently being produced. Figure 11-53 illustrates five degrees of muscling combined as follows: The current thick (superior) muscling includes only those carcasses previously classed as very thick. Current average muscling includes the
previous thick and moderately thick degrees, and the current thin (inferior) muscling includes the previous slightly thin and thin degrees.

The grade is determined by calculating a preliminary grade according to the
schedule shown in Table 11-10 and adjusting up or down one grade for thick or
thin muscling respectively. Alternatively, the equation may be used.

Table 11-10. Preliminary Carcass Grade Based on Backfat Thickness over the Last Rib

Preliminary Grade	Backfat Thickness Range
U.S. No. 1	Less than 1.00 inch
U.S. No. 2	1.00 to 1.24 inches
U.S. No. 3	1.25 to 1.49 inches
U.S. No. 4	1.50 inches and over[1]

[1]Carcasses with last rib backfat thickness of 1.75 inches or over cannot be graded U.S. No. 3, even with thick muscling.

Carcasses with average muscling will be graded according to their backfat thickness over the last rib. Carcasses with thin muscling will be graded one grade lower than indicated by their backfat thickness over the last rib. Carcasses with thick muscling will be graded one grade higher than indicated by their backfat thickness over the last rib except that carcasses with 1.75 inches or greater backfat thickness over the last rib must remain in the U.S. No. 4 grade. *Equation:* Carcass grade = (4.0 × backfat, in.) − (1.0 × muscling score [thin = 1, average = 2, thick = 3]).

Specifications for Grades of Barrow and Gilt Carcasses

U.S. No. 1

Barrow and gilt carcasses in this grade have an acceptable quality of lean and belly thickness and a high expected yield (60.4 percent and over) of four lean cuts. U.S. No. 1 barrow and gilt carcasses must have less than average backfat thickness over the last rib with average muscling or average backfat thickness over the last rib coupled with thick muscling.

Barrow and gilt carcasses with average muscling may be graded U.S. No. 1 if their backfat thickness over the last rib is less than 1 inch. Carcasses with thick muscling may be graded U.S. No. 1 if their backfat thickness over the last rib is less than 1.25 inches. Carcasses with thin muscling may not be graded U.S. No. 1.

U.S. No. 2

Barrow and gilt carcasses in this grade have an acceptable quality of lean and belly thickness and an average expected yield (57.4 to 60.3 percent) of four lean cuts. Carcasses with average backfat thickness over the last rib and average muscling, less than average backfat thickness over the last rib and thin muscling, and greater than average backfat thickness over the last rib and thick muscling will qualify for this grade.

Pork carcasses with average muscling may be graded U.S. No. 2 if their backfat thickness over the last rib is 1 to 1.24 inches. Carcasses with thick muscling may be graded U.S. No. 2 if their backfat thickness over the last rib is 1.25

to 1.49 inches. Carcasses with thin muscling must have less than 1 inch of backfat thickness over the last rib to be graded U.S. No. 2.

U.S. No. 3

Barrow and gilt carcasses in this grade have an acceptable quality of lean and belly thickness and a slightly low expected yield (54.4 to 57.3 percent) of four lean cuts. Carcasses with average muscling and more than average backfat thickness over the last rib, thin muscling and average backfat thickness over the last rib, and thick muscling and much greater than average backfat thickness over the last rib will qualify for this grade.

Barrow and gilt carcasses with average muscling will be graded U.S. No. 3 if their backfat thickness over the last rib is 1.25 to 1.49 inches. Carcasses with thick muscling will be graded U.S. No. 3 if their backfat thickness over the last rib is 1.5 to 1.74 inches. Carcasses with 1.75 inches or greater backfat thickness over the last rib cannot grade U.S. No. 3. Carcasses with thin muscling will be graded U.S. No. 3 if their backfat thickness over the last rib is 1 to 1.24 inches.

U.S. No. 4

Barrow and gilt carcasses in this grade have an acceptable quality of lean and belly thickness and low yield expected (less than 54.4 percent) of four lean cuts. Carcasses in U.S. No. 4 grade always have more than average backfat thickness over the last rib and thick, average, or thin muscling, depending on the degree to which the backfat thickness over the last rib exceeds the average.

Barrow and gilt carcasses with average muscling will be graded U.S. No. 4 if their backfat thickness over the last rib is 1.5 inches or greater. Carcasses with thick muscling will be graded U.S. No. 4 with backfat thickness over the last rib of 1.75 inches or greater, and those with thin muscling will be graded U.S. No. 4 with 1.25 inches or greater backfat thickness over the last rib.

U.S. Utility

All carcasses with unacceptable quality of lean and belly thickness will be graded U.S. Utility, regardless of their degree of muscling or backfat thickness over the last rib. Also, all carcasses which are soft and/or oily, or are pale, soft, and exudative (PSE), will be graded U.S. Utility.

In 1983, 1,880,000 pounds of federally inspected pork carcasses were graded by the USDA.

Live and Carcass Illustrations of Pork Carcass Grades

Illustrations depicting live and carcass representatives of the various pork carcass grades are provided by South Dakota State University. Photos were taken by Kevin W. Jones (Figures 11-54 through 11-62).

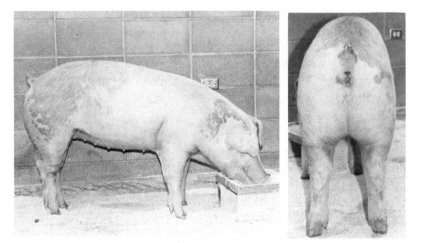

Fig. 11-54. A U.S. No. 1 grade gilt weighing 246 pounds. She is very trim in her jowls, shoulders, and middle. When she moved, her shoulder blades were clearly evident, indicating trimness. From the rear she is wider through the center of her hams than any other place on her body, indicating superior muscling. Over her top, her loins are rounded and smooth, further evidences of muscling and trimness. She is free of evidence of excess fat at the root of the tail and in the crotch.

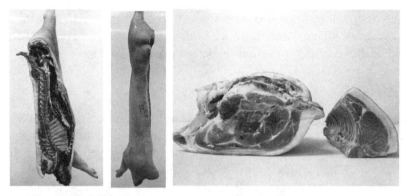

Fig. 11-55. A U.S. No. 1 grade carcass that has 0.95 inch of backfat at the last rib. Average to thick muscling is viewed over the back. The untrimmed ham and loin show a lack of excess fatness. The blade end of the loin exposed at the 10th rib shows 0.65 inch of fat cover and 5.80 square inches of loin-eye area.

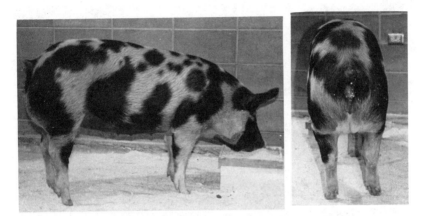

Fig. 11-56. A U.S. No. 2 grade gilt weighing 244 pounds. She shows evidence of having slightly more fat in her jowls, shoulders, and middle than the No. 1 grade gilt. From the rear some evidence of fat at the base of the tail and in the crotch is present. A significant amount of her thickness is due to fat, as indicated by the somewhat flattened turn of her top as compared to the No. 1 hog.

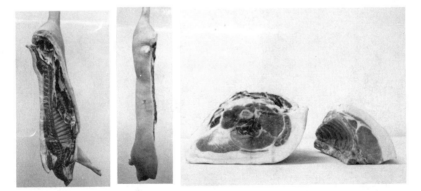

Fig. 11-57. A U.S. No. 2 grade carcass that has 1.22 inches of backfat at the last rib. Average muscling is viewed over the back. The untrimmed ham and loin show evidence of excessive fatness. The blade end of the loin exposed at the 10th rib shows 1.25 inches of fat cover and 4.50 square inches of loin-eye area.

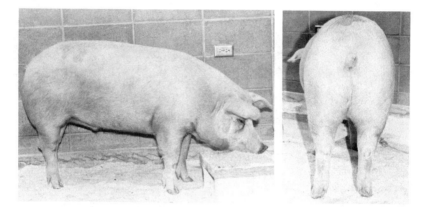

Fig. 11-58. A U.S. No. 3 grade barrow weighing 246 pounds. From the side he shows evidence of excess fat in his jowls and shoulders, over the edge of his loin, and in his belly. From the rear he shows some evidence of adequate muscling and a desirable rounded turn over his loins. His rear leg placement is somewhat narrow, indicating a lack of muscling.

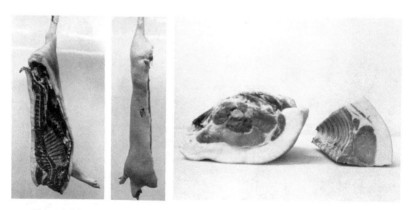

Fig. 11-59. A U.S. No. 3 grade carcass that has 1.35 inches of backfat at the last rib. Average muscling is viewed over the back. The untrimmed ham and loin show excessive fat. The blade end of the loin exposed at the 10th rib shows 1.45 inches of fat cover and 5.35 square inches of loin-eye area.

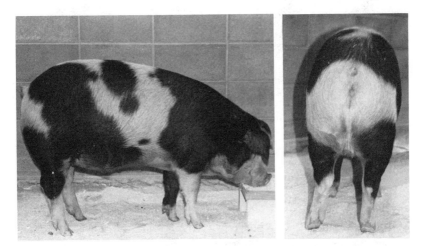

Fig. 11-60. A U.S. No. 4 grade gilt weighing 225 pounds. In the side view she clearly shows evidence of excess fatness in her jowls, shoulder, belly, and flanks. From the rear, excess fat is evident in the crotch. A serious lack of muscling is apparent from the rear, since her middle is wider than her hams.

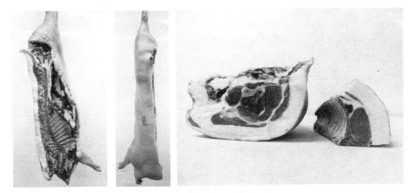

Fig. 11-61. A U.S. No. 4 grade carcass that has 1.30 inches of backfat at the last rib. Because of its thin muscling, one degree less than standard, the fat thickness is adjusted to 1.55, thus causing the carcass to grade No. 4. Noting the fatness all along the back, it can be seen that this carcass is leanest at the last rib. The untrimmed ham and loin clearly show excessive fat. The blade end of the loin exposed at the 10th rib shows 1.80 inches of fat cover and 4.50 square inches of loin-eye area. This is definitely a No. 4 carcass.

Fig. 11-62. Left to right. The blade end of loins exposed at the 10th rib from U.S. No. 1, No. 2, No. 3, and No. 4 grades pictured in Figures 11-55, 11-57, 11-59, and 11-61. Fat depth (in inches at the three-fourths loin-eye distance from the chine bone) and loin-eye area (in square inches), respectively: 0.65, 5.80; 1.25, 4.50; 1.45, 5.35; 1.80, 4.50.

Grade Standards for Sow Carcasses

The establishment of U.S. standards for grades of slaughter sows and sow carcasses became effective September, 1, 1956.

The grades are based on differences in yields of lean cuts and fat cuts and differences in quality of pork. In the development of the standards, sow carcass data were studied to establish relationships between measurements and carcass differences important in grading. As a result, these standards include average backfat thickness measurements as objective guides to grade. These principles are similar to those which have received widespread acceptance in the standards for barrows and gilts, adopted in 1952.

The five grades for sows and sow carcasses are U.S. No. 1, U.S. No. 2, U.S. No. 3, Medium, and Cull. The U.S. No. 1 grade includes sows and carcasses with about the minimum finish required to produce pork cuts of acceptable palatability. The U.S. No. 2 and U.S. No. 3 grades represent overfinish with resulting lower yields of lean and higher yields of fat. Medium and Cull are underfinished grades producing pork with low palatability.

The backfat measurements which qualify sow carcasses for the various grades are listed in Table 11-11.

Table 11-11. Backfat Measurement Guides to Grades
for Sow Carcasses

Average Backfat Thickness	Grade
(in.)	
1.5 to 1.9	U.S. No. 1
1.9 to 2.3	U.S. No. 2
2.3 or more	U.S. No. 3
1.1. to 1.5	Medium
Less than 1.1	Cull

What Makes Progress?

We must continue to improve by openly comparing our top-performing animals with established standards and with those owned by others. The USDA Agricultural Marketing Service, through the Federal Meat Grading and Certification Branch of its Livestock Division, gives us a set of uniform standards on which to make comparisons and which have been instrumental in fostering improvement. For the meat animal industry to continue to flourish, it must encourage the development of animals that produce high-quality edible product most efficiently.

By comparing the description of hogs and carcasses in this text with those in the 1974 and 1977 editions, one can readily see that progress has been made in increasing the number of lean, meaty hogs in the United States.

USDA'S CERTIFICATION SERVICE AND PRODUCT EXAMINATION SERVICE

In addition to grading carcasses and cuts, the meat graders of the Livestock Division of the USDA Agricultural Marketing Service provide two more services to the meat industry. Both services are paid for by the segment of the meat industry using them, and both are becoming more widely used in the United States.

Certification Service for Meat and Meat Products[13]

Since 1923, the Federal Meat Grading Service has assisted organizations such as government agencies, private institutions, and other purveyors of meats in their meat procurement programs. This service involves (1) assisting the purchaser in the development of specifications to assure accurate and uniform interpretation and (2) examining the product to assure its compliance with the specifications.

The Meat Certification Service is based on USDA approved Institutional Meat Purchase Specifications, commonly called IMPS. IMP Specifications are available for fresh beef, fresh lamb and mutton, fresh veal and calf, fresh pork, cured pork, cured beef, edible by-products, sausage products, and portion-cut meat products. (See Chapter 12.) Each item is numbered and belongs to a series according to the product—Series 100, for instance, is fresh beef, and Item 104 is an oven-ready rib. IMP Specifications can be requested from the Standardization and Review Branch of the Livestock Division.

Purchasers, be they hospitals, schools, restaurants, hotels, air lines, or

---

[13]*Marketing Bulletin No. 47.* Agricultural Marketing Service, USDA.

steamship lines, may ask suppliers to submit bids on products based on IMP Specifications.

When the purchaser requests delivery, the supplier asks the nearest USDA meat grading office to have a grader examine the product. The meat grader is responsible for accepting the product and certifying that it is in compliance with specifications.

The federal grader stamps each acceptable meat item, or the sealed carton in which it is contained, with a shield-shaped stamp (as illustrated below) bearing the words "USDA accepted as specified." This assures the purchaser that all products delivered met the requirements of the specifications at the time of acceptance.

This method of meat procurement assures the purchaser of a wholesome product (only meat that has passed inspection for wholesomeness will be examined for "acceptance") of the grade, trim, weight, and other options requested. This system also encourages competitive bidding and usually results in overall lower costs, permits long-range meal planning, and eliminates controversies between the buyer and seller over compliance of product.

In 1981, the USDA certified 1,839,356,000 pounds of meat, including 359,935,000 pounds for commercial firms, 878,933,000 pounds for the Department of Defense, and 600,488,000 pounds for federal purchases.[14]

Product Examination Service[15]

If a business involves meat shipments, there may be times when it is necessary to have an impartial expert officially establish a shipment's physical condition. This service is available to meat packers, wholesalers, brokers, carriers, and their insurance agents—in short, to anyone with a financial interest in the meat shipment who may want to substantiate a damage claim, protect against one, or save the time and expense of having to personally examine it. Product examinations are made on request, and a fee is charged for the service. The examination may be performed wherever the meat shipment is located—in packing plants, warehouses, trucks, or railroad cars—provided the meat is accessible to the meat grader.

---

[14]*Agricultural Outlook.* Dec. 1983. Economic Research Service, USDA.
[15]*Marketing Bulletin No. 55.* Agricultural Marketing Service, USDA.

During an examination, the grader impartially documents the facts about the physical condition of a meat or meat product shipment. The meat involved may be fresh or frozen carcasses or wholesale cuts of beef, pork, lamb, veal, or calf; sausage; smoked or cured meats; etc. The grader issues an official certificate attesting to the physical condition of the shipment. Such certificates are accepted as *prima facie* evidence by all federal and most state courts.

In a product examination, the meat grader reports only on physical conditions that can be accurately determined and described. For example, meat or meat products are examined for extent and kind of damage, freezer burn, thawing and refreezing, cleanliness, freshness, temperature, weight ranges, and fat thickness. Also, the temperature of the conveyance and the extent of container damage are determined. Management should specify the factors to be determined by the examination and whether it is to be conducted by random sample or by 100 percent examination of the product.

☞ NOTE: Meat that is unwholesome is not eligible for product examina- ☜ tion. If there is any question about the meat being fit for human consumption, the meat grader will refer the problem to USDA's Food Safety and Inspection Service (FSIS) or, if appropriate, to state and local authorities having jurisdiction over the wholesomeness of meat.

THE MEAT GRADER

Government meat graders are appointed from a list of eligibles submitted by the Civil Service Commission. To qualify, a candidate must have had at least three years of suitable practical experience in wholesale meat marketing and grading. A maximum of three years of appropriate college training (courses in meat and livestock judging) can be substituted for practical experience at the rate of one year of study for nine months of required experience. Before being given a permanent appointment as graders, the candidates must serve a probationary period of one year, during which time they are given intensive training in the application of standards, and their work is reviewed very carefully by a grading supervisor to ascertain their ability to do the job in strict conformance with the federal standards. The government takes precautions to obtain competent men and women of high integrity.

At the end of 1984, there were 7 GS-4 cooperative training students; 350 federal meat graders listed in the salary range of GS-5 to GS-9; and 65 supervisory grading personnel at the GS-11 through GS-15 levels, totaling 415 personnel.

In 1984, cooperative training students GS-4 received $12,427 per year, while trainees were employed at the GS-5 grade with an annual salary of $13,903. They participate in an intensive training program for five to six months before assuming a grading or reporting assignment. That training is accomplished at group meetings and assignments at two different field offices. Trainees are paid travel and subsistence allowances in addition to salary for approximately three months of the training period—while attending meetings and training at the first field office. They are eligible for promotion to the GS-7 ($17,221) a year after their original employment. They are eligible for promotion to the GS-9 grade ($21,066), which is the journeyperson level for most grader and reporter positions, after a year of satisfactory performance at the GS-7 grade. Higher-grade positions usually involve supervisory duties; selections for these openings are based on merit. Every grader's work is identified by a code in his or her grade stamp by means of a combination of letters of the alphabet. (See Figure 11-51.)

COST OF THE GRADING SERVICE

- The rate for grading service, periodically increasing due to inflation, is presently $26.40 per hour. The fees for service are based on the time required to render the service, including the time required for the preparation of certificates and travel of the official grader in connection with the performance of the service.

- Fees for grading performed on a weekly contract basis are $1,000 per calendar week (less any allowable credits) to cover up to 40 hours of weekly grading service, and at the regular rate prescribed in the first paragraph of this section for grading time in excess of 40 hours per week between the hours of 6:00 A.M. and 6:00 P.M. Monday through Friday. The fee increases to $34.40 per hour from 6:00 A.M. to 6:00 P.M. on Saturday or Sunday and to $52.80 per hour on a federal legal holiday.

THE STAMP INK IS HARMLESS

The ink used in stamping meat, whether it is the inspection stamp or the grade stamp, is made from a vegetable dye that is absolutely harmless and need not be trimmed off the meat.

The following inks are FSIS approved for grading and inspection stamps on beef, veal, pork, lamb, mutton, and chevon: A.C.M.I. Violet 31, Purple Marking Ink 98P, Violet Marking Ink 2-504, Purple Meat Branding Inks GL-31 and GL-35/75, and Meat Marking Violet Inks 3-90-1 and 3-90-4.

Ring Bologna
Knackwurst
Polish Sausage
Beerwurst
Blood and
  Tongue
Hamburger
  Loaf
Pickle and
  Pimento Loaf
Thuringer
Liverwurst
Headcheese
Fresh Pork
  Sausages

Bratwurst
Prasky
Mettwurst
Minced Ham
Bologna
Kiszka
Frankfurters
Mortadella
Minced Ham
Frankfurters
Pepper Loaf
Smoked
  Sausage Links

Sausages

Courtesy National Live Stock and Meat Board

Pork Color and Marbling Scores

(Illustrations framed in black are considered UNACCEPTABLE)

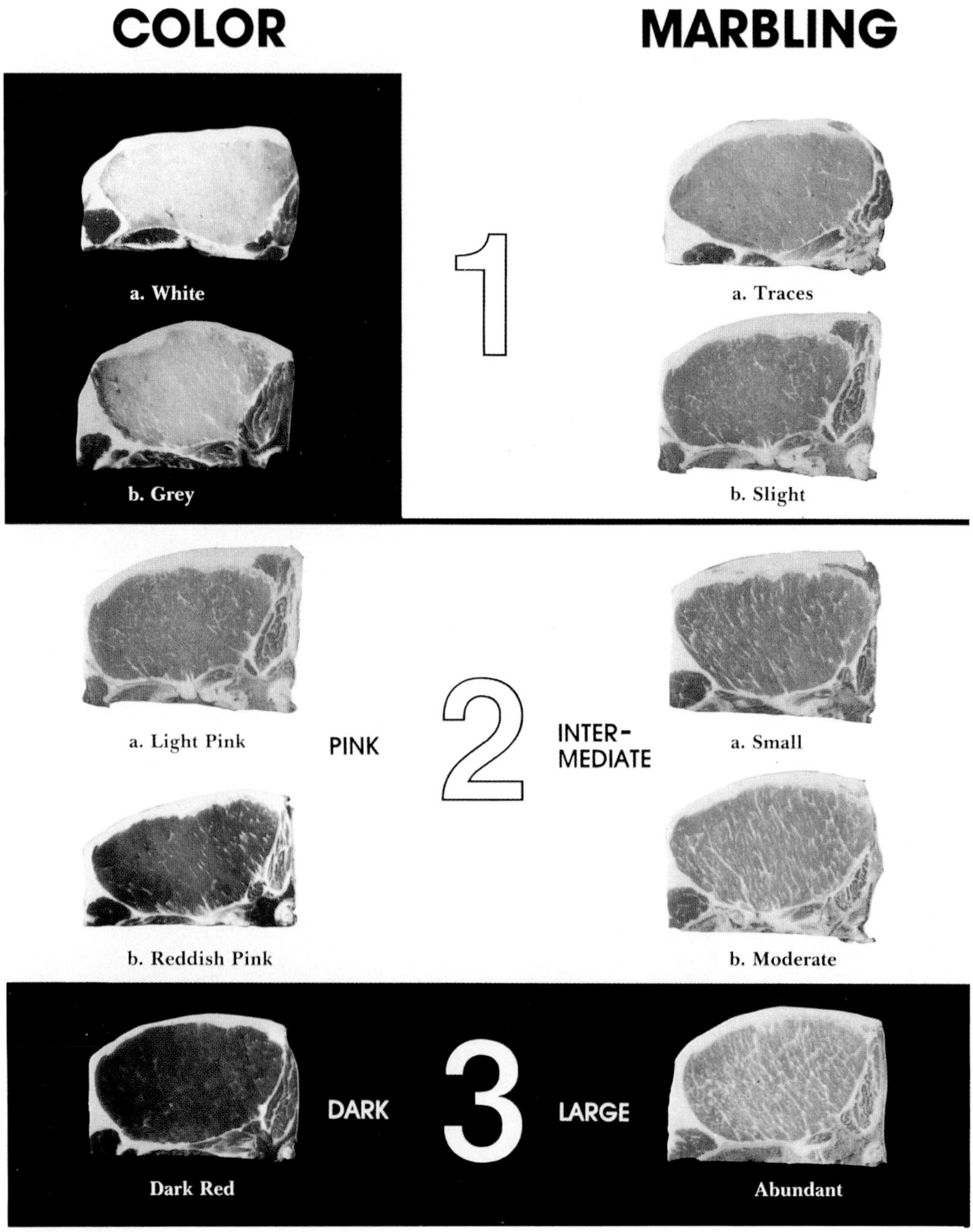

COLOR
MARBLING
a. White
b. Grey
1
a. Traces
b. Slight
a. Light Pink
b. Reddish Pink
PINK
2
INTER-
MEDIATE
a. Small
b. Moderate
Dark Red
DARK
3
LARGE
Abundant

Courtesy National Pork Producers Council, Des Moines, Iowa

Illustrations of Beef Marbling

Since marbling is such an important factor in judging beef, the following pictures illustrate the lower limits of six marbling degrees: Moderately Abundant, Slightly Abundant, Moderate, Modest, Small, and Slight.

It should be noted that there are nine degrees of marbling referred to in the Official United States Standards for Grades of Carcass Beef. These color photographs have been developed to assist government, industry, and academia in the proper application of official grade standards.

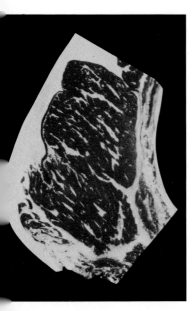

Moderately Abundant

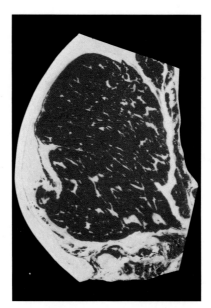

Slightly Abundant

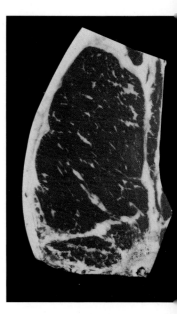

Moderate

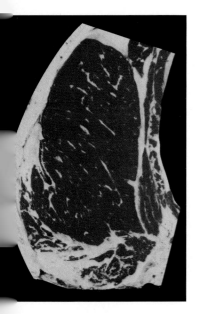

Modest

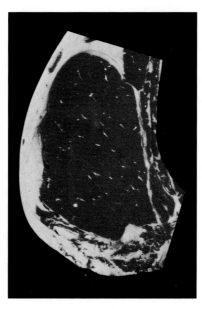

Small

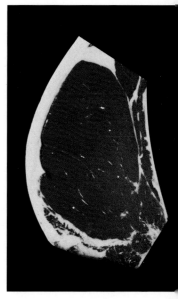

Slight

The above illustrations are reduced reproductions of the Official USDA Marbling Photographs prepared by the National Live Stock and Meat Board for the U.S. Department of Agriculture. For accuracy in evaluation, it is recommended that the official set be used.

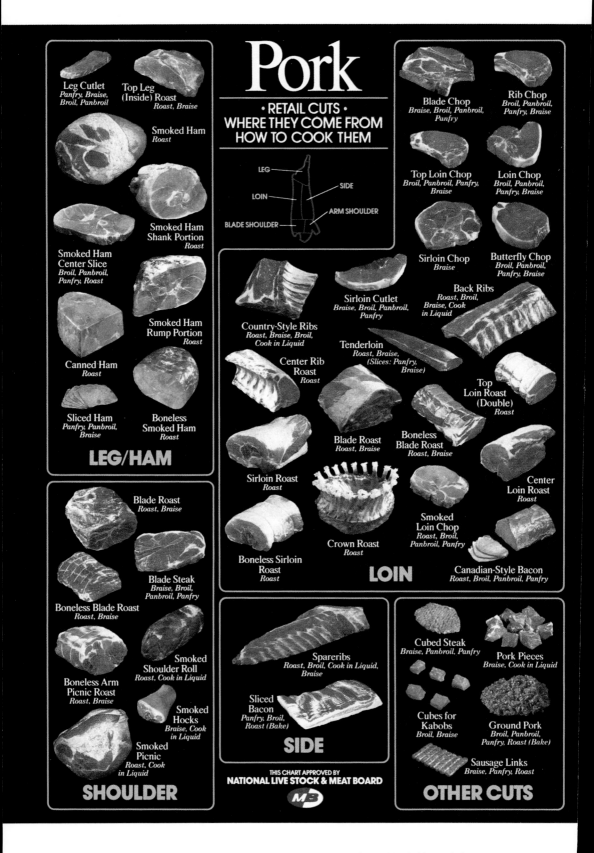

Pork

• RETAIL CUTS •
WHERE THEY COME FROM
HOW TO COOK THEM

LEG — SIDE
LOIN
ARM SHOULDER
BLADE SHOULDER

LEG/HAM

Leg Cutlet
Panfry, Braise, Broil, Panbroil

Top Leg (Inside) Roast
Roast, Braise

Smoked Ham
Roast

Smoked Ham Shank Portion
Roast

Smoked Ham Center Slice
Broil, Panbroil, Panfry, Roast

Smoked Ham Rump Portion
Roast

Canned Ham
Roast

Sliced Ham
Panfry, Panbroil, Braise

Boneless Smoked Ham
Roast

SHOULDER

Blade Roast
Roast, Braise

Blade Steak
Braise, Broil, Panbroil, Panfry

Boneless Blade Roast
Roast, Braise

Smoked Shoulder Roll
Roast, Cook in Liquid

Boneless Arm Picnic Roast
Roast, Braise

Smoked Hocks
Braise, Cook in Liquid

Smoked Picnic
Roast, Cook in Liquid

LOIN

Country-Style Ribs
Roast, Braise, Broil, Cook in Liquid

Center Rib Roast
Roast

Sirloin Cutlet
Braise, Broil, Panbroil, Panfry

Tenderloin
Roast, Braise, (Slices: Panfry, Braise)

Back Ribs
Roast, Broil, Braise, Cook in Liquid

Blade Chop
Braise, Broil, Panbroil, Panfry

Rib Chop
Broil, Panbroil, Panfry, Braise

Top Loin Chop
Broil, Panbroil, Panfry, Braise

Loin Chop
Broil, Panbroil, Panfry, Braise

Sirloin Chop
Braise

Butterfly Chop
Broil, Panbroil, Panfry, Braise

Blade Roast
Roast, Braise

Boneless Blade Roast
Roast, Braise

Top Loin Roast (Double)
Roast

Sirloin Roast
Roast

Boneless Sirloin Roast
Roast

Crown Roast
Roast

Smoked Loin Chop
Roast, Broil, Panbroil, Panfry

Center Loin Roast
Roast

Canadian-Style Bacon
Roast, Broil, Panbroil, Panfry

SIDE

Spareribs
Roast, Broil, Cook in Liquid, Braise

Sliced Bacon
Panfry, Broil, Roast (Bake)

OTHER CUTS

Cubed Steak
Braise, Panbroil, Panfry

Pork Pieces
Braise, Cook in Liquid

Cubes for Kabobs
Broil, Braise

Ground Pork
Broil, Panbroil, Panfry, Roast (Bake)

Sausage Links
Braise, Panfry, Roast

THIS CHART APPROVED BY
NATIONAL LIVE STOCK & MEAT BOARD

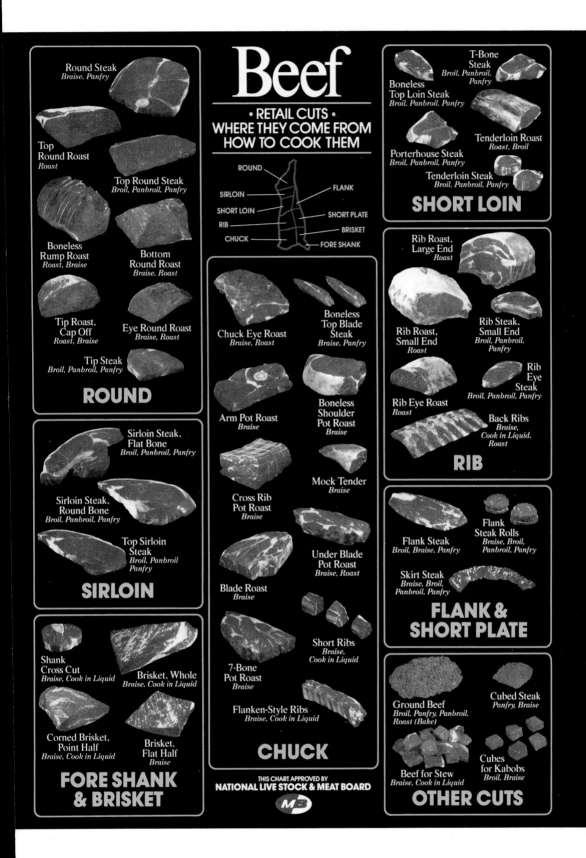

Beef

• RETAIL CUTS •
WHERE THEY COME FROM
HOW TO COOK THEM

ROUND
SIRLOIN
SHORT LOIN
RIB
CHUCK

FLANK
SHORT PLATE
BRISKET
FORE SHANK

ROUND

Round Steak
Braise, Panfry

Top Round Roast
Roast

Top Round Steak
Broil, Panbroil, Panfry

Boneless Rump Roast
Roast, Braise

Bottom Round Roast
Braise, Roast

Tip Roast, Cap Off
Roast, Braise

Eye Round Roast
Braise, Roast

Tip Steak
Broil, Panbroil, Panfry

SIRLOIN

Sirloin Steak, Flat Bone
Broil, Panbroil, Panfry

Sirloin Steak, Round Bone
Broil, Panbroil, Panfry

Top Sirloin Steak
Broil, Panbroil Panfry

FORE SHANK & BRISKET

Shank Cross Cut
Braise, Cook in Liquid

Brisket, Whole
Braise, Cook in Liquid

Corned Brisket, Point Half
Braise, Cook in Liquid

Brisket, Flat Half
Braise

CHUCK

Chuck Eye Roast
Braise, Roast

Boneless Top Blade Steak
Braise, Panfry

Arm Pot Roast
Braise

Boneless Shoulder Pot Roast
Braise

Mock Tender
Braise

Cross Rib Pot Roast
Braise

Under Blade Pot Roast
Braise, Roast

Blade Roast
Braise

Short Ribs
Braise, Cook in Liquid

7-Bone Pot Roast
Braise

Flanken-Style Ribs
Braise, Cook in Liquid

THIS CHART APPROVED BY
NATIONAL LIVE STOCK & MEAT BOARD

SHORT LOIN

T-Bone Steak
Broil, Panbroil, Panfry

Boneless Top Loin Steak
Broil, Panbroil, Panfry

Porterhouse Steak
Broil, Panbroil, Panfry

Tenderloin Roast
Roast, Broil

Tenderloin Steak
Broil, Panbroil, Panfry

RIB

Rib Roast, Large End
Roast

Rib Roast, Small End
Roast

Rib Steak, Small End
Broil, Panbroil, Panfry

Rib Eye Roast
Roast

Rib Eye Steak
Broil, Panbroil, Panfry

Back Ribs
Braise, Cook in Liquid, Roast

FLANK & SHORT PLATE

Flank Steak
Broil, Braise, Panfry

Flank Steak Rolls
Braise, Broil, Panbroil, Panfry

Skirt Steak
Braise, Broil, Panbroil, Panfry

OTHER CUTS

Ground Beef
Broil, Panfry, Panbroil, Roast (Bake)

Cubed Steak
Panfry, Braise

Beef for Stew
Braise, Cook in Liquid

Cubes for Kabobs
Broil, Braise

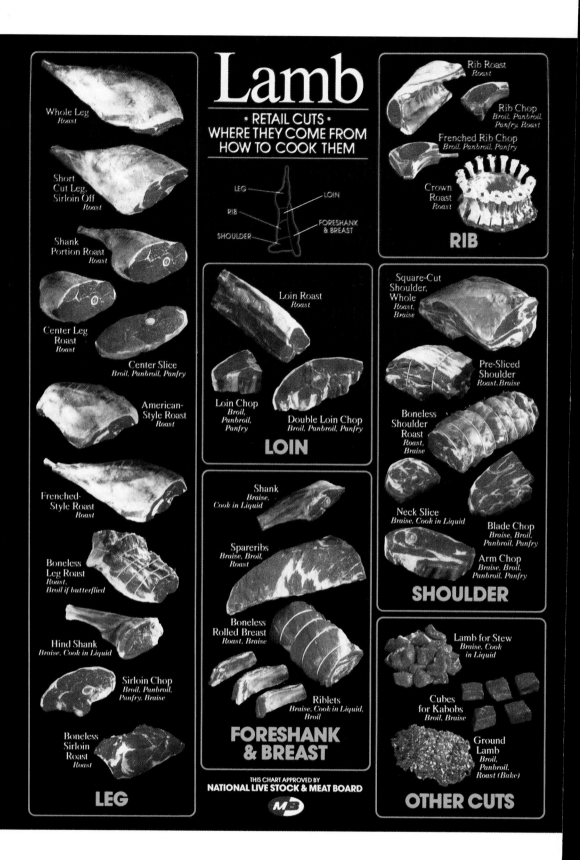

Lamb
· RETAIL CUTS ·
WHERE THEY COME FROM HOW TO COOK THEM

LEG
RIB
SHOULDER
LOIN
FORESHANK & BREAST

LEG

Whole Leg
Roast

Short Cut Leg, Sirloin Off
Roast

Shank Portion Roast
Roast

Center Leg Roast
Roast

Center Slice
Broil, Panbroil, Panfry

American-Style Roast
Roast

Frenched-Style Roast
Roast

Boneless Leg Roast
Roast, Broil if butterflied

Hind Shank
Braise, Cook in Liquid

Sirloin Chop
Broil, Panbroil, Panfry, Braise

Boneless Sirloin Roast
Roast

LOIN

Loin Roast
Roast

Loin Chop
Broil, Panbroil, Panfry

Double Loin Chop
Broil, Panbroil, Panfry

FORESHANK & BREAST

Shank
Braise, Cook in Liquid

Spareribs
Braise, Broil, Roast

Boneless Rolled Breast
Roast, Braise

Riblets
Braise, Cook in Liquid, Broil

RIB

Rib Roast
Roast

Rib Chop
Broil, Panbroil, Panfry, Roast

Frenched Rib Chop
Broil, Panbroil, Panfry

Crown Roast
Roast

SHOULDER

Square-Cut Shoulder, Whole
Roast, Braise

Pre-Sliced Shoulder
Roast, Braise

Boneless Shoulder Roast
Roast, Braise

Neck Slice
Braise, Cook in Liquid

Blade Chop
Braise, Broil, Panbroil, Panfry

Arm Chop
Braise, Broil, Panbroil, Panfry

OTHER CUTS

Lamb for Stew
Braise, Cook in Liquid

Cubes for Kabobs
Broil, Braise

Ground Lamb
Broil, Panbroil, Roast (Bake)

THIS CHART APPROVED BY
NATIONAL LIVE STOCK & MEAT BOARD

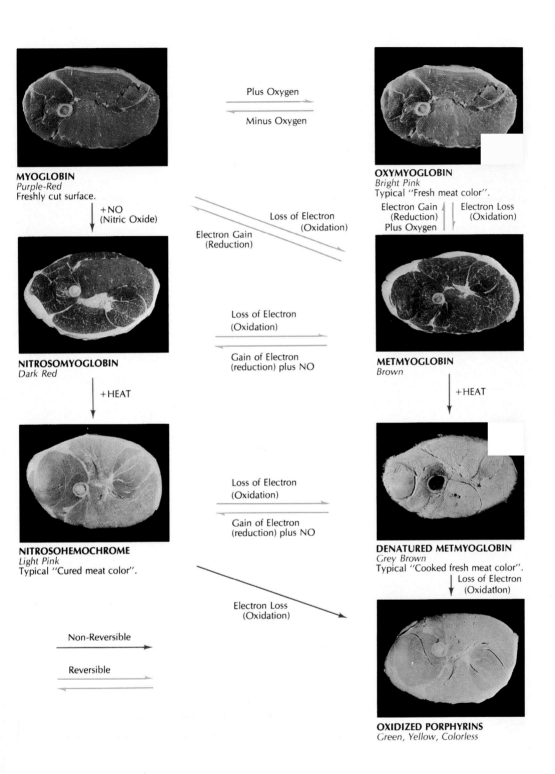

MYOGLOBIN
Purple-Red
Freshly cut surface.

Plus Oxygen

Minus Oxygen

OXYMYOGLOBIN
Bright Pink
Typical "Fresh meat color".

+NO
(Nitric Oxide)

Loss of Electron
(Oxidation)

Electron Gain
(Reduction)

Electron Gain
(Reduction)
Plus Oxygen

Electron Loss
(Oxidation)

NITROSOMYOGLOBIN
Dark Red

Loss of Electron
(Oxidation)

Gain of Electron
(reduction) plus NO

METMYOGLOBIN
Brown

+HEAT

+HEAT

Loss of Electron
(Oxidation)

Gain of Electron
(reduction) plus NO

NITROSOHEMOCHROME
Light Pink
Typical "Cured meat color".

DENATURED METMYOGLOBIN
Grey Brown
Typical "Cooked fresh meat color".
Loss of Electron
(Oxidation)

Electron Loss
(Oxidation)

Non-Reversible

Reversible

OXIDIZED PORPHYRINS
Green, Yellow, Colorless

Diagrammatic Table of Meat Color
Using Actual Photos of the Color Steps

The Meat Board
Beef Steak Color Guide

Degrees of Doneness

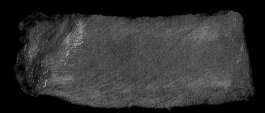

Very Rare
130° F. Approx. 55°C

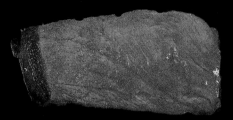

Rare
140° F. Approx. 60°C

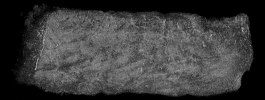

Medium Rare
150° F. Approx. 65°C

Medium
160° F. Approx. 70°C

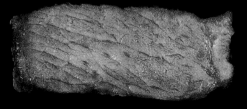

Well Done
170° F. Approx. 75°C

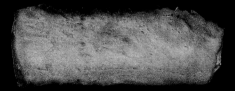

Very Well Done
180° F. Approx. 80°C

Official Guide Published by
National Live Stock & Meat Board
in cooperation with
Texas Agricultural Experiment Station

Chapter 12

Meat Merchandising

THE SOURCE OF SUPPLY IN THE UNITED STATES

MARKETING practices have changed considerably from the days of John Pynchon who is reputed to have been the first meat packer in the United States (Springfield, Massachusetts, 1641). The packing industry moved westward with the railroads. Cities such as Cincinnati, Chicago, and Kansas City became large packing centers. Herds of livestock were driven on foot for long distances to packing centers or to railroad loading points to be shipped to market. Much of the glamorized history of our West centers around railroads, Indians, settlers, gold, cowboys, and cattle drives.

Large stockyards were constructed to receive live shipments. Commission houses and banking facilities were necessary adjuncts for the buying and selling operation. Meat packers had their buyers at the large centers, or they operated through brokers. It was to govern illegal practices in the buying and selling of livestock that the Packers and Stockyards Act was passed. Contract buying became popular, under which livestock came direct to the packer holding pens, thus eliminating an intermediate handler. As trucking livestock took over, buyers went directly to the farmer or feeder, or to the auction which had become popular in livestock communities the country over. In recent years, the practice of dealing in futures and options has taken hold.

As livestock production moved westward from eastern population centers, distribution became a problem that was solved in 1880 by the invention of the refrigerator car. Unlike the mechanically refrigerated cars of today, these cars were cooled from ice bunkers (ice and salt) built into each end of the cars. These bunkers had to be refilled at different points along the route. A number of cars derailed because of the swaying of the suspended quarters of beef inside. Today, the refrigerator car has been supplemented and displaced by modern, mechanically refrigerated trucks.

Packing houses which originally stood on the outskirts of towns found themselves practically in the centers of cities. Many plants had become obsolete; traffic conditions were slow; taxes increased with land values; and, to keep the business profitable, old plants were dismantled and modern ones were erected closer to the areas of livestock production. This had an added advantage in that it cut down transportation costs. That is the story as of now.

IMPORTED MEAT

The Meat Import Act of 1979 establishes a maximum tonnage of meat that can be imported into the United States from other countries. The act also gives the president the power to adjust that maximum allowable imported tonnage as domestic meat production rises and falls. When U.S. production is high, imports are held to a lower level, when domestic production levels are lower, imports are still restricted, but at a higher than normal level. This "countercyclical" law is designed to minimize economic problems for U.S. meat producers and meat consumers.

During 1983, slightly more than 93 percent of the total red meat handled by the U.S. meat industry was produced in the United States, and less than 7 percent was imported from other countries. USDA data[1] indicate that approximately 3.1 percent more total pounds of red meat were imported in 1983 than in 1982. Slightly more than 73 percent of the total imported tonnage consisted of beef and veal. Australia and New Zealand are major sources of fresh beef, the bulk of which is manufacturing beef. Canada, Central America, and South America are other fresh beef sources. Argentina and Brazil provide much of the imported cured, corned, or other cooked beef.

Of the imported pork, which comprised 26 percent of total meat imports in 1983, nearly half arrived in the United States as cured or processed product, principally ham. Denmark, Poland, Yugoslavia, and the Netherlands are major sources of cured pork products, whereas Canada is the most common source of imported fresh pork. New Zealand and Australia provide most of the lamb entering the U.S. market. All imported meats must be slaughtered, processed, and shipped under conditions approved by and monitored by the U.S. Food Safety and Inspection Service. Ever since the inception of meat inspection, all imported meats have to be passed at the port of entry by FSIS before the importer can permit the product to be sold or used in the United States. (See Chapter 3.)

MEAT INDUSTRY BUSINESSES

Several types of business operations make up the industry that converts livestock to steaks, chops, roasts, and processed meat products. Conversion of the inedible portion of the livestock to useful products is also accomplished to a large extent by members of the meat industry. (See Chapter 10.) Delivery of the edible and inedible products to the ultimate consumer and user completes the function of the industry.

The era of John Pynchon established the designation *meat packer* or *packing house* for the commercial enterprise that starts with live animals and performs the necessary steps to make edible and convenient products for the con-

---

[1]As published by *Meat Facts*. 1984. American Meat Institute. See also *Agricultural Statistics*. 1983. USDA.

sumer. Those early wholesale provisioners packed the heavily salted chunks of meat into strong, tightly sealed barrels. Modern packers buy livestock and may perform the slaughtering, chilling, cutting, and processing procedures necessary to sell carcasses; primal (wholesale), sub-primal, and retail cuts; processed meats (cured meat, sausage, etc.); as well as all forms of by-products.

Large, multiplant packers may have distribution centers, branch houses, or complete packing houses throughout the United States. The larger packers sell in volume to wholesalers, large retail operators, processors, boners, breakers, and hotel-restaurant operators and suppliers. The large plants ship product long distances from the plants in livestock-production areas to the urban-population centers.

Other types of meat packers include those plants described as *kill* and *chill* operations which sell only carcasses and fresh or processed by-products. Smaller, independent packers are found in most areas of the United States, buying livestock and selling products within their own regions, often specializing in products which are local favorites. The smallest packers are frequently family operations that are localized in a single community and provide custom slaughtering and meat processing for livestock owners in the area. Frequently these operations sell fresh and frozen retail cuts and produce a number of processed meat items for retail sale. Though many of them no longer provide the frozen food locker rental service, they are still often referred to as *locker plants*.

The meat industry includes businesses called *boners*, which reduce carcasses and cuts to boneless product for further processing. Some boners purchase and slaughter mature cows or other lower-grade animals, whereas other boners purchase youthful, high-quality carcasses and wholesale cuts to provide raw material for their boning lines. Boners may sell some primal and sub-primal cuts, but serve principally as suppliers of boneless trim for further processing.

Many packers are also meat processors, but the term *processors* refers to operators who purchase carcasses, primal or sub-primal cuts, trimmings, variety meat items, and nonmeat ingredients. By applying meat-processing technology to the ingredients, the processor is able to sell the resulting products at a profit. Some processors buy all of these raw materials and sell a wide variety of processed products. More often, processors specialize in one operation, such as sausage making, curing operations, or frozen-food processing. There are single-product processors, an example of which would be the Kosher corned beef processor who purchases only beef briskets. Processors may sell at all wholesale levels or through their own specialty retail stores.

The meat industry also includes *wholesalers* and *distributors* who purchase meat products in large wholesale volume. The distributor's warehouse or delivery vehicle subsequently provides the smaller retail operation with a wide variety of products from which to order on a relatively small wholesale-volume basis. Wholesalers and distributors provide access to segments of the retail market that are more difficult for packers and processors to serve.

An important and growing segment of the meat industry are those operators who serve the special needs of the hotel-restaurant-institutional (HRI) meat buyers. The HRI meat *purveyors* buy carcasses, primals, sub-primals, boneless processing meats, and variety meats in the wholesale trade. Steaks, chops, roasts, patties, cubes for stew or kabobs, and many processed items are produced in compliance with rigid specifications for grade, weight, size, quality, and uniformity. The strict weight/size specifications are achieved by skilled "portion control" personnel in the HRI purveyors' cutting rooms. Meat patties formulated within tight compositional specifications are produced in various weights and shapes by the HRI purveyors. By purchasing from skilled and reputable HRI purveyors, the management of the foodservice unit of a hotel or plant cafeteria is assured that, with proper preparation and serving, each customer will receive a meat portion that is appetizing to the eye and satisfying to the palate.

MEAT RETAILING

Meat retailers are the final link in the chain between livestock producers and meat consumers. The consumer buys meat to prepare at home from a variety of retail operations or buys meat cooked and ready-to-eat from an equally wide variety of foodservice units.

Meat consumers in the United States make most of their purchases from the self-service meat counter of a supermarket. Fresh (unfrozen) meat cuts are displayed in a clear film packaging material that is sealed to a rigid back board. Each package displays a label which has the product identity (generally including species, wholesale cut, and retail cut); net weight; price per pound; and total value of the package clearly indicated. Meat department personnel may be available to answer questions and prepare special orders upon request in self-service meat departments. Some supermarkets, and many retail meat markets, have a "service" meat counter run by sales personnel skilled in meat merchandising. The focal point of a service meat department is a glass-fronted refrigerated case displaying a wide assortment of fresh and processed meats. The meat buyer may select cuts from the display or may request special orders, such as extra-thick steaks. The meat cutter then removes the selected cuts from the case or prepares the special order, weighs, wraps, and labels the packaged product with the cut identification and the value while the customer waits. The person behind a service counter is able to assist meat shoppers with suggestions concerning type of cut, amount required, and preparation techniques for specific situations.

The food freezers in a large number of households are re-supplied periodically with volume meat purchases. The freezer owner may choose to buy a whole carcass, a side, a quarter, a bundle, or several of each of a number of particular cuts. The retailer cuts, trims, wraps, labels, and freezes the purchase. The buyer may specify steak or chop thickness, roast size and weight, or number of items per package. Meat retailers often provide the freezer meat service as well as the self-service counter or service department. Some retail meat outlets serve the freezer meat trade exclusively. Packaged, frozen meat cuts are available in many retail outlets from supermarkets to convenience stores, but the majority of U.S. meat shoppers continue to prefer fresh meat cuts. Marketing of fresh and frozen meat items through vending machines may be one of the meat-retailing techniques of the future.

Specialty meat shops may not handle fresh meats but rather specialize in various brands of sausage (dry and semi-dry), cured smoked meats, domestic and foreign varieties of cheese, meat supplements, meat sauces, tenderizers, condiments, and various items that are relevant but different. Most specialty shops bear names indicative of the products handled; for example, "Smoke House," "Sausage Shop," "Gourmet Cave," "Missing Link," etc. These shops are selective in their buying and are always searching for the unusual. They experience some difficulty in getting some items, because they do not buy in large quantities unless they are a chain. Some specialty shops furnish their own recipes to a processor who makes the product for them.

Recent USDA[2] sources estimate that American consumers spent 28 percent of their household food dollars on meals and snacks away from home in the second quarter of 1984. When business meals were included, 34.5 percent of the food expenditures in the United States were spent in the nonhome foodservice setting in 1983. Cafeterias, fast-food counters, and all other eating establishments, therefore, retail a significant amount of meat and meat products. A 1979 USDA-sponsored[3] in-depth study of the foodservice industry credited food purchases of 49 billion pounds to the foodservice industry for resale to its customers. Of the 49 billion pounds, up 12 billion from the previous 1969 study, nearly 11 percent (5 billion pounds) was made up of red meat, variety meats, and meat products. Poultry represented approximately 3 percent and fish 2 percent of the total food purchases.

The U.S. meat industry offers the consumer a wide variety of purchase options. The consumer can select the option that provides the combination of product variety, product selection, convenience, service, and economy that best fits the individual situation. The competition among the segments of the industry assures the availability of quality product and good service at reasonable cost.

Meat Business Management

The management of meat operations at all levels of the industry is complex. Skillful buying, excellent and consistent quality of work, temperature control, sanitation, competitive pricing, and innovative merchandising are essentials of a

---

[2]*National Food Review.* Fall 1983. Vol. 24:27, and Fall 1984. Vol. 27:21, Economic Research Service, USDA.

[3]Michael G. Van Dress. *The Food Service Industry: Structure, Organization and Use of Food, Equipment and Supplies.* USDA, ERS Stat. Bul. No. 609.

good meat operation. Whereas most major industries are engaged in assembling sub-units into finished products such as appliances, furniture pieces, stereos, computers, or vehicles, the meat industry starts with intact animals and disassembles them into a wide variety of consumer products with an even wider range of individual values.

Most meat businesses have a range of purchase options. Slaughter operations may be selective of livestock type, weight, maturity, and grade. Processors and retailers may choose to buy carcasses, primal (wholesale), and sub-primal cuts (see meat identification section of this chapter) of various grades. (See Chapter 11.) By varying the degree of processing, the cutting procedure, or packaging technique, the processor or retailer may influence the value or the price of the unit sold. Close trimming increases the convenience and edible yield to the ultimate user, but involves higher investment in a lesser weight of saleable merchandise. The willingness of the buyer to pay a higher price for the added value must be adequate to compensate for the extra investment.

Inventory control is a constant problem in the meat industry, particularly with fresh meat. In a cutting operation, parts of the purchased unit are divided and sold to a variety of customers, some utilizing only one or two of the portions removed. Each different cut may have a different value. The demand for some cuts may be much greater than for other cuts; however, when a steer is killed, all of the parts are generated. Basic logic says that before another steer is killed, all of the parts of the preceding steer should be sold. How are all of the least desirable cuts of a steer sold in the same time interval as the most desirable portions? Pricing is the tool used to "control the flow" of the different-valued items. By pricing the items with a high-demand level relatively high and the low-demand items at a low level to encourage their purchase, the packers hope to match sales with supply availability. Accumulation of any sub-unit due to lack of a market or low demand can be economically disastrous in a meat operation.

Meat retail displays, processing areas, and personnel must be neat, clean, and sanitary, not only to instill confidence in the consumer but also to maximize shelf life and product appearance. Fresh meat in a self-service case must be sold within three to five days. If not, the color and overall appearance of the product will deteriorate, making it unsaleable.

Management and employees of a retail meat department should be aware of the income level, the ethnic background, the number of persons per living unit, and regional preferences of their customers. Meat-portion sizes should vary to correspond to the levels of physical activity and the ages of the meat consumers who purchase from a self-service case. Each different retail cut has unique palatability traits and edible portion yields which influence its relative value.

The retail meat manager must price the product in the case to operate profitably, to correspond with differences in product value, to account for demand differences unique to the customers, and to result in sales volume that corresponds to the rate of production in the operation.

Meat Pricing

Retail meat prices are the culmination of a series of marketing functions. Ultimately, meat pricing is determined by supply and demand, the availability of meat animals, and the relative availability of consumer disposable income for the purchase of meat. Variables which may influence one or both of the major price determinants include livestock production costs such as feed availability, interest rates, energy costs, and labor costs. Some of the same variables influence processing charges. The state of the economy has a major influence on demand, but the price of food options which are competitors for the consumers' protein dollars also influences the demand situation. Competition among different meat sources—beef, pork, lamb, poultry, and fish—may have a significant influence on meat pricing.

Meat-pricing Terms

Meat-pricing techniques utilize a number of yield factors and value adjustments. Terminology associated with those meat industry variables must be clearly understood in order to follow the development of meat prices. The following terms and definitions should be helpful to persons unfamiliar with meat-pricing principles.

Live Weight

Live weight refers to the weight of the live animal at the time of purchase by the slaughterer or at the time of slaughter. Sometimes it is referred to as *slaughter weight*.

Carcass Weight

Carcass weight is the weight of the carcass after all slaughter procedures are completed.

Hot Carcass Weight

Hot carcass weight is obtained immediately after slaughter and just before the carcass enters the chilling cooler. Most beef and lamb carcasses in commercial meat packer coolers are tagged with the hot carcass weight.

Chilled Carcass Weight

Chilled carcass weight can be obtained at any time after the carcass has completed the post-mortem chill period and before subdivision of the carcass is initiated. Chilled carcass weight is less than hot carcass weight due to moisture loss during cooling.

Dressing Percent

Dressing percent is that proportion of the live weight that remains in the carcass or carcasses of a given animal or group of animals. Mathematically, dressing percent = (carcass weight ÷ live weight) 100. Many persons in the meat and livestock industry refer to dressing percent as *yield*.

Grades

USDA and company (house) grades communicate expected value in the meat trade. Quality grades, yield grades, and carcass grades are discussed in detail in Chapter 11.

Cutting Yield

Cutting yield is the proportion of the weight of an original unit (carcass, side, primal cut, etc.) that is saleable product after the original unit has been trimmed or processed for sale.

By-product

By-product is a broad term that refers to those portions of a meat animal that are not composed mainly of skeletal muscle. Fat trimmings and bone are by-products of the retail cutting operation. Offal items are by-products of meat-animal slaughter.

Offal

Offal refers to all parts of the animal that are not a part of the carcass as it leaves the slaughter area. Edible offal items include heart, tongue, liver, head meat, tripe, etc. (See Chapter 10.) Inedible offal include hides, hair, digestive tract contents, and other items.

Drop Credit (Hide and Offal Value)

Drop credit is the value of the offal portion of the slaughter animal. The U.S. Crop and Livestock Reporting Service reports a value based on the prices of selected beef offal items. The weekly value is quoted as a marketing guideline in dollars per 100 pounds of live weight. The actual value of the offal to a specific packer will vary widely, depending on the degree of offal processing accomplished, the volume, and the access to available markets.

Slaughter Cost (Kill Cost)

Slaughter cost, or kill cost, is the total cost of the buying process, including transportation to the slaughter facility, slaughtering, chilling, transferring from the cooler, and handling the by-products of a slaughtered animal. Slaughter cost does not include the purchase value of the animal. Slaughter cost is often quoted on a per head basis, but may be based on weight in some cases.

Margin

Margin designates the portion of sales income that is utilized to operate the selling unit. Margin is the difference in product value between buying and selling. Margin is used to compensate for operating costs such as rents, labor, utilities, equipment purchases and maintenance, profit, and other miscellaneous expenses. Margin may be quoted in total dollars, in dollars per some base unit of sales (dollars/$1,000 sales), or as a percentage of sales income.

Markup

Markup is a portion of wholesale cost that is added to the wholesale cost to cover operating expenses. Markup differs from margin in that margin is based on sales value, and markup is based on purchase cost. That is an important difference when either item is quoted as a percentage or as a rate per $1,000.

Shrinkage

Shrinkage refers to the weight losses which may occur throughout the processing sequence. Cooler shrinkage = hot carcass weight − cold carcass weight. Cooler shrinkage, transportation shrinkage, and storage shrinkage are due, in large part, to moisture loss from fresh and processed product. Cutting shrinkage, trimming shrinkage, and cooking shrinkage, or losses, are reductions in tissue mass as well as water losses. The losses may be accounted for or identified by the complementary term *yield* in each case.

Yield

Yield is the portion of the original weight which remains following any processing or handling procedure in the meat-merchandising sequence. Dressing percent = slaughter yield. Yield is usually quoted in percent and may be cited for each of the steps that were discussed as shrinkages.

Price and Value

The pricing discussion in this chapter will use the term *price* on a per unit basis (cents/pound), whereas the term *value* refers to total cost (worth) of a cut, of carcasses, of a load, etc. (price × weight = value).

Tests (Cutting, Yield, Price, Shrink, Labor)

The purpose of tests in meat operations is to provide actual information under working conditions that will aid in making management decisions. Because testing is an information-gathering process, a report form is necessary to record data. The data then can be evaluated and used to compare sources of supply, raw material options, effects of purchase cost variation, processing procedures, product mix, worker efficiency, pricing procedures, and volume effects. Biological variation of animals, the wide range of cut types possible, and the constant fluctuation of prices at every level of the meat industry make testing on a regular basis essential to all meat operations. Continual updating of test data is required to monitor the effectiveness of computer-assisted meat industry management.

Forms for recording test data for three meat animal species are provided in Figures 12-1, 12-2, and 12-3. The forms include information blanks which relate to live animals, carcasses, wholesale cuts, and retail cuts. Margin and markup calculations are also illustrated. A packer, processor, purveyor, or retailer may use only portions of the forms shown that pertain to the data that are gathered and recorded at certain points in the marketing sequence.

Where possible, tests should report data on several of each wholesale unit being tested in order to reflect the variation in wholesale product. Tests should be designed and conducted to account for variations in technique and the ability of employees who are, or will be, involved in actual production. Pricing or decision errors may result from tests which are not true indicators of production conditions.

Tests are extremely important when procedural changes, equipment changes, pricing changes, or personnel changes are made. If repeat testing develops predictable patterns, testing frequency may be reduced. Testing requires time and effort; therefore, its effectiveness should be reviewed in relation to the cost of testing.

Cutting tests not only provide information about operations but also are used by management to determine prices. For example, a retail meat department manager may want to determine the influence on retail prices of an increase in wholesale meat cost. If the meat department operated on a 20 percent margin, by the earlier definition, 20 percent of total retail sales income was used to pay for in-department costs, expenses, and profits. Therefore, 80 percent of the total sales income could be used to buy wholesale meat.

The increase in wholesale costs to the meat department in the example may be an increase in beef rib wholesale price from $150 per 100 pounds to $165 per

BEEF CUTTING TEST

Cutting date _____ Live animal ID _____ Carcass ID _____

Breed _____ Sex _____

Live animal weight _____ Price _____ Sales value _____

By-products: List _____ Estimated value _____

Slaughter date _____ Hot carcass weight R _____ L _____ Total _____

Dressing percent _____

GRADING INFORMATION

Maturity _____ Marbling_____ Quality grade _____

Fat thickness _____ Kidney, pelvic, and heart fat weight _____ Percent _____

Rib-eye area _____ Yield grade _____

I. WHOLESALE

Item	Weight	Percent of Side	Price	Value	Remarks
Side					
Forequarter					
Hindquarter					
Heart fat					
Kidney & pelvic fat only					
Kidney					
Round ⎱ may be combined					
Rump[a] ⎰ for pricing					
Loin					
Rib					
Chuck[b]					
Flank					
Plate					
Brisket					
Foreshank[b]					
Totals					
Average price					
Cutting loss or gain					

Fig. 12-1

Beef Markup and Margin Calculations

(Based on One Side of Beef [Half Animal])

Live Animal to Carcass

Sales value of a hanging side	$ _____
Live cost × 0.5 (half animal)	$ _____
Margin per side (sales value − live cost)	$ _____
By-product value per animal × 0.5 (half animal)	$ _____
Adjusted margin/side (margin + by-product value)	$ _____
Percent markup	
(Adjusted margin ÷ live cost of half animal) 100	_____ %
Markup per pound	
(Adjusted margin ÷ pounds of half animal purchased)	$ _____
Percent margin	
(Adjusted margin ÷ sales value of side) 100	_____ %
Margin per pound	
(Adjusted margin ÷ pounds of one side sold)	$ _____

Carcass to Wholesale Cuts

Cost (value) of hanging side	$ _____
Summed value of all wholesale cuts	$ _____
Margin (wholesale cut value − side cost)	$ _____
Percent markup per side	
(Margin ÷ cost of side) 100	_____ %
Markup per pound	
(Margin ÷ side weight purchased)	$ _____
Percent margin per side	
(Margin ÷ wholesale cut value) 100	_____ %
Margin per pound	
(Margin ÷ pounds of wholesale cuts sold)	$ _____

II. RETAIL (PRIMAL CUTS)

Item	Weight	Percent of Wholesale Cut	Price	Value	Item	Weight	Percent of Wholesale Cut	Price	Value
Round (wholesale)					Bone				
Top					Totals				
Bottom					Rump (wholesale)[a]				
Tip					Boneless roast				
Heel									
Boneless rump roast[a]					Lean				
					Fat				
Lean					Bone				
Fat					Totals				

Fig. 12-1 (continued)

II. RETAIL (PRIMAL CUTS) (Continued)

Item	Weight	Percent of Wholesale Cut	Price	Value	Item	Weight	Percent of Wholesale Cut	Price	Value
Loin (wholesale)					Fat				
Steaks					Bone				
Wedge bone					Totals				
Round bone					Chuck, square cut or arm (wholesale)[b]				
Flat bone (double)					Cross rib				
Pin bone					Blade roasts				
Top-sirloin boneless					Arm roasts				
Porterhouse					Seven-bone roasts				
T-bone					Shoulder pot roasts (boneless)				
Top-loin bone-in					Chuck eye				
Top-loin boneless					Back ribs				
Tenderloin					Soup bones				
Lean					Cross-cuts foreshank[b]				
Fat					Stew				
Bone					Lean				
Totals					Fat				
Rib (wholesale)					Bone				
Short standing rib roast					Totals				
Rib steaks					Primal summary		(% of side)		
Small end					Roasts & steaks				
Large end					Other retail cuts				
Boneless					Lean				
Rib-eye steaks					(Retail) Subtotal				
Back ribs					Fat				
Short ribs					Bone				
Lean					Totals				

Fig. 12-1 (continued)

II. RETAIL (ROUGH CUTS) (Continued)

Item	Weight	Percent of Wholesale Cut	Price	Value
Flank (wholesale)				
Flank steak				
Lean				
Fat				
Bone				
Totals				
Brisket (wholesale)				
Boneless roast[c]				
Lean				
Fat				
Bone				
Totals				
Plate (wholesale)				
Short ribs				
Pastrami piece				
Boneless plate				
Boiling beef				
Lean				
Fat				
Bone				

Item	Weight	Percent of Wholesale Cut	Price	Value
Totals				
Foreshank (wholesale)[b]				
Cross-cuts				
Lean				
Fat				
Bone				
Totals				

III. SUMMARY (TOTALS FOR SIDE)

Item	Weight	Percent of Side	Price	Value
Roasts & steaks				
Other retail cuts				
Lean				
(Retail) Subtotal				
Fat (incl. KPH)				
Bone				
Totals				
Error		(compared to side)		

Fig. 12-1 (continued)

Beef Retail Markup and Margin Calculations

(Based on One Side of Beef [Half Animal])

Live Animal to Retail Product

Retail value	$ _____
Live cost × 0.5 (half animal)	$ _____
Margin per side (retail value − live cost)	$ _____
By-product value per animal × 0.5 (half animal)	$ _____
Adjusted margin per side (margin + by-product value)	$ _____
Percent markup	
(Adjusted margin ÷ live cost of half animal) 100	_____ %
Percent margin	
(Adjusted margin ÷ retail value of one side) 100	_____ %
Markup per lb.	
(Adjusted margin ÷ pounds of half animal purchased)	$ _____
Margin per lb.	
(Adjusted margin ÷ pounds of retail product sold from half carcass)	$ _____

Hanging Side to Retail Product

Retail value	$ _____
Cost of hanging side	$ _____
Margin (retail value − cost of side)	$ _____
Percent markup	
(Margin ÷ side cost) 100	_____ %
Percent margin	
(Margin ÷ retail sales value) 100	_____ %
Markup per lb.	
(Margin ÷ pounds purchased)	$ _____
Margin per lb.	
(Margin ÷ pounds sold)	$ _____

Footnotes for Figure 12-1.

[a]The rump is usually part of the round but may be handled as separate wholesale cuts.

[b]Chucks are traded as arm chucks, with the foreshank on or as square-cut chucks, the foreshank being a separate wholesale cut. Square-cut chuck and foreshank may be summed to determine the wholesale price of the foreshank based on arm-chuck price, by subtracting out the value of the square-cut chuck, leaving the balance for foreshank value.

[c]Curing and smoking charges to produce corned beef from the brisket or other cuts are:

_____ ¢/pound.

Carcass fabricator _____

Data taker _____

Sources of all prices: Live _____

Wholesale _____

Retail _____

Fig. 12-1 (continued)

PORK CUTTING TEST

Cutting date _____ Live animal ID _____ Carcass ID _____

Breed _____ Sex _____

Live animal weight _____ Price _____ Sales value _____

By-products: List _____ Estimated value _____

Slaughter date _____ Hot carcass weight R _____ L _____ Total _____

Dressing percent _____

GRADING INFORMATION

Backfat: Last lumbar _____ Last rib _____ First rib _____ Total _____ Average _____

Fat depth at 10th rib _____ Muscling score _____ Carcass length _____

Loin-eye area _____ Carcass grade _____

I. WHOLESALE

Item	Weight	Percent of Side	Price	Value	Remarks
Side					
Ham					
Loin					
Boston shoulder					
Picnic shoulder					
Belly					
Spareribs					
Neck bones					
Jowl					
Front feet					
Hind feet					
Tail					
Fatback					
Clear plate					
Kidney fat					
Other fat					
Lean trim 50/50 (belly)					
Lean trim 90/10					
Totals					
Average price					
Cutting loss or gain					

Fig. 12-2

Pork Markup and Margin Calculations

Live to Wholesale Cuts from One Side
  Summed wholesale cuts sales value $ _____
  Live cost × 0.5 (half animal) $ _____
  Margin (wholesale value − live cost) $ _____
  By-product value × 0.5 (half animal) $ _____
  Adjusted margin (margin + by-product value) $ _____
  Percent markup
    (Adjusted margin ÷ live cost of half animal) 100   _____ %
  Markup per pound
    (Adjusted margin ÷ pounds of half animal purchased) $ _____
  Percent margin
    (Adjusted margin ÷ sales value of wholesale cuts) 100   _____ %
  Margin per pound
    (Adjusted margin ÷ pounds of wholesale cuts from side) $ _____

II. RETAIL FROM LEAN CUTS

Item	Weight	Percent of Wholesale Cut	Price	Value	Item	Weight	Percent of Wholesale Cut	Price	Value
Ham (wholesale)					Sirloin chops				
Ham (fresh, bone-in)[a]					Boneless sirloin roast				
Rump portion					Boneless blade-end roast				
Shank portion					Butterfly chops				
Center slices					Tenderloin				
Boneless					Pocket chops				
Lean[a]					Country ribs				
Fat					Back ribs				
Skin					Lean[a]				
Bone					Fat				
Totals					Bone				
Loin (wholesale)					Totals				
Center-cut loin chops					Boston shoulder (wholesale)[a]				
Center-cut rib chops					Blade steaks				
Blade chops					Bone-in roast				
Blade roast bone-in					Boneless roast				

Fig. 12-2 (continued)

II. RETAIL FROM LEAN CUTS (Continued)

Item	Weight	Percent of Wholesale Cut	Price	Value
Shish kabobs				
Lean[a]				
Fat				
Bone				
Totals				
Picnic shoulder (wholesale)[a]				
Arm steaks				
Pork hocks				
Bone-in roast				
Boneless roast				
Sausage trim[a]				
Lean[a]				
Fat				
Skin				
Bone				
Totals				
Lean cut summary		(% of side)		
Roast				
Steaks				
Chops				
Other retail cuts				
Lean				
(Retail) Subtotal				
Fat				
Bone				
Totals				

III. BELLY AND MINOR CUTS

Item	Weight	Percent of Wholesale Cut	Price	Value
Belly (wholesale)[a]				
Fresh side pork				
Salt pork				
Slab bacon				
Sliced bacon				
Totals				
Spareribs				
Neck bones				
Jowl[a]				
Front feet				
Hind feet				
Tail				
Fatback[b]				
Clear plate[b]				
Salt pork				
All other fat				

IV. SUMMARY (TOTALS FOR SIDE)

Item	Weight	Percent of Side	Price	Value
Totals: Roasts, chops, &/or steaks				
Other retail cuts				
Lean[a]				
(Retail) Subtotal				
Fat				
Bone				
Grand Total				
Cutting loss or gain				

Fig. 12-2 (continued)

Pork Markup and Margin Calculations

(Based on One Side of Pork [Half Animal])

Live to Retail

Retail value (summed from one side) $ _____
Live cost × 0.5 (half animal) $ _____
Margin (retail value − live cost) $ _____
By-product value × 0.5 (half animal) $ _____
Adjusted margin (margin + by-product value) $ _____
Percent markup
  (Adjusted margin ÷ cost of half animal) 100   _____ %
Percent margin
  (Adjusted margin ÷ retail value of one side)   _____ %
Markup per pound
  (Adjusted margin ÷ pounds of half animal purchased) $ _____
Margin per pound
  (Adjusted margin ÷ pounds of retail product) $ _____

Wholesale to Retail

Retail value (summed from one side) $ _____
Cost of summed wholesale cuts (one side) $ _____
Margin (retail value − wholesale cost) $ _____
Percent markup
  (Margin ÷ cost of wholesale cuts) 100   _____ %
Percent margin
  (Margin ÷ retail value) 100   _____ %
Markup per pound
  (Margin ÷ pounds of wholesale cuts) $ _____
Margin per pound
  (Margin ÷ pounds of retail product) 100 $ _____

Footnotes for Figure 12-2.

[a]Processing from the fresh state costs: curing and smoking _____ ¢/pound, grinding and seasoning for sausage, _____ ¢/pounds.
[b]Rendering pork fat into lard costs _____ ¢/pound plus the rendering shrink which approximates 25 percent.

Carcass fabricator _____

Data taker _____

Sources of all prices: Live _____ Wholesale _____ Retail _____

Fig. 12-2 (continued)

LAMB CUTTING TEST

Cutting date _____ Live animal ID _____ Carcass ID _____

Breed _____ Sex _____

Live animal weight _____ Price _____ Sales value (cost) _____

By-products: List _____ Estimated value _____

Slaughter date _____ Hot carcass weight _____ Dressing percent _____

GRADING INFORMATION

Maturity _____ Flank streaking _____ Flank fullness & firmness _____

Conformation _____ Quality grade _____ Leg conformation_____

Fat thickness: Top loin _____ Lower rib _____ Rib-eye area _____

Kidney and pelvic fat weight _____ Percent _____ Yield grade _____

I. WHOLESALE

Item	Weight	Percent of Side or Carcass	Price	Value	Remarks
Side or carcass					
Kidney & pelvic fat only					
Kidney					
Leg					
Loin					
Rib (rack)					
Shoulder					
Flank ⎫ may be combined Breast ⎬ for pricing shank ⎭					
Spleen					
Totals			▓		
Average price	▓	▓	▓	▓	
Cutting loss or gain					

Fig. 12-3

Lamb Markup and Margin Calculations

Live to Wholesale

Sales value of hanging side or carcass	$ _____
Live cost*	$ _____
Margin (sales value − live cost)	$ _____
By-product value*	$ _____
Adjusted margin (margin + by-product value)	$ _____
Percent markup (Adjusted margin ÷ live cost) 100*	_____ %
Markup per pound (Adjusted margin ÷ pounds of animal)*	$ _____
Percent margin (Adjusted margin ÷ sales value of carcass) 100*	_____ %
Margin per pound (Adjusted margin ÷ pounds of carcass sold)*	$ _____

*If carcasses are split and sold as sides, multiply values, cost, or weights for the animal by 0.5.

II. RETAIL (PRIMAL CUTS)

Item	Weight	Percent of Wholesale Cut	Price	Value	Item	Weight	Percent of Wholesale Cut	Price	Value
Kidney & pelvic fat only					Fat				
Kidney					Bone				
Leg					Totals				
Frenched leg					Rib (rack)				
American leg					Rib chops				
Boneless leg					Rib roast (crown)				
Leg steaks					Lean				
Shish kabobs					Fat				
Lean					Bone				
Fat					Totals				
Bone					Shoulder				
Totals					Blade chops				
Loin					Arm chops				
Loin chops					Boneless shoulder				
Lean									

Fig. 12-3 (continued)

II. RETAIL (PRIMAL CUTS) (Continued)

Item	Weight	Percent of Wholesale Cut	Price	Value
Saratoga chops				
Shish kabobs				
Lean				
Fat				
Bone				
Totals				

II. RETAIL (ROUGH CUTS) (Continued)

Item	Weight	Percent of Wholesale Cut	Price	Value
Flank				
Lean				
Fat				
Bone				
Totals				
Breast				
Spareribs				
Riblets				
Boneless breast				
Lean				
Fat				
Bone				
Totals				
Foreshank				
Lamb shanks				

Item	Weight	Percent of Wholesale Cut	Price	Value
Lean				
Fat				
Bone				
Totals				
Neck				
Neck slices				
Lean				
Fat				
Bone				
Totals				

III. SUMMARY (TOTALS FOR SIDE)

Item	Weight	Percent of Side	Price	Value
Roasts, chops, &/or steaks				
Other retail cuts				
Lean				
(Retail) Subtotal				
Fat (incl. KPH)				
Bone				
Totals				
Error (compared to side or carcass)				

Fig. 12-3 (continued)

Lamb Markup and Margin Calculations

Live to Retail

Retail value summed from carcass or side $ _____

Live cost* $ _____

Margin (retail value − live cost) $ _____

By-product value* $ _____

Adjusted margin (margin + by-product value) $ _____

Percent markup
    (Adjusted margin ÷ live cost) 100*   _____ %

Markup per pound
    (Adjusted margin ÷ pounds of animal purchased)* $ _____

Percent margin
    (Adjusted margin ÷ retail value) 100*   _____ %

Margin per pound
    (Adjusted margin ÷ pounds of retail product) $ _____

Wholesale to Retail

Retail value summed from all product* $ _____

Wholesale cost of hanging carcass or side $ _____

Margin (retail value − wholesale cost) $ _____

Percent markup
    (Margin ÷ cost of carcass or side) 100   _____ %

Markup per pound
    (Margin ÷ pounds of carcass or side) $ _____

Percent margin
    (Margin ÷ retail value, summed) 100*   _____ %

Margin per pound
    (Margin ÷ pounds of retail product) $ _____

---

*If carcasses are split and sold as sides, multiply values, cost, or weights for the animal by 0.5.

Carcass fabricator _____

Data taker _____

Sources of all prices: Live _____

                 Wholesale _____

                 Retail _____

Fig. 12-3 (continued)

100 pounds, due to a combination of season change and decreased beef supplies. Since the wholesale cost represented 80 percent of the expected retail value of the rib, the manager determined that the retail value of a 30-pound beef rib must be: (30 × $1.65) ÷ 0.80 = $61.88.

Figure 12-4 is an adaptation of the "Rib (wholesale)" section of Figure 12-1 and illustrates the results of a cutting test the meat department manager performed on one rib. The retail prices were determined by combining relative desirability of the retail cuts, prices in competitive stores, and what the manager thought consumers would be willing to pay for the various retail cuts. If the retail prices necessary to achieve the required margin were too high, the manager would have had to make adjustments. For instance, if competitors in the area were selling small-end steaks for $2.85 per pound, it is unlikely that large volumes of steak would have moved at $2.92 per pound. If small-end steaks had been reduced to $2.85 per pound, the retail value of the rib would have been reduced $0.42 to $61.54, subsequently decreasing the margin to 19.54 percent. The manager could have either re-priced one of the other retail portions of the rib to make up the difference or increased the price of some other item in the retail case to maintain the level of margin. Modification of the cutting procedure by cutting the roast into steaks or by cutting a portion of the rib into boneless steaks may have resulted in a greater retail return. Often consumers are willing to pay more for a high-yielding, "classy," boneless cut.

The manager would have reduced the display of beef rib retail cuts if beef rib wholesale prices had reached the level that made it impossible to achieve margin requirements at the maximum retail prices the normal consumer was willing to pay. The space normally occupied by rib cuts would have been filled with alternatives at a price the consumer was willing to pay. The result would have been a decrease in wholesale demand for beef ribs and probably downward pressure on the wholesale rib price. At the same time, the retailer would have been attempting to maintain total sales volume by merchandising more competitively priced meat items.

The effect of margin and yield on pricing can be demonstrated using the cutting test above. The wholesale price increase of $15/100 pounds ($150/100 lb. to $165/100 lb.) may be adjusted to account for the margin: $15 ÷ (1−margin) = $15 ÷ .80 = $18.75/100 pounds. Assuming the total 30 pounds of wholesale rib is saleable, the retail price of rib cuts would be increased an average of 18¾¢ per pound. However, in many retail situations, the fat and bone (3.5 pounds) have little or no value. Saleable product, therefore, accounts for 88.3 percent of the original weight (26.5 lb./30 lb.). Correcting the retail price change for yield: $18.75/100 pounds ÷ yield = 18.75 ÷ .883 = $21.23/100 pounds. Each pound of retail product from the rib must produce 21¼¢ more return to the retailer if the wholesale cost of ribs increases $15 per 100 pounds (15¢ per pound). It is unlikely that the retail price of ground beef or short ribs can be immediately increased 21¢ per pound. Therefore, the higher-priced steaks and roasts must be increased more than the average 21¢ per pound to achieve the required retail val-

RETAIL CUTTING TEST FOR BEEF RIBS

Cut by _____ WJ _____

Item _____ 103 rib _____

No. of item _____ 1 _____

Wholesale price ___ $165 / 100 pounds ___

Expected retail value (20% margin) ___ $61.88 ___

Date _____ 7/2/84 _____

Source ___ Z Packing Co. ___

Weight _____ 30 pounds _____

Wholesale cost ___ $49.50 ___

	Weight	Percent of Wholesale Cut	Retail Price	Value
Rib (wholesale)	30 lb.	100%	$1.65	$49.50
Short standing rib roast	4.0	13.3	2.62	10.48
Rib steaks				
Small end	6.0	20.0	2.92	17.52
Large end	10.0	33.3	2.77	27.70
Boneless				
Rib-eye steaks				
Back ribs				
Short ribs	5.5	18.3	.85	4.67
Lean	1.0	3.0	1.59	1.59
Fat	1.5	5.0	–	
Bone	2.0	6.7	–	
Totals	30.0	99.6	–	61.96

Actual retail value _____ $61.96 _____

Wholesale cost _____ $49.50 _____

Margin _____ $12.46 _____

Margin as percent of sales _____ 20.11% _____

Fig. 12-4

ue. The cutting test would be a more effective tool for adjusting prices than attempting to add on to prices as discussed in this paragraph.

Animal-Carcass—Retail Price Relationship

Urban consumers may not understand why steak prices may be $3.69 per pound when the farmer is selling market beef cattle for less than $0.75 per pound. Many persons in the livestock and meat industry may not be able to fully explain the seemingly wide discrepancy between live prices and retail prices.

The data in Table 12-1 illustrate some reasons for the differences in meat prices at different points in the meat-marketing process. A 1,000-pound Choice, Yield Grade 3 steer priced at $65.75 per hundred pounds cost the packer $657.50. The USDA price sources indicated that the carcass of that steer was priced at $103.75 per 100 pounds. Assuming that the steer had a dressing percentage of 62 percent, the 620-pound carcass had a wholesale value of $643.25. The by-product value, quoted by the USDA source, was $5.85 per 100 pounds of live weight. The 1,000-pound steer yielded by-product valued at $58.50, which, when added to the carcass value, totaled $701.75. The meat packer sold the carcass and by-product of the steer for $44.25 more than the cost of the live steer. The packer may have decided to sell the carcass as quarters. If so, the value of the carcass would have been $657.20. When the by-product value was added,

Table 12-1. Beef Price, Weight, and Value Relationships

Item	Price[1]	Weight	Value
	($/100 lb.)	(lb.)	($)
Steer, Choice, YG 3	65.75	1,000	657.50
Carcass, steer, Choice, YG 3	103.75	620	643.25
By-product value	5.85[2]	1,000	58.50
Carcass + by-product			701.75
Margin to packer as carcass			44.25
Forequarters	91.00	322.4[3]	293.38
Hindquarters	122.25	297.6[3]	363.82
Carcass as quarters			657.20
Carcass as quarters + by-product			715.70
Margin to packer as quarters			58.20
Carcass, retail value (20% margin)		620	804.06[4]
Carcass, retail weight (75% yield)[5]		465	804.06
Carcass, retail weight (60% yield)[5]		372	804.06
Hindquarters, wholesale	122.25	620[6]	757.95
Hindquarters, retail (20% margin)		620	947.44[4]
Hindquarters, retail weight (60% yield)[5]		372	947.44

[1]Mean of four weekly averages, USDA *Market News*. June and July 1983.

[2]USDA quoted on basis of per hundred pounds live weight.

[3]Forequarters 52 percent, hindquarters 48 percent of carcass weight.

[4]Retail value = wholesale value ÷ (1 − margin).

[5]Approximate cutting yields: 75 percent bone-in, 60 percent boneless.

[6]Equals carcass weight to demonstrate value differences.

the total value of the steer was $715.70, as quarters. The packer's margin was increased to $58.20 by choosing to sell that carcass as four separate units rather than as one unit.

The retailer who purchased the steer carcass would have paid $643.25 for it, providing there were no shipping costs. If the retail unit operated on a 20 percent margin, the wholesale value, $643.25, represented 80 percent of the total value of retail cuts from that carcass. Therefore, the retail value was $804.06 ($643.25/0.80). To produce attractive, saleable retail cuts, the retailer trimmed away excess bone, fat, and connective tissue. The total weight of saleable retail cuts from a beef carcass will vary from approximately 60 percent to 75 percent of the carcass weight. The degree of trimming and the proportion of boneless cuts will determine yield. Therefore, the original 620-pound carcass produced from 372 to 465 pounds of retail cuts, which had a total value of $804.06.

A cutting test would have been used to establish specific retail-cut prices and to determine optimum cutting procedures for the particular retail situation. Some beef cuts, such as pot roasts, are not popular during the summer. The less-desired cuts were economically priced to encourage consumers to purchase them. To compensate, the more highly demanded steaks were marked with relatively high prices in order to achieve the $804.06 retail value.

The difference between the retail value, $804.06, and the live-steer value, $657.50, was used by the packers and the retailer to pay the costs of slaughter, chilling, cutting, packaging, transportation, and other costs. The value increase, coupled with the weight decrease from 1,000 pounds of steer to as low as 372 pounds of meat after the nonmeat parts of the animal were removed, resulted in large price per pound increases. The average retail price of beef from the carcass is $2.16 per pound ($804.06/372 lb.).

The retailer may have elected to purchase hindquarters only in order to increase the proportion of higher-valued steak cuts in the retail case. Purchasing an equal amount, 620 pounds, of hindquarters only would also have resulted in an equal number of pounds of retail product, 370 to 460 pounds, depending upon trimming and boning procedures. The retail value of the hindquarter cuts would have been $947.44 [(620 lb. × $122.25/100 lb.) ÷ 0.80]. To achieve an average price of $2.55 per pound ($947/372 lb.), some steaks would have been priced in excess of $3.50 per pound to compensate for ground beef and other lower-value cuts.

Few specific retail prices have been quoted in the examples above. Wide variation exists in meat retail pricing as a result of store-to-store differences in relative demand, cutting procedures, margins, volume, by-product value, and many other factors. Meat prices are constantly changing at each level in the meat-marketing system. Successful managers at all levels must be prepared to adjust to the changes.

Livestock price increases usually result in carcass and wholesale prices moving to higher levels, and, depending on the demand situation, meat retail prices will reflect the upward trend. Similarly, downward movements in live price may gradually travel through the supply chain to the consumer. Demand reductions due to price increases, income loss, or diet changes result in volume reductions at the retail meat case which are soon detected in decreased demand and strong downward pressure on prices at the wholesale level. The slaughterer generally passes the effects of demand reduction on to the livestock producer as lower prices.

Fresh meat is a perishable item and must be moved through marketing channels rather quickly. Negative demand trends fill the "pipeline" quickly and often result in rapid price decreases to salvage the product or clear out the system. The perishable food industry has an old adage: "Sell it or smell it."

Seasonal variations in price relationships may occur due to variation in livestock and meat supply; however, consumer preferences do have major influences on the seasonal changes in meat prices. Table 12-2 demonstrates the beef-price relationship changes that occur between winter and summer. The summer prices are the prices used in Table 12-1 and the example discussed previously. The winter prices were reported by the USDA as averages for the week ending

Table 12-2. Beef Price, Weight, and Value Seasonal Relationships

Item	Weight	Summer		Winter	
		Price[1]	Value	Price[2]	Value
	(lb.)	*($/100 lb.)*	*($)*	*($/100 lb.)*	*($)*
Steer, live	1,000	65.75	657.50	64.00	640.00
Carcass	620	103.75	643.25	103.75	643.25
By-product	–	5.85[3]	58.50	5.75	57.50
Carcass + by-product	–	–	701.75	–	700.75
Margin to packer	–	–	44.25	–	60.75
Forequarters	322.4[4]	91.00	293.38	98.50	317.56
Hindquarters	297.6[4]	122.25	363.82	115.50	343.73
Carcass (quarters)	620	–	657.20	–	661.29
Carcass + by-product	–	–	715.70	–	718.79
Margin to packer/carcass	–	–	58.20	–	78.79
Rounds	138.9[5]	111.20	154.46	122.50	170.15
Loins	106.6	170.60	181.86	130.00	138.58
Flanks	32.2	53.10	17.10	56.00	18.03
Arm chucks	184.8	86.50	159.85	101.00	186.65
Ribs	59.5	163.75	97.43	137.50	81.81
Plates	51.5	53.25	27.42	57.50	29.61
Briskets	23.5	57.75	13.57	64.00	15.04
Fats, kidneys, etc.	22.3	10.00[6]	2.23	10.00[6]	2.23
Carcass (primals)	619.3	–	653.92	–	642.10
Margin to packer[7]	–	–	54.92	–	59.60

[1]Mean of four weekly averages, USDA *Market News*. June and July 1983.

[2]Weekly averages, USDA *Market News*. Jan. 16, 1982.

[3]USDA quoted on basis of per hundred pounds live weight.

[4]Forequarters 52 percent, hindquarters 48 percent of carcass weight.

[5]Wholesale cut weights calculated from percents quoted by the National Live Stock and Meat Board in *A Steer Is Not All Steak*.

[6]Estimated price.

[7]Carcass value (primals) + by-product value – live value = margin to packer.

January 16, 1982, and were selected because the Choice, Yield Grade 3 steer-carcass price on that report was the same as the summer carcass price, $103.75 per 100 pounds. This comparison illustrates that carcass prices and livestock prices vary somewhat independently, since the live prices in this case differed $1.75 per 100 pounds while the carcass prices were identical.

Summer is steak season. In the winter, beef roasts become more popular meat items in many households. The $30 spread between forequarters and hindquarters in summer 1983 compares to a $17 spread in winter 1982, while carcass prices were identical. Wholesale loins from identically priced carcasses were $40 per 100 pounds more valuable, and wholesale ribs were $26 per 100 pounds more valuable during the summer steak season than during the winter. Primal (wholesale) loins and ribs together represented 42.7 percent of the carcass value during the summer, whereas in winter, the two cuts accounted for only 34.3 percent of the total carcass value. The prices increased for chucks ($15 per 100 pounds) and for rounds ($11 per 100 pounds) in the winter, the roast season.

MEAT CUTS

Wholesale and Retail Cuts

The meat industry in the United States has adopted a system of standardized *wholesale*, or *primal*, cuts for each species. Wholesale cuts result from the subdivision of the carcass into more easily handled units. Guidelines long used by the meat industry in making both wholesale and retail cuts are:

- To separate tender portions of the carcass from less-tender areas.
- To separate lean areas from the portions having greater amounts of fat.
- To separate thicker, more heavily muscled portions of the carcass from the thin-muscled areas.

Wholesale or primal cutting separates the legs, made up of large locomotion-muscle systems, from the back, composed of large support-muscle systems, from the thinner body-wall sections of the carcass. Subdivision of the carcass into wholesale cuts permits the purchase of only the part(s) of the carcass which may be best adapted to the buyer's purpose; for example, corned beef processors purchase only beef briskets. Wholesale cuts have been defined as large subdivisions of the carcass which are traded in volume by segments of the meat industry. Relatively rigid standardization of wholesale cuts has been established for many years to provide efficient communication between high-volume buyers and sellers.

Retail cuts can be defined as subdivisions of wholesale cuts or carcasses which are sold to the consumer in a ready-to-cook or ready-to-eat form. A large selection of different retail cuts are found in American retail-meat cases due to the innovative nature of meat merchandisers. The consumer can quite easily learn to identify cuts from different species by differences in cut size, color, and fat characteristics. Beef cuts are large and have a cherry red color and a white, firm fat. Pork cuts are intermediate in size, tend to be a grayish pink color, and have the softest fat. Lamb cuts will be small in size, will be dark pink to light red in color, and will have hard, white fat. Consumers are also able to classify most of the bone-in cuts into one of seven types based on muscle/bone shape and size relationships. Familiarity with the seven types will enable the meat buyer to identify the source location in the carcass for meat cuts, therefore making it possible to predict relative palatability as well as to identify the retail-cut name (Figure 12-5).

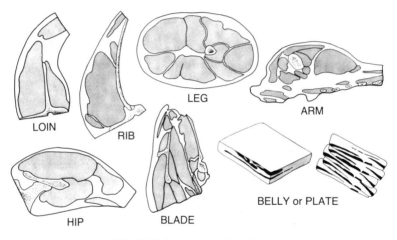

LEG

ARM

LOIN

RIB

BELLY or PLATE

HIP

BLADE

Fig. 12-5. Seven types of retail cuts.

The seven retail-cut types include two cuts which have a cross section of the *longissimus* or "eye" muscle of the back as the major muscle component. *Loin* cuts often contain a section of the tenderloin muscle *(psoas major)* in addition to the *longissimus*. The *lumbar* vertebra (loin) has the typical T-bone configuration formed by the lateral process, the body and the dorsal process of the vertebra. *Rib* cuts have no tenderloin section and often exhibit a section of rib bone.

The cross-section of a round bone is typical of two cut types. The *leg* muscle configuration of top *(semimembranosus)*, bottom *(biceps femoris)*, eye *(semitendinosus)*, and sirloin tip *(quadriceps)* is easily distinguished from the more diffuse arrangement of small muscles in the *arm* cuts.

Two types of cuts include flat or irregularly shaped bones. The *hip* cuts are composed of a small number of relatively large, parallel muscles. The *blade* cuts display numerous small, non-parallel muscles typical of the foreleg system.

The seventh cut type includes those cuts that are best recognized by the alternating layers of fat, lean, and rib bones which make up the body wall. Those *belly* or *plate* cuts would include bacon, spareribs, and beef brisket.

If the identification system outlined above is utilized, an experienced meat consumer will develop the ability to recognize muscle size and shape relationships that will aid in the identification and carcass location of many boneless cuts. It should be noted that the seven types of retail cuts do not include all of the retail cuts that are available. (See color photos of cuts and Chapters 13 thru 16.)

MEAT CUT STANDARDS

Uniform Retail Meat Identity Standards

Presently in the United States, there are estimated to be more than 1,000 names to identify the approximately 300 fresh cuts of beef, pork, lamb, and veal offered for sale at the retail level. Because of this situation, the Industrial Cooperative Meat Identification Standards Committee (ICMISC) was formed in mid 1972, composed of representatives of the entire meat industry, including meat packers and processors and all different sizes of grocery and supermarket operations. Livestock producers were represented through the National Live Stock and Meat Board.

This committee worked diligently to produce a most needed document for the U.S. meat industry. It is entitled *Uniform Retail Meat Identity Standards* and is published and distributed at cost by:

Department of Merchandising
National Live Stock and Meat Board
444 North Michigan Avenue
Chicago, Illinois 60611

This document includes a master list of 314 identifications for retail cuts of all species and is designed to serve all marketing areas in the United States.

The meat label information recommended by the ICMISC includes:

- The kind (species) of meat—beef, pork, lamb, or veal. It is listed first on every label.
- The primal (wholesale) cut—loin, leg (ham), shoulder, chuck, etc.—tells where the meat comes from on the animal.
- The retail cut—blade roast, spareribs, loin chops, etc.—tells what part of the primal cut the meat comes from.
- Ground beef—not less than X percent lean—tells what percentage of lean in ground beef the retailer will guarantee. "X" in this case may be any figure the retailer wishes to put there. According to federal law, ground beef and hamburger must be at least 70 percent lean, but the area between 70 and 100 percent is entirely up to the retailer.

If desired, the retailer may put any familiar "fanciful" name on a label after any of these four items. Further, if cooking instructions are given, chances for the consumer to "go wrong" are significantly lessened.

This uniform identity system is strictly voluntary, not government controlled in any way. Some states already had identification laws on the books before these standards were published. In order for the system to be truly "uniform" and thus serve *all* the people, regardless of where they move or relocate in the United States, it must be widely used.

In this text, the Uniform Retail Meat Identity Standards are used. Also some of the more common "fanciful" names are added where appropriate in the following chapters. The retail cuts in the colored illustrations of beef, lamb, and pork near the center of this text are identified by these standards.

Institutional Meat Purchase Specifications

Wholesale cuts are often trimmed or portioned into intermediate units larger than conventional retail cuts. The resulting *sub-primal* cuts are in demand by foodservice units, institutions, fresh meat retailers, and meat processors. The boxed-meat programs used by much of the fresh meat industry are often based on the production, vacuum packaging, and boxing of sub-primal cuts.

The meat industry recognized the marketing problems inherent in the virtually unlimited options for shape, composition, and degree of trim introduced by the sub-primal and boxed-meat programs. The industry has adopted and is using a set of guidelines to standardize carcasses, primal cuts, sub-primal cuts, and retail cuts.

The Agricultural Marketing Service, USDA, has prepared a series of Institutional Meat Purchase Specifications (IMPS), as follows:

- *Series 100* (fresh beef)
- *Series 200* (fresh lamb and mutton)
- *Series 300* (fresh veal and calf)
- *Series 400* (fresh pork)

- *Series 500* (cured, cured and smoked, and fully cooked pork products)
- *Series 600* (cured and dried and smoked beef products)
- *Series 700* (edible by-products)

- *Series 800* (sausage products)
- *Series 1,000* (portion-cut meat products)

These specifications contain detailed indexing and descriptions of the various products customarily purchased in volume. Copies of the specifications for these products may be purchased from the Superintendent of Documents.

The National Association of Meat Purveyors, Tucson, Arizona, has published *The Meat Buyers Guide*, which contains excellent descriptive color pictures and accompanying discussion for the *Series 100, 200, 300, 400,* and *1,000* USDA specifications.

Universal Product Code, Retail Meat Cuts

A new concept in retail meat standardization has been developed by the National Live Stock and Meat Board and the Food Marketing Institute in cooperation with several retail food chains and equipment manufacturers. The system is an adaptation of the Universal Product Code (UPC) system to retail meat cuts. The UPC system is a series of lines and spaces printed on many uniform weight and labeled products. Random-weight products have presented problems in the development of scanning technology in the UPC system. The Meat Board and the Food Marketing Institute have encouraged scale and scanning equipment manufacturers to develop and to make available to food chains the equipment necessary to handle random-weight coding. The coded area of the package label will inform the in-store computer of the species; the primal, sub-primal, and retail cuts; the packer source; and other items of information the retailer may choose to code into the system. Meat-cut labels are then read by the same type of scanner equipment used by store personnel at the checkout, during case inventory checks, and for monitoring movement of product through the whole marketing sequence. Not only does the UPC system make the checkout procedure more accurate and efficient, but also it provides retail managers with data related to product flow, consumer preferences, and meat department efficiency.

The National Live Stock and Meat Board produces many useful resources which illustrate meat cuts and recommended preparation and cooking procedures. Examples of these sources have been used throughout this text. Copies of materials may be obtained by contacting the National Live Stock and Meat Board at the address listed earlier in this chapter.

Chapter 13

Pork Identification and Fabrication

PORK does not normally leave the packing plant in carcass form but is fabricated at the plant. Historically this procedure differed from that used in beef, veal, and lamb distribution where whole carcasses were shipped from the packing plant to distribution centers. Now more fabrication of carcasses of all these species, as well as pork, is taking place right in the plant where the animals are slaughtered.

Why is this chapter necessary if most commercially slaughtered pork is cut into wholesale and retail cuts before it leaves the plant? Pork has been, and will continue to be, the traditional "farmer's" meat. The pig is a simple-stomached creature which could not be considered a fussy eater. Many hogs have been, and probably will continue to be, raised in small pens and outbuildings throughout the United States to provide pork for the family. The pig is small enough to make home slaughter feasible with minimum facilities. Home raised or locally purchased hogs are the primary reason for the existence of a large number of local lockers or slaughter plants in many communities. This chapter will make the consumers of the home raised pork chops and hams aware of how the pork carcass can be properly processed into chops and roasts either at home or in a federally or state-inspected processing facility. This chapter will also make the reader aware of the relationship between structure and function of the various sections of the animal and the palatability of the product. Students and consumers will find some excellent meat identification tips in the next four chapters.

Most pork carcasses are chilled overnight, and the chilled carcasses move from the cooler to a continuous chain cutting operation. The cutting room must have a capacity equal to the slaughter rate in the plant and may process more than 1,000 pork carcasses per hour. Mechanization, specialization, automation, and skill are as important on the pork-disassembly lines as they are on the vehicle-assembly lines. Two thirds of the fresh pork cuts and trimmings produced by the carcass-disassembly process are transferred to other areas of the plant for further processing. Hams, shoulders, bellies, jowls, spareribs, and neck bones may be cured, smoked, and cooked. Lean trim is incorporated into many sausage items. Fat trim is rendered into lard and cracklings. Because processed pork products have been such an important part of the American diet, and because all of the processing procedures can be accomplished more efficiently and with greater uniformity on a large scale, the slaughter plant has cut the pork carcass into parts since the beginning of the meat packing industry in the United States. (See Chapter 12.)

Although most of the pork processed today is chilled before cutting, hot processing of pork carcasses is currently occurring in some plants. Cutting the pork carcass before the animal heat has been removed saves the energy required to cool and reheat those portions of the carcass that are heat processed. (See Chapter 18.) Cured and smoked meats are cooked, lard is heated during rendering, and many sausage products are smoked and cooked. Those products that are sold fresh must be chilled immediately after the carcass has been subdivided. Most plants are designed for the chilling period following slaughter today, but the hot processing of pork and other carcasses may be adopted by more of the industry in the future.

The method of cutting pork is similar in all sections of the United States, even though there may be differences in regional preferences for pork cuts. Please refer to the section on meat cuts in Chapter 12 for information on general cut types and sources of information about meat cuts. The following material will detail specific cutting procedures for pork.

FABRICATION

The skeletal structure of the porcine in relation to an outline of the wholesale cuts is shown in Figure 13-1, and a more detailed chart of wholesale and retail cuts and recommended cooking methods appears as Figure 13-2. Use these two figures for reference as the discussion moves through cutting a pork carcass. Refer to the color photos of the retail cuts of pork found near the center of the text. Figures 13-3 through 13-67 appear through the courtesy of the University of Illinois and South Dakota State University.

The authors and Miss Teann Garnant, an SDSU student and meat lab employee, demonstrate the fabrication steps in the following figures.

Generally, when hogs are finished for market they will weigh *approximately* 240 pounds. After slaughter, the resulting carcass represents approximately 70 percent of the live weight. Therefore, a 240-pound live hog will produce a carcass weighing approximately 168 pounds.

PORK CHART
LOCATION, STRUCTURE
AND NAMES OF BONES

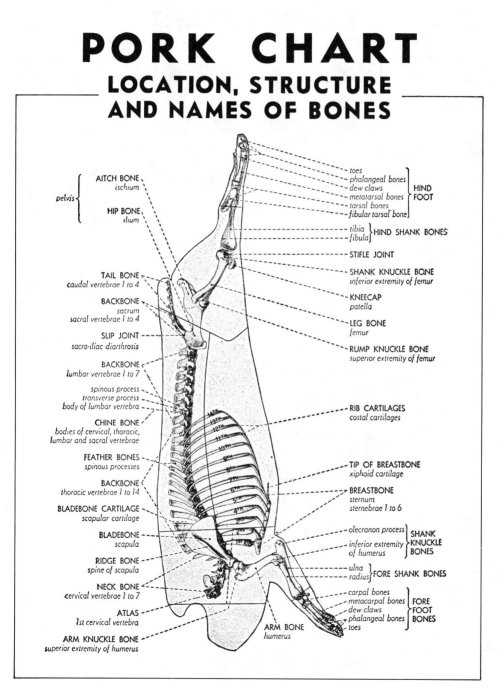

Fig. 13-1. Porcine anatomy

PORK CHART

RETAIL CUTS OF PORK — WHERE THEY COME FROM AND HOW TO COOK THEM

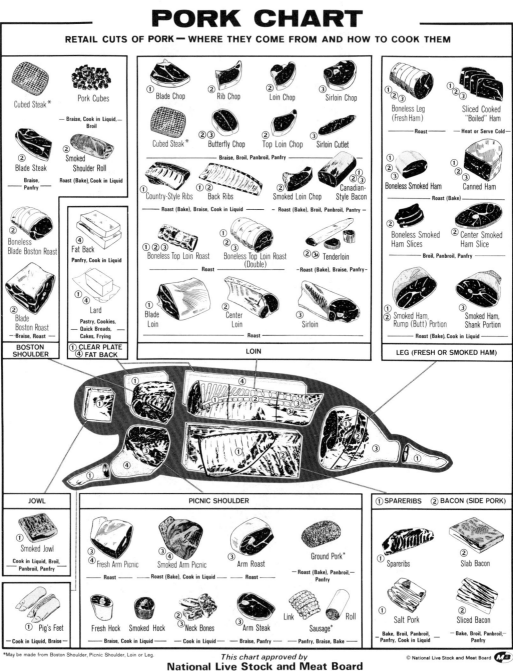

This chart approved by
National Live Stock and Meat Board

Fig. 13-2

*May be made from Boston Shoulder, Picnic Shoulder, Loin or Leg.

© National Live Stock and Meat Board

Figure 13-3. Before leaving the slaughter floor, hog carcasses are generally split into right and left sides of approximately equal weight.

In many packing houses, hog carcasses are not completely split but are hung on a single hook or gambrel, and the sides are attached by the fatback only. Generally, the hams are faced; that is, the collar fat or the fat about the inside of the ham is trimmed away on the kill floor. Note the difference between a faced ham (right) and an unfaced ham (left). The kidney fat, which is called leaf fat in pork, is removed on the kill floor. The left side indicates where the leaf fat is located. This procedure differs from lamb and beef slaughter, in which the kidneys and kidney fat are normally left intact in the carcass. Both kidneys are left in the left side to demonstrate their relative size and location. When the leaf fat is removed, the kidneys would also be removed from the pork carcass before it enters the chilling cooler.

Fig. 13-3

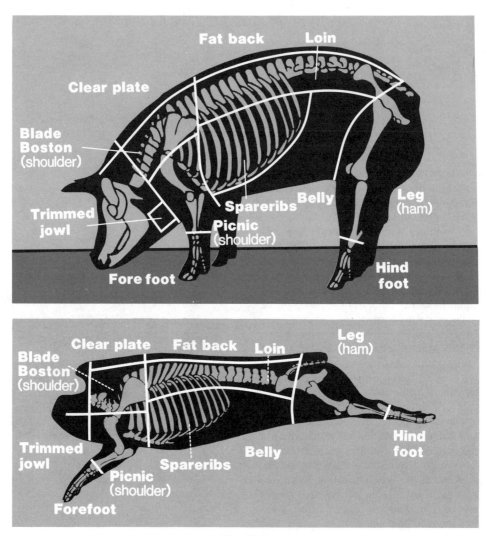

Fig. 13-4

Figure 13-4. The wholesale cuts of pork sometimes double as retail cuts. The leg (ham) and the loin are the two most valuable lean primal cuts, with the blade Boston (Boston shoulder or butt) and the picnic shoulder being of less value. The belly is the fifth primal cut. Therefore, the five primal cuts are the leg (ham), loin, blade Boston (Boston shoulder or butt), picnic shoulder, and belly, the first four being further recognized as the four lean cuts. Other cuts may be known as minor cuts. The clear plate and backfat (fatback) are termed *fat cuts*, and the neck bones and spareribs are termed *bone cuts* of the pork carcass The jowl is the fifth minor cut.

Figure 13-5. The first step in fabricating a pork carcass is to remove the hindfoot. The point of separation may vary, depending on the projected use of the ham. If a long-term, country-cure process is intended, be sure that the separation is made through the middle of the hock joint where the bone is solid to prevent bacterial contamination from invading the hollow, marrow space of the ham shank bone.

The hindfoot is generally not used for human consumption, since it does contain an exceptionally high proportion of bone and very little edible muscle. In addition, the hindfoot must be opened on the kill floor for insertion of the gambrel or hook in the tendon to hang the carcass. This also detracts from its desirability as a human food product.

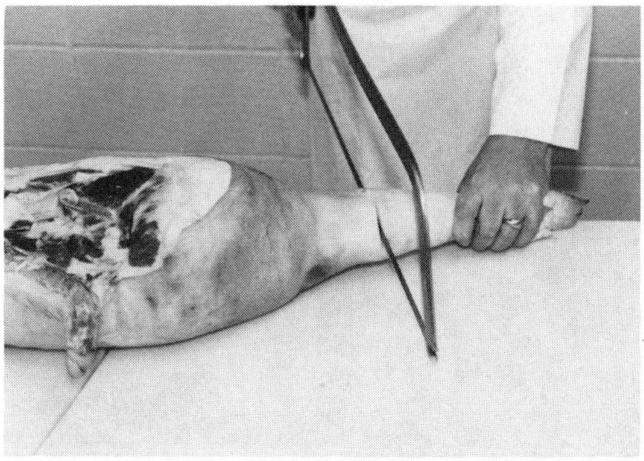

Fig. 13-5

Figure 13-6. The next step is the removal of the forefoot at the junction between the foreshank bone and the forefoot bone. The forefoot, unlike the hindfoot, has a larger percentage of edible tissue composed largely of muscle and a lower percentage of tendon and bone. The forefoot is generally easier to clean and process and is therefore utilized for human consumption as pickled pigs' feet or sometimes as a raw material for sausage.

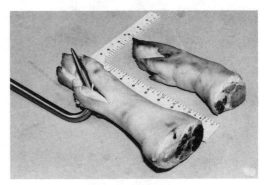

Fig. 13-6

The leg (ham) is the largest single cut in the pork carcass comprising, on the average, 16 percent of the live weight and 22 percent of the carcass weight. Note its location on the live animal and the carcass (Figure 13-4). Note the skeletal designations for future reference to Figure 13-1.

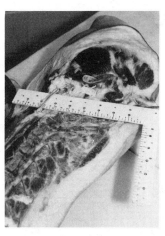

Fig. 13-7

Figure 13-7. Caution should be used in removing the ham so that the *femoral* artery which provides a natural method for applying cure is not destroyed. Previous steps must be taken on the slaughter floor to remove the kidney and pelvic fat with care so that the *femoral* artery is not destroyed. When removing the ham from the carcass, be careful not to cut the *femoral* artery too short. (Note Chapter 19.)

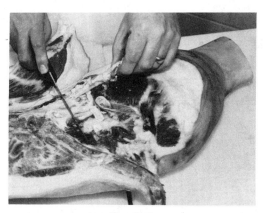

Fig. 13-8

Figure 13-8. The ham is removed from the carcass at a point approximately 2½ inches cranially from (forward from) the tip of the aitch bone. The cut is made perpendicularly to the long axis of the ham, and usually between the third and fourth *sacral* vertebrae.

After the bone has been severed with a saw, the knife is used to complete the removal of the ham. Note the *lumbar* lean which may be used as an indication of the overall leanness and muscling of intact carcasses.

If the flank is trimmed away from the anterior portion of the ham before the separation is made, the flank is left on the belly or side.

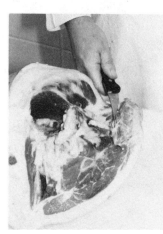

Figure 13-9. To further trim the ham once it is removed, first remove the tail bone. Then trim the flank side and very carefully remove the lymph glands.

Fig. 13-9

Figure 13-10. The most common ham is a ham which is skinned approximately three fourths of the distance from the rump (butt) face and is called a skinned ham. The skin, as well as a certain amount of the fat, is removed.

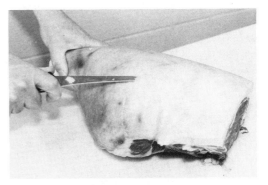

Fig. 13-10

Figure 13-11. The idea is to bevel the fat in a uniform manner so that the fat will measure approximately ¼ inch in thickness at the rump (butt) face. Different amounts of fat are trimmed, according to the market potential for the pork leg. All legs (hams) are not trimmed this closely.

Fig. 13-11

Figure 13-12. If the ham was not faced during the slaughter process, the collar fat over the cushion area of the ham should be trimmed off. Note the location of the aitch, or pelvic, bone (a) as well as the shaft of the *ilium* (b) which is the round portion of the pelvic bone. Do not confuse this round bone with the round ham bone, or *femur*, which is located caudally to (to the rear of) the aitch bone. The large muscle in the face of every ham is the *gluteus medius*, or top sirloin (c).

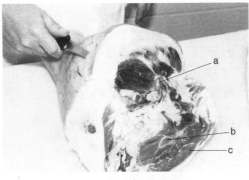

Fig. 13-12

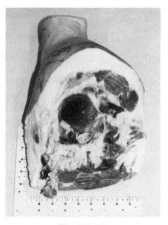

Fig. 13-13

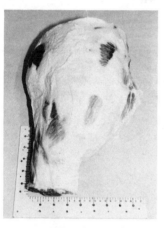

Fig. 13-14

Fig. 13-15

Figure 13-13. A pork leg (fresh ham) (IMPS 402) is composed of a shank portion, a rump (butt) portion, and the center section and may be sold whole, either fresh or cured and smoked. The center section is the most valuable portion of the ham and often center slices are removed for sale.

Nowadays few consumers care to purchase a whole ham, except for very special occasions, due to its size and total cost. Also, the large pelvic bone present in a bone-in leg may cause considerable grief for the modern consumer in terms of carving and serving. Thus, more processors and packers are removing the bone before it is merchandised, that is, fabricating a boneless leg (ham). This is a very simple task for the processor, but usually a nearly impossible one for the consumer. Furthermore, a boneless leg (ham) may be easily merchandised in smaller portions.

Figure 13-14. Before a ham is boned, the skin must be completely removed. The fat should be trimmed relatively close, as fat tends to be more evident in boneless meat cuts.

Figure 13-15. The pelvic (aitch) bone is removed from the butt portion of the leg (ham). Note the socket which must be released from the *femur* ball.

Figure 13-16. Although the shank meat may be left on the boneless ham by removing the meat from the shank before the bone is removed, a more desirable boneless ham results if the shank meat is removed with the bone and utilized as pork trimming in sausage. The shank could be used as a soup bone as well. The shank is removed by working the knife through the stifle joint between the *femur* (thigh bone) and the *tibia* and *fibula* (shank bones).

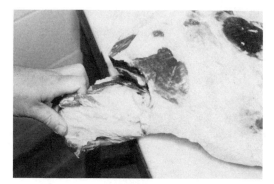

Fig. 13-16

Figure 13-17. Note the ball of the *femur*, or ham bone, which has been exposed. Loosening the meat surrounding the *femur* without cutting to the outside is known as "tunnel boning." Some processors are now completely opening up the ham at the natural seams in order to remove the fat deposits between muscles. That is necessary to produce the "90 + percent fat free" cured and smoked boneless hams currently displayed in some retail meat cases.

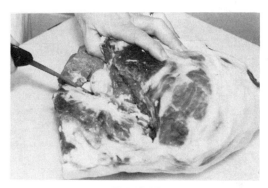

Fig. 13-17

Figure 13-18. Since the *femur*, or ham bone, had been loosened from the rump end of the ham, it is a rather simple matter to remove the ham bone intact after loosening this shank end.

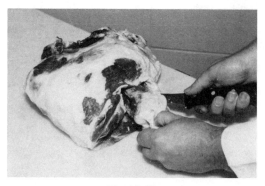

Fig. 13-18

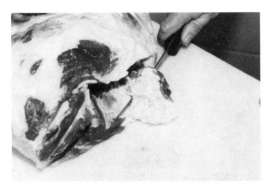

Fig. 13-19

Figure 13-19. The knee cap *(patella)* must be removed from the boneless ham. This can be accomplished during removal of the *femur* or separately, as shown here.

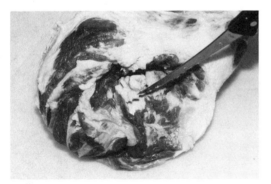

Fig. 13-20

Figure 13-20. Removing the shank, as has been illustrated, makes it easier to locate and remove the *popliteal* lymph node (see Figure 14-65) from the boneless ham. It is located in the seam between the *semitendinosus* and *biceps femoris* muscles of the bottom or outside leg muscle system.

Fig. 13-21

Figure 13-21. The tunnel-boned ham has been separated from the pelvic (aitch), leg *(femur)*, shank *(tibia-fibula)*, and knee cap *(patella)* bones as well as the popliteal lymph node and its surrounding fat deposit. The ham could be opened further to remove additional intermuscular (seam) fat.

Figure 13-22. The introduction of a jet netter and the elastic net has made the tying of roasts with the string and butcher's knot outdated. Using the jet netter, it is a simple and quick operation to tie any boneless roast. Boneless hams completely separated into muscle systems for bone and fat removal can be restructured into easily sliceable intact roasts by a technique known as tumbling, or massaging. When muscle systems containing as much as 2 percent salt and 0.5 percent phosphate are introduced into the physical action, myosin (muscle fiber protein) is released and promotes the binding together of the different muscle systems. (See Chapter 19.)

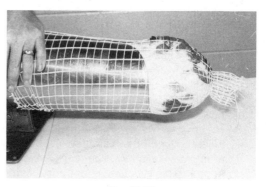

Fig. 13-22

Figure 13-23. A boneless fresh ham (IMPS 402B) may be somewhat difficult to identify; however, its main identifying feature is its size, for it is much larger than most other boneless pork cuts. A boneless whole shoulder may be similar in size but tends to be longer and less plump when netted. The fresh ham on the right is accompanied here by a similarly prepared, cured and smoked boneless ham. The excess netting on the fresh ham will be used to hang the boneless roast in the smokehouse. See Chapter 19 for further information concerning meat curing.

Fig. 13-23

Figure 13-24. Although most hams are boneless today, the ham is also handled as a bone-in roast. Because of its large size, a bone-in ham, either cured and smoked or fresh (as shown), is often subdivided into butt (rump) portions, center portions, and shank portions. When it is cut into two parts, the terminology becomes *butt half* and *shank half*. The center portion is usually cut into slices (fresh or cured) if it is utilized separately.

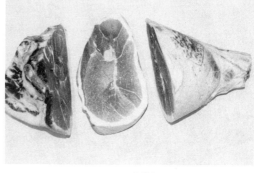

Fig. 13-24

You can see the desirability of a center slice, with only a small cross section of the ham bone *(femur)* present. The shank portion is less desirable from a palatability standpoint, since there are more connective tissues in that lower portion of the ham. The rump (butt) portion is more palatable, but the large pelvic bone present in this portion makes carving difficult. If the word *portion* is used in describing either the rump or the shank part, this indicates that at least part of the center slices have been removed. However, if the parts are designated shank half or rump half, the center section must be included in the half you buy.

The lean primal cut which is usually the most valuable cut in the pork carcass is the loin. The loin comprises approximately 11 percent of the live weight of a hog and 16 percent of its carcass weight. These percentage figures are strictly averages, and the more muscular, minimally finished hogs will have a considerably higher percentage of their carcass or live weight represented by the ham and the loin. The pork loin encompasses a longer area of the carcass than does the beef or lamb loin. The pork loin includes the area of the ribs, so in fact, you may have a retail cut properly named a rib chop from the loin in pork. This is definitely not possible in beef and lamb. The pork loin encompasses the *sacral, lumbar*, and a good portion of the *thoracic* areas of the vertebral column.

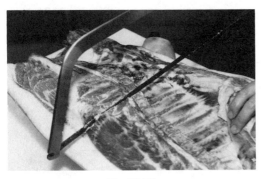

Fig. 13-25

Figure 13-25. The first step in removing the loin from the pork carcass is to separate it from the shoulder. This is often done by cutting across the third rib, although variations are made (note later section in this chapter).

Fig. 13-26

Figure 13-26. After the bone has been cut with a saw, a final severance is made with a knife. Note that this separation is made at right angles to the long axis of the carcass.

Figure 13-27. The spareribs and belly are separated from the loin, which at this point still has the fatback remaining intact. Note in the skeletal diagrams (Figures 13-1 and 13-4), that the line indicating where the ribs will be severed separates the loin from the spareribs. Orient yourself with the blade bone or *scapula* on the front or cranial end of the carcass and the sirloin end to the rear. The saw cut can be made first, or, as shown here, a knife cut can be made. The cut is made beginning from a point lateral to the tenderloin muscle (a) on the sirloin end, cranially to a point on the last rib 1 to 2 inches lateral to the tenderloin muscle. Some modification of these locations will occur with variations in cutting procedures.

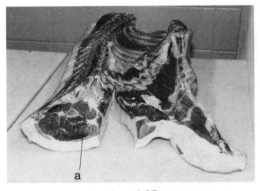

Fig. 13-27

Figure 13-28. When sawing through the ribs from the blade end, place the saw against the third *thoracic* vertebra with the distal end of the saw moving in the space resulting from the knife cut described above. After cutting through the ribs, complete the separation of the rib end of the loin and the belly with the knife.

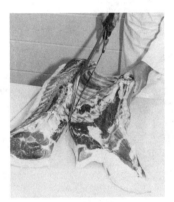

Fig. 13-28

Figure 13-29. The fatback is removed from the loin. In trimming the blade end of the loin down to the point where only ¼ inch of outside fat remains, you expose the false lean or *trapezius* muscle. In order to accomplish the proper job of trimming, you must expose this particular muscle.

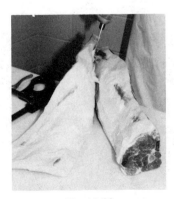

Fig. 13-29

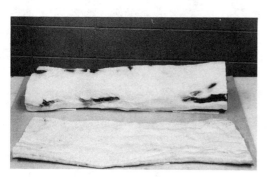

Fig. 13-30

Figure 13-30. In commercial practice, the loin is pulled from the belly and spareribs with a U-shaped loin knife. When loins are removed under packing house conditions with a loin-pulling knife, sometimes the *longissimus* muscle is scored; that is, instead of only removing the fat, the knife penetrates into the loin muscle itself. This is very undesirable, since it destroys a part of the most valuable edible portion of the carcass. Under the laboratory conditions shown in this figure, the fatback is removed very carefully without scoring the loin but exposing the muscle in a few places to demonstrate the close trim. The fatback is processed into lard, although some areas still prefer *chunks* of fatback to be used in certain types of cooking, but the demand at present is very small for such use. This fat makes very high-quality rendered lard.

Fig. 13-31

Figure 13-31. The trimmed pork loin (IMPS 410) has a higher-valued center portion and two slightly lower-valued ends. The ends include muscles of the leg (ham) and shoulder systems as well as the *longissimus*, which makes up most of the musculature of the center loin.

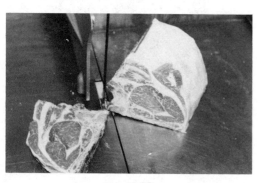

Fig. 13-32

Figure 13-32. Beginning at the cranial (front) portion of the loin, a retail chop called a blade chop is removed. The blade chop is identified by the presence of the blade bone and the attached cartilage.

Figure 13-33. The center-cut rib chop in this figure came from the tenth rib, but the area from the fifth rib through the last rib would be the area for the center-cut rib chops. Notice the prevalent use of the term *chop*, which originated from the use of a cleaver to chop the meat into small pieces. Now we use modern saws to make chops, but previously anything small enough to be chopped with a cleaver was designated a chop, while anything larger that had to be separated with a saw was called a steak. Thus, today we have pork and lamb chops but beef steaks.

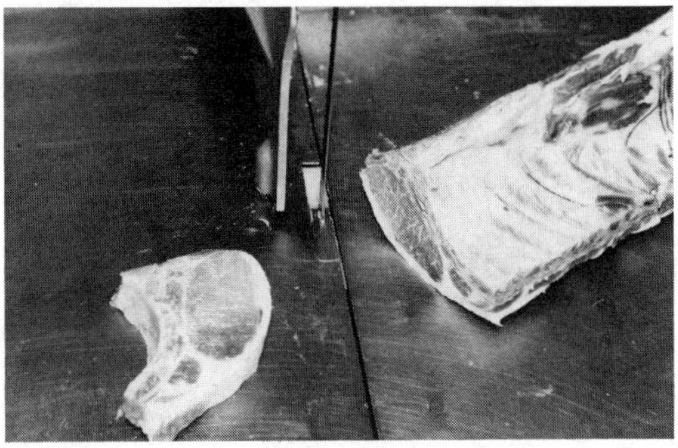

Fig. 13-33

Figure 13-34. The center-cut loin chop in this example has been removed from the area of the seventh (rearmost) *lumbar* vertebra. All chops removed cranially to (to the front of) this seventh *lumbar* vertebra up to the last rib are called center-cut loin chops and are identified by the presence of the tenderloin muscle and by the absence of a rib. In other words, all chops removed from the *lumbar* area of the vertebral column would be termed *center-cut loin chops*.

Fig. 13-34

Figure 13-35. From the caudal (rear) end of the loin, the rearmost chop removed is termed a *sirloin chop*. This chop is removed from the face, which was directly connected to the rump (butt) face of the ham. Sirloin chops include the eye (*longissimus*) and the tenderloin (*psoas*) in addition to the top sirloin (*gluteus*) muscles. Chops containing sacral vertebrae and portions of the hipbone (*ilium*) are sirloin chops.

Fig. 13-35

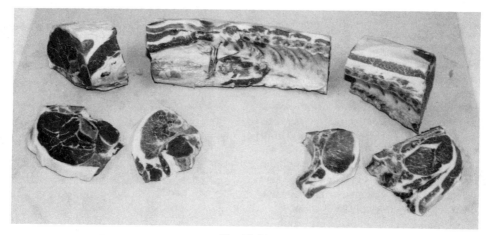

Fig. 13-36

Figure 13-36. In review, here are the component retail cuts (left to right) from the loin: a sirloin chop, a center-cut loin chop, a center-cut rib chop, and a blade chop. The pork loin chop contains a bonus, the tenderloin, which is indeed a tasty morsel! The sirloin, center, and blade roasts from the loin are also very desirable cuts.

Fig. 13-37

Figure 13-37. Several innovations can be made with pork chops, one of them being pocket chops. Ordinarily, regular bone-in pork chops are cut approximately ¾ to 1 inch thick, but for pocket chops, they may be cut slightly thicker. Immediately adjacent to the rib, in the loin-eye muscle, a slit is made which can be filled with dressing. The use of the rib chop is most appropriate for a pocket chop, since the slit can be made under the rib and is hardly noticeable when the chops are served.

The pork loin may be boned, although it is definitely more widely merchandised as bone-in chops and roasts. Boning the main portion of the loin does not normally present a problem if you first carefully remove the tenderloin (composed of the *psoas major* and *psoas minor*). The small minor is removed from the major for pork trim. From the blade end, it is an easy matter to remove the remaining *scapular* cartilage by keeping the knife flat and close to the cartilage. By using the same technique, remove the backbone and remaining portion of the ribs. After the chine (body) and feather bones (dorsal processes) have been removed, only the back ribs remain, an excellent cut for barbecuing.

Figure 13-38. The one somewhat difficult portion of the loin to bone is the sirloin end, because it contains a portion of the pelvic bone. Start by entering from the ham side and staying close to the bone. Continuing close to the bone, follow around in the same initial direction. It can be separated from the backbone through the *sacroiliac* joint, and each can be removed separately or removed together as one bone.

Fig. 13-38

Figure 13-39. A popular chop made from a boneless loin is a butterfly chop. This is in reality a double chop, since a normal ¾-inch thickness of cut is made but not completed. Again, a second normal thickness beyond that is made, and the cut is completed.

Fig. 13-39

Figure 13-40. The double chop thus removed is folded out and becomes a very desirable boneless butterfly chop.

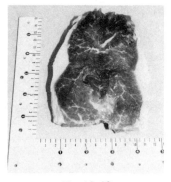

Fig. 13-40

Fig. 13-41

Figure 13-41. The blade end of the pork loin can be fabricated into meaty country-style ribs by cutting through the rib bones laterally to the vertebrae, as shown. The cut shown is continued back through the muscle but not completely through the roast.

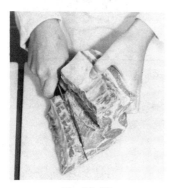

Fig. 13-42

Figure 13-42. Cutting back under the rib ends and the backbone to form an M-shaped cut opens up the roast.

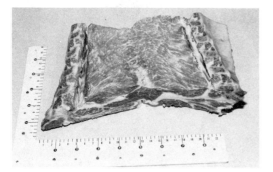

Fig. 13-43

Figure 13-43. The resulting country-style ribs produce some tender, meaty portions for barbecuing.

Figure 13-44. The blade portion of the pork loin may be boned to produce a boneless roast and back ribs. The roast can be cut into butterfly chops, tied or netted with another boneless blade or sirloin to produce a larger double roast, or used as a single small pork roast.

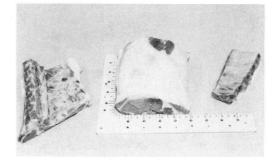

Fig. 13-44

In summary, there are a multitude of retail cuts from the pork loin (see Figure 13-36). Remember, the most forward portion of the loin is termed *blade end*. From the blade end, the blade chops may be removed, or the blade end may be boned, and the result is a boneless blade-end roast. The roast may be cut into boneless chops, which may be single or doubled by butterflying them. The blade end may also be utilized as meaty country-style ribs.

The most variety arises from the center section of the loin. The center itself may be utilized as a roast, but more often it is cut into chops, center-cut loin chops, and the center-cut rib chops. Also rib chops may be made into pocket chops. The boneless loin may be made into butterfly chops or simply left as a boneless loin. The boneless loin from large packing sows, that is, those weighing over 360 pounds alive, is oftentimes cured and utilized as Canadian bacon. When the loin is boned out, a portion of the bone which is removed is very desirable for barbecuing. This is known as the back ribs.

The center section, as well as the whole pork loin, may be cured and smoked to be cut into cured and smoked pork chops, generally bone-in. A thick, 1½ inches plus, smoked pork chop, properly prepared, is elegant eating at its best! *Windsor loins* or *Windsor loin chops* is the name sometimes applied to cured and smoked loins and chops.

Finally, the sirloin end may be utilized intact as a bone-in roast or may be cut into chops, but more desirably, it is boned and two of them are put together for a boneless sirloin roast.

Fig. 13-45

Figure 13-45. When the spareribs are removed from the belly, the idea is to leave as small an amount of meat on the spareribs as is possible; thus, the name *spareribs*. Also, one must be certain to remove all portions of the bone and cartilage including the *sternum* with the spareribs, since these make very unpalatable bacon. The loin side of the belly is that side which was attached to the loin. The length of the ribs themselves on the spareribs is a function of where the cut was made in separating the loin and spareribs. Recall our earlier discussion concerning separating the loin from the spareribs and belly (Figures 13-27 and 13-28). Modern cutting methods dictate that the largest portion of the rib cage be left on the spareribs. Actually, pricewise, spareribs sometimes exceed loin in retail price. This is a matter of supply and demand, since loins account for approximately 16 percent of the carcass weight, while spareribs account for only 3 percent of the carcass weight. The remaining portion, after the removal of the spareribs, is the fifth primal cut, called the belly. The belly comprises 9 percent of the live weight and approximately 13 percent of the carcass weight.

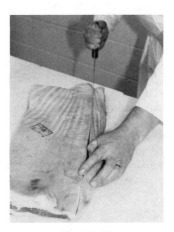

Fig. 13-46

Figure 13-46. Once the spareribs have been removed, all that remains to be done to the belly is to simply square it up. First of all, the flank end is squared by cutting through the center of the length of the flank pocket. The flank side of the belly can be cut 1 inch longer than the loin side, due to differential shrinkage which occurs during the smoking process in the *rectus abdominus* muscle, a long, flat muscle in the flank side of the belly. The loin side is trimmed, and finally the teat line is removed so that no rudimentary mammary glands are left in the bacon slab. Less expensive bacon is not trimmed so closely, and thus these mammary "seeds" may be found in such bacon (hence the term *seedy bacon*). Note the state inspection stamp on the skin surface of the belly.

Figure 13-47. The most common way to use the belly (IMPS 408) is to cure and smoke it, which results in a cured and smoked slab. The most popular form of bacon is, of course, sliced bacon. Some people do prefer fresh pork belly or fresh side pork. This also may be sliced. Others prefer the belly dry-cured or salted, which is called salt pork.

The spareribs (IMPS 416) may be cured and smoked also but are more often used fresh and barbecued. For best results, the spareribs are cut into individual portions, each containing two to six ribs, and cooked in liquid prior to the final heating on a grill. Spareribs enjoy a high demand in many parts of the United States and are a favored picnic item.

Fig. 13-47

The shoulder area of the pork carcass, that area anterior to the third rib in this discussion, is composed of two primal cuts and three minor wholesale cuts. As has been true earlier, the wholesale cuts are often retail cuts as well. The two primal cuts, and lean cuts as well, are the Boston shoulder (butt) and the picnic shoulder. Often the blade Boston (Boston shoulder or butt) and the picnic are nearly equal in weight, although the picnic is usually slightly heavier than the blade Boston. The skeleton plays a very important role in identifying retail cuts from the Boston shoulder (butt) and the picnic, as indicated in Figures 13-1 and 13-4. The three minor wholesale cuts originating from the anterior portion of the pork carcass are the neck bones, the jowl, and the clear plate.

Figure 13-48. The first step in fabricating the rough shoulder is to remove the neck bones. The neck bones should be removed with as little lean as possible, but at the same time leaving no cartilage or bone chips in the shoulder.

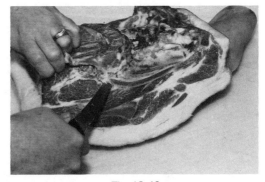

Fig. 13-48

Fig. 13-49

Figure 13-49. Neck bones (IMPS 421) make somewhat desirable, but less expensive, "spareribs." Utilized as such, they can be very delicious barbecued. Some packers cure and smoke neck bones. They are an excellent source of stock for soups and sauces as well as for flavoring pieces in vegetables.

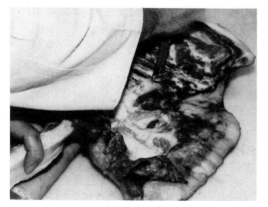

Fig. 13-50

Figure 13-50. The jowl is removed by beginning with the fat collar immediately above the foreshank and continuing straight across the cranial portion of the shoulder parallel with the cut which separated the shoulder from the loin, belly, and spareribs. The cut should be made posterior to the "notch" which results from the cut made behind the ear when the head is removed at slaughter.

Fig. 13-51

Figure 13-51. The rough-cut jowl, after removal, is trimmed into the square-cut jowl. The square-cut jowl is utilized as cured and smoked jowl bacon squares. The bulk is utilized in sausage and loaf manufacture.

Figure 13-52. The clear plate, a fat cut much like the backfat, is removed from the shoulder. The remaining skin should not cover more than 25 percent of the area between the elbow and the butt edge. An alternate method is to remove the clear plate from the Boston shoulder after the lean cuts have been separated. False lean of the *trapezius* muscle should be exposed. ("False lean" refers to a small or thin muscle near the skin surface which may have larger muscles underlying it.)

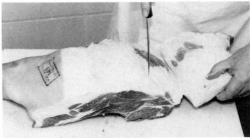

Fig. 13-52

Figure 13-53. The remaining portion, after removal of the neck bones and jowl, is called the fresh pork shoulder (IMPS 404), or New York shoulder. In this form, it is a wholesale cut. Although it may be cured and smoked or used fresh as a retail roast, the whole shoulder is a larger cut than most of today's households can use efficiently. The extensive bone distribution (see Figures 13-1 and 13-4) makes the shoulder a very difficult cut to carve.

Fig. 13-53

Figure 13-54. At the same time, however, it is composed of two other wholesale cuts, the blade Boston (Boston shoulder or butt) and the picnic. The Boston butt is separated from the picnic by cutting at right angles to the long axis of the shoulder at a distance approximately 1 inch below the exposed surface of the blade bone. Figures 13-1 and 13-4 show the bone structures that must be severed to make the separation. A cut is made first with a knife, then with a saw to sever the bone.

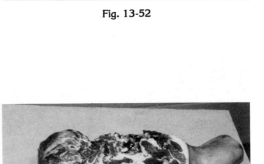

Fig. 13-54

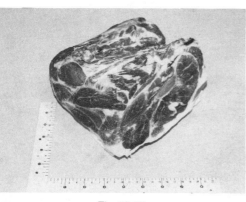

Fig. 13-55

Figure 13-55. The blade Boston (IMPS 406) is a wholesale cut and also a retail cut. Note that the blade bone is now exposed on two adjacent surfaces of the Boston shoulder (butt). The Boston shoulder can be utilized as a bone-in roast or cut into blade steaks. It makes an excellent boneless roast, which can be used fresh or cured and smoked.

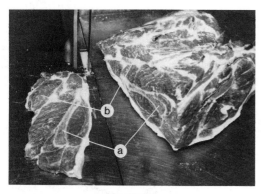

Fig. 13-56

Figure 13-56. The most popular retail cut of the blade Boston (Boston shoulder or butt) is the pork shoulder blade steak, which is removed with a saw across the edge which contains the blade bone. The identifying feature of the pork shoulder blade steak is the blade bone or *scapula*. (a). Observe also the extension of the rib eye or the *longissimus* muscle (b). Note that we now refer to the pork shoulder *steak* or blade *steak*. This is a larger portion than a rib or blade chop and must be removed with a saw; thus, the term *steak* is used.

Figure 13-57. The blade chop from the loin is very easily confused with the blade steak from the Boston shoulder (butt). The key to identifying them is the presence of a rib and vertebra on the blade chop from the loin. Since the neck bones and ribs are removed in fabricating a pork shoulder, they cannot be present on a blade steak. Although some variation exists in where the shoulder and loin are separated, there will usually be a large amount of cartilage in the blade chop from the loin. This results from the chop being removed farther down on the blade bone, where the cartilage is more pronounced. The blade bone in the blade steak is much heavier, and a very small portion of cartilage is present. Also, the overall size of the blade steak is usually greater than that of the blade chop.

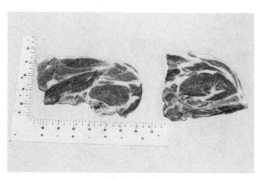

Fig. 13-57

Figure 13-58. Of all the boning operations, perhaps boning the blade Boston
(Boston shoulder or butt) is the easiest.
The one bone in the Boston butt is the
blade bone or *scapula*. On the upper side
of the blade bone is the spine of the
scapula protruding from the broad, flat
area of the bone.

The boneless Boston butt is highly
desirable, especially when cured and
smoked. In this form it often rivals the
boneless, cured, and smoked ham and
will generally be sold at a lower price.

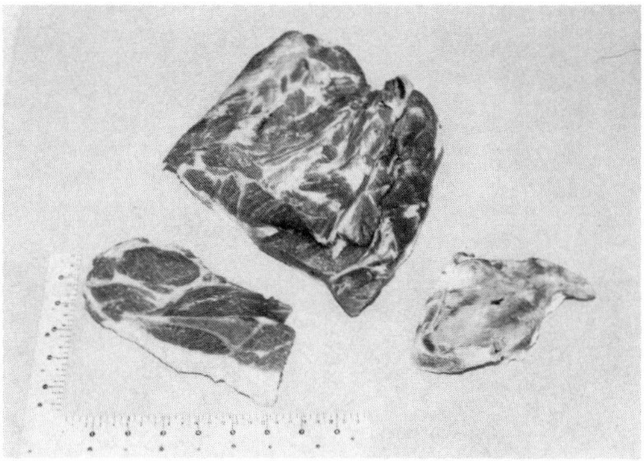

Fig. 13-58

Figure 13-59. The remaining portion
of the shoulder is the one other lean cut
called the pork shoulder picnic (IMPS
405). In some parts of the country, it is
called the *cala* because this type of cut
originated in California. Most modern
fabricators do remove approximately one
third of the skin, if the clear plate was not
removed earlier. The fresh picnic
shoulder also doubles as a wholesale cut
and a retail cut.

Fig. 13-59

Figure 13-60. The shoulder hock
(IMPS 417) may be removed from the
picnic shoulder, and, as such, it makes a
separate retail cut. It may be sold fresh,
or it may be cured and smoked. Often it
is boned and used in sausage production.

Fig. 13-60

Fig. 13-61

Figure 13-61. The remainder of the picnic shoulder may be cut into arm roasts or arm steaks. Pork arm steaks are usually removed from the center of the picnic, the roasts being the portions left at either end. If you cut arm steaks from the center, two roasts will result, one from each end. Many times the picnic will be sold only as a roast, either fresh or cured and smoked. The picnic shoulder is located lower on the live animal where smaller muscles, held together with large amounts of connective tissue, are required to work more as the animal moves about. The more that muscles are used in the live animal, the less palatable they are.

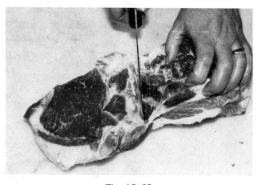

Fig. 13-62

Figure 13-62. A major portion of the bone in the shoulder is located in the picnic. The bones are the foreshank bone, which is the *radius* and *ulna* fused together, and the arm bone or *humerus*. The shank has been removed in the illustration, since it is relatively high in connective tissue. The arm bone is then removed. Note the size of the arm bone. Tunnel boning is not employed here but rather a cut is made from the outside to the bones. Because of the large amount of bone and connective tissue in a wholesale picnic as contrasted with the Boston butt, the picnic has less value.

Figure 13-63. Because of the large proportion of bone (the arm and shank bones) in the picnic shoulder, much more convenience is gained in boning the picnic than is gained by boning the Boston shoulder (butt), which contains only the blade bone. Often, picnics are boned out to obtain high-quality sausage material.

Fig. 13-63

Note the intermuscular connective tissue and fat seams in the arm steak. The picnic shoulder itself in fresh or cured and smoked form is a retail cut. The cured and smoked picnic gains its symmetrical shape from being hung in a stockinet as it is heated and smoked in the smokehouse. The shoulder hock can be removed either from the fresh picnic or from the cured, smoked picnic. The arm steak, arm roast, and boneless picnic may also be utilized as cured and smoked product.

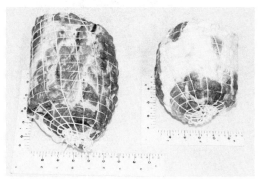

Figure 13-64. The boneless, rolled, and netted Boston shoulder (left) and picnic shoulder (right) shown here are both compact, uniformly shaped roasts that cook well in either the fresh or the cured form. They are much easier to slice and serve than the bone-in versions.

Fig. 13-64

Discussion to this point has covered the four lean cuts—the leg (ham), loin, blade Boston (Boston shoulder or butt), and picnic shoulder, as well as the other primal cut—the belly. Foremost among the minor cuts would be the spareribs. Also of great importance is the lean trim. Pork trimmings fulfill a vital role in the fabrication of processed meats, such as the multitude of sausage products and sandwich loaves, frankfurters, etc. (see Chapter 20). Fat trim is generally processed into lard. The neck bones serve as a substitute for spareribs, and the jowl serves as a rather inexpensive form of slab bacon.

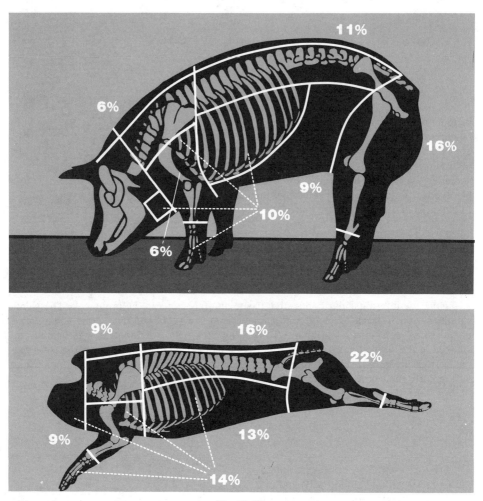

Fig. 13-65

Figure 13-65. In summary, the percentages listed adjacent to the cuts of the live animal and carcass represent the percentage of live and carcass weight, respectively, found in each wholesale cut. It must be stressed that these are average figures taken from a so-called average hog. This hog would probably have approximately 1.25 inches of fat at the last rib, be approximately 30 inches long, and have a loin-eye area of approximately 4½ square inches.

Table 13-1 further summarizes the values shown in Figure 13-65. By adding 16 and 11 percent for the leg (ham) and loin respectively in the live animal or 22 and 16 percent respectively in the carcass, the ham and loin percentages are 27 and 38 percent respectively of the live weight and carcass weight. Live and carcass yields of the four lean cuts and primal cuts of the carcass in Figure 13-65 are listed in Table 13-1, along with percentages of fat trim and minor cuts.

Table 13-1. Percent of Live Hog and Pork Carcass Weights in Primal Cuts, Lean Cuts, Ham and Loin, Fat Trim, and Minor Cuts

Portion of Carcass	Live Weight	Carcass Weight
	.... (percent)	
Five primal cuts	48	69
Four lean cuts	39	56
Ham and loin	27	38
Minor cuts	10	14
Fat trim	11	16

One should be aware that yield data vary with differences in carcasses, cutting procedures, and other factors. Table 13-2 demonstrates the variation in pork carcass yield data as derived from only three sources. Each source has reported the information as it has acquired it. None of the data are in error but they do differ. The next portion of this chapter demonstrates how variations in data may occur.

Table 13-2. Percent of Pork Carcass Weight in Pork Cuts

Cut	Source of Data		
	NLSMB[1]	AMI[2]	MWE[3]
	.... (percent)		
Trimmed ham	21.8	16.5	22
Trimmed loin	14.2	18.7	16
Bacon side	13.2	11.3	13
Spareribs	3.1	3.9	3
Boston + trim	8.9	7.3	9
Picnic + trim	6.25	6.9	9
Jowls, feet, etc.	11.0	6.3	11
Retail pork	79.5	76.5	83
Fat, bone waste	20.5	23.5	17

[1]National Live Stock and Meat Board. *A Hog's Not All Chops.*
[2]American Meat Institute. *Meat Facts.* 1984.
[3]*The Meat We Eat.* 1985.

THE IMPACT OF CUTTING METHOD AND USDA GRADE ON LEAN CUT YIELD

Almost all of the progress in producing the modern, meaty hog dates from the time breeders began using carcass data on related hogs in selecting herd replacements. Today, nearly every hog producer uses cut-out information at one time or another in selecting replacements.

There is one problem with that, though. Many people think that once a hog is dead, the carcass is measured and that's it. But *there are big differences in the way hogs are evaluated for slaughter data*, depending on which plant is doing the work. How the cuts are taken and how they are trimmed can make 5 or even 7 percent difference in cut-out from identical hogs.

While one way is not necessarily *better* than the other, you should know how each packer collects the data if you are to derive the maximum value from the information.

The differences crop up in the way the packer cuts up carcasses and trims the cuts. A given packer may alter the cutting procedure from day to day, depending on the relative prices of the various cuts. In general, those that trim closely will usually operate that way; those with a market that doesn't mind more fat usually leave more on.

Backfat, length, and loin eye will generally be consistent with either method. *But weights and percentages can vary tremendously*—never compare those figures from two different packers or processors. To illustrate this point, two hogs of quite different types were slaughtered and processed.

One hog was an overfat No. 4 barrow weighing 208 pounds. The other pig, a gilt, weighed 234 pounds and was a meaty USDA No. 1, with excellent ham development and a groove down her back, indicating a high degree of muscling with only moderate finish or fat.

After slaughter, both carcasses were carefully split so that, as nearly as possible, each half weighed the same. The overfat carcass was 30.1 inches long, had an average backfat thickness of 1.87 inches, and had slightly thin muscling (by 1985 standards, thin muscling). The meaty hog had a carcass length of 30.2 inches with an average backfat thickness of 1.10 inches and displayed very thick muscling (by 1985 standards, thick muscling). The fat carcass weighed 160.5 pounds, and the meaty carcass weighed 179 pounds. After the loins were cut at the tenth rib, the loin-eye area in the fat carcass was 3.95 square inches and was 7.40 square inches in the meaty carcass.

The carcass halves were cut up, using two rather sharply contrasting methods. Both methods are used by packing plants. The first method, termed the *maximum-trim method*, means that the maximum amount of fat is trimmed away and results in lower percentages (based on carcass weight) of the four lean cuts (leg [ham], loin, Boston shoulder [butt], and picnic shoulder). The second method, termed the *minimum-trim method*, means that very little fat is trimmed away, which results in higher percentage figures. The impact of cutting methods on the lean yield of the two different types of pigs should be the same, or should it?

Some specific differences between the maximum-trim method (resulting in higher lean to fat relationships) and the minimum-trim method should be identified. The maximum-trim method resulted in cutting a longer foot off both the ham and the picnic, leaving a short shank. In the maximum-trim method, the ham was removed a bit anterior to the line shown in Figures 13-1 and 13-4, whereas in the minimum-trim method, that line was moved slightly to the rear, or caudally, leaving more weight in the loin. Correspondingly, the shoulder/loin separation in the maximum-trim method was on the third rib, as indicated in Figure 13-25, but the minimum-trim line for separating the shoulder was over the first rib, permitting meat that would normally be in the blade Boston to be merchandised as loin.

The maximum-trim method further reduced the weight of the loin by separating the loin and the sparerib/belly section closer to the vertebra than the procedure described for Figures 13-27 and 13-28. The minimum-trim method separated the loin and belly on a line more distant from the backbone than indicated in Figures 13-27 and 13-28. The result was long rib ends left on the pork chops from the loin and narrower bellies and spareribs. The maximum-trim method removed all accessible surface fat to no more than ¼ inch thick, whereas the minimum-trim method removed the skin and only a minimum amount of fat in trimming the wholesale cuts.

The right side of each of the two hogs was cut using the maximum-trim method, and the left side was cut using the minimum-trim technique. It should be noted that the right ham of each carcass had been faced on the kill floor as well (see Figure 13-3). The weights of the trimmed wholesale cuts from the fat and meaty carcasses are listed in Table 13-3.

Table 13-3. Trimmed Cut Weights as Influenced by Carcass Type and Cutting Method

Cut	Overfat Carcass		Meaty Carcass	
	Minimum Trim	Maximum Trim	Minimum Trim	Maximum Trim
		(pounds)		
Ham	16.5	14.4	20.7	20.7
Loin	15.8	10.4	23.0	16.2
Ham and loin	32.3	24.8	43.7	36.9
Boston butt	6.3	5.9	8.5	8.9
Picnic	5.6	5.4	8.5	8.3
Four lean cuts	44.2	36.1	60.7	54.1
Fat trim	14.0	22.5	5.5	10.0
Chilled side	80.0	80.0	88.5	87.7

Maximum-trim cutting reduced the ham weight in the overfat carcass 2 pounds more than the minimum-trim method, due to the greater amount of fat and skin removed. Much less fat was removed from the meaty ham, using the maximum-trim method of cutting. Therefore, in the meaty carcass, the extra fat removed in the maximum-trim technique was compensated for by the shorter length of ham in the minimum-cutting technique. Therefore, the meaty hams were equal in weight.

Adding a small portion of what would normally have been ham on the posterior and two ribs to the anterior end of the loin resulted in longer, heavier loins in the minimum-trimmed group. Longer rib ends, more outside fat, and the additional length resulted in loins that were approximately 50 percent heavier than those produced by the maximum-trim method.

The minimum-trim Boston shoulder (butt) actually weighed less than the maximum trim from the meaty hog because of the fact that the cut separating the shoulder from the loin was made a distance of 1½ ribs farther forward, thus subtracting from the weight of the Boston shoulder (butt) and picnic shoulder, while adding to the weight of the loin. However, with the overfat hog, the difference in fatness overshadowed this difference in cutting location on the weight of the Boston shoulder (butt).

In looking at the picnic shoulder, that cut which is the lower portion of the shoulder, the influence of the number of ribs left on the loin does not appear to be as great as was the case with the Boston shoulder (butt), since the minimum-trim picnic weighs slightly more in both cases than the maximum-trim picnic shoulder. This difference is due largely to differences in amount of fat and skin trimmed; considerably more is trimmed in the maximum-trim method.

Observe the total amount of fat trimmed from each carcass during the complete processing of the carcass into its five primal cuts (including the belly) and the minor cuts, such as spareribs, neck bones, jowl, etc. This is total fat trim in the carcass. As would be expected with a maximum trim, there are larger amounts of fat trim as compared to the minimum trim, which leaves more fat on the four lean cuts. It is certainly an eye opener to see the tremendous difference in fat trim, especially in the overfat hog.

The comparison of cutting methods may be reviewed by observing the percentages of the carcass weight found in various carcass components listed in Table 13-4. Recall again the definitions of *minimum trim* and *maximum trim*. Minimum trim is a fabrication method that leaves as much fat as the market will bear on the leg (ham), loin, blade Boston (Boston shoulder or butt), and picnic shoulder. Maximum trim removes the maximum amount of fat, leaving only that which is necessary to insure palatability for the consumer. The fabrication meth-

Table 13-4. Percent of Carcass Weight Found in Lean Cuts of Fat and Meaty Pork Carcasses Cut Using Minimum- and Maximum-Trim Methods

Cut	Overfat Carcass		Meaty Carcass	
	Minimum Trim	Maximum Trim	Minimum Trim	Maximum Trim
	. . . . *(percent of carcass)*			
Ham	20.6	18.0	23.4	23.6
Loin	19.8	13.0	26.0	18.5
Ham and loin	40.4	31.0	49.4	42.1
Boston butt	7.9	7.4	9.6	10.1
Picnic	7.0	6.7	9.6	9.5
Four lean cuts	55.3	45.1	68.6	61.7
Fat trim	17.5	28.1	6.2	11.4

ods were compared on one overfat pork carcass and one trim, meaty pork carcass.

The percent of the carcass weight in the ham was greater in the meaty hog than in the overfat carcass. Cutting method was not related to large differences in ham percentages, especially in the meaty carcass, because the longer length of the closer-trimmed ham compensated for the greater amount of fat left on the surface of the minimum-trimmed ham. The more closely trimmed (maximum) ham from the overfat carcass, however, lost more fat in trimming and, even though slightly longer, represented 2.6 percent less of the carcass weight than the shorter, less closely trimmed ham from the other side of the overfat pig. Loin percentages vary most widely not only because of the differences in degree of fat trimmed but also in the amount of rib and belly left on the loin and in the overall loin length. Ham and loin percent was a widely used measure of carcass merit and is still a good indicator of relative pork-carcass values. Ham and loin percentage differs 9.4 percent between cutting methods in the overfat pig, whereas in the meaty carcass, the difference in ham and loin percentage is only 7.3 percent between the two methods. This is evidence that a fat hog carcass can benefit more from the use of a minimum-trim method than can a properly finished, meaty pork carcass. Percent of carcass found in the four lean cuts has also been used as a method of evaluating and ranking pork carcasses. Those values parallel the ham and loin percentages in Table 13-4. Fat-trim percentages show similar but opposite numerical trends and emphasize the difference between the two carcasses better than the weights in Table 13-3, due to the weight adjusting or weight equalizing effect of reporting percentages.

Since these differences do exist, how can you evaluate a set of carcass figures and tell what kind of a cutting method was used in securing the data? Here are some guidelines to keep in mind.

- The percentage of ham and loin or four lean cuts should increase as backfat declines and decrease as backfat increases. If this is not generally true, that is, hogs that have 1.3 to 1.4 inches of backfat have about as high a percentage of ham and loin as hogs with 1.0 or 1.1 inches, a maximum-trim method was *not* used.

- The general level of the percentage ham and loin that is secured will tell you something. If a maximum trim is being used, hogs that have 1.0 to 1.1 inches of backfat and have approximately 5.5 square inches of loin eye will have 40, 41, or possibly 42 percent ham and loin. If such hogs cut 43 to 46 percent ham and loin, a minimum-type trim method was used.

- Finally, remember three things when checking carcass data: *First*, never compare percentage figures secured at different packing plants. *Second*, don't attach too much significance to small differences between hogs. If we could get a 7 to 9 percent difference between two sides of the same hog due to cutting methods, differences of up to 1 percent between different hogs can easily exist due to differences in how the carcasses were split, cut, and trimmed within the same plant, even if every effort was put forth to be consistent. Remember, large commercial operations cut as many as 17 hogs per *minute* (1,020/hour) when in normal production. A slip of the knife can come quite easily under such conditions. *Third*, in spite of all this, carcass data properly evaluated have been, and will continue to be, the basis of much of the future improvement in the hog business.

Carcass Merit and Production Efficiency

The pork-carcass–evaluation procedure discussed on the preceding pages is a time-consuming process which requires that the carcass be cut before a comparison can be made. It should be noted that an evaluation procedure which results in actual weights of saleable product is the most accurate method of evaluation. However, in many instances, it is important to be able to evaluate the merits of a carcass without the expenditure of time and effort required to cut it into wholesale cuts. In commercial pork-processing plants that cut 1,000 carcasses or more per hour, maintaining carcass identity on each cut and obtaining accurate cut weights from even a small number of carcasses are difficult at best and normally impossible.

The most popular current method of pork-carcass evaluation is recommended and published by the National Pork Producers Council (NPPC). *The Procedures to Evaluate Market Hog Performance*[1] emphasizes both the carcass merit and the production efficiency of the animal. Each hog and its carcass are evaluated and can be compared with other hogs on the basis of "pounds of acceptable quality lean pork gain per day on test." That value is predicted by inserting adjusted hot carcass weight, loin-muscle area, tenth rib fat depth, days on test, and weight on test into a prediction equation.

---

[1]Second ed. 1983. National Pork Producers Council, P.O. Box 10383, Des Moines, IA 50306.

Figure 13-66. As has been noted earlier in this chapter, pork-carcass evaluation often involves measurement of the size of the loin-eye muscle, that is, the *longissimus dorsi*. This is done by breaking the loin between the tenth and eleventh ribs at right angles to the backbone. If possible, the loin should be broken (cut) between the tenth and eleventh ribs before the backfat has been removed for two reasons. *First*, some pork-carcass–evaluation procedures utilize a fat depth or thickness measurement taken at three fourths of the length of the eye muscle on this tenth rib-eye surface. *Second*, if the backfat is intact, the loin-eye muscle maintains its natural size and shape, while the area is determined by using a grid or by making a tracing. In most packing house situations, it is necessary to obtain tracings after the loins have been pulled away from the backfat with a U-shaped loin knife. Care must be taken to prevent distortion when the supporting fat has been removed. It is important that the loin is cut perpendicularly to its axis to assure accurate measurement of the muscle area. Normally, pork loins are not broken at the tenth rib in regular packing house operations.

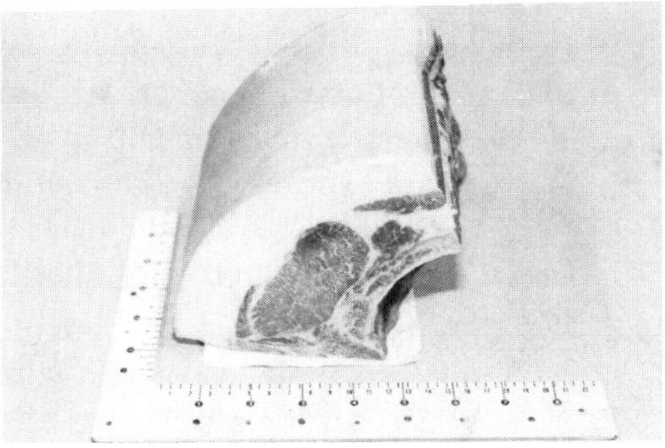

Fig. 13-66

Figure 13-67. Some pork-carcass evaluators (NPPC recommended procedures) break or rib the loin while the carcass is hanging, somewhat similar to the procedure in ribbing a beef carcass (Chapter 14). In using this technique, avoid damaging the belly by cutting into it from the side near the loin and maintain a cut across the loin muscle exactly perpendicular to its long axis, thus measuring its true size and assuring an accurate fat-thickness measurement.

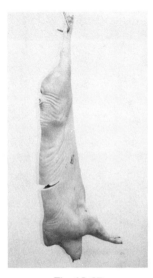

Fig. 13-67

One of the alternate equations in the NPPC's publication is adaptable to the data available for the overfat and meaty pork carcasses identified in Tables 13-3 and 13-4. The adjustment procedure of Boggs and Merkel[2] was used to convert average backfat thickness to tenth rib fat depth values. Then it was possible to use the equation designed to predict the "pounds of acceptable quality lean pork adjusted to a 160-pound carcass." The equation predicted that the overfat carcass would produce approximately 69 pounds of lean pork, whereas the meaty carcass would produce approximately 95 pounds of lean pork, if adjusted to a 160-pound carcass. The 26-pound difference in pounds of lean pork represents 16.25 percent of a 160-pound carcass. It is interesting to note that the difference in actual yield of four lean cuts, using the maximum-trim method of cutting, was 16.6 percent. Therefore, although measuring and estimating different parameters, the NPPC's system did evaluate the relative carcass merit of the two examples in parallel with the actual yield of cuts obtained from the two carcasses.

---

[2]Donald L. Boggs and Robert A. Merkel. *Live Animal–Carcass Evaluation and Selection Manual.* 1979. Kendall/Hunt Publishing Co., 2460 Kerper Boulevard, Dubuque, IA 52001.

Chapter 14

Beef Identification and Fabrication

IDENTIFICATION

THE *Uniform Retail Meat Identity Standards*[1] (see Chapter 12) will be used as the basis for identification in this chapter.

FABRICATION

The skeletal structure of the bovine in relation to an outline of the wholesale cuts is shown in Figure 14-1, and a more detailed chart of wholesale and retail cuts and recommended cooking methods appears as Figure 14-2. Use these two figures for

reference as the discussion moves through cutting a beef carcass. The center section of this text includes color photos of beef retail cuts.

Figures 14-3 through 14-105 appear through the courtesy of the University of Illinois and South Dakota State University. Figures 14-106 through 14-112 were made available for this text by the IBP Corp., Dakota City, Nebraska.

The First Steps

For demonstration purposes let us assume that generally, when beef animals are finished for market, they will weigh approximately 1,100 pounds. After slaughter, the resulting carcasses represent approximately 60 percent of their live weight. Therefore, a 1,100-pound live animal will have a carcass weighing approximately 660 pounds. Before leaving the slaughter floor, beef carcasses are split into right and left sides, each side then weighing about 330 pounds.

---

[1]National Live Stock and Meat Board.

BEEF CHART
LOCATION, STRUCTURE
AND NAMES OF BONES

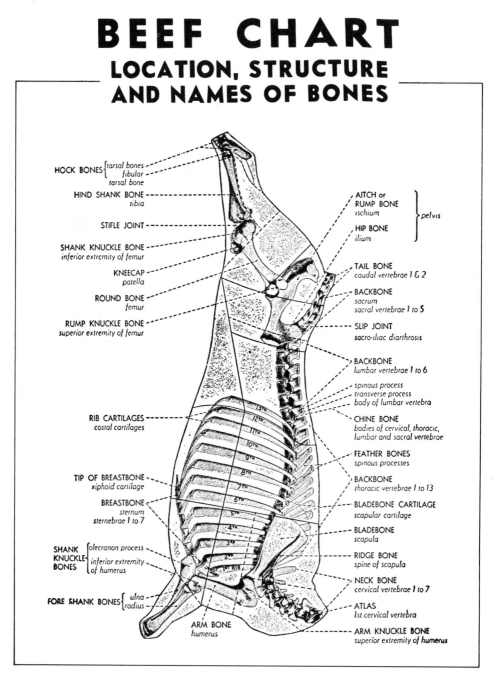

Fig. 14-1. Bovine anatomy

BEEF CHART

RETAIL CUTS OF BEEF — WHERE THEY COME FROM AND HOW TO COOK THEM

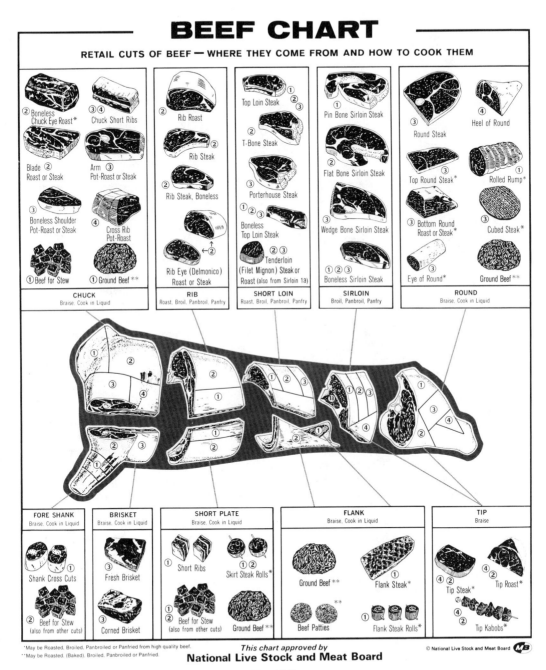

CHUCK
Braise, Cook in Liquid

RIB
Roast, Broil, Panbroil, Panfry

SHORT LOIN
Roast, Broil, Panbroil, Panfry

SIRLOIN
Broil, Panbroil, Panfry

ROUND
Braise, Cook in Liquid

② Boneless Chuck Eye Roast*
③④ Chuck Short Ribs
Blade ② Roast or Steak
Arm ③ Pot-Roast or Steak
③ Boneless Shoulder Pot-Roast or Steak
④ Cross Rib Pot-Roast
① Beef for Stew
① Ground Beef**

② Rib Roast
② Rib Steak
② Rib Steak, Boneless
Rib Eye (Delmonico) Roast or Steak ←②

Top Loin Steak ①②③
T-Bone Steak ②
Porterhouse Steak ③
①②③ Boneless Top Loin Steak
②③ Tenderloin (Filet Mignon) Steak or Roast (also from Sirloin 1a)

① Pin Bone Sirloin Steak
② Flat Bone Sirloin Steak
③ Wedge Bone Sirloin Steak
①②③ Boneless Sirloin Steak

③ Round Steak ①
④ Heel of Round
③ Top Round Steak*
① Rolled Rump*
③ Bottom Round Roast or Steak*
③ Cubed Steak*
③ Eye of Round*
Ground Beef**

FORE SHANK
Braise, Cook in Liquid

BRISKET
Braise, Cook in Liquid

SHORT PLATE
Braise, Cook in Liquid

FLANK
Braise, Cook in Liquid

TIP
Braise

① Shank Cross Cuts
② Beef for Stew (also from other cuts)

③ Fresh Brisket
③ Corned Brisket

① Short Ribs
①② Skirt Steak Rolls*
①② Beef for Stew (also from other cuts)
Ground Beef**

Ground Beef**
① Flank Steak*
Beef Patties**
①② Flank Steak Rolls*

④② Tip Steak*
④② Tip Roast*
④② Tip Kabobs*

*May be Roasted, Broiled, Panbroiled or Panfried from high quality beef.
**May be Roasted, (Baked), Broiled, Panbroiled or Panfried.

This chart approved by
National Live Stock and Meat Board

© National Live Stock and Meat Board **MB**

Fig. 14-2

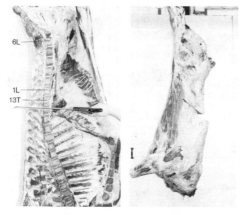

Fig. 14-3

Figure 14-3. To be properly evaluated and merchandised, beef sides must be separated into forequarters and hindquarters, a practice called ribbing or quartering (left). This is normally done by severing the half carcass between the twelfth and thirteenth ribs, although some local areas modify the number of ribs left on the hindquarter. In actual practice, rather than counting the ribs in a posterior direction from front to rear to find the twelfth rib, count the exposed bodies of the *lumbar* and *thoracic* vertebrae in an anterior direction or from the rear to the front. Counting off six *lumbar,* Nos. 6L through 1L, and 1½ *thoracic* vertebrae, Nos. 13T and 12T, a total of 7½ vertebrae, you arrive at a point midway between the twelfth and thirteenth ribs. These exposed vertebrae are much easier to locate and to count than are the ribs. A saw is used to sever the backbone, and a knife is used to complete the cut, leaving a portion of the flank attached to the plate in order that the entire half carcass may still hang from a rail (right). For shipping and further processing, the separation into forequarters and hindquarters is completed by easily severing this plate-flank attachment with a knife. The ribbing procedure may be reversed by inserting the knife at the desired spot below the rib-eye muscle and making a smooth cut toward the backbone. The first method is preferred for making a most nearly perpendicular cut across the rib-eye muscle in order to measure its true size. See page 322.

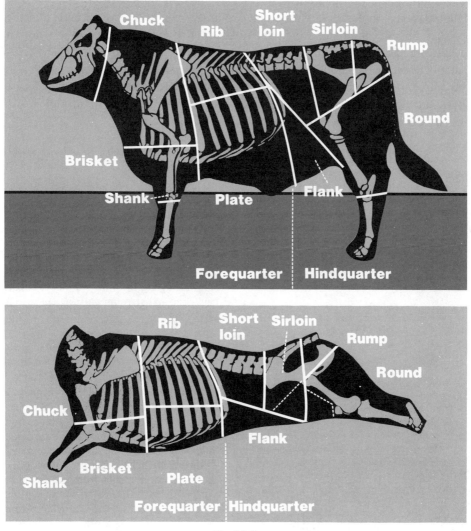

Fig. 14-4

Figure 14-4. Each carcass side or half is divided into quarters. The forequarter, composed of five wholesale cuts (the primal chuck and rib and the rough brisket, plate, and shank), is usually the heavier quarter. The hindquarter, composed of three wholesale cuts (the primal round, loin, and rough flank), is the most valuable quarter, based on retail pricing.

Forequarter

The forequarter represents approximately 52 percent of the carcass weight and approximately 31 percent of the live weight. The forequarter is composed of the following wholesale cuts: the chuck and rib, which are primal cuts, and the brisket, shank, and plate, which are rough cuts. If the carcass or side is not all cut at the same time, it may be advisable to cut the forequarter first. The forequarter produces cuts which are best utilized as roasts or other forms which will be cooked and prepared in a manner to reduce the effects of connective tissue on palatability. The hindquarter, however, is a primary source of high-value steaks and would benefit more from additional aging time.

Fig. 14-5

Figure 14-5. The rib and plate are severed from the chuck, brisket, and shank by making a cut between the fifth and sixth ribs. The ribs are counted, and the cut is marked from the inside before the front quarter is removed from the rail. The cut can then be made from the outside, following the fifth rib closely. The backbone and sternum must be cut with a saw.

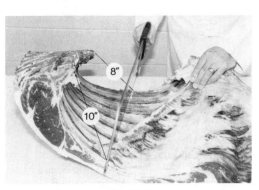

Fig. 14-6

Figure 14-6. The wholesale plate is separated from the wholesale primal rib by a cut which is no more than 10 inches from the chine on the large (blade) end and 10 inches from the chine at the small (loin) end (IMPS 103). The cut illustrated was made 10 inches from the chine on the small end and 8 inches from the chine on the blade end. The blade end is thus called because the blade cartilage from the blade bone is exposed, while the loin end is thus named because it is adjacent to the loin from the hindquarter. The completion of the separation of the rib and plate is done with a knife.

The two primal ribs in wholesale form represent 5 percent of live weight and 9 percent of the carcass weight. Steaks and roasts trimmed for retail trade from the two wholesale ribs (right and left) amount to approximately 3 percent of the live weight and 5 percent of the carcass weight.

Figure 14-7. Fabrication of the rib is begun by removing, with a saw, the ends of all rib bones on a line approximately 1 inch from the lateral edge of the *longissimus dorsi* (rib eye) at the loin end. The location of this cut depends on market demand, so it may be located at a greater distance from the eye, giving more weight, but at the same time more waste, to the rib roast. USDA specifications state that this separation should be not more than 3 inches from the eye on the loin end (small end) and not more than 4 inches from the eye on the blade end (large end). In a typical blade-end view of a wholesale rib, note the blade cartilage (a). It has been separated from the blade bone itself, which remains on the chuck. The resulting rib ends, which are removed, may be made into short ribs (Figures 14-43 and 14-44).

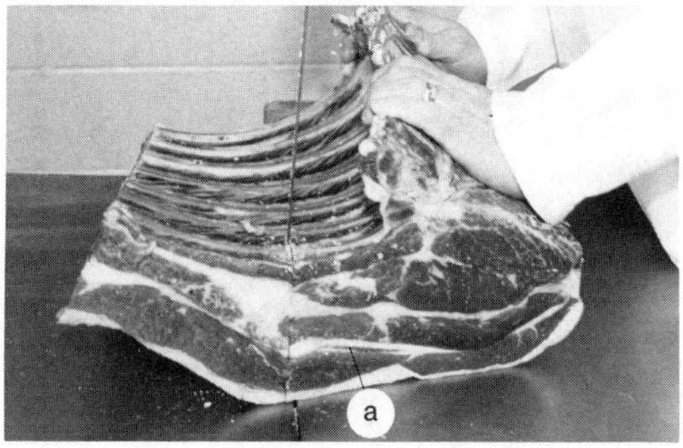

Fig. 14-7

Figure 14-8. The bodies of the vertebrae (chine bones) are removed on the saw (see Figure 14-11).

Fig. 14-8

Figure 14-9. In further processing this rib, some of which is optional, the feather bones or *dorsal* processes of the *thoracic* vertebrae are removed. Since the main portion or the bodies of the *thoracic* vertebrae have already been removed, the only bones remaining are the ribs themselves and the blade bone and cartilage.

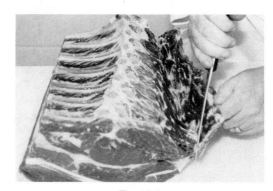

Fig. 14-9

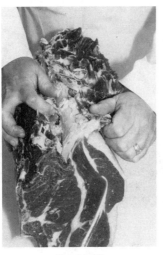

Fig. 14-10

Figure 14-10. It is essential that the neck leader or backstrap *(ligamentum nuchae)* be removed, since it is composed almost entirely of elastin connective tissue, which is absolutely impossible to make palatable by means of cookery.

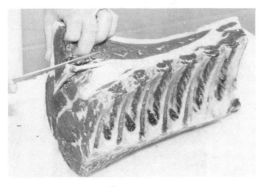

Fig. 14-11

Figure 14-11. The blade bone and/or its cartilage is removed. In this view of the standing rib roast, note the natural rack formed by the ribs which remain in the roast. This roast would be placed in the oven inverted on this natural rack for roasting. This rib plus the feather bones and backstrap could be identified as a 107 rib, oven prepared, by IMPS.

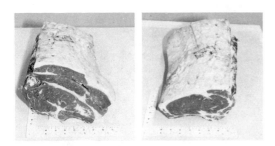

Fig. 14-12

Figure 14-12. Finally, outside fat in excess of 0.25 inch is removed, resulting in a finished short standing rib roast, so named for its short length of ribs remaining. Officially this would be labeled ''extra trim.'' The components which were removed from the wholesale rib to get this standing rib roast were the body and *dorsal* processes of the *thoracic* vertebrae (bone), the neck leader, the blade, the rib ends, and the surface fat trim. The roast on the left illustrates the blade (large) end, and the roast on the right illustrates the loin (small) end.

Figure 14-13. Rib steaks removed from this standing rib roast from the loin (small) end are called rib steaks, small end. Note the "cap" muscle (*spinalis dorsi*), which aids in rib steak identification (a).

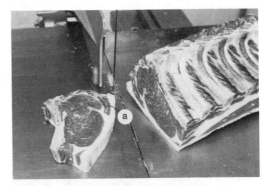

Fig. 14-13

Figure 14-14. Rib steaks removed from the blade (large) end are called rib steaks, large end. Note the number of accessory muscles in the large end which make it less palatable as a steak source. The large end is often effectively utilized as a roast source.

Fig. 14-14

Figure 14-15. Steaks from the rib do vary in value. In this close comparison between a steak from the loin (small) end and one from the blade (large) end, the steak from the loin end is more desirable in that it has a larger rib-eye muscle and less intermuscular (seam) fat. Both of these cuts are called rib steaks, but now the uniform labeling standards identify them for their value differences.

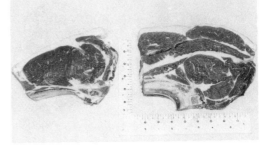

Fig. 14-15

Figure 14-16. The ribs may be removed from the trimmed rib to make an excellent boneless rib roast or a source of very desirable boneless rib steaks.

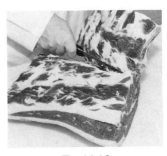

Fig. 14-16

Fig. 14-17

Figure 14-17. The removal of the cap muscles which cover the blade (cranial) end of the rib produces a rib-eye roast (*Delmonico* roast) from the boneless rib.

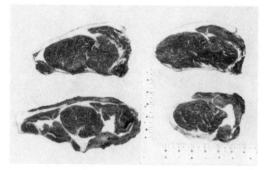

Fig. 14-18

Figure 14-18. The boneless rib steaks on the left demonstrate end to end differences in the rib similar to those shown in Figure 14-15. The top row has loin-end steaks and the bottom row has blade-end steaks. The rib-eye steaks on the right differ less, because all cap muscles except the *spinalis dorsi* have been removed resulting in rib-eye (*Delmonico*) steaks. The bone-in retail roasts and steaks from the rib equal only about 3 percent of the live weight and 5 percent of the carcass weight of a beef animal. Boneless yields would be somewhat less.

Figure 14-19. Returning to that portion of the forequarter remaining after the removal of the rib and plate, the primal chuck is separated from the brisket and shank by a cut parallel to the top side of the chuck which severs the distal (far) end of the *humerus* (arm bone). If this cut is made just slightly (½″ to 1″) above the point where the knuckle bone can be felt from the outside (top), a butterfly-shaped cut surface on the bone will appear, signifying that a correct cut has been made. This cut must be completed with a saw (bottom), as the *humerus* must be severed. The chuck is the largest wholesale primal cut in the beef animal and the combined right and left chucks represent 16 percent of the live weight and 26 percent of the carcass weight. The semi-boneless blade and arm retail roasts represent 9 percent of the live weight and 15 percent of the carcass weight. However, the boneless retail roasts secured from the chucks represent only 7 percent of the live weight and 11 percent of the carcass weight. This does not mean that the portion of the chuck not represented by roasts is all waste, since there are considerable amounts of lean trim and some miscellaneous cuts secured from the chuck in addition to these roasts.

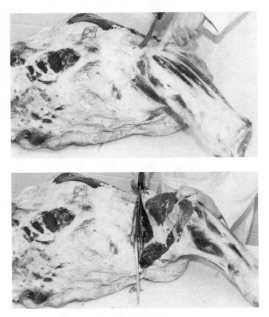

Fig. 14-19

Figure 14-20. This is a chuck (IMPS 113) from the animal's left side. Visualize the rib having been removed from the blade face (a) and the brisket and shank removed from the arm face (b). The chuck may be processed by several methods. It may be cut bone-in on a band saw. The arm and blade roasts and steaks are cut by alternating cuts into the joint of the *scapula* and *humerus*. The rib bones may be trimmed from arm steaks and roasts but are usually left on blade cuts. A method in general use leaves the *humerus* bone and the *scapula* (blade) in the roasts, resulting in about a 4 percent higher roast yield on a carcass basis or raising the 11 percent quoted earlier (Figure 14-19) to 15 percent. With the second method, an institutional method, all bone is removed from the roasts, resulting in inside (chuck eye) and outside (shoulder) chuck roasts.

Fig. 14-20

Fig. 14-21

Figure 14-21. The square corner of the square-cut chuck is called the cross-rib pot roast, formerly known as the English cut. The cross-rib may be cut from the chuck when any of the cutting methods discussed here are used. A bone-in cross-rib roast is shown. The cross-rib may also be merchandised as a boneless roast. This is a low-quality roast with obvious fat seams even in this trim chuck. It should be cooked slowly with moist heat.

Fig. 14-22

Figure 14-22. The rib and neck bones are removed from the chuck by trimming them out as cleanly as possible (top). The task can be simplified by sawing across the rib bones near the backbone, or by sawing across the *dorsal* processes of the vertebrae. The leverage against the boner is reduced by making the saw cuts (bottom).

Fig. 14-23

Figure 14-23. The neck leader (*ligamentum nuchae*) is removed. This piece of connective tissue (seen earlier in the rib) is sometimes called the backstrap and is responsible for holding the animal's head erect. As discussed earlier, it is impossible to make this piece of connective tissue palatable.

Figure 14-24. For the most commonly used bone-in method, the square-cut chuck thus prepared is placed on a saw and two or three blade roasts are removed. The blade bone (*scapula*) appears in these roasts, thus the name *blade roast*. Note that the rib eye still remains in the chuck, although it is somewhat small in size at this point. In many retail meat cases, these roasts would contain vertebrae and ribs in addition to the *scapula*.

Fig. 14-24

Figure 14-25. Next, the chuck is turned 90°, or a right angle, and several arm roasts are cut. The arm bone (*humerus*) is very evident in these cuts, thus the name *arm roasts*. Note that there are fewer and larger muscles in the arm roasts and thus less seam fat than occurs in the blade roasts. The main muscle systems of the arm roasts are the *triceps brachii,* long head (a), and the *triceps brachii,* lateral head (b).

Fig. 14-25

Figure 14-26. After the removal of two or three arm roasts, the remaining portion of the chuck is returned to its original position and more blade roasts are removed. Proceeding forward (cranially), the spine of the *scapula* (blade) bone (a) becomes evident in the blade roast. The spine of the *scapula* forms, in fact, a characteristic seven configuration, and these roasts are thus known as seven-bone roasts. A neck pot roast may be cut from the remainder of the chuck by making the cut in the same plane as the seven-bone roasts. The *pre-scapular* lymph node and surrounding fat deposit should be removed from the neck roast, a rather low-quality roast, which may be better utilized as lean trim for ground beef.

Fig. 14-26

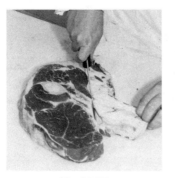

Fig. 14-27

Figure 14-27. The bone-in roasts from the chuck may require that fat seams be trimmed before packaging for the freezer or retail case.

Fig. 14-28

Figure 14-28. Because a roast should be at least 1½ inches thick, to cook properly without excessive drying, the bone-in chuck roasts are larger than most consumers prefer. Therefore the roasts are often subdivided. This is advisable in all cases, since it permits the roasts to be cut thicker than 1½ inches.

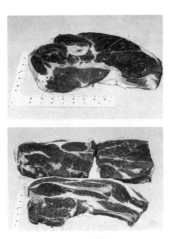

Fig. 14-29

Figure 14-29. The bone-in chuck produces arm roasts (top) and blade roasts (bottom) as illustrated. The cross-rib roast (Figure 14-21) and a neck roast may also result from bone-in cutting of the chuck. The remainder of the chuck is used for lean for stew, ground beef, or other products. These bone-in retail cuts equal approximately 9 percent of the live weight of the animal and 15 percent of the carcass.

Figure 14-30. An alternate method of fabricating this primal is to muscle bone the chuck, resulting in two large boneless roasts. This is sometimes termed as the *Institutional Method,* since the boneless roasts are ideal for use in large institutions such as hospitals and dormitories where uniform servings in terms of size and quality are essential. One roast, the shoulder pot roast, was formerly called the outside or arm chuck. Clod was also a name in popular use for this sub-primal, but since the term *clod* is quite nondescriptive, ICMISC recommends that it not be used. The shoulder pot roast (outside chuck) is the large muscle group which lies behind the lower end of the arm bone and below the spine of the *scapula.*

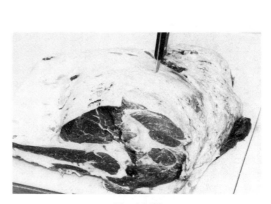

Fig. 14-30

In removing the boneless shoulder (outside) enter caudally to (behind) the *humerus* and follow this bone closely to its juncture with the *scapula.* Then follow the spine of the *scapula* toward the adjacent side. Then, by riding the knife against the spine of the *scapula,* on the lower, or arm, side of the spine, follow the spine to the blade face of the chuck.

Figure 14-31. As the outside chuck is trimmed away, the *scapula* (blade bone) is exposed. One method of removing the outside chuck recommends that a cut be made down (toward the table) at the lower (ventral) edge of the *scapula.* That cut should expose a fat seam, which is then followed out to the arm face to remove the outside chuck roast.

Fig. 14-31

Figure 14-32. The boneless chuck thus removed is trimmed of excess intermuscular (seam) fat so that the lean tissue on the inner surface is exposed. The edges of the roast are squared up. The thin or upper end is trimmed considerably, resulting in a more uniform roast. This roast is what a foodservice manager would expect to see when a box of 114A Beef Shoulder Clod Roasts (IMPS) is opened. External fat may be trimmed to approximately 0.25 of an inch, depending on specifications.

Fig. 14-32

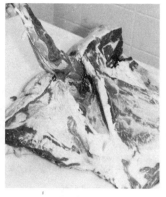

Fig. 14-33

Figure 14-33. The small muscle remaining above the spine of the *scapula* is the *supraspinatus* and is known as the mock, or Scotch, tender. It has a significant amount of connective tissue; therefore, it may often be tenderized mechanically to be made into minute (cubed) steaks or to be used as lean trim.

Fig. 14-34

Figure 14-34. After removing the thin muscle (*trapezius*) on the surface of the remaining chuck and the arm (*humerus*) and the blade (*scapula*) bones, separate the second boneless chuck roast from the chuck. The inside chuck or chuck roll is an extension of the rib-eye muscle with surrounding muscles included. Using the silver-dollar–shaped fat deposit in the center of the blade face of the boneless chuck as a starting place, cut through the chuck to the table, parallel with the back line, to the indentation where the joint between the arm and blade was removed.

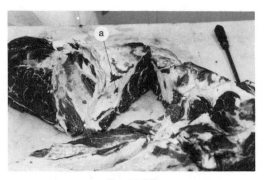

Fig. 14-35

Figure 14-35. At the indentation, a right-angle cut, which passes out through the back (dorsal side) of the chuck, removes the large inside chuck roast. If the *pre-scapular* lymph node with its large fat deposit has not been removed, this cut should bisect the lymph node as shown (a). The remainder of the chuck is used for lean trim, stew, or similar purposes.

Figure 14-36. The shoulder pot roast (outside chuck roast) is trimmed by removing seam fat from the inner surface and subcutaneous fat from the outer surface. Thin muscles, particularly on the upper, or dorsal, edge, also are trimmed to square the shape of the roast and improve its uniformity. The small muscle remaining on the surface of the roast, which was over the spine of the *scapula* (blade bone), is the *cutaneous omobrachialis*. This thin external muscle is often called the false lean, or rose, muscle. The fat covering this muscle is sometimes called the "frosting over the rose."

Fig. 14-36

Figure 14-37. In this view of the finished inside chuck roll, from the caudal (rib) end, the roast is turned over so the inside surface which was removed from the first five ribs is exposed, and you can see where the ribs were once located. This serves as a useful key in identifying this inside chuck roll (116A Chuck Roll, IMPS). Yield of roasts with this boneless method of cutting is somewhat less than that using the retail bone-in method and approximates 7 percent of the live weight and 11 percent of the carcass weight.

The roast shown in this figure could be trimmed more closely to include the "eye" muscle and a few of the immediately adjacent muscles. The roast shown not only increases the roast yield but also provides foodservice units with a large unit with greater cooking and serving potential.

Fig. 14-37

Figure 14-38. The inside chuck roast can be tied or netted (see Figure 13-22, page 421) to produce an even more desirable institutional roast. Consumer-sized roasts can be produced by cutting both the outside chuck and the inside chuck roasts into sections.

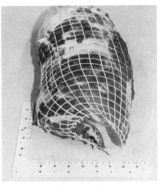

Fig. 14-38

Fig. 14-39

Figure 14-39. The *supraspinatus* (chuck tender) may also be used as a small roast but is probably better utilized as steaks which are passed through a tenderizer or cuber (Figure 14-67). The remainder of the boneless chuck is utilized for trim as ground beef, stew beef, or raw material for processed meat.

Fig. 14-40

Figure 14-40. The chuck neck bones may be cut into sections to use for excellent soup bones to make stock or for flavoring pieces. Deboning equipment is available and in use which will salvage significant amounts of mechanically separated beef from these and other bones. The resulting tissue is a sausage raw material. (See Chapter 20.)

The shank, brisket, and plate of the forequarter and the flank from the hindquarter are termed *rough cuts*. Wholesale percentages of these four cuts in total equal approximately 14 percent of the live weight and 24 percent of the carcass weight. Various retail cuts may be fabricated from these rough cuts, although through the use of modern retailing methods, much of this product is utilized as ground beef or stew beef. Retail yield in this form approximates 7 percent of live weight and 12 percent of carcass weight.

Fig. 14-41

Figure 14-41. The wholesale plate is directly ventral to the rib. The diaphragm may be removed from the plate after the membrane which covers the interior surface of the plate posterior to and covering the diaphragm muscle has been pulled out. This muscle is commonly called the skirt in this location of the carcass.

Figure 14-42. The thin, flat skirt muscle can be rolled up and skewered. The skirt will thus produce two skirt-steak rolls (pin-wheel steaks). These skirt-steak rolls look quite desirable, and when they come from high-quality beef and are properly prepared, they are. When the muscle is thin, the diaphragm, or skirt, may be best used as lean trim.

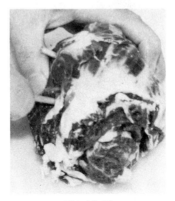

Fig. 14-42

Figure 14-43. Short ribs can be removed from ribs 6, 7, 8, and 9 (counting from the front). The heavy muscle (*serratus ventralis*) is known as the pastrami piece (a) the piece usually cured, smoked, and spiced for pastrami. These short ribs are adjacent to but ventral to those short ribs removed from the wholesale rib. Other options from the plate are boiling beef, beef spareribs, and a boneless plate. The boneless plate may be cured and sold as beef bacon, rolled and sold fresh, or used for ground beef.

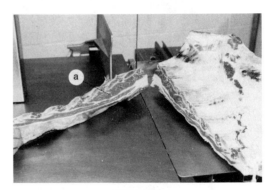

Fig. 14-43

Figure 14-44. The plate cuts are often economical cuts which, with proper preparation, make excellent family meals. The skirt steaks, short ribs, and plate ribs shown here add variety to the beef diet.

Fig. 14-44

Figure 14-45. The shank is separated from the brisket by being cut through the natural seam which separates them.

Fig. 14-45

Figure 14-46. The brisket may yield a boneless brisket roast (119 Brisket, Boneless, Deckle On, IMPS) if the sternum is removed. This boneless roast may be processed into corned beef following removal of the deckle—the muscle layer and fat on the top surface.

Fig. 14-46

Figure 14-47. The wholesale foreshank is easily identified. Shank cross cuts may be obtained from the foreshank. The prominent shank bones are the *radius* and the *ulna*. The shank, being very lean, is often boned and placed into lean trim for ground beef.

Fig. 14-47

Hindquarter

The hindquarter, composed of three wholesale cuts (the round, loin, and rough flank), represents approximately 48 percent of the carcass weight and 29 percent of the live weight.

Figure 14-48. The removal of the flank is one of the earliest procedures necessary in breaking the beef hind-quarter down into wholesale cuts. The wholesale flanks represent 4 percent of the live weight of an animal and 7 percent of its carcass weight. The flank is removed by following the contour of the round and removing the cod, or udder fat, with the flank. Towards the front, the cut is marked laterally to the loin-eye muscle not more than 10 inches from the backbone, where a saw must be used to sever the thirteenth rib. In the illustration, the thirteenth rib was cut about 1 inch past the lateral edge of the loin eye. The same procedures can be performed on the table but are illustrated here on the rail.

Fig. 14-48

Figure 14-49. Although the kidney and pelvic fat can be removed earlier, it is sometimes easier to trim and pull it free after the flank has been trimmed away from its attachments with the round. Care should be taken not to score into the tenderloin muscle which lies under the fat. The kidney fat just above the diaphragm is more obvious in the left side than in the right. This is caused by the alternate filling and emptying of the rumen in the live animal, leaving an open cavity for the kidney and surrounding fat to drop down into on the left side when the rumen is empty. For this reason, the right side is called the tight side, while the left is called the loose side in a beef carcass.

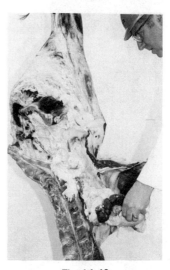

Fig. 14-49

Fig. 14-50

Figure 14-50. Separation of the flank is completed with a knife. The hanging tenderloin (a) is a portion of the diaphragm muscle attached to the back region of the last rib and contracts with each breath in the live animal. Because of this, it may not be as tender as some other, less active muscles, so it is best utilized as lean trim. The hanging tenderloin may be removed when the kidney and kidney fat are removed, after the flank has been cut away, or, later, when the wholesale loin is trimmed. The true tenderloin (*psoas*) should not be cut into when the diaphragm is removed.

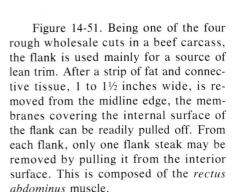

Fig. 14-51

Figure 14-51. Being one of the four rough wholesale cuts in a beef carcass, the flank is used mainly for a source of lean trim. After a strip of fat and connective tissue, 1 to 1½ inches wide, is removed from the midline edge, the membranes covering the internal surface of the flank can be readily pulled off. From each flank, only one flank steak may be removed by pulling it from the interior surface. This is composed of the *rectus abdominus* muscle.

Fig. 14-52

Figure 14-52. After its removal, the flank steak is usually scored by cutting the surface lightly at right angles. The remainder of the flank is trimmed of waste fat and utilized for ground beef.

   This text has indicated in several places that there are several methods used
to break specific wholesale units down into retail units. Local preferences, personal choices, and merchandising possibilities, as well as many other factors,
may determine the technique which best serves a situation. Cutting hindquarters
is an area in which a variety of options are used by meat cutters.

   There are several possible methods of removing the primal round from the
hindquarter. The meat industry generally acknowledges three types of beef
rounds that vary in the manner of cutting the sirloin tip. The sirloin tip is that
portion of the leg musculature that is anterior (forward) to the *femur* (thigh bone)
in a standing animal. It is often referred to as the knuckle in the meat industry.

   The Chicago round is the round described in the Institutional Meat Purchase
Specifications (IMPS) and is the most common round available in the wholesale
trade. The Chicago round is removed by a cut which begins at the junction of the
last sacral vertebra and the first caudal vertebra and passes through a point anterior to the prominence of the aitch bone. This cut is continued on a straight line
out through the flank, or ventral, portion of the hindquarter. The ball of the
femur is exposed by the properly executed cut. The wholesale rounds in Figures
14-1 and 14-2 are Chicago rounds. Note also the outlines in Figure 14-4. The Chicago round cutting style can be criticized because it divides the sirloin tip muscle
system, leaving one portion on the sirloin end of the wholesale loin and the other
portion of the sirloin tip on the wholesale round. A further problem is the fact
that the cutting angle reduces the potential value of both portions of the tip. One
can visualize the problem by observing Figure 14-1.

   An alternative method results in a Diamond round; named for its diamond
shape, which includes the portion of the sirloin tip that is left on the loin using
the Chicago style. The Diamond round is cut on the same angle through the sac-
ral/caudal junction and exposes the ball of the *femur*, but does not continue the
cut through the anterior, or flank side, of the round. The sirloin tip is separated
by making a cut which continues the angle of the aitch bone through the flank
side of the round. Figure 14-74, later in this chapter, illustrates the first step in
separation of the Diamond round, and Figure 14-4 shows the angle for cutting
the anterior face of the sirloin tip area. Although the Diamond round results in
the most useful wholesale round, it is difficult to produce without using a power
saw to separate the loin and round on the rail.

   The third option is a compromise which results from the separation of the
sirloin tip muscle system from the hindquarter before separating the round and
loin. The New York round is cut through the *sacrum* and *femur* socket at the
same angle and produces the equivalent of either a Chicago round or a Diamond
round with the sirloin tip removed (see diagram below).

The Chicago round will represent approximately 22.5 percent of the carcass or 14 percent of the live weight. The New York round and the Diamond round, including the sirloin tip in each case, will approximate 25 percent of the carcass weight and 15 percent of the live weight. The boneless retail yield of the round in any of these forms will represent 60 to 64 percent of the wholesale round weight.

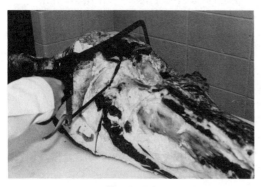

Fig. 14-53

Figure 14-53. Another method of separating the round cuts from the hindquarters without subdividing the sirloin tip is illustrated. The first cut is made caudally to (to the rear of) the *ischium* of the pelvic bone (commonly referred to as the aitch bone). By cutting parallel to and just behind the aitch bone, one must saw through the *femur*. The rump portion of the round must be removed from the loin separately, resulting in two pieces, the equivalent of a Diamond round (see Figure 14-69).

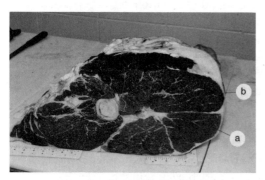

Fig. 14-54

Figure 14-54. The nomenclature of the retail cuts from the round can be explained very logically if one knows the origin of the names. The outside round is that portion which is toward the outside of the animal and is down on the table (a). The inside round (b) is that portion of the round toward the animal's midline. The outside round became known as the bottom round because in cutting, the round is usually placed on the table or block with the outside next to the table. Therefore, it was called bottom round. The inside, being on top, was thus called top round. The eye of the round is the lower right (*semitendinosus*) muscle, and the remainder of the bottom round is the *biceps femoris* muscle. The muscle system to the left (cranial) of the *femur* is the sirloin tip, discussed extensively above, made up of four muscles called collectively the *quadriceps*. The top round is made up of the *semimembranosus* and the *adductor* muscles. The top and the tip are more tender than the bottom and the eye.

Figure 14-55. The *quadriceps* muscles which compose the round tip and lie to the cranial (front) side of the *femur* are removed by entering through the natural seam which separates the round tip from the top or inside round on one side and from the bottom or outside round on the other. From this dorsal view, you can see the cross section of the *femur*.

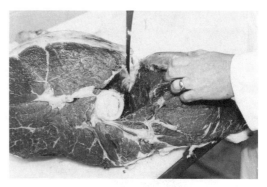

Fig. 14-55

Figure 14-56. Entering from the stifle joint near the *patella,* at the shank knuckle bone, which is the caudal or distal end of the *femur*, you can loosen the connections sufficiently so that it is possible to grasp the tip in one hand and the knuckle of the *femur* in the other to separate the two. Final separation is made with a knife. From the rough tip (IMPS 167), the *patella* (a), or knee cap, is removed.

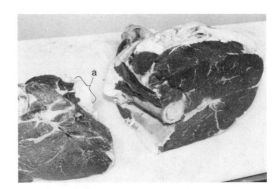

Fig. 14-56

Figure 14-57. The outside, or cap, muscles may be removed to make a highly desirable retail cut, the cap off ("bald") round (sirloin) tip roast or steak. This cut is located at the junction of the sirloin tip and the round tip but is officially named round tip, since it is a subprimal of the primal round. The term *sirloin tip* is well recognized by industry but often confusing to consumers.

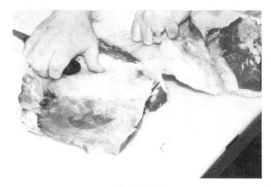

Fig. 14-57

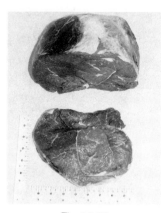

Fig. 14-58

Figure 14-58. The sirloin tip makes an excellent roast and may be cut into steaks if the animal has sufficient quality. The tip is usually the most tender roast from the round, due to its location in the live animal, in front of the large round bone (*femur*). Also in this position, the muscle fibers are stretched when the carcass is hung up in the conventional manner, thus making the roast more tender (see Chapter 21). The sirloin tip cuts can be identified by the horseshoe- or oval-shaped connective tissue line in the center of each cut.

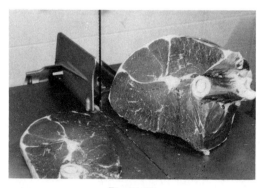

Fig. 14-59

Figure 14-59. Full round (bone-in) steaks are cut after the sirloin tip has been removed. To lead the cut through the saw with the bone it was necessary to cut this left round with the outside round at the top. Modern meat-cutting techniques have reduced the number of full round steaks in the retail case. Tenderness differences between the top and bottom round make it difficult to effectively use the full round steak as a single unit. In fabricating the remainder of the round into sub-primals, a technique known as muscle boning will be employed. In so doing, the muscles and muscle groups are separated into sub-primal and retail cuts by following the natural seams as much as possible. The muscles are taken apart in much the same manner as nature put them together. Muscles grouped naturally together for function in the live animal are more uniform in palatability and thus form much more desirable retail cuts.

Figure 14-60. The next step in muscle boning the round is the removal of the hind shank bone, or *tibia*. Muscles of the lower round are anchored to this bone through strong connective tissues, which must be severed. The large tendon on the back of the shank is the *Achilles* tendon in which was inserted a hook on a trolley to suspend the carcass on a rail. The *tibia*, or shank bone, is followed with a knife up into the round to the stifle joint. This is the knee joint of the animal where the *tibia* joins the large bone of the round, the *femur*. When this joint is severed with a knife, the remaining connective tissue attachments to the *tibia* are loosened, and the shank bone is removed.

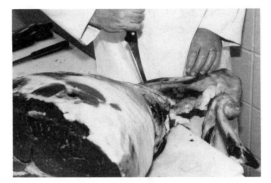

Fig. 14-60

Figure 14-61. From the shank knuckle or stifle joint, the large *femur* bone is being removed. With older methods of cutting, this large bone was left in the retail round steaks. Noting the size of this bone, don't you agree that it is unwise to leave it in the round and later have it occupy valuable freezer space?

Fig. 14-61

Figure 14-62. The *femur* and the shank bones, along with the *patella*, represent a significant amount of the weight of the round.

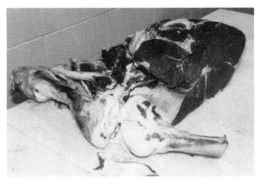

Fig. 14-62

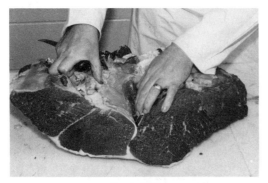

Fig. 14-63

Figure 14-63. Separate the bottom round on your left from the top round on your right by following the natural seam which separates them. The top round sub-primal is trimmed on the newly separated face as well as on the inside or top surface, where there is usually a portion of the lean that has been darkened due to exposure to air for several days during marketing or aging.

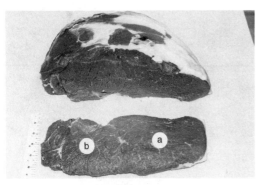

Fig. 14-64

Figure 14-64. Steaks, preferably for moist-heat cookery (braising), can be cut from the top round. The top round is composed of two major muscles, the *semimembranosus* (a) and the *adductor* (b). The top-round roasts and steaks appear very solid and homogeneous; thus, the muscle separation is difficult to see. The top round and the sirloin tip are generally considered to be the more tender round roasts.

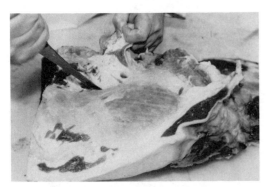

Fig. 14-65

Figure 14-65. Especially important in trimming the bottom round sub-primal is the removal of the *popliteal* lymph node and its surrounding fat deposit, which is located on the center top of the roast between the *semitendinosus,* or eye, and the *biceps femoris* muscles. The heel of the round is separated from the lower portion of the bottom round.

Figure 14-66. Steaks may also be cut from the bottom round. They may be cut thicker than the top-round steaks and make ideal material for swissing. Steaks and roasts from the bottom round can be most easily identified by the distinct eye of the round, the *semitendinosus* muscle (a). The other large muscle in the bottom round, sometimes alone called the bottom round, is the *biceps femoris* (b). The eye and bottom round roasts may be separated as well for smaller portions or for ease in slicing.

Fig. 14-66

Figure 14-67. Bottom-round steaks from lower-grade cattle may be utilized as minute (cubed) steaks by being passed through a mechanical tenderizer, a desirable method of merchandising the lower-quality bottom-round steaks.

Fig. 14-67

Figure 14-68. The heel, or *Pikes Peak,* roast should be utilized with moist-heat cookery, since it contains high amounts of connective tissues because of its location in the live animal—the lower portion of the round. The main muscle of the heel roast is the *gastrocmemius* muscle (a). The heel of the round may be used as lean trim for ground beef along with the rest of the shank meat and trimmings.

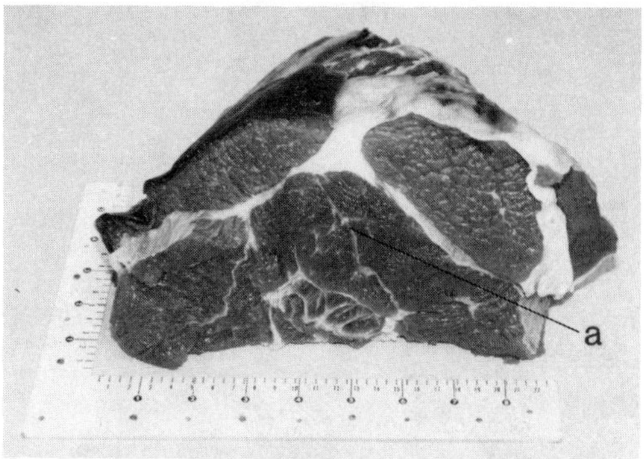

Fig. 14-68

The method for removing the round from the hind illustrated in Figure 14-53 resulted in an intact sirloin tip but left the rump portion of the round still attached to the loin. However, in industry, the rump is considered a part of the round. Thus, when rounds are quoted and traded, it is done with the assumption that the rump is a part of the round. Here we consider it a separate cut for purposes of clarity. The rumps, when considered wholesale cuts, represent 2 percent of the live weight and 4 percent of the carcass weight, while in retail form, the boneless roasts represent 1 percent of the live weight and 2 percent of the carcass weight.

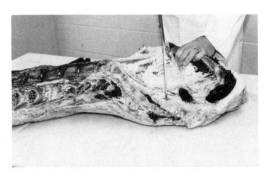

Fig. 14-69

Figure 14-69. The rump is separated from the loin by a cut made on an imaginary line connecting two points: (1) a point on the backbone between the fifth *sacral* vertebra and the first *coccygeal* (caudal, or tail) vertebra and (2) the anterior (front) tip of the proximal (inward) end of the *femur*. This cut is identical to the Chicago-style separation described earlier and indicated in Figures 14-1 and 14-2, except that now the sirloin tip has been removed and will not be cut in two by this separation. A piece of bone from the *femur* knuckle, resembling a silver dollar in size, results when a proper cut has been made. The Institutional Meat Purchase Specifications (IMPS) indicate that the cut is anterior to the *femur* knuckle but exposes the *femur*.

Fig. 14-70

Figure 14-70. The beef round rump roast (standing rump) normally has the knuckle from the *femur* removed, but still contains the large pelvic bone, which creates difficulty in carving and serving for the consumer. Note in Figure 14-1, the large portion of the pelvic bone contained in the rump area.

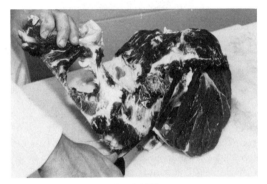

Fig. 14-71

Figure 14-71. It is more appropriate for the butcher or meat cutter to bone the rump and ultimately much more satisfying to the consumer to be able to effortlessly carve an exquisite roast for company. The pelvic bone (aitch bone) is "cleared" by following closely on either side with the knife. Note the size of the pelvic bone as its last attachments are severed.

Figure 14-72. The boneless rump (left) is rather diffuse, so it must be tied for ease of cooking and slicing. Any boneless roast can be very easily and quickly passed through the tube of a jet netter to become closely wrapped in this elastic net (right). The jet net is left on during roasting and should be removed before the roast is served.

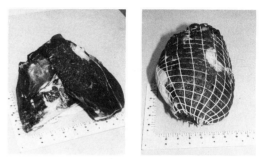

Fig. 14-72

Figure 14-73. Another approach to cutting the hindquarter on the rail begins after the flank has been removed (see Figures 14-48 through 14-50). The sirloin tip (a) is removed prior to making the round/loin separation in the following procedure.

Fig. 14-73

Figure 14-74. A cut beginning in front of the aitch-bone prominence (a) is made through the round to the flank or lower side of the hind. The cut is made at the same angle as the aitch-bone angle.

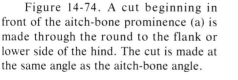

Fig. 14-74

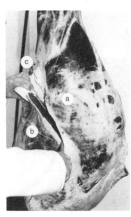

Fig. 14-75

Figure 14-75. Following the seams separating the outside round (a) (shown) and the inside round (on the opposite side) from the sirloin tip (b) system, cut down from the stifle joint (c) to the cut described in Figure 14-74.

Fig. 14-76

Figure 14-76. The front of the *femur* is exposed as the sirloin tip is pulled out and separated near the *femur*/pelvic joint.

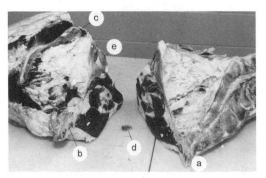

Fig. 14-77

Figure 14-77. After the sirloin tip is removed, the round with the rump on is separated from the loin on a line beginning at the fifth *sacral* vertebra (a) junction with the first *caudal* vertebra (b) to a point just anterior to the aitch-bone prominence (c). Note the small portion (d) of the *femur* ball (e) that was removed in making the cut correctly.

Figure 14-78. The aitch bone (pelvic bone) is removed from the round. Some butchers push the steel (for aligning knife edges) through the *obturator foramen* (hole through the pelvic bone) and use it as a lever to aid in removing the bone. Note the ball of the *femur*, which must be released from its socket in the pelvic bone. After the shank bone is removed (Figure 14-60), the *femur* is removed, as in Figure 14-61. Because the rump is on the round, the *femur* will be longer than in Figure 14-62, since it will include the ball head on its proximal end. The top round and bottom round are separated on the seam between them, as in Figure 14-63. Trimming is simplified by leaving the seam fat on the bottom round. To stay on the proper seam, the top round should be made as small as possible.

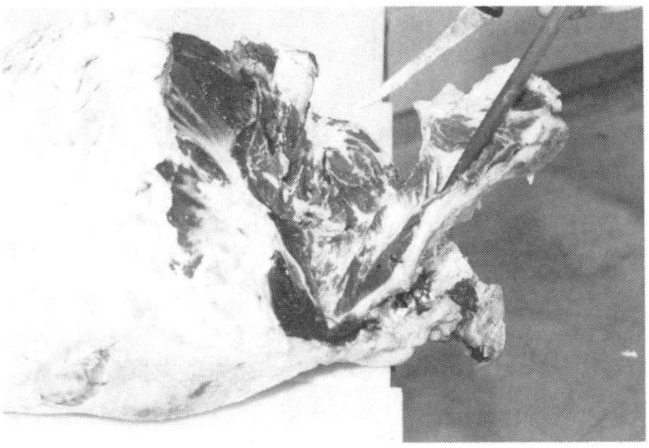

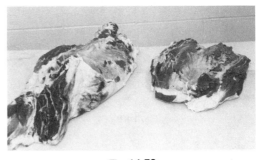

Fig. 14-78

Figure 14-79. The top round (IMPS 168) (right) can be trimmed of surface fat to no more than ¼ to ⅛ inch, and the dry and knife-scored areas can be trimmed away to produce a very desirable roast, representing approximately 15 percent of the round weight. The bottom round (IMPS 170) and shank area (left) will yield three roasts.

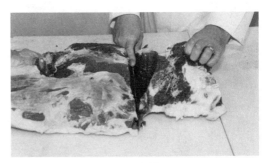

Fig. 14-79

Figure 14-80. After the fat is trimmed from the inside surface of the outside round, and after the *popliteal* lymph node and fat deposit surrounding it are carefully removed (Figure 14-65), the rump is separated. The cut is made at the cranial end of the large boneless cut, behind the point where the rear of the pelvic bone was removed and perpendicular to the axis of the roast.

Fig. 14-80

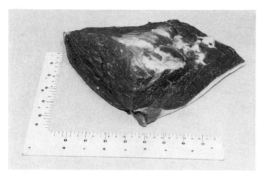

Fig. 14-81

Figure 14-81. The rump roast can be trimmed even more closely than illustrated in Figure 14-72. The finished small roast is one major muscle system comprising about 4 percent of the wholesale round.

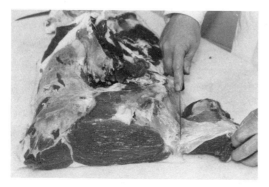

Fig. 14-82

Figure 14-82. When the round muscles are separated and trimmed, the heavy membranes of the *epimysium* connective tissue that surround some muscles should be trimmed away. (Note membrane to the left of the knife blade.)

Fig. 14-83

Figure 14-83. The shank (heel of round) is separated cranially to the tapering of the heel roast. The small round muscle in the center of the face of the bottom round should be trimmed from the roast for ground beef because of its connective tissue content. The trimmed bottom round roast as shown (right) approximates 17 percent of the wholesale round. The heel roast is trimmed, as shown in Figure 14-68.

Figure 14-84. The wholesale primal loin (IMPS 172) is composed of two sub-primals—the sirloin (IMPS 181) and the shortloin (IMPS 173). Normally, when full loins are fabricated into bone-in steaks, the sirloin and shortloin will not be separated, but rather the whole loin will be sawed into steaks from this sirloin end. The sirloin sub-primals wholesale weight approximates 5 percent of the live weight and 8 percent of the carcass weight. Steak yield from the sirloins equals about 3 percent of the live weight and 5 percent of the carcass weight.

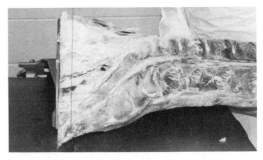

Fig. 14-84

The shortloin is that sub-primal of the primal loin remaining after the removal of the sirloin. The shortloin is composed of the *lumbar* section of the hindquarter. As wholesale sub-primal cuts the two shortloins from a carcass represent 4 percent of the live weight and 7 percent of the carcass weight. Steak yield from the shortloin will equal approximately 2 percent of the live weight and 4 percent of the carcass weight. Note that one rib remains on the cranial end of the shortloin. Since steak yield (retail yield) is 4 percent of the carcass weight and wholesale sub-primal yield is 7 percent of the carcass weight, the yield of the wholesale sub-primal in steaks is four-sevenths, or 57 percent. By using this same relationship, the retail yield of any wholesale cut can be determined. (See Figure 14-99 for separation of the shortloin and sirloin.)

Figure 14-85. The sirloin is cut on a band saw into steaks which may differ considerably from one another in their value due to the amount of bone they contain. The ICMISC recommends four names for the various sirloin steaks, based on the shape of the pelvic bone each contains. Beginning at the caudal (rear) end of the sirloin, this steak and one or two more cranially from (toward the front of) this steak are called wedge-bone sirloin steaks, because that portion of the *ilium* (pelvic bone) has the characteristic wedge shape. The first steak removed is sometimes called a butt-bone sirloin steak, because the pelvic bone has a depression where the *femur* (round bone) engaged the pelvic bone. Note the top-sirloin, or *gluteus medius,* muscle. This muscle carries throughout the whole sirloin and may be removed and sold as a boneless top-sirloin roast or cut into top-sirloin steaks. Note on the remaining portion of the sirloin, the wedge bone, the *ilium,* and the *sacrum.*

Fig. 14-85

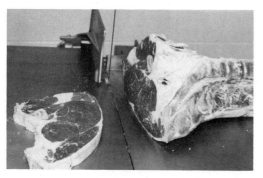

Fig. 14-86

Figure 14-86. Proceeding further in a cranial direction, the next steak is the round-bone sirloin steak which receives its name from the fact that the shaft of the *ilium* is almost round at this point. Note again in the remaining portion of the sirloin, the round bone and the *sacrum*.

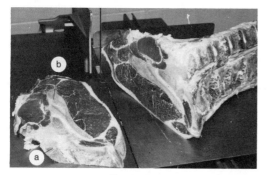

Fig. 14-87

Figure 14-87. Proceeding further in a cranial direction, we approach that portion of the sirloin which contains the backbone or *sacrum* (a) joined to the pelvic bone or *ilium* (b), thus known as the *sacroiliac* joint. Since there are two bones in this particular steak, it has received the name double-bone sirloin steak.

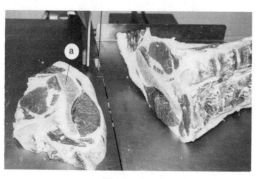

Fig. 14-88

Figure 14-88. The most cranial steak of the sirloin is the pin-bone sirloin steak. The pin bone (a) or hipbone (*tuber coxae*) is quite prominent at this location, as can be evidenced in the steak on the saw. There is usually one pin-bone sirloin steak and the steak immediately caudal (to the rear) to it is a (double-bone) flat-boned sirloin steak. The tenderloin (*psoas*) muscle appears quite prominently in the pin-bone sirloin steak. Note the pin-bone structure remaining on the loin. The next steak will include the junction between the sirloin and the short loin, resulting in the other face of the steak having the anatomy of the porterhouse steak.

Figure 14-89. Of all the steaks of the sirloin, the round-bone steak is the meatiest, due to the small amount of bone and few fat seams. Trimming the *sacral* vertebra (top) makes the round-bone sirloin even more merchandisable.

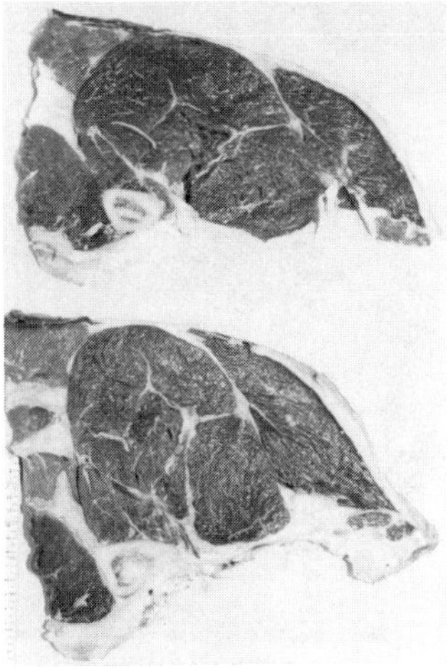

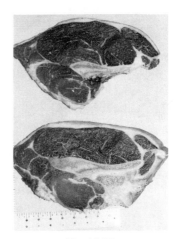

Fig. 14-89

Figure 14-90. The double-bone sirloin steak includes a significant amount of bone. In modern methods of merchandising, the backbone, or *sacrum*, is removed, leaving only the *ilium* present, thus a flat-bone sirloin steak (top). Note the top-sirloin, or *gluteus medius*, muscle which persists throughout the sirloin.

Fig. 14-90

Figure 14-91. The hipbone, or pin bone, separates the top sirloin, *gluteus medius* (a) from the tenderloin, *psoas major* (b), in the pin-bone sirloin. The pin-bone sirloin has an extremely large amount of bone in relation to the amount of meat, thus there is considerable value difference between types of sirloin steaks.

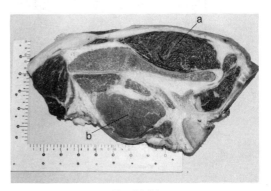

Fig. 14-91

Fig. 14-92

Figure 14-92. Beginning to saw from the sirloin or caudal end, the first steaks off the shortloin are the porterhouse steaks. Porterhouse steaks may be identified by the large size of the tenderloin (*psoas*) muscle (a) ventral to (below) the *transverse* (T) vertebral processes. The real key to identifying the porterhouse steak is, however, the presence of an additional muscle dorsal (above) to the *longissimus,* which is actually an extension of the top sirloin, the *gluteus medius* muscle (b). The Uniform Retail Meat Identity Standards indicate that a porterhouse steak has a tenderloin muscle which exceeds 1¼ inches in diameter.

Fig. 14-93

Figure 14-93. Moving forward in the shortloin, as the tenderloin decreases in size and the *gluteus medius* disappears, the steaks become T-bone steaks. The characteristic T is formed by the *transverse* processes of the *lumbar* vertebrae.

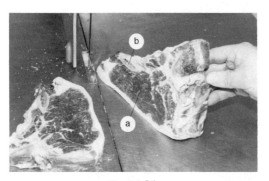

Fig. 14-94

Figure 14-94. Proceeding forward, where the tenderloin disappears and a rib or a portion of an obliquely split rib appears, the steaks are properly named top-loin steaks. Previously this particular steak was called a club steak, but the term *club* was so widely used, oftentimes inappropriately, for many other cuts from other locations on the carcass that the ICMISC decided to use the universal descriptive term *top-loin steak*. Note the large loin-eye (*longissimus*) muscle (a) and the portion of the split rib (b). Note also that the cap muscle observed on rib steaks (Figure 14-13) is not present.

Figure 14-95. The bone-in shortloin yields three kinds of steaks. The most caudally is a porterhouse steak (upper left), since it includes a small portion of the *gluteus medius* muscle (see Figure 14-92). However, the upper-right steak, technically a T-bone steak, can be merchandised by IMPS as a porterhouse steak, since the tenderloin diameter exceeds 1¼ inches. A T-bone steak, by all definitions, is shown lower left, and the lower-right steak is a top loin with the identifying bias-cut thirteenth rib exposed on the lower-left area of the steak (a). The large, meaty loin cut for the illustrations demonstrates the nomenclature discrepancies that may exist. When steaks

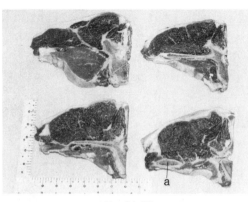

Fig. 14-95

are cut 1 inch thick, and porterhouse steaks are identified by the presence of a portion of the *gluteus medius*, and top-loin steaks by the presence of a portion of the thirteenth rib, this shortloin yields 3 porterhouse, 10 T-bone, and 3 top-loin steaks. Use of the IMPS changes the identities of the steaks. Porterhouse steaks are identified as having 1¼-inch or more tenderloin diameter, and T-bone steaks are identified as having ½-inch or less tenderloin diameter, resulting in 10 porterhouse steaks, 5 T-bone steaks, and 1 top-loin steak from the same shortloin. The number of top-loin steaks decreased from three to one because two of the steaks which contained a bias-cut section of the thirteenth rib also exhibited a portion of tenderloin that exceeded ½ inch in diameter, thereby qualifying them for classification as T-bones.

Figure 14-96. In an alternate method of fabricating the entire loin, the tenderloin muscle, *not* the hanging tenderloin, may be removed and processed as a fillet. Care must be taken in removing this tenderloin muscle.

Fig. 14-96

Figure 14-97. The tenderloin (IMPS 190) is trimmed of accessory muscles (*psoas minor*) and membranous fat layers to produce a trimmed tenderloin from which tenderloin fillet steaks are cut. The popular *filet mignon* is a tenderloin steak wrapped with a slice of bacon.

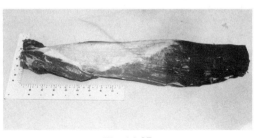

Fig. 14-97

Fig. 14-98

Figure 14-98. The tenderloin fillet is the most palatable muscle in the beef carcass but may be easily confused with the eye of the round which is quite similar in appearance. Recall that the eye of the round is the *semitendinosus* muscle from the bottom, or outside, round. During the animal's life, this *semitendinosus* muscle is used every time the animal takes a step as it moves from place to place. On the other hand, the *psoas* muscle of which the tenderloin fillet is composed, lying on the inside of the *transverse* processes of the loin, does very little work in the animal other than perhaps aid in some small way the maintenance of the animal's posture or aid it as it turns its body to a slight degree. The eye of the round is considerably less tender than the tenderloin fillet, and this fact is related to the function of the respective muscles within the live animal. To differentiate these two steaks, note the coarse structure of the eye of the round as compared to the fine texture of the tenderloin fillet. Here is a simple lesson in muscle structure. The muscle bundles are much larger in the eye of the round, thus causing the coarser texture. Another identifying hint is the fact that the eye of the round displays a strip of external fat, seen in the upper portion of this figure, which is normal, since this cut lies to the outside of the animal. The tenderloin fillet, on the other hand, from the inside of the animal, has no surface exhibiting subcutaneous fat.

Fig. 14-99

Figure 14-99. To continue the alternate loin fabrication procedure, the sirloin is separated from the shortloin by a cut made between the fifth and sixth *lumbar* vertebrae which nicks the cranial or front portion of the pelvic girdle (pin bone). After the backbone has been separated with a saw, the cut is completed with a knife.

Figure 14-100. The pelvic bone and *sacral* vertebra are removed from the sirloin. The bone should be carefully followed with the tip of the knife to avoid scoring into the muscle. Note the size of the bone removed (left hand).

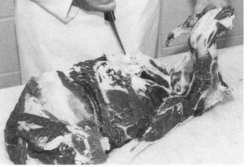

Fig. 14-100

Figure 14-101. The boneless sirloin (IMPS 182) is trimmed by removing a 1- to 1½-inch strip along the back line to remove the connective tissue attachments to the vertebrae. The ventral tail of the sirloin is removed by following a hard-to-find seam between the top sirloin (*gluteus medius*) and the tail muscle (*tensor fasciae latae*). The front (cranial) and rear (caudal) ends of the roast should be squared. The trimmings can be used for kabobs or ground beef.

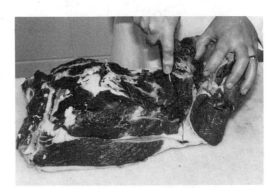

Fig. 14-101

Figure 14-102. Very desirable boneless sirloin steaks can be cut from the boneless sirloin roast. The meat cutter must examine the roast carefully to assure that steaks are cut "across the grain" (perpendicular to the muscle-fiber direction). This steak is often called a top-butt steak or a boneless top-sirloin steak.

Fig. 14-102

Fig. 14-103

Figure 14-103. Further fabrication of the shortloin by this alternative method involves removing the chine bones or body of the *lumbar* vertebrae of the backbone, trimming the "tail" of the shortloin to approximately 1 inch lateral to the loin-eye muscle (the shortloin tail is adjacent to the flank in the intact carcass), and trimming away the external fat in excess of 0.25 to 0.3 of an inch. The above trimming is subject to market demand and specifications. The resulting subprimal then is called a top-loin or striploin, bone-in or New York strip (IMPS 179).

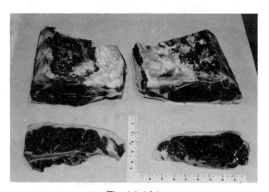

Fig. 14-104

Figure 14-104. Any steak from this top-loin, bone-in, or striploin would be officially called a top-loin steak. It may be called a bone-in strip steak or a New York steak unofficially. Note the *transverse* processes of the *lumbar* vertebrae. The *dorsal* processes of these vertebrae may also remain on the steak or roast. This particular steak is actually a porterhouse steak without the bone and tenderloin muscle. If the remaining bone is removed (the *transverse* process and that remaining portion of the *dorsal* process), the resulting steak would be officially a top-loin steak, boneless or unofficially a boneless strip steak, sometimes called a Kansas City steak.

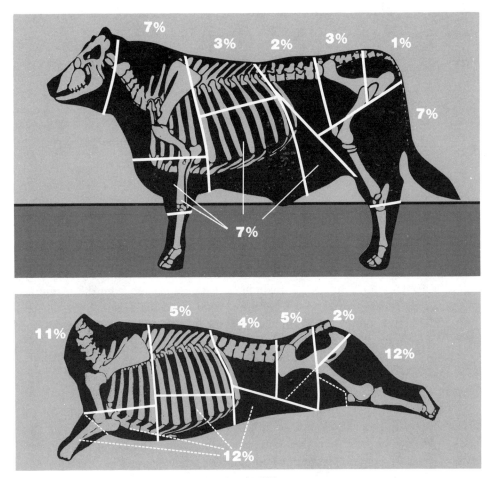

Fig. 14-105

Figure 14-105. Total retail roasts and steaks fabricated and closely trimmed as shown previously in this presentation make up approximately 23 percent of the live weight of the animal (Table 14-1). This 23 percent figure is obtained by adding the percentages listed in the six primal cuts—7 percent in the round, 1 percent in the rump, 3 percent in the sirloin, 2 percent in the shortloin, 3 percent in the rib, and 7 percent in the chuck. Identical figures were presented as each wholesale cut was fabricated. Similar figures may be added in the carcass diagram, summing to a total of 39 percent of the carcass weight in the closely trimmed steaks and roasts (Table 14-1). Now note in the live animal diagram the figure 7 percent, which is directed toward all four rough cuts—the shank, brisket, plate, and flank. This 7 percent means that 7 percent of the live weight of the animal is represented in retail yield or, in this case, lean trim from these four rough cuts. Likewise, the 12 percent figure in the carcass diagram has the same connotation.

Table 14-1. Percent of Beef Live and Carcass Weights Expected in
Roasts and Steaks, Lean Trim, Fat Trim, and Bone

	Percent of	
Component	Live Weight	Carcass Weight
Roasts and steaks	23	39
Lean trim	15	25
Total retail meat[1]	38	64
Fat trim	14	24
Bone	7	12
Total[2]	59	100

[1]Sum of roasts and steaks and lean trim.
[2]Total is sum of retail meat, bone, and fat trim (carcass).

Table 14-1 indicates that lean trim accounted for approximately 15 percent of the live weight. Of this, 7 percent was obtained from the rough cuts, and the remaining 8 percent resulted from the trimming of the primal cuts to produce desirable roasts and steaks. Likewise in the carcass column, the 25 percent figure for total lean trim is made up of the 12 percent of the carcass weight obtained from the rough cuts, while the remaining 13 percent comes from the trimmings of the primal roasts and steaks. Therefore, total retail yield, which would be a sum of these two figures, approximates 38 percent of the live weight of the beef animal or 64 percent of its carcass weight.

Note further in the table that the remaining components of the carcass of a beef animal are fat trim, which approximates 14 percent of the live weight and 24 percent of the carcass weight, and bone trim, which represents 7 percent of the live weight and 12 percent of the carcass weight. Thus, we can account for the gross composition of the carcass produced by this 1,100-pound live animal. Perhaps it will not be so difficult for you now to rationalize the retail price of beef cuts in comparison to the price of live animals, when you realize the relatively small percentage of the live animal which is represented in high-quality edible meat.

It should be stressed that these are average yield figures and may vary considerably, due to methods of fabrication and the makeup of the animal itself, in terms of muscling and fat. These average figures are the result of the fabrication methods shown here. An animal yielding this percentage of retail product might very well be our 1,100-pound live animal with a 660-pound carcass that might have approximately 11 to 11.6 square inches of rib eye at the twelfth rib; possess about 3.5 percent of its carcass weight in kidney, pelvic, and heart fat; and carry about 0.6 of an inch of subcutaneous fat over the rib eye at the twelfth rib. What is the USDA yield grade of this steer? Check Chapter 11 if you do not remember how to calculate this yield grade.

Animals differing in fatness and muscling may cause the 38 percent retail yield in the live animal to vary from 33 to 45 percent, while the retail yield from carcass weight may vary from 55 to 75 percent. The data reported here included

Table 14-2. Percent of Carcass Weight in Roasts and Steaks from Each Primal Cut and in Total Roasts and Steaks, Lean Trim and Fat, and Bone, as Estimated by Three Sources

Product Type and Wholesale Cut	Percent of Carcass Weight		
	NLSMB[1]	AMI[2]	MWE[3]
Roast and steaks			
Chuck	16.6	16.8	11.0
Rib	6.0	5.9	5.0
Loin	12.2	12.0	9.0
Round	9.6	9.7	14.0
Total	44.4	45.0	39.0
Lean trim	25.3	23.4	25.0
Fat and bone	29.8	31.0	36.0

[1]National Live Stock and Meat Board. *A Steer's Not All Steak.*
[2]American Meat Institute. *Meat Facts.* 1984.
[3]*The Meat We Eat.* 1985.

some bone in the loin and rib steaks. The rest of the roasts and steaks and rough cuts were processed to the boneless forms indicated in the discussion.

Table 14-2 lists data in Table 14-1 and in Figure 14-105 again and compares it to similar information from other sources. Although similar, there are variations in the data. That variability is one reason why cutting tests, as discussed in Chapter 12, are so important in meat merchandising. A wide discrepancy exists between this text and the other sources in the roast and steak yield from the chuck. The data reported in *The Meat We Eat* are based on boneless chuck roasts, whereas the other sources cut blade, arm, and cross-rib roasts, leaving the bone and some of the seam fat in the roasts. Closer trimming of the rib and loin in the procedures described in this book results in steak cuts with less fat, bone, and "tail" on them than may be true in most cases. Correspondingly, more of the chuck, rib, and loin are found in the fat and bone portions. Much of the 4 percent advantage in roast and steak yield from the round is due to the salvage of the sirloin tip as an intact roast in procedures described earlier in this chapter. Utilizing the heel of round as a roast, rather than lean trim, contributes to the roast-yield advantage in the round as well. The higher proportion of fat and bone listed in *The Meat We Eat* may be due to a combination of more boneless cuts, closer trimming of surface and seam fat, as well as differences in carcass fat content.

NEW DEVELOPMENTS IN BEEF HANDLING AND DISTRIBUTION

Prior to 1966, most beef was distributed in carcass form; that is, generally trucked or shipped by rail from the packing house in forequarters and hindquarters. The quarters were eventually cut into wholesale and retail cuts in some wholesale house or in the "back room" (a meat processing room) of a retail store.

In 1976, more than two thirds of the beef shipped to supermarkets was centrally fabricated, either at the packing house or at a meat warehouse, and the prefabricated primal and sub-primal cuts were delivered to stores in boxes instead of in the conventional carcass form.

Thus the term *boxed beef*, or *beef in a box*, was born. In 1982, 79 percent of the beef in the wholesale market was "in a box," and there are some estimates that boxed beef will represent 82 percent of wholesale beef in 1985.[2]

Modern beef plants have been designed to process beef into the box. Extensive cutting, vacuum packaging, boxing, and product handling systems have been developed to process 200 to 400 beef carcasses per hour. Most of the modern operations are based on moving rails and conveyor systems, which bring the product to the meat cutters, who each perform one specific step in the preparation of a sub-primal cut. Figures 14-106 through 14-112 (courtesy of IBP Corp.) illustrate some aspects of carcass disassembly and the boxing operation.

Fig. 14-106. After approximately 20 hours in a chill cooler (30° to 34° F), the beef carcasses are ribbed and displayed to USDA meat graders on a moving rail. Note excellent lighting and plant employee applying the grade roll. (See Chapter 11.)

[2]*1982 Retail Boxed Beef Survey*. Cryovac Division, W. R. Grace & Co.

Fig. 14-107. Inventory information, including weight and grade, is entered into the plant computers immediately after the carcass has been graded.

Fig. 14-108. Carcasses undergo some of the initial breakdown while on the moving rail. Chuck and rib separation is an early step.

Fig. 14-109. Still on the rail, the sirloin tip is pulled from the round and dropped on the conveyor below. (Note empty round hooks returning in background.)

Fig. 14-110. As product passes down the conveyor, further trimming is accomplished. Conveyor in foreground carries chucks, whereas ribs are sent down the one on the left.

Figure 14-111. Finished cuts are bagged, and the bags are evacuated and sealed under the hoods of vacuum-packaging machines.

Fig. 14-112. The bagged product is placed in boxes. In this plant, the boxes are automatically sorted, using the Universal Product Code System (see box labels on left); palletized; and warehoused by computerized automation. The automatic robot crane can be seen at the end of the aisle moving a pallet of product about halfway up on the beam.

Chapter 15

Lamb Identification and Fabrication

ALTHOUGH the per capita lamb consumption in the United States was only 1.7 pounds, carcass weight (1.5 pounds, retail weight) in 1983, those persons who eat lamb probably eat it regularly and thus consume a considerable amount. Evidence that lamb consumers are truly a small segment of the population is found in the comparison of lamb consumption with the per capita beef consumption of 106.5 pounds, carcass weight (78.8 pounds, retail) and per capita pork consumption of 66.2 pounds, carcass weight (62.3 pounds, retail) in 1983. Based on consumption data, it may be assumed that the majority of meat consumers are not aware of the many attributes of lamb and do not include it as an option when planning menus for  the family or when ordering in a restaurant. In effect, many consumers cannot consider lamb as an option, because it does not appear as a restaurant-menu item or in the retail-meat cases in some areas of this country.

The purpose of this chapter is to share with the reader information about the identity, relative value, and palatability of lamb wholesale cuts. Livestock producers may utilize the product knowledge in breeding animal selection and in planning lamb-feeding programs to enhance consumer acceptance. Members of the meat industry may be introduced to new ideas that will improve the processing, distribution, demand, and retailing in the lamb business. Educators at all levels have less access to information about lamb than other meat species. This chapter is designed to help fill that void. Each reader is a food consumer, who, after reading this chapter, will perhaps find increased variety and enjoyment derived from adding lamb to his or her meat diet.

Lamb has long been noted for its delicacy of flavor and for its tenderness. Starting in biblical times, and continuing through to this modern age, reference has continually been made to the desirability of lamb meat. It is featured frequently for gourmet dining at home and in hotels and restaurants. However, many people have had unpleasant experiences with lamb or mutton, due to a lack of knowledge on their part, or, perhaps, even on the part of those responsible for the processing, fabrication, merchandising, preparation, or serving of the ovine meat, be it lamb or mutton.

IDENTIFICATION

The *Uniform Retail Meat Identity Standards*[1] (see Chapter 12) will be used as the basis for identification in this chapter.

FABRICATION

The skeletal structure of the ovine in relation to an outline of the wholesale cuts is shown in Figure 15-1, and a more detailed chart of wholesale and retail cuts and how to cook them appears as Figure 15-2. Use these two figures for reference as the discussion moves through cutting a lamb carcass. Observe the lamb retail cut photos in the center section of this text for further aid in visualizing the many lamb cuts discussed here.

Figures 15-3 through 15-64 appear through the courtesy of the University of Illinois and South Dakota State University.

More Detailed References

In addition to those references indicated in Chapter 12, the following is especially appropriate for lamb: *How to Cut Today's New Lamb for Greater Sales and Profits,* published by the American Lamb Council, 200 Clayton Street, Denver, CO 80206, and the National Live Stock and Meat Board Lamb Committee. The publication can be obtained from the American Lamb Council.

----

[1]National Live Stock and Meat Board.

LAMB CHART
LOCATION, STRUCTURE
AND NAMES OF BONES

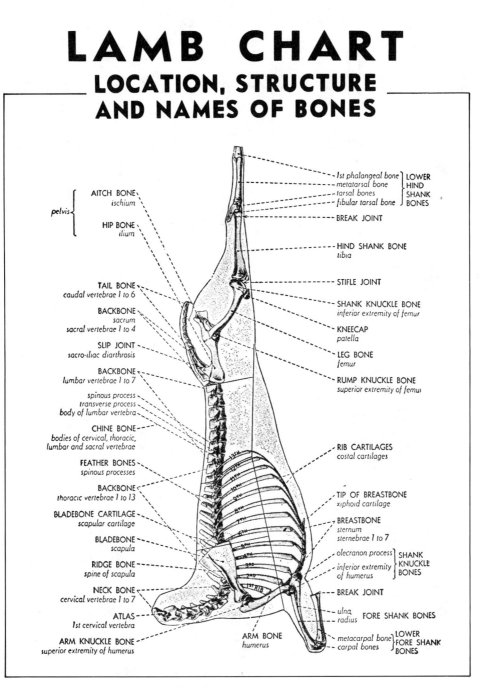

Fig. 15-1. Ovine anatomy

LAMB CHART

RETAIL CUTS OF LAMB — WHERE THEY COME FROM AND HOW TO COOK THEM

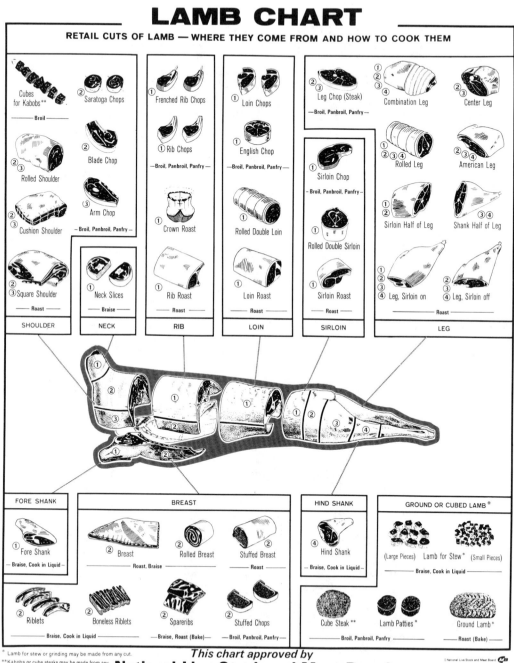

This chart approved by
National Live Stock and Meat Board

* Lamb for stew or grinding may be made from any cut.
**Kabobs or cube steaks may be made from any thick solid piece of boneless Lamb.

Fig. 15-2

The First Steps

For demonstration purposes, let us assume that generally, when lambs are finished for market, they will weigh approximately 100 pounds. Perhaps in the future, we will see lambs ready for market at 150 pounds, due to the current trend toward larger lambs. After slaughter, the resulting carcass represents approximately 50 percent of the live weight. Therefore, the 100-pound live animal will have a carcass weight of approximately 50 pounds.

Figure 15-3. Unlike beef or pork carcasses, lamb carcasses are not split before leaving the slaughter floor. One reason for this is that lamb carcasses, being light in weight, can be cooled and handled as whole carcasses very readily. In addition, some cuts from the loin and rib may be double cuts, if the carcass is unsplit.

Fig. 15-3

The lamb-packing industry does not rib (separate foresaddle and hindsaddle) lamb carcasses under normal marketing procedures. Shipping ribbed carcasses results in shrinkage and dehydration of the exposed rib-eye surface, and the lamb carcass need not be divided for ease of handling, as noted previously; however, in order for a lamb carcass to be properly evaluated for subcutaneous fat thickness and rib-eye area in classroom, laboratory, or carcass-contest situations, the carcasses must be ribbed. When this is done, the foresaddles and hindsaddles are separated between the twelfth and thirteenth ribs. Since the lamb carcass is not split longitudinally, the backbone cannot be seen, so the ribber must count ribs from the inside to find the twelfth rib.

Normally, the knife is inserted toward the outside to mark between the twelfth and thirteenth ribs. A cut is then made in a manner so as to bisect the loin-eye muscle (*longissimus dorsi*) at right angles to its long axis so that there is no distortion in its cross-sectional size. One skilled with a knife may cut through the muscle and the cartilaginous connection between *thoracic* vertebrae, thus completing the separation without a saw. However, the use of a saw has proven very beneficial in neatly severing the backbone.

After the carcass has been ribbed, the rib eye and fat cover over the rib eye are easily measured. The terms *hindsaddle* and *foresaddle* are used because the carcass has not been split. For fabrication, the hindsaddle is easily separated from the foresaddle by severing the attachment between the two, which is the flank muscle.

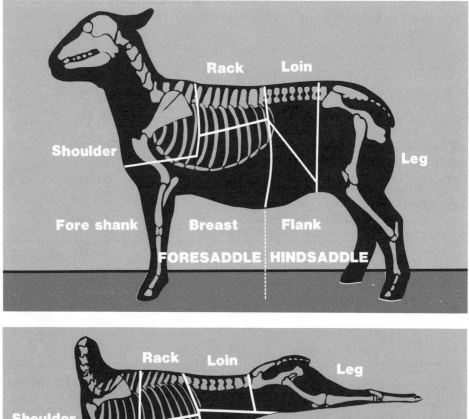

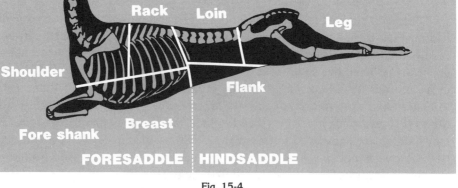

Fig. 15-4

Figure 15-4. The foresaddle comprises slightly more than half of the carcass weight as is the case with the beef forequarter. The foresaddle, composed of the shoulder, rack, foreshank, and breast, comprises 51 percent of the carcass and 25.5 percent of the live animal, while the hindsaddle, composed of the loin, leg, and flank, comprises 49 percent of the carcass weight and 24.5 percent of the live weight. The leg, loin, rack, and shoulder are the primal cuts of lamb.

Hindsaddle

The first step in the fabrication of the hindsaddle is the removal of the one rough cut from the hindsaddle, the flank. The flank in wholesale form comprises 1.5 percent of the live weight or 3 percent of the carcass weight. Possible retail cuts from the flank will be discussed later.

Figure 15-5. There are several modifications in the way in which the flank is removed. The modern method is to remove as much of the flank from the loin as possible, since the loin is much more palatable than the flank. As shown, a measurement may be made laterally or away from the end of the loin eye, a distance equal to one half the width of the loin eye itself. In industry, a 3-inch distance from the eye is common, but this leaves a rather long, undesirable "tail" on each loin chop. The flank muscles are just not palatable compared to the loin eye itself. Thus, a close trim here means a more satisfied consumer.

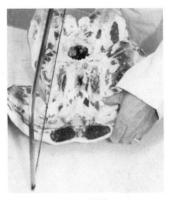

Fig. 15-5

When a double or kidney chop is made, much more flank must be left on the loin, and the separation must be made farther into the flank. This flank can be completely removed with a knife because of the wide flank left on the loin, allowing the knife to go outside or laterally to the end of the thirteenth rib. The *longissimus dorsi* is called loin eye because it is located in the loin. It is called the rib eye when it is viewed on the rib (rack) of the foresaddle. The largest, flat, straight muscle in the flank is called the *rectus abdominis*. In beef, this muscle becomes the flank steak, but in lamb its size, texture, and flavor dictate its use as ground lamb. Flanks are sometimes rolled and pinned to be roasted.

Kidneys are enclosed in the cranial (front) portion of the kidney fat. The kidneys and kidney fat together are sometimes called the kidney knob. Before cutting the thirteenth rib, remove the kidney knob by carefully trimming and pulling it away from the underlying tenderloin.

The largest wholesale and retail cut in the lamb carcass is the leg. In wholesale form, it comprises 33 percent of the carcass and 16.5 percent of the live lamb. Note that these carcass and live percentage figures have a relationship of 2 to 1, since the carcass weight is one half of the live weight. The retail leg of lamb, fabricated with the bone in, comprises about 24 percent of the carcass and 12 percent of the live weight. A boneless leg would account for a smaller proportion of the carcass and live weight. A retail cut is ready for the consumer, while a wholesale primal cut may contain several sub-primals, or retail cuts, and, as such, usually requires additional fabrication before being ready for the consumer. Exact percentages are unimportant, but relative weights are meaningful to establish a concept of live animal and carcass composition in terms of wholesale and retail cuts.

When the leg is severed from the loin, the sirloin is normally included with the leg. The resulting primal cut may be termed a *long-cut leg;* however, simply *leg-o-lamb,* or *leg of lamb,* implies the whole leg with sirloin. This lamb *pelvic* limb fabrication differs from beef fabrication where the round, including the rump, and the sirloin are at least two separate cuts, and pork fabrication, where the sirloin is left with the loin rather than with the leg (ham). The separation in lamb is made at the seventh or last *lumbar* vertebra. Skeletal landmarks which are keys to cut identification are *sacral* and *coccygeal* (tail) sections of the vertebral column, the pelvic bone, and the leg bone *(femur).* (See Figure 15-1.)

Figure 15-6. The leg of lamb and loin separation may be achieved by cutting through the muscle and connective tissue on each side of and perpendicular to the backbone at the point where the *femoral* arteries branch from the large artery (dorsal aorta), which lies ventrally to (under) the spinal column. The cut should graze or pass immediately cranially to (front of) the hipbones.

Fig. 15-6

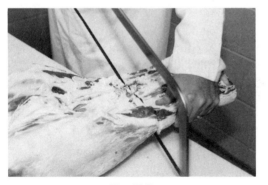

Figure 15-7. Cutting through the vertebral column with a saw results in a flat surface on the face of the loin (IMPS 232) and the leg (IMPS 233). Complete separation can be made easily with a power saw.

Fig. 15-7

Figure 15-8. The cranial (front) face of the leg is made up of muscle and bone sections which will be identified by starting at the dorsal surface and moving ventrally on the leg face. The muscles immediately below the subcutaneous fat on the back are the top sirloins (a) *(gluteus medius)*. The cartilage tip of the anterior end of the pin bone (b) *(ilium)* is shown as the whiter structure in the fat on the lower right edge of the right *gluteus medius*. Ventral to the *gluteus medius* on each side are the loin-eye muscles (c) *(longissimus dorsi)*, which are not only the largest muscles on this cut surface but also the heaviest muscles in the whole

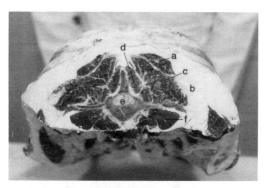

Fig. 15-8

carcass. The small muscles adjacent to the *dorsal* processes of the vertebra are the (d) *multifidus dorsi*. The large, rounded portion of the bone is the body of the seventh (e) *lumbar* vertebra with its *dorsal* process extending upward between the two *multifidus dorsi* and the *transverse* processes or T-bones extending ventrally to (under) the *longissimus*. The small muscle system ventral to the T-bones is the tenderloin muscles. These are the most tender muscles in the carcass. The larger muscle is the (f) *psoas major* and is associated with a much smaller *psoas minor*.

The Latin names are universal throughout the world, thus their inclusion. Common names can vary even between sections of our own country. The ISMISC standards should eliminate this variation.

Figure 15-9. In order to fabricate this pair of legs, it is necessary to split them. The aitch bone, or *pubic* bone, can normally be split with a knife in a young animal, since it is joined by cartilage. Once the aitch bone has been split, the remaining muscles can also be split with a knife, exposing the cut surfaces of the aitch bone. Then the *sacral* and *coccygeal* portions of the vertebral column must be severed with a saw to complete the separation of the pair of legs. A power saw completes the whole process in a single pass.

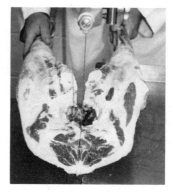

Fig. 15-9

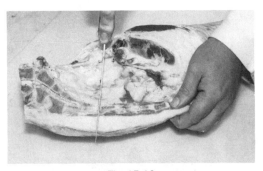

Fig. 15-10

Fig. 15-11

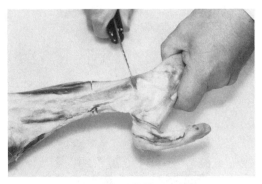

Fig. 15-12

Figure 15-10. The first fabrication step in any case is the removal of the tail bone, composed of a number of the *coccygeal* vertebrae, the actual number depending on how long the tail was in the live lamb. Ordinarily, three tail vertebrae are left on the leg.

A lamb carcass usually has a small amount of pelvic fat in the pelvic cavity located between the aitch bone and the backbone, and it is removed when a leg of lamb is fabricated.

Figure 15-11. Outside fat is trimmed where needed until the remaining fat cover on the leg does not exceed ¼ inch. The fell, or thin membrane, separating the pelt from the subcutaneous fat is not removed unless absolutely necessary in trimming, since this fell holds the shape of the leg and helps retain moisture and juices during cooking. When the flank side of the leg is trimmed, special care must be taken to remove the *prefemoral* lymph node.

Figure 15-12. The large tendon at the rear of the leg, the *Achilles* tendon, is severed at its origin in preparation for the hock removal. There are three common types of lamb legs: Frenched, American, and boneless.

In a Frenched leg, after the tendon is loosened, the muscle is severed about 1½ inches above the hock joint and cleared away from the bone. Slightly above the hock joint is the break joint. It is scored with a knife and may be broken across the edge of the cutting table. The hock bone, or trotter, is thus removed.

The break joint or *epiphyseal* plate is the area of growth in the long bones, so in young animals it is cartilaginous and easily broken, while in older animals where the bones have ceased growing, the *epiphyseal* plate has turned completely to bone and is tightly fused. The break joint is located above the hock joint, that joint which allows the animal to flex its legs and walk.

Figure 15-13. The exposed shank bone is composed of the *tibia* and the *fibula* which are fused together in the ovine. The Frenched leg is so named because the shank bone *(tibia-fibula)* is bare and exposed. When any bone is exposed, it is termed *Frenched.*

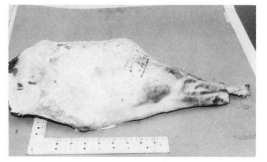

Fig. 15-13

Figure 15-14. In fabrication of the American leg, after the shank muscle and *Achilles* tendon are loosened from the shank bone, as was done in the fabrication of the Frenched leg, the stifle joint is entered from the caudal (tail) side. This is the knee joint which joins the shank bone with the large leg bone, the *femur.*

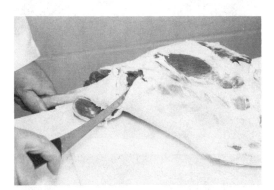

Fig. 15-14

Figure 15-15. The shank muscle (largely *gastrocnemius*) may be left attached to the American leg after the shank bone *(tibia-fibula)* has been removed. The knee cap, or *patella,* is normally left in an American leg. The shank muscle can be folded over the small cavity next to the *femur,* created when the shank bone was removed at the stifle joint and pinned in place with tooth picks or skewers. The main portion of the American leg is identical to the Frenched leg. The reason an American leg is more valuable than the Frenched leg is that more bone has been removed; that is, the shank bone as well as the lower shank bone.

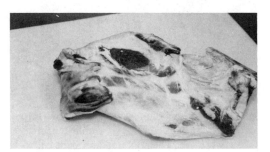

Fig. 15-15

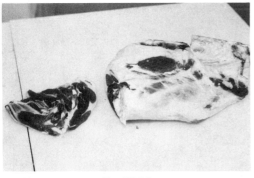

Fig. 15-16

Figure 15-16. Note the abundant amount of connective tissue in the shank muscle, as indicated by the white or silvery streaks. Therefore, the authors prefer to cut away and separate the shank muscle from the leg and utilize it as ground lamb, since it does contain large amounts of connective tissue and is considerably less palatable than the leg itself. This modification of the American leg results in a more uniformly palatable bone-in leg roast.

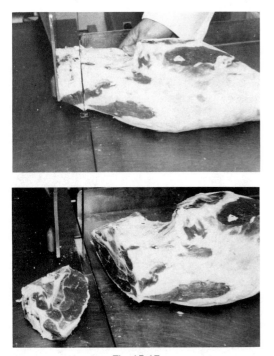

Figure 15-17. The sirloin is normally left on the wholesale primal leg of lamb. Thus, sirloin chops can be removed from a leg of lamb. (Top) The location of the chop removal from the leg is indicated here. The largest muscle in a sirloin chop (bottom) is the top sirloin or *gluteus medius* muscle. Note the presence of the flat pelvic bone in this sirloin chop.

Fig. 15-17

The whole leg may be sawed into chops and slices (steaks) starting at this cranial (front) end. The result would be four to six sirloin chops, depending on how thick they were cut. These chops are identical to beef sirloin steaks in bone and muscle structure, differing of course in size and color. A wedge-shaped area corresponding to the rump in beef and containing the main portion of the pelvic bone most logically should be removed before the rest of the leg can be cut into leg slices (steaks).

The rump section cannot be sliced or steaked because of the pelvic bone but makes excellent kabobs after the bone is removed. The center roast and most of the shank half can then be sliced.

Figure 15-18. The sirloin may be removed from the wholesale leg to produce a lamb leg, short cut, and retail roast. The sirloin chops may be merchandised as well as this popular form of the leg.

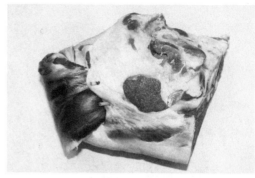

Fig. 15-18

Figure 15-19. The lamb leg may be separated into a sirloin half (left) and a shank half (right) as shown. A sirloin portion or half, a center roast, and a shank portion are other possible breakdowns of the leg. The rump section can be included with the sirloin as shown, with the shank half, or with a center portion.

Fig. 15-19

Figure 15-20. The bone-in lamb leg offers as much versatility as any wholesale cut can. In addition to the sirloin chops and roasts, a center-leg slice (steak) is illustrated between the sirloin roast and the shank roast. Center-leg slices (steaks) contain the full cross section of the leg composed of three muscle systems, which may differ considerably in palatability: the top (left), bottom (right), and tip (near rule).

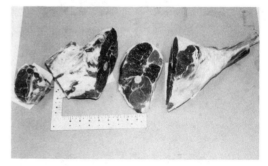

Fig. 15-20

The leg can be boned out rather than steaked. The boneless leg would be roasted and, as such, is almost certain to be highly palatable. There is a considerable amount of bone (the whole pelvic bone and the *femur*) left in the Frenched and American legs. These bones can cause inconvenience to the modern consumer when such a roast is being served. By boning the leg and jet netting the resulting boneless roast before offering it to the consumer, you can eliminate any grief connected in serving a leg of lamb.

Fig. 15-21

Figure 15-21. The pelvic bone is removed when the slip joint *(sacroiliac)* is entered, the pelvic bone *(ilium)* is separated from the backbone *(sacrum)*, and the *sacrum* is removed.

Fig. 15-22

Figure 15-22. The pelvic bone is separated from the surface of the leg and from the ball of the *femur* and removed.

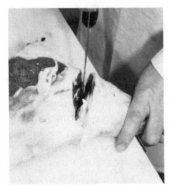

Fig. 15-23

Figure 15-23. The shank is removed at the stifle joint.

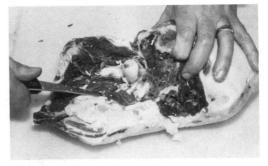

Figure 15-24. The proximal (inward) end of the *femur* is loosened. . . .

Fig. 15-24

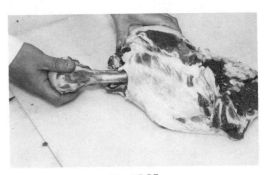

Figure 15-25. . . . so the *femur* can be pulled out from the shank end without cutting through any muscles in the leg. This technique is called tunnel boning. After all bone is removed, the sirloin end is trimmed uniformly. The femur may be more easily removed by opening the lean on the top (medial) surface of the leg from one end of the *femur* to the other. The *femur* is then freed of muscle and connective tissue and lifted out.

Fig. 15-25

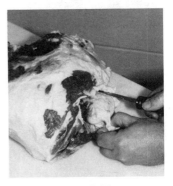

Figure 15-26. The *popliteal* lymph node and surrounding fat should be removed from the seam between the bottom *(biceps femoris)* and eye *(semitendinosus)* muscles at the distal (rear) end of the boneless leg.

Fig. 15-26

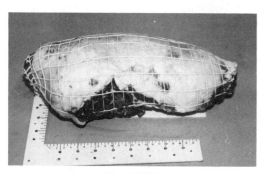

Fig. 15-27

Figure 15-27. The boneless leg (IMPS 233B), as any boneless roast, needs to be tied so that it can be roasted in a uniform manner and easily sliced after cooking. The introduction of a jet netter and the elastic net has made the tying of roasts with the old-fashioned string and butcher's knot outdated. Using the jet netter is a simple and quick operation to tie any boneless roast (see Figure 13-22, page 421). A boneless leg of lamb is identified by its size and shape. Since it is a long-cut leg containing the sirloin, you can recognize the boneless leg by its length. Keys to species identification are cut size and color. Lamb fat is much harder than beef or pork fat. Also, the fell will be present on lamb roasts. No such membrane is present on beef or pork cuts.

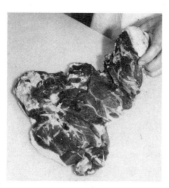

Fig. 15-28

Figure 15-28. By following the intermuscular seam between the inside *(semimembranosus)* and the outside *(biceps femoris)* muscles of a tunnel-boned leg, you can open up the leg to form a large portion, which, due to its thickness (1 or 2 inches), could be charcoal grilled or broiled.

Figure 15-29. The loin is the most valuable wholesale and retail cut in a lamb carcass, since it contains the most tender muscles of the carcass. The main reason for the high retail price of lamb loin chops is that there are so few of them from each animal. Only 4 percent of the live animal ends up as tender, juicy loin chops. Surface fat should be removed carefully, using a shaving action to prevent scoring the surface of the lean of this valuable wholesale cut.

Fig. 15-29

The composition of the loin varies slightly from end to end. At the seventh *lumbar* vertebra (the caudal end) the *longissimus dorsi,* or loin eye, is oval in shape; the tenderloin is at its maximum size; and the *gluteus medius* (top sirloin muscle) is present. From a cranial view (not shown), the loin eye is obviously larger and more symmetrically shaped. The tenderloin is not present, since it originates at the last rib as a very thin muscle and gets progressively larger until it reaches its maximum size at the seventh *lumbar* vertebra. The hanging tenderloin is not to be confused with the actual tenderloin. Since the hanging tenderloin is a part of the diaphragm muscle, it is considerably less tender than the actual tenderloin muscle because of its constant activity in the live animal.

The fell membrane, separating the pelt from the subcutaneous fat, is removed from the loin except for the specific area containing inspection and grading stamps. Since the loin will be made into chops which will be broiled, that is, cooked with very high heat for a short time, the fell will shrink and distort the shape of the chop. The fell is left intact on leg roasts to hold in the juices during the slower roasting process. The subcutaneous fat on the surface of the loins should be reduced to ⅛ to ¼ inch thick when the fell is removed. The trimmed loin may be used as a roast but is most often cut into chops.

Fig. 15-30

Figure 15-30. Double loin chops may be fabricated, since the lamb carcass is not split for chilling. The advantage of cutting double chops is the increased portion size. If the kidneys are left in the loin, they are called kidney chops (not shown); another name is English chops. A 2- to 3-inch chop is cut from the rib (cranial) end of the loin before the double loin is split. After cutting the muscle with a steak knife, complete the separation with a saw. The resulting double or kidney chop is in demand mainly because of its uniqueness, rather than for its utility, since it does contain large amounts of subcutaneous and kidney fat. Also the kidneys and flank muscles are present, neither of which compares favorably at all with the quality of the loin-eye muscle itself.

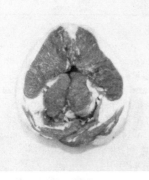

Fig. 15-31

Figure 15-31. Removing the *lumbar* vertebra produces a boneless double loin chop. The tail, or flank end, of the chop should be cut 1½ to 2 inches longer than shown in Figure 15-30 to produce the round configuration illustrated here.

Fig. 15-32

Figure 15-32. The double loin may be split on a band saw. The *transverse* processes of the *lumbar* vertebrae form the characteristic T of loin chops, just as they form the T in the T-bone steak of beef. Each *lumbar* vertebra is split exactly in the middle, resulting in two loin roasts, each the mirror image of the other.

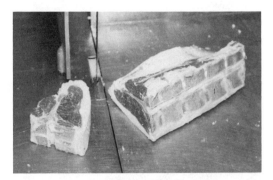

Figure 15-33. The most popular method for fabricating the loin is to make individual single loin chops.

Fig. 15-33

The term *chop* originally was used for any piece of meat which was fabricated by the use of the cleaver; that is, it was chopped off with a cleaver. Only relatively small cuts could thus be fabricated, while larger cuts called steaks were removed with a saw. This is why all lamb cuts, except leg slices (steaks), are called chops; for instance, loin chops, rack chops, arm chops, and blade chops. In modern times, even chops are fabricated by an electric band saw. The main identification key for the loin chop is the presence of the tenderloin muscle.

Foresaddle

The foresaddle, which represents slightly more than one half of the carcass weight, is composed of the primal rib (rack), shoulder, and the rough cuts, the foreshank and breast. Recall that the separation between the foresaddle and the hindsaddle is made between the twelfth and thirteenth ribs. (See Figure 15-4.)

Figure 15-34. While breasts and shanks are still in saddle form, separate them from the shoulders by sawing across the arm bone or *humerus* at a point slightly above the knuckle, that is, the junction of the arm bone to the foreshank bone, which is the fused *radius* and *ulna*. The ribs and body wall are cut parallel to the back-line.

Fig. 15-34

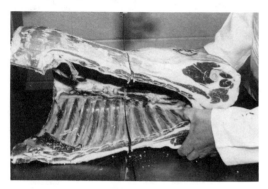

Fig. 15-35

Figure 15-35. The separation of the foresaddle half into the portion containing the shoulder and the portion containing the rib (rack) is made between the fifth and sixth ribs, leaving a seven-rib rack just as the beef chuck is separated from the beef rib. Some lamb fabricators cut an eight-rib rack, that is, cut between the fourth and fifth ribs (IMPS) to make a longer rack. However, the cranial portion of the rack near the shoulder is less desirable than the caudal portion near the loin, since it contains more connective tissue, and the *longissimus dorsi,* or rib-eye muscle, is much smaller. This front chop should be left on the shoulder to insure uniform palatability within each cut.

The largest wholesale cut in the foresaddle, and second only in size to the leg in the whole lamb carcass, is the shoulder. The shoulder contains a number of bones, making it a difficult cut to carve and slice. Shoulder cuts can be priced very economically and, if fabricated and cooked correctly, can provide delightful dining. Skeletal notations are especially helpful in shoulder-cut identification, so note them carefully on the carcass diagram (Figure 15-1).

Fig. 15-36

Figure 15-36. Remove the neck from the shoulder by cutting in a straight line extending from the back. IMPS indicates that the cut is perpendicular to the neck, leaving no more than 1 inch of neck on the shoulder.

Fig. 15-37

Figure 15-37. The pair of shoulders can be separated exactly down the middle of the backbone with a band saw.

The wholesale primal shoulder is often called the square-cut shoulder because it usually fits the dimensions of a square. The blade face has a portion of blade cartilage exposed, as well as the eye *(longissimus)* muscle.

Figure 15-38. If chops are removed from the shoulder, those from the blade face are called blade chops because of the presence of the blade bone *(scapula)*. The rib eye, or *longissimus dorsi,* extends through this area.

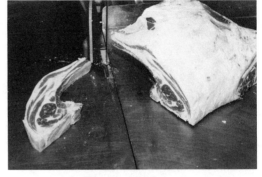

Fig. 15-38

Figure 15-39. At right angles to the blade face is the arm face. Thus, chops here are called arm chops. As in beef, the arm side of the shoulder is more muscular than the blade side, but less tender, since the *triceps brachii* and deep *pectoral* muscles do a considerable amount of work in providing locomotion for the live animal. Generally, the more use a muscle receives throughout life, the less palatable it will be.

Fig. 15-39

Figure 15-40. Shoulder chops, be they blade (top) or arm (bottom) chops, are less palatable than rack or loin chops, because the shoulder contains a considerable amount of connective tissue surrounding numerous small muscle systems. Arm chops from the lamb shoulder normally include sections of the ribs as shown in Figure 15-39, but when fat seams are large, the chops may be more closely trimmed, as shown here.

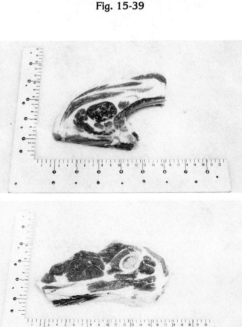

Fig. 15-40

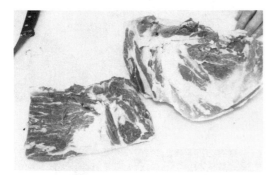

Fig. 15-41

Figure 15-41. There are several alternatives for shoulder fabrication which are preferred to making chops, because the resulting products are more likely to please the consumer. One alternative is the boneless blade (Saratoga) roll, which is an extension of the rib-eye muscle into the shoulder. The first step in its fabrication is the removal of the rib cage. Next, the eye and all muscles medially to (lying above) the blade bone are removed, following the natural seam.

The eye is rolled tightly within the adjacent muscles and held in the tight Saratoga roll by wooden skewers.

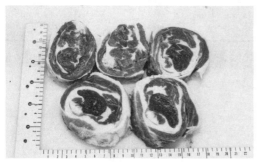

Fig. 15-42

Figure 15-42. The Saratoga roll is cut between skewers to form boneless blade (Saratoga) chops. The rib-eye muscle, or *longissimus dorsi,* begins here in the shoulder and extends all the way to the leg. It is the largest muscle in the lamb carcass but also one of the most palatable. Thus, boneless Saratoga chops composed largely of this eye muscle are very tender and juicy and the type of lamb cut that will encourage the consumer to repeat lamb purchases.

Fig. 15-43

Figure 15-43. The remainder of the shoulder may be diced for lamb stew or kabobs (shish kabobs) after the blade and arm bones have been removed. Outdoor cookery is becoming increasingly popular, since people have more leisure time. Shish kabobs rank high on the list of desirable meats for outdoor cookery.

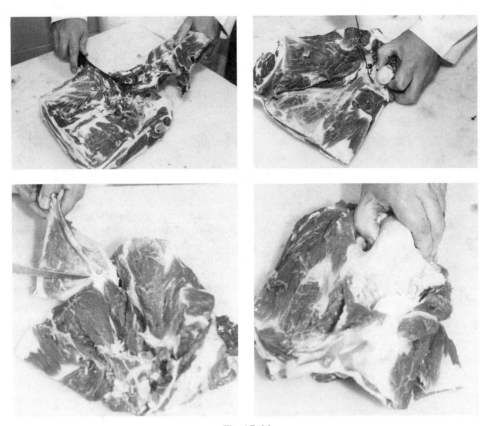

Fig. 15-44

Figure 15-44. A third alternative for shoulder fabrication is to completely bone out the square-cut shoulder and utilize it as a boneless roast.

In boning the shoulder, remove the muscle along the inside of the neck bone *(longus colli)* and trim away any evidence of dried blood deposits remaining from slaughter. The rib cage is removed, leaving as little muscle as possible on the bones (top left). The neck leader *(ligamentum nuchae)* is very unpalatable, since it is composed of elastin, which will not break down during cooking and must be removed (top right).

The large, flat blade bone *(scapula)* and the round arm bone *(humerus)* are separated at their junction, and the blade bone with its protruding spine is pulled free (bottom left). The arm bone is then removed.

The *pre-scapular* lymph node is located in the shoulder in front of the *scapula* (bottom right) and is surrounded by a large fat deposit. It must be removed, since it would detract from the desirability of the roast. Finally, outside fat cover is removed where it exceeds ¼ inch in thickness. The fell is left intact if possible.

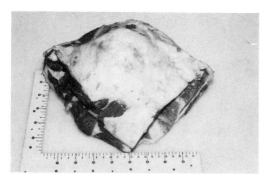

Fig. 15-45

Figure 15-45. The boneless shoulder maintains its original square shape quite well. It may be stuffed with dressing and sewn together at the blade and arm-face seams to make a rather desirable cushion-shoulder roast.

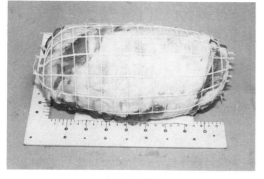

Fig. 15-46

Figure 15-46. The jet netted boneless shoulder roast may be the most preferred method of merchandising the shoulder. The roast is rolled around the extension of the rib eye before being placed in the netting. This netting remains in place throughout roasting. After the roast has reached the desired degree of doneness, the net is removed before serving.

Figure 15-47. When cutting through the center of this boneless shoulder (left), even after boning and trimming rather carefully, you will note the boneless shoulder still contains some undesirable seam fat and connective tissues. Yet when properly roasted, this boneless roast will please far more consumers than would be pleased with improperly broiled arm or blade chops from the bone-in shoulder.

It is useful to compare the boneless leg (right) and the boneless shoulder. Certainly the leg is more desirable, since it has larger muscle systems with less seam fat and fewer connective tissues. However, if we put a price differential between these two cuts, perhaps the shoulder would seem more desirable. The point is, the boneless shoulder can provide some very delicious and economical eating. Certainly the leg would be preferred for "special" occasions, but the shoulder in this form is special for "regular" occasions.

Fig. 15-47

Figure 15-48. The neck, which was removed from the paired shoulders, may be sawed into neck slices, a retail cut that lacks wide consumer demand. These are probably consumed only by those people who really appreciate lamb and enjoy the economy of such a meal.

Fig. 15-48

A wide variety of retail cuts originate in the lamb shoulder. The square-cut shoulder may be merchandised as a retail cut, but usually arm chops and blade chops are removed from it. Alternately, the rib-eye muscle may be utilized as Saratoga chops and the remainder for kabobs. A third method is to bone and roll the shoulder and make a highly desirable boneless shoulder roast.

The smallest wholesale and retail cut of lamb is the rib (rack), since on the average only 3.5 percent of the live lamb weight is actually retail rib (rack) chops. The term *rib* is sometimes used synonymously with the term *rack* and is now the recommended ICMISC name. A small portion of the blade bone cartilage appears in rack chops from the cranial end of the rack. (See Figure 15-1.)

Figure 15-49. The commonly accepted cutting technique for lamb (Figure 15-1) retains much of the rib end within the wholesale rib. To make acceptable retail rib chops, you must remove the "tails" of those ribs at a point approximately one half the width of the rib eye away from the eye.

Fig. 15-49

Figure 15-50. The fell and some fat are removed from the rack roast, since like the loin, its most popular use is as rib (rack) chops, which are usually broiled. If the wholesale rack possesses more than ¼ inch of outside fat cover, it is trimmed to that thickness. When trimming federally graded overfat lambs, it may be necessary to remove the grade stamp, but usually care is taken to leave the stamp on the rib so that it can be seen on the chops in the self-service meat case.

Fig. 15-50

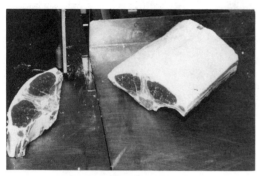

Fig. 15-51

Figure 15-51. Use of the unsplit, trimmed rib to produce the double chops discussed earlier would result in larger portions. However, the rack is more often split down the center of the backbone into two single lamb ribs.

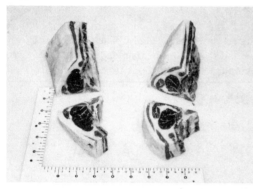

Fig. 15-52

Figure 15-52. The wholesale primal rib differs in composition from end to end. Note the blade-bone cartilage and intermuscular fat seams in the cranial end rib chops (right). More merchandisable rib (rack) chops may be produced if the bodies of the *thoracic* vertebrae are trimmed away with the power saw before the chops are cut. (See Figure 14-8.)

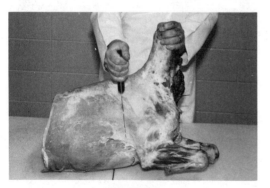

Fig. 15-53

Figure 15-53. Although a crown roast can be fabricated from a foresaddle cut as described (Figures 15-34 and 15-35), some modifications of cutting procedure may result in a larger, more easily shaped roast. In the example (Figure 15-58), the breast and shank were not removed from the foresaddle, and the rib was separated from the shoulder at the third rib.

Figure 15-54. Normally, for a close trim, the breast is separated from the rib (rack) by cutting laterally to the eye, a distance equal to one half the width of the rib eye. However, to make a crown roast, which is a rather prestigious cut, leave much longer rib ends on the rack to form the crown.

The diaphragm is termed *the skirt* as it appears on the foresaddle. Remember on the hindsaddle it was termed *the hanging tenderloin*. The rib eyes are large and symmetrical when viewed from this loin or caudal end.

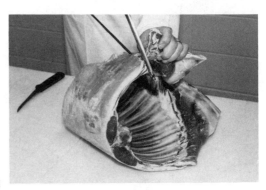

Fig. 15-54

Figure 15-55. The fabrication of a crown roast is not something that is done every day by a meat retailer; in fact, the consumer would undoubtedly have to make a special request for one. It does take valuable time to fabricate this roast and therefore the retail price will be quite high. However, it is truly *the* roast for a special occasion.

Using the double rib (rack), remove the breast, leaving as much rib as possible on the rib (rack). After the ribs are sawed, the separation is completed with a knife.

The chine bones or bodies of the *thoracic* vertebrae are loosened by severing the rib connections with a saw and removed with the feather bones (*dorsal spinous processes*) attached. These feather bones are split in a beef carcass and the cartilaginous ends are examined closely in the evaluation of carcass maturity. The blade bone and its cartilage are prominent in this cranial (shoulder) end of the rib.

Fig. 15-55

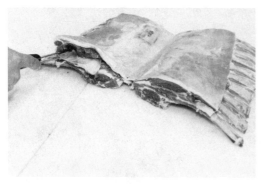

Fig. 15-56

Figure 15-56. The fell and outside finish were not broken as the chine bones were removed. The rib bones are Frenched, that is, trimmed of most muscle and fatback from the rib ends, a distance of 2½ to 3 inches.

The portion being trimmed away, which ordinarily belongs on the breast, is composed of thin layers of muscle interspersed with layers of fat and therefore lacks the high degree of palatability which the rib eye possesses. The muscles between the ribs, called rib fingers or *intercostal* muscles, are removed so that the rib ends are completely bare, that is, Frenched. The blade-bone cartilage in the cranial end is removed.

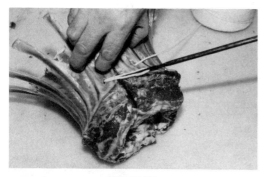

Fig. 15-57

Figure 15-57. Using a tying needle to penetrate the tissues, tie each end of the rib together with heavy butcher's string.

Fig. 15-58

Figure 15-58. The finished crown roast may be filled with dressing. Lean ground lamb may be used to fill the center as well. The Frenched rib ends can be decorated with small colorful paper or aluminum foil collars for serving. The properly prepared crown roast takes on a true "royal" appearance.

The roast is very easy to serve, since all that needs to be done to obtain an individual serving is to cut between the ribs with a knife, giving each person one rib and the accompanying tissues. (See Chapter 22.)

The rough cuts of lamb are the flank of the hindsaddle and the foreshank and breast of the foresaddle. All told, in retail form, they amount to 10 percent of the carcass weight.

The most proper use of lamb flanks is to trim away excess fat and grind the remainder for lamb patties.

Figure 15-59. The portion of the breast which corresponds to the plate in beef contains rib bones and the *sternum*. For the several alternate fabrication methods, the diaphragm, or skirt, is removed. The breast may be cut into riblets by merely cutting between each rib of the breast. An alternative is to remove the *sternum* and rib cage, leaving a boneless breast composed of layers of thin muscles separated by layers of intermuscular fat. The rib cage and *sternum* are called lamb spareribs and may be prepared by bar-

Fig. 15-59

becuing much as pork spareribs are prepared. The boneless breast is usually rolled. Only thin muscles are included in the rolled breast, which may be confused with the boneless blade (Saratoga) roll. There is a vast difference in the value of a rolled breast and a boneless blade (Saratoga) roll, since the Saratoga roll contains the very tender rib-eye muscle. By closely examining the two, you can quickly determine the presence of the rib eye in the Saratoga roll. The rolled breast is identified by the alternate layers of thin muscle and fat in the cross section of the roll.

That portion of the breast which corresponds to the brisket in beef does contain a rather thick muscle. This may be separated from the *sternum* and utilized for cubing, much as the beef bottom-round steak is utilized.

To fabricate a breast for stuffing (Scotch roast), the meat portion of the breast is peeled away from the bone structure, leaving a natural hinge along the full length of one side. This pocket then may be stuffed with ground lamb or lamb dressing and sewed, jet-netted, or tied for roasting.

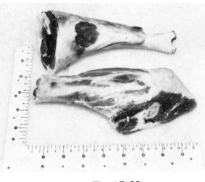

Figure 15-60. Lamb foreshanks (bottom) may be utilized as mock duck or simply lamb shanks for braising. Rear shanks removed from legs can be used in a similar manner.

Fig. 15-60

The rough cuts may be many and varied. Shish kabobs or lamb for stewing can originate from large, thick pieces of rough cut lean. (See Figure 15-43.)

Perhaps the most highly justified use for all of the rough cuts would be to process them into ground lamb. Ground lamb has very many delightful uses, while some of these variations of rough cut fabrication can only lead to consumer dissatisfaction, due to improper cooking or simply unfulfilled expectations of good eating.

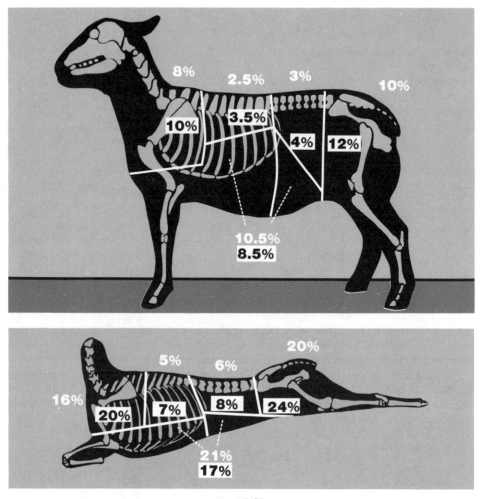

Fig. 15-61

Figure 15-61. By studying this figure and Table 15-1, you will be able to answer the question: How much retail product does one lamb (or lamb carcass) produce?

First of all, solidify your thoughts concerning the relation between live and carcass weights. With lambs it is simple, since a lamb carcass weighs approximately 50 percent as much as the live lamb. Thus the bone-in retail leg represents 12 percent of the live lamb (shown in black figures in a white box on the live lamb) while this same bone-in leg represents 24 percent of the carcass (shown in black on the carcass diagram). The carcass weighs one half as much as the live animal, so the percent of leg must be two times as great.

Bone-in figures (in black) and boneless figures (in white) are shown, since in this chapter, we have discussed bone-in and boneless fabrication for the two largest primal cuts, the leg and the shoulder. We have not shown the fabrication of a boneless loin or rack because it is so seldom done in industry. However, the boneless yield figures mean that *all* cuts are boneless, including the rough cuts.

Examine the bone-in figures first. Table 15-1 indicates that roasts and chops comprise 29.5 percent of the live animal. Check these percentages on the live animal by

Table 15-1. Percent of Live and Carcass Weights Expected in Bone-in and Boneless Products of Lamb Cutting

	Percent of			
	Live Weight		Carcass Weight	
Product	Bone-in	Boneless	Bone-in	Boneless
Roasts and chops	29.5	23.5	59	47
Lean trim	8.5	10.5	17	21
Fat trim	6.5	7.5	13	15
Bone	5.5	8.5	11	17

adding 12 percent from the leg, 4 percent from the loin, 3.5 percent from the rack, and 10 percent from the shoulder to total 29.5 percent. This same system works for carcass and live bone-in and boneless cuts. The figures placed on each cut indicate the percent which that particular retail cut represents of the carcass or live weight.

In the table, lean trim represents the trimmed boneless rough cuts as well as lean trim produced when the wholesale primal cuts were fabricated into retail cuts. With the live lamb, lean trim from bone-in fabrication amounts to 8.5 percent of the live weight. Thus total bone-in *retail* yield amounts to 38 percent of the live lamb, that is, 29.5 percent + 8.5 percent.

This same system works for bone-in and boneless retail yield based on live or carcass weights.

Check yourself:

What percent of the carcass is bone-in retail product?
Answer—76 percent determined by adding 59 percent for roasts and chops to 17 percent for lean trim.

What percent of live weight is boneless retail product?
Answer—34 percent; 23.5 percent + 10.5 percent

What percent of the carcass weight is total boneless retail product?
Answer—68 percent which is 2 × 34 percent, since the carcass = ½ the live weight, or is also 47 percent + 21 percent from the carcass chart.

All percentage figures used here are based on averages of the detailed carcass analysis of approximately 175 lambs evaluated in the University of Illinois Live Animal and Carcass Evaluation Course during the period from 1965 through 1974. This so-called average lamb possessed the following "vital statistics":

Leg conformation	High Choice
Loin-eye area	2.35 square inches
Fat over twelfth rib (top)	0.23 inch
Fat over lower twelfth rib	0.67 inch
Average fat cover	0.45 inch
Percent of kidney fat	3.8 percent
USDA Yield Grade	3.6

Since all percentage figures quoted here are averages, they can vary according to the methods of cutting and the fatness and muscling of the lambs.

DEVELOPMENTS IN MERCHANDISING LAMB

Historically, lamb merchandising procedures have been similar to beef marketing in that lamb carcasses were cut into wholesale (primal) and retail cuts by the retailer. Some trading in foresaddles and hindsaddles and limited merchandising of untrimmed paper- or plastic-wrapped wholesale cuts occurred in the foodservice business. Cutting lamb carcasses and selling portions of the carcass made the less desirable portions of the carcass very difficult to merchandise.

The strong movement to boxed beef has been accompanied by a similar trend in lamb packer sales. Major packers who are lamb processors report that 30 to 50 percent (possibly as high as 70 percent) of the lamb carcasses they produce are cut into primal or sub-primal cuts, vacuum bagged, and boxed. Boxes may contain three-piece carcasses (leg, back, shoulder) or may be filled with one primal or sub-primal cut (leg, loin, rack, etc.). Streamlined carcasses which have been trimmed of the thin, less valuable portions of the carcass (breast, flank, shanks, kidney) are available from some suppliers, boxed or hanging.

The fat and bone content of the thin (rough) cuts of the lamb carcass and the amount of labor required to obtain the usable lean product from those cuts have made them economic liabilities in the meat trade. The development of mechanical de-boning equipment and the acceptance of mechanically de-boned meat as a raw material have improved the utility of some portions of the ovine carcass significantly. Reducing the cost of processing those thin cuts may reduce the retail prices of the more desirable lamb cuts. Centralized cutting, boxing, and streamlining have resulted in the availability to foodservice or other retailers of bulk ground lamb for loaf or burger raw material.

Vacuum packaging and boxing lamb cuts is expensive because the cuts are smaller than beef cuts. The retail consumer must pay for this service, or it must be compensated for by reducing other losses. Major advantages of the vacuumized boxed lamb are the maintenance of freshness for relatively long periods; the reduction of shipping and storage yield losses; and a more efficient utilization of thin cuts, trimmed fat, and bone at a large processing plant.

Preservation of freshness would not be a concern for the retail sales outlet in a large metropolitan area (Los Angeles) that has a daily, local, fresh-slaughtered lamb source. The added cost of bagging and boxing may only inflate the retail value of lamb in that situation. Contrastingly, the mid-American lamb processors with limited local markets may realize significant benefits from boxing 100 percent of the lamb which must be shipped to population centers on the coasts.

The National Lamb Council is planning to introduce new cutting methods for lamb carcasses, which are designed to serve the needs of the HRI customer more adequately than the conventional cuts have in the past. The lamb processing industry recognizes that the low volume of lamb produced in the United States results in high unit costs. An increased demand in the HRI segment of the market may minimize distribution problems.

Chapter 16

Veal and Calf Identification
and Fabrication

THE VEAL carcass has very little fat covering, is high in moisture, and does not lend itself to aging or ripening. It is necessary, therefore, to move veal into retail channels without delay. (Note the protective film wrap on the carcass in Figure 16-4.)

Beef and veal cuts, aside from their water, fat, and ash content, differ mainly in size and terminology. Veal is tender by nature because of its age. Calf carcasses fall between the veal and beef stage and are usually considered inferior to both, grade for grade. This is because the flesh of calf carcasses has developed beef characteristics without the accompanying fat covering and marbling that enhance beef qualities.

WHOLESALE CUTS

The size of the carcass will determine the method of cutting. The larger calf carcasses are generally halved and then quartered, whereas the smaller calf carcasses and practically all veal carcasses are cut into foresaddles and hindsaddles. A *foresaddle* (IMPS 304) is the two unsplit forequarters, anterior to the twelfth rib. A *hindsaddle* (IMPS 330) consists of the two unsplit hindquarters posterior to the twelfth rib. Other wholesale cuts of veal are *long saddle*—two unsplit hindquarters with loin and nine ribs attached; *shoulders or veal chucks*— split or unsplit shoulders of four ribs with briskets and foreshanks attached; *legs*—single or unsplit (IMPS 334), cut in front of the hips; *veal backs*—single or unsplit, and cut from the fourth rib to the hipbone (including loin and nine ribs); *rattles*—the unsplit shoulders with breast and shanks attached; and *rib backs*— the unsplit ribs (eight ribs on each side). Variation in wholesale cutting procedures is greater in veal and calf than with other species. (See Figures 16-1, 16-2, and 16-3.)

A veal carcass is generally sold with the liver and sweetbread (thymus) attached. The demand for veal liver is so great that retailers invariably demand it. The average liver in a 90-pound veal carcass weighs about 3½ pounds. At $3.00

VEAL CHART
LOCATION, STRUCTURE
AND NAMES OF BONES

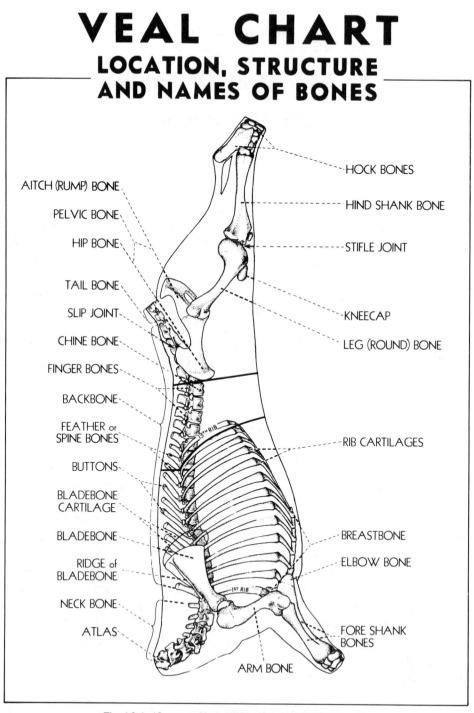

Fig. 16-1. (Courtesy National Live Stock and Meat Board)

VEAL CHART

RETAIL CUTS OF VEAL — WHERE THEY COME FROM AND HOW TO COOK THEM

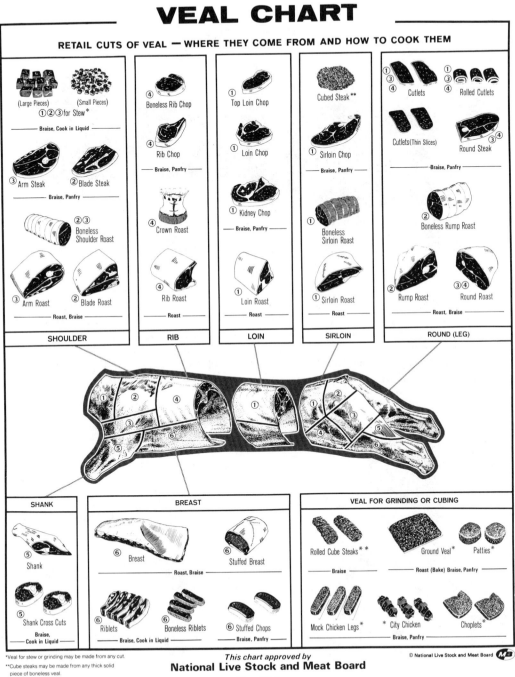

Fig. 16-2

*Veal for stew or grinding may be made from any cut.

**Cube steaks may be made from any thick solid piece of boneless veal.

This chart approved by
National Live Stock and Meat Board

© National Live Stock and Meat Board

per pound (summer 1983), it represents a value of $10.50, or approximately $0.12 per pound on the 90-pound carcass.

RETAIL CUTS

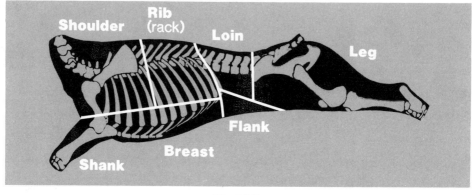

Fig. 16-3

Figure 16-3. The method of cutting veal follows the same pattern that is employed in cutting beef (Chicago style), with a few exceptions. One of these exceptions is one in which the boned neck and brisket are rolled in with the shoulder of light veal carcasses when the shoulder is rolled. Another variation is one in which a style of cutting is followed similar to cutting lamb, where the sirloin is left on the leg. (See Figures 16-1 and 16-2.)

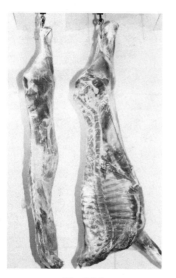

Figure 16-4. The calf carcass used in the following discussion was split into two sides. The carcass is wrapped in film to prevent dehydration in the cooler. Note the minimal fat outside and inside the carcass.

Fig. 16-4

Figure 16-5. Veal and calf carcasses are separated into forequarters and hindquarters or saddles, either between the twelfth and thirteenth ribs or between the eleventh and twelfth ribs. This separation was made between the twelfth and thirteenth ribs.

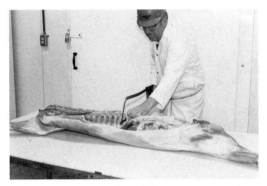

Fig. 16-5

Figure 16-6. The breast and shank are separated from the shoulder when a cut is made across the arm bone *(humerus)* just above (dorsal to) the joint between the *humerus* and the shank bones *(radius* and *ulna)*. The cut shown was made parallel to the back-line through the length of the forequarter.

Fig. 16-6

Figure 16-7. The veal shoulder and rib are separated between the fifth and sixth ribs, as shown, or between the fourth and fifth ribs.

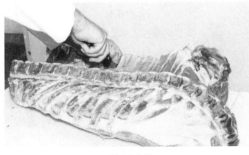

Fig. 16-7

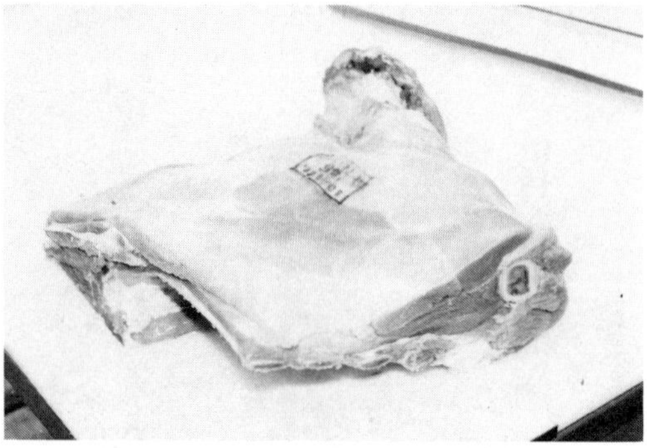

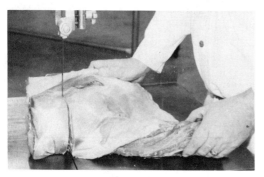

Fig. 16-8

Figure 16-8. The term *veal chuck* is just as appropriate as the term *veal shoulder*. Small veal shoulders are often boned and rolled by the same method as is followed in boning a shoulder of lamb. The chucks of calf carcasses can be made into top and bottom shoulder rolls, which are more nearly the size desired by the trade. Although a five-rib chuck is standard, in some regions of the United States the shoulder is cut from the carcass between the third and fourth ribs to make a three-rib chuck; however, in other regions, it is cut as a four-rib chuck.

Fig. 16-9

Figure 16-9. Slices cut parallel to the arm face are termed *veal arm steaks*. An arm roast of the veal shoulder is being cut on the saw in the illustration. Unscrupulous dealers have been known to misrepresent veal arm steak as veal round steak (veal cutlets).

Fig. 16-10

Figure 16-10. Turning the shoulder 90° results in the production of blade roasts and steaks of the veal shoulder. Note the absence of seam fat in the blade roast as compared to other blade cuts in Chapters 13, 14, and 15. Large quantities of veal shoulder are used in making veal stew and city chicken. The latter generally consists of 1-inch squares of veal cut ½ inch thick and placed on a 5-inch skewer with alternate layers of pork.

Veal shanks and breasts usually are boned, and the meat is diced or ground. *Ground veal* is commonly used in combination with pork for veal loaf (20 percent pork), and for mock chicken. When used for the latter, the mixture should be seasoned by the retailer before it is molded and placed on skewers. The seasoning for veal loaf is added by the consumers when it is prepared. The same combination can also be molded into patties, each patty bound by a slice of cured bacon. A mixture of 80 percent pork and 20 percent veal makes an excellent sausage. Veal breasts can also be boned and made into breast rolls, boiled, and served as jelled cuts. Sausage breast rolls are made by rolling a layer of sausage into the boned breast or by making a pocket between the ribs and the meat and stuffing it with sausage. A five-rib shoulder is 23 to 25 percent of the weight of the carcass. The neck, shank, and breast represent another 16 to 18 percent.

Figure 16-11. The veal rib, veal rack, or hotel rack may be double (unsplit) or split down the back-line as shown. The rib ends are trimmed and may be boned out and used, as indicated for breasts. The separation may be made 1 inch laterally to the rib eye as shown. If the rib is to be used for a crown roast (see Figures 15-53 through 15-58), the "tails" must be cut longer. Removing the meat from the end of the rib for a distance of 1½ to 2 inches is called *Frenching*. When ribs are to be Frenched, the "tails" must be cut to accommodate that procedure.

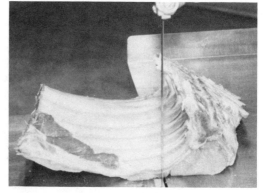

Fig. 16-11

Figure 16-12. The chine bones (central axes of the vertebrae or bodies of the vertebrae) are trimmed away prior to cutting steaks.

Fig. 16-12

Figure 16-13. Veal rib chops may be single, as shown, or cut as double chops.

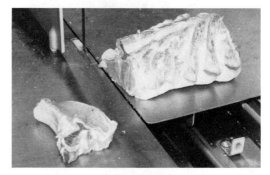

Fig. 16-13

A cut which is the result of chopping with a cleaver is termed a *chop*. Only soft bones and the solid bones of young animals are adapted to the use of the cleaver. Round or hollow bones and the bones of older animals will splinter and should be sawed. The ribs of veal when cut into slices for braising are called veal rib chops. When a rib chop is cut from ¾ to 1 inch thick and a pocket is made in the eye muscle, the chop is called a bird or a chop for stuffing. This practice is followed in both veal and pork. If the opening for the pocket is made from the inside surface of the rib, the stuffing can be inserted and will remain so without pinning the opening. On a loin chop, the opening can be made on the flank side of the loin-eye muscle (see Figure 13-37). Ribs from heavy veal carcasses can be prepared for roasts either as standing ribs (bone-in) or as rolled veal ribs (boneless). A seven-rib saddle represents 6.5 to 7.0 percent of the weight of the carcass.

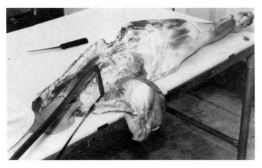

Fig. 16-14

Figure 16-14. The flank is removed by separating it from the leg at the rear and cutting along the lower edge of the loin. The thirteenth rib may be ossified sufficiently to make the use of a saw necessary, as shown. Flank meat is usually diced or ground and may be utilized as indicated earlier in the discussion concerning ground veal.

Fig. 16-15

Figure 16-15. From the last rib to the hipbone (shortloin in beef is the region) of the *lumbar* vertebrae from which loin veal chops (porterhouse and T-bone steak in beef) are cut. If the chop includes a slice of kidney imbedded in the kidney fat, the cut is known as a kidney veal chop. The veal fillet or tenderloin which lies on the underside of the vertebrae is seldom removed as a separate cut. The loin is more suitable as a roast if it is boned and rolled, preferably taking the loin saddle which includes both sides. The loin of veal with kidney, suet, and flank represents about 17 percent of the carcass weight. The front of the last (sixth) *lumbar* vertebra is the reference point used to make the cut between the loin and the leg.

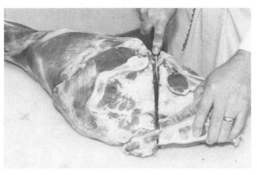

Fig. 16-16

Figure 16-16. The veal round is removed in the same manner as a leg of lamb. As in lamb, the leg includes the sirloin roast and steaks. The sirloin section can be removed by cutting on a line from the joint between the last (fifth) *sacral* vertebra and the first tail vertebra through a point exposing and just anterior to the ball of the *femur*. (See Chapter 14 for a detailed discussion of the bones of the sirloin.)

Figure 16-17. The rump roast can be removed by cutting just caudally to and parallel with the aitch (pelvic) bone. The roast can be utilized as a bone-in roast or as a boneless tied roast. A larger bone-in or boneless roast results if the sirloin and rump are removed as a single unit resulting in a sirloin half of the leg.

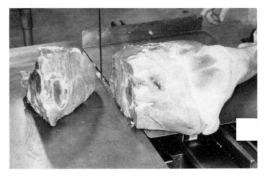

Fig. 16-17

Figure 16-18. The greatest demand is for veal round steak or veal cutlet. Considering that the leg with the rump off represents 27 percent of the carcass weight and that only 50 to 60 percent of this can be cut into cutlets (round steak in beef), it must be evident that this cut is the most expensive in a veal carcass. Cutlets may be made in thicknesses of ½ to 1½ inches, depending upon the use to which they are put. A cutlet for breading is cut ½ inch thick unless the customer specifies otherwise. The first three to four slices should be priced higher than the remainder of the cuts. Slicing ceases when the stifle joint (shank knuckle bone) is reached. The meaty part on the back of the shank can be cut for a small heel of veal round pot roast.

Fig. 16-18

Chapter 17

Fresh Meat Processing

MEAT PROCESSING may be defined as any mechanical, chemical, or enzymatic treatment of meat which alters the form from which it originally occurs. *Further processing* is a term often used in the industry which implies that the meat undergoes particle-size reduction (that is, grinding) and/or other processes. Processing of meat serves a number of functions which may include any one or more of the following:

- Preservation and/or shelf-life extension.
- Tenderization (mechanical, enzymatic, or other means).
- Meat cookery.
- Manipulation and control of composition (protein, fat, and moisture content).
- Portion control (size, weight, and shape).
- Improvement of consumer convenience.

Hence, by definition, virtually all meat is processed in one way or another prior to consumption.

Preservation and shelf-life extension (improvement) are the most important processes that meat undergoes, since it is a highly perishable commodity. The methods of meat preservation include freezing, heat pasteurization, heat sterilization, curing and smoking, dehydration, and irradiation. These important topics are covered in Chapters 18 and 19. The shelf life of meat may be defined as the length of time before the meat or meat product becomes unpalatable for human consumption, due to microbial spoilage and/or rancidity development (see Chapter 18). Mechanical refrigeration is the most useful means of extending the shelf life of fresh meat. Without mechanical refrigeration, the shelf life of fresh meat would be limited to hours instead of days or weeks (when the ambient temperature is over 40° F).

Since preservation is covered in the following two chapters, this chapter is devoted to the methods and reasons for fresh meat processing. The broad topic of fresh meat processing has gained an increased importance in our society during the past 20 years, due to changes in consumer attitudes, habits, and life styles.

Consumer attitudes about meat, health, and physical fitness have resulted in changes in the way meat is processed and merchandised. For example, consumers today are concerned more than ever about their health and physical fitness. This has brought about changes in eating habits and has placed an increased demand on leaner meat products. Genetically, the production of leaner meat-type animals is a slow process. To compound the problem, the marketing and slaughtering segments of the livestock industry are often confused or do not understand what the consumer really wants. Presently, there appears to be a trade-off between meat quality (tenderness, flavor, juiciness) that is predicted by marbling (see Chapter 11) and the production of leaner meat-type animals. Since marbling levels tend to decrease with less total fat in the meat animal's carcass, two opposing forces exist between meat quality and the production of leaner meat. These opposing forces have created some confusion in the direction in which livestock production should move in the future. It is interesting to note that while numerous consumer groups, human nutritionists, meat scientists, and even many retailers are calling for the production of leaner beef, the packer often pays the top market price for Yield Grade 3 cattle which are moderately fat. As long as this marketing situation exists, there is little incentive for the producer to market leaner cattle. Similar trends are evident in the hog industry; however, the factors of growth and reproductive efficiency play much greater roles in determining the type of hog marketed.

THE FOODSERVICE INDUSTRY—IMPORTANCE OF PORTION AND COMPOSITIONAL CONTROL

What does all of this have to do with fresh meat processing? Since there is such wide variation in the fat composition of the livestock slaughtered, compositional uniformity has become a major problem for the meat processor. Compositional uniformity is important in the production of leaner products, and it is even more important in the large volume of meat products marketed through the foodservice industry (for restaurants, cafeterias, fast-food chains, etc.).

Changing consumer habits and life styles have greatly increased the demand for more convenient meat products which require less preparation time. Table 17-1 lists the percentage of families with two wage earners in the working force since 1960. The number of households having both husband and wife working

Table 17-1. Percentage of Family Units with Both Husband and Wife Employed: 1960-1981[1]

Year	1960	1965	1970	1975	1977	1978	1979	1980	1981
Percent[2]	25.5	29.4	34.3	38.9	40.9	41.8	43.9	44.8	45.6

[1]*Statistical Abstracts of the United States National Data Book and Guide to Sources.* 103rd ed. 1982-83. U.S. Bureau of the Census, Washington, D.C., p. 382.

[2]Percent is based upon the total number of employed husbands.

has greatly increased the demand for convenient-food items. When both husband and wife work, neither feels like spending hours in the kitchen preparing a meal. The increased demand for convenience foods is further substantiated by the "boom" in microwave oven sales that has occurred during the last 10 years.

In addition to the convenience factor, this shift in the working force has closely coincided with an increase in the number of meals eaten away from the home. This trend of an increased percentage of meals consumed outside the home has caused tremendous growth in the hotel-restaurant-institutional (HRI) trade. The fast-food industry, a major segment of the HRI market, has experienced the greatest growth and has become an extremely competitive industry. The fast-food industry has greatly increased the demand for exact compositional control, portion control (size, weight, and shape), and rapid preparation convenience. The major fast-food franchises are characterized by very large volumes and relatively low-profit margins. Obviously, it is *meat* that is the principal drawing card for customers and accounts for the largest percentage of sales. A 1979 USDA-sponsored in-depth study of the foodservice industry revealed that its consumers purchased approximately 5 billion pounds of red meat products and 1.5 billion pounds of poultry products.[1] This accounted for a combined (red meat and poultry) per capita consumption (retail weight) of almost 30 pounds per person.[2]

Due to the extreme competitiveness, high-volume, and low-profit margins, very small variations in fat content or portion weight can account for hundreds of thousands of dollars in net profit differences annually for the larger fast-food corporations. Many specialized meat-processing plants cater strictly to meeting the needs of the foodservice industry. Suppliers of the foodservice industry must meet the rigid portion size, quality, and cost specifications of the retailing corporation that purchases their products or another supplier will be quickly found.

TENDERIZATION—MECHANICAL AND ENZYMATIC

Another segment of the HRI trade has specialized in offering intact muscle steaks (not ground or formed) at economy prices. Family steakhouses achieve economy by utilizing lower-quality carcasses (usually from the Standard, Commercial, and Utility grades [see Chapter 11]) and assuring uniform tenderness with mechanical and/or enzymatic tenderization.

Mechanical tenderization is accomplished by passing the steaks (or boneless sub-primals) through a bank of needles or through a rotary steak mascerator (cuber). The former method is referred to as needle tenderization (or sometimes blade tenderization) and is illustrated in Figure 17-1. The bank of needles passes through the meat, severing connective tissues and muscle fibers, making the

---

[1]Michael G. Van Dress. *The Food Service Industry: Structure, Organization and Use of Food, Equipment and Supplies.* 1979. USDA, ERS Stat. Bul. No. 609.

[2]Based on 1980 population census of 220 million people.

Fig. 17-1. (Top) Needle (or blade) tenderization of boneless sub-primals is one method of improving the palatability of less-tender cuts. A Ross® tenderizer (Ross Industries, Midland, Virginia) is illustrated in this example. (Bottom) A close-up photo of the bank of needles. (Photographs courtesy South Dakota State University)

meat more palatable. A variable speed conveyor is used to advance the meat while the up-and-down motion of the needle bank penetrates the muscle tissue. Best results are obtained when boneless sub-primals such as boneless rib eye and boneless top sirloin are used. It is desirable to needle tenderize sub-primals several times (two or three passes) through the tenderizing machine when the source of sub-primals is from more mature beef (Commercial and Utility grades).

The steak mascerator (Figure 17-2) can be used for tenderizing individual loin, sirloin, and round steaks; however, more often it is used to make cubed steaks from less-tender portions of the carcass such as the chuck (or shoulder). Steak trimmings and end pieces are sometimes fused together in a mascerator to produce high-quality cubed steaks rather than utilizing these pieces for stew or kabob meat.

Enzymatic tenderization is another means of improving the tenderness of steaks from mature cattle. Three proteolytic enzymes are commonly used in the tenderization of meat—papain, bromelin, and ficin. These enzymes, derived from certain tropical plants (papaya, pineapple, and fig respectively), degrade muscle fibers and connective tissue to different degrees. Of the three, papain is the most widely used; however, it has the least degradative effect on collagen (the most abundant connective tissue protein). Bromelin exhibits strong degradative properties for collagen and has the least affinity for the muscle fibers and the connective tissue elastin (found in blood vessels and the sarcolemma or cell membrane). Ficin exhibits the greatest degradative action on collagen and elas-

Fig. 17-2. (Top) A rotary blade tenderizer such as the Hobart® tenderizer (Hobart, Troy, Ohio) pictured here is used for making cubed steaks. (Bottom) The two mascerating blade assemblies rotate in opposite directions, pulling the meat through the center. (Photographs courtesy Hy-Vee Supermarket, Brookings, South Dakota)

tin; however, it also has the greatest degradative effect on the myofibrillar proteins of the muscle fibers. Care must be taken not to overtenderize and turn the meat to mush, particularly when the enzyme ficin is used. The commercial tenderizers used by many of the family steakhouses are usually a blend of the three enzymes. The tenderizer is usually applied by spraying or dipping the steaks with a solution that may also contain salt and phosphate.

FROZEN CONVENIENCE FOODS

Frozen convenience foods, the term given to the new generation of TV dinners, have had a substantial impact in both the retail food industry (supermarkets) and the foodservice industry. Frozen convenience meals of today emphasize variety, quality, and microwavable convenience with menu offerings such as chicken cordon bleu, beef stroganoff, veal parmesan, and many other delicacies. The aluminum trays have been replaced by attractively designed plastic dishes which allow microwave cooking. The gourmet image, more rigid quality specifications, attractive packaging graphics, and nutritional information have also contributed to their recent upsurge in sales at the retail supermarkets.

Frozen convenience meals have made a big hit in the airline foodservice market where the quality meal image is a big factor for airline companies in attracting customers. One company (Bil Mar Foods, Zeeland, Michigan) has developed a business with past receipts exceeding $10 million annually with the airline foodservice industry.[3] Numerous other companies, such as Armour and Co. with its Dinner Classics® line of frozen entrées and Stouffers with its Lean Cuisine® line for calorie-conscious consumers, have also experienced phenomenal market success.

Industry experts project that the frozen convenience-meal market will be one of the hottest growth areas for the meat industry throughout the 1980s and 1990s, particularly as new processing technologies for automation are developed.

The key to quality in processing frozen convenience meals is in carefully monitored sanitation programs and rapid freezing processes. Meat, pasta, and vegetable items are precooked (vegetables are sometimes only blanched), loaded onto plastic dishes moving down a conveyor line, frozen in a cryogenic tunnel, and sealed with a thin plastic film over the top before the lid is applied. The end result is a complete gourmet-style hot meal that can go from freezer to plate in less than 10 minutes—that's convenience.

THE BEEF PATTY INDUSTRY

The largest segment of the fresh meat processing industry is unquestionably the highly specialized beef patty industry. The hamburger continues to be the mainstay for the fast-food industry with 1983 ground beef per capita consumption figures estimated at 18.8 pounds per person.[4] Although not all of this figure represents hamburgers consumed from fast-food franchises, it is nonetheless a significant portion.

The ground beef segment of the meat industry is subject to specific compositional and labeling regulations by the USDA. The Code of Federal Regulation (CFR) (sec. 319.15) stipulates that: "The terms *ground beef* and *chopped beef* are synonymous. Products so labeled must be made with fresh and/or frozen beef with or without seasoning and without the addition of fat as such and shall contain no more than 30 percent fat. It may not contain added water, binders or extenders. It may contain beef cheek meat not to exceed 25 percent. If the name is qualified by the name of a particular cut, such as *ground beef round* or *beef chuck, ground* the product must consist entirely of meat from that particular cut or part. Product labeled *ground beef, beef fat added* may have beef fat added, however the total fat may not exceed 30 percent."

Hamburger is defined in the CFR (sec. 319.15b) as: "Chopped fresh and/or frozen beef, with or without added beef fat and/or seasonings. Shall not contain more than 30 percent fat, and shall not contain added water, binders or extenders. Beef cheek meat may be used up to 25 percent of the meat formulation."

If ground beef or hamburger contains soy products or other extenders, it must be descriptively labeled as such. For example, *ground beef and texturized*

---

[3]*Meat Processing*. March 1984. Vol. 23 (3):32.

[4]*Meat Facts*. 1984. American Meat Institute.

vegetable protein is the required label when one of the latter ingredients is added.

If a nutritionally inferior product is used, then the product must be labeled *imitation ground beef* or more often *beef patties (beef patty mix)*. The CFR (sec. 319.15c) defines *beef patties* as: "Chopped fresh and/or frozen beef with or without the addition of beef fat as such and/or seasonings. Binders or extenders and/or partially defatted beef fatty tissue may be used without added water or with added water only in amounts such that the product's characteristics are essentially that of a beef patty."

Ground beef used in the production of patties may take on a variety of shapes and sizes from modern patty-making machines. The most common shape is round; however, square patties and oval-shaped patties are not uncommon. The most common sizes of patties are 6/lb., 4/lb., and 3/lb. (2.6, 4, and 5.3 ounces respectively).

The sources of raw materials used in the production of ground beef varies widely between processors and in the quality of ground beef patty produced. Typical operations will use boneless cow meat, choice beef plates, or flanks, as well as a variety of trimmings from various portions of the carcass, depending on supply and price. Some ground beef patty makers use imported frozen beef when supplies and prices are favorable. Quality patty operations limit the use of frozen trimmings to a minimum, since higher cooking losses are generally associated with the use of frozen trimmings. In typical ground beef operations, a "leaner blend" and a "fat blend" are produced in a ¼- to ⅜-inch grind. Fat content of ground beef is routinely estimated using rapid infrared scanning techniques (see Figure 17-8). The lean and fat blends are combined in the proper ratios to achieve the desired target fat content according to the company specifications. After a short blending period, the ground beef is typically reground through a ⅛-inch plate with a bone collection apparatus (Figure 17-3) used to remove any bone chips or fragments from the finished product. Following the final grind, the meat is transported (by bulk bin or conveyor) to a patty-forming machine (Figure 17-4) and the subsequent patties are conveyed into a cryogenic freezing tunnel or spirulator (Figures 17-5 and 17-6). Rapid cryogenic freezing systems have been shown to improve patty quality by minimizing cooking losses and preserving flavor for longer periods of time. Frozen patties are boxed, generally with the aid of a patty-stacking machine (Figure 17-7), and moved to a holding freezer for shipment.

Larger patty plants run at production capacities of 10,000 pounds per hour. They utilize a continuous system of grinding, blending, forming, freezing, and packaging such that product flow never stops once grinding commences. In such operations, the lean and fat blends are checked almost constantly for fat content, using an infrared spectral analysis machine such as the Anyl-Ray® machine pictured in Figure 17-8. Estimates of fat content are obtained in seconds, allowing for continual on-line adjustments of lean and fat blends to keep the product in the company's acceptable fat range.

Fig. 17-3. Bone chip collectors are used by ground beef manufacturers to eliminate cartilage and bone chips from the final product. The bone chips and cartilage do not pass through a fine grinder plate but rather come out the center tube shown in this photo. A special grinder plate that has shallow grooves radiating to the center forces any bone chips to the exit tube. (Photo courtesy Howard Beef, Howard, South Dakota)

Fig. 17-4. A Formax® patty machine (Formax, Inc., Mokena, Illinois) is shown making "4 to the pound" (¼-pounder) round patties. (Photo courtesy Howard Beef, Howard, South Dakota)

Fig. 17-5. Before going into the spirulating freezer, patties are perforated with serator blades for a more rapid and uniform freeze. (Photo courtesy Howard Beef, Howard, South Dakota)

Fig. 17-6. Patties are frozen to 0° F in 10 minutes inside this −80° F spirulator. The patties enter on the bottom level of the stainless steel conveyor and exit on the top level. (Photo courtesy Howard Beef, Howard, South Dakota)

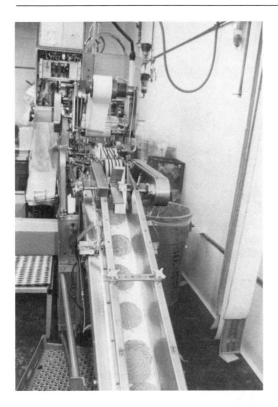

Fig. 17-7. Frozen patties are conveyed up to a patty-stacking machine, which reduces labor costs in boxing. Note the perforations on the patties from the serators shown in Figure 17-5. (Photo courtesy Howard Beef, Howard, South Dakota)

Fig. 17-8. Fat content can be determined in seconds by using one of several infrared spectral analysis machines. The Anyl-Ray® machine (Kartidg Pak Co., Davenport, Iowa) shown here requires approximately 13 pounds of ground meat tightly packed in the round canister and is a non-destructive fat determination method. (Photo courtesy Howard Beef, Howard, South Dakota)

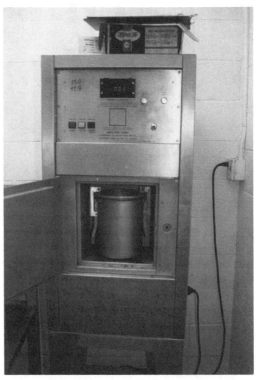

RESTRUCTURED MEAT PRODUCTS

Restructured meat is the term often applied to a new class of products which have been ground, flaked, or chopped and formed into steak- or roast-like products that have a texture that is closer to that of an intact steak rather than ground meat.

Roger W. Mandigo, a University of Nebraska scientist and one of the leading experts in the field of restructured meat products, summarized the need for this processing technology very effectively in the following quote:[5]

> With today's ever-increasing demand for fast food, meals away from home, convenience, individual meals, smaller portion size and less time for food preparation, the demand for boneless steak meat cannot be met from existing supplies at prices perceived reasonable by many consumers. Manufacturing steaks from roast meat and meat trimmings is an often expressed desire of processors. Keeping beef, pork and lamb competitive and in high demand can be improved through these steak-making techniques. At the same time the finished products can be priced more competitively to other more inexpensive main meal entrée items.

Depending on the method of particle-size reduction, restructured products are also referred to as *flaked and formed, chunked and formed, sectioned and formed,* or *chopped and formed.* Unlike ground beef, pork, or lamb, the meat is mixed after comminution (particle reduction) with some agent (usually salt, phosphate, and/or protein binders) that will either directly or indirectly bind the particles back together. The meat is then formed into the desired shape and, upon being cooked, will maintain that approximate shape.

A number of processing considerations are important in the production of restructured products and include raw material source and quality, comminution method, particle size, mixing time, processing temperature, forming method, nonmeat ingredients, processing equipment, and packaging system(s).

Raw Material Source and Quality

A wide range of muscle tissue and meat by-products can be used in the formulation of restructured products, depending on the desired cost and quality of the finished product. The choice of raw materials should be largely influenced by the market segment the processor intends to enter and whether the product as formulated will offer a competitive value alternative to the prospective consumer. For example, if the intended consumer is within an institutional market, it often becomes more important to minimize the raw material costs providing that the quality specifications for that market are met.

---

[5]R. W. Mandigo. *Restructured Meats.* 1982. Proceedings of the Fourth Annual Sausage and Processed Meats Short Course, Iowa State University.

Raw materials that have the greatest economic value for restructuring processes are those which have a lower wholesale value or are lower-valued by-products of the processing industry. Such materials allow greater room for the higher production costs associated with restructuring (which is necessary to upgrade the ultimate quality of the finished product) and yet maintain an economically priced item. Raw material items that can be effectively used include primal cuts, boneless primal cuts, selected trimmings, non-selected trimmings, mechanically deboned tissue, and/or partially defatted tissue. Primal cuts that may be used are generally limited to the chuck (or shoulder) and round (or leg). The loin meat is generally considered infeasible for two reasons. First, it is too costly as a raw material, and second, restructuring would not generally result in quality improvement of the product, since the loin is more tender and contains less connective tissue than lower-valued cuts such as the beef chuck.

Raw materials should be as fresh as possible to prevent problems caused by bacterial, enzymatic, and oxidative degradation which result in off-flavors, off-odors, and a more rapid deterioration of desirable color. Ideally, meat from pork and poultry carcasses should be boned within 24 hours post-mortem and further processed within 48 hours post-mortem. Beef and lamb should be further processed within seven days post-mortem.

Methods of Particle-Size Reduction

The method and degree of comminution is largely determined by available equipment and the connective tissue content of the raw materials being used. Generally, meats containing moderate to high levels of connective tissue, such as certain muscle systems of the shoulder and shank regions, should be reduced to a fine particle size to insure uniform palatability. Raw materials that have lower concentrations of connective tissue are palatable in larger particle sizes and contribute a more desirable texture to the finished product. It is important to remove the dense connective tissue membranes from larger muscle systems if a large particle size is used in the restructured product. Removal of these connective tissue membranes is termed *denuding* and results in extremely high-quality raw material.

Particle-size reduction may be accomplished by sectioning, chunking, slicing, flaking, grinding, or chopping. Each method contributes different textural properties to the final restructured product.

Sectioning

Sectioning involves the separation of entire muscles (or muscle systems) by seaming between the muscles with a knife. Sectioning is a widely used practice in the commerical ham industry for producing *sectioned and formed hams*. It is desirable to denude the sectioned muscles where heavy connective tissue membranes exist. The tenderness of the individual muscle and the amount of internal connective tissue and fat will affect the usefulness of the sectioned portions.

Chunking

Chunking is a term used to describe the process of making coarse particle sizes (but smaller than sectioned muscles). Meat can be made into chunks with a very coarse grinder plate, a meat dicer, a bowl chopper, or an ordinary knife. Coarse grinding and dicing are the two most common methods of producing desirable-sized chunks.

Slicing

Slicing can be used as a method of particle reduction by thin-slicing frozen and tempered meat on a high-speed slicer or cleaver. Slicing is particularly useful for tissues such as the beef plate and Boston butt, which are often relatively high in fat. Thin-slicing these cuts and formulating with a lean fraction results in a restructured product that has an appearance very similar to natural marbling.

Flaking

Flaking is the process of reducing particle size with the Urschel Comitrol® (Valparaiso, Indiana) or similar equipment. An example of a laboratory-sized Comitrol® and a flaking head used in particle-size reduction are shown in Figure 17-9. Meat should be frozen and tempered (24° to 26° F) prior to flaking in a Comitrol® to insure uniform flake sizes. Flaking temperature of lean and fat fractions is also critical to insure good protein extraction necessary for product bind as well as dispersion of the fat throughout the product. A wide variety of flaking heads that result in flake sizes ranging from very coarse to very fine are available.

Grinding

Grinding is a more restrictive term than *chunking* and is accomplished by passing meat through a grinder plate and cutting the meat with a knife that fits snugly against the plate. Grinder plates are available with a large variety of hole sizes ranging from $\frac{1}{16}$ inch to large kidney plates which have dimensions of $1\frac{1}{2}$ by 3 inches per hole or larger. As with most comminution equipment, the importance of sharp knives cannot be overstressed. Sharp knives are critical for efficient operation, particle definition, and the prevention of fat smearing.

Fig. 17-9. (Top) A laboratory-sized Comitrol® (Urschel Laboratories, Inc., Valparaiso, Indiana) machine used for flaking meat in restructured products. (Bottom) An example of a flaking head used by this machine. A rotating impeller (not shown) with knives forces frozen meat through the openings on the flaking heads which are available in many sizes. (Photographs courtesy South Dakota State University)

Chopping

Chopping is the method of particle reduction often used to describe processing in a bowl chopper or silent cutter. These machines are available with vacuum domes, and some models have the capabilities for simultaneous chopping and cooking. A large bowl chopper is illustrated in Chapter 20 (Figure 20-11) and consists of a rotating bowl which forces the meat to pass through a series of rotating knives that are closely adjusted to the surface of the bowl. Extreme caution must be exercised when chunking meat in a bowl chopper such that batch-to-batch variation in chunk size is minimized. Nonetheless, the bowl chopper has the greatest versatility in creating different particle sizes (large to small) without requiring a change of parts (that is, plates, cutting head, dicing head, etc.).

Nonmeat Ingredients

Certain nonmeat ingredients are considered essential for the production of acceptable restructured meat products. Of these, salt (NaCl) is presently the most widely used nonmeat ingredient. Salt functions to solubilize proteins which form a tacky exudate that, when heated, acts to bind the meat particles together in a continuous mass. Salt also functions as a flavor enhancer at the levels commonly used in restructured products. Research conducted at the University of

Nebraska[6] showed that salt levels of 0.5 to 0.75 percent are required to achieve an adequate bind in restructured products. Since salt acts as a pro-oxidant leading to the development of oxidative rancidity (see page 562), it is extremely important that high purity grade salt be used in manufacturing restructured products. Common table salt typically contains small amounts of copper, iron, and chromium impurities, which can act as catalysts in the development of oxidative rancidity.

Polyphosphates are another group of compounds which are commonly added to restructured products at levels of 0.15 to 0.40 percent. Polyphosphates function to improve the protein extraction of the myofibrillar proteins necessary for good product bind and also to decrease the shrinkage that occurs during cooking. Phosphates are also reported to improve the fresh color of restructured products (which is often a problem) and inhibit oxidative rancidity. For a more detailed discussion of phosphates, see Chapter 20.

Other ingredients that may be used to add variety to restructured products or reduce the cost include seasonings, binders, extenders, and nitrites. These nonmeat ingredients are also discussed in Chapter 20 and serve the same functions that are outlined in sausage ingredients.

Processing Procedures for Restructured Products

The National Live Stock and Meat Board has published an excellent source of information on manufacturing practices for restructured products.[7] The interested reader should refer to this booklet for more detailed information on the subject matter. The authors of *The Meat We Eat* are again grateful to the National Live Stock and Meat Board for much of the information that is given on restructured meat products.

Following raw material selection, particle-size reduction and selection of nonmeat ingredients to be used, the processor is ready to begin manufacturing restructured products. The next processing step is to combine raw materials with salt, phosphates, etc., in a mixer (blender, massager, or tumbler). Fresh restructured products are typically blended for 5 to 15 minutes, depending on meat temperature, particle size, and blending action to get the necessary protein extraction required for binding. Low temperatures (approximately 30° to 32° F) are critical in preserving a desirable oxygenated fresh meat color. After blending, the product is typically stuffed in a polyethylene bag or other low-cost casing and tempered to a temperature of approximately 26° F. Hydraulic meat presses can then form the restructured product to virtually any shape desired. A uniform temperature of the product is extremely critical to minimize pressing losses (if

---

[6]*Ibid.*

[7]*Manufacturing Guidelines for Restructured Pork Products.* 1985. National Live Stock and Meat Board, Chicago.

too warm) or insure that the product does not break apart later when cleaving into individual portion sizes (if too cold).

Alternative forming methods may become more practical in the future with the introduction of new and better equipment. There is considerable interest in eliminating the critical tempering times required for hydraulic pressing as a forming method. One such method is to form the product into individual portion sizes much like hamburger patties are presently made. Using such a modified patty-forming machine would eliminate not only the lengthy and critical tempering times (to 26° F) but also the need for cleaving into individual portions. Such a system would be less labor and energy intensive.

Another forming method that has been used is known as the *extrusion* process. The restructured product is extruded from specially designed stuffing horns onto a conveyor, subsequently frozen or tempered and sliced or cleaved into individual portions.

Chapter 18

Preservation and Storage of Meat

PRESERVATION of red meat, fish, and poultry products is accomplished by creating an unfavorable environment for the growth of spoilage organisms (bacteria, yeasts, molds, and parasites) and the prevention of chemical oxidation of lipids which leads to rancidity. Various methods of preserving meat have evolved throughout history and include drying, smoking, salting, freezing, canning, freeze-drying, and irradiation or combinations thereof. Throughout recorded history, salting or curing has played the most significant role in keeping meat safe for later consumption.

In modern societies, refrigeration and sub-zero storage have replaced salt to a large degree as a means of preservation. Due to mechanical refrigeration, cured and smoked products of today are generally not dependent upon high salt concentrations for preservation as their earlier predecessors were. Mildness and brightness (appearance) play a larger role in appetite and sales appeal. Yield is a factor that has become more critical to the processor than in bygone days when haste and waste were not so important. Speed has become a necessary part of our economy. Fast, mild cures and light smokes have largely replaced long, heavy cures. Today, cured meats add variety and a distinctive flavor to products that must in turn be preserved by refrigeration.

The broad topic of meat curing is presently so vast that it merits a separate chapter for discussion purposes. Therefore, the reader is referred to Chapter 19 for detailed coverage of this subject.

WHY PRESERVATION?—MICROBIOLOGY

Animal tissues are sterile, or nearly so, except for lymph nodes in the living and growing animal. But as soon as the process begins which converts that living tissue into human food, it becomes subject to degradation by chemical, physical, and biological reactions. If the knife used to sever the jugular vein and carotid arteries is not sterilized before use, an infection is inoculated into all of the animal tissues, edible and non-edible, via the circulatory system. Chapter 2 covered sanitation and stressed its extreme importance to the meat industry. Proper sanitation is the first big step for successful meat preservation.

A means of preservation should be practical and usable, should not make the product unpalatable or worsen its appearance, and it must be safe, that is, not harmful in any way to those who consume the meat. Meat is preserved from microorganisms which thrive on its rich supply of nutrients, just as humans thrive on meat.

A book about meat must have a section on meat microbiology, since it is indeed an integral part of the story of meat. Yet, this book can only touch on the very most important points in meat microbiology. There are very excellent whole books published on the subject.

Meat is an ideal culture medium for microbes; that is, they like it and thrive on it because meat is:

- High in moisture.
- Rich in nitrogenous foods of various degrees of complexity.
- Plentifully supplied with minerals and accessory growth factors.
- Sufficient in fermentable carbohydrates.
- At a fairly favorable pH (~ 5.6).

Factors Affecting Microorganism Growth on or in Meat

Temperature

Some microbes grow well at temperatures of 32° to 68° F (0° to 20° C). They are known as *psychrophiles*. Examples are *pseudomonas* bacteria and some yeasts and molds. Others called *mesophiles* grow well at temperatures between 68° and 113° F (20° and 45° C). Most bacteria belong to this group. *Staphylococcus aureus* is an example. A few microbes called *thermophiles* grow at higher temperatures between 113° and 150° F (45° and 65° C). An example is *Clostridium botulinum*.

Moisture

Moisture level can be expressed as percent water, and when expressed that way, most bacteria require at least 18 percent water to grow. Molds grow in media (tissue) that can be as low as 13 percent water. A more definitive measure is water activity (a_w) which is defined as:

$$a_w = \frac{\text{vapor pressure of solution}}{\text{vapor pressure of pure solvent (H}_2\text{O)}}$$

Most spoilage organisms require an a_w of 0.91 or greater for growth. Since fresh meat has an a_w of 0.99 or higher, it provides an ideal environment for growth. Some spoilage organisms can grow at a_w levels below 0.91 and are noted in Table 18-3.

Oxygen

Some microbes need oxygen in order to grow. These are called *aerobic* and include many bacteria and the yeasts and molds.

Those that cannot grow where oxygen is present are called *anaerobic*. Examples are *Clostridium* and the putrifiers (those which degrade proteins and form very strong smelling gasses).

A third group is called *facultative*, meaning that these microbes grow in either aerobic or anaerobic conditions.

Degree of Acidity or Alkalinity (pH)

Most microorganisms thrive at a pH near neutrality (pH of 7.0). Most meat products fall in the pH range of 4.8 to 6.8. In general, microorganisms grow slower and less efficiently at a pH of 5.0 or lower.

Physical Properties of Meat

If the meat is in a form that provides a large surface area exposed to oxygen (air), the microbes have more room to grow. For example, a whole carcass has the minimum amount of surface area exposed. When it is cut into wholesale cuts, more area is exposed; when it is cut into retail cuts (steaks, etc.) still more area is exposed. When meat is ground, the maximum total surface area is exposed. Thus, the further meat is processed, the more vulnerable it is to microbial action.

Meat processors can control the factors that affect microbial growth and thus give meat a longer shelf life, meaning that the meat stays in top wholesome condition longer. For example, beef carcasses *can* hang in a 33° to 35° F cooler for perhaps three weeks to age (*not* a standard practice), and, depending on the relative humidity of the cooler, perhaps only a few molds and some harmless bacteria such as *pseudomonas* would grow on limited areas of the carcass. But cut the carcass into its cuts, or grind it, or raise the temperature and/or relative humidity of the cooler, and the above factors are affected such that microbiological growth could flourish.

New processing techniques involving the use of vacuum-packaging allow carcasses to be cut into primal, sub-primal, or even retail cuts, after which each cut is placed in an airtight envelope. In such a wrap, at the proper temperatures, cuts may be held for extended time periods during marketing without the occurrence of damaging microbial growth.

Organisms of Concern in Meat

Clostridium botulinum

Botulism, a food-borne *intoxication* (ingestion of a food which contains a microbial toxin) is caused by various strains of *C. botulinum* which produce toxins types A, B, and E affecting the nervous system of humans. The organism is identified as a gram-positive, anaerobic, spore-forming rod, able to grow under relatively thermophilic conditions. The toxin, which is destroyed by boiling for 15 to 20 minutes, is probably the most deadly poison known. One gram properly distributed (an impossibility) would be enough to kill up to 10 million people. Clostridia are fairly widely distributed in soil and will grow well on protein, low-acid, and improperly canned foods. Canned vegetables, meat, and fish are most often incriminated. About 24 to 48 hours after consumption of the contaminated food, the symptoms begin to be noticeable; for example, muscular weakness and loss of those functions dependent on nerve action. About two thirds of the cases are fatal. The nitrite component of a curing mixture has been shown to control the growth of *C. botulinum* in cured meats (see page 592).

Staphylococcus aureus

Staphylococcal food poisoning (intoxication) is caused, in most cases, by *Staphylococcus aureus* which produces a very heat-stable toxin capable of causing severe gastrointestinal upsets. These may occur two to six hours after ingestion of the toxin. *S. aureus* is a typical gram-positive staphylococcus occurring in masses like clusters of grapes or in pairs and short chains. All enterotoxin-producing *S. aureus* cultures are beta-hemolytic and coagulase-positive (coagulating oxalated blood plasma). They are facultative in their oxygen requirements in a complex glucose medium, but not all coagulase-positive staphylococci are necessarily toxigenic. Some of the toxigenic cocci are tolerant and grow in salt concentrations approaching saturation and also tolerate nitrites fairly well. The cocci also can grow in lower pH solutions aerobically, below 5.0, than anaerobically. The microorganism itself is somewhat heat-resistant, sometimes being able to withstand pasteurization temperatures in milk, but the enterotoxin is *very* heat-resistant.

This organism most often causes a problem after a roast, perhaps cured and smoked or even fresh, has been properly cooked and served. If the leftover roast remains at room temperature for a period of time, before being refrigerated, the *S. aureus*, if present, has a good opportunity to grow at favorable temperature conditions. Also the previous cooking may have destroyed any competitive "friendly" microorganisms, thus giving *S. aureus*, with its slight temperature tolerance, a chance to grow. Even if the leftover roast is thoroughly re-cooked before being eaten, any toxin formed during the period at room temperature will not be destroyed.

Salmonella

A food-borne infection results from ingesting a high number of potentially infectious microorganisms or by allowing their growth and multiplication in the digestive tract. The bacteria must be living to cause infection.

Salmonella infection is caused by any one of a large number of species of that genus. The salmonellae are gram-negative, nonspore-forming rods that ferment glucose, usually with gas, but not lactose or sucrose. They grow best in nonacid foods at an optimum temperature of about 98.5° F but will grow well at room temperatures. They are killed by pasteurization equivalent to that given milk (161° F for 15 seconds or 143° F for 30 minutes). After ingestion, the incubation period will normally be 12 to 24 hours before digestive upsets occur.

Clostridium perfringens

Food poisoning from *C. perfringens* is caused by the release of an enterotoxin during sporulation. Based upon their ability to produce various exotoxins, six types of *C. perfringens* are recognized: types A, B, C, D, E, and F. The food-poisoning strains belong mainly to the type A and C groups. Most perfringens food-poisoning cases on the North American continent are produced from the type A strain and cause a mild case of gastroenteritis, which is rarely fatal. The type C strains have been mainly isolated in Europe and are much more virulent. A 35 to 40 percent mortality rate is associated with food poisoning contracted from the type C strains.

C. perfringens are anaerobic spore formers that can withstand the adverse environmental conditions of drying, heating, and certain toxic compounds. Different strains display wide variations in their resistance to heat. These organisms may find their way into meats either directly from the slaughtered animals or from subsequent contamination from containers, handling, or dust. The lowest reported temperature for growth is 68° F, while sporulation requires higher temperatures (98° to 99° F). Sporulation (and subsequent toxin production) occurs in the gastrointestinal tract after the ingestion of *C. perfringens* spores. Symptoms of acute abdominal pain and diarrhea develop 8 to 22 hours after ingestion, with most cases being attributed to canned meat, gravy, or soups. Most contamination occurs after the product is cooked, thus sanitation and refrigeration are important.

Other Sources of Food-borne Infections and Food Poisoning That Are Less Common

Numerous other microorganisms are involved in food-poisoning outbreaks but occur less frequently. Streptococcal food-poisoning outbreaks have been associated with cheese, turkey, turkey dressing, Vienna sausage, and barbecued beef. Specific organisms of concern include *Streptococcus faecalis, Streptococcus viridans,* and *Streptococcus pyogenes.*

Bacillus cereus causes a food-poisoning reaction similar to *C. perfringens;* however, the documented cases of food poisoning from *B. cereus* are quite low in the United States. Meat dishes that have been incriminated in outbreaks include meat loaf, minced meat, liver sausage, and soups.

Virulent strains of *Escherichia coli* have been reported to be the causative agent in some food-poisoning outbreaks. *E. coli* is one of the most widespread microorganisms in nature; however, few virulent strains exist. Contamination of *E. coli* usually occurs with unsanitary handling of the cooked product or improper handling and cooking of the raw product.

Vibrio parahaemolyticus food poisoning is contracted almost exclusively from seafoods. Symptoms consist of diarrhea, abdominal pain, nausea, vomiting, and chills. While its incidence is quite low in the United States, it is the leading cause of food poisoning in Japan.

THE FREEZING OF MEAT

In developed countries, freezing is the most common method of preserving fresh (uncured) meats for extended periods of time. Even though the bulk of beef, pork, and lamb is purchased unfrozen in retail meat counters, it is often subsequently frozen at home. As consumer habits and buying patterns gradually change, we will undoubtedly see a gradual trend towards the purchase of frozen meat. A rise in frozen meat sales has been predicted since the 1950s, but it has been slow to come. Consumers are still suspicious of frozen meat at the retail level. Unfortunately, they would rather buy fresh retail cuts and freeze them at home, often in an inadequate freezer and in an inadequate wrap.

In an extensive study conducted by Kansas State University,[1] the distribution costs, acceptance, cooking, and eating qualities of frozen meat were investigated. Processing steps which met with the most success included fabricating retail cuts at least 48 hours post-mortem in a room no warmer than 55° F; allowing a 30-minute period for bloom to develop before freezing or packaging; freezing initially at −70° F, storing in the dark at −15° F or lower, and displaying at case temperatures of −20° F or lower; packaging in a film with moderate- to high-oxygen permeability to maintain color; and using a 5- to 6-second water dip at 90° to 95° F to remove color bleaching.

The least costly method of distribution was to ship the carcass to a central processor to be cut into retail cuts and then shipped to retail stores. The most costly was to ship the carcass through a central distributor and then to a retail store to be cut and wrapped.

Results of comparative cooking tests of frozen versus unfrozen steaks and chops showed that freezing did not impair the palatability of beef steaks, but that fresh, unfrozen pork chops were preferred to frozen chops, even though the frozen chops were acceptable. Frozen chops and beef steaks generally had higher cooking losses than their fresh counterparts. In the retail market, most of the customers who purchased the frozen meat liked its palatability, but not all; 23

---

[1]*Research Publication 166*. Sept. 1973.

percent were dissatisfied with tenderness, juiciness, or texture. More than 85 percent of the nonpurchasers indicated that the frozen meat was unappealing and unappetizing or appeared artificial.

Physical Effects of Freezing

Freezing acts as a preservation method by the almost complete inactivation of meat enzymes and growth inhibition of spoilage organisms. The lower the temperature, the greater the inhibitory action and the longer the period of satisfactory storage. To secure and maintain very low temperatures requires expensive construction and entails high operating costs. The industry has been utilizing temperatures ranging from 0° to −32° F.

Lean meat averages from 60 to 65 percent water, which expands at both high and low temperatures. The actual point at which meat juices will freeze solid is not 32° F but 28° to 29° F. The rate of crystallization and the size of the crystals formed are dependent upon the temperature. Slow freezing causes the water to separate from the tissue into pools that form large crystals. These stretch and rupture some of the surrounding tissue. Rapid freezing results in very little water separation, and the crystals are therefore small and less expansive. Because there is practically no pool crystallization in very low-temperature freezing, the drip is considerably less than on meats frozen at higher temperatures.

To prevent the growth of spoilage-producing bacteria deep in carcass tissues or in the center of containers of warm meat, meat temperatures must be brought down to 40° F within a 16-hour period. If hot meat goes directly to the freezer, it must reach 0° F within 72 hours to prevent the growth of putrefactive bacteria. For large packs of hot meat, a freezer temperature of −5° F with air velocities of 500 to 1,000 feet per minute is recommended. For some freezers, the amount of meat frozen at one time should not exceed 2 pounds per cubic foot of freezer space. More than this amount raises the freezer temperature and slows down the freezing process. The key rule is "fast to 0° F or below."

Low temperatures do not destroy vitamins. Most of the vitamin loss is caused by heat or light or is lost in the juices that escape.

Experimental work to date has indicated that best results are obtained when:

- An animal is physically sound (in good health).
- An animal is properly bled (fiery carcasses do not keep well).
- The animal is properly chilled (chill room temperature of 29° to 36° F).
- The aging period is restricted.
- The wrapping material is of good quality.
- The holding temperature is 0° F or lower.

This is predicated on the assumption that proper sanitary precautions have been observed.

Cutting Method for Frozen Storage Meats

Tests conducted at Kansas and Michigan experiment stations show that boning meat has no effect on the flavor or juiciness of the cooked meat and that packaging boneless meat is easier, causes less damage to wrappers, and saves up to 35 percent of frozen storage space. The expense of boning adds to the labor charge over the bone-in method, which is absorbed in part by less necessary rental space, smaller amount of paper required, and the ease and satisfaction in cooking and carving.

One of the best methods for breaking down a carcass of beef for subsequent boning is the method explained in this text. Regardless of the method used, the cuts should be made ready for the oven and in such sizes as will best meet the needs of the family. Some cuts other than the standard ones, such as the top round muscle sliced into chipped steak (about ¼ inch on the slicing machine), make for variety and aid in menu making. Giving the top round a slight freeze (not solid) will make it slice evenly.

Oxidative Rancidity

Since the growth of spoilage microorganisms is inhibited at freezer temperatures, oxidative rancidity is the principal factor limiting the storage life of frozen meats (refer to Table 18-1 for the maximum recommended storage times for various frozen meats).

The development of rancidity in animal fats depends upon their ability to absorb oxygen from the air. This weakness for oxygen varies with the basic chemical structure of the fat involved. Any fat that has one or more double bonds in the carbon chain will be vulnerable to a cleavage caused by the oxygen taking the place of the double bond and forming aldehydes and fatty acids. These products, so formed, generally are no longer pleasing in taste or odor. As a result, they affect the palatibility of the fat and the adjoining lean.

Since pork fat is fairly high in unsaturated fatty acids (*e.g.*, oleic acid—one double bond; linoleic acid—two double bonds; linolenic acid—three double bonds) which have the ability to absorb oxygen, it follows that its storage life is lessened considerably. Beef and lamb, on the other hand, have a higher proportion of saturated fatty acids with no double bonds in the carbon chain and therefore are less susceptible to oxygen absorption and oxidative rancidity, with a subsequently longer storage life.

The obvious ways to combat oxidative rancidity are to eliminate the air and to use anti-oxidants. The elimination of air can be done in several ways, the most practical of which is to use a wrapping material that is airtight, moistureproof, and properly applied. The loss of moisture from meat or any other food is usually termed *shrink* or *dehydration*. The loss of moisture from the frozen surface of meat has been dubbed *freezer burn*. A good wrapping material will serve to reduce both oxidative rancidity and freezer burn.

Australian workers have shown that freezer burn occurs less when meat is frozen at $-4°$ F compared to $14°$ F. This is of practical use for the storage of large cuts and carcasses which are too unwieldy to wrap properly.

Factors Which Stimulate Oxygen Absorption

Increased temperatures accelerate oxygen absorption, as has been mentioned previously. Ultraviolet light used in the sterile lamps that are part of the equipment of some coolers accelerates oxidation. The minerals copper, iron, manganese, cobalt, and lead are also guilty. Salt (NaCl) increases the susceptibility of fats to oxidation.

Aging

Experiments show that the length of the holding (aging or ripening) period has a direct bearing on storage life because it permits oxygen absorption by the exposed fat. This raises the question whether meat that is to be frozen should be aged. Aged meat has shown higher peroxide values and a shorter storage life than 48-hour chilled meat. Aging also showed that, although the ripened meat was slightly more tender during the first month of storage, this advantage disappeared in the subsequent months, the fresh and aged meats being on a par for tenderness. These things being true, aging meat for the development of flavor, aside from its tenderizing effect, becomes a questionable practice for meat that is to be held in zero storage for more than six months.

Trimming Fat

What about the fat on meat? Again the results indicate that it is advisable to trim closely before freezing. The fat probably won't be eaten, even if it is palatable; it will taint the lean if it oxidizes, and it will take up much more storage space. In the case of pork, the nature of the fat makes it more vulnerable to oxidation and therefore lowers its storage life below that of beef, veal, and lamb. It is very important, therefore, to trim closely or to freeze only those cuts that are quite lean. For example, tests on sausage of different degrees of fatness showed that the lean sausages had longer storage life than those containing more fat. It has also been demonstrated that pork that was frozen after 48 hours of chill had longer storage life than pork that was chilled for 7 to 14 days before it was frozen. The same was true of the sausage made from such pork.

Packaging Meat for the Freezer

It is highly important that cuts of meat be compact and as nearly square or rectangular in shape as possible. There should be no sharp edges of bone protruding to puncture the paper. The paper should be pressed tightly to the meat to exclude all the air possible and make the entire job practically airtight. A double layer of waxed paper should be placed between cuts, if several are wrapped in the same package. An adhesive tape made especially for low temperatures is used in securing the package. It has an added advantage in that the mark of identification can be placed on it.

The most desirable method from the standpoint of maximum air seal is the apothecary or drugstore wrap, although work at Kansas showed as good results with the use of the butcher's wrap, a quicker and more rugged method, which gives the package a double thickness of paper. With a little practice, most people become quite proficient and speedy with the drugstore wrap. It does not pay to economize on paper, either in quality or in quantity.

Vacuum packaging in an abrasion-resistant, moisture-impermeable film provides the best packaging system for shelf-life longevity; however, this requires the use of specialized equipment generally not available to most consumers.

Temperature, Length of Storage, and Thawing

The lower the temperature, the longer the period of successful storage. Recommendations are definitely for zero or lower, the limiting factor being the cost of the equipment and the cost of maintaining the lower temperatures.

The length of the storage period should not be over 12 months, for economic reasons if for no other. With a proper wrap in good-quality paper and a zero temperature, practically all lean meat will keep well for six to eight months, with some exceptions. These exceptions have to do with products that contain salt, such as seasoned sausage, liver pudding, scrapple, sliced ham, and bacon slices. Table 18-1 summarizes the recommended storage times for red meat.

W. L. Sulzbacher, former bacteriologist for the USDA, reports: "There is no indication that frozen meat becomes more perishable after thawing than fresh meat."

Repeated tests made at the Pennsylvania station showed that:

- Meat which was thawed in the unopened package exceeded the keeping quality of unwrapped fresh meat.
- Meat which was alternately thawed and refrozen as many as three times before being unwrapped was the equal of the meat used after one thawing.
- Meat which was unwrapped and thawed and then rewrapped and refrozen was not materially changed in palatability other than that it was slightly drier because of the juices it had lost.

Table 18-1. Storage Time Chart[1] (Maximum Storage Time
Recommendations for Fresh, Cooked, and
Processed Meats[2])

Meat	Refrigerator (38° to 40° F)	Freezer (at 0° F or lower)
Beef (fresh)	2 to 4 days	6 to 12 months
Veal (fresh)	2 to 4 days	6 to 9 months
Pork (fresh)	2 to 4 days	3 to 6 months
Lamb (fresh)	2 to 4 days	6 to 9 months
Poultry (fresh)	2 to 3 days	3 to 6 months
Ground beef, veal, and lamb	1 to 2 days	3 to 4 months
Ground pork	1 to 2 days	1 to 3 months
Variety meats	1 to 2 days	3 to 4 months
Luncheon meats	1 week	not recommended
Sausage, fresh pork	1 week	2 months
Sausage, smoked	3 to 7 days	1 month
Frankfurters	4 to 5 days	1 month
Bacon	5 to 7 days	1 month
Smoked ham, whole	1 week	2 months
Smoked ham, slices	3 to 4 days	1 month
Beef, corned	1 week	2 weeks
Leftover cooked meat	4 to 5 days	2 to 3 months
Frozen combination foods		
Meat pies (cooked)	—	3 months
Swiss steak (cooked)	—	3 months
Stews (cooked)	—	3 to 4 months
Prepared meat dinners	—	2 to 6 months

[1]*Lessons on Meat.* National Live Stock and Meat Board.

[2]The range in time reflects recommendations for maximum storage time from several authorities. For top quality, fresh meats should be used in two or three days; ground meat and variety meats should be used in 24 hours.

There is no reason to hesitate to refreeze meat when occasion demands, but it should be done within the day. If it is to be used the following day, it should be placed in the rear of the refrigerator rather than refrozen for that short period. Thawing may be accomplished in various ways to suit the conditions, or the meat may be cooked in the frozen state, in which case it will require a slightly longer cooking period (see Chapter 22).

Freezer Storage of Seasoned Meats

Salt affects the rate of fat oxidation, causing cured meats or meat products seasoned with salt to acquire a flat, rancid taste in a shorter time than the unseasoned product. Whole hams, picnics, or butts, properly wrapped in a good grade of locker paper, will maintain the original flavor for two months or possibly more; however, half-hams or sliced bacon will lose flavor within the month.

The freezer-storage life of fresh sausage can be lengthened by omitting the seasoning and then adding the seasoning after the sausage is thawed. The addition of an antioxidant such as BHA, BHT, or propyl gallate (see Chapter 20) inhibits oxidation and may be added to the sausage at the second grinding. Pork trimmings can be frozen for future sausage making, but the holding period should not exceed one month. Smoked sausage has a longer storage life than fresh sausage.

Preparing and Freezing Poultry

Broilers and fryers are cut into halves or quarters, and fowl can be left whole for roasting, cut into stewing joints, or boned. Cut-up fowl is popular because second joints, drum sticks, and white meat can be packed separately from the less desirable wings, backs, and necks. It has the added advantage of compactness, eliminating the large body cavity which traps considerable air, thus requiring less storage space. Roasters frozen whole should have the excess internal fat removed as it will oxidize and become rancid far more rapidly than the rest of the fowl. The giblets are wrapped in cellophane or foil and placed inside the bird.

In large-scale operations, poultry, turkeys, ducks, geese, and other fowl are usually vacuum packed. The fowl are placed in CRYOVAC® bags and the air is exhausted with a vacuum pump. The end of each bag is then made airtight by a metal clamp placed around the opening. It is then immersed in hot water to give it a skintight shrink. Storage at 0° F will keep properly wrapped fowl edible for from four to six months.

Freezing Eggs

Eggs may be frozen whole or as separate whites and yolks, but they must be stirred enough to break the membranes, but not enough to cause foaming. Only strictly fresh eggs should be frozen. Cooked eggs should not be frozen, as they become tough and rubbery. Each egg should be broken individually into a cup to check for soundness before it is placed in the container with the good eggs. Eight broken eggs are required to fill a pint cup to weigh approximately 1 pound. One tablespoonful of corn syrup, honey, or sugar or 1 teaspoonful of salt (depending upon how the eggs are to be used) is added to and gently stirred into each cup of broken whole eggs or yolks. This is necessary to prevent the eggs from becoming gummy. If a good, liquid-tight container is used, the eggs can be held at 0° F for one year. They should be thawed in the container before they are used.

Freezing Fish

Fish, including shell fish, can be divided into two main groups, based on the oil content of the flesh. The non-oily fish (less than 3 percent oil) store their oil in the liver rather than in the flesh and are represented by the cod, haddock, halibut, and swordfish, to mention a few. The oily fish (over 3 percent oil) have the oil distributed throughout the flesh. Some representatives of this group are herring, mackerel, and salmon.

The chief type of spoilage in frozen fish, as in warmblooded animals, is caused by the oxidation of the fats, resulting in rancidity. The action of bacteria and enzymes is inhibited by low temperatures, but air must be excluded if oxidation is to be held to a minimum. It is necessary, therefore, to wrap the eviscerated fish, either in the round (unsplit) or the fillet, in plastic film, excluding as much air as possible, and freezing it at 0° F or below. Fish that are too large to be wrapped should be quick-frozen and dipped in cold water several times to cover them with a glaze of ice. The ice glaze will evaporate within several months unless the humidity of the holding room is very high. Reglazing or wrapping in moisture-vapor–proof material is then necessary. Frozen fish that have been well wrapped can be stored with other foods without imparting or transferring any odor or flavor to them. Table 18-2 lists maximum recommended frozen-storage periods for some common species of fish.

Table 18-2. Storage Period for Frozen Fish

Species	Round or Headed and Gutted	Wrapped, Packaged
Croaker	6-8 mo.	8-10 mo.
Grouper	6-8 mo.	8-10 mo.
Lake herring	6-8 mo.	8-10 mo.
Ling cod	6-8 mo.	8-10 mo.
Mackerel (Spanish and Boston)	6-8 mo.	8-10 mo.
Mullet	6-8 mo.	8-10 mo.
Red snapper	6-8 mo.	8-10 mo.
Rockfish	6-8 mo.	8-10 mo.
Rosefish (ocean perch)	6-8 mo.	8-10 mo.
Sablefish	6-8 mo.	8-10 mo.
Salmon	6-8 mo.	8-10 mo.
Sea trout	6-8 mo.	8-10 mo.
Shrimp	6-8 mo.	8-10 mo.
Cod	8-10 mo.	10-12 mo.
Flounder (sole)	8-10 mo.	10-12 mo.
Haddock	8-10 mo.	10-12 mo.
Halibut	8-10 mo.	10-12 mo.
Pike (all species)	8-10 mo.	10-12 mo.
Pollock	8-10 mo.	10-12 mo.
Porgie (scup)	8-10 mo.	10-12 mo.
Sole	8-10 mo.	10-12 mo.
Whiting	8-10 mo.	10-12 mo.
Smelt	8-10 mo.	8-10 mo.
Whitefish	8-10 mo.	8-10 mo.

Home Freezer Storage Units

Home freezer units are made in three popular styles referred to as:

• The vertical, upright, or side-door type.

- The horizontal, top-door, or cabinet type.
- The combination refrigerator-freezer with separate doors for refrigerating and freezing compartments.

The last one far outnumbers the other two types, having been universally adopted by manufacturers of all the popular makes of household refrigerators.

The side-door type has the outward appearance of the regular household refrigerator, but the freezing element circulates through pipes between the walls, and in some models the pipes are in the shelves, making the shelf a freezer plate. The shelves are at a convenient height for easy access and visibility; the depth of the shelves does not require long arms or the use of tongs, and the space can be utilized efficiently. This model has been criticized by some for door leakage and spilling of cold air whenever the door is opened.

The top-door (dunk-in) type has the advantage of less door leakage and practically no spillage of cold air but has the disadvantage of requiring reaching and stooping, and the food is rather inaccessible without considerable rearrangement. It also utilizes more floor space than the side-door type. Most of them are designed with a $-10°$ F sharp-freeze compartment and a zero storage compartment. Both side- and top-door types are made in popular sizes ranging from 10- to 50-cubic foot capacity at prices varying from \$25 to \$60 per cubic foot.

Walk-in Storage Units

Manufactured units of this type usually are prefabricated at the plant and assembled on the owner's premises. The more popular practice is to buy the refrigerating unit and have local labor construct the refrigerator. This permits the owner a wider choice in capacity, and generally it is cheaper. The most popular insulation consists of 4 or 6 inches of styrofoam, sandwiched between laminated steel or plastic panels. The cooler should be capable of operating at $35°$ F. A $0°$ to $-10°$ F walk-in freezer may be built within the cooler to economize on insulation materials. Single refrigerating machinery operated by a ½- to 1-horsepower motor (depending on cooler/freezer size) is satisfactory, but it is advisable to have a separate machine for each box. This will save closing down the plant in case one unit goes bad. Freon-12 gas is probably the most satisfactory refrigerant to use, and forced-air cooling units that are self-defrosting are in equal favor with gravity units. Sizes of 400- to 1,200-cubic foot capacity best suit the needs of farm families.

Cryogenic Freezing

Methods of rapidly freezing meat are termed *cryogenic*. Various systems of producing cold are utilized.

The freezing system most closely resembling a conventional refrigerator or freezer employs a brine ($CaCl_2$ solution) of $-40°$ F circulating in a tank below and through plate fin coils above a stainless steel belt which carries the meat product through a chamber. Moist product (normally steaks, chops, or patties) freezes instantly to the belt as it makes contact and later pops off at the end where the belt turns under to begin to return. Freezing capacities of up to 3,600 pounds of 0° F product per hour are available.

Liquid nitrogen is more widely used as the refrigerant in similar conveyor—tunnel—chamber arrangements. Liquid nitrogen itself has a temperature of −320° F, but very seldom is meat immersed directly into the liquid; rather, the liquid is placed under pressure of 15 to 22 psi (pounds per square inch) and sprayed through nozzles over the product. The closer the product is to the nozzle, the nearer to −320° F is the temperature. Normally, freezing temperatures range from −100° to −200° F.

Liquid carbon dioxide (CO_2) is also used as a refrigerant in cryogenic systems, using the same principles that were described for liquid nitrogen. Liquid CO_2 has a temperature of −180° F.

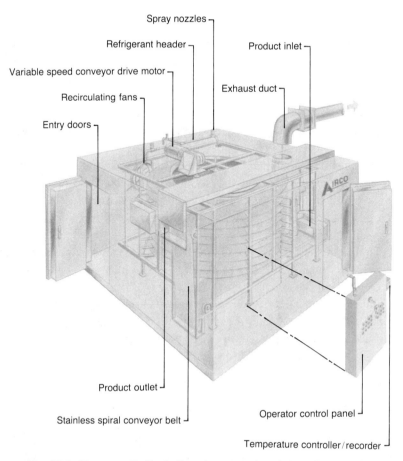

Fig. 18-1. Diagrammatic illustration of a cryogenic spirulator freezer. (Diagram courtesy Airco Kryofoods, Murray Hill, New Jersey)

Cryogenic spirulators have gained widespread acceptance in recent years as a rapid-freezing technique in the beef patty industry and other specialized industries where large freezing volumes are needed but where only limited floor space is available. Spirulators have conveyors shaped like large corkscrews and can be purchased as self-contained units as small as 8- by 8-foot square, or as larger models with freezing capacities over 12,000 pounds per hour. A diagrammatic illustration of an Airco® spirulator (Murray Hill, New Jersey) is shown in Figure 18-1.

THE CANNING OF MEAT

The sterilization of food products by canning is an efficient method of food preservation whose discovery in 1809 is attributed to a Frenchman, Nicolas Appert. A year later, Peter Durland, an Englishman, invented the tin can, which led to the first commercial canning operations. With the invention of the tin can came the term *canning*, which was broadened in later years to include preservation in other forms of containers.

The canning of meat represents the second most common preservation method of meat products for extended periods of time (freezing is the most common). Canned-meat products may be conveniently grouped into two categories: sterilized products and pasteurized products. Sterilized products are shelf stable (require no refrigeration), while pasteurized products require refrigeration to inhibit spoilage.

Canned foods are preserved by hermetically sealing (sealed to prevent the escape or entry of air) the product in a container and destroying, through application of heat, microorganisms capable of producing spoilage. Spoilage microorganisms are universally present throughout our environment and include bacteria, yeasts, and molds. The organisms of greatest concern in the canning industry are certain anaerobic bacteria that are capable of forming protective endospores. Some genera of bacteria form spores which are capable of resisting adverse conditions such as high temperatures, low moisture, and antiseptics. In fact, considerable time at 240° F is required to kill some spores whose vegetative cells can cause food poisoning in humans. When sufficiently adverse conditions are encountered, the vegetative bacterial forms are killed, leaving behind the spore forms. When spores again meet favorable conditions, they may slowly return to the actively growing (vegetative) form and then multiply very rapidly.

Sterilization

A food substance becomes absolutely sterile only when no vegetative microorganisms or spores are present, either because there were none originally or because they have all been killed. Absolute sterility seldom exists in commercially canned products. Commercial sterility refers to the destruction of spoilage organisms and their spores such that a product will not undergo spoilage under indefinite storage periods. The principal method of assuring commercial sterility is by heat treatment in a sealed container. Sanitation is extremely important in meat-canning plants, since the fewer the number of organisms originally present, the more effective is the sterilization process.

A time and temperature relationship is required for the destruction of most microorganisms. That is to say that simply reaching a specified internal temperature is often not sufficient to assure destruction of certain spores. The product must be held at this minimum temperature for a specified period of time to achieve destruction of spores of concern. This concept is known as the F_o value, which is defined as the equivalent time in minutes at 250° F required to destroy vegetative and spore forms of a given organism. The usual practice is to calculate this in terms of the destruction of *Clostridium botulinum* for an equivalent number of minutes at 250° F. While the F_o value is a somewhat arbitrary way of expressing the effectiveness of a process, it does provide a useful tool for the determination of safe, commercial sterilization levels. A safe cook is considered to have an F_o value of 2.78 or more. This means a minimum of 2.78 minutes at 250° F. Many use 3 minutes as the minimum time at 250° F for a safe cook.

Canning under the sterilizing temperature (*i.e.*, 250° F) necessitates processing the cans in a *retort cooker* in order to achieve the needed temperature. A retort cooker is simply a giant pressure cooker that operates under pressures of 12 to 15 pounds per square inch (psi). This enables the cooking-water temperature to rise above the normal boiling point of 212° F (at sea level). *Retort processing* simply refers to cooking under pressure.

The advantages of canning meat products using sterilization temperatures are numerous. Perhaps the greatest advantage is the indefinite storage life over a wide temperature range. Canned-meat products, opened after many years of storage, have still maintained an edible quality (although flavor deterioration had occurred). Canning provides an effective preservation method for high-quality food in times of catastrophes (natural or human-made) which result in the disruption of power and thus mechanical refrigeration. Other advantages include:

- Energy is saved (because no refrigeration is required).
- Merchandising speed is less critical.
- Product spoilage during merchandising is minimal (some "leakers" do occur).
- The stabilizing effect canned goods can have on market prices when raw material supplies become low.
- Convenience is an additional advantage of canned meat, since no cooking is required after the can is opened.

The principal disadvantage of utilizing sterilization temperatures in the canning of meat and meat products is the slight deterioration in food quality that occurs.

Examples of popular canned-meat products that have undergone commercial sterilization include the potted-meat products, Vienna-style wieners, corned beef hash, roast beef, corned beef, beef stews, pickled pigs' feet, and numerous others.

Pasteurization

Pasteurized meat products have undergone a less-severe heat treatment (generally 155° to 165° F internal temperature) and, as such, have not destroyed the thermophilic spore-forming microorganisms. However, under good sanitation practices and proper refrigerated storage conditions (less than 40° F), these canned products will typically have a shelf life of six months. This relatively long shelf life (compared to other cooked products) is primarily because no recontamination of the product can occur while it's in a sealed container. Since it was cooked in this container to a temperature of at least 155° F, virtually all vegetative forms of microorganisms have been destroyed. Refrigeration temperatures of 40° F effectively inhibit the outgrowth of bacterial spores that survived the heat treatment.

Examples of pasteurized products include most canned hams. Cured meats, such as ham, often pick up an undesirable metallic flavor when cooked to sterilization temperatures and thus are generally pasteurized. The consumer should be very cautious about purchasing a canned ham that is not refrigerated. Pasteurized products, by law, must display in bold type on the principal display panel of the can the words "Keep Refrigerated."

Retortable Pouches

Laminated foil pouches have become a popular and more economical means of canning in recent years. Although nonmeat foods are the most common contents of the retortable pouch, some meat products are being preserved in this manner. Foil-laminated retort pouches are in widespread use in the food industry for producing shelf-stable, sterilized products. The advantages of the foil pouch are lower container costs and generally faster processing (retorting) times due to the flatter configuration of a pouch compared to a can. Figure 18-2 illustrates several meat items offered in the U.S. military's MRE (Meal, Ready to Eat) ration. The lighter weight of the retortable pouch was a major factor in the military's decision to switch from the can to this type of package. Figure 18-3 illustrates how flexible retortable pouches are used in the entire MRE ration.

Cook-in Processing

Cook-in processing is a recent innovation that has gained widespread acceptance among larger processors in the meat industry. The cook-in process refers to the cooking of products (usually boneless hams and boneless turkey breasts) in a plastic, formable film and using this same film as the merchandising container. Actually two types of film—a forming film and a non-forming film for the lid—are used. Both plastic films are composed of multiple layers of materials designed to meet their specific functions. A modified roll stock packaging machine handles the two rolls of film.

Fig. 18-2. Examples of flexible retort packaging of meat products used in the U.S. military's MRE (Meal, Ready to Eat) rations. Meat products pictured here are shelf-stable, heat-sterilized products. (Photograph courtesy U.S. Army Natick Research Center, Natick, Massachusetts)

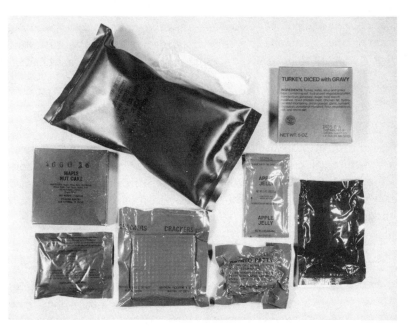

Fig. 18-3. Example of an entire MRE (Meal, Ready to Eat) ration illustrating the extensive use of flexible packaging. The elimination of the tin can in military rations has decreased the cost and weight of the ration. (Photograph courtesy U.S. Army Natick Research Center, Natick, Massachusetts)

The forming film is fed through a heated chamber which forms a cavity in the film which is filled with product. The non-forming film is then conveyed over the top of the product-filled cavities, a vacuum is drawn, and the two films are sealed together to form an airtight container ready for water or steam cooking. The end result is a canned ham without the can. Cook-in films are available in both transparent and opaque forms. Cook-in hams are generally placed in stainless steel molds prior to and during cooking to give them the desired shapes.

Advantages of the cook-in ham-processing system over conventional water-cooking systems include elimination of cooking losses, a twofold to threefold increase in shelf life, greater efficiency in packaging costs and labor, and a reduction of sewage-treatment costs. Cooking losses are eliminated, since the product is merchandised in the cooking film. The skintight evacuated package allows little or no moisture purge between the products and the plastic film. Shelf life is increased because there is no recontamination of the product after cooking by packaging lines and handling of the product. Labor savings are also claimed, since there is no repackaging after cooking. Furthermore, the task of cleaning the cooking molds is greatly simplified when a sealed plastic film is used. Finally, sewage-treatment costs are significantly reduced, because the rendering of fat and the leaching of meat proteins into the cooking water are eliminated.

When compared to conventionally canned hams, considerable cost savings are also observed. Obviously, the container costs are substantially lower for the cook-in process. Equipment costs are also greatly reduced with the elimination of the specialized equipment required for handling cans and the retorting chamber; however, the canned ham will have a longer shelf life with its sturdy container that is less apt to become punctured or damaged during merchandising.

PRESERVATION OF MEAT BY DRYING

The removal of moisture from meat is an effective means of preservation, because moisture is a critical element required by all microorganisms for growth. This method of preserving meats lowers the moisture content of the product to a point at which the activity of food-spoilage and food-poisoning microorganisms is inhibited.

Moisture requirements for microbial growth should be defined in terms of water activity (a_w) in the environment (meat) of the microorganisms. Water activity is defined as the ratio of the water-vapor pressure of the food substrate (P) over the vapor pressure of pure water (P_o) at the same temperature (*e.g.*, a_w = P/P_o). The a_w of fresh meat is above 0.99. The minimum a_w values required for the growth of various classes or genera of microorganisms are reported in Table 18-3.

Foods preserved by dehydration are conveniently divided into two groups, low-moisture foods and intermediate-moisture foods based on a_w levels. Low-

Table 18-3. Approximate Minimum a_w Levels Required for
Growth of Food-Spoilage Microorganisms[1]

Microorganisms	Minimum a_w
Groups	
Most spoilage bacteria	0.91
Most spoilage yeasts	0.88
Most spoilage molds	0.80
Holophilic bacteria	0.75
Xerophilic molds	0.65
Osmophilic yeasts	0.60
Specific organisms	
Acinetobacter	0.96
Enterobacter aerogenes	0.95
Bacillus subtillis	0.95
Clostridium botulinum	0.95
Escherichia coli	0.96
Pseudomonas	0.97
Staphylococcus aureus	0.86
Saccharomyces rouxii	0.62

[1]James M. Jay. *Modern Food Microbiology*. 1978. Van Nostrand Reinhold Co., New York.

moisture foods are defined as those having an a_w of less than 0.60 and containing less than 25 percent moisture. Intermediate-moisture foods have an a_w between 0.60 and 0.85 and contain less than 50 percent moisture.

Moisture removal may be accomplished by low-temperature drying (less than 120° F), high-temperature drying (greater than 200° F), freeze-drying, and/or salting.

Low-moisture meat products include freeze-dried foods produced by specialty companies for camping and backpacking and for the military.

Meat powders for soup stock and bouillon are also low-moisture foods but are generally produced by high-temperature drying procedures such as spray or drum drying. Beef jerky sometimes falls into the upper end of the low-moisture category with a_w values typically ranging from 0.55 to 0.70.

The U.S. military has developed low-moisture compressed food bars designed for long-range patrol operations. "Food Packet, Assault" rations, as pictured in Figure 18-4, are designed to be a lightweight, high-density meal with a relatively long shelf life, suitable for both cold and hot climates. Foods in the assault packet are all low-moisture or intermediate-moisture and can be consumed dry or rehydrated.

The true dry sausages fall into the intermediate-moisture category with a_w values typically ranging from 0.65 to 0.80. As such they require no refrigeration; however, they are often subject to external mold growth unless treated with a mold inhibitor such as potassium sorbate (see Chapter 20).

The a_w of meat products may be lowered by the addition of salt; however, the addition of salt alone requires extremely high levels to reach the a_w level required for the preservation of intermediate-moisture foods. Table 18-4 gives the

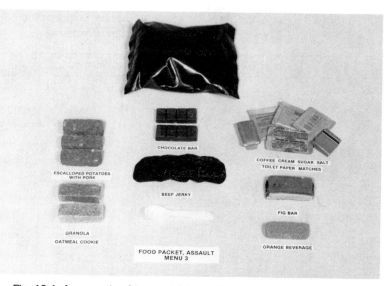

Fig. 18-4. An example of low- and intermediate-moisture foods is illustrated in the U.S. military's Food Packet, Assault menu. (Photograph courtesy U.S. Army Natick Research Center, Natick, Massachusetts)

Table 18-4. Relationship Between a_w and Concentration of Salt Solutions[1]

	Sodium Chloride Concentration	
Water Activity	Molal	Percent (w/v)[2]
0.995	0.15	0.9
0.99	0.30	1.7
0.98	0.61	3.5
0.94	1.77	10.0
0.90	2.83	16.0
0.86	3.81	22.0

[1]*The Science of Meat and Meat Products*. 1960. American Meat Institute Foundation, W. H. Freeman & Co., Publishers, San Francisco.

[2]Weight/volume (*e.g.*, 3.5% = 3.5 pounds NaCl per 100 pounds water).

effects of salt concentrations on a_w. For this reason, salting is generally accompanied by drying to produce intermediate-moisture foods.

PRESERVATION OF MEAT BY IRRADIATION

Background and Current Status

The preservation of meat or other food products by exposure to ionizing radiation dates back to a 1909 French patent by O. Wust. However, it was not until the early 1950s that significant progress was made utilizing irradiation as a practical method of food preservation. Since the late 1950s, the U.S. government has sponsored much of the research related to this method of preservation. The U.S. Army has probably played the most significant role in the development and testing of the process. Governmental support for irradiation research was cut significantly after 1968, when the FDA refused approval of a petition to produce irradiated ham for public consumption. Nonetheless, international interests in the irradiation process have continued to be strong.

A recent recommendation by the Joint Expert Committee on the Wholesomeness of Irradiated Foods, convened by the World Health Organization (WHO), has sparked new interest in the irradiation process.[2]

> After reviewing approximately 30 years of safety and toxicology testing of irradiated foods, the committee concluded that any food irradiated to a dose of 1.0 Mrad[3] or less is a wholesome product for human consumption and therefore should be approved without further testing.

The WHO Joint Expert Committee deferred general recommendations for foods irradiated at levels higher than 1 Mrad, until data are available from numerous long-term ongoing studies.

Following the recommendations of the WHO Joint Expert Committee, the FDA is considering the approval of slightly more conservative legislation. A current pending petition to the FDA is asking for approval of 0.1 Mrad levels of irradiation to most foods. Action on the irradiation process at the 0.1 Mrad level is expected by the FDA in 1986.

Pasteurization (Radurization)

Radiation levels of 0.1 to 1.0 Mrad are considered sufficient for food pasteurization. Irradiation to pasteurizing dosages is commonly referred to as *radurization* and results in a product shelf life comparable to heat-pasteurized food (*e.g.*, canned hams). Like heat pasteurization, radurization also requires refrigerated storage of the products or spoilage will occur. Radurization is designed to kill or inactivate food-spoilage organisms without the use of heat. This gives the process tremendous potential applications in merchandising fresh meat, fish, and poultry products. For example, fresh (unfrozen) meats could be effectively exported to any country in the world and still maintain a fresh, acceptably long shelf life upon arrival. Poultry products often create merchandising problems for retailers due to their short display period before bacterial spoilage occurs. Fresh seafoods from coastal regions could be shipped to the interior portions of larger countries without the expense of rapid air freight, the method by which shipments are now made. The shelf life of other products, such as fresh pork sausage, bacon, corned beef, frankfurters, and luncheon meats, could

---

[2]WHO, 1981. *Wholesomeness of Irradiated Food.* Report of a Joint FAE/IAEA/WHO Expert Committee, Oct. 27 to Nov. 3, 1980, World Health Organization Technical Report Series, No. 659, Geneva, Switzerland.

[3]Unit for measuring radiation energy.

be extended from 3 to 10 times their present safe-storage period. Savings from product spoilage would be passed on to the consumer, since the retailers must include in their operating margin an allowance for perishable products that spoil before being sold.

Radiation pasteurization of poultry meat provides a very effective means of controlling contamination with salmonella. Salmonella contamination can be serious enough that some health authorities have suggested that radurization should be required for poultry products to bring the problem under control. An estimated 2 million cases of salmonella food poisoning plague this country each year, resulting in estimated economic losses of $1.5 billion annually (from food-spoilage losses, medical expenses, law suits, etc.).[4]

Sterilization (Radappertization)

Sterilization by irradiation is commonly referred to as *radappertization*. Depending on the food, spoilage and disease organisms are killed at radiation doses of 1 to 5 Mrad. Similar to their heat-sterilized (canned) counterparts, radappertized meat products may be stored for years without refrigeration (providing the sealed container is not punctured or corroded).

Before radappertization processing, meat must be cooked to 140° to 150° F in order to inactivate any autolytic enzymes that could produce off-flavors during prolonged storage at room temperature. After cooking, the meat is frozen to a −40° F and irradiated while frozen, which further minimizes flavor changes that would otherwise occur at the higher radiation doses (1 to 5 Mrad). Cooking is *not* required for radurization (pasteurization) processing because the meat must be stored under refrigerated conditions, and the shelf life is still limited.

Radappertized ham was consumed by the Apollo astronauts who landed on the moon and by the American astronauts and the Soviet cosmonauts who participated in the joint American-Soviet *Apollo-Soyuz* space flight in 1975. During the *Columbia* space shuttle flights, American astronauts dined on radappertized beef, pork, smoked turkey, and corned beef. Radiation-sterilized food products are approved for use in special hospital meals needed by patients who must have sterile foods because their immunity to disease has been altered by medical treatments, such as chemotherapy. These special diets are used primarily by terminally ill cancer patients and have shown no effect on the rate of cancer progression. Some examples of food preservation applications by irradiation are presented in Table 18-5.

Methods of Irradiation Processing

Two types of irradiation processes, gamma radiation and electron radiation,

---

[4]A. Brynjolfsson. 1980. *Food Irradiation in the United States*. Proceedings of the 26th European Meeting of Meat Research Workers, Vol. I, Colorado Springs, American Meat Science Assn., p. 172.

Table 18-5. Applications of Radiation Preservation to Food[1]

Dosage	Food	Main Objective	Means of Attaining Objective
Above 1 Mrad	Meat, poultry, fish, and many other highly perishable foods	Safe long-term preservation without refrigerated storage	Destruction of spoilage organisms and any pathogens present, particularly *C. botulinum*
	Spices and other special food ingredients	Minimization of contamination of food to which the ingredients are added	Reduction of population of microbes in special ingredient
Below 1 Mrad	Meat, poultry, fish, and many other highly perishable foods	Extension of refrigerated storage below 37° F	Reduction of population of microorganisms capable of growth at these temperatures
	Frozen meat, poultry, eggs, and other foods, including animal feeds, liable to contamination with pathogens	Prevention of food poisoning	Destruction of salmonellae
	Meat and other foods carrying pathogenic parasites	Prevention of parasitic disease transmitted through food	Destruction of parasites such as *Trichinella spiralis* and *Taenia saginata*
	Cereals, flour, fresh and dried fruit and other products liable to infestation	Prevention of loss of stored food or spread of pests	Killing or sexual sterilization of insects
	Fruit and certain vegetables	Improvement of keeping properties	Reduction of population of molds and yeasts and/or in some instances delay of maturation
	Tubers (*e.g.*, potatoes), bulbs (*e.g.*, onions), and other underground organs of plants	Extension of storage life	Inhibition of sprouting

[1]*Food Technology*. Feb. 1983. Vol. 37 (2):55. "Radiation Preservation of Foods."

are in commercial use today. The gamma radiation facilities utilize the isotopes cobalt-60 and cesium-137. Gamma radiation has a very short wavelength of the same nature as short X-rays. Electron radiation is created by large electron accelerators that range in power from 4 to 10 Mev (million electron volts). Both processing methods achieve the same result and *do not* result in the accumulation of radiation in the food product. Irradiation-preservation methods leave no more radiation in a product than what was originally there by nature (from the sun and environment).

Substantial energy savings are incurred for radappertized products during processing, merchandising, and storing when compared to other methods. Table 18-6 compares several preservation methods and the amount of energy required to maintain wholesomeness of the product. As indicated in this table, heat sterilization (canning) requires over five times as much energy as radappertization to achieve sterility. Storing frozen meat at −13° F for 25 days requires over 30 times the energy required to radappertize a product.

Table 18-6. Typical Energy Values Used in the Processing of Food[1]

Preservation Method	kj/kg[2]
Radurization (with 0.25 Mrad)	21
Radappertization (with 3 Mrad)	157
Heat sterilization	918
Blast freezing chicken meat from 40° to −10° F (4.4° to −23.2° C)	7,552
Storing product at −13° F (−25° C) for 25 days	5,149
Refrigeration storage for 5.5 days at 32° F (0° C)	318
Refrigeration storage for 10.5 days at 32° F (0° C)	396

[1]A. Brynjolfsson. 1980. *Food Irradiation in the United States*. Proceedings of the 26th European Meeting of Meat Research Workers, Vol. I, Colorado Springs, American Meat Science Assn., p. 172.

[2]Kilojoules per kilogram. Standard International (SI) units of measure for energy and mass (weight).

As one would expect from these energy-usage values, the preservation cost per pound of irradiated products is economical. Although large initial capital investments are required for irradiation facilities ($2 to $10 million), the volume of product that can be processed makes it economical on a per pound basis. At a cost of $10 million, such a facility would have a capacity to handle 100 million pounds of product per year.[5] Table 18-7 gives cost estimates of irradiation by gamma rays and electron radiation on a per pound basis.

Table 18-7. Cost of Radappertization of Bacon[1, 2]

Radiation Source[3]	Five-Year Plant Depreciation Costs	Operational Costs	Total Costs
	.... *(cents per lb.)*		
Co-60	2.03	1.2	3.23
Cs-137	2.03	0.32	2.38
10-Mev accelerator	0.49	0.43	0.92
4-Mev accelerator	0.36	0.40	0.76

[1]Brynjolfsson, A. 1980. *Food Irradiation in the United States*. Proceedings of the 26th European Meeting of Meat Research Workers, Vol. I, Colorado Springs, American Meat Science Assn., p. 172.

[2]Reflects a plant size capable of irradiating 100 million pounds annually.

[3]Radiation sources: Co-60 (cobalt-60); Cs-137 (cesium-137); 10-Mev (10-million electron volt) accelerator; 4-Mev (4-million electron volt) accelerator.

---

[5]*Ibid.*

PACKAGING MATERIALS

Wrapping materials suitable for meats come under six general groups:

- Wax- or paraffin-treated kraft papers.
- Cellophane (of increasingly lessening importance).
- Aluminum foil.
- Laminated foils.
- The films (polyethylene and pure or mixed polymers or copolymers of vinyl chloride, vinyl acetate, or vinylidine chloride).
- Food-grade "plastic" dips (Dermatex®).

The characteristics of these wrapping materials differ, and their suitability for packaging meats, vegetables, and fruits depends upon how closely they come to meeting the following criteria:

- Have low moisture-vapor transmission.
- Have differing oxygen permeability, depending on use.
- Have good tensile strength and are puncture resistant.
- Are pliable.
- Will maintain pliability and tensile strength at sub-zero temperatures.
- Are non-toxic.
- Are odorless.
- Provide for ease of marking for identification.
- Have good stripping qualities (will peel from meat when frozen).
- Are greaseproof and stainproof.
- Provide good sealing properties.

The moisture loss, or shrinkage, of any food during freezer storage must be held to a minimum. An excess of 8 percent shrink in meats and 3 percent in fowl is considered to make them unacceptable as fresh meat. This loss in weight is easily measured, and the changes in color, aroma, flavor, and texture are in about the same proportion as the loss in weight. To hold dehydration to a minimum necessitates the use of a material that has a low moisture-vapor transmission at low temperatures.

Wax- or Paraffin-treated Kraft Papers

The packaging materials in the first group are basically wood pulp papers. Kraft (German meaning *strong*) is probably the wrapping material most often used in the home and by local meat processors for freezing meat. These materials come in many forms having different qualities that give protection against oil, grease, chemicals, molds, moisture-vapor and oxygen transmission, and water.

Another wood-pulp paper that is used extensively is vegetable parchment, which has many uses, depending upon its treatment. Some grades are impervious to oxygen, carbon dioxide, and nitrogen. The coated parchment is used for freezer-wrapped meats. The newest type is the silicone-treated vegetable parchment which has anti-sticking properties that make it particularly useful as dividers for frozen meat cuts and for hamburger, sausage, and lamb patties.

Cellophane

Cellophane, the original transparent film, has lost much of its retail packaging market to the polyvinyl chloride copolymer films. Many types of cellophane materials are manufactured with varying properties. The one most prominent for packaging fresh meats is coated on one side with nitrocellulose. When the film is dry, it has a low oxygen-transmission rate; however, when in contact with a cut meat surface, the moisture from within the meat causes a significant increase in the oxygen-transmission properties of the film. This phenomenon helps keep the meat in retail counters a desirable bright red color through surface oxygenation of the myoglobin pigment protein (see discussion on meat color, page 614). Cellophane *is not* a good packaging material for the freezing of meat because of its high oxygen-transmission properties which will shorten the frozen shelf life of the product (see discussion on lipid oxidation, page 562). Therefore, when a consumer intends to freeze meat, any meat purchased with a cellophane wrap should be rewrapped tightly in an oxygen-barrier paper or film before freezing.

Aluminum Foil

Aluminum foil is used primarily by consumers for home freezing because it is readily available, is easy to work with, and has the additional advantage in that it can be used as a cooking wrap as well as a storage wrap. The primary disadvantages of aluminum foil are that it is easily torn, has a low resistance to punctures from bones or other sharp objects, and cannot be used in a microwave oven.

Laminated Foils

Laminated foils have become popular packaging material for intermediate-moisture foods such as beef sticks, jerky, and meat products that are heat-sterilized using laminated foil retortable pouches. The laminated foils offer high resistance to puncture and abrasion, making them an ideal economic means of offering sterilized, shelf-stable products.

Films

The greatest technological advancements in the packaging industry have occurred with the copolymer films. Meat-packaging films are generally multilayer structures using a variety of polymer resins (long-chain structural chemical compounds) and coatings to meet the specific needs of various products in the industry. They usually contain three or more coextruded layers, with each layer contributing a specific property to the structure (Figure 18-5).

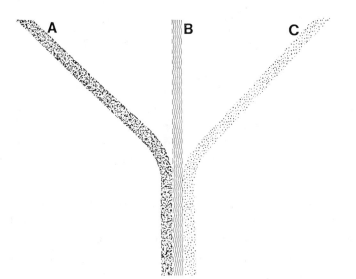

Fig. 18-5. Composite polymer film.

Layer A. The outside layer must be scuff- and abrasive-resistant.
Layer B. The middle layer generally provides the barrier to oxygen.
Layer C. The sealant layer must be capable of being melted and
welded under pressure to the sealant layer of the other film
to make a gastight seal.

Most processed meat is vacuum packaged and sometimes flushed with an inert gas such as nitrogen to reduce the speed of lipid oxidation, which causes oxidative rancidity. The packaging film must keep atmospheric oxygen from getting back into the package. The ability of different films to slow down the transmission of oxygen is dependent on a number of factors, such as the type of polymer, thickness, ambient temperature, and humidity. Some films are designed as oxygen barriers, while other films are designed for high oxygen transmission to give fresh meats a desirable, oxygenated surface color. Oxygen-transmission values are available for most films and are generally reported as milliliters (cubic centimeters) of oxygen transmitted through 100 square inches (ml O_2/100 square inches) of film in 24 hours under standardized conditions. Typical oxygen-transmission values for some commonly used films are reported in Table 18-8.

Table 18-8. Typical Oxygen-Transmission Values[1]

	ml O_2/100 sq. in/ 24 hrs., 73° F, 0% RH
Ethylene vinyl alcohol (EVOH)	0.1
Saran (PVDC) coatings	1.5-1.0
1 ml nylon	2-3
0.5 mil polyester	9
0.75 mil oriented polypropylene	200
2 mil sealant (EVA or Surlyn)	300

[1]Willard Gehrke. 1983. *Film Properties Required for Thermo-formed and Thermal-processed Meat Packages.* In Proceedings of the 36th Reciprocal Meats Conference, National Live Stock and Meat Board, Chicago, p. 55.

The authors are grateful to Willard Gehrke, an expert in the meat-packaging film industry from which most of the following information was obtained.[6]

According to this packaging expert, the oxygen transmission of the finished multilayered film is a complicated subject, and a number of factors must be considered. A few of the factors that influence the oxygen barrier of the finished package are:

- Films for packaging processed meat are generally a combination of various polymer films. Each layer in the finished film will contribute something to the barrier properties of the finished structure.

- Some of the polymer layers in the film structure will vary in initial thickness.

- The forming film will be considerably thinner after forming than it was originally. This thinning is not uniform and varies with the size and shape of the cavity, as well as the technique used to do the thermo-forming. Thin films permit more oxygen to pass through and enter the package. Each layer in the packaging film will be thinner after forming.

- The entire package will be handled in packaging, boxing, shipping, and merchandising. Any abuse will tend to harm the barrier properties.

- The amount of moisture in the air can change the rate of oxygen transmission for certain packaging films.

- Processed meats are generally stored at refrigerated temperatures. Less oxygen is transmitted through all polymer films as the temperature is reduced.

Films are manufactured for many different purposes which depend upon the product to be packaged and the equipment on which the packaging is accomplished. A good seal is paramount in the production of a vacuum-packaged prod-

---

[6]Willard Gehrke. 1983. *Film Properties Required for Thermo-formed and Thermal-processed Meat Packages.* In Proceedings of the 36th Reciprocal Meats Conference, National Live Stock and Meat Board, Chicago, p. 55.

uct. The sealant is usually the thickest layer of packaging film and is generally composed of one of several compounds, each having slightly different properties to meet different packaging needs.

The four sealants most commonly used in meat films are:

1. Low-density polyethylene (poly)
2. Ethylene-vinyl acetate (EVA)
3. Ionomer (Surlyn)
4. Linear low-density poly (LLDPE)

Many processed meat products are now packaged on machines that require two different types of laminated (fused together) films, a non-forming film and a forming film.

The packaging system generally employs a machine which performs most of the following steps:

1. Heats the forming film.
2. Forms it into a cavity of the desired size.
3. Provides a loading area—so product can be placed in the formed cavity.
4. Brings the non-formed film in place over the formed film containing the product.
5. Pulls a vacuum to remove air inside the package.
6. Introduces an inert gas, such as nitrogen, into the package, if desired.
7. Seals the package. This is done with a heat-seal bar that melts the two sealant layers and welds them together under pressure.
8. Cuts or trims the films to provide individual packages.

The most widely used non-forming film is a structure consisting of oriented polyester/saran/sealant.

The polyester layer is usually ½ mil (0.0005″) in thickness. The sealant layer is usually 2 to 3 mils (0.002-0.003″) in thickness.

The non-forming film is frequently printed. Any printing is generally buried in the structure—that is, placed between the saran and the sealant.

Each layer in the structure has a specific purpose.

• The polyester is resistant to the temperature needed to melt the sealant and is scuff- and abrasion-resistant.
• The saran layer provides the barrier to oxygen.
• The sealant must be capable of being melted at a relatively low temperature to make a hermetic bond with the sealant layer of the formed film.

The forming film structure used for the largest number of processed meat-packaging applications is nylon/saran/sealant.

The gauge of the nylon and the gauge of the sealant are adjusted, depending on the size and shape of the cavity the film must be thermo-formed into to accommodate the product. The total thickness will vary from approximately 2.75 mils for products such as bacon to 11 mils or more for items such as large hams.

There is another popular class of forming films that are referred to in the industry as semi-rigid films. Semi-rigid films were developed for use on finished packages of processed meat which are displayed in stores by hanging on pegs. For these applications, a stiffer structure is used—generally consisting of a 7.5- to 12-mil semi-rigid polyester sheet combined with a barrier layer and a sealant.

In order to allow for the thinning that takes place in thermo-forming, the starting film must have sufficient material to insure that the side walls and corners of the shaped package have adequate thickness to provide the desired protection.

Each layer in the formed web structure has its specific purpose.

- The nylon or semi-rigid film must be capable of being softened by heat and formed into a cavity of the desired size and shape; yet be resistant to temperatures needed to melt the sealant. It must also be scuff- and abrasion-resistant.
- The barrier layer provides the main resistance to oxygen transmission.
- The sealant must be capable of being melted at a relatively low temperature to make a hermetic bond with the sealant layer of the non-formed film.

The process of heating and forming a film into a cavity produces a shaped film that is considerably thinner than the original flat film. The degree of thinning and uniformity of the shaped film is a complicated subject and will depend on a number of factors, including:

- The type and original thickness of the forming film.
- The size of the area being formed.
- The depth and contour of the cavity.
- The procedure used to heat and draw or push the heated film into the cavity.

Forming Films for Thermal Processing

Perhaps the most demanding requirement for forming films is for those applications where the thermal processing (or cooking) is done in the same package used for shipping and merchandising. There are about 30 cook-in lines for hams in the United States and about 14 cook-in lines for poultry products (1983).

These applications are some of the most demanding for forming films, because the film must shrink with the product as the product is cooked and cooled. In addition, the film must have a good appearance and be tough enough to stand abuse in shipping and handling.

The most important characteristic of these films is their ability to shrink. Films formed on a conventional processed meat-packaging machine by vacuum alone do not shrink to an acceptable degree. To have shrink in the films, they must be formed at lower temperatures. To do this forming at the critical temperature, the packaging machines are designed with plugs to assist in getting the film in the cavity properly.

Food-Grade Dermatex® (Acytelated Monoglycerides)

An innovation in packaging cooked, processed products consists of dipping the product in food-grade acytelated monoglycerides (trade name, Dermatex®). Dermatex® is approved by the USDA for use as packaging film in cooked sausage products such as summer sausage, liver sausage, and other "chub" items. This compound is made from highly distilled monoglycerides from animal fats. Dermatex® is marketed in wax-like blocks which are subsequently melted in a dipping tank. Product is submerged into the dipping tank whose temperature largely controls the coating thickness. Dermatex® is available in a number of colors in addition to the transparent one which appears most popular. The principal advantage of this innovation is that it provides a more economical alternative to vacuum packaging with shelf-life properties similar to vacuum-packaged products.

Chapter 19

Meat Curing and Smoking

HISTORICAL ORIGINS

THE SALTING, or curing, of meat is the oldest known form of preservation for this perishable commodity. The ancient Sumerian culture that flourished around 3000 B.C. is believed to have been the first to make use of salted meats and fish as a form of preservation. Between the years 3000 B.C. and 1200 B.C., historical evidence indicates that the Jewish people gathered salt from the Dead Sea for use in the preservation of various foods. The Chinese and Greeks are also reported to have used salted fish in their diets and are credited with passing this practice on to the Romans, who included pickled meats in their diets.[1]

The following was written by the Roman scholar Cato,[2] who lived in the third century B.C.

Salting of hams and of the small pieces such as are put up at Puteoli

Hams should be salted in the large storage jars or in the smaller jars in this way: (1) When you buy the hams cut off the feet; (take) for each ham a half-*modius*[3] of Roman salt ground in a mill, sprinkle the bottom of the large jar or of the smaller one with the salt, then put in a ham, skin side down, and cover completely with salt. (2) Then put a second on top, cover the same way, take care that meat does not touch meat. Cover them all the same way. When you have placed them all, cover above with salt so that the meat will not show. Make the salt level. When they have been five days in the salt, take them all out, salt and all. Place on the bottom those that were on top, and cover and arrange in the same way. (3) After twelve days in all, take the hams out, wipe off all salt and hang them in the wind for two days. On the third day wipe them off well with a sponge, rub thoroughly with olive oil and vinegar mixed, and hang them up on the meat rack. Neither moths nor worms will touch them.

Just how dry salting originated is not definitely known. It is quite likely that the use of salt for preserving meat was entirely accidental. Since saltpeter was

---

[1]James Jay. *Modern Food Microbiology*. 1978. D. Van Nostrand Reinhold Co., New York.
[2]Ernest Brehaut. *Cato the Censor on Farming*. 1933. Columbia University Press, pp. 145-146.
[3]A unit of volume of Cato's time, approximately equal to ½ U.S. gallon.

probably an impurity in the salt that was used, it remained for the chemist to develop this color-retaining agent in its pure state. The salted meats of the ancients were very unevenly cured and were objectionably dry and salty. The latter part of the eighteenth century marked the beginning of the salt curing of meat on a scientific basis.

Historical information of the smoking of meat is more scanty; however, it is quite likely that it evolved as a result of cooking meat on open fires. On the North American continent, it is known that the Indians often hung strips of meat in the tops of their tepees out of reach of dogs. As the strips dried, the meat took on a smoky flavor from the campfire smoke inside the tepee. The dried strips, which were very hard and inflexible, were powdered by being beaten with stones or wooden mallets and then mixed with dried fruits and vegetables to make *pemmican*. In this form, the dried meat was transported in skin sacks or bladders and was the principal food whenever the tribes were migrating.

Modern meat curing and smoking practices are employed principally to create flavor and appearance variations in our food supply and to inhibit the outgrowth of *C. botulinum* spores which cause the lethal food poisoning botulism (see Chapter 18). The function of preservation as the principal reason for curing and smoking meat has become less significant in modern times with the advent of mechanical refrigeration. The result is that modern-day hams are considerably more mild (less salty) than those produced in earlier historical periods.

Country-cured meats that employ dry-curing methods are a major processing industry, particularly in the southern regions of the United States. Preservation by salting and dehydration of country-cured ham and bacon is of major importance for these products, since they are often not refrigerated during storage. Preservation by salting requires much higher salt levels than are present in regular commercial hams and other cured products that are not country-cured. Many people outside the southern regions of the United States consider the country-cured hams to be too salty and unpalatable unless they are boiled for long periods to remove the high salt content from the meat.

FUNCTIONS OF CURING AND SMOKING

The following functions of meat curing and/or smoking processes are considered to be the most important economically:

- Color development (internal and external)
- Flavor development
- Preservation
- Shelf-life extension

The characteristic pink cured-meat color associated with the lean of cured-meat products provides important aesthetic qualities. Nitrite, the preservative chemical responsible for the development of the pink color in cooked cured meats is discussed in detail later in the chapter. Without the addition of nitrite (or nitrate), cured meats as we know them would not exist. Rather, ham or other products would have an unattractive grayish-green color and a flat, salty flavor. Generations of consumers have become accustomed to the bright pink cured-meat color displayed by cured products. The chemistry of cured-meat color development is discussed later in the chapter and is illustrated in the color section near the center of the book. The external color of smoked cured-meat products is primarily determined by the amount of smoke deposition (natural or artificial) on the product surface during thermal processing. A moderately intense chestnut color is considered most desirable by many.

Meat curing is considered a bacteriostatic process, which creates an unfavorable environment for microbial growth. Most bacteria, yeasts, and molds that cause food spoilage have a relatively low tolerance to salt (NaCl). Salt acts as a dehydrating agent, osmotically lowering the water content of bacterial organisms and thus limiting their ability to thrive and reproduce. However, there are numerous spoilage and pathogenic (disease-causing) bacteria that can tolerate the salt levels present in most cured products (2 to 4 percent). These organisms are collectively referred to as *halophilic* (salt-loving) organisms.

The microbiological safety that is assured from nitrite addition in cured meats is perhaps the most important function of the curing process. Nitrite is the most effective antibotulinal agent known to humans in the prevention of the food poisoning botulism (see page 557).

The extension of product shelf life is another function of the curing process that is related to the bacteriostatic properties of salt. Since most cured-meat products are fully cooked prior to sale (except country-cured hams, bacon, and dry sausage) the thermal processing also acts in the extension of product shelf life before spoilage occurs.

MEAT-CURING INGREDIENTS

Salt

The most important of the curing ingredients is salt (NaCl). It makes up the bulk of the curing mixture, not only because it is a good preservative, but also because it provides the most desirable flavor. Its diffusion in meat is by the process of osmosis. When salt is added to a meat product using water as a solvent and carrier, the solution is called a pickle, in which case the water makes up the bulk of this curing pickle.

Sugar

Sugar, a secondary ingredient in the curing formula, counteracts the astringent quality (harshness) of the salt, enhances the flavor of the product, and aids in lowering the pH of the cure. Its role in color development and color stability under present commercial curing practices has been found to be negligible. The sugars most frequently used are sucrose and dextrose and their derivatives.

Nitrite and Nitrate

Sodium nitrate is a naturally occurring substance in vegetables, water, soil, and even the air. Originally discovered as an impurity in salt, it has been used in small amounts for thousands of years to cure meats.

Sodium nitrite, also used to cure processed meats, is a derivative of sodium nitrate. When nitrate is used to cure meats, it is readily converted to nitrite by microbes or by a reducing agent such as ascorbate. Nitrite is the active ingredient. Both nitrate and nitrite usage is allowed under the Meat Inspection Act (see Chapter 3).

Nitrite is considered essential in cured meats because it performs several vital functions:

- It prevents botulism and has other bacteriostatic properties to provide a safeguard against the mishandling of meat by manufacturers, distributors, retailers, or consumers (*i.e.,* failure to refrigerate properly because of mechanical malfunction, negligence, or ignorance). A recent USDA study concluded that 63 percent of the 2,500 households surveyed ran a high risk of food-borne illness because of a lack of awareness of basic safe food-handling practices.[4]
- It is responsible for color fixation of the meat pigments which result in the characteristic cured pink color.
- It retards lipid oxidation which otherwise causes an undesirable (warmed-over) flavor.
- It provides cured meats with their characteristic cured flavor.

Nitrite is the only substance that will do all these things. No substitute has been found, even though more than 700 substances have been tested as possible replacements.

FSIS regulations (CFR 318.7b, 318.7c) permit the use of sodium or potassium *nitrate* only in dry-cured country ham, dry-cured country bacon, and dry sausages. Approved usage levels for nitrate in dry-cured country ham and bacon are 3½ ounces for each 100 pounds of meat in dry salt or dry cure and 2¾ ounces in 100 pounds of chopped meat and/or meat by-product.

Current FSIS regulations permit the use of sodium or potassium *nitrite* in all products except pumped bacon at the following levels:

2 pounds in 100 gallons of pickle at 10 percent pump level
1 ounce for each 100 pounds of meat in dry salt or dry cure
¼ ounce in 100 pounds of chopped meat and/or meat by-product

[4]*Food Safety: Homemakers' Attitudes and Practices.* Jan. 1977. USDA Agricultural Economic Report #360.

The use of nitrites or nitrates, or combinations thereof, is not permitted to result in more than 200 parts per million (ppm) of nitrite, calculated as sodium nitrite in the finished products other than bacon.

In pumped bacon, ingoing nitrite levels may not exceed 120 ppm and must be accompanied by at least 550 ppm of sodium erythorbate or ascorbate as a cure accelerator. Residual nitrite levels in the finished bacon are not permitted to exceed 40 ppm.

The addition of nitrite to meat indirectly induces the development of the characteristic cured-meat color. This color change from the purple-red of fresh meat to the pink of cured meats involves the protein myoglobin which is the vehicle for oxygen storage in the muscle. In meat curing, nitrite (NO_2) is broken down to nitric oxide (NO) which binds to the myoglobin molecules to induce the color change (see meat-color chemistry section, page 612).

Sodium Erythorbate and Ascorbate

The time element has posed some problems in color development and retention in cured meats and particularly in emulsion-type products that are heat processed immediately. It was found that ascorbic acid and erythorbic acid and their salts hastened color production, due to a chemical reaction, with the nitrite either producing more nitric oxide or reducing metmyoglobin to myoglobin.

Federal regulations permit the addition of 75 ounces of ascorbic acid or 87½ ounces of sodium ascorbate or erythorbate to 100 gallons of pickle and ¾ ounce of ascorbic acid or ⅞ ounce of sodium ascorbate to each 100 pounds of chopped meat (550 ppm).

Spraying the surface of cured cuts prior to packaging with 5 to 10 percent ascorbic acid or ascorbate solution, the use of which will not result in a significant addition of water to the product, has been found to deter color fading caused by light. The ability of ascorbic-treated cured meats to resist fading is ascribed to residual ascorbic acid maintaining reducing conditions on the exposed cured-meat surface.

Ascorbate is commonly known as vitamin C, and in addition to its beneficial effects on color, it has been shown to inhibit the formation of nitrosamines in cured meats. Nitrosamines are a known carcinogen (cancer-causing agent) and are discussed more thoroughly later in this chapter (see "The Nitrite Controversy").

Alkaline Phosphates

The primary purpose of alkaline phosphates is to decrease the shrinkage in smoked meat and meat products and the "cook-out" in canned meat products. The ability to increase the water-binding quality of meat results in increased yields of up to 10 percent.

The following phosphates are approved for use: disodium phosphate, mono-sodium phosphate, sodium metaphosphate, sodium polyphosphate, sodium tri-polyphosphate, sodium pyrophosphate, sodium acid pyrophosphate, di-potassium phosphate, monopotassium phosphate, potassium tripolyphosphate, and potassium pyrophosphate. Phosphates may be added to the pickle for hams, bacon, pork shoulders, picnics, Boston butts, boneless butts, and pork loins. The pickle shall not contain more than 5 percent of approved phosphate, and the finished product shall contain no more than 0.5 percent phosphate.

Spices and Flavorings

A number of spices or flavorings are often used to season ham and other cured-meat products to give them unique flavor characteristics. Soluble spices, usually on a dextrose or salt carrier, may be dissolved in the pumping pickle for incorporation into the cured product. Alternatively, some hams such as prosciut-ti (see page 626) are rubbed with spices on the surface. Spices commonly used in seasoning various hams include pepper, cinnamon, clove, nutmeg, and allspice (see page 637, Table 20-3).

Water

Water is an often overlooked curing ingredient used in pumping- or immer-sion-curing techniques. Water's primary functions in modern meat-curing prac-tices are:

- To act as a dispersing medium for salt, nitrite, sugar, phosphates, and other curing ingredients.
- To assist in maintaining a moist, juicy end product.
- To compensate for moisture loss during thermal processing.
- To reduce product cost by merchandising products which have water lev-els greater than were naturally present in the raw meat.

The quantity of water contained in processed-meat (including poultry) prod-ucts that is in excess of the normal water content in the unprocessed form is termed *added water*. A major new USDA regulation which is designed to control the added water in cured and cooked pork and poultry products has been ap-proved and became effective on April 15, 1985. In the past, added-water catego-ries were established to categorize cured products as follows: A ham containing up to 10 percent added water was labeled "ham—water added." A ham contain-ing less than 20 percent added water (but more than 10 percent) was labeled "ham and 20 percent added water." Of course, hams containing no added water have no restrictive labeling requirements for added water. The formula for deter-mining the quantity of added water (AW) in the past was:

AW = (percent of moisture in final product) − (water factor × protein content of product)

where the water factor = 3.79 smoked ham
                           3.83 canned ham
                           3.93 smoked pork shoulder picnics
                           4.0 all other products

Table 19-1 gives the allowable AW levels permitted for various classes of processed meats.

Table 19-1. Allowable Added Water (AW) for Various Classes of Processed Meat Products[1]

Product Class	AW Permitted	Regulations[2]
	(percent)	*(section)*
Fresh sausage	3	319.140
Cooked sausage	10	319.140
Chopped ham	3	319.105
Luncheon meat	3	319.260
Meat loaf	3	319.261
"Water-added" cured pork[3]	10	319.104
Cured pork	0	319.104

[1]*Encyclopedia of Labeling Meat and Poultry Products.* 1983. Published by *Meat Plant Magazine.*

[2]Code of Federal Regulations, part and subparts (meat inspection regulations).

[3]Subject to change. See text for specifics.

Under the new regulation, which was published in the *Federal Register* on April 13, 1984, AW is calculated differently. The new method and final rule for calculating AW is based on the PFF (protein fat-free) method and replaces the previous standards which limited added water and other substances. PFF regulations set the standards for measuring minimum meat-protein content in cured-pork products on a fat-free basis without considering the water content. The revised rules are accompanied by changes in labeling and have new procedures for determining compliance. Cured-pork products which do not measure up to the prescribed PFF levels shall be labeled to reflect this and are called "non-traditional" products. The formula for determining PFF values is given here:

$$\frac{\text{PFF}}{\text{value}} = \frac{\text{percent of meat protein}}{100 - \text{percent of fat}} \times 100$$

☞   Note: Moisture is no longer a part of the formula for determining AW.   ☜
The labeling changes reflected in the PFF regulations are listed in Table 19-2.

Table 19-2. Protein Fat-free (PFF) Values for Determination of Added Water in Cured-Pork Products and Labeling Implications[1]

Type of Cured-Pork Product	Minimum Meat PFF Percentage[2]	Product Name and Qualifying Statements
Cooked ham, loin[3]	20.5	(Common and usual)
	18.5	(Common and usual) with natural juices
	17.0	(Common and usual) water added
	<17.0	(Common and usual) and water product–X% of weight is added ingredients[4]
Cooked shoulder, butt, picnic[3]	20.0	(Common and usual)
	18.0	(Common and usual) with natural juices
	16.5	(Common and usual) water added
	<16.5	(Common and usual) and water product–X% of weight is added ingredients[4]
Uncooked cured ham, loin	18.0	Uncooked (common and usual)
	<18.0	Uncooked (common and usual) and water product–X% of weight is added ingredients[4]
Uncooked cured shoulder, butt, picnic	17.5	Uncooked (common and usual)
	<17.5	Uncooked (common and usual) and water product–X% of weight is added ingredients[4]
"Ham patties," "chopped ham," "pressed ham," and "spiced ham"	19.5	(Common and usual)
"Ham patties," "chopped ham," "pressed ham," and "spiced ham"	17.5	(Common and usual) with natural juices
"Ham patties," "chopped ham," "pressed ham," and "spiced ham"	16.0	(Common and usual) water added
"Ham patties," "chopped ham," "pressed ham," and "spiced ham"	<16.0	(Common and usual) and water product–X% of weight is added ingredients[4]

[1]*Federal Register.* Vol. 49, No. 73, Friday, April 13, 1984. "Rules and Regulations."

[2]The minimum meat PFF percentage shall be the minimum meat protein which is indigenous to the raw, unprocessed pork expressed as a percent of the nonfat portion of the finished product; and compliance shall be determined under §318.19 of the Code of Federal Regulations.

[3]The term *cooked* is not appropriate for use on labels of cured-pork products heated only for the purpose of destruction of possible live trichinae. Cooked products should be heated to a minimum of 148° F.

[4]Processors may immediately follow this qualifying statement with a list of the ingredients in descending order of predominance rather than having the traditional ingredients statement. In any case, the maximum percent of added substances in the finished product on a total weight percentage basis would be inserted as the X value; *e.g.*, Ham and Water Product–20% of Weight Is Added Ingredients. A prerequisite for label approval of these products is a quality control program approved by the USDA under §318.4 of the Code of Federal Regulations.

THE NITRITE CONTROVERSY

The food industry and government regulatory agencies make every effort to insure that the risks associated with our food supply are kept as low as possible. However, attempts to simplify scientific concerns in order to write effective legislation have led to some dilemmas. One such case occurred in 1959, when the U.S. Congress enacted the Food Additives Amendment to the federal Food, Drug and Cosmetic Act. This amendment, commonly referred to as the Delaney Clause, provides *"that no additive shall be deemed safe if it is found to induce cancer when ingested by man or animal, or if it is found after tests which are appropriate for the evaluation of the safety of food additives, to induce cancer in man or animal.''* The Delaney Clause flatly states that no cancer-causing substance or *carcinogen* shall be added in any concentration to the food supply.

Most scientists would probably agree that the passage of the Delaney Clause was beneficial in 1959, considering the state of the development of the science of toxicology at that time. Since the enactment of this legislation, our technological capabilities have greatly improved in the detection of minute traces of substances. This has brought about interpretational controversies of the Delaney Clause because ''no'' was intended in the legislation to refer to ''zero,'' a finite entity. However, we now realize that nature does not follow finite rules. If a proven carcinogen is formed from a food additive at ''parts per billion'' or ''parts per trillion'' levels (which are now detectable), does it pose a threat to our food supply? Should it be banned outright as a food additive, or should the benefits the compound provides be weighed against the potential risks it may pose? Such has been the case with sodium nitrite, which was very nearly banned in 1980.

Events leading to the nitrite controversy began in the early 1970s when several parts per billion (ppb) of N-nitrosopyrrolidine (NPYR), a known carcinogen in laboratory animals, were detected in some samples of bacon fried at high temperatures (350° to 400° F). Subsequent research found traces (less than 5 ppb) of two other nitrosamines, dimethylnitrosamine (DMNA) and N-nitrosomorpholine (NMOR), present in fried bacon. Nitrosamine formation occurs as a reaction between nitrite (NO_2) and secondary amines (protein breakdown products). Both nitrite and secondary amines occur naturally at varying levels in most food substances. These early findings brought a flurry of intensive research on nitrite and cured-meat products and their subsequent safety for consumption by the public.

The Risk-Benefit Concept

The problem of defining the carcinogenic risk of a very low level of a chemical in foods provides a dilemma for regulatory officials. The present ''yes'' or ''no'' legal framework for accepting or rejecting makes the decision process relatively easy. However, many scientists wonder if it is really in the public's best interest, since there is no room for scientific judgment as to the degree of risk versus the overall potential benefit.

Scientific and historical evidence have effectively proven that the minute, unknown risks of cancer posed by the inclusion of nitrite from meat products in the diet are far outweighed by the important benefits this preservative provides to our society. In addition, nitrites have a proven safety record of thousands of years of use in cured-meat products (see chapter introduction). Most important, nitrite is the only known chemical substance that can be added to food that prevents the highly lethal food poisoning botulism (see Chapter 18). Since *C. botulinum* produces a highly heat-stable spore, it survives the cooking temperatures employed for most meat products. The deadly toxin is produced only by vegetative (growing) forms of the cell. Nitrite prevents the outgrowth of the microorganism's spores and thus the subsequent possibility of toxin production.

The importance of this inhibitory action of nitrite on *C. botulinum* is often not recognized by the consuming public. Without the use of nitrite, there are few doubts that there would be a tremendous increase in the number of deaths due to botulism. Just think of the points at which meat products are temperature-abused, either by consumers or during the distribution of meat through the merchandising channels. The brown-bag lunches consumed in our public schools and many places of employment often sit for hours without refrigeration. These could provide prime incubation periods for *C. botulinum* growth. Pasteurized canned hams and other meats are even a greater concern. One never knows if these canned products have been temperature-abused (held unrefrigerated for periods of time) during distribution, which would allow growth of *C. botulinum*, if no nitrite were present.

It should be clear, however, that "zero risk" or "absolute safety" of food is an unattainable goal. Nonetheless, absolute safety is a worthy goal, and some industries have approached it. For example, more than 800 billion units of commercially canned foods have been produced in North America since 1940, with only five deaths attributable to botulism.[5] None of these five deaths were from canned-meat products which contained nitrite. This indeed represents a remarkable safety record!

Other Sources of Nitrite

Fred Deatherage, a renowned biochemist from The Ohio State University, reported that the total dietary intake of nitrite or nitrate from cured-meat products represents only 2 to 4 percent of the nitrite getting into our bodies.[6] He reported that certain vegetables have nitrate levels many times that found in cured meats. Celery has 1,600 to 2,600 ppm, radishes 2,400 to 3,000 ppm, lettuce 100 to 1,400 ppm, zucchini squash 600 ppm, and potatoes 120 ppm. Deatherage also reported that the salivary glands and intestines are significant production sites (from bacterial nitrification processes) of nitrite and nitrate in the human body. The average person ingests 8 milligrams of nitrite daily from salivation—more than twice the amount consumed from cured meats. Intestinal nitrite production has been reported to be approximately 80 to 130 milligrams per day, *which leads some experts to believe that nitrite might represent a natural chemical defense system protecting us from botulism.* For some reason, *C. botulinum* spores do not germinate to active vegetative cells in the gut. Deatherage further stated that "evidence suggests the nitrite produced by the gut (and that ingested from

---

[5]*The Risk/Benefit Concept as Applied to Food.* 1978. Summary of the Expert Panel on Food Safety and Nutrition, published by Institute Food Technologists.

[6]Fred Deatherage. *Man and His Food and Nitrite.* 1978. National Livestock and Meat Board, Chicago.

foods) inhibits germination of *C. botulinum* spores in the gut just as it does in cured meats.''

Most people have heard of sudden infant-death syndrome, in which babies die in their cribs for no apparent reason. It has been shown that many of these fatalities are due to the outgrowth, in the infant's intestine, of *C. botulinum* spores to active vegetative bacteria, which produce enough toxin to kill the baby. It is reasonable then to consider that normal bacterial nitrification processes producing nitrite have not been sufficiently developed in the first months of life to protect some babies.[7] This idea needs further study. Nevertheless, it has sufficient merit to cause the Sioux Honey Association to issue a notice of warning that honey should not be fed to infants less than one year old. Honey is well laced with *C. botulinum* spores.[8]

In response to the finding of nitrosamines in fried bacon, the meat-processing industry immediately set out to investigate corrective-processing methods. A task force was set up to examine bacon-processing methods and determine how these methods may be altered to eliminate nitrosamine formation after the product is cooked. It was determined that by reducing ingoing nitrite levels in the bacon to 120 ppm and by achieving residual levels (after processing) of less than 40 ppm, nitrosamine formation was reduced to non-detectible levels or at least trace levels (less than 5 ppb). It was found that the 40 ppm residual levels could be attained by adding the maximum amount (550 ppm) of ascorbate, a vitamin C derivative, which acts as a reducing agent. Also, by lengthening cooking and processing times, the residual nitrite level could be further reduced. These levels (120 ppm ingoing) were still sufficient for controlling outgrowth of *C. botulinum* spores. Most of the bacon-processing industry was using the lower nitrite levels long before they were enacted into law.

At the present time, nitrosamine formation has not been detected (at problem levels) in other cured-meat products. Trace levels have been detected in country dry-cured hams. Higher levels have been detected in country dry-cured bacon, which is currently under investigation for regulative changes.

METHODS OF MEAT CURING

Prolonged exposure of meat to salt action results in excessive shrinkage and a high salt content. It is therefore important in any of these methods that quantitative measurements and time schedules be observed.

Dry Salt Cure

The dry salt cure was the original method employed by our ancestors who practically had to pick the salt out of their teeth. It involved the rubbing and packing of meat in salt for considerable periods of time. The only use made of

---

[7]*Ibid.*
[8]*Ibid.*

this method today is in the production of salt pork in which fatbacks, heavy jowls, and occasionally heavy sow bellies are packed or rubbed with dry salt. Salt pork finds favor in the South where it is used as "seasoning meat" with greens. It is well to add 10 ounces of saltpeter (sodium or potassium nitrate) to each 100 pounds of salt and use 10 pounds of cure per 100 pounds of pork as it is layered. Allow it to cure for two to eight weeks.

Dry Sugar Cure (Dry Country Cure)

The dry sugar cure has proven to be the safest method for farmers and operators who do not have refrigerated curing rooms or the equipment for injection curing. Its chief advantages are that:

- The rate of cure is faster than the immersion cure because the curing ingredients are applied directly to the meat surface in their full concentration.
- The curing can be conducted safely at higher and wider temperature variations than is possible in immersion curing.
- The time schedule is not exacting.
- There is less spoilage in the hands of the novice or under unfavorable curing conditions.

A simple and time-tested formula consists of mixing 8 pounds of table or curing salt, 3 pounds of cane sugar, 2 ounces of nitrate (saltpeter), and 1 ounce of sodium or potassium nitrite. This is commonly referred to as the 8-3-2-1 formula. Precautions regarding the use of nitrate and nitrite are discussed in Chapter 20.

For each pound of pork 1 ounce of cure is used. This will require three separate rubbings for hams at three- to five-day intervals; two rubbings for picnics and butts; and one thorough rubbing for bacon, with a light sprinkling over the flesh side of each bacon after it is rubbed. For heavy hams (over 20 pounds), 1½ ounces of cure per pound of ham or four rubbings are used. The rubbed meats are placed in boxes, on shelves, or on non-metal tables to cure; however, they should not be placed in tight boxes or barrels where they will rest in their own brine.

The length of the curing period is seven days per inch of thickness. Since most hams weighing 12 to 15 pounds measure 5 inches through the cushion, they will cure in 35 days; a bacon 2 inches thick will cure in 14 days. If the cured cuts remain in cure for a longer period of time, they cannot become any saltier, and that makes it possible to smoke them all at the same time.

If some salt is forced into the aitch bone joint to guard against bone souring, the curing can be done at a higher temperature (50° F) and in a shorter period of time. This is because salt absorption is more rapid at the higher temperatures.

Box Curing (Pressure)

The pressure method of box curing was a popular method before curing became so highly mechanized and is applicable only to bacon. It is practiced by packers who produce a mild-cured product. The amount of cure used is about ¾ ounce to each pound of bacon. Tight boxes and in many cases ordinary curing vats are used. The size of the box should be such that the bacon will fit in snugly without overlapping, regardless of the number of rows. A lid fits loosely inside the box or vat, upon which considerable pressure can be exerted either by a dead weight or by a screw jack on a crossbar attached to the sides of the container.

The pressure on the bacon causes the brine that is formed to rise to the top and cover the meat, thereby sealing it from the air. With just the right amount of curing having been added, the length of time to cure the bacon is of minor concern because it cannot become too salty. Packers have used this method of curing bacon as a means of storage, allowing it to remain in cure for as long as 90 days.

Hot Salt Cure

A practice followed in some communities and tested at the Pennsylvania Experiment Station with success is as follows:

Rub the cushion side and butt of the ham with saltpeter (1 ounce). Follow immediately with a rubbing of granulated or brown sugar over the entire ham. Allow the ham to absorb these ingredients for several hours. Then, heat sufficient salt so that it is uncomfortable to the hands (wear cotton gloves). Place the ham in the hot salt and cover for five minutes to get the ham in a soft condition. Take a clean, round, pencil-size stick, and force the hot salt into the aitch bone joint. Give the ham a thorough rubbing with the hot salt. An accurate measure would be an increase in weight of ¾ ounce per pound of ham.

Allow the hams to absorb the cure for five to seven days, rub with black pepper, and smoke.

Sweet Pickle Cure

A combination of salt and water is called a brine or pickle. A brine that has preservative qualities, such as sodium chloride (common salt) and water, is used for curing meat. The addition of sugar to a sodium chloride brine is called a sweet pickle. The proportion of salt to water determines the strength or salinity of brine or pickle. In the meat industry, two different types of brine—the pickling brines and the refrigerating brines—are common. The latter are a combination of calcium chloride and water and are used to carry cold to refrigerated boxes operating under the indirect expansion system.

Sweet pickle with a salimeter reading of 75° to 85° is recommended for farm curing. Table 19-3 gives the amounts of the different ingredients and the water necessary to make such pickles. A good practice to follow is to heat the water to the boiling point and dissolve the ingredients in it. Allow the pickle to chill before pouring it over the meat. A salimeter is a necessary piece of equipment when making up curing solutions of different salinities.

Table 19-3. Sweet Pickle Formulations

Salt	Sugar	Sodium Nitrite	Cold Water		Degree of Pickle by Salimeter 40° F
.... (lb.)		(oz.)	(gal.)	(lb.)	
10	3	¼	4	33⅓	95
9	3	¼	4	33⅓	90
10	3	¼	5	41⅔	85
8	3	¼	4	33⅓	85
8	3	¼	5	41⅔	75
6	3	¼	4	33⅓	70
7	3	¼	5	41⅔	65
6	3	¼	5	41⅔	60

Cold water weighs 8.33 pounds per gallon, hot water 8 pounds per gallon.
Seven pints of salt weigh 8 pounds.
One quart of syrup weighs 3 pounds.
If salimeter is calibrated for reading at 60° F, subtract .116 per degree below 60°.

Containers

Plastic or fiberglass barrels, large stone crocks, and stainless-steel containers make suitable curing receptacles. Clean lids and weights are necessary. Barrels in which any sort of spray material has been mixed must never be used. Metal containers that will corrode are also unthinkable.

Temperature

The best temperature for the curing room is from 35° to 40° F. Unless curing vessels have been previously contaminated with ham-souring bacteria, very little spoilage is experienced at these temperatures. Successful sweet pickle curing can be done at temperatures ranging from 40° to 50° F, which is the usual cellar temperature on farms during the winter. Temperatures of 50° F and over are too high for safe pickle curing. The brine will sour and become ropy, indicating bacterial growth, and the hams will develop an off-flavor or will sour around the bone.

Length of Cure

Hams should be measured through the cushion back of the aitch bone to determine their thickness. Hams measuring the greatest thickness should be placed in the bottom of the vat with the lighter hams placed on top. The thickness of each layer should be recorded on a card or in a book so that the date when they are to be taken out can be determined. The following is the curing schedule for the different strengths of sweet pickle:

85° pickle cure . . . . 9 days per inch
75° pickle cure . . . . 11 days per inch
60° pickle cure . . . . 13 days per inch

Procedure on the Farm

The chilled hams, shoulders, and bacon are packed in the barrel or vat in the order named, and sufficient cold pickle is poured over the pack so that it will be covered when the lid is weighted down. Four gallons of pickle will cover 100 pounds of closely packed meat, but 4½ to 5 gallons are necessary if the meat is loosely packed. Overhauling (moving and turning) meat once or twice during the curing period is desirable to permit the pickle to reach all parts of the meat.

It is necessary to follow rather closely the length of cure prescribed for the different strengths of pickle. As an example, consider a barrel of pork consisting of hams 5½ inches thick, on top of which are shoulders 3½ inches thick, over which are bacons 2 inches thick, curing in an 85° pickle (see Table 19-3). According to the schedule of 9 days per inch in cure, the bacons must come out in 18 days, the shoulders in 30 days, and the hams in 50 days. If 5 gallons of water are used to dissolve the 8-3-2-1 formula, making a 75° pickle, it will be necessary to cure at the rate of 11 days per inch. A 75° pickle is preferable, if the curing room temperature does not rise above 45° F.

A large part of the pork spoilage occurring on the farm could be eliminated if from 4 to 8 ounces of pickle were pumped into the center of the ham and around the hip joint soon after slaughter. A good syringe-type pickle pump will cost approximately $50.

Shrinkage

There is very little difference in the shrinkage of pork cured by the dry and sweet pickle methods at the end of a 60-day aging period. Sweet pickle–cured hams will gain an average of 5 percent in cure as compared to a loss of 5 to 7 percent for hams in dry cure. Sweet pickle–cured hams lose about 5 percent during smoking as compared to 2 percent for dry-cured hams. Bacon will show about 5 percent higher shrink than hams because of the large surface area.

Soaking

All farm-cured pork, whether dry or sweet pickle cured, should be rinsed or soaked in cold water before it is placed in smoke. The soaking removes the excess salt on the outside and eliminates the formation of salt streaks on the meat when exposed to the heat of the smokehouse. Quick-cured meats are rinsed but not soaked.

Soaking pickled bacon and shoulders one hour and ham about two hours is sufficient for farm purposes. Dry sugar–cured meats cured with 1 ounce of the cure per pound of meat need not be soaked. Rinsing is sufficient.

Curing Terms Defined

Artery Cure. Injection of cure into *femoral* artery of ham and artery in the shoulder (Figures 19-2 and 19-3).

Cover Pickle. The pickle in which hams cure.

Immersion Cure. Curing in a cover pickle.

Injection Cure. Stitch pumping or needling (Figure 19-1).

Overhaul. The rehandling or repacking of ham during the pickling period to permit a more uniform distribution of pickle.

Pumping. The forcible introduction of pickle into a ham by means of a ham pump.

Quick Cure. A term applied to a pickle containing sodium nitrite.

Salimeter (Salometer or Salinometer). A ballasted glass vacuum tube graduated in degrees and used for testing the strength or salinity of pickle.

Stitch or Gun. A single insertion of the needle in the pump method. The number of stitches or guns given each ham varies with the size of the ham and the strength of the pump pickle. It is more desirable to stitch small amounts in numerous locations than large amounts in few locations.

The Combination Cure

A successful method of curing hams on the farm is the use of the dry cure and the sweet pickle cure in combination. The quicker the salt gets to the center of the ham, the less danger there is of loss from spoilage. Experiments show that when 1½ to 2 pounds of salt per 100 pounds of pork is rubbed into hams 24 to 48 hours before they are placed in a 75° pickle, they will cure in nine days per inch. This one rubbing of salt is all absorbed overnight and has more rapid penetrating qualities because it is neither mixed with other ingredients nor dissolved in water.

Pump Pickling (Stitch or Spray Pumping)

To hasten the introduction of the cure to the center of the ham, processors can use the practice known as pumping. This consists of forcing the curing pickle into the center of the ham through a needle attached to a plunger-type syringe or by means of mechanically operated pumps (Figures 19-1 and 19-4). Pumping pickle may be the same strength as the cover pickle, although a 5° to 10° stronger pickle is generally used. The curing of a ham is hastened considerably under this method, as it takes place from the inside as well as from the outside. Hams

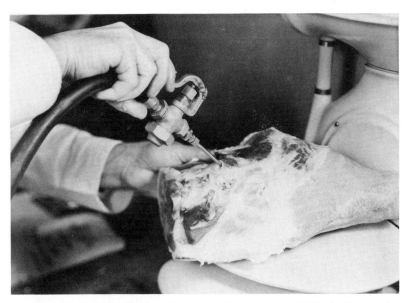

Fig. 19-1. Giving the ham a stitch at the hip joint (pump pickling). Give remaining stitches in cushion, stifle joint, hock end, and butt end.

should not be stitch-pumped with more than 8 percent of their weight of 85° pickle. That would mean that 1½ pounds of 85° pickle is pumped into a 20-pound ham, and the ham is then cured in a 75° cover pickle for 7 to 14 days (total).

When using a 70° to 75° pickle, stitch-pump the ham with 10 percent and cure in a 75° cover pickle for 7 to 14 days.

Hams have been cured very successfully, experimentally, by using a combination of pumping and dry curing. The procedure is to use an 85° pickle at the rate of 8 percent of the weight of the ham and then rub with one half its usual application or ½ ounce of the recommended formula per pound of ham and cure it 2½ days per pound.

Advantages of Pump Pickling

- The salt is introduced to the center of the ham before spoilage has a chance to take place.
- The curing period is shortened by almost one third.
- A quicker turnover is effected.

The Artery Cure

Artery curing consists of forcing a pickle into the *femoral* artery on the inside butt end of the ham by means of a small needle attached to a hose and connected to a pump that exerts a pressure of 40 to 50 pounds (Figure 19-2 and 19-3). The artery-pumped hams are either rubbed with the dry mix or placed in a

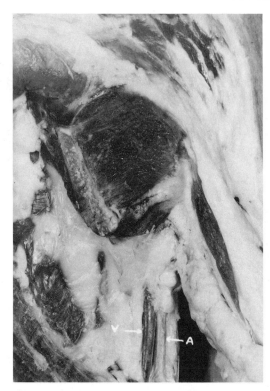

Fig. 19-2. Prior to removing the ham from the pork carcass, carefully loosen the leaf fat over the inside of the ham butt and separate the artery "A" from the fat. Cut the artery long. It differs from the vein "V" in that the artery is strong and elastic.

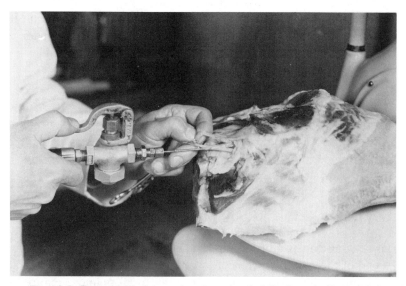

Fig. 19-3. Pump pickle into main artery ahead of the branch. If artery is too short, pump each branch separately.

pickle of similar strength for five to seven days to complete the curing. The salinity of the pickle to use will depend upon how long and under what conditions the ham is to be stored. A pickle strength of 60° to 65° on the salimeter is desirable when hams are to be held under refrigeration (less than 40° F).

Repeated tests, using the formula recommended in this text, produced higher flavor and better keeping quality when hams were artery-pumped with 10 percent of their weight using a 75° pickle. These artery-pumped hams were given a light rubbing of the dry cure mixture and left to shelf-cure for several days before being smoked. These hams had a longer storage life, and their flavor was more acceptable to the trade than were the hams cured with 60° or 65° pickle. For farm curing, the use of the 85° pickle (10 percent by weight) and one thorough rubbing of the dry cure produced a ham of high flavor and good holding quality.

The advantages of artery curing are speed and uniform flavor. It is particularly adapted to pork processors and those doing custom curing.

To be labeled as "ham" under federal requirements, the meat, while in the smokehouse, must be shrunk back to its original fresh weight before the curing solution is injected. If they do not come down to their fresh weight, the hams are returned for further heating and shrinking or must be labeled "ham, water added" or "ham and X% water added." (See earlier discussion.)

Mechanized Pumping

Both artery pumping and stitch or spray pumping have been mechanized to a high degree in industry. Figure 19-4 shows the Comcure®;[9] a machine designed to eliminate problems of control, yield, and production in ham processing. Each of the four stations is independent and may be turned off without affecting the others. The desired percentage is dialed on the front of each station and reads down to tenths of a percent. The operator has no decisions to make regarding the correct amount of pickle. He or she merely places the ham on the platform, clamps the gun onto the artery, pushes the programming start button, and proceeds to the next station.

Should a ham be an extremely bad leaker and not hold the correct percentage, it will sit on the platform, and the red fill light on the panel will flash on and off, indicating that the ham should be removed for stitch pumping. A production figure of 240 per hour is a conservative one to allow for situations involving heavy hams at high percentages. Pumping hams in the 12- to 16-pound range at percentages of 10 to 15 percent will exceed the 240/hour figure easily.

Figure 19-5 shows an InjectoMat®,[10] a machine using the spray-pumping principle with the exception that the needles inject the brine at hundreds of points. The pressure of the brine forces relatively uniform brine distribution throughout the meat.

---

[9]Vogt, Inc., Clawson, Michigan.
[10]Koch Supplies, Inc., Kansas City, Missouri.

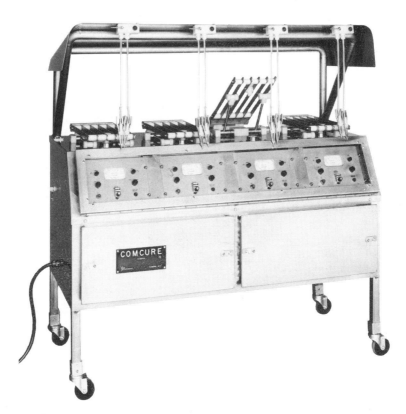

Fig. 19-4. An automatic artery pumping machine which is electronically controlled. (Courtesy Vogt, Inc.)

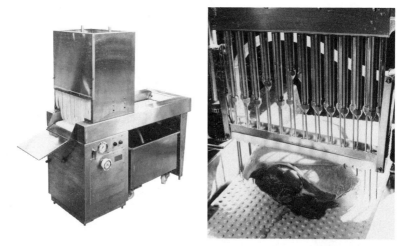

Fig. 19-5. An automatic bone-in pickle injector. (Left) An overall view showing volume and stroke adjustment controls. (Right) A close-up of InjectoMat® injecting a ham.

The InjectoMat® is fully automatic. The operator loads the conveyor belt with the product—bone-in, semi-boned, or boneless—which discharges into a barrel or truck after injection.

Needles operate in unison until they strike an obstruction; *spring-loading* protects each needle when it strikes a bone. A control bar aligns each needle and holds the product in position during the instant of the injection stroke. Quantity of pickle is adjusted by belt speed, *volume,* and *strokes per minute.* Hams are pumped at the rate of 6 to 12 per minute, up to 6,000 pounds per hour. Smaller product units, like briskets, may be loaded onto the conveyor belt two wide for more than 12 per minute production speed.

Boneless Hams

Most commercial hams produced today are boneless. Boneless hams are produced in two forms: whole boneless hams and sectioned and formed hams. Traditional whole boneless hams are made from one ham only and are often tunnel-boned. Tunnel-boning is a process of removing the *femur* by tunneling around it with a knife and leaving the ham in one piece, with all muscles attached together. This boning procedure is illustrated in Figures 19-6 through 19-9. The ham may be boned before or after it has been sufficiently cured, using approved curing methods. If it is boned after curing, it is important to have a market or method of utilizing the miscellaneous cured trimmings which are created but not used in the ham (*e.g.,* the shank meat, etc.).

A high percentage of the commercial boneless hams that are produced are the sectioned and formed variety. This includes the majority of the hams which are labeled "water added" or "with natural juices." Hams which are manufactured in this manner are generally boned by seaming out the three major muscle groups of the pork leg, the knuckle, the inside and outside muscle groups which correspond to the sirloin tip, and the top and bottom round sub-primals illustrated in the beef fabrication chapter (Chapter 14). Grouping like muscle groups produces a uniformly colored sectioned and formed ham. The boneless muscle chunks are typically stitch-pumped in a high-speed pumping machine similar to the one illustrated in Figure 19-5. After pumping, the muscles are tumbled or massaged to extract sufficient quantities of meat proteins which function to bind the muscles back together. Massaging and tumbling of boneless hams both accomplish the same function, but they utilize slightly different principles. Figure 19-10 illustrates an example of a small laboratory vacuum tumbler. Tumbling relies on gravitational impact (from meat pieces dropping from top to bottom of the drum as it rotates) and abrasion against other meat pieces to extract the essential myofibrillar proteins necessary for binding meat sections together. A meat massager relies mainly on abrasion of meat against meat or meat against stainless steel rotating paddles. Most massagers have a heavy, blunt paddle positioned vertically in a square vat. The paddle rotates slowly to gently massage the meat while minimizing tearing of muscle tissue and particle-size reduction. Both

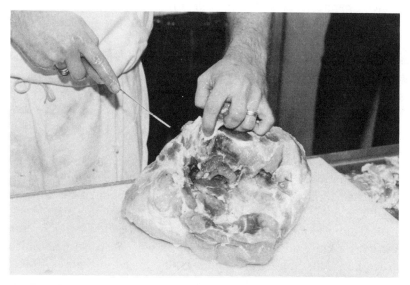

Fig. 19-6. The first step in tunnel-boning a ham is to remove the aitch bone (shown here) and the ham's shank through the stifle joint (not shown).

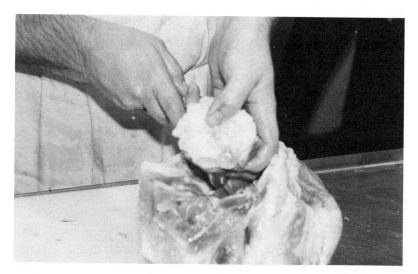

Fig. 19-7. The *femur* is removed by closely encircling it with a boning knife.

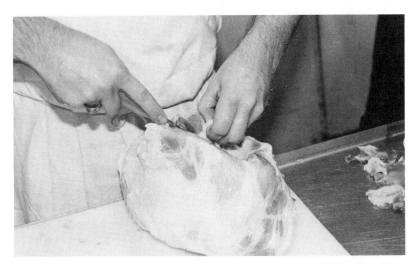

Fig. 19-8. Internal seam fat should be removed to produce a high-quality ham, using the tunnel-boning method. Note the boneless ham is in one piece and has not been cut open.

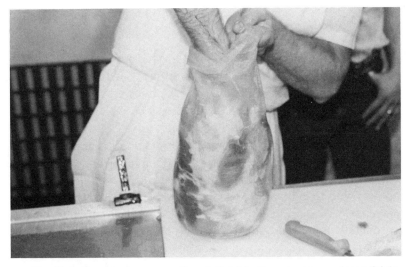

Fig. 19-9. Boneless hams are often stuffed into fibrous casings, clipped tightly, and placed in spring-loaded molds to flatten the ham (not shown). This prevents the occurrence of air pockets in the center region of the ham where the *femur* was removed.

Fig. 19-10. This small Vortron® vacuum tumbler (Vortron, Inc., Beloit, Wisconsin) can handle up to 250 pounds of product. The presence of vacuum is reported to reduce tumbling time required for sufficient protein extraction. (Photograph courtesy South Dakota State University)

tumbling and massaging have a tenderizing effect on the ham muscle and allow for greater pickup and retention of moisture by the meat.

Chemistry of Cured-Meat Color

The color of ham and other cured-meat products is one of the principal factors that affect their selection from the retail meat display case in the supermarket. A properly cured ham should have a uniform bright pink color that is free of uncured gray spots or other color defects and which, under proper packaging methods, will maintain this desirable color.

Numerous factors affect the ultimate color of cured-meat products and their ability to retain this color. These factors include myoglobin content of the muscle, pH, amount and dispersed uniformity of nitrite in the muscle, condition of the meat, exposure to light and oxygen after processing, and the type of packaging employed.

To properly understand how these factors affect the color of cured meat, one must understand the chemical reactions that are involved in curing processes.

Myoglobin is the meat protein principally responsible for the color of all meat. The myoglobin content of muscle tissue varies with the species, sex, and chronological age of the animal and the particular muscle. The oxygen needs of a muscle largely determine its myoglobin content. For example, the heart muscle, the most vascularized and hardest-working muscle in the body, has the highest myoglobin content. The myoglobin content of lean muscle tissue in veal and pork ranges from 1 to 3 milligrams per gram of wet tissue. Young beef contains 4 to 10 milligrams of myoglobin per gram of wet tissue, while mature beef contains 16 to 20 milligrams per gram. The myoglobin content of poultry white meat is generally less than 0.5 milligram per gram, while the dark meat ranges from 2 to 4 milligrams per gram. Lean-muscle tissue from lamb and mutton contains 10 to 22 milligrams of myoglobin per gram of wet tissue.

Chemical Structure of Myoglobin

Myoglobin is a very dynamic protein which can readily undergo changes in color, depending on its immediate environment. An explanation of how and why these changes occur is difficult without first understanding the structure of myoglobin, illustrated in Figures 19-11, and 19-12. Myoglobin is a complex protein that, in addition to its protein portion, contains another moiety, non-peptide in nature, complexed to the peptide chain (long chain of amino acids). This non-peptide portion is called a "heme group" and is composed of two parts, an iron atom and a porphyrin ring (Figure 19-11). For this reason, myoglobin is often called a "heme protein" and is the source of dietary iron in meat.

Examination of Figure 19-11 reveals that there are six binding sites for the iron (Fe) atom in the heme group. Four of these sites are used in stabilizing the porphyrin ring, one binding site is used to link the protein portion (the globin) to the heme group, and the sixth site is free to interact with a number of chemical elements.

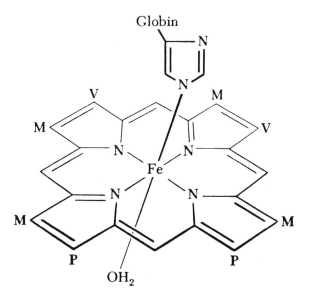

Fig. 19-11. A simplified schematic illustration of the heme group moiety of the myoglobin molecule. M, V, and P represent methyl, vinyl, and propyl groups respectively.

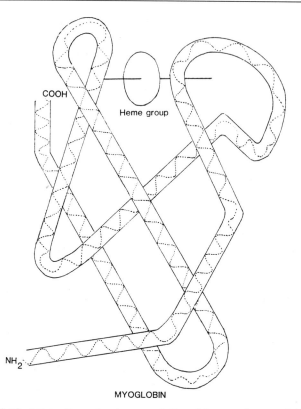

COOH

Heme group

NH$_2$

MYOGLOBIN

Fig. 19-12. A three-dimensional representation of the complex myoglobin molecule and orientation of the heme group within the molecule. Note the small size (proportionally) of the heme group which contributes the important color properties to meat. The actual myoglobin molecule is much more tightly compacted, and the heme group is held in a stationary position by hydrophobic interactions. The dotted line represents the approximate configuration of the polypeptide chain (chain of amino acid residues composing the protein).

The color state of meat is largely determined by the chemical compound present at this sixth site and the redox state of the iron atom (Fe^{+2} or Fe^{+3}). The physical state of the globin also influences the color complex that is formed. Heat denaturation (from cooking) of the globin results in a brown to gray color of the meat, unless nitric oxide is present at the sixth binding site. Table 19-4 lists the most common pigments (colors) of the myoglobin molecule in meat, the redox state of the iron atom, and the complexed element at the sixth binding site. The reader is also referred to the myoglobin figure in the color section near the center of the book.

Chemistry of Meat Curing

When nitrite (NO_2) is combined with meat, it will eventually be reduced to nitric oxide (NO). Reduction to NO is often accelerated by the use of ascorbate or erythorbate, both of which act as strong reducing agents. Nitric oxide has a

Table 19-4. Heme Pigments of Muscle

Pigment Forms	Fe Redox State	Element at Sixth Bind Site	State of Globin	State of Heme Group	Color
Myoglobin	Fe^{+2}	H_2O	Native	Intact	Purplish red
Oxymyoglobin	Fe^{+2}	O_2	Native	Intact	Bright red
Metmyoglobin	Fe^{+3}	H_2O	Native	Intact	Brown
Denatured myoglobin	Fe^{+2}	H_2O	Denatured	Intact	Brown
Nitrosomyoglobin	Fe^{+2}	NO	Native	Intact	Bright red
Nitrosohemochrome	Fe^{+2}	NO	Denatured	Intact	Bright red (pink)
Sulfmyoglobin	Fe^{+3}	H_2S	Denatured	Intact but reduced	Green
Cholemyoglobin	Fe^{+2} or Fe^{+3}	H_2O_2	Denatured	Intact but reduced	Green
Free and oxidized porphyrins	Fe absent		Absent	Ring structure destroyed— open chains	Yellow Colorless

greater affinity than water for the sixth bind site of the heme group. Hence, the myoglobin pigment is readily converted to *nitrosomyoglobin*, which, in reality, is a relatively unstable intermediate in the curing process. When the globin (protein portion of myoglobin) is denatured by heat, it has a stabilizing effect on the NO bound to the heme group. The subsequent pigment is termed *nitrosohemochrome* (previously known as nitric oxide hemochromagen) and results in the typical bright pink color associated with cured meats. The general reaction responsible for cured-meat color is given here:

$$myoglobin \ + \ \underset{H_2O}{\overset{NO_2}{NO}} \longrightarrow nitrosomyoglobin + heat \longrightarrow nitrosohemochrome$$

The stability of the nitrosohemochrome pigments is influenced by oxygen tension, light, and other factors. Fading of the desirable bright pink color will occur when cured meats are exposed to light and oxygen. For this reason, ham, bacon, and other cured-meat products are generally vacuum-packaged to exclude exposure to atmospheric oxygen during storage and merchandising. Center-ham slices are sometimes merchandised in an oxygen-permeable film and are turned face down in the display counter to minimize exposure to light and subsequent color fading. However, the recommended practice is to vacuum-package cured-meat products when surface color is critical to merchandising. Vacuum-packaging is less critical in whole hams and bellies (bacon), since the interior of the product is not exposed to light and is only exposed to minimal oxygen.

SMOKING

Prior to the application of smoke, it is critical that the surface conditions of the product be slightly moist. If the product's surface is too wet, the product will have a streaked appearance after it has been smoked. If the surface is too dry, the product will be pale in color due to poor adherence of the smoke volatiles to the product. Therefore, products are generally exposed to a short drying period prior to smoking.

Natural wood smoke is generally produced from sawdust, woodchips, or logs from hardwoods. Although the wood of the hickory tree is the most popular source of smoke, other hardwoods, such as oak, maple, ash, mesquite, apple, cherry, and other fruitwoods, are commonly used. Pine and other coniferous trees should be avoided as a source of smoke due to their high tar content and bitter flavor. The smoke may be produced from an elaborate electronically controlled smoke generator or from a variety of much simpler versions, ranging from log burning to manually controlled smoke generators.

Smoke may be applied to the surface of products in one of two forms: *natural* or *liquid*. Natural wood smoke is composed of three principal phases: (1) *solids*—fly ash and tar, (2) *noncondensibles*—air and combustion gases, and (3) *condensibles*—acids, carbonyls, phenolics, and polycyclic hydrocarbons.

The solid and noncondensible phases do not contribute significantly to the flavor, color, aroma, or preservative properties of smoked products. Dr. Hugo Wistreich[11] of B. Heller and Co., Chicago, has identified the phenolic fraction as the primary source of the smoky aroma and flavor. The phenolic fraction also is thought to be responsible for the preservative properties of wood smoke. The carbonyl fraction has been demonstrated to be the source of the desirable amber-brown color generated during smoking processes.

Liquid smoke (sometimes inappropriately called "artificial smoke") is manufactured from the condensible fraction of natural wood smoke through a water permeation-distillation process. Through further processing, the essential volatiles are refined and marketed in various forms (*e.g.,* oil based, water soluble, or dry on a dextrose or salt carrier). The application of liquid smoke in meat processing is a well-established practice. The principal methods of application are: (1) *drenching* or *dipping,* (2) *atomizing,* and (3) *regenerating.* High-volume continuous frankfurter lines often dip the franks into an acidic bath containing liquid smoke. In addition to imparting flavor, the acidic liquid smoke bath acts to "set the skin" prior to thermal processing, thus greatly aiding the peelability properties of the finished product (see Chapter 20). Atomizing the liquid smoke into very fine particles in a modern smokehouse is also a common method of application. Regenerated smoke is applied by superheating the atomized liquid smoke prior to venting it into the modern smokehouse.

Liquid smoke or granular dry smoke on a dextrose carrier may be added directly into the formulation of processed meat products to impart a smoky flavor uniformly throughout. When this latter practice is employed, not more than ⅛ ounce liquid smoke per 100 pounds of meat should be added. Liquid smoke is extremely purified and potent (particularly the oil-based type) and as such will impart an objectionable bitter flavor at higher levels.

---

[11]Hugo E. Wistreich. 1977. *Smoking of Meats.* Proceedings of Meat Industry Research Conference, Chicago, p. 37.

Chapter 20

Sausages

THE CONTEMPORARY role of sausages fits conveniently into modern life styles as an elegant entrée for entertaining as well as the main course in "quick and easy" meals. Sausage is also one of the oldest forms of processed food and has become highly prized throughout various stages of history. The word *sausage* is derived from the Latin word *salsus* meaning "preserved," or literally, "salted."

The earliest known recorded reference to sausage was in Homer's *Odyssey* written in the ninth century, B.C., over 2,800 years ago. There is strong historical evidence that the ancient Babylonian society produced and consumed sausages almost 3,500 years ago. Sausage became so popular during the wild festivals in Rome that, as the Christian Era began, it was banned. Strong public protest eventually forced a repeal of the ban.

By the Middle Ages, many varieties of sausages were being produced throughout Europe. Each variety became unique, dependent on the climate of the region and the availability of various spices. The warmer climates of Italy, southern Spain, and southern France led to the development of dry and semi-dry sausages. The cooler climates of Germany, Austria, and Denmark produced fresh and cooked sausages because preservation was less of a problem.

It was during this period (the Middle Ages) that spices became very important and valuable commodities. Black pepper was in such demand by sausage makers that it was traded as a form of currency. The discovery of the western world was largely due to the popularity of sausage throughout Europe. History tells us that when Columbus made his eventful voyage, he was searching for a shorter route to the Spice Islands of the Far East to obtain spices used in the production of sausage. It is remarkable and ironic that he should land in the area of what is now known as Jamaica, the one place in this hemisphere where spices flourish.

Partly a result of the increasing availability of different spices from the Far East, sausage makers, or wurstmachers, became very skilled at creating distinctive types of sausages throughout Europe. Many sausages became known by the town from which they originated. Specific examples include Genoa and Milano salamis from Genoa and Milan, Italy; bologna from Bologna, Italy; frankfurters from Frankfurt, Germany; and braunschweiger from Brunswick (Braunschweig), Germany.

Early American Indians in this country produced a form of sausage by combining dried berries and meat into a cake and then smoking or sun drying them. However, the American sausage industry did not see great development until during and after the Civil War, when the Industrial Revolution began. With the Industrial Revolution came many jobs and many European immigrants. During this time period, a great influx of German, Italian, Polish, Dutch, and Danish people brought with them Old World sausage recipes and skills to satisfy their ethnic tastes.

The modern sausage industry is extremely diversified in the size and types of plants in operation and represents a unique blend of art and science. Most of the major sausage processing plants are highly mechanized and automated to handle large volumes of products with great efficiency.

The art of making sausage developed slowly throughout the centuries with little or no understanding of the scientific principles underlying their manufacture. Today, science plays an integral role in sausage production by minimizing the batch-to-batch variation that once occurred and virtually eliminating the unsuccessful batches that had to be thrown away. During the past 50 years, a phenomenal amount of scientific data and literature has been compiled on the topic of sausage production. The modern meat processor must have a basic understanding of the scientific principles responsible for the desired textural, water binding, flavor, color, and shelf-life characteristics of the finished product.

In 1983, approximately 2,100 sausage processors in the United States produced 6.28 billion pounds of sausage representing over 200 different varieties. This figure can be further broken down into the following statistics:[1]

Variety	*Billion Pounds*
Fresh sausage	1.140
Dry and semi-dry sausage	0.333
Franks and wieners	1.396
Bologna	0.679
Sliced luncheon meats	0.980
Liver sausage, specialty items, and other sausages	1.432
Vienna and other canned sausages	0.319

USDA statistics indicate that revenue from the export of processed meats from the United States was approximately equal to the dollars imported in 1983 ($19.3 and $25.6 million, respectively). The most recent available estimate for per capita consumption of sausage products in the United States is 26.9 pounds per person.[2] In 1973, approximately 4.4 billion pounds of sausage was produced in the United States, corresponding to a per capita consumption of 20.6 pounds.

---

[1]*Meat Facts*. 1983. American Meat Institute.

[2]Based on the 1982 U.S. population estimate of 233 million people. From *National Food Review*. 1983. No. 23, Economic Research Service, USDA.

The production of sausage thus constitutes an important and growing segment of the meat industry. Through sausage production, virtually unlimited variety is created in our meat supply.

SAUSAGE IDENTIFICATION

Sausages and ready-to-serve meats are generally grouped according to the processing method used. The following classifications should be helpful in the identification, selection, and care of these products.

Fresh Sausage

Fresh sausage is made from selected cuts of fresh meat (not cooked or cured) and must be stored in a refrigerated (or frozen) state prior to consumption. Fresh sausages should not be held in a refrigerator for more than three days prior to consumption and should be cooked thoroughly before serving. Examples of products in this classification include fresh pork sausage (bulk, patty, or link); country-style pork sausage; fresh kielbasa (Polish); Italian sausage; bratwurst; bockwurst; chorizos (fresh); and thuringer (fresh).

Uncooked Smoked Sausages

Sausages in this class may be cured (contain nitrite) or fresh. These sausages are smoked but not cooked prior to sale. They must be held under refrigerated conditions and for no longer than seven days. These sausages should be cooked thoroughly before serving. Examples of products in this classification include smoked pork sausage, kielbasa, mettwurst, and smoked country-style sausage.

Cooked Sausages

These sausages are usually made from fresh meats which are cured during processing, fully cooked, and smoked. This class constitutes the greatest tonnage of sausage produced in the United States. These products should be refrigerated until consumed. Since many are vacuum-packaged, one should check the freshness date for storage life. Generally, these sausages will keep seven days, under refrigerated conditions, after a package is opened. Since they are fully cooked, these sausages are ready to eat (although some are generally served hot). Examples of sausages in this class include frankfurters (wieners), bologna, beerwurst (beef salami), New England sausage (berliner), mettwurst, cotto salami, German-style mortadella, kielbasa, knackwurst, smoked thuringer links, teawurst, and Vienna sausage.

Dry and Semi-dry Sausages

The dry and semi-dry sausages are made from fresh meats which are cured during processing and may or may not be smoked. A carefully controlled bacterial fermentation causes these products to have a lower pH (4.7 to 5.3), which aids in the preservation and produces the tangy flavors associated with this class of products. The dry sausages are generally not cooked and include the Italian salamis and pepperoni. Dry sausages require long-drying periods (generally from 21 to 90 days, depending on product diameter), whereas semi-dry sausages are often fermented and cooked in a smokehouse. Both are ready to eat. Cool storage is recommended for dry sausages, while semi-dry sausages should be refrigerated. Examples of dry and semi-dry sausages include summer sausage; cervelat (many variations); thuringer; the salamis (Genoa, Milano, Sicilian, B. C., D'Arles, and hundreds more); chorizos; frizzes; Lebanon bologna; pepperoni; mortadella; lyono; landjaegar; and sopressata.

Specialty Meats (Luncheon Meats)

These products are made from fresh meats which are cured during processing, are fully cooked, and may be smoked (although usually they are not). Luncheon meats may be cooked in loaf pans or casings or water-cooked in stainless steel molds. Specialty meats are often sliced and vacuum-packaged at the processing plant. These products are ready to eat and should be refrigerated and may be held approximately seven days after a package is opened. There is usually a freshness date on the package. Examples of products in this class include loaves, such as Dutch, ham and cheese, honey, jellied tongue, old-fashioned, olive, pepper, pickle and pimento, headcheese, chopped ham, and many more; scrapple; and Vienna sausage.

SAUSAGE TERMINOLOGY

The matter of sausage terminology is a tangle of Gordian complexity, a strange and untidy knotting of history, tradition, secrecy, myth, religion, politics, and stubborn pride. Considerable confusion has arisen from the meshing of Old World and New World traditions in regard to the naming of sausage products. It is hoped that the glossary will illuminate a bit of that tradition through a descriptive list of selected sausage types and styles. This list is by no means complete, but rather a representation of some of the more common products and styles of products on the market in this country. An anonymous author in *Meat Industry* magazine[3] summed up this tumultuous task of compiling a glossary rather well:

---

[3]Anonymous. 1983. "A Glossary of Old World Sausage," *Meat Industry*, 29 (6):46.

Even a brief glossary such as this reveals the new world translations of old world products that plague accurate sausage identification. Take a word like "cervelat," for instance: "Cervelat" properly refers to a family of semi-dry sausages, to which summer sausage and thuringer belong. But some U.S. processors make a summer sausage and call it thuringer, others make a thuringer and call it summer sausage, and still others make both and call it cervelat.

Sorting out salamis would take a team of Italian geographers, and even they wouldn't be able to solve the problem of how to spell the word. "Salami" or "Salame"? San Francisco Bay Area processors use the latter; everyone else uses the former. In Italy, it's spelled both ways.

Then there's kielbasa. It would take pages to talk about kielbasa. Suffice it to say that in Poland, frankfurters and salamis are kielbasa, but here kielbasa is Polish sausage. But not always—sometimes Polish sausage is a little less spicy than kielbasa. Sometimes Polish sausage is called kielbasa anyway. But salami is never kielbasa—except in Poland.

In 1938, Paul Aldrich[4] stated:

Some standardization in types and uniformity in the use of names appears to be necessary if dry sausage is to achieve real popularity in the United States. The word *frankfurter* carries with it a fairly definite idea of shape, color, flavor and texture of the product named; *salami* or *cervelat* mean little to the housewife [consumer] because these terms are so loosely used and so indefinite.

Unfortunately, this observation still holds true today—almost 50 years later.

Sausage Glossary

Alpina Salami. A spicy Italian-style salami of U.S. origin.

B. C. Salami. A dry Italian-style salami stuffed into a natural beef casing.

Balleron. A mildly seasoned bologna-type product made of finely chopped veal and pork with bits of smoked beef tongue and pistachio nuts added for flavor. A sandwich meat, stuffed in a dark casing.

Bangers. English breakfast sausage, though technically not allowed to be called "sausage" by USDA standards, because the formula calls for bread crumbs to be mixed with the pork emulsion. A plump product—usually about 4 inches long and 1½ inches thick.

Beerwurst. One of the big sausages—usually 2½ to 3½ inches in diameter—stuffed into distinctive veined natural casings or vein-decorated artificial casings. It is basically a coarse-ground German-style salami, made of beef and pork, spiced with garlic.

Blood Sausage. Known as blatwurst in Germany. Primary meat constituent is diced pork or pork fat, mixed with beef blood and mild spices. Very dark in appearance and often available in both chubs and rings.

---

[4]P. I. Aldrich. 1938. *Sausage and Meat Specialties, The Packers Encyclopedia*, Part III. *The National Provisioner*, Chicago.

Blood-and-Tongue Sausage. Distinguished from blood sausage by its main ingredient—chunks of cured, cooked tongue in various sizes and shapes. This sausage is also available in a variety of conformations: ring, stick, and loaf are typical. Sometimes it is spiced with a small amount of ground clove.

Bockwurst. The traditional way to sell bockwurst is in a fresh state, though several cooked bockwurst products are now available at retail. Fresh bockwurst, like the cooked varieties, is a veal and pork sausage with a taste and bite similar to that of frankfurters. Styles include the coarse-ground farm-style bockwurst.

Bockwurst (cooked). Until recently, most bockwursts were sold fresh. The guarantee of increased shelf life has brought about the retailing of cooked bockwurst. White bockwursts are similar to frankfurters in flavor, white in color, and generally thicker in diameter.

Veal is the primary meat constituent, supplemented by pork. Other ingredients include milk, eggs, and, often, chopped parsley. The casings tend to be very thin, and cooked bockwursts are usually boiled at the meat plant, not smoked. Regional bockwurst products in the United States include farm-style (similar to bratwurst), Swiss-style (very mildly seasoned), and thuringer bockwurst (coarse texture with a speckled appearance).

Bologna. The second most popular sausage in the United States, bologna products account for nearly 20 percent of all the sausage sold in the United States, second only to frankfurters. In generic terms, bologna is a fully cooked, mildly seasoned sausage made from beef and/or pork. But several styles and shapes of bologna are produced in volume, among them:

- **Berliner** or **New England Sausage.** A product made of coarse-ground pork with pieces of ham or chopped beef interspersed in the emulsion. When berliner is in small chub form, it is sometimes called Leona bologna; when in stick form, it is known as jagdwurst. Generally it is stuffed into large casings, similar to beerwurst.
- **German.** A light-colored bologna with medium-coarse texture. Meat ingredients include beef and pork, and spices are dominated by garlic. Also stuffed into large-diameter casings.
- **Ham-Style.** According to regulation, the dominant ingredient in ham-style bologna must be cured ham pieces. These pieces are suspended in a fine emulsion, and the product is both smoked and cooked.
- **Vienna.** Sometimes called fleischwurst. Another large-diameter sausage with a creamy meat texture. Pistachio nuts are used for seasoning in place of garlic. Vienna sausage is produced in both rings and chubs.

Bratwurst or **White Hots.** Produced both fresh and cooked. In cooked form these sausages are known regionally as white hots and are made in the same way as frankfurters, except that nitrite is omitted (hence the white color). Main meat ingredients include pork and/or veal. Fresh bratwurst (uncooked) is prevalent in much of the upper Midwest with German-style, Wisconsin-style, and Swiss-style the predominating names. Each is coarse ground, combined with various seasoning combinations, and stuffed into hog casings.

Braunschweiger. Basically, a creamy-textured liver sausage that has been smoked. See **Liverwurst**.

Buendnerfleisch. A pressed and dried-beef round cured with wine—not a sausage.

Cappicola. Made from cured boneless pork butts which have been lightly rubbed with pepper and other seasonings. The boneless butts are then smoked and dried to take on a unique aged flavor.

Caserta Pepperoni. An Italian product consisting of 75 percent pork and 25 percent beef, stuffed into hog casings and linked in pairs (12 ounces to each piece). Generically, pepperoni means red pepper pods, and Caserta is a city in southern Italy.

Cervela. See **Garlic Sausage**.

Cervelat. *Cervelat*, like *salami*, is one of those sausage terms encompassing a whole family of products. It is used in the United States interchangeably with summer sausage and thuringer, making accurate identification confusing at best. In general, cervelat is a mildly flavored semi-dry sausage, somewhat softer in consistency than the Italian salamis. For the purposes of this glossary, thuringer, in its semi-dry form, is included as a member of the cervelat family; summer sausage falls into the farm-style group of the cervelat family.

Within the definition of cervelat are many sausage styles and types. These include (in alphabetical order):

- **Farm-Style.** Made of equal parts of beef and pork, rough chopped and mildly seasoned, and stuffed into 1½-inch diameter casings of varying lengths.
- **Goettinger.** A distinctively flavored hard cervelat.
- **Göteborg.** A salty and heavily smoked, coarsely chopped cervelat.
- **Gothaer.** Made of fine-chopped lean pork.
- **Holsteiner.** Similar to farm-style except that it is packed in rings.
- **Landjaegar.** A flattened and smoked cervelat about the size of a frankfurter.
- **Thuringer.** A tangy semi-dry sausage, mildly seasoned and characterized by a relatively soft consistency.

Chorizo. A Spanish dry sausage made from coarse-cut pork. About the size of a large frank, chorizo is a highly spiced, hot sausage.

Coney Island Red Hots. High-quality frankfurters, which were originally formulated to be roasted (not boiled) on hot plates by sandwich stand vendors. The red hots originated in New York City, were dyed red, and are considered to be the namesake of the hot dog.

Cooked Salami. The cooked salamis include beer, cooked, and Kosher sausages. Each is produced from spice-enhanced cured meats or fresh meats to which cure and spices are added. Generally, these are beef-dominated salamis (as opposed to the dry salamis, which are usually pork-dominant), stuffed into natural or fibrous casings, and smoked to a deep red.

D'Arles Salami. A very coarse-textured dry French salami that is stuffed into 18-inch hog bungs and wrapped with cord designed to show a diamond-shaped pattern.

Farmer Sausage. A mildly seasoned dry or semi-dry sausage. Farmer sausage originated with the farmers of northern Europe and is made of 65 percent beef and 35 percent pork. It is chopped medium fine, seasoned, stuffed into beef middles, and heavily smoked. Each piece weighs from 1 to 2 pounds.

Frankfurter. In its various forms, the frankfurter accounts for 25 percent of all sausage sold in this country, easily making it the most popular sausage on this side of the Atlantic. Nearly 95 percent of frankfurters are sold skinless, and the most popular size is 1.6 ounces, or 10 to a pound. Most frankfurters in the United States are a blend of beef, pork, and/or poultry meat, mildly seasoned with paprika and other spices and smoked. In recent years, a substantial volume of poultry frankfurters have entered the marketplace. Poultry meat is often combined with red meats to produce frank-furters at a lower production cost. German-style frankfurters are made of veal, and franks called "old fashioned" or "old style" are almost always stuffed into natural casings. Cocktail franks are the same as regular franks except that they are much smaller and are used as an appetizer. Vienna-style sausages (frankfurters) are cooked and canned.

Frizzes. Another highly seasoned dry sausage, made from lean, cured pork that has been rough-chopped and stuffed into hog middles.

Galician Sausage. A smoked beef and pork sausage that is seasoned with garlic and stuffed into beef rounds 17 to 20 inches long. The casing is broken in the middle, and the ends are tied together. Galician is favored by the Polish and by the people of the western U.S.S.R. around Easter time.

Garlic Sausage. Also known as cervela and knoblauch. Similar to frankfurters, except that garlic is a prime spice ingredient. The shape is also distinctive: very short and squat, typically about 5 inches long and 2 inches thick. Usually stuffed into natural casings.

Gelbwurst. A luncheon meat similar to the mildly flavored Swiss-style bockwurst. Made from veal, pork, milk, and eggs. A typical gelbwurst chub is 24 inches long and about 2½ inches thick.

Genoa Salami. A moderately spiced dry salami of Italian origin. Traditionally, this salami is stuffed into hog bungs, 20 to 24 inches long and wrapped vertically and horizontally with flax twine, starting at the small end, with loops 2 inches apart joined with slip-hitch knots (Figure 20-1).

German Salami. A mildly spiced dry salami that is heavily smoked. Traditionally this salami is stuffed into calf bladders or beef middles and tied with flax twine.

Göteborg, or Swedish, Sausage. A salty, heavily smoked cervelat that is somewhat coarsely chopped. Göteborg is traditionally soaked in a salt brine before smoking. This sausage takes its name from the Swedish city of Göteborg.

Gothaer. A summer sausage reported to be impossible to manufacture in any but the winter months (believe it or not).

Headcheese. Available in the United States in two styles: French and German. French headcheese consists of cured pork pieces suspended in vinegar-flavored gelatin, with bits of pickles and pimentos added for flavor. There is no casing. German head-cheese, or Suelze, also is composed of cured pork pieces, plus some beef, but the gelatin is produced by heating to resemble more of an emulsion. German headcheese requires a casing; usually natural casings are used.

Holsteiner. Similar to farmer sausage, except that the ends are tied together (sometimes called horse shoe sausage). Dried and smoked, it appears on the market in pieces weighing about 1 pound each.

Hungarian Salami. Similar to German salami. Mild in flavor but heavily smoked.

Italian Salami. Characterized by a unique "tang" in the flavor. Regional specialties include Genoa, Milano, Sicilian, and southern Italian, among hundreds more. In this country, Italian-style salami produced in the San Francisco Bay Area is allowed to develop an outside coating of white mold; in other parts of the United States, the mold is brushed off the casing. It should also be noted that the Bay Area salami manufacturers call their product "salame." The basic meat ingredients of Italian salami are coarse-chopped cured pork and fine-chopped lean beef. Drying time is typically three months.

Jagdwurst. Berliner bologna in stick form. See **Bologna.**

Jaternice. A ring liver pudding of Bohemian origin. Made of liver, head meat, bread crumbs, and cornmeal or other fillers. Traditionally contains too much filler to be classified as a sausage in the United States.

Kielbasa. Some sausage producers call kielbasa "Polish sausage"; others who make both products give a little spicier taste to kielbasa. It is a coarse-ground beef and pork sausage, usually smoked to a medium red. See **Polish Sausage.**

Kiszka. A smoked beef and pork sausage from Hungary that also contains rice and pork fat pieces. Primary spice is garlic.

Knackwurst. Also known in Switzerland as schueblig. A coarse-ground beef, pork, and veal sausage smoked to a reddish-brown. Garlic or onion dominates the several spices used in knackwurst. The product is typically stuffed into a natural casing.

Knoblauch. See **Garlic Sausage.**

Landjaegar. A Danish "hunter's sausage" made of beef and pork. This unique dry to semi-dry sausage is stuffed into hog casings, cut into double links, and pressed flat between weighted boards to give it a square appearance. Garlic and caraway seeds give it a unique flavor profile.

Lebanon Bologna. Not actually an Old World sausage but a development of the Pennsylvania Dutch culture. This semi-dry bologna is an all-beef product, characterized by a very heavy smoke and a very acid flavor.

Linguisa. In cooked form, this Portuguese sausage is made of coarse-ground, highly spiced pork. Typically smoked to a medium reddish-orange.

Liverwurst. A very fine-ground cooked sausage seasoned with onions, sometimes pistachios, and other spices. Meat ingredients include pork and liver. Regional varieties include French-style (a coarser mixture) slightly darker in color; country-style (flavored with bacon bits); and others. By regulation, liverwurst must contain at least 30 percent liver, and if a pink color is desired, the meats must be cured.

Lombardia Salami. This coarse-cut Italian salami has a moderately high fat content and contains brandy. Like most of the Italian salamis, Lombardia has its own characteristic twine wrap consisting of a 1-inch square pattern.

Luncheon Meats. This large group of loaf products is primarily of American origin and does not require a casing for cooking. Condiments are added to fine-ground or comminuted meat batters to create infinite variety. Examples include pickle and pimento loaf, olive loaf, ham and cheese loaf, etc.

Lyons Sausage. A French all-pork sausage, seasoned with garlic and other spices and stuffed into large-diameter casings. Traditionally, the pork is portioned in parts as follows: four parts of fine-chopped lean pork and one or two parts of diced pork fat.

Mettwurst. *Mettwurst* and *teawurst* are closely related, with the terms often interchanged. Some producers call their pork, veal, and garlic spreadable chub sausages teawurst, while at the same time producing the same product in a ring and calling it mettwurst. The sausages are of German origin and designed for sandwich spreading.

Milano Salami. A beef and pork dry sausage of Italian origin. Milano is generally stuffed into large hog bungs and wrapped with twine, three coils per inch.

Mortadella. Looks very similar to common bologna, except that cubes of pork fat are suspended in the emulsion. It is a delicately spiced sausage made from both pork and beef, very finely chopped, and usually smoked briefly at a high temperature prior to air drying. It is nearly always stuffed into large-diameter beef bladders. Mortadella originated in Bologna, Italy.

Pepperoni. In some regions of Europe it is known as kabanossi. A hotly spiced, coarse-ground pork and beef sausage with a very distinctive flavor, making it ideal as a pizza topping (its overwhelming dominant use in the United States). Ground red pepper is the major flavor ingredient.

Polish Sausage. Similar to kielbasa, another Polish-style coarse-ground sausage. Pork is the main meat ingredient, and the flavor profile is accented toward garlic. The sausages—usually 4 to 5 inches long, 1½ inches in diameter—are smoked thoroughly. *Kielbasa* is a generic word in Poland meaning "sausage." It is used in the United States exclusively to describe Polish or Polish-style sausages.

Polka Sausage. A sausage with Old World predecessors, but in the United States strictly a West Coast specialty. Made of pork, beef, and onions, blended to make the product an ideal complement to beer. Generally stuffed in a natural casing and smoked to a dark red.

Pork and **Potato Ring,** or **Korv.** A sausage of Swedish origin, still popular there at Christmas. The basic meat is pork, to which potatoes, onions, and spices are added. It is usually uncooked and sold frozen and contains no nitrite.

Pork Sausage. A fresh, uncooked sausage made entirely from pork and seasoned with salt, pepper, and sage. Pork sausage is sold in bulk 1- or 2-pound chubs or in link or patty form.

Prasky. Sausage of Polish or Czechoslovakian origin. Made from pork trimmings cured with salt and seasoned with sugar, pepper, and garlic. Prasky is traditionally stuffed into beef weasand casings.

Prosciutti (pronounced proshooti). A dry-cured Italian ham that is highly seasoned with pepper and other spices (including nutmeg).

Salami. Like cervelat, salami refers to a family of sausage products, mostly of Italian origin but also from other parts of Europe as well. Most dry salamis are made from pork and have a coarse-ground texture; garlic is the main spice, enhanced with others. The mild Italian winters are perfect for the extended drying periods required for high-quality dry salami, and hundreds of regional salami specialties have developed in cities from Brindisi to Venice. While most of these salamis go back several hundred years, it wasn't known until after World War II that the operative factor in the development of dry salami flavor and quality was bacteria (specifically, lactobacilli). Traditional salamis are characterized with different variations of twining or cording to provide a distinctive decorative effect and give support to the hanging sausage during fermentation and drying. B. C. salami, for example, has only a few vertical and horizontal cordings, whereas Genoa salami has many wrappings of twine, both vertical and horizontal, in a basket-weave effect (Figure 20-1). Some of the Salami sausages on the market are as follows: B. C. salami, H. C. salami, Genoa, Savonna Genoa, Milano, D'Arles, Nola, Sicilian or Sicani, Lola, cooked salami, Liguria salami, Lombardia, Kosher salami, Alessandria, Lazio, Novaro, Catania, Bobbio, Sorrento, Ancona, Capri, Corti, and many others.

Schinkenwurst. An Old World sausage closely related to berliner bologna. Ham pieces and diced pistachios are interspersed in a fine emulsion of pork and veal. Schinkenwurst is distinguished from berliner by the fineness of the emulsion and a no-garlic flavor profile.

Schueblig. See **Knackwurst.**

Scrapple. An American product of Pennsylvania Dutch origin that is not a sausage, but rather is a cornmeal-based suspension of head meat and/or pork sausage. Scrapple is sliced cold and fried before serving—usually for breakfast.

Sopressata. In Italy, any sausage stuffed into the wrinkled casings made from hog middles. In the United States, a sausage made from coarse-chopped pork and seasoned, in part, with whole black pepper.

Souse. A loaf product similar to headcheese that is bound together with gelatin. Souse typically will contain cured and cooked head meat, tongues, hog lips and hog snouts, and pickles and pimentos for an attractive color.

Suelze. See **Headcheese**.

Teawurst. See **Mettwurst**.

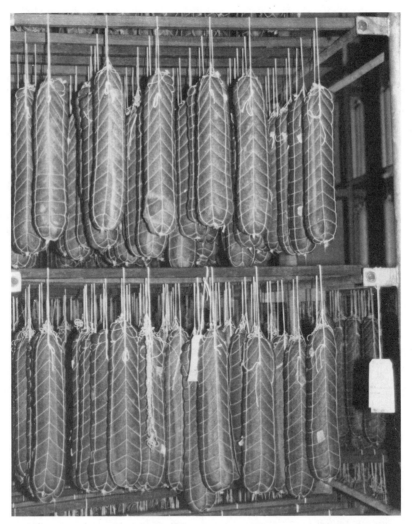

Fig. 20-1. Genoa salami stuffed in natural casings and hanging in a "drying room." Each of the traditional dry Italian salamis has a unique and different type of "cording." Twine is wrapped vertically and horizontally for Genoa salami, as shown above.

Thuringer. Some thuringers are fermented semi-dry sausages; while others are not fermented, but sold fresh. In fact, considerable confusion surrounds the definition of thuringer, since in certain parts of the United States, the name is used interchangeably with summer sausage. Other sausage manufacturers produce both a thuringer and a summer sausage. For this glossary, cooked thuringer sausage is defined as being made mostly from pork (sometimes beef and veal are added in small quantities), with similar flavor used in fresh pork breakfast links, though sage is usually not used to season thuringer. In years past, this sausage was sold fresh, but—as in the case of bockwurst—many processors are offering thuringer in cooked form for better shelf life.

Vienna Sausage (Wieners). Vienna wieners are a popular canned sausage, similar to frankfurters, although usually more bland in flavor and smaller in diameter. Vienna wieners are generally not linked, but are cut into uniform lengths prior to canning.

Information for this glossary was obtained from the following sources:

Heller's Secrets of Meat Curing and Sausage Making. 1929. Published by B. Heller & Co. Chicago.

Sausage and Meat Specialties. 1938. (Part III, *The Packers Encyclopedia.*) Published by *The National Provisioner.* Chicago.

Meat Industry. 1983. Vol. 29:6.

Encylopedia of Labeling Meat and Poultry Products. 1983. Published by *Meat Plant* magazine. St. Louis, Missouri.

SAUSAGE INGREDIENTS

A basic understanding of the properties and/or function(s) of raw meat materials, and nonmeat ingredients and the interaction of the two is essential in order for the reader to comprehend why manufacturing practices are different from one product to the next.

Raw Meat Materials

Large numbers of different raw meat materials are used in the production of all types of sausage, with each contributing particular properties to the finished product. Raw materials vary considerably in their proximate composition (percentage protein, moisture, fat, and ash), color, connective tissue (collagen) content, and binding ability. This fact is illustrated in Table 20-1 in which 27 of the more common raw materials for sausage production are listed. The values represented in this table serve only as a guide for average values. Actual values may vary considerably from lot to lot of the same raw material. Other raw meat materials not listed in this table but that are in widespread use include mechanically separated beef, pork, and poultry tissue (formerly known as mechanically deboned meat), and mechanically desinewed meat.

Values for protein, moisture, fat, collagen, color, and bind are heavily relied on by larger processors who utilize preblending practices and computer formulations (see discussion later in chapter).

In recent years the use of poultry meat has had a considerable impact on the

Table 20-1. Meats Most Often Used in Commercial Sausage Production with All Bones and Bone Fragments, Heavy Sinew, and Gristle Removed[1]

Meat	Protein	Moisture	Fat	Collagen[2]	Color[3]	Bind[4]
	. . . . (percent)				. . . . . (index)	
Bull meat, full carcass	20	68	11	20	100	100
Cow meat, full carcass	19	70	10	21	95	100
Beef shank meat	19	73	7	66	90	80
Beef chucks	18	61	20	30	85	85
Beef trimmings, 90% lean	17	72	10	30	90	85
Beef trimmings, 75% lean	15	59	25	38	85	80
Beef plates	15	34	50	–	–	–
Beef flanks	13	43	42	–	55	50
Beef head meat	17	68	14	73	60	85
Beef cheeks, trimmed	17	68	14	59	10	85
Beef tissue (partially defatted)	20	59	20	–	30	25
Veal trimmings, 90% lean	18	70	10	–	70	80
Mutton	19	65	15	–	85	85
Poultry meat (dark)	19	67	12	–	80	90
Pork trimmings, 50% lean	10	39	50	34	35	55
Pork trimmings, 80% lean	16	63	20	24	57	58
Pork blade, 95% lean	19	75	05	23	80	95
Picnic trimmings, 85% lean	17	67	15	24	60	85
Pork jowls	06	22	72	43	20	35
Pork cheeks, trimmed	17	67	15	72	65	75
Pork tissue (partially defatted)	14	50	35	–	15	20
Pork hearts	16	69	14	27	85	30
Pork tripe	10	74	15	–	20	05
Beef hearts	15	64	20	27	90	30
Beef tripe	12	75	12	–	05	10
Beef lips	15	60	24	–	05	20
Beef weasand meat	14	75	11	–	75	80

  – = No data available.

  [1]John C. Hickey and A. Wade Brant. *Sausage Makers Handbook*. Dept. of Food Science & Technology, University of California, Davis.

  [2]Values represent the percentage of the protein content that is collagen.

  [3]Color values are expressed as an index in which 100 represents a raw material with the most intense red color and 0 (zero) represents a raw material with virtually no contribution of red color.

  [4]Bind values are expressed as an index in which 100 represents a raw material which contributes the maximum bind and 0 (zero) represents a raw material which contributes little or no bind.

production of sausage. The use of turkey meat to produce turkey frankfurters, bologna, and breakfast sausage has undergone tremendous growth during the last five years. This growth can be attributed to two primary factors. First, processed poultry products are generally cheaper than their red meat counterparts. Second, processed poultry products have been perceived to be leaner and more nutritious than the traditional sausage counterparts by a significant segment of this country's consumers.

The variety meats most commonly used in sausage production include hearts, tongues, livers, kidneys, tripe (beef stomachs), and pork stomachs. The amount and type of variety meat that may be used is dependent on the product being produced and on the quality of the product. Certain products such as braunschweiger and liverwurst contain predominantly liver (a variety meat). Other products such as frankfurters and bologna may contain no variety meats or may contain as much as 30 to 40 percent of variety meats. Extreme caution must be used when formulations contain low-bind variety meats such as tripe or pork stomachs. These low-binding meats should be limited to 15 percent or less of the formulation, or product instability may occur. Hearts provide a relatively good raw material source for products such as summer sausage, where a dark red color is desired. Hearts are also commonly used in emulsion-type products such as bologna. Tongues that have been cured are the sole meat source in jellied-tongue loaf and a major constituent of blood-and-tongue loaf. Tongues are also used in numerous other sausage products, depending upon the price and the supply. Kidneys are the least used of the variety meats in sausage production. The vast majority of the kidneys in the United States are either exported or used in pet food.

The proper selection of meat ingredients is essential for the production of sausage of uniform quality. This does not mean that only the more costly meat ingredients should be used in all situations, but rather the meat ingredients should be combined to meet pre-defined standards of fat content, color, binding properties, and other characteristics. The raw materials should be as fresh as possible and in good bacteriological condition. Inventory control of raw meat materials should be set up on the FIFO (first in, first out) principle to insure a constant and uniform turnover of raw materials from either the holding cooler or the freezer. The intelligent meat processors will avoid "bargain" raw materials, since there is usually a reason for the bargain, such as temperature abuse resulting in a high-microbial load. When it comes to the use of meat ingredients, there's an old saying that is very true: "Garbage in, garbage out." The end product can only be as good as the quality of the raw materials going into its production.

Meat ingredients will vary considerably in their moisture-to-protein ratio, in their lean-to-fat ratio, in their water-binding properties, and in their relative amounts of pigment (red color). Binding refers to the ability of the meat ingredients to hold and entrap fat and water to produce a stable emulsion (see meat emulsions, page 664). Binding also refers to the ability of lean meat particles to be held or glued together. The specific factors responsible for variations in the binding properties of meat ingredients used for sausage are very complex and are discussed in more detail later in this chapter.

Raw meat materials are classified according to their binding ability as *binder* or *filler* meats. Binder meats have been further subdivided into high, medium, and low, based on their ability to bind the product. Skeletal meat from the beef animal is considered to have the highest binding properties. Examples include bull meat, shank meat, boneless chuck, and cow meat. Meat ingredients having an intermediate binding ability (medium) include head meat, cheek meat, and lean pork trimmings. Low-binding meats usually contain a large proportion of fat or are non-skeletal meats and include regular pork trimmings such as jowls, ham fat, beef briskets, hearts, hanging tenders, and tongue trimmings. Meats with little or no binding ability are called filler meats and include tripe, pork stomachs, lips, snouts, skin, and partially defatted pork and beef tissue. While these meat ingredients are nutritionally acceptable, their use in sausage products must be severely limited, if a high-quality product is to be produced. Perhaps the most desirable means of using these last named ingredients is through the jellied-loaf products (*e.g.,* souse and headcheese), because the binding ability of the meat is not important in these products.

The moisture-protein ratios of the various raw meat materials are important to the processor in predicting the composition of the final product. FSIS meat inspection regulations dictate that the moisture content of the final product cannot exceed four times the protein content plus 10 percent (4 P + 10). When using meat with a low moisture-protein ratio (such as bull meat), processors can add more water to the meat during mincing than with meat containing a high moisture-protein (M:P) ratio (beef hearts). However, economics dictates the formulation of ingredients to a large part, and meats having a high M:P ratio are generally more expensive—even though slightly more water may be added. Table 20-2 lists the M:P ratio of selected meats used in sausage formulations. It should be recognized that these values reflect averages of the various ingredients listed. Actual M:P ratios will vary somewhat from sample to sample, depending on the condition of the meat and the fat content.

Table 20-2. Percentage of Approximate Moisture and Protein Content and M:P Ratios of Some Meats Used in Sausage Making

Sample	Moisture	Protein	M:P Ratio
Regular pork trimmings	29.0	7.0	4.1
Cheek meat (pork)	71.7	19.6	3.7
Head meat (pork)	63.1	16.4	3.8
Pork fat	6.2	1.33	4.6
Chucks	72.0	20.0	3.6
Bull meat	73.6	21.2	3.4
Beef tripe	72.8	15.2	4.7
Beef trimmings	71.1	19.8	3.5
Beef flanks	59.2	15.4	3.8
Beef plates	70.6	19.2	3.7
Beef hearts	79.0	16.0	4.9

Nonmeat Ingredients

There are a number of nonmeat ingredients used in the production of sausage to provide certain functional properties, to create variations in flavor and appearance, to prolong the shelf life of the product, and to insure microbiological safety of the final product. Some of these ingredients, such as salt, cannot be omitted. Others are adjuncts whose primary function is to reduce the cost of the products (as in cereal grain fillers).

The following ingredients are ones most commonly used in sausage production. The regulatory limits of certain ingredients are always subject to change. The limits expressed in this text are current for January 1985.

Salt

Salt is the most critical ingredient in sausage manufacturing other than meat. Without salt, sausage, as we know it today, could not be made. Salt has three primary functions: preservation, flavor enhancement, and protein extraction to create product bind. In modern sausage production, the preservative effect of salt is not very important (except in dry sausages). To be an effective preservation method, brine concentrations in the product must be around 17 percent. The percentage of brine concentration in the product may be calculated as follows:

$$\frac{\% \text{ salt added to meat}}{\% \text{ water in meat } + \% \text{ salt}} \times 100 = \% \text{ brine concentration}$$

Most sausages will have a brine concentration of 4 to 6 percent, which corresponds to 2 to 3 percent added salt.

The two most important functions of salt are to impart flavor and enhance the binding properties of the meat proteins actin and myosin (see Chapter 21). Salt acts to solubilize these meat proteins, enabling them to entrap fat and bind water. This action stabilizes the sausage batter so that the fat does not coalesce into large fat pockets or migrate to the surface to form fat caps during cooking.

Sodium chloride is the most common salt used in sausage manufacturing, because it has the most desirable flavor, is readily available, and has the greatest protein solubilization properties. However, sodium has been linked to hypertension which can lead to coronary heart disease in some people. For this reason, patients suffering from high blood pressure have often been placed on reduced sodium diets. In recent years, the major meat processors have responded with low-sodium luncheon meats and other sausages by replacing a portion of the sodium chloride with potassium chloride, calcium chloride, or other salts. However, these salts are generally more bitter than sodium chloride. Intensive research is ongoing in this area.

Water

Water is often an overlooked ingredient in sausage formulations; nonetheless, it is an extremely important nonmeat ingredient. Emulsion products would be very dry and unpalatable if they contained only the moisture inherent in the meat ingredients. Water is often added in the form of ice during the chopping or mincing process to hold the temperature of the meat batter down, thus insuring stability of the batter. Moisture functions to make the final product juicier and works with the salt in helping to solubilize the meat proteins for binding fat.

Nitrite and Nitrate

Historically, nitrate was present as a naturally occurring contaminant in the salt used for sausage production. Early sausage makers found that salts from certain areas of Europe produced superior sausages; however, they did not recognize that this was due to a contaminant. Chemists later isolated the compound nitrate, and it was then added intentionally in the form of saltpeter (potassium nitrate). Eventually chemists recognized that nitrite (NO_2) and not nitrate (NO_3) was responsible for the beneficial color and flavor properties which enhanced sausage products. Nitrate is actually broken down by bacteria to nitrite during long curing processes. With current rapid-processing techniques in use, nitrate is seldom used except in dry-sausage production. Nitrite is added directly to the sausage batter—usually in the form of sodium nitrite. Potassium nitrite can also be used.

Nitrite has four primary functions in cured meat and sausage products. These functions are: (1) development of the characteristic pink cured-meat color, (2) flavor improvement, (3) powerful antioxidant properties, and (4) bacteriostatic properties. The development of the cured-meat color was discussed in Chapter 19 and is illustrated in the color section near the center of this text. About 40 parts per million (ppm) of nitrite in the finished product is considered necessary for the formation of the cured color. Nitrite also inhibits the oxidation of the lipids (fat) in meat which would otherwise lead to the development of oxidative rancidity. Without nitrite a stale, warmed-over flavor would exist in most products. Oxidative rancidity is often a problem in fresh sausages because they do not contain nitrite. The bacteriostatic properties of nitrite are extemely important in the thermally processed, vacuum-packaged products such as frankfurters and luncheon meats. Without nitrite, the safety of these products would be jeopardized by the organism *Clostridium botulinum*, the bacteria that cause the lethal food poisoning botulism. These organisms are not destroyed at normal meat processing temperatures, but rather form protective spores. Nitrite prevents the outgrowth of *C. botulinum* spores and the subsequent production of one of the world's most deadly toxins.

Nitrite and nitrate are the most regulated and most controversial of the sausage ingredients. Sodium or potassium *nitrite* may be used at a level not to exceed ¼ ounce per 100 pounds of meat (156 ppm) in sausage products. Sodium or potassium *nitrate* may be used at a level not to exceed 2¾ ounce per 100 pounds of meat. Extreme caution must be exercised in adding nitrite or nitrate to meat, since too much of either ingredient can be toxic to humans. Since such small quantities are difficult to weigh on most available scales, *it is strongly recommended that a commercial premixed cure be used when nitrite and/or nitrate is called for in the formulation.*

Ascorbates and Erythorbates

These vitamin C derivatives are also known as cure accelerators, since they act to speed the curing reaction. Ascorbate and erythorbate are strong reducing agents which accelerate the conversion of metmyoglobin and nitrite to myoglobin and nitric oxide (see Chapter 19, pages 614-616). Residual amounts of these compounds present in the finished product also add stability to the cured color by reducing the deterioration rate of the nitrosohemochrome pigment. A further beneficial function appears to be that ascorbates and erythorbates seem to inhibit the formation of nitrosoamines (see Chapter 19). Examples of specific compounds approved for use as cure accelerators include ascorbic acid, erythorbic acid, sodium erythorbate, sodium ascorbate, citric acid, sodium citrate, sodium acid pyrophosphate, and glucono delta lactone (GDL).

Sugars

A variety of sugars are commonly used in different sausage products, ranging from sucrose (cane or beet sugar) to dextrose (corn sugar). Included in this latter group are corn syrup, corn syrup solids, and sorbitol. Sugars are used primarily for flavoring to counteract the salt-flavor intensity and to provide food for microbial fermentation in the fermented sausages. Most sugars (except sorbitol) increase the browning of the meat during cooking, which may or may not be desirable, depending on the product. Dextrose is essential in fermented sausages because fermentation bacteria require a simple sugar to produce lactic acid. Dextrose is usually added at the 0.5 to 1 percent level (of the meat weight) in the formulation for fermented sausages. Sorbitol has been credited with reducing the charring of frankfurters on the grill.

Antioxidants

Several compounds may be added to fresh and dry sausage to retard the development of oxidative rancidity. The most common of these compounds are BHA (butylated hydroxyanisole), BHT (butylated hydroxytoluene), and propyl gallate. In fresh sausage, the allowable level that can be used is 0.01 percent of the fat content for any one above or 0.02 percent for any two or more used in combination. In dry sausage, the allowable level is 0.003 percent of the meat block weight for any one or 0.006 percent for two or more used in combination. Regulations of allowable levels are based upon fat content of fresh sausages and total meat block weight for dry sausages.

Phosphates

On March 12, 1982, the Food Safety and Inspection Service (FSIS) approved the *direct addition* of phosphates to cooked sausage products. Prior to 1982, phosphates could be present in cooked sausage products, but only when added through cured pork trimmings which had been cured with phosphate in the brine.

Phosphates function to improve the water-binding ability of the meat, solubilize proteins, act as an antioxidant, and help protect and stabilize the flavor and color of the finished product. Through the use of phosphates, processors can attain a longer product shelf life and improve the smokehouse yield.

Phosphates are approved at a level not to exceed 0.5 percent in the finished product. There is about 0.1 percent naturally occurring phosphate in muscle tissue which must be considered in the analysis when phosphates are added. The following phosphates are approved for use: disodium phosphate, monosodium phosphate, sodium meta-phosphate, sodium polyphosphate, sodium tri-polyphosphate, sodium pyrophosphate, sodium acid pyrophosphate, dipotassium phosphate, monopotassium phosphate, potassium tripolyphosphate, and potassium pyrophosphate.

Mold Inhibitors

Mold growth is a common problem in the production of dry sausage. To inhibit mold growth, sausages may be dipped in a 2.5 percent solution of potassium sorbate or a 3.5 percent solution of propylparaben (propyl-p-hydroxy benzoate).

Glucono Delta Lactone (GDL)

This compound serves as a cure accelerator and also produces an acid tang similar to that produced by natural fermentation. It is used at the 0.5 percent level in some fermented sausages, with the exception of Genoa salami, for which use is permitted at the 1 percent level.

Extenders

A number of ingredients are used as extenders to reduce the cost of the product and to provide certain functional properties related to product bind, texture, and flavor. The extenders are discussed in detail in the following section of this chapter. Examples of extenders include nonfat dry milk, sodium caseinate, cereal flours, and soy protein, which comes in a variety of forms.

Spices, Seasonings, and Flavorings

There are many different spices, seasonings, and flavorings used in sausage products. Their use levels are primarily dictated by product-identity standards and personal flavor preferences. Combining different levels of the various spices, seasonings, and flavorings available creates infinite variety in our sausage supply. This topic is sufficiently vast that it merits its own section. For information on individual spices, refer to that section of this chapter.

By definition, *spices* refer to any aromatic vegetable substances in whole, broken, or ground form whose function as ingredients in foods is seasoning rather than nutrition and from which none of the flavoring principle has been removed. *Flavorings* refer to extractives containing flavoring constituents from fruits, vegetables, herbs, roots, meat, seafood, poultry, eggs, and dairy products whose function is flavoring rather than nutrition. *Seasoning* is a comprehensive term that can be applied to any ingredient which improves the flavor of the product in question.

Spices vary greatly in composition, but the aromatic and pungent properties which render them valuable probably reside in volatile oils, resins, or oleoresins. The active principles are usually very small proportions of the spice as a whole. Some success has been achieved in separating this active principle and using it for flavor instead of using the spice from which it is obtained. In addition to having an agreeable effect upon the organs of taste and smell, the principles of spices stimulate the flow of digestive juices (gastric and salivary).

Spices may be added as natural spice or as spice extracts. In the latter case, they must be labeled as "flavorings." Actually, the spice extracts offer the advantages of being easier to control for flavor intensity and of not being visible in the product as spice particles would be. Spice extracts result in fewer microbial contamination problems in the product because the oil-based extracts do not provide a source of spoilage microorganisms like their natural counterparts. The extracts are also easier to store because they are less bulky than natural spices.

Table 20-3 is a handy reference chart of major spices available to food processors. Data include physical description, flavor characteristics, and examples of use.

Extenders and Binders

Certain sausage products may contain extenders and binders such as nonfat dry milk, dried milk, dried whey, reduced lactose whey, whey protein concentrate, calcium lactate, cereal flours, soy flour, soy protein concentrate, isolated soy protein, and/or vegetable starch at a level up to 3.5 percent of the finished product (alone or in combination). Isolated soy protein is an exception to the regulations and may be used up to a level of 2 percent in the finished product. If these ingredients are used in the formulation, their presence must be reflected in the product name label, for example, "frankfurter, cereal added" or "bologna, soy protein concentrate and nonfat dry milk added."

The following discussion of various extenders gives some insight on their functional properties and appropriate usage.

Milk Powder

Nonfat dry milk and similar dried products of milk origin function primarily as extenders; however, some feel there is an improvement in product flavor—probably due to a sweetening effect. The calcium-reduced form of nonfat dry milk is most commonly used, since high-calcium levels interfere with protein sol-

Table 20-3. Spices at a Glance[1]

Spice and Sources	Description	Flavor	Typical Uses[2]

Allspice
  Jamaica
  Honduras
  Mexico
	Reddish-brown pimento berries, nearly globular; ⅛- to 5⁄16-in. diameter.
	Available: whole and ground.
		Pungent, clovelike odor and taste.	Bologna, pork sausage, frankfurters, hamburger, mince meat, potato sausage, headcheese, and many other meat food products.

Anise seed
  Spain
  Netherlands
  Mexico
	Greenish-brown, ovoid-shaped seeds; 3⁄16 in. long.
	Available: whole and ground.
		Pleasant, licorice-like odor and taste.	Dry sausage: mortadella and pepperoni.

Basil
  California
  Hungary
  France
  Yugoslavia
	As marketed, small bits of green leaves.
	Available: whole and ground.
		Aromatic; faintly anise-like, mildly pungent taste.	Pizza sausage, certain poultry products.

Bay leaves
  Turkey
  Portugal
	Elliptical leaves; up to 3 in. long; deep green upper surface, paler underneath.
	Available: whole and ground.
		Fragrant, sweetly aromatic; slightly bitter taste.	Pickling spice for corned beef, beef tongue, lamb tongue, pork tongue, and pigs' feet.

Caraway seed
  Netherlands
  Poland
	Curved, tapered brown seeds; up to ¼ in. long.
	Available: whole.
		Characteristic odor; warm, slightly sharp taste.	Polish sausage.

Cardamon seed
  Guatemala
  India
	Small, angular, reddish-brown seeds; often marketed in their pods—greenish- or buff-colored (blanched).
	Available: whole, decorticated, and ground.
		Pleasantly fragrant odor; warm, slightly sharp taste.	Bologna, frankfurters, similar products.

Celery flakes California	Medium to dark green flakes; about ⅜-in. diameter. Available: flakes, granulated, and powdered.	Sweet, strong, typical celery odor and taste.	Chicken and turkey products.

Celery seed
  India
  France
	Grayish-brown seeds; up to 1⁄16-in. diameter.
	Available: whole, ground, and as salt.
		Warm, slightly bitter celery odor and taste.	Beef stews, meat loaf, chicken and turkey products.

Chili powder California	Red to very dark red powder. Contains chili pepper, cumin, oregano, garlic, salt, and sometimes other spices.	Characteristic aromatic odor, with varying levels of heat or pungency.	Chili con carne, taco meat, some Spanish and Mexican sausages.

Cinnamon
  Indonesia
  Seychelles
  Taiwan
  Ceylon
	Tan to reddish-brown quills (stocks) of rolled bark, varying lengths.
	Available: whole and ground.
		Agreeably aromatic, with sweet, pungent taste.	Ham loaf, other pork loaves, pastrami rub, and sometimes in bologna, mortadella, and blood sausage.

(Continued)

Table 20-3 (Continued)

Spice and Sources	Description	Flavor	Typical Uses[2]

Cloves
    Madagascar
    Indonesia
    Tanzania
	Reddish-brown; ½ to ¾ in. long.
	Available: whole and ground.
		Strong, pungent, sweet odor and taste.	Bologna, frankfurters, headcheese, liver sausage, corned beef, and pastrami. Whole cloves are often stuck into hams when the hams are baked.

Coriander seed
    Morocco
    Rumania
    Argentina
	Yellowish-brown nearly globular seeds; ⅛- to 3/16- in. diameter.
	Available: whole and ground.
		Distinctively fragrant; lemon-like taste.	Frankfurters, bologna, knackwurst, Polish sausage, and many other cooked sausages.

Cumin seed
    Iran
    India
    Lebanon
	Yellowish-brown elongated oval seeds; ⅛- to ¼-in. diameter.
	Available: whole and ground.
		Strong, aromatic, somewhat bitter.	Chorizos, chili con carne, and other Mexican and Italian sausages.
			Its principal use is in making curry powder.

Dill seed India	Light brown oval seeds; 3/32 to 3/16 in. long. Available: whole and ground.	Clean, aromatic odor; warm, caraway-like taste.	Headcheese, souse, jellied tongue loaf, and similar products.

Fennel seed
    India
    Argentina
	Green to yellowish-brown oblong or oval seeds; 5/32 to 5/16 in. long.
	Available: whole and ground.
		Warm, sweet, anise-like odor and taste.	Italian sausages, pizza sausage, pizza salami.

Garlic, dried California	White material, ranging in standard particle size from powdered, granulated, ground, minced, chopped, large chopped, sliced, to large sliced.	Strong, characteristic odor; extremely pungent taste.	Polish sausage, most beef sausages, salamis; subtle amounts in bologna, frankfurters, and similar products.

Ginger
    Nigeria
    Sierra Leone
    Jamaica
	Irregularly shaped pieces ("hands") 2½ to 4 in. long; brownish- to buff-colored (when peeled and bleached).
	Available: whole, ground, and cracked.
		Pungent, spicy-sweet odor; clean, hot taste.	Pork sausage, frankfurters, knackwurst, and numerous other cooked sausages.

Mace
    Indonesia
    Grenada
	Flat, brittle pieces of lacy material, yellow to brownish-orange in color.
	Available: whole and ground.
		See nutmeg; but somewhat stronger, less delicate.	Bologna, mortadella, bratwurst, bockwurst, and many other sausages—both fresh and cooked.

Marjoram
    France
    Portugal
    Greece
    Rumania
	As marketed, small pieces of grayish-green leaves.
	Available: whole and ground.
		Warm, aromatic, pleasantly bitter, slightly camphoraceous.	Braunschweiger, liverwurst, headcheese, and Polish sausage.

(Continued)

Table 20-3 (Continued)

Spice and Sources	Description	Flavor	Typical Uses[2]

Mustard
  Denmark
  Canada
  U.K.
  U.S.
	Tiny, smooth, nearly globular seeds, yellowish- or reddish-brown. Available: whole and ground.	Yellow: no odor, but sharp, pungent taste when water is added. Brown: with water added, sharp, irritating odor; pungent taste.	Bologna, frankfurters, salami, summer sausage, and similar products.

Nutmeg
  Indonesia
  Grenada
	Large, brown, ovular seeds; up to 1¼ in. long. Available: whole and ground.	Characteristic sweet, warm odor and taste.	Frankfurters, bologna, knackwurst, minced ham sausages, liver sausage, and headcheese.

Onion, dried California	White material ranging in particle size from powdered, granulated, ground, minced, chopped, large chopped, sliced, to large sliced.	Sweetly pungent onion odor and taste.	Braunschweiger, liver sausage, headcheese, and baked luncheon loaves, including Dutch loaf and old-fashioned loaf.

Oregano
  Greece
  Mexico
  Japan
	As marketed, small pieces of green leaves. Available: whole and ground.	Strong, pleasant, somewhat camphoraceous odor and taste.	Most Mexican and Spanish sausages, fresh Italian sausage; sometimes in frankfurters and bologna.

Paprika
  California
  Spain
  Bulgaria
  Morocco
	Powder, ranging in color from bright rich red to brick-red, depending on variety and handling.	Slightly sweet odor and taste; may have moderate bite.	Frankfurters, bologna, and many other cooked and smoked sausage products; also fresh Italian sausage.

Pepper, black
  Indonesia
  Brazil
  India
  Malaysia
	From green (unripe) pepper berries, brownish-black, wrinkled berries; up to ⅛-in. diameter. Available: whole, ground, cracked, and decorticated.	Characteristic, penetrating odor; hot, biting taste.	Most used of any spice. Frankfurters, bologna, pork sausage, summer sausage, salamis, liver sausage, loaf products, and *most* other sausages.

Pepper, red
  Japan
  Mexico
  Turkey
  U.S.
	Elongated and oblate-shaped red pods of varying sizes; from ⅜ to ½ in., depending on variety. Available: whole and ground.	Characteristic odor, with heat levels mildly to intensely pungent.	Chorizos, smoked country sausage, Italian sausage, pepperoni, fresh pork sausage, and many others.

Pepper, white
  Indonesia
  Brazil
  Malaysia
	From ripe, yellowish-gray pepper seeds; up to 3⁄32-in. diameter. Available: whole and ground.	Like black pepper, but less pungent.	Used where black pepper specks are not desired; *e.g.,* pork sausage and deviled ham.

(Continued)

Table 20-3 (Continued)

Spice and Sources	Description	Flavor	Typical Uses[2]
Rosemary France Spain Portugal California	Bits of pine, needle-like green leaves. Available: whole and ground.	Agreeable, aromatic odor; fresh, bittersweet taste. Somewhat like sage in flavor.	Chicken stews and a few other poultry products.
Saffron Spain Portugal	Orange and yellow strands; approximately ½ to ¾ in. long. Available: whole and ground.	Strong, somewhat medicinal odor; bitter taste.	The most expensive of all spices. Used primarily for color, but in very few sausages.
Sage Yugoslavia Albania	Oblanceolate-shaped leaves, grayish-green; about 3 in. long. Available: whole, cut, rubbed, ground.	Highly aromatic, with strong, warm, slightly bitter taste.	Pork sausage, pizza sausage, breakfast sausage, and old-fashioned loaf.
Savory France Spain	As marketed, bits of dried, greenish-brown leaves. Available: whole and ground.	Fragrant, aromatic odor.	Used primarily in pork sausage, but is good in many sausages.
Thyme Spain France	As marketed, bits of gray to greenish-brown leaves. Available: whole and ground.	Fragrant, aromatic odor; warm, quite pungent taste.	Pork sausage, liver sausage products, headcheese, and bockwurst.
Turmeric India Jamaica	Fibrous roots, orange-yellow in color; ⅓ in. long. Available: ground.	Characteristic odor, reminiscent of pepper; slightly bitter taste.	Used more for color than flavor. Constituent of curry powder.

[1]Adapted from information made available from the American Spice Trade Ass., New York City.
[2]The seasoning in most sausage and meat products is a blend of several spices.

ubility. Dried milk powders and whey powders exhibit limited binding properties.

Cereal Flours

Cereal flours are composed principally of starch; however, they may serve the function of both extender and binder. Their function varies with the source, which may be wheat, rice, oats, corn, etc. The cereal flours are generally added to the lower-quality products for economic reasons. Some of the cereal flours are reported to improve binding qualities, cooking yields, and slicing characteristics.

Soy Flour, Grits, and Texturized Soy Protein

These soy protein ingredients contain 50 percent protein and are used to boost the protein content and help bind water. Flour, grits, and texturized soy differ primarily in their particle size and texture. Soy grits have a larger particle size than soy flour. Texturized soy is very similar to grits except that the texture is changed to more closely duplicate the texture of ground meat. The principal use of soy flour is in nonspecific loaf products, whereas grits are often used in pizza toppings, chili, and sloppy Joe mixes. Texturized soy protein is commonly used in meat patties, meat loaf, and similar items.

Soy Protein Concentrate

This is a 70 percent protein product that is available as a flour or in a coarse granular form similar to grits. Soy concentrates are generally used in an emulsion-type sausage in which they function to bind water.

Soy Protein Isolate

This 90 percent protein product is useful as both a binder and an emulsifier. It is the only soy product that functions similarly to meat in forming an emulsion. The soy protein isolate should not be considered equal in functional quality to the contractile meat proteins (e.g., actomyosin), but nonetheless is useful in creating more stable emulsions in marginal formulations.

One problem sometimes encountered with soy concentrates, grits, or flours is the development of a beany flavor. This usually occurs after storage of the product and is more pronounced when higher concentrations are used. Modern soy processing technology has substantially reduced the beany-flavor problem.

PRINCIPLES OF SAUSAGE PRODUCTION

The topic of sausage production is such a vast subject that it cannot be adequately covered in a single chapter. Individuals who seek additional information on the principles and modern practices of sausage production should refer to the book entitled *Sausage and Processed Meats Manufacturing*, written by Professor Robert E. Rust of Iowa State University.[5] This excellent book is probably the most complete compilation of information related to sausage manufacturing practices ever published in this country.

Fundamental processing methods which include particle reduction (chopping, grinding, emulsifying, or flaking), mixing or blending, stuffing, smoking, peeling, and packaging are important in the successful production of sausage. These manufacturing processes are discussed throughout the chapter.

Manufacturing Procedures for Fresh Sausage

Pork sausage comprises the vast majority of fresh sausage produced

---

[5]Robert E. Rust. *Sausage and Processed Meats Manufacturing.* 1975. AMI Center for Continuing Education, American Meat Institute, Chicago.

throughout the world, whether it be in bulk, link, or patty form. Two sources of raw materials are used to produce fresh pork sausage, trimmings from pork-cutting operations and boneless pork derived from boning entire hog carcasses (usually sows). By-products or variety meats may be used in fresh sausage with proper labeling. Most recently, fresh turkey sausage has emerged on the marketplace and appears to have established a steadily growing market share. Fresh turkey sausage is generally made from boneless drumsticks, mechanically separated turkey meat, and/or miscellaneous trimmings from the boning lines. Other common fresh sausages include fresh thuringer, fresh bratwurst, fresh bockwurst, and country-style sausage.

The term *pork sausage* is generally implied to mean the ground and seasoned fresh pork product. There are, however, several kinds or ''styles'' of fresh pork sausage which differ in texture, seasoning, and meat content.

Country-Style

Country-style usually contains from 10 to 20 percent of beef ground with the fresh pork; it is coarsely ground, using the ³⁄₁₆-inch plate, and does not contain sage as a seasoning. It is stuffed into hog casings or regenerated collagen casings of different sizes and is unlinked. It is also sold loose (unstuffed).

The term *farm* or *country* cannot be used on labels in connection with products unless such products are actually prepared on the farm or in the country. However, if the product is prepared in the same way as on the farm or in the country, these terms, if qualified by the word *style* in the same size and style of lettering, may be used. Further, the term *farm* may be used as part of a brand designation when qualified by the word *brand* in the same size and style of lettering and followed with a statement identifying the locality in which the product is prepared. Sausage containing cereal grains, soy or vegetable proteins cannot be labeled ''farm-style'' or ''country-style.''

Breakfast-Style

Breakfast-style is an all-pork sausage that is finely ground and seasoned with sage, salt, and pepper. It is stuffed into narrow or medium sheep casings, narrow hog casings, or regenerated collagen casings which are then linked to make the various-sized sausages.

Whole Hog Sausage

Whole hog sausage is sausage prepared with fresh and/or frozen meat from swine in such proportions as are normal to a single animal and may be seasoned with condimental substances as permitted in any sausage product. It cannot be made with any lot of product which, in the aggregate, contains more than 50 percent trimmable fat, that is, fat which can be removed by thorough practicable trimming and sorting. To facilitate chopping or mixing, water or ice may be used in an amount not to exceed 3 percent of the total ingredients used.

Modern whole hog sausage is made from 240- to 250-pound butcher hogs. Depending on the fatness (grade) of the hogs, leaf lard may be pulled out of the raw material for sausage or, conversely, certain lean muscles (ham, loin, tenderloin) may be pulled out if the hogs are too lean. Of course, the carcasses must be boned out before being ground into sausage.

The Code of Federal Regulations (CFR 319.141) stipulates that fresh pork sausage may not contain more than 50 percent fat. However, optimum palatability appears to be found around 30 percent fat. The ratio of fat to lean has a pronounced effect upon flavor, texture, and tenderness of the sausage. Up to a point, a fatter pork sausage will exhibit more desirable palatability characteristics; however, consumer preferences are creating a greater demand for leaner sausage. Producing leaner pork sausage requires quality raw materials. Pork blade meat trimmings work very well. The finished product should have a bright red appearance of the lean pork mottled with the white of the fat. Good lean and fat particle distinction are paramount in a high-quality fresh pork sausage.

Grinding

The first step in fresh sausage processing is grinding. Prior to grinding, the meat should be chilled to 32° F to minimize smearing of the product. The pork trimmings are ground through a 3/16-inch grinder plate with a bone-chip collection attachment, if available. Some processors prefer to grind the lean pork through a 1/8-inch plate and the fatter pork through a 3/16-inch plate before blending together. It is extremely important that grinder knives always be kept sharp and that meat temperature be maintained at close to freezing. Grinding prepares the meat for easy mixing with the spices.

Fresh sausage prepared in a silent chopper usually has a slightly brighter color that is retained longer; however, the chopper has no effective means of screening out bone or cartilage chips that were missed in the boning operation.

Mixing

The ground pork trimmings are placed in a mixer with salt, pepper, and sage (or any other spices which may be used) and are thoroughly mixed with the seasoning ingredients as well as any anti-oxidants which may be used.

An alternative method preferred by many processors is to grind the pork through a coarse plate (1/2 to 3/4 inch) and then mix in a blender with the seasonings. After thorough mixing, the sausage is reground through a 3/16- or 1/8-inch grinder plate. A combination mixer/grinder lends itself to this method very well and minimizes the cleaning of equipment when the job is done.

Excessive mixing after the salt has been added can cause too much extraction of the soluble proteins and can result in a tough, rubbery end product. If fat begins to adhere to the inside of the mixer, it is a sure sign of overmixing.

Stuffing

The majority of the pork sausage produced is stuffed into 1- or 2-pound chubs (bulk-style) in opaque plastic bags. When pork sausage is stuffed into casings and linked, the appearance is much more critical. Many processors prefer to stuff at 28° to 30° F to minimize smearing and to get a good "bloom" on the product. Pork sausage is generally stuffed into narrow or medium sheep casings, narrow pork casings, or collagen casings. Natural casings should be flushed with water prior to use, which also enables them to be placed on the stuffing horn with greater ease. During stuffing, the casing should be held firmly between the fingers on the stuffing horn, with sufficient pressure applied so that the casing fills to maximum capacity (or slightly less) with no air pockets.

Seasoning Pork Sausage

The farm practice of tasting raw pork sausage to determine the proper amount of seasoning caters to only one person's desires and can be detrimental to one's health. The most satisfactory way to get the seasoning nearly right for the greatest number of people is to weigh the meat and add the following mixture for each 100 pounds of ground pork:

> 28 to 30 oz. table salt
> 6 oz. black pepper
> 2 oz. ground sage

☞   NOTE: Small quantities are more accurately weighed in grams (1 oz.   ☜
     = 28 g).

This imparts an excellent flavor, and the different batches of sausage will always be seasoned the same. Many prefer butcher's pepper (coarsely ground black pepper) instead of table pepper.

Those desiring a more highly seasoned sausage might try the following formula:

Highly Seasoned Pork Sausage

2 lb. salt	⅜ oz. (10 g) Jamaica ginger
6 oz. dextrose (corn sugar)	⅜ oz. (10 g) ground mace
⅜ oz. (10 g) red pepper	⅜ oz. (10 g) thyme
3 oz. white pepper	2 oz. rubbed sage

Add and mix with each 100 pounds of ground pork.

Bratwurst

100 lb. 70% to 80% lean pork trimmings	3.2 oz. (90 g) black pepper
	1.6 oz. (45 g) mace
1.9 lb. salt	1.6 oz. (45 g) coriander
1.9 lb. nonfat dry milk	1.6 oz. (45 g) hot mustard
1.9 lb. water	

Grind pork through a ⅜-inch grinder plate, mix ingredients uniformly, re-grind through a ³⁄₁₆-inch plate, and stuff into hog casings.

Manufacturing Procedures for Cooked and Smoked Sausages

There are many varieties of cooked and smoked sausages, each requiring slightly different processing procedures. Products such as frankfurters, bologna, and many of the loaf products are made from finely comminuted meat emulsions. Other products, such as smoked pork sausage, Polish sausage, and smoked thuringer, are coarser cut in texture and do not require the use of a silent chopper or an emulsion mill. Since frankfurters and bologna are the most popular of all sausages, their production will be discussed first.

Frankfurters and Bologna

The manufacture of frankfurters (wieners or hot dogs) is more complex with more critical steps than the production of fresh sausage. For optimal stability of emulsion products, a lean-meat source and a fat-meat source should be handled separately.

GRINDING

The first operation in the production of frankfurters is grinding. Grinding procedures vary with processors. Some prefer to grind the lean trimmings of the formula through a ¾- to 1-inch plate and depend upon the silent cutter to achieve the necessary fineness before stuffing. Others prefer to grind the lean meat through a finer plate (*e.g.*, ⅛ to ¼ inch) and use an emulsion mill after blending to achieve the desired emulsion texture. The first procedure increases the capacity of the grinder, while the latter procedure is more readily adapted to continuous processing lines. Usually the lean meats (generally beef) are chopped or ground first and to a finer consistency than the fat meats (generally pork).

CHOPPING

After the lean meat has been ground, it is placed in a bowl chopper (silent cutter) and the salt, curing ingredients, and one half of the water (ice) are added. The temperature of the meat will rise during chopping due to friction of the knives passing through the meat. The lean meat should be chopped to a temperature of 42° to 45° F before the remaining water, fat meat, and spices are added. Chopping should continue until a temperature of 62 to 63° F is reached. For optimal stability of the final product, temperature is more critical than chopping time. When an emulsion mill is used following chopping, the meat should be chopped to a temperature of 52° to 55° F. In cases where "hard" fats are used as the fat-meat source (*e.g.,* beef or mutton) the end-point chopping temperature should be 2° to 3° F higher (65° to 66° F) in order to make the fat plastic enough to produce a good emulsion.

Other ingredients, particularly nonmeat proteins such as cereal flours and nonfat dry milk, are added with the last ingredients, since they readily absorb water and make it less available for soluble protein extraction. It is important that *half* the water (generally in the form of ice) be added initially to produce a brine concentration of sufficient strength to solubilize the meat proteins, which are essential for a stable emulsion. If nonskeletal meats which are high in collagen (such as tripe) are used in the formulation, they should be added with the fat-meat portion.

BLENDING

Most larger meat processors employ a practice known as "preblending" in the production of frankfurters and other emulsion products. Preblending refers to the blending of the salt, cure, and a portion of the water with coarse-ground meats prior to use. Preblends are generally held 24 hours before chopping. The reasons and advantages for preblending are discussed later in this chapter (see page 692). Some processors use a grinder, a mixer, and an emulsion mill to produce emulsion products—skipping the use of the bowl chopper. It is necessary to grind both the lean and the fat trim finer (⅛ inch and ¼ inch respectively) if this procedure is followed. The lean meat is then blended with the salt, cure, and one half of the water for several minutes before the remaining ingredients are added. The meat batter is then emulsified in an emulsion mill.

STUFFING AND LINKING

The chopped meat is placed in a stuffer which may be the "piston type" or a "continuous vacuum stuffer." Frankfurters and similar products are stuffed into cellulose casings (for skinless wieners) or sheep, hog, or edible collagen casings of the appropriate diameter. Bologna is generally stuffed into 3½- to 4-inch diameter fibrous casings (except for ring bologna which is stuffed into 1¼- to 1¾-inch cellulose, collagen, or natural beef round casings [small intestine of beef]).

Cleaned and salted hog or sheep casings should be flushed with water prior to being placed upon the stuffing horn. The water should be expelled from the casing by being stripped between the operator's fingers before it is stuffed. As the casing fills with meat, it is allowed to work off the stuffing horn. It should be held firmly between the fingers of the operator to insure that it is filled to maximum capacity and to eliminate air pockets.

There are available linking machines that link natural casings, collagen casings, and cellulose casings to produce a uniform link portion. Otherwise, natural and collagen casings may be linked manually, using one of two methods. The first method known as *braiding* is accomplished by taking a 5- to 6-foot section of stuffed product and folding it in half. Pinch the casing about 5 inches from the folded end on both halves. Fold one side over the pinched spot, and loop the other end through the loop that is created. Repeat this process to the ends of the casing (see Figure 20-2). The second hand-linking method is accomplished by twisting every other link in opposite directions (see Figure 20-3). More care is required in hanging the product on smokehouse trucks using this second method, as the links can become untwisted. However, this latter method is faster than the former method.

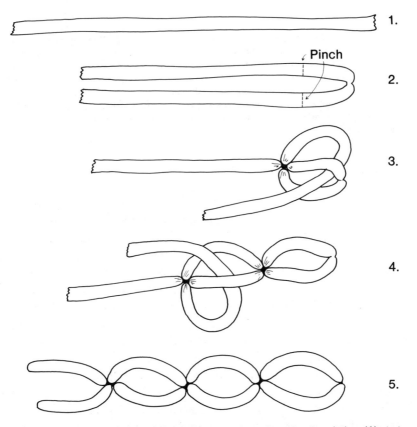

Fig. 20-2. The "braiding" method of hand linking. Stepwise description: (1) start with a 5- to 6-foot section of stuffed casing; (2) fold it in half, and pinch the casing on both halves about 5 inches from the fold; (3) fold one side over the pinched area, and work the end back through the loop that has just been formed; and (4) repeat the process until the ends of the casing have been reached (5).

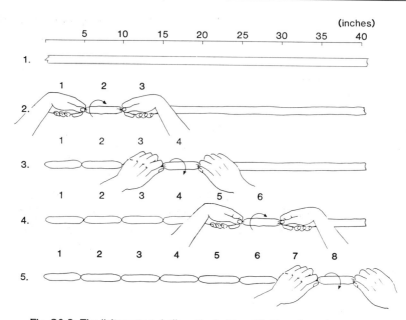

Fig. 20-3. The "alternate twist" method of hand linking. Stepwise description: (1) the casing to be linked can be virtually any length; (2) start on one end, and pinch the casing between thumb and forefinger at points 5 and 10 inches from the end, using both hands, twisting the link between your hands away from yourself; (3) continue in the same fashion, pinching the casing at points 15 and 20 inches from the same end, except this time twist the link between your hands towards yourself; and (4 and 5) continue twisting every other link in the opposite direction as the preceding twist until the end of the casing is reached.

SMOKING THE SAUSAGE

Smoke was originally applied to meat products as a form of preservation. However, today smoking serves two primary functions: it provides flavor and it acts as a surface-coloring agent. Two forms of smoke, natural wood smoke and liquid smoke, are applied to products. The smoke should be applied to a surface that is not wet, but rather sticky or slightly moist. A short-drying cycle is generally completed in the smokehouse before application of smoke is initiated. This drying cycle conditions the meat surface for optimal adherence of smoke particles to the product. The length of time a product is smoked is dependent upon the smoke density inside the smokehouse, the surface condition of the product, and the air velocity inside the smoking chamber.

In the modern closed system, thermal processing chambers, a good smoke application can be achieved in 30 minutes. Liquid smoke can be uniformly applied in three to five minutes by a pulsing atomization system.

If the product surface is too wet during smoking, an unattractive streaking of smoke will occur on the product. If the product surface is too dry during smoking, little smoke will adhere to the surface, and the sausage will be pale in color. It is also important that the product be loaded onto a smokehouse truck evenly, with product not touching other product. This is essential for uniformity of surface color.

THERMAL PROCESSING

Thermal processing is the last critical step in producing a stable emulsion product. The highest-quality meat formulations can produce an inferior product when an improper thermal processing schedule is followed. The modern thermal processing chamber serves many functions including (1) drying, (2) smoking, (3) heating, (4) cooling, and (5) adding humidity to the chamber environment. Each of these functions can be carefully controlled independently.

Processed meat products may be processed in "batch-type" or "continuous" thermal processing chambers. The continuous smokehouses are primarily used for small-diameter sausages with continuous frankfurter processing lines capable of producing in excess of 12,000 pounds of franks per hour.

In a typical system, the frankfurter emulsion is stuffed into a cellulose casing and linked with a high-speed mechanical linker. The links are then treated with either one or a combination of coagulants (natural smoke, vinegar, citric acid, or malic acid mixed with liquid smoke). The coagulants precipitate the meat proteins at the inner surface of the casing (to aid peelability) and to contribute to color and flavor.

After coagulation, franks are cooked in high-velocity, high–relative humidity cooking zones in a very large cooker. The cooked frankfurters are then chilled, peeled, and packaged. After the franks are stuffed and linked, the remaining processing time is usually 60 to 80 minutes. Chilling can be accomplished in about 10 minutes with a 60 percent brine solution refrigerated at about 28° F. Product is chilled to 38° to 40° F.

High-speed peeling machines can peel franks at a rate in excess of 5,000 pounds per hour with one operator. A knife edge slits the moistened casing as the strand of frankfurters is moved through the machine. An air jet assists in blowing the casing off the frankfurters (Figure 20-4).

To get the characteristic cured-meat color in such a short processing time, ascorbate is essential, and the product must remain additional time in the oven *after* the internal meat temperature has reached 150° to 155° F. This extra heating to develop color takes about 20 minutes in a high-velocity 200° F oven.

In batch systems, the frankfurters are generally processed in a graduated, stepwise processing schedule. A typical smokehouse schedule for frankfurters is listed below:

Time	Temperature	Percent Relative Humidity	Function	Damper
30 min.	125° F	25	drying	open
1 hr.	140° F	35	smoking	closed
1 hr.	165° F	35	cooking	closed
10 min.	180° F	100	steam cooking	closed

Fig. 20-4. After cooking and cooling, the cellulose casing is removed by this high-speed peeler to make "skinless wieners." A closely adjusted knife slits the casing, and an air jet blows the casing off the wieners.

Formulations for Cooked and Smoked Sausages

Quality Deli Franks

50 lb. fresh regular pork trimmings
   (50% lean)
20 lb. fresh veal trimmings
30 lb. bull or cow trimmings
   (90% lean)
26 lb. crushed ice
2 lb. 4 oz. salt
5 oz. dextrose
4 oz. white pepper

4 oz. paprika
¾ oz. (22 g) mace
¾ oz. (22 g) ginger
½ oz. (14 g) allspice
1½ oz. onion powder
⅞ oz. (24 g) monosodium glutamate
2 oz. phosphate
4 oz. certified cure (6% nitrite)

☞   NOTE: Small quantities are more accurately weighed in grams (1 oz.   ✎
= 28 g).

Grind bull or cow trimmings through a ¼-inch grinder plate; veal and pork trimmings through a ⅜-inch grinder plate. Place bull (or cow) trimmings in a bowl chopper, and add one half (10 pounds) of the moisture (ice), the salt, and the cure. Chop to a temperature of 42° to 45° F. Add the remaining meat ingredients, ice, and spices to the chopper while chopping continues. Chop the emulsion to 62° to 63° F. Stuff into narrow or medium sheep casings, and link at 5-inch intervals.

Bologna

40 lb. regular pork trimmings (50% lean)	1 oz. coriander
60 lb. cow meat (85% to 90% lean)	¾ oz. (22 g) mace
20 lb. crushed ice	¾ oz. (22 g) ginger
2 lb. salt	½ oz. (14 g) cloves
5 oz. dextrose	1 oz. onion powder
3 lb. nonfat dry milk	¼ oz. (7 g) garlic powder
4 oz. white pepper	1 oz. monosodium glutamate
2 oz. paprika	2 oz. phosphate
	4 oz. certified cure (6% nitrite)

☞ NOTE: Small quantities are more accurately weighed in grams (1 oz. ✎ = 28 g).

Follow the procedures for frankfurters, except stuff into 3½- to 4-inch fibrous casings. If ring bologna is to be made, stuff into 1¼- to 1¾-inch collagen casings or beef rounds 18 inches in length, and tie into a loop.

Polish Sausage

20 lb. regular pork trimmings (50% lean)	3 oz. black pepper
50 lb. special pork trimmings (80% lean)	2 oz. white pepper
	4 oz. paprika
30 lb. boneless cow trimmings (85% lean)	¾ oz. (22 g) mace
	¾ oz. (22 g) allspice
15 lb. cold water (crushed ice if a chopper is used)	1 oz. onion powder
	½ oz. (14 g) garlic powder
5 oz. dextrose	1 oz. monosodium glutamate
2 lb. 8 oz. salt	4 oz. certified cure (6% nitrite)

☞ NOTE: Small quantities are more accurately weighed in grams (1 oz. ✎ = 28 g).

Grind lean beef and pork trimmings through a ¼-inch grinder plate; regular pork trimmings through a ⅜-inch grinder plate. Add lean meat to blender with the salt, cure, and water. Mix until water is absorbed by the meat. Add remaining meat, dextrose, and spices to blender, and mix for 2 minutes. Regrind through a ⅛- or ³⁄₁₆-inch grinder plate, and stuff into hog casings. Linking should be done at 5-inch intervals.

Smoked (Country-Style) Pork Sausage

100 lb. lean pork (20% to 25% fat)	4 oz. red pepper
15 lb. water (ice if chopper is used)	4 oz. paprika
2 lb. 8 oz. salt	2 oz. coriander
5 oz. dextrose	1 oz. nutmeg
4 oz. certified cure (6% nitrite)	1 oz. sage

☞   NOTE: Small quantities are more accurately weighed in grams (1 oz.   ✑
= 28 g).

Grind meat through a ⅜-inch grinder plate. Add salt, cure, and water, and blend until water is absorbed by the meat. Add remaining ingredients, and blend until adequate tackiness, or bind, is achieved (usually two additional minutes). Regrind sausage through a fine (⅛- or ³⁄₁₆-inch) grinder plate, and stuff into hog casings. Stuffed sausage should be coiled around smokesticks in 12- to 18-inch loops. Do not link.

Braunschweiger

40 lb. pork liver (semi-frozen)	8 oz. white pepper
20 lb. special pork trimmings	2 oz. marjoram
(80% lean)	2 oz. nutmeg
10 lb. bacon end pieces	2 oz. ginger
30 lb. pork jowls	1 oz. allspice
2 lb. 8 oz. salt	1 oz. cloves
3 lb. 8 oz. nonfat dry milk	⅞ oz. (24 g) sodium erythorbate
8 oz. dextrose	4 oz. certified cure (6% nitrite)
4 lb. peeled onions (or 3-4 oz.	½ oz. (14 g) liquid smoke (optional)
onion powder)	

☞   NOTE: Small quantities are more accurately weighed in grams (1 oz.   ✑
= 28 g).

Grind the liver, pork jowls, bacon ends, and onions separately through a ⅜-inch grinder plate. Place ground liver, salt, and cure in chopper, and chop until bubbles appear on surface (usually one to three minutes). Add ground pork jowls, bacon ends, and spices, and continue to chop until temperature reaches 50° F. At approximately 50° F, the nonfat milk should be added, and the final chopping temperature should not exceed 65° F. Stuff into moistureproof casings (about 3 inches in diameter) or sewed hog bungs. Steam or cook in hot water (180° F) until internal temperature reaches 155° F. Immediately chill in an ice bath to 40° F. If the product is to be smoked, use a cool smoke *after* the product has been chilled.

Principles and Manufacturing Practices
for Dry and Semi-dry Sausages

In modern practices, the dry and semi-dry sausages are the most difficult and time consuming to produce. Ironically, these types of sausages are thought to have been the first produced in early history, dating back to the Babylonian culture around 1500 B.C. (approximately 3,500 years ago).

The key to successful production of dry and semi-dry sausages hinges on *controlled bacterial fermentation* of the sausage during processing. This fermentation was traditionally accomplished by chance contaminants that originated from the meat itself or from the sausage production equipment. As a result, successful products were often achieved in one plant, but not in another, when apparently identical procedures were followed. Today, most of the dry and semi-dry sausages are produced using a starter culture which inoculates the sausage batch with desirable lactic acid–producing bacteria. These bacteria function to lower the pH of the sausage (by producing lactic acid), giving the product its characteristic tangy flavor. Also, reducing the pH of the sausage creates an unfavorable environment for spoilage organisms, thus aiding in the preservation of the product.

Bacterial Fermentation

Fermentation of the product is achieved by the use of one of three methods: starter cultures, backslop, or natural fermentation.

STARTER CULTURES

The use of starter cultures is by far the most common fermentation method used in this country. Starter cultures are available from commercial sources in two forms: as a frozen concentrate or in a lyophilized (freeze-dried) form. Both forms have been demonstrated to produce acceptable results. Starter cultures typically contain a blend of two or more different organisms and sometimes different strains of the same organism. The organisms most commonly used belong to the genera *Pediococcus, Micrococcus*, and *Lactobacillus*. Specific species include: *P. cerevisiae, P. acidilactici, M. aurantiacus*, and *L. plantarum*, the last also occurring as the most common organism in natural fermentation methods.

Some dry sausage processors use a custom-developed starter culture whose microorganism content and ratio is a trade secret. This enables their product to have a unique flavor, different from other products on the market.

The primary advantages for using starter cultures over the other two fermentation methods are as follows:

- A more uniform product is produced from batch to batch.
- The risk of a batch undergoing spoilage is greatly reduced.

- Fermentation is more controlled, and the time to desired end point pH can be more readily predicted.
- The fermentation takes place faster.

BACKSLOP

The backslop fermentation method is the second most common of the three fermentation methods. Backslopping refers to setting aside a small quantity of one day's production for use the next day. Generally, processors who routinely practice backslopping will keep isolated cultures in frozen storage and then periodically revert to their original culture. However, day-to-day production will depend on the backslopping method. There is a greater chance of undesirable organisms predominating during the fermentation process, sometimes resulting in product spoilage.

NATURAL FERMENTATION

When there is no inoculation of a batch with microflora, it is called natural fermentation. This method is the least practiced, primarily because it produces the most inconsistent results. The end product may display little or no fermentation, it may spoil, or it may occasionally produce an outstanding batch. The success or failure of this method is dependent upon the types of bacteria that happen to be present within the meat and anything with which the meat comes in contact. If there are insufficient lactic acid–producing organisms present, spoilage organisms have the opportunity to grow and flourish, due to the lack of a pH decline.

In order for fermentation to occur, it is necessary for the bacteria to have food to grown on. Lactic acid–producing bacteria require simple sugars for optimal growth. The most common simple sugar used in sausage formulations is dextrose or corn sugar. Varying the level of dextrose in the formulation can alter the end point pH. More dextrose is used for lower pH products. Typically, dry and semi-dry sausages will contain 0.5 to 2 percent dextrose.

The actual fermentation process requires specific environmental conditions for optimal product quality. These conditions may be achieved in a greening room or in a smokehouse. Larger dry sausage processors typically use a greening room to hang product during fermentation. This is a room where temperature, humidity, and air velocity can be carefully controlled. Dry sausage is typically fermented at 70° to 75° F, 75 to 80 percent relative humidity (RH), and a slow, steady movement of air until the desired pH is attained (usually one to three days). Semi-dry sausages are normally fermented in a smokehouse for 8 to 12 hours at slightly higher temperatures, 85° to 100° F.

The final pH of fermented sausages will typically range from about 4.8 to 5.4, depending upon the tanginess desired and the product identity characteristics.

Drying and Cooking

Dry sausages such as the Italian salamis are not cooked and do not require refrigeration after manufacture (except for prolonged storage). These sausages are preserved by a low-moisture content of 30 to 40 percent and must have a moisture-protein ratio of 1.9:1 or less. Because these sausages are not generally cooked, the USDA has established procedures for certifying the use of dried products containing pork in the Code of Federal Regulations (CFR 318.10c). The sausage must be held at specified temperatures for specified times, which are determined by the sausage diameter. It is preferable to use certified trichinae-free trimmings when possible (see Chapter 3).

Carefully controlled environmental conditions are also critical during the drying process of dry sausage. If the product dries too fast, a problem known as *case hardening* will occur. Case hardening refers to a condition in which the outside of the sausage is hard and dry, inhibiting further moisture migration out of the interior part of the sausage, which is still wet. Sausages that have undergone case hardening are prone to internal spoilage by anaerobic bacteria. This problem is caused by too low humidity during drying. On the other hand, excessive humidity causes the product to dry too slowly and will often result in excessive mold growth, yeast growth, and bacterial surface slime on the product surface.

Theoretically, the drying rate at the surface of the sausage should be only slightly greater than that required to remove the moisture which migrates from the inside of the sausage. These conditions are approached by various combinations of temperature, humidity, and air velocity. There is little agreement as to the most effective combination of these three variables. As a general recommendation, the temperature of the drying room should be maintained between 45° and 55° F at a relative humidity of 70 to 72 percent. Air velocity in the room should be between 15 and 25 air changes per hour.

The drying time will vary considerably, depending on the diameter of the product and the size and type of product. Most sausages will require between 10 and 120 days of drying time.

Semi-dry sausages are generally cooked in a smokehouse to a temperature of 145° to 148° F, especially if they contain uncertified pork (pork that has not undergone certification as trichinae free). Semi-dry sausages will generally contain approximately 50 percent moisture and thus require refrigeration to prevent spoilage. Most semi-dry sausages are smoked, whereas many of the dry sausages are not.

Manufacturing Practices

Maximum particle definition is an important factor in the attractiveness of dry sausages. Particle definition refers to a clear demarcation between fat and lean particles. To achieve the best particle definition, the raw meat materials should be ground (or chopped) at very low temperatures (20 to 25° F). Mixing and unnecessary handling of the meat should be minimized as much as possible. The cold meat temperatures act to reduce protein extraction and to help reduce the deformation in particle shapes. This is important in the reduction of "smearing" of the product. Mixing time should be only enough to allow uniform distribution of spices and other ingredients.

The traditional dry sausages such as the Italian salamis are generally stuffed into hog middles, hog bungs, or sewed beef middles (see "Sausage Casings"). Some processors have switched to artificial casings due to the cost of natural casings. Artificial casings used in dry sausage are the "fibrous" type made from a special paper base impregnated with cellulose.

Dry sausages are the most expensive sausages to produce (and to buy), due to the lengthy production time and subsequent product shrinkage.

Semi-dry sausages such as thuringer, cervelat, and Lebanon bologna are generally fermented and cooked in a smokehouse and subsequently can be produced in 12 to 18 hours. This shorter process dramatically reduces production costs. Particle definition is less critical in these sausages, and a moderate bind is often desirable. Therefore, the meat in semi-dry sausages is not ground at subfreezing temperatures. The meat batter should also be mixed for a longer period of time in order for adequate protein extraction to occur. Without adequate protein extraction, the sausages may "fat out" during thermal processing. Fat out occurs when there is insufficient emulsifying protein available to entrap and stabilize fat particles. The fat will migrate during cooking to form "pockets" of fat inside the product or to the surface, forming undesirable "fat caps."

Formulations for Dry and Semi-dry Sausages

Genoa Salami (Dry)

40 lb. regular pork trimmings (50% lean, semi-frozen)	6 oz. white pepper
	1 oz. fresh garlic (peeled)
20 lb. pork shoulder butts (80% lean, semi-frozen)	⅛ oz. (3.5 g) cloves
	¼ oz. (7 g) nitrate
40 lb. boneless chuck	⅛ oz. (3.5 g) nitrite
3 lb. 6 oz. salt	starter culture (follow directions
6 oz. sugar (sucrose)	for recommended amount)
6 oz. dextrose	8 oz. red wine (optional)

☞   NOTE: Small quantities are more accurately weighed in grams (1 oz.   ☜
    = 28 g).

Grind pork trimmings and shoulder butts through a ½-inch grinder plate. Grind beef through a ¼-inch grinder plate. Place all ingredients except salt in a bowl chopper, and chop for one to two minutes. Add the salt during the last 30 seconds of chopping. Stuff sausage into hog bung casings, or if hog bung casings are unavailable, use fibrous casings of similar size. Place the sausage in a greening room for 36 hours (75° F and 75 percent RH). Genoa is traditionally wrapped tightly with No. 9 Italian hemp 2-ply cord. When wrapping, make a hitch of the twine every ½ inch the entire length of the sausage (see Figure 20-1). The salami is not smoked, but should be dried for 9 to 10 weeks in a drying room (50° F and 70 percent RH).

Pepperoni

50 lb. special pork trimmings
   (80% lean, semi-frozen)
20 lb. regular pork trimmings
   (50% lean, semi-frozen)
30 lb. boneless chuck (85% lean)
3 lb. 6 oz. salt
6 oz. sugar
6 oz. dextrose

8 oz. cayenne pepper
8 oz. pimento
4 oz. crushed red pepper
1 oz. whole anise seed
½ oz. (14 g) fresh garlic (peeled)
¼ oz. (7 g) nitrate
⅛ oz. (3.5 g) nitrite

☞   NOTE: Small quantities are more accurately weighed in grams (1 oz.   ☜
    = 28 g).

Grind all meat ingredients through a ⅛-inch grinder plate, and mix with non-meat ingredients in a blender for two minutes. The meat is then cured in 6-inch deep pans for 48 hours at a temperature of 40° F. After this curing period, remix meat in blender for approximately one minute, and stuff into narrow or medium hog casings. Break the casing into 21- to 22-inch lengths, making allowance for enough casing to loop over each end where casing is broken. Twist casings in the center to form twin links about 10 inches long. The pepperoni should be held in a greening room for 24 hours (75° F and 75 percent RH) before it is ready for drying. If the pepperoni is to be smoked, apply a cold smoke *after* the sausage has come out of the greening room, but *before* it starts to dry. Dry at 50° F (70 percent RH) for 21 days.

Summer Sausage (Semi-dry)

40 lb. special pork trimmings
   (80% lean)
20 lb. regular pork trimmings
   (50% lean)
30 lb. boneless chuck (80% lean)
10 lb. beef hearts
2 lb. 8 oz. salt
1 lb. dextrose (corn sugar)
2 oz. coarse-ground black pepper
2 oz. intermediate grind black pepper

3 oz. paprika
1 oz. red pepper
¾ oz. (22 g) coriander
1 oz. whole mustard seed
½ oz. (14 g) mace
½ oz. (14 g) allspice
¼ oz. (7 g) fresh garlic (peeled)
4 oz. certified cure (6% nitrite)
starter culture (follow directions
    for recommended amount)

☞   NOTE: Small quantities are more accurately weighed in grams (1 oz.   ☜
    = 28 g).

Grind meats through a ⅜-inch grinder plate, and mix thoroughly with non-meat ingredients; however, do not overmix. Three to four minutes is generally sufficient, depending on mixer's speed. Regrind meat through a ⅛-inch grinder plate, and stuff into appropriate natural or fibrous casings, usually 2 to 3 inches in diameter.

The following smokehouse schedule should be used with a starter culture in the production of summer sausage:

- Ferment for 12 hours at 100° F (75 to 85 percent RH).
- Smoke for one to two hours near end of fermentation cycle.
- Cook for 1½ to 3 hours at 155° F (60 percent RH).
- Cook until done (148° F internal temperature) at 165° F (45 percent RH).
- Shower with cold water for three minutes, and chill immediately in cooler.

SAUSAGE CASINGS

Sausage casings may be categorized as two types: natural and manufactured.

Natural casings (also called animal casings) are made from the stomachs, intestines, and bladders of hogs, sheep, and cattle. Packing houses that save casings will flush them thoroughly with water and pack them in salt, whereupon they are generally purchased by casing processors. The casing processors do the final cleaning, scraping, sorting, grading, and salting of the casings. Most animal casings are sold on the basis of a *tierce* (55-gallon barrel). The number of pieces of casing packed per tierce will depend on the size and capacity of the casing.

Sometimes, locker operators will clean their own hog and sheep casings for sausage production. This practice is generally limited to the small intestines of hogs and sheep. The intestine is first inverted by turning one end inside out and running water into the lip that is created. The weight of the water will easily force the casing to be completely inverted. The inverted casing should then be thoroughly washed in a dilute chlorine solution (0.5 percent) and brushed with a soft-bristled brush. Next, the casing is rinsed in clean water (inside and out) and inverted back into its original form and packed in a saturated salt solution for storage. It is best to store the casings in a freezer for long-term storage (more than two weeks). The casings will not freeze in the saturated salt solution. Halophilic (salt loving) bacteria thus do not have an opportunity to grow as they might if stored in a cooler. The saturated salt solution inhibits the growth of other bacteria that survived the dilute chlorine solution wash.

Hog Casings

Natural hog casings used in sausage production include bungs, stomachs, middles, small casings, and bladders.

Hog Bungs

Hog bungs are straight sections of the large intestine, generally cut into 32-inch lengths. They are graded according to size and may hold 1½ to 4 pounds per each 32-inch cut piece (depending on the grade) after stuffing. The grade names from largest to smallest are export, large prime, medium prime, special prime, small prime, and skip, with a diameter range of 1⅛ inches to over 2 inches. Specially selected grades of hog bungs are often used to manufacture sewed hog bungs for stuffing liver sausage, thuringer, and Genoa salami. Sewed hog bungs are manufactured to specified sizes and are generally made of double or triple thicknesses for additional strength (see Figure 20-5).

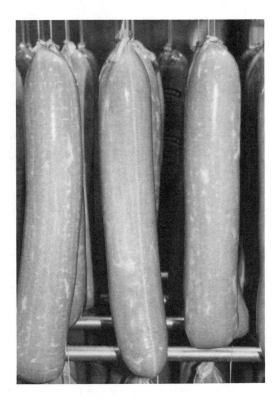

Fig. 20-5. An example of sewed hog bungs used for liver sausage production.

Hog Stomachs

Hog stomachs are principally used for stuffing headcheese and will hold between 5 and 8 pounds. Three hundred pieces are packed to a tierce.

Hog Middles

Hog middles are made from the caecum, or blind gut, of the hog. They are used for stuffing frizzes, Italian type sausages, and liver sausage.

They are put up according to three specifications:

- *Hog middle caps* are the closed end of the hog middles and are sold by the piece in tierces packed to capacity.
- *Cap-on hog middles* are the total length of the middle containing the cap end and the open end. Cap-on middles measure 8 to 10 feet in length with a diameter of 3 to 4½ inches. There are 150 cap-on middles packed to a tierce.
- *Cap-off hog middles* have the cap end removed and thus are open on both ends. Cap-off middles measure 8 to 10 feet in length and are packed 180 pieces to a tierce.

Small Hog Casings

Small hog casings are made from the small intestines and have three commonly accepted grades: narrow, medium, and wide. These grades may be further subdivided by some of the casings processors, depending on the demands of their customers. These casings are put up in "hanks" with 290 to 355 hanks per tierce, depending on the grade (size). A hank is the equivalent of 100 yards of casing tied up in a bundle. The stuffing capacity of narrow hog casings is approximately 90 pounds/hank; medium hog casings, 115 pounds/hank; and wide hog casings, 135 pounds/hank. Small hog casings are used for stuffing products such as Polish sausage, smoked pork sausage, bratwurst, and many other coarse-cut, small-diameter sausage products.

Hog Bladders

Hog bladders are sold in both a salted and a dried form and are used primarily as a casing for minced luncheon meats. Hog bladders are more fragile than beef bladders, are round in shape, and have a stuffing capacity of 2 to 6 pounds. They are most commonly put up salted and packed in tierces to capacity.

Sheep Casings

Sheep casings are among the most valuable of the animal casings and are produced from the small intestine. The other portions of the digestive tract of sheep are not generally used as a sausage casing. The length, diameter, and thinness of tissue make sheep casings ideal for most of the small-diameter sausages. Sheep casings are used primarily for stuffing frankfurters and pork sausages. After cleaning, they are graded for quality, measured for diameter and length, and salted and packed according to the specifications required by the trade.

North American sheep casings are put up in 100-yard hanks and graded according to size. The available sizes and their stuffing capacities are as follows: 16 to 18 mm (37 lbs.), 18 to 20 mm (46 lbs.), 20 to 22 mmm (55 lbs.), 22 to 24 mm (64 lbs.), 24 to 26 mm (72 lbs.), and 26 mm and over (77 lbs.).

Sheep casings of South American origin are generally taken from older animals and thus are a stronger casing with a slightly higher stuffing capacity. They also are graded into 2 mm size increments starting at 17 to 19 mm for the smallest available and ranging to 27 mm and over for the largest grade of casing.

Other countries which supply sheep casings to the United States include New Zealand, Iran, Iraq, Syria, and the Soviet Union. Of these, New Zealand is the largest supplier of sheep casings to the United States.

Beef Casings

Beef casings which are used in sausage production include the *weasand* (esophagus), *beef rounds* (small intestine), *beef middles* (large intestine), *beef bungs* (caecum), and *bladders*.

Beef Weasand

Beef weasand casings are made from the tubes leading from the throat to the first stomach (rumen). These tubes are prepared by removing the weasand meat from the casing, inspecting them for grubs, and, if the casings are free of grubs, inflating the weasand with air and drying it. At this point, the casings are graded for quality and size. Weasand casings are used for stuffing bologna and salami.

Beef Rounds

Beef rounds, made from the small intestine, are used for stuffing ring bologna, holsteiner, and mettwurst. Beef rounds are put up in "sets" containing a minimum of 100 feet and no more than five pieces with a length less than 5 feet. They are separated into "export" (those that have a clear surface) and "domestic" (those that have small nodules of scar tissue left by intestinal parasites) grades and further graded according to size. Exports have four size grades: narrow (29 to 36 mm), medium (36 to 38 mm), wide (38 to 44 mm), and extra wide (over 44 mm). Export rounds have a stuffing capacity of 55 to 90 pounds (per 100-foot set), depending on the size grade. Domestic rounds are available in two sizes: wide (over 38 mm) and regular (under 38 mm). Domestic rounds have a stuffing capacity of 65 to 75 pounds (per 100-foot set), depending on the size grade.

Beef Middles

Beef middles obtained from the large intestine are used for stuffing bologna, cervelat, and salamis. Middles are put up in "sets" containing a minimum of 57 feet and in no more than five pieces comprising the set. They are graded according to the following sizes: medium (45 to 50 mm), regular (50 to 60 mm), wide (60 to 70 mm), and extra wide (over 70 mm). Stuffing capacity ranges from 55 to 95 pounds (per 57-foot set), depending on the graded size.

Beef Bungs

Beef bungs obtained from the caecum or appendix of cattle are used for stuffing bologna and cappicola. Beef bungs are put up by the piece in tierces which may contain 350 to 500 pieces, depending on the size. The diameter of bungs will range from 3 inches to over 5 inches, representing eight different grades. Stuffing capacity per bung will range from 12 to over 20 pounds.

Beef Bladders

Beef bladders are sold in both a salted and a dried form and are used to stuff mortadella and various luncheon meat specialities. Bladders have a diameter range of less than 5 inches to over 7½ inches and a stuffing capacity of 3 to 10 pounds.

Stuffing capacities given in the preceding discussion are *approximate*. Actual stuffing capacity will vary, depending on the tightness of the stuff, whether the sausage is linked, and the natural size variations that occur between animal casings. For more information on natural casings, the reader is referred to the book *Sausage and Processed Meats Manufacturing*[6] from which much of this information was obtained.

Manufactured Casings

Manufactured casings presently account for the vast majority of casings used to stuff sausage products. The principal advantages of manufactured casings over their natural counterparts are in price, uniformity, and versatility. There are three classifications of manufactured casings: *cellulose, collagen*, and *co-extruded casings*.

Cellulose Casings

Cellulose casings are inedible, provide high-structural strength, and display good permeability to moisture and smoke when the casings are moist. There are three types of cellulose casings manufactured: small cellulose, large cellulose, and fibrous casings. Small and large cellulose casings are produced from cotton linters (the fuzz from cotton seeds) which are dissolved and regenerated into a casing. Fibrous casings are the toughest of all casings and are made from a special paper pulp base which is impregnated with cellulose.

---

[6]*Ibid.*

Small cellulose casings are used to produce skinless wieners and similar skinless, small-diameter products. They are available as clear casings or in a variety of tinted colors which impart a food-grade dye on the product's surface. Small cellulose casings are designed to be easily peeled from the product's surface (see Figure 20-4). They are available in sizes ranging from 15 to 50 mm in diameter and 70 to 160 feet in length. They are generally produced in the "shirred" form, which means they are crimped into short strands. For example, a shirred casing that is 8 inches in length may stuff out at 100 feet.

Large cellulose casings are used for stuffing products such as bologna and braunschweiger. They are often colored but do not impart dye onto the product's surface. They are available as moisture permeable or impermeable. Impermeable large cellulose casings are used for water-cooking processes (as in braunschweiger).

Fibrous casings exhibit the greatest structural strength and thus insure uniform product diameter from end to end. This is an important factor in portion-controlled luncheon meat slicing where uniform slices are essential in automated slicing/packaging lines. Fibrous casings are commonly used for luncheon meat specialities, summer sausage, bologna, and many other products.

Collagen Casings

Collagen casings were developed as an edible casing replacement for natural casings, but at the same time, they have the uniformity of a manufactured product. Collagen casings are made from the corium layer of split beef hides. The collagen from the corium layer is ground, swelled in acid, sieved, filtered, and finally extruded. These casings are generally limited to smaller-diameter products because of their lower structural strength. Palatability of the casing is a function of casing diameter (and subsequent membrane thickness). Larger-diameter collagen casings (which are heavier casings) are generally less palatable than smaller-diameter casings. Collagen casings should not be soaked in water prior to stuffing as large cellulose and fibrous casings are. As the product is stuffed into the casing, the casing absorbs moisture from within the meat, making it pliable for linking. Product stuffed into collagen casings may be smoked immediately upon entering the smokehouse (no drying period needed). Collagen casings are available in shirred form or as regular flat casings. Small-diameter shirred casings tend to become brittle with age, resulting in breakage problems during stuffing. It is best to store these casings in a 40° F cooler and use within six months.

Co-extruded Casings

Co-extruded collagen casings are the newest class of casings to be used in the sausage industry. The casing is actually formed from a collagen paste which is co-extruded over the sausage batter during stuffing. This co-extrusion process requires very specialized equipment and a very large capital investment. It is designed exclusively for high-volume operations. At the time of this writing, there are only a handful of these systems in the country.

MEAT EMULSIONS

Finely comminuted sausage batters are often referred to as *meat emulsions*. However, by strictest definition, they are not true emulsions and hence need defining. An appropriate definition follows: "A meat emulsion is a finely comminuted dispersion of lean and fat particles into a two-phase system which consists of a dispersed phase (fat droplets) and a complex continuous phase composed of water, solubilized proteins, cellular components, and miscellaneous spices and seasonings." A true emulsion, such as mayonnaise, is a heterogeneous mixture of two immiscible liquids (fat and water) stabilized by an emulsifying agent such as the protein casein. The result is a stable colloidal suspension of the two liquids which do not readily separate because the emulsifying agent acts as a physical barrier between the two phases but is miscible with both. This phenomenon occurs because the emulsifying agent (the protein) will undergo a conformational change in its structure toward a point of maximum stability with the immediate environment. The protein will unfold, orienting *hydrophobic (water hating)* portions of the protein molecules toward the lipid (fat) phase and *hydrophilic (water loving)* portions of the protein molecules toward the continuous phase. This is the mechanism of membrane formation around lipid droplets.

A schematic illustration of the structural components present in a typical meat emulsion is shown in Figure 20-6. The scanning electron micrograph illustrated in Figure 20-7 depicts a three-dimensional view of fat droplet entrapment in the emulsion matrix (continuous phase).

In order for a protein to be a good emulsifier of fat and water, it must pos-

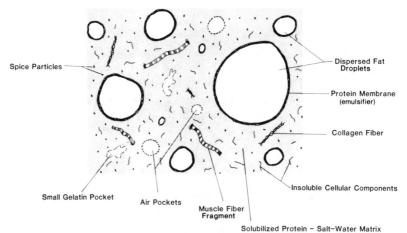

Fig. 20-6. Structural components of a typical meat emulsion.

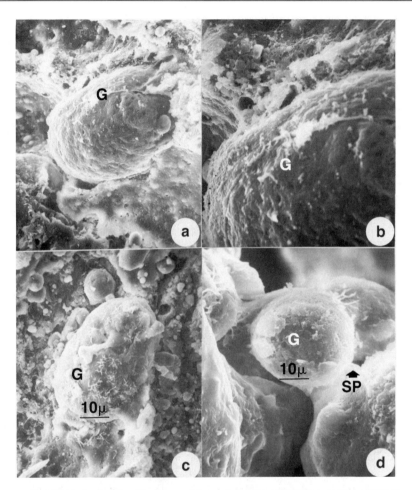

Fig. 20-7. Cooked frankfurter emulsions chopped to 60° F (a and b) and 72° F (c and d). Large encapsulated fat globules are labeled "G." Solubilized proteins binding fat particles together are represented by "SP." Magnifications: (a) 625×, (b) 1250×, (c) 1250×, and (d) 2500×. Scale bars represent 10 micrometers. (Scanning Electron Micrographs by Kevin W. Jones)

sess both hydrophobic and hydrophilic properties. Of the meat proteins, myosin demonstrates the best emulsifying properties and is readily abundant, comprising approximately 45 percent of the myofibrillar proteins in skeletal muscle. The protein collagen exhibits little or no emulsifying properties due to its unique triple-helical structure, which is extremely stable and not readily solubilized. For this reason, meat emulsions containing a high percentage of the protein as collagen are very unstable (phase separation will readily occur).

There are several factors which must be critically controlled in order for the sausage processor to produce a stable emulsified product. Meat temperature during chopping, particle size, protein quality, fat source, fat condition, and thermal processing conditions are extremely important. Control of chopping temperature is critical in the extraction of the contractile proteins within muscle fibers. The contractile protein myosin is known to be the principal emulsifier in comminuted meat systems. Optimal extraction temperatures of myosin have been demonstrated to occur between 40° and 45° F.

Particle size of the fat droplets can also have a profound effect on the ultimate stability of a meat emulsion, especially if the emulsifying proteins are of marginal quality (*i.e.,* moderate collagen levels). Particle size is primarily influenced by chopping time or by the degree of comminution that is attained in high-speed emulsion mills. A reduction in particle size of the fat droplets greatly increases the exposed surface area of lipid droplets and, therefore, also increases the amount of emulsifier needed to stabilize the lipid droplets. When the surface area of the lipid droplets exceeds the level of available emulsifying agent, the emulsion will short out, or fat out. Shorting out refers to the separation of phases and subsequent coalescence of large pools of fat. This is seldom a problem in meat emulsions made with high-quality raw meat materials.

The selection of raw materials is important to the sausage processor in determining the desired quality and price of the finished product. For stable emulsions, the processor should limit the use of low-binding raw materials such as tripe to no more than 15 percent of the formulation. Low-binding meats often contain a large percentage of their total protein content in the form of collagen rather than contractile proteins. Collagen may bind considerable amounts of water during chopping; however, this water is lost during thermal processing. Collagen molecules will shrink up to one third of their original length when heated to 150° F and will be converted to gelatin at higher temperatures. When collagen, as an emulsifier, surrounds a fat droplet, shrinkage and/or gelatinization of the collagen will occur upon thermal processing. This allows fat droplets to migrate and coalesce into fat pockets. Gelatin will also migrate into gelatin pockets. In severe shorting out of the emulsion, the fat droplets will migrate to the surface to form fat caps. This problem will often occur in poor-quality formulations with a high fat content and a low ratio of contractile proteins to collagen.

The source of fat, as well as the condition of the fat, can influence the stability of meat emulsions to lesser degrees. Rendered fats such as lard will produce unstable emulsions (probably due to a large surface area of fine fat droplets). Beef, pork, and lamb or mutton fats will respond differently, due to differences in their plasticization properties. The harder (more saturated) fats from the kidney knob are much less desirable than the softer intermuscular fats from the carcass.

When emulsion products are heated too rapidly, some fatting out will often occur. This problem is generally associated with weaker formulations and is characterized by a greasy coating on the product surface or a small pocket of fat in the smokehouse stick mark. Slower-processing schedules allow a better skin to be formed on the product surface which minimizes this problem.

PRODUCT DEFECTS

· There are a number of problems which plague sausage manufacturers, often resulting in unacceptable or undesirable finished products. These problems are often related to poor sanitation practices or the use of poor-quality raw materials.

Table 20-4 is a useful trouble-shooting source of information when product defects or manufacturing problems arise.

The data contained in Table 20-4 and in Tables 20-6 through 20-24 are courtesy of Least Cost Formulations, Ltd., 827 Hialeah Drive, Virginia Beach, VA 23464.

Table 20-4. Sausage Product Defects, Causes, and Remedies

Defect	Description	Cause(s)	Remedy(ies)
Ruptured casings	Casings rupture or split during cooking	Stuffed too tightly Bacterial-gas producer	Stuff less tightly Check sanitation Change sanitizer
Color problems (external)	Streaking Poor smoke color	Smoked while too wet Smoked while too dry	Increase drying time Shorten drying time
Color problems (internal)			
Pale color	Lacks desirable color	Insufficient curing time	Use preblends
Greening	Uniform greenish color	Use of too much NO_2 (nitrite burn)	Reduce NO_2 Check sanitation
Green cores	Green center	Bacterial-*Lactobacillus* viridescence	Check cook cycle (153° F critical temperature)
Green rings	Concentric green rings	Bacterial-*Lactobacillus* viridescence	Check cook cycle (153° F critical temperature)
Fading	Light or chemical fading	Display lights Sloppy rinse of sanitizers	Gas flush Vacuum package Use amber film Rinse equipment better
Case hardening	Hard, dry surface Wet interior	Humidity too low Drying too fast	Increase humidity Increase drying time
Fatting out	Separation of fat	Too much fat in formulation	Reduce fat level
Fat caps	Fat on surface	Chopping temperature too high	Chop to 60° to 62° F
Fat caps	Fat on surface between casing and product	Cooked too fast	Longer cook cycle
Fat pockets	Fat pockets throughout product	Insufficient contractile proteins Excessive chopping Use of too little salt or salt omitted	Decrease level of high-collagen meats Sharpen chopper knives, reduce chopping time Check salt addition

(Continued)

Table 20-4 (Continued)

Defect	Description	Cause(s)	Remedy(ies)
		Overworking or pumping long distances	Change production layout
		Use of too much rework (previously cooked product)	Reduce rework level
		Insufficient mixing (coarse-cut products)	Increase mixing time
Gassiness	Vacuum packages bloat	Bacterial gas produced	Check sanitation, thermal-processing schedule
Gelatin pockets	Gelatin pockets	Too much connective tissue in formulation	Reduce connective-tissue level, preblend
Mold growth	Fuzz on surface	Poor sanitation (outside air in plant)	Check sanitation Use filters in ventilation system Maintain positive room air pressure
Particle definition poor	Coarse-cut sausages lack clean fat/lean distinction	Dull chopper-grinder blades Chopping temperatures too warm	Sharpen plates and knives Chop at 28° to 32° F
Peelability poor	Inedible casings won't peel easily	Surface drying too severe Poor skin formation	Operate smokehouse at higher humidity Steam cook last 10 minutes of cycle Reduce collagen level in formulation
Putrefaction	Undesirable odor	Putrefactive bacteria	Check thermal-processing temperature Check sanitation
Smearing	Fat smearing, leaving undesirable appearance in fresh sausages	Overworking Meat temperature too warm	Minimize mixing Stuff at 32° F

LEAST COST SAUSAGE FORMULATIONS

The best sausage formulation is one that meets the predetermined quality specifications at the lowest possible cost of production. This is a problem that is not easily determined without the assistance of computer formulations; and even then, the results are only as good as the information entered into the computer. The dimensions of the best formulation problem vary widely from company to company, depending on such factors as size of the firm, ingredient availability, operating capacity, storage capacity, and product diversity. For most companies, the ultimate formulation of the product is also influenced by raw material procurement policies, inventory policies, production scheduling, and price strategies.

Much of the following information pertaining to least cost computer formulations and preblending applications was obtained from George C. Selfridge,

Jr., and Robert A. LaBudde, Least Cost Formulations, Ltd.[7, 8] The authors are grateful to these two individuals for providing the valuable information that has been compiled on the following pages. Tables 20-6 through 20-24 are copies of the printouts received from the Least Cost Formulator®, a computer software system designed by the aforementioned company. The information in these tables is an abbreviated form for computer input simplicity reasons. The authors are also indebted to Professor Robert Rust, Iowa State University, for his useful suggestions and the valuable information that has been published in the Iowa State Sausage Short Course Proceedings.

Why Computers and Least Cost Formulation?

In the last 20 years, the advances in electronics have brought the cost of computers down from millions of dollars to where small-business systems are available for under $10,000. Along with the drastic drop in prices has come ease-of-use. The user does not need an engineering degree to use the computer. Though the art of sausage making is thousands of years old, it is only recently that formulating has begun to make use of twentieth-century technology. With the systems available today, even the smallest family-owned meat processor can take advantage of sophisticated techniques to produce consistent quality products and increase the bottom line profit. Systems have been designed assuming the user has not had any computer experience or training and has no familiarity with the intricacies of least cost formulation.

Least cost formulation is a mathematical technique for determining the best use (optimum allocation) of available resources when there are many possible uses for these resources. The limited resources are the raw materials which are purchased or are available from a slaughtering operation from which finished products are made. Sausage manufacturers face the problem of manufacturing many products each day, with fixed amounts and types of raw materials whose chemical analyses change with each lot. Moreover, the finished products must meet product specifications. Control of the many necessary variables cannot be done by hand in a practical time frame, but can be done in a matter of seconds on a computer.

Misconceptions About Least Cost Formulation

Today, linear programming is usually referred to as least cost formulation. With this nickname has come the misconception that only inferior products can be made using least cost formulation—after all they are the "cheapest." Least

---

[7]G. C. Selfridge, Jr. and R. A. LaBudde. 1982. *The Use of Least Cost Formulation in Sausage Formulation.* In Proceedings of the Fourth Annual Sausage and Processed Meats Short Course, Iowa State University.

[8]R. A. LaBudde and G. C. Selfridge, Jr. 1982. *Preblending and Least Cost Formulations.* In Proceedings of the Fourth Annual Sausage and Processed Meats Short Course, Iowa State University.

cost formulation can better be described as *yielding the least expensive formulation that will satisfy the product specifications and make use of the available raw materials.*

The mathematical technique first finds the raw materials that achieve the product specifications, and only then does it find substitutions that still meet the product specifications but reduce the price. Therefore, any quality-level product can be least cost formulated and produced consistently, using either the same or the varying raw materials, providing that the specifications are well defined and the raw materials accurately characterized. To prove this technique works on different product-quality levels, we shall consider the manufacture of four types of franks on the following pages, using least cost formulation:

- Regular frank (#1)
- By-product frank (#2)
- Regular frank with isolated soy protein (ISP)
- By-product frank with ISP

The Problems with Standard Formulas

It is common practice for meat processors to have a standard formula for making a product. For example, the regular frank might have the following:

Quantity (lb.)	Raw Material	Short Name
30	Beef plates	B-plates
70	Pork trimmings (80%)	P-80
25	Water	Water
2¾	Salt	Salt
1	Corn syrup	I-corn
¾	Frank spice and cure	S-frank

Raw meat materials used in manufacturing processes have a wide range of variability that make it impossible to produce a consistent finished product using a standard formula. The fat content may vary by 5 to 10 percent and the protein content may vary by 2 to 3 percent between suppliers. Hence, only by chance can one repeatedly make the same finished product when using fixed amounts of the raw materials. Table 20-5 shows the analysis of four regular frank batches made using the standard formula and different lots of raw materials. It is these typical variations that are responsible for product failure, as well as the wide variation in product cost.

Creating Product Standards

Let's examine five categories that affect product specifications and see what their contributions are to the product standards for the regular frank.

Table 20-5. Standard Formula Analysis

| Attribute | Regular Frank Sample | | | |
	#1	#2	#3	#4
	.... (percent)			
Fat	27.0	23.5	30.8	29.0
Protein	12.5	12.4	12.1	11.6
Moisture	56.3	59.9	52.9	55.2
USDA-AW[1]	6.3	10.3	4.5	8.6

[1]AW (Added water). The level of water that can be legally added to formulations according to USDA regulations is 4 × protein content + 10 percent. USDA-AW values must therefore be 10 percent or less to remain in regulatory compliance. Calculations of USDA-AW from moisture and protein analyses indicate that the Number 2 batch is out of compliance by 0.3 percent for added water. Batch Numbers 1, 3, and 4 could have had additional water added to a level not exceeding 10 percent. (Refer to discussion on page 595.)

USDA Regulations for Franks

The USDA specifies what may and may not be included in the product as well as the upper limit of many product attributes: (1) fat content, 30 percent finished weight; (2) USDA added water, 10 percent finished weight; (3) isolated soy protein, 2 percent finished weight; and (4) nonfat dry milk, 3.5 percent finished weight.

Product Label

The choice of the product label determines the quantity of each raw material and its relation to other raw materials. Hence a label that reads "pork, beef, water, salt, corn syrup, spices, and nitrite" would have to have more pounds of pork than pounds of beef, more pounds of beef than pounds of water, and so on.

Plant Equipment

The type and size of equipment in the plant will have an influence on the product as will the care employees take in carrying out the manufacturing steps. A successful formulation in one plant may fail in another plant in which equipment and technique vary.

Equipment will affect product specifications in several ways. The first deals with how the batches are handled. The size of each batch will be a function of equipment size. The use of frozen meat and its quantity will depend on freezer space and the ability to temper and/or flake the boxes of meat. The ability to accurately control the smokehouse will affect the product shrink. The available cooler storage area could determine how many different raw materials are on hand to use in the formulations, as well as the ability to make preblends. The importance of preblends will be covered later.

Market Requirements

Input from salespersons and customers and knowledge of the marketplace will influence the kind and amount of spices in the products as well as the salt content, the color level, and the bite (bind). The method of consumer cooking, the use of a roller grill for instance, will influence the dextrose level and the percent and kind of fat (pork vs. beef). All this must also take into account the price for which the product may be sold.

Company Standards

Company reputation is maintained by choice of labels and the quality level (Type #1 vs. #2) of the products produced. For each kind and type, it must be consistent and at a price that will allow a profit.

The use of a least cost formulation system allows all these parameters to be taken into account, thus producing the identical product, at the specified quality level, day after day.

Getting Started Using Least Cost Formulation

The overall process of setting up a least cost formulation system is flowcharted in Figure 20-8.

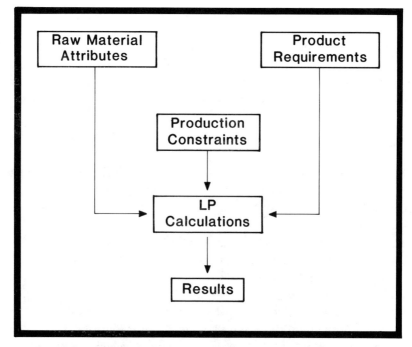

Fig. 20-8. Least cost formulation procedure. Linear programming (LP) calculations provide the final results.

Raw Material Attributes

Table 20-6 is an example listing of raw materials which might be available and the attributes to control in the production.

To begin, five attributes are generally sufficient: fat, moisture, protein, bind, and color, but there really is no limit to the possible attributes to measure and control. A comprehensive raw material attribute table is included in Table 20-24. The data are based upon the averages for each material collected over several years from dozens of installed least cost formulation users. The color and bind values are constants by meat type. The values shown were obtained from Professor John Carpenter at the University of Georgia.

The raw material parameter table in Table 20-7 shows the prices and quantity of each raw material available to the formulations.

Least Cost Formulation: Programming Product Standards

To create product standards that will be used in least cost formulation, each attribute must be expressed as a constraint with a low limit and a high limit. A standard formula has been designed so that the values of its attributes would most likely fall into allowable ranges required by the USDA, the product label, and so on, but specific values were never given as targets.

The following are the steps required to establish the least cost formulation low and high limits on both the raw materials and the product attributes. The four types of limits that can be chosen are: finished weight (FW), green weight (GW), batch weight (BW), and pounds (lb.).

Raw Material Limits

For each product, determine what raw materials will be offered to the formulation. For illustration purposes, assume that the local availability of raw materials has led to the selection of 10 items from the raw material table (Table 20-24). These materials and their limits are shown in Table 20-8.

The majority of the raw materials will have a low limit of zero and a high limit of BIG. In mathematical terms, this means that the least cost formulation system may use as few or as many of the raw materials as it desires to achieve the desired product specifications. Hence, in Table 20-8, the 10 raw materials offered all have low limits of zero and high limits of BIG with the limit type of green weight (GW). I-salt has a low limit of zero and a high limit of 3 percent of the green weight.

I-corn has identical values for both the low and the high limits, which require *exactly* 1 percent of the block weight of corn syrup. Similarly, S-Frank limits will cause exactly ¾ pound of spices and cure to be included in the batch.

Table 20-6. Short Attribute Level Table for Least Cost Formulations

SEQ NO.	SHORT NAME	RAW MATERIAL	GROUP TYPES	MOISTURE	FAT	PROTEIN	SALT	USDA-AW	BIND	COLOR	COLLAGEN
1.	B-BULL	BF FC BULL MEAT	B	0.7000	0.0950	0.2000	0.0000	-0.1000	6.0000	9.4000	0.0400
2.	B-CHEEK	BF CHEEK MEAT	BK	0.6300	0.1950	0.1700	0.0000	-0.0500	2.3800	8.1600	0.1003
3.	B-CHUCK	BF BNLS CHUCKS	B	0.5700	0.2600	0.1600	0.0000	-0.0700	3.8400	13.6000	0.0480
4.	B-COW	BF FC COW MEAT	B	0.6650	0.1500	0.1800	0.0000	-0.0550	4.4100	7.0200	0.0378
5.	B-HEAD	BF HEAD MEAT	BK	0.6100	0.2300	0.1550	0.0000	-0.0100	1.2400	4.0300	0.1131
6.	B-HEART	BF HEARTS	O	0.6650	0.1800	0.1500	0.0000	0.0650	0.9000	6.0000	0.0405
7.	B-LIP	BF LIPS	O	0.6200	0.2200	0.1550	0.0000	0.0000	0.0465	0.0620	0.1395
8.	B-NAVEL	BF BNLS NAVELS	B	0.3800	0.5150	0.1000	0.0000	-0.0200	1.2500	1.9000	0.0420
9.	B-PLATE	BF PLATES	B	0.4000	0.4800	0.1100	0.0000	-0.0400	1.7600	2.7500	0.0462
10.	B-SHANK	BF SHANK MEAT	BK	0.7100	0.1000	0.1850	0.0000	-0.0300	5.1800	8.5100	0.0610
11.	B-50	BF 50% TRIMMINGS	B	0.3750	0.5100	0.1050	0.0000	-0.0450	1.2600	1.9950	0.0441
12.	B-60	BF 60% TRIMMINGS	B	0.4500	0.4100	0.1300	0.0000	-0.0700	2.0800	3.2500	0.0526
13.	B-65	BF 65% TRIMMINGS	B	0.4900	0.3600	0.1500	0.0000	-0.1100	2.7000	4.2000	0.0600
14.	B-75	BF 75% TRIMMINGS	B	0.5700	0.2700	0.1550	0.0000	-0.0500	3.4100	5.2700	0.0589
15.	B-85	BF 85% TRIMMINGS	B	0.6550	0.1550	0.1800	0.0000	-0.0650	4.3200	7.0200	0.0504
16.	B-90	BF 90% TRIMMINGS	B	0.6950	0.1100	0.1850	0.0000	-0.0450	4.8100	7.4000	0.0518
17.	P-BACEND	PK BACON ENDS	OR	0.3050	0.6000	0.0900	0.0000	-0.0550	0.0090	0.2250	0.0774
18.	P-BELLY	PK BELLY STRIPS	P	0.3400	0.5650	0.0900	0.0000	-0.0200	0.5400	0.4500	0.0315
19.	P-BLADE	P-BLADE MEAT	P	0.7150	0.0900	0.1900	0.0000	-0.0450	4.5600	3.8000	0.0437
20.	P-CHEEK	PK CHEEK MEAT	PKZ	0.6900	0.1400	0.1600	0.0000	0.0500	1.4400	4.6400	0.1152
21.	P-DIAPH	PK DIAPHRAGM MEAT	P	0.7000	0.1500	0.1450	0.0000	0.1200	2.3200	2.6100	0.1015
22.	P-FATBCK	PK SKND FATBACKS	P	0.1500	0.8000	0.0400	0.0000	-0.0100	0.1200	0.0400	0.0240
23.	P-HAM	PK FRESH HAM TRIM	P	0.5650	0.3000	0.1300	0.0000	0.0450	2.7300	2.4700	0.0325
24.	P-HAMCUR	PK CURED HAM TRIM	PR	0.5700	0.3000	0.1200	0.0000	0.0900	2.1600	1.2000	0.0360
25.	P-HAMFAT	PK HAM FAT	P	0.0400	0.9500	0.0100	0.0000	0.0000	0.0100	0.0100	0.0040
26.	P-HEAD	PK HEAD MEAT	PK	0.6100	0.2200	0.1600	0.0000	-0.0300	1.2000	2.5600	0.1120
27.	P-HEART	PK HEARTS	OZ	0.7000	0.1400	0.1550	0.0000	0.0800	0.9300	5.1150	0.0418
28.	P-JOWL	PK SKND JOWLS	P	0.3000	0.6150	0.0750	0.0000	0.0000	0.3450	0.1275	0.0322
29.	P-LIP	PK LIPS	O	0.6200	0.2200	0.1550	0.0000	0.0000	0.1550	0.0775	0.1395
30.	P-LIVER	PK LIVERS	OL	0.7400	0.0600	0.1900	0.0000	-0.0200	0.3800	9.3100	0.1805
31.	P-PIC	PK BNLS PICNICS	P	0.6000	0.2400	0.1550	0.0000	-0.0200	3.1000	2.4800	0.0356
32.	P-NECK	PK NECKBONE TRIM	P	0.6000	0.2300	0.1600	0.0000	-0.0400	3.0400	2.5600	0.0400
33.	P-PICHRT	PK PICNIC HEARTS	P	0.7000	0.1100	0.1800	0.0000	-0.0200	4.1400	3.4200	0.0396
34.	P-SNOUT	PK SNOUTS	O	0.5500	0.3000	0.1450	0.0000	-0.0300	0.3625	0.0725	0.1160
35.	P-SPLEEN	PK SPLEENS	O	0.7000	0.1450	0.1550	0.0000	0.0800	0.1085	8.6800	0.1395
36.	P-STOM	PK STOMACHS	O	0.0000	0.0000	0.0000	0.0000	0.0000	0.0000	0.0000	0.0000
37.	P-TNGTRM	PK TONGUE TRIMMINGS	O	0.0000	0.0000	0.0000	0.0000	0.0000	0.0000	0.0000	0.0000
38.	P-TONGUE	PK TONGUES	O	0.0000	0.0000	0.0000	0.0000	0.0000	0.0000	0.0000	0.0000
39.	P-50	PK 50% TRIMMINGS	P	0.3200	0.5900	0.0850	0.0000	-0.0200	1.0200	0.7650	0.0289
40.	P-80	PK 80% TRIMMINGS	P	0.5600	0.2800	0.1500	0.0000	-0.0400	2.8500	2.5500	0.0360
41.	P-95	PK 95% VL TRIMMINGS	P	0.0000	0.0000	0.0000	0.0000	0.0000	0.0000	0.0000	0.0000
42.	C-MDB	CK COMMINUTED MEAT	C	0.6600	0.2050	0.1250	0.0000	0.1600	1.8750	0.6250	0.0437
43.	T-MDB	TK MDB COMMINUTED MT	C	0.0000	0.0000	0.0000	0.0000	0.0000	0.0000	0.0000	0.0000
44.	V-BNLS	VEAL BNLS	V	0.0000	0.0000	0.0000	0.0000	0.0000	0.0000	0.0000	0.0000
45.	S-BEBR	BEBRECZINER SPICE		0.0500	0.0000	0.0100	0.0000	0.0100	0.0000	0.0000	0.0000
46.	S-BOLO	BOLOGNA SPICE		0.0000	0.0000	0.0000	0.0000	0.0000	0.0000	0.0000	0.0000
47.	S-FRANK	FRANK SPICE		0.0800	0.0000	0.1500	0.0000	-0.5200	3.0000	1.5000	0.0000
48.	S-SALAMI	SALAMI SPICE		0.0000	0.0000	0.0000	0.0000	0.0000	0.0000	0.0000	0.0000
49.	S-SAUS	FR SAUSAGE SPICE		0.0000	0.0000	0.0000	0.0000	0.0000	0.0000	0.0000	0.0000
50.	I-CORN	CORN SYRUP		0.2600	0.0000	0.0000	0.0000	0.2600	0.0000	0.0000	0.0000
51.	I-CURE	DRY CURE (W/SALT)		0.0000	0.0000	0.0000	0.0000	0.0000	0.0000	0.0000	0.0000
52.	I-SALT	SALT		0.0000	0.0000	0.0000	1.0000	0.0000	0.0000	0.0000	0.0000
53.	X-CEREAL	CEREAL FLOUR	X	0.0000	0.0000	0.0000	0.0000	0.0000	0.0000	0.0000	0.0000
54.	X-ISP	ISOLATED SOY PROTEIN	X	0.2900	0.0000	0.7100	0.0000	-2.5500	31.9500	0.0000	0.0000
55.	X-NFDM	NON-FAT DRY MILK	X	0.0000	0.0000	0.0000	0.0000	0.0000	0.0000	0.0000	0.0000
56.	X-PREMUL	CASEINATE PRE-EMULS.		0.5030	0.4190	0.0620	0.0100	0.2550	12.3380	0.0000	0.0000
57.	X-SPC	SOY PROTEIN CONCENC.	X	0.2000	0.0000	0.7000	0.0000	-2.6000	28.0000	0.0000	0.0000
58.	X-TVP	TEXTURED VEG PROTEIN	X	0.0000	0.0000	0.5000	0.0000	-2.0000	20.0000	0.0000	0.0000
59.	X-TSPC	TEXTURED SOY PROTEIN	X	0.0000	0.0000	0.7000	0.0000	-2.8000	28.0000	0.0000	0.0000
60.	WATER	WATER		1.0000	0.0000	0.0000	0.0000	1.0000	0.0000	0.0000	0.0000

Table 20-7. Raw Material Parameter Table: Example Listing

SEQ NO.	SHORT NAME	RAW MATERIAL	MAT. TYPE	GROUP TYPES	COST #1	COST #2	COST #3	QUANTITY #1 (LBS)	QUANTITY #2 (LBS)	ROUNDING UNIT	MATERIAL USAGE CURRENT	TO-DATE
1.	B-BULL	BF FC BULL MEAT	B		1.5400	1.5600	0.0000	BIG	BIG	0.000	0.	0.
2.	B-CHEEK	BF CHEEK MEAT	BK		0.9600	0.9800	0.0000	60.0	BIG	1.000	0.	0.
3.	B-CHUCK	BF BNLS CHUCKS	B		1.4400	1.4500	0.0000	BIG	BIG	0.000	0.	0.
4.	B-COW	BF FC COW MEAT	B		1.4700	1.4500	0.0000	600.0	BIG	1.000	0.	0.
5.	B-HEAD	BF HEAD MEAT	BK		0.8100	0.8100	0.0000	BIG	BIG	0.000	0.	0.
6.	B-HEART	BF HEARTS	0		0.4900	0.4900	0.0000	60.0	BIG	0.000	0.	0.
7.	B-LIP	BF LIPS	0		0.5500	0.5500	0.0000	BIG	BIG	0.000	0.	0.
8.	B-NAVEL	BF BNLS NAVELS	B		0.6200	0.6300	0.0000	BIG	BIG	0.000	0.	0.
9.	B-PLATE	BF PLATES	B		0.5200	0.5200	0.0000	.0	BIG	1.000	0.	0.
10.	B-SHANK	BF SHANK MEAT	BK		1.2100	1.2200	0.0000	BIG	BIG	0.000	0.	0.
11.	B-50	BF 50% TRIMMINGS	B		0.6400	0.6450	0.0000	.0	BIG	1.000	0.	0.
12.	B-60	BF 60% TRIMMINGS	B		0.8000	0.8000	0.0000	BIG	BIG	0.000	0.	0.
13.	B-65	BF 65% TRIMMINGS	B		0.8700	0.8700	0.0000	BIG	BIG	0.000	0.	0.
14.	B-75	BF 75% TRIMMINGS	B		1.0200	1.0300	0.0000	BIG	BIG	0.000	0.	0.
15.	B-85	BF 85% TRIMMINGS	B		1.1500	1.1700	0.0000	.0	BIG	1.000	0.	0.
16.	B-90	BF 90% TRIMMINGS	B		1.3000	1.3100	0.0000	BIG	BIG	0.000	0.	0.
17.	P-BACEND	PK BACON ENDS	OR		0.2400	0.2400	0.0000	BIG	BIG	0.000	0.	0.
18.	P-BELLY	PK BELLY STRIPS	P		0.4100	0.4100	0.0000	BIG	BIG	0.000	0.	0.
19.	P-BLADE	P-BLADE MEAT	P		1.1000	1.2000	0.0000	BIG	BIG	0.000	0.	0.
20.	P-CHEEK	PK CHEEK MEAT	PKZ		0.8900	0.9000	0.0000	60.0	BIG	0.000	0.	0.
21.	P-DIAPH	PK DIAPHRAGM MEAT	P		0.6300	0.6300	0.0000	BIG	BIG	0.000	0.	0.
22.	P-FATBCK	PK SKND FATBACKS	P		0.2700	0.2750	0.0000	.0	BIG	1.000	0.	0.
23.	P-HAM	PK FRESH HAM TRIM	P		0.7000	0.7100	0.0000	BIG	BIG	0.000	0.	0.
24.	P-HAMCUR	PK CURED HAM TRIM	PR		0.5700	0.5700	0.0000	BIG	BIG	0.000	0.	0.
25.	P-HAMFAT	PK HAM FAT	P		0.0800	0.0800	0.0000	10000.0	BIG	1.000	0.	0.
26.	P-HEAD	PK HEAD MEAT	PK		0.6600	0.6600	0.0000	BIG	BIG	0.000	0.	0.
27.	P-HEART	PK HEARTS	0Z		0.4800	0.4800	0.0000	.0	BIG	1.000	0.	0.
28.	P-JOWL	PK SKND JOWLS	P		0.3600	0.3650	0.0000	17.0	BIG	1.000	0.	0.
29.	P-LIP	PK LIPS	0		0.3400	0.3500	0.0000	75.0	BIG	0.000	0.	0.
30.	P-LIVER	PK LIVERS	OL		0.1750	0.1700	0.0000	BIG	BIG	0.000	0.	0.
31.	P-PIC	PK BNLS PICNICS	P		0.8300	0.8300	0.0000	BIG	BIG	0.000	0.	0.
32.	P-NECK	PK NECKBONE TRIM	P		0.8100	0.8150	0.0000	BIG	BIG	0.000	0.	0.
33.	P-PICHRT	PK PICNIC HEARTS	P		1.1500	1.1600	0.0000	BIG	BIG	0.000	0.	0.
34.	P-SNOUT	PK SNOUTS	0		0.3100	0.3100	0.0000	25.0	BIG	1.000	0.	0.
35.	P-SPLEEN	PK SPLEENS	0		0.1200	0.1200	0.0000	BIG	BIG	0.000	0.	0.
36.	P-STOM	PK STOMACHS	0		0.2200	0.2100	0.0000	BIG	BIG	0.000	0.	0.
37.	P-TNGTRM	PK TONGUE TRIMMINGS	0		0.2000	0.2000	0.0000	BIG	BIG	0.000	0.	0.
38.	P-TONGUE	PK TONGUES	0		0.5800	0.5800	0.0000	BIG	BIG	0.000	0.	0.
39.	P-50	PK 50% TRIMMINGS	P		0.4200	0.4300	0.0000	375.0	BIG	1.000	0.	0.
40.	P-80	PK 80% TRIMMINGS	P		0.8000	0.8200	0.0000	365.0	BIG	1.000	0.	0.
41.	P-95	PK 95% VL TRIMMINGS	P		1.1000	1.1300	0.0000	BIG	BIG	0.000	0.	0.
42.	C-MDB	CK COMMINUTED MEAT	C		0.2400	0.2400	0.0000	BIG	BIG	0.000	0.	0.
43.	T-MDB	TK MDB COMMINUTED MT	C		0.3200	0.3200	0.0000	BIG	BIG	0.000	0.	0.
44.	V-BNLS	VEAL BNLS	V		1.3500	1.3600	0.0000	BIG	BIG	0.000	0.	0.
45.	S-BEBR	BEBRECZINER SPICE			0.3000	0.3000	0.0000	BIG	BIG	0.000	0.	0.
46.	S-BOLO	BOLOGNA SPICE			0.4500	0.4500	0.0000	BIG	BIG	0.000	0.	0.
47.	S-FRANK	FRANK SPICE			0.3400	0.3400	0.0000	BIG	BIG	0.000	0.	0.
48.	S-SALAMI	SALAMI SPICE			0.5500	0.5500	0.0000	BIG	BIG	0.000	0.	0.
49.	S-SAUS	FR SAUSAGE SPICE			0.3200	0.3200	0.0000	BIG	BIG	0.000	0.	0.
50.	I-CORN	CORN SYRUP			0.3000	0.3000	0.0000	BIG	BIG	0.000	0.	0.
51.	I-CURE	DRY CURE (W/SALT)			0.2500	0.2500	0.0000	BIG	BIG	0.000	0.	0.
52.	I-SALT	SALT			0.0550	0.0550	0.0000	BIG	BIG	0.000	0.	0.
53.	X-CEREAL	CEREAL FLOUR		X	0.2200	0.2200	0.0000	BIG	BIG	0.000	0.	0.
54.	X-ISP	ISOLATED SOY PROTEIN P		X	1.0100	1.0100	0.0000	BIG	BIG	0.000	0.	0.
55.	X-NFDM	NON-FAT DRY MILK		X	1.0200	1.0200	0.0000	BIG	BIG	0.000	0.	0.
56.	X-PREMUL	CASEINATE PRE-EMULS.			0.1620	0.1620	0.0000	BIG	BIG	0.000	0.	0.
57.	X-SPC	SOY PROTEIN CONCENC.		X	0.5000	0.5000	0.0000	BIG	BIG	0.000	0.	0.
58.	X-TVP	TEXTURED VEG PROTEIN		X	0.3500	0.3500	0.0000	BIG	BIG	0.000	0.	0.
59.	X-TSPC	TEXTURED SOY PROTEIN		X	0.5400	0.5400	0.0000	BIG	BIG	0.000	0.	0.
60.	WATER	WATER			0.0000	0.0000	0.0000	BIG	BIG	0.000	0.	0.

Attribute Limits

From analysis of the four batches shown in Table 20-5, one can determine a range of limits that the standard formulation yields. Using these data as a base, combined with the USDA regulations and experience, one can set target limits for the attributes of the regular frank. This is shown in Table 20-8.

For fat, one may wish to approach the high limit of 30 percent of the finished weight (FW) allowed by the USDA regulations. To give a margin of safety, set the high limit as 29 percent FW. The low limit for fat has been set at 25 percent FW. The least cost formulation system will use as much fat as possible, because it is an inexpensive attribute.

Table 20-8. Product Specifications for Regular Frank

```
PRODUCT SHORT NAME:  ISU-REG          AS OF:        7/13/81
PRODUCT LONG NAME:   ISU REGULAR FRANK   PRODUCT TYPE:         GROUPS:

LAST COSTS:    $     0.6556     $    0.6671    $    0.0000
LAST CALCULATION USED COST # 1       DATE LAST CALCULATION: 7/13/81    CALC. #    3
MIN OR MAX:          MIN

NO. RAW MATERIALS:      14          PRODUCT BLOCK SIZE:    100.0000 (SCALED ON BW)
NO. REQUIREMENTS:       10          SHRINKAGE RATE:          8.0000 %
PRODUCT DISPLAY UNITS:  LBS
```

<--- RAW MATERIALS IN PRODUCT --->　　(NAMES FROM TABLE: ISU MEAT COURSE)

SEQ NO.	SHORT NAME	RAW MATERIAL DESCRIPTION	IN BLOCK	ACCEPTABLE RANGE OF VALUES LO LIMIT	HI LIMIT	LIMIT TYPE	CURRENT VALUE	CURRENT STATUS
1.	B-CHEEK	BF CHEEK MEAT	Y	.0000	BIG	% GW	16.3674	BETWEEN
2.	B-COW	BF FC COW MEAT	Y	.0000	BIG	% GW	8.5269	BETWEEN
3.	B-PLATE	BF PLATES	Y	.0000	BIG	% GW	0.0000	HI-LIMIT
4.	B-50	BF 50% TRIMMINGS	Y	.0000	BIG	% GW	0.0000	HI-LIMIT
5.	B-85	BF 85% TRIMMINGS	Y	.0000	BIG	% GW	0.0000	HI-LIMIT
6.	P-FATBCK	PK SKND FATBACKS	Y	.0000	BIG	% GW	0.0000	HI-LIMIT
7.	P-JOWL	PK SKND JOWLS	Y	.0000	BIG	% GW	0.0000	LO-LIMIT
8.	P-50	PK 50% TRIMMINGS	Y	.0000	BIG	% GW	29.0407	BETWEEN
9.	P-80	PK 80% TRIMMINGS	Y	.0000	BIG	% GW	46.0651	BETWEEN
10.	I-CORN	CORN SYRUP		1.0000	1.0000	% BW	1.0000	BETWEEN
11.	I-SALT	SALT		.0000	3.0000	% GW	2.6769	BETWEEN
12.	S-FRANK	FRANK SPICE		.7500	.7500	LBS	0.7500	BETWEEN
13.	WATER	WATER		.0000	BIG	% GW	24.8942	BETWEEN
14.	P-CHEEK	PK CHEEK MEAT	Y	.0000	BIG	% GW	0.0000	LO-LIMIT

<--- REQUIREMENTS IN PRODUCT --->

SEQ NO.	<-- REQUIREMENT --> ITEM-1	ITEM-2	REQ. TYPE	ACCEPTABLE RANGE OF VALUES LO LIMIT	HI LIMIT	LIMIT TYPE	LAST CALC. VALUE	CURRENT STATUS
1	MOISTURE		A	-BIG	BIG	% FW	55.4230	BETWEEN
2	PROTEIN		A	-BIG	BIG	% FW	11.6058	BETWEEN
3	FAT		A	25.0000	29.0000	% FW	29.0000	HI-LIMIT
4	BIND		A	180.0000	BIG	% FW	201.4825	BETWEEN
5	COLOR		A	190.0000	BIG	% FW	280.9178	BETWEEN
6	SALT		A	2.2500	2.2500	% FW	2.2500	HI-LIMIT
7	COLLAGEN		A	-BIG	3.7500	% FW	3.7500	HI-LIMIT
8	USDA-AW		A	-BIG	9.0000	% FW	9.0000	HI-LIMIT
9	P	> B	GG	.0000	BIG	% GW	38.8270	BETWEEN
10	B	> WATER	GM	.0000	BIG	% GW	-0.0000	LO-LIMIT

For protein, there are really no requirements. The least cost formulation system will use as little protein as possible, because it is an expensive attribute. From the analysis of the standard formula, set the low limit at 10 percent finished weight (FW), and leave the high limit unbounded at BIG.

For moisture, leave it completely unbounded, because it will be controlled by the USDA added water (AW).

For USDA-AW, the USDA regulation limits the maximum amount to 10 percent finished weight (FW).

Because water is an inexpensive ingredient, the least cost formulation system will include as much as possible. To give a margin of safety, set the high limit at 9 percent finished weight (FW).

Bind and color are dependent on the protein content of the raw materials. The least cost formulation system will use as little protein as possible, because it is an expensive attribute. Therefore, the bind and color will be close to their low limits, and the high limits are unbounded at BIG.

The values 180 for bind and 190 for color are based upon the analysis of many hundreds of formulations.

The identical limits for salt will ensure that exactly 2¼ pounds of salt will be used. This requirement comes from both the choice of label and the marketplace preference.

The limits for collagen were chosen to prevent excessive gelatin formulation in the product. An upper limit of 3.75 percent of finished weight was used.

At this point we have taken care of the attributes we wish to control in the regular frank. The next step is to insure that product label constraints are met.

One of the unique features of the least cost formulator is its group capability. The group parameter allows the handling of a number of raw materials as a single item. For example, all beef raw materials may be designated by Group B. In Table 20-6, all the raw materials have been categorized into one or more groups.

To ensure that the batch meets the requirements of the chosen label, specify that all the pounds of pork raw materials (P) are greater than the pounds of beef raw materials (B) by Group P > B. Similarly, specify beef greater than water (W) by Group B > W. The other items in the label (salt, corn syrup, and spices) have been fixed by specifying their individual limits. In Table 20-8, the complete product specifications are shown for the regular frank.

Least Cost Formulation: Final Reports

Product Formulation Material Usage Report

Table 20-9 shows the calculated cost of the regular frank and the quantities of raw materials required to achieve these results. Of 10 meat raw materials offered, only 4 were used. This resulted in a raw material price for regular franks

Table 20-9. Product Formulation Material Usage Report for Regular Frank

```
        PRODUCT NAMES - SHORT: ISU-REG      LONG: ISU REGULAR FRANK    TYPE: MIN
        LAST UPDATED:          7/13/81      LAST CALCULATED:  7/13/81

                        CALC. #   3 BASED ON COST #   1

                        <- CALCULATED COST ($ PER LB) ->
                        COST #1   COST #2   COST #3
                        ------------------------------------
        BLOCK WEIGHT:    0.7800    0.7937   0.0000
        GROSS WEIGHT:    0.6031    0.6137   0.0000
        FINISHED WEIGHT: 0.6556    0.6671   0.0000
```

SEQ NO.	MATERIAL NAME SHORT	LONG	USAGE	COST # 1	EXTENDED COST # 1	BW	GW	FW	LO-LIMIT	HI-LIMIT	TYPE
1.	B-CHEEK	BF CHEEK MEAT	16.4	0.9600	15.71	16.37	12.66	13.76	.0000	BIG	% GW
2.	B-COW	BF FC COW MEAT	8.5	1.4700	12.53	8.53	6.59	7.17	.0000	BIG	% GW
3.	P-50	PK 50% TRIMMINGS	29.0	0.4200	12.20	29.04	22.46	24.41	.0000	BIG	% GW
4.	P-80	PK 80% TRIMMINGS	46.1	0.8000	36.85	46.07	35.62	38.72	.0000	BIG	% GW
		BLOCK WEIGHT (BW):	100.0 LBS		77.30	100.00	77.33	84.05			
5.	I-CORN	CORN SYRUP	1.0	0.3000	0.30	1.00	0.77	0.84	1.0000	1.0000	% BW
6.	I-SALT	SALT	2.7	0.0550	0.15	2.68	2.07	2.25	.0000	3.0000	% GW
7.	S-FRANK	FRANK SPICE	0.7	0.3400	0.25	0.75	0.58	0.63	.7500	.7500	LBS
8.	WATER	WATER	24.9	0.0000	0.00	24.89	19.25	20.92	.0000	BIG	% GW
		GROSS WEIGHT (GW):	129.3 LBS		78.00	129.32	100.00	108.70			
		SHRINKAGE LOSS:	10.3			10.35	8.00	8.70			
		FINISHED WEIGHT:	119.0 LBS		78.00	118.98	92.00	100.00			

of 66¢/pound (total cost ÷ finished weight = 78/119 = $.66). When the standard formula was used, the cost per pound was calculated at 69¢, using the same day's raw material prices. This represents a savings of 3¢/pound. In large operations which produce over 100,000 pounds of franks weekly, this can amount to a savings of $3,000/week for this example.

Product Formulation Requirements Report

This report displays the predicted value of the attributes for the product and is shown in Table 20-10 in the column labeled "Value." All calculations fall between the low and high limits. As expected, the predicted values for fat and USDA-AW are at the high limits, while the values for protein, bind, and color are near the low limits.

Product Formulation Price-ranging Report

Table 20-11 is divided into two groups of raw materials: those that were used in the formulation and those that were excluded. Those included offered the right combination of attributes for a given price. For example, item #2, B-Cow, costs $1.47/pound; if this raw material were to drop to $1.21/pound, more of it would be included; if B-Cow rose in price to $4.31/pound, less of it would be used. Notice that P-50 is very close to the price where less would be

Table 20-10. Product Formulation Requirements Report for Regular Frank

```
PRODUCT NAMES - SHORT: ISU-REG      LONG: ISU REGULAR FRANK     TYPE: MIN
LAST UPDATED:          7/13/81      LAST CALCULATED:   7/13/81
```

			LAST CALCULATION # 3 USING COST # 1				
SEQ		(------ REQUIREMENT ------)	VALUE	(----- LIMITS ON RANGE ----)			PENALTY
NO.	TYPE	DESCRIPTION	VALUE	LOW-LIMIT	HI-LIMIT	TYPE	COST
1	A	MOISTURE	55.4230	-BIG	BIG	% FW	0.0000
2	A	PROTEIN	11.6058	-BIG	BIG	% FW	0.0000
3	A	FAT	29.0000	25.0000	29.0000	% FW	0.0159
4	A	BIND	201.4825	180.0000	BIG	% FW	0.0000
5	A	COLOR	280.9178	190.0000	BIG	% FW	0.0000
6	A	SALT	2.2500	2.2500	2.2500	% FW	0.0157
7	A	COLLAGEN	3.7500	-BIG	3.7500	% FW	0.0848
8	A	USDA-AW	9.0000	-BIG	9.0000	% FW	0.0105
9	GG	P > B	38.8270	.0000	BIG	% GW	0.0000
10	GM	B > WATER	-0.0000	.0000	BIG	% GW	0.0064

```
NOTE: "PENALTY-COST" IS THE AMOUNT IN $ PER LB BY WHICH THE FORMULATION COST COULD BE REDUCED
      IF THE REQUIREMENT LIMIT WAS CHANGED BY 1.0 AS GIVEN. "VALUE" IS GIVEN IN THE SAME UNITS
      AS THE LIMITS.
```

Table 20-11. Product Formulation Price-ranging Report for Regular Frank

```
PRODUCT NAMES - SHORT: ISU-REG   LONG: ISU REGULAR FRANK     TYPE: MIN
LAST UPDATED:          7/13/81   LAST CALCULATED:   7/13/81

                          CALC. #   3 ... USING COST # 1

                          (-- CALC. COST ($ PER LB) --)
                          COST #1   COST #2   COST #3
                          -------------------------------
          BLOCK WEIGHT:    0.7800    0.7937    0.0000
          GROSS WEIGHT:    0.6031    0.6137    0.0000
          FINISHED WEIGHT: 0.6556    0.6671    0.0000
```

SEQ		(----- MATERIAL NAME -----)	USAGE	(-- COST RANGE --)		(---- CURRENT COSTS ----)		
NO.	SHORT	LONG	(LBS)	USE-MORE	USE-LESS	COST-1	COST-2	COST-3
1.	B-CHEEK	BF CHEEK MEAT	16.367	-20.8668	1.1611	0.9600	0.9800	0.0000
2.	B-COW	BF FC COW MEAT	8.527	1.2068	4.3138	1.4700	1.4500	0.0000
3.	P-50	PK 50% TRIMMINGS	29.041	-.6438	.4335	0.4200	0.4300	0.0000
4.	P-80	PK 80% TRIMMINGS	46.065	.6837	1.1456	0.8000	0.8200	0.0000
5.	I-CORN	CORN SYRUP	1.000	-86.3779	2187.7686	0.3000	0.3000	0.0000
6.	I-SALT	SALT	2.677	-BIG	BIG	0.0550	0.0550	0.0000
7.	S-FRANK	FRANK SPICE	0.750	-115.2305	2916.9653	0.3400	0.3400	0.0000
8.	WATER	WATER	24.894	-21.4985	.8519	0.0000	0.0000	0.0000
9.	B-PLATE	BF PLATES	0.000	-BIG	.9572	0.5200	0.5200	0.0000
10.	B-50	BF 50% TRIMMINGS	0.000	-BIG	.9366	0.6400	0.6450	0.0000
11.	B-85	BF 85% TRIMMINGS	0.000	-BIG	1.3823	1.1500	1.1700	0.0000
12.	P-FATBCK	PK SKND FATBACKS	0.000	-BIG	.3737	0.2700	0.2750	0.0000
13.	P-JOWL	PK SKND JOWLS	0.000	.3452	BIG	0.3600	0.3650	0.0000
14.	P-CHEEK	PK CHEEK MEAT	0.000	.3427	BIG	0.8900	0.9000	0.0000

```
NOTE: "COST-TO-USE-MORE" IS THE PRICE AT WHICH THE FORMULATION WILL START TO USE MORE OF THE MATERIAL.
      "COST-TO-USE-LESS" IS THE PRICE AT WHICH LESS OF THE MATERIAL WILL BE USED IN THE FORMULATION.
```

used. However, the negative number in the "Use-More" column means that no additional P-50 would be included in the formulation regardless of a lower price, due to other constraints (fat content).

For those raw materials (#9 through #14) that were not included in the formulation, the "Use-More" column represents the price at which those raw materials will enter the formulation. Hence P-Jowl that costs 36¢/pound will enter at 34½¢/pound. Other than P-Cheek, all the rest of the raw materials will not enter regardless of a lower price, due to lack of availability.

With these price ranges, it is possible to improve the advance purchase of the raw materials. If a slaughter operation is involved, it is possible to decide if the raw material is better used in the products or sold in the open market, by comparing *The Yellow Sheet*[9] or other pricing services with the "Use-More" prices.

Other Reports

The Least Cost Formulator® provides a wide range of additional reports that keep a running inventory of raw materials and finished product on hand; raw material usage by day, week, month, and year-to-date; and production formulation tickets.

Other Frank Formulations

For each of the other products to be made, additional raw materials were offered, but the product specifications for moisture, protein, and fat remained the same. The identical reports for each product are shown following their description.

Regular Frank with By-products

To produce this #2 regular frank, the formulation includes two additional raw materials: beef hearts and pork lips. The label will read "pork, water, pork lips, beef hearts, beef, salt, corn syrup, spices, and nitrite."

Except for bind and collagen, the low and high limits of the attributes for this product are identical to those of the #1 regular frank. The reports for this by-product (BP) frank are shown in Tables 20-12 through 20-15.

Regular Frank with Isolated Soy Protein

To produce this frank, we will force into the formulation isolated soy protein (ISP) in the amount of 2 percent of the finished weight of the product. The label will read "pork, beef, water, salt, ISP, corn syrup, spices, and nitrite."

---

[9]*The Yellow Sheet. The National Provisioner* Daily Market Service, 15 West Huron Street, Chicago, IL 60610.

Table 20-12. Product Formulation Material Usage Report for BP Frank

PRODUCT NAMES - SHORT: ISU-BP   LONG: ISU BYPRODUCT FRANK   TYPE: MIN
LAST UPDATED: 7/13/81   LAST CALCULATED: 7/13/81

CALC. # 3 BASED ON COST # 1

<- CALCULATED COST ($ PER LB) ->
COST #1   COST #2   COST #3

BLOCK WEIGHT: 0.6172   0.6293   0.0000
GROSS WEIGHT: 0.4873   0.4970   0.0000
FINISHED WEIGHT: 0.5297   0.5402   0.0000

SEQ NO.	<----- MATERIAL NAME -----> SHORT	LONG	USAGE	COST # 1	EXTENDED COST # 1	<--- PERCENTAGE OF ---> BW	GW	FW	<----- LIMITS ON USE -----> LO-LIMIT	HI-LIMIT	TYPE
1.	B-COW	BF FC COW MEAT	2.6	1.4700	3.85	2.62	2.07	2.25	.0000	BIG	% GW
2.	P-50	PK 50% TRIMMINGS	26.8	0.4200	11.27	26.83	21.19	23.03	.0000	BIG	% GW
3.	P-80	PK 80% TRIMMINGS	43.2	0.8000	34.53	43.16	34.08	37.05	.0000	BIG	% GW
4.	B-HEART	BF HEARTS	13.7	0.4900	6.71	13.69	10.81	11.75	.0000	BIG	% GW
5.	P-LIP	PK LIPS	13.7	0.3400	4.66	13.69	10.81	11.75	.0000	BIG	% GW
		BLOCK WEIGHT (BW):	100.0	LBS	61.02	100.00	78.97	85.83			
6.	I-CORN	CORN SYRUP	1.0	0.3000	0.30	1.00	0.79	0.86	1.0000	1.0000	% BW
7.	I-SALT	SALT	2.6	0.0550	0.14	2.62	2.07	2.25	.0000	3.0000	% GW
8.	S-FRANK	FRANK SPICE	0.8	0.3400	0.25	0.75	0.59	0.64	.7500	.7500	LBS
9.	WATER	WATER	22.3	0.0000	0.00	22.26	17.58	19.11	.0000	BIG	% GW
		GROSS WEIGHT (GW):	126.6	LBS	61.72	126.63	100.00	108.70			
		SHRINKAGE LOSS:	10.1			10.13	8.00	8.70			
		FINISHED WEIGHT:	116.5	LBS	61.72	116.50	92.00	100.00			

Table 20-13. Product Specifications for BP Frank

```
PRODUCT SHORT NAME:  ISU-BP              AS OF:        7/13/81
PRODUCT LONG NAME:   ISU BYPRODUCT FRANK  PRODUCT TYPE:    F      GROUPS:

LAST COSTS:    $    0.5297      $    0.5402      $    0.0000
LAST CALCULATION USED COST # 1          DATE LAST CALCULATION: 7/13/81    CALC. #   3
MIN OR MAX:         MIN

NO. RAW MATERIALS:      16        PRODUCT BLOCK SIZE:     100.0000 (SCALED ON BW)
NO. REQUIREMENTS:       13        SHRINKAGE RATE:           8.0000 %
PRODUCT DISPLAY UNITS: LBS
```

<--- RAW MATERIALS IN PRODUCT --->   (NAMES FROM TABLE: ISU MEAT COURSE)

SEQ NO.	SHORT NAME	RAW MATERIAL DESCRIPTION	IN BLOCK	ACCEPTABLE RANGE OF VALUES LO LIMIT	HI LIMIT	LIMIT TYPE	CURRENT VALUE	CURRENT STATUS
1.	B-CHEEK	BF CHEEK MEAT	Y	.0000	BIG	% GW	0.0000	LO-LIMIT
2.	B-COW	BF FC COW MEAT	Y	.0000	BIG	% GW	2.6213	BETWEEN
3.	B-PLATE	BF PLATES	Y	.0000	BIG	% GW	0.0000	HI-LIMIT
4.	B-50	BF 50% TRIMMINGS	Y	.0000	BIG	% GW	0.0000	HI-LIMIT
5.	B-85	BF 85% TRIMMINGS	Y	.0000	BIG	% GW	0.0000	HI-LIMIT
6.	P-FATBCK	PK SKND FATBACKS	Y	.0000	BIG	% GW	0.0000	HI-LIMIT
7.	P-JOWL	PK SKND JOWLS	Y	.0000	BIG	% GW	0.0000	LO-LIMIT
8.	P-50	PK 50% TRIMMINGS	Y	.0000	BIG	% GW	26.8319	BETWEEN
9.	P-80	PK 80% TRIMMINGS	Y	.0000	BIG	% GW	43.1589	BETWEEN
10.	I-CORN	CORN SYRUP		1.0000	1.0000	% BW	1.0000	BETWEEN
11.	I-SALT	SALT		.0000	3.0000	% GW	2.6213	BETWEEN
12.	S-FRANK	FRANK SPICE		.7500	.7500	LBS	0.7500	BETWEEN
13.	WATER	WATER		.0000	BIG	% GW	22.2632	BETWEEN
14.	B-HEART	BF HEARTS	Y	.0000	BIG	% GW	13.6939	BETWEEN
15.	P-LIP	PK LIPS	Y	.0000	BIG	% GW	13.6939	BETWEEN
16.	P-CHEEK	PK CHEEK MEAT	Y	.0000	BIG	% GW	0.0000	LO-LIMIT

<--- REQUIREMENTS IN PRODUCT --->

SEQ NO.	REQUIREMENT ITEM-1	ITEM-2	REQ. TYPE	ACCEPTABLE RANGE OF VALUES LO LIMIT	HI LIMIT	LIMIT TYPE	LAST CALC. VALUE	CURRENT STATUS
1	MOISTURE		A	-BIG	BIG	% FW	55.4038	BETWEEN
2	PROTEIN		A	-BIG	BIG	% FW	11.6009	BETWEEN
3	FAT		A	25.0000	29.0000	% FW	29.0000	BETWEEN
4	BIND		A	130.0000	BIG	% FW	153.3243	BETWEEN
5	COLOR		A	190.0000	BIG	% FW	200.2796	BETWEEN
6	SALT		A	2.2500	2.2500	% FW	2.2500	HI-LIMIT
7	COLLAGEN		A	-BIG	4.2000	% FW	4.2000	BETWEEN
8	USDA-AW		A	-BIG	9.0000	% FW	9.0000	BETWEEN
9	P	> WATER	GM	.0000	BIG	% GW	37.6893	BETWEEN
10	WATER	> P-LIP	MM	.0000	BIG	% GW	6.7669	BETWEEN
11	P-LIP	> B-HEART	MM	.0000	BIG	% GW	0.0000	LO-LIMIT
12	B-HEART	> B	MG	.0000	BIG	% GW	8.7437	BETWEEN

<--- REQUIREMENTS IN PRODUCT --->

SEQ NO.	REQUIREMENT ITEM-1	ITEM-2	REQ. TYPE	ACCEPTABLE RANGE OF VALUES LO LIMIT	HI LIMIT	LIMIT TYPE	LAST CALC. VALUE	CURRENT STATUS
13	B	> SALT	GA	.0000	BIG	% GW	0.0000	LO-LIMIT

Table 20-14. Product Formulation Requirements Report for BP Frank

```
PRODUCT NAMES - SHORT: ISU-BP      LONG: ISU BYPRODUCT FRANK   TYPE: MIN
LAST UPDATED:          7/13/81      LAST CALCULATED:  7/13/81
```

				LAST CALCULATION #	3	USING COST #	1	

SEQ NO.	TYPE	REQUIREMENT DESCRIPTION		VALUE	LIMITS ON RANGE LOW-LIMIT	HI-LIMIT	TYPE	PENALTY COST
1	A	MOISTURE		55.4038	-BIG	BIG	% FW	0.0000
2	A	PROTEIN		11.6009	-BIG	BIG	% FW	0.0000
3	A	FAT		29.0000	25.0000	29.0000	% FW	0.0152
4	A	BIND		153.3243	130.0000	BIG	% FW	0.0000
5	A	COLOR		200.2796	190.0000	BIG	% FW	0.0000
6	A	SALT		2.2500	2.2500	2.2500	% FW	0.0094
7	A	COLLAGEN		4.2000	-BIG	4.2000	% FW	0.0844
8	A	USDA-AW		9.0000	-BIG	9.0000	% FW	0.0158
9	GM	P >	WATER	37.6893	.0000	BIG	% GW	0.0000
10	MM	WATER >	P-LIP	6.7669	.0000	BIG	% GW	0.0000
11	MM	P-LIP >	B-HEART	0.0000	.0000	BIG	% GW	0.0034
12	MG	B-HEART >	B	8.7437	.0000	BIG	% GW	0.0000
13	GA	B >	SALT	0.0000	.0000	BIG	% GW	0.0062

NOTE: "PENALTY-COST" IS THE AMOUNT IN $ PER LB BY WHICH THE FORMULATION COST COULD BE REDUCED
      IF THE REQUIREMENT LIMIT WAS CHANGED BY 1.0 AS GIVEN. "VALUE" IS GIVEN IN THE SAME UNITS
      AS THE LIMITS.

Table 20-15. Product Formulation Price-ranging Report for BP Frank

```
PRODUCT NAMES - SHORT: ISU-BP   LONG: ISU BYPRODUCT FRANK   TYPE: MIN
LAST UPDATED:        7/13/81   LAST CALCULATED:  7/13/81
```

```
CALC. #    3 ... USING COST # 1
```

			CALC. COST ($ PER LB)		
			COST #1	COST #2	COST #3
	BLOCK WEIGHT:		0.6172	0.6293	0.0000
	GROSS WEIGHT:		0.4873	0.4970	0.0000
	FINISHED WEIGHT:		0.5297	0.5402	0.0000

SEQ NO.	MATERIAL NAME SHORT	LONG	USAGE (LBS)	COST RANGE USE-MORE	USE-LESS	CURRENT COSTS COST-1	COST-2	COST-3
1.	B-COW	BF FC COW MEAT	2.621	.9768	1.4784	1.4700	1.4500	0.0000
2.	P-50	PK 50% TRIMMINGS	26.832	-.7271	.4415	0.4200	0.4300	0.0000
3.	P-80	PK 80% TRIMMINGS	43.159	.7942	1.1377	0.8000	0.8200	0.0000
4.	I-CORN	CORN SYRUP	1.000	-153.8512	10.6458	0.3000	0.3000	0.0000
5.	I-SALT	SALT	2.621	-BIG	BIG	0.0550	0.0550	0.0000
6.	S-FRANK	FRANK SPICE	0.750	-205.1950	14.1344	0.3400	0.3400	0.0000
7.	WATER	WATER	22.263	-.1017	1.5150	0.0000	0.0000	0.0000
8.	B-HEART	BF HEARTS	13.694	-37.3937	.5046	0.4900	0.4900	0.0000
9.	P-LIP	PK LIPS	13.694	-4.1098	.3546	0.3400	0.3500	0.0000
10.	B-CHEEK	BF CHEEK MEAT	0.000	.9516	BIG	0.9600	0.9800	0.0000
11.	B-PLATE	BF PLATES	0.000	-BIG	.9583	0.5200	0.5200	0.0000
12.	B-50	BF 50% TRIMMINGS	0.000	-BIG	.9412	0.6400	0.6450	0.0000
13.	B-85	BF 85% TRIMMINGS	0.000	-BIG	1.3857	1.1500	1.1700	0.0000
14.	P-FATBCK	PK SKND FATBACKS	0.000	-BIG	.3720	0.2700	0.2750	0.0000
15.	P-JOWL	PK SKND JOWLS	0.000	.3361	BIG	0.3600	0.3650	0.0000
16.	P-CHEEK	PK CHEEK MEAT	0.000	.2870	BIG	0.8900	0.9000	0.0000

NOTE: "COST-TO-USE-MORE" IS THE PRICE AT WHICH THE FORMULATION WILL START TO USE MORE OF THE MATERIAL.
      "COST-TO-USE-LESS" IS THE PRICE AT WHICH LESS OF THE MATERIAL WILL BE USED IN THE FORMULATION.

The limits for the attributes for this product are identical to those of the #1 regular frank shown in Table 20-8. The reports for this product are shown in Tables 20-16 through 20-19.

Regular Frank with By-products and Isolated Soy Protein

To produce this frank, we will include all the raw materials that were used in both the second and the third formulations. The label will read "pork, water, pork lips, beef heart, beef, salt, ISP, corn syrup, spices, and nitrite." The limits for the attributes on this product are identical to those of the #2 regular frank with by-products shown in Table 20-13. The reports for this product are shown in Tables 20-20 through 20-23.

When one briefly reviews the results of the least cost formulation examples, it can be seen that considerable cost variations are encountered for the four frankfurter formulations. These cost variations are summarized here.

Regular frankfurter (Table 20-9)	$65.56/cwt.
By-product (BP) frankfurter (Table 20-12)	$52.97/cwt.
Isolated soy protein (ISP) frankfurter (Table 20-16)	$60.35/cwt.
BP-ISP frankfurter (Table 20-20)	$46.82/cwt.

Table 20-16. Product Formulation Material Usage Report for ISP Frank

```
PRODUCT NAMES - SHORT: ISU-ISP      LONG: ISU FRANK WITH ISP    TYPE: MIN
LAST UPDATED:         7/13/81       LAST CALCULATED:   7/13/81

                     CALC. #    2 BASED ON COST #    1

                     <- CALCULATED COST ($ PER LB) ->
                     COST #1   COST #2   COST #3
                     -----------------------------------
     BLOCK WEIGHT:    0.7793    0.7914    0.0000
     GROSS WEIGHT:    0.5552    0.5638    0.0000
     FINISHED WEIGHT: 0.6035    0.6128    0.0000
```

SEQ NO.	MATERIAL NAME SHORT	LONG	USAGE	COST #1	EXTENDED COST #1	BW	GW	FW	LO-LIMIT	HI-LIMIT	TYPE
1.	B-CHEEK	BF CHEEK MEAT	23.6	0.9600	22.64	23.58	16.80	18.26	.0000	BIG	% GW
2.	B-COW	BF FC COW MEAT	9.5	1.4700	14.03	9.55	6.80	7.39	.0000	BIG	% GW
3.	P-50	PK 50% TRIMMINGS	41.0	0.4200	17.20	40.95	29.18	31.71	.0000	BIG	% GW
4.	P-80	PK 80% TRIMMINGS	25.9	0.8000	20.74	25.92	18.47	20.07	.0000	BIG	% GW
		BLOCK WEIGHT (BW):	100.0 LBS		74.61	100.00	71.24	77.44			
5.	I-CORN	CORN SYRUP	1.0	0.3000	0.30	1.00	0.71	0.77	1.0000	1.0000	% BW
6.	I-SALT	SALT	2.9	0.0550	0.16	2.91	2.07	2.25	.0000	3.0000	% GW
7.	S-FRANK	FRANK SPICE	0.7	0.3400	0.25	0.75	0.53	0.58	.7500	.7500	LBS
8.	WATER	WATER	33.1	0.0000	0.00	33.13	23.60	25.65	.0000	BIG	% GW
9.	I-ISP	ISOLATED SOY PROTEIN	2.6	1.0100	2.61	2.58	1.84	2.00	2.0000	2.0000	% FW
		GROSS WEIGHT (GW):	140.4 LBS		77.93	140.37	100.00	108.70			
		SHRINKAGE LOSS:	11.2			11.23	8.00	8.70			
		FINISHED WEIGHT:	129.1 LBS		77.93	129.14	92.00	100.00			

Table 20-17. Product Specifications for ISP Frank

```
PRODUCT SHORT NAME:  ISU-ISP              AS OF:        7/13/81
PRODUCT LONG NAME:   ISU FRANK WITH ISP   PRODUCT TYPE:  F       GROUPS:

LAST COSTS:     $    0.6035      $    0.6128     $    0.0000
LAST CALCULATION USED COST # 1        DATE LAST CALCULATION: 7/13/81    CALC. #   2
MIN OR MAX:          MIN

NO. RAW MATERIALS:       15       PRODUCT BLOCK SIZE:    100.0000 (SCALED ON BW)
NO. REQUIREMENTS:        10       SHRINKAGE RATE:          8.0000 %
PRODUCT DISPLAY UNITS: LBS
```

<--- RAW MATERIALS IN PRODUCT --->   (NAMES FROM TABLE: ISU MEAT COURSE)

SEQ NO.	SHORT NAME	RAW MATERIAL DESCRIPTION	IN BLOCK	ACCEPTABLE RANGE OF VALUES LO LIMIT	HI LIMIT	LIMIT TYPE	CURRENT VALUE	CURRENT STATUS
1.	B-CHEEK	BF CHEEK MEAT	Y	.0000	BIG	% GW	23.5800	BETWEEN
2.	B-COW	BF FC COW MEAT	Y	.0000	BIG	% GW	9.5475	BETWEEN
3.	B-PLATE	BF PLATES	Y	.0000	BIG	% GW	0.0000	HI-LIMIT
4.	B-50	BF 50% TRIMMINGS	Y	.0000	BIG	% GW	0.0000	HI-LIMIT
5.	B-85	BF 85% TRIMMINGS	Y	.0000	BIG	% GW	0.0000	HI-LIMIT
6.	P-FATBCK	PK SKND FATBACKS	Y	.0000	BIG	% GW	0.0000	HI-LIMIT
7.	P-JOWL	PK SKND JOWLS	Y	.0000	BIG	% GW	0.0000	LO-LIMIT
8.	P-50	PK 50% TRIMMINGS	Y	.0000	BIG	% GW	40.9518	BETWEEN
9.	P-80	PK 80% TRIMMINGS	Y	.0000	BIG	% GW	25.9207	BETWEEN
10.	I-CORN	CORN SYRUP		1.0000	1.0000	% BW	1.0000	BETWEEN
11.	I-SALT	SALT		.0000	3.0000	% GW	2.9056	BETWEEN
12.	S-FRANK	FRANK SPICE		.7500	.7500	LBS	0.7500	BETWEEN
13.	WATER	WATER		.0000	BIG	% GW	33.1275	BETWEEN
14.	P-CHEEK	PK CHEEK MEAT	Y	.0000	BIG	% GW	0.0000	LO-LIMIT
15.	X-ISP	ISOLATED SOY PROTEIN		2.0000	2.0000	% FW	2.5827	HI-LIMIT

<--- REQUIREMENTS IN PRODUCT --->

SEQ NO.	<-- REQUIREMENT --> ITEM-1	ITEM-2	REQ. TYPE	ACCEPTABLE RANGE OF VALUES LO LIMIT	HI LIMIT	LIMIT TYPE	LAST CALC. VALUE	CURRENT STATUS
1	MOISTURE		A	-BIG	BIG	% FW	55.5938	BETWEEN
2	PROTEIN		A	-BIG	BIG	% FW	11.6484	BETWEEN
3	FAT		A	25.0000	29.0000	% FW	29.0000	HI-LIMIT
4	BIND		A	180.0000	BIG	% FW	231.2576	BETWEEN
5	COLOR		A	190.0000	BIG	% FW	277.2159	BETWEEN
6	SALT		A	2.2500	2.2500	% FW	2.2500	HI-LIMIT
7	COLLAGEN		A	-BIG	3.7500	% FW	3.7500	HI-LIMIT
8	USDA-AW		A	-BIG	9.0000	% FW	9.0000	HI-LIMIT
9	P	> B	GG	.0000	BIG	% GW	24.0408	BETWEEN
10	B	> WATER	GW	.0000	BIG	% GW	0.0000	LO-LIMIT

Table 20-18. Product Formulation Requirements Report for ISP Frank

```
PRODUCT NAMES - SHORT: ISU-ISP      LONG: ISU FRANK WITH ISP    TYPE: MIN
LAST UPDATED:           7/13/81      LAST CALCULATED:   7/13/81


                            LAST CALCULATION #   2  USING COST #   1
SEQ  <------ REQUIREMENT ------>             <----- LIMITS ON RANGE ----->    PENALTY
NO.  TYPE       DESCRIPTION        VALUE    LOW-LIMIT     HI-LIMIT   TYPE      COST
-------------------------------------------------------------------------------------
  1  A    MOISTURE                55.5938      -BIG          BIG     % FW     0.0000
  2  A    PROTEIN                 11.6484      -BIG          BIG     % FW     0.0000
  3  A    FAT                     29.0000    25.0000      29.0000    % FW     0.0172
  4  A    BIND                   231.2576   180.0000          BIG    % FW     0.0000
  5  A    COLOR                  277.2159   190.0000          BIG    % FW     0.0000
  6  A    SALT                     2.2500     2.2500       2.2500    % FW     0.0170
  7  A    COLLAGEN                 3.7500      -BIG          3.7500   % FW     0.0921
  8  A    USDA-AW                  9.0000      -BIG          9.0000   % FW     0.0113
  9  GG   P       )    B          24.0408      .0000          BIG    % GW     0.0000
 10  GM   B       )    WATER       0.0000      .0000          BIG    % GW     0.0070
```

NOTE: "PENALTY-COST" IS THE AMOUNT IN $ PER LB BY WHICH THE FORMULATION COST COULD BE REDUCED
      IF THE REQUIREMENT LIMIT WAS CHANGED BY 1.0 AS GIVEN. "VALUE" IS GIVEN IN THE SAME UNITS
      AS THE LIMITS.

Table 20-19. Production Formulation Price-ranging Report for ISP Frank

```
PRODUCT NAMES - SHORT: ISU-ISP   LONG: ISU FRANK WITH ISP    TYPE: MIN
LAST UPDATED:           7/13/81   LAST CALCULATED:   7/13/81


                       CALC. #    2 ... USING COST # 1

                       <-- CALC. COST ($ PER LB) -->
                       COST #1    COST #2    COST #3
                       ------------------------------
              BLOCK WEIGHT:    0.7793    0.7914    0.0000
              GROSS WEIGHT:    0.5552    0.5638    0.0000
              FINISHED WEIGHT: 0.6035    0.6128    0.0000
SEQ  <----- MATERIAL NAME ----->   USAGE   <-- COST RANGE -->   <---- CURRENT COSTS ---->
NO.  SHORT     LONG               (LBS )  USE-MORE   USE-LESS   COST-1    COST-2   COST-3
------------------------------------------------------------------------------------------
 1.  B-CHEEK  BF CHEEK MEAT       23.580  -20.8668    1.1611    0.9600    0.9800   0.0000
 2.  B-COW    BF FC COW MEAT       9.548    1.2068    4.3138    1.4700    1.4500   0.0000
 3.  P-50     PK 50% TRIMMINGS    40.952    -.6438     .4335    0.4200    0.4300   0.0000
 4.  P-80     PK 80% TRIMMINGS    25.921     .6837    1.1456    0.8000    0.8200   0.0000
 5.  I-CORN   CORN SYRUP           1.000  -86.3779 2187.7686    0.3000    0.3000   0.0000
 6.  I-SALT   SALT                 2.906      -BIG     .0550    0.0550    0.0550   0.0000
 7.  S-FRANK  FRANK SPICE          0.750 -115.2305 2916.9653    0.3400    0.3400   0.0000
 8.  WATER    WATER               33.127  -21.4985    .8519    0.0000    0.0000   0.0000
 9.  X-ISP    ISOLATED SOY PROTEIN 2.583      -BIG    3.6153    1.0100    1.0100   0.0000

10.  B-PLATE  BF PLATES            0.000      -BIG     .9572    0.5200    0.5200   0.0000
11.  B-50     BF 50% TRIMMINGS     0.000      -BIG     .9366    0.6400    0.6450   0.0000
12.  B-85     BF 85% TRIMMINGS     0.000      -BIG    1.3823    1.1500    1.1700   0.0000
13.  P-FATBCK PK SKND FATBACKS     0.000      -BIG     .3737    0.2700    0.2750   0.0000
14.  P-JOWL   PK SKND JOWLS        0.000     .3452      BIG     0.3600    0.3650   0.0000
15.  P-CHEEK  PK CHEEK MEAT        0.000     .3427      BIG     0.8900    0.9000   0.0000
```

NOTE: "COST-TO-USE-MORE" IS THE PRICE AT WHICH THE FORMULATION WILL START TO USE MORE OF THE MATERIAL.
      "COST-TO-USE-LESS" IS THE PRICE AT WHICH LESS OF THE MATERIAL WILL BE USED IN THE FORMULATION.

Table 20-20. Product Formulation Material Usage Report for BP-ISP Frank

```
PRODUCT NAMES - SHORT: ISU-BPI    LONG: ISU BYPRODUCT + ISP   TYPE: MIN
LAST UPDATED:         7/13/81      LAST CALCULATED:   7/13/81

                    CALC. #    2 BASED ON COST #    1

                        (- CALCULATED COST ($ PER LB) -)
                        COST #1   COST #2   COST #3
                        -----------------------------------
            BLOCK WEIGHT:    0.5912    0.6032    0.0000
            GROSS WEIGHT:    0.4307    0.4394    0.0000
            FINISHED WEIGHT: 0.4682    0.4776    0.0000
```

SEQ NO.	SHORT	LONG	USAGE	COST #1	EXTENDED COST #1	BW	GW	FW	LO-LIMIT	HI-LIMIT	TYPE
1.	B-CHEEK	BF CHEEK MEAT	6.4	0.9600	6.12	6.38	4.65	5.05	.0000	BIG	% GW
2.	P-50	PK 50% TRIMMINGS	37.0	0.4200	15.54	37.01	26.96	29.30	.0000	BIG	% GW
3.	P-80	PK 80% TRIMMINGS	27.8	0.8000	22.23	27.79	20.24	22.00	.0000	BIG	% GW
4.	B-HEART	BF HEARTS	14.4	0.4900	7.06	14.41	10.50	11.41	.0000	BIG	% GW
5.	P-LIP	PK LIPS	14.4	0.3400	4.90	14.41	10.50	11.41	.0000	BIG	% GW
		BLOCK WEIGHT (BW):	100.0 LBS		55.86	100.00	72.85	79.18			
6.	I-CORN	CORN SYRUP	1.0	0.3000	0.30	1.00	0.73	0.79	1.0000	1.0000	% BW
7.	I-SALT	SALT	2.8	0.0550	0.16	2.84	2.07	2.25	.0000	3.0000	% GW
8.	S-FRANK	FRANK SPICE	0.8	0.3400	0.25	0.75	0.55	0.59	.7500	.7500	LBS
9.	WATER	WATER	30.2	0.0000	0.00	30.15	21.97	23.88	.0000	BIG	% GW
10.	X-ISP	ISOLATED SOY PROTEIN	2.5	1.0100	2.55	2.53	1.84	2.00	2.0000	2.0000	% FW
		GROSS WEIGHT (GW):	137.3 LBS		59.12	137.27	100.00	108.70			
		SHRINKAGE LOSS:	11.0			10.98	8.00	8.70			
		FINISHED WEIGHT:	126.3 LBS		59.12	126.29	92.00	100.00			

Table 20-21. Product Specifications for BP-ISP Frank

```
PRODUCT SHORT NAME:  ISU-BPI          AS OF:        7/13/81
PRODUCT LONG NAME:   ISU BYPRODUCT + ISP   PRODUCT TYPE:   F       GROUPS:

LAST COSTS:     $      0.4682      $     0.4776     $     0.0000
LAST CALCULATION USED COST # 1          DATE LAST CALCULATION:  7/13/81    CALC. #   2
MIN OR MAX:          MIN

NO. RAW MATERIALS:       17      PRODUCT BLOCK SIZE:      100.0000 (SCALED ON BW)
NO. REQUIREMENTS:        13      SHRINKAGE RATE:            8.0000 %
PRODUCT DISPLAY UNITS:  LBS
```

<--- RAW MATERIALS IN PRODUCT --->   (NAMES FROM TABLE: ISU MEAT COURSE)

SEQ NO.	SHORT NAME	RAW MATERIAL DESCRIPTION	IN BLOCK	ACCEPTABLE RANGE OF VALUES LO LIMIT	HI LIMIT	LIMIT TYPE	CURRENT VALUE	CURRENT STATUS
1.	B-CHEEK	BF CHEEK MEAT	Y	.0000	BIG	% GW	6.3792	BETWEEN
2.	B-COW	BF FC COW MEAT	Y	.0000	BIG	% GW	0.0000	LO-LIMIT
3.	B-PLATE	BF PLATES	Y	.0000	BIG	% GW	0.0000	HI-LIMIT
4.	B-50	BF 50% TRIMMINGS	Y	.0000	BIG	% GW	0.0000	HI-LIMIT
5.	B-85	BF 85% TRIMMINGS	Y	.0000	BIG	% GW	0.0000	HI-LIMIT
6.	P-FATBCK	PK SKND FATBACKS	Y	.0000	BIG	% GW	0.0000	HI-LIMIT
7.	P-JOWL	PK SKND JOWLS	Y	.0000	BIG	% GW	0.0000	LO-LIMIT
8.	P-50	PK 50% TRIMMINGS	Y	.0000	BIG	% GW	37.0056	BETWEEN
9.	P-80	PK 80% TRIMMINGS	Y	.0000	BIG	% GW	27.7887	BETWEEN
10.	I-CORN	CORN SYRUP		1.0000	1.0000	% BW	1.0000	BETWEEN
11.	I-SALT	SALT		.0000	3.0000	% GW	2.8415	BETWEEN
12.	S-FRANK	FRANK SPICE		.7500	.7500	LBS	0.7500	BETWEEN
13.	WATER	WATER		.0000	BIG	% GW	30.1518	BETWEEN
14.	B-HEART	BF HEARTS	Y	.0000	BIG	% GW	14.4133	BETWEEN
15.	P-LIP	PK LIPS	Y	.0000	BIG	% GW	14.4133	BETWEEN
16.	P-CHEEK	PK CHEEK MEAT	Y	.0000	BIG	% GW	0.0000	LO-LIMIT
17.	X-ISP	ISOLATED SOY PROTEIN		2.0000	2.0000	% FW	2.5258	HI-LIMIT

<--- REQUIREMENTS IN PRODUCT --->

SEQ NO.	<-- REQUIREMENT --> ITEM-1	ITEM-2	REQ. TYPE	ACCEPTABLE RANGE OF VALUES LO LIMIT	HI LIMIT	LIMIT TYPE	LAST CALC. VALUE	CURRENT STATUS
1	MOISTURE		A	-BIG	BIG	% FW	55.5606	BETWEEN
2	PROTEIN		A	-BIG	BIG	% FW	11.6402	BETWEEN
3	FAT		A	25.0000	29.0000	% FW	29.0000	BETWEEN
4	BIND		A	130.0000	BIG	% FW	182.3456	BETWEEN
5	COLOR		A	190.0000	BIG	% FW	190.0000	LO-LIMIT
6	SALT		A	2.2500	2.2500	% FW	2.2500	HI-LIMIT
7	COLLAGEN		A	-BIG	4.2000	% FW	4.2000	HI-LIMIT
8	USDA-AW		A	-BIG	9.0000	% FW	9.0000	HI-LIMIT
9	P	> WATER	GM	.0000	BIG	% GW	25.2369	BETWEEN
10	WATER	> P-LIP	MM	.0000	BIG	% GW	11.4655	BETWEEN
11	P-LIP	> B-HEART	MM	.0000	BIG	% GW	0.0000	LO-LIMIT

<--- REQUIREMENTS IN PRODUCT --->

SEQ NO.	<-- REQUIREMENT --> ITEM-1	ITEM-2	REQ. TYPE	ACCEPTABLE RANGE OF VALUES LO LIMIT	HI LIMIT	LIMIT TYPE	LAST CALC. VALUE	CURRENT STATUS
12	B-HEART	> B	MG	.0000	BIG	% GW	5.8528	BETWEEN
13	B	> SALT	GA	.0000	BIG	% GW	2.5772	BETWEEN

Table 20-22. Product Formulation Requirements Report for BP-ISP Frank

```
PRODUCT NAMES - SHORT: ISU-BPI    LONG: ISU BYPRODUCT + ISP   TYPE: MIN
LAST UPDATED:            7/13/81   LAST CALCULATED:   7/13/81
```

LAST CALCULATION #   2 USING COST #   1

SEQ		REQUIREMENT			LIMITS ON RANGE			PENALTY
NO.	TYPE	DESCRIPTION		VALUE	LOW-LIMIT	HI-LIMIT	TYPE	COST
1	A	MOISTURE		55.5606	-BIG	BIG	% FW	0.0000
2	A	PROTEIN		11.6402	-BIG	BIG	% FW	0.0000
3	A	FAT		29.0000	25.0000	29.0000	% FW	0.0093
4	A	BIND		182.3456	130.0000	BIG	% FW	0.0000
5	A	COLOR		190.0000	190.0000	BIG	% FW	0.0013
6	A	SALT		2.2500	2.2500	2.2500	% FW	0.0115
7	A	COLLAGEN		4.2000	-BIG	4.2000	% FW	0.0995
8	A	USDA-AW		9.0000	-BIG	9.0000	% FW	0.0122
9	GM	P	>	WATER	25.2369	.0000	BIG	% GW	0.0000
10	MM	WATER	>	P-LIP	11.4655	.0000	BIG	% GW	0.0000
11	MM	P-LIP	>	B-HEART	0.0000	.0000	BIG	% GW	0.0084
12	MG	B-HEART	>	B	5.8528	.0000	BIG	% GW	0.0000
13	GA	B	>	SALT	2.5772	.0000	BIG	% GW	0.0000

NOTE: "PENALTY-COST" IS THE AMOUNT IN $ PER LB BY WHICH THE FORMULATION COST COULD BE REDUCED
IF THE REQUIREMENT LIMIT WAS CHANGED BY 1.0 AS GIVEN. "VALUE" IS GIVEN IN THE SAME UNITS
AS THE LIMITS.

Table 20-23. Product Formulation Price-ranging Report for BP-ISP Frank

```
PRODUCT NAMES - SHORT: ISU-BPI   LONG: ISU BYPRODUCT + ISP   TYPE: MIN
LAST UPDATED:           7/13/81   LAST CALCULATED:   7/13/81
```

CALC. #   2 ... USING COST # 1

<-- CALC. COST ($ PER LB) -->

	COST #1	COST #2	COST #3
BLOCK WEIGHT:	0.5912	0.6032	0.0000
GROSS WEIGHT:	0.4307	0.4394	0.0000
FINISHED WEIGHT:	0.4682	0.4776	0.0000

SEQ	MATERIAL NAME		USAGE	COST RANGE		CURRENT COSTS		
NO.	SHORT	LONG	(LBS)	USE-MORE	USE-LESS	COST-1	COST-2	COST-3
1.	B-CHEEK	BF CHEEK MEAT	6.379	.4584	1.0885	0.9600	0.9800	0.0000
2.	P-50	PK 50% TRIMMINGS	37.006	.1570	.4847	0.4200	0.4300	0.0000
3.	P-80	PK 80% TRIMMINGS	27.789	.5932	.8696	0.8000	0.8200	0.0000
4.	I-CORN	CORN SYRUP	1.000	-102.1871	398.2238	0.3000	0.3000	0.0000
5.	I-SALT	SALT	2.841	-BIG	BIG	0.0550	0.0550	0.0000
6.	S-FRANK	FRANK SPICE	0.750	-136.3094	530.9052	0.3400	0.3400	0.0000
7.	WATER	WATER	30.152	-3.9108	1.0072	0.0000	0.0000	0.0000
8.	B-HEART	BF HEARTS	14.413	.2683	.9506	0.4900	0.4900	0.0000
9.	P-LIP	PK LIPS	14.413	.1183	1.0893	0.3400	0.3500	0.0000
10.	X-ISP	ISOLATED SOY PROTEIN	2.526	-BIG	3.4305	1.0100	1.0100	0.0000
11.	B-COW	BF FC COW MEAT	0.000	1.3694	BIG	1.4700	1.4500	0.0000
12.	B-PLATE	BF PLATES	0.000	-BIG	.5941	0.5200	0.5200	0.0000
13.	B-50	BF 50% TRIMMINGS	0.000	-BIG	.7665	0.6400	0.6450	0.0000
14.	B-85	BF 85% TRIMMINGS	0.000	-BIG	1.2762	1.1500	1.1700	0.0000
15.	P-FATBCK	PK SKND FATBACKS	0.000	-BIG	.3220	0.2700	0.2750	0.0000
16.	P-JOWL	PK SKND JOWLS	0.000	.2884	BIG	0.3600	0.3650	0.0000
17.	P-CHEEK	PK CHEEK MEAT	0.000	.4133	BIG	0.8900	0.9000	0.0000

NOTE: "COST-TO-USE-MORE" IS THE PRICE AT WHICH THE FORMULATION WILL START TO USE MORE OF THE MATERIAL.
"COST-TO-USE-LESS" IS THE PRICE AT WHICH LESS OF THE MATERIAL WILL BE USED IN THE FORMULATION.

Table 20-24. Material Attribute Level Table

SEQ NO.	SHORT NAME	ATTRIBUTE LONG NAME	ATT. TYPE	B-BULL	B-CHEEK	B-CHUCK	B-COW	B-HEAD	B-HEART	B-LIP	B-NAVEL	B-PLATE
1.	MOISTURE	MOISTURE		0.7000	0.6300	0.5700	0.6650	0.6100	0.6650	0.6200	0.3800	0.4000
2.	FAT	FAT		0.0950	0.1950	0.2600	0.1500	0.2300	0.1800	0.2200	0.5150	0.4800
3.	PROTEIN	PROTEIN (TOTAL)		0.2000	0.1700	0.1600	0.1800	0.1550	0.1500	0.1550	0.1000	0.1100
4.	SALT	SALT		0.0000	0.0000	0.0000	0.0000	0.0000	0.0000	0.0000	0.0000	0.0000
5.	USDA-AW	USDA ADDED WATER	C	-0.1000	-0.0500	-0.0700	-0.0550	-0.0100	0.0650	0.0000	-0.0200	-0.0400
6.	BIND	CALCULATED BIND	C	6.0000	2.3800	3.8400	4.4100	1.2400	0.9000	0.0465	1.2500	1.7600
7.	COLOR	CALCULATED COLOR	C	9.4000	8.1600	13.6000	7.0200	4.0300	6.0000	0.0620	1.9000	2.7500
8.	COLLAGEN	CALC COLLAGEN	C	0.0400	0.0400	0.0480	0.0378	0.1131	0.0405	0.1395	0.0420	0.0462
9.	M+P+F	MOISTURE+PROTEIN+FAT	C	0.9950	0.9950	0.9900	0.9950	0.9950	0.9950	0.9950	0.9950	0.9900
10.	BIND-NX	UNIV-GA BIND INDEX		30.0000	14.0000	24.0000	24.5000	8.0000	6.0000	0.3000	12.5000	16.0000
11.	COLOR-NX	UNIV-GA COLOR INDEX		47.0000	48.0000	B5.0000	39.0000	26.0000	40.0000	0.4000	19.0000	25.0000
12.	COLL-NX	COLLAGEN/PROT RATIO		0.2000	0.5900	0.3000	0.2100	0.7300	0.2700	0.9000	0.4200	0.4200

SEQ NO.	SHORT NAME	ATTRIBUTE LONG NAME	ATT. TYPE	B-SHANK	B-50	B-60	B-65	B-75	B-85	B-90	P-BACEND	P-BELLY
1.	MOISTURE	MOISTURE		0.7100	0.3750	0.4500	0.4900	0.5700	0.6550	0.6950	0.3050	0.3400
2.	FAT	FAT		0.1000	0.5100	0.4100	0.3600	0.2700	0.1550	0.1100	0.6000	0.5650
3.	PROTEIN	PROTEIN (TOTAL)		0.1850	0.1050	0.1300	0.1500	0.1550	0.1800	0.1850	0.0900	0.0900
4.	SALT	SALT		0.0000	0.0000	0.0000	0.0000	0.0000	0.0000	0.0000	0.0000	0.0000
5.	USDA-AW	USDA ADDED WATER	C	-0.0300	-0.0450	-0.0700	-0.1100	-0.0500	-0.0650	-0.0450	-0.0550	-0.0200
6.	BIND	CALCULATED BIND	C	5.1800	1.2600	2.0800	2.7000	3.4100	4.3200	4.8100	0.0090	0.5400
7.	COLOR	CALCULATED COLOR	C	8.5100	1.9950	3.2500	4.2000	5.2700	7.0200	7.4000	0.2250	0.4500
8.	COLLAGEN	CALC COLLAGEN	C	0.0610	0.0441	0.0526	0.0600	0.0589	0.0504	0.0518	0.0774	0.0315
9.	M+P+F	MOISTURE+PROTEIN+FAT	C	0.9950	0.9900	0.9900	1.0000	0.9950	0.9900	0.9900	0.9950	0.9950
10.	BIND-NX	UNIV-GA BIND INDEX		28.0000	12.0000	16.0000	18.0000	22.0000	24.0000	26.0000	0.!000	6.0000
11.	COLOR-NX	UNIV-GA COLOR INDEX		46.0000	19.0000	25.0000	28.0000	34.0000	39.0000	40.0000	2.5000	5.0000
12.	COLL-NX	COLLAGEN/PROT RATIO		0.3300	0.4200	0.4050	0.4000	0.3800	0.2800	0.2800	0.8600	0.3500

SEQ NO.	SHORT NAME	ATTRIBUTE LONG NAME	ATT. TYPE	P-BLADE	P-CHEEK	P-DIAPH	P-FATBCK	P-HAM	P-HAMCUR	P-HAMFAT	P-HEAD	P-HEART
1.	MOISTURE	MOISTURE		0.7150	0.6900	0.7000	0.1500	0.5650	0.5700	0.0400	0.6100	0.7000
2.	FAT	FAT		0.0900	0.1400	0.1500	0.8000	0.3000	0.3000	0.9500	0.2200	0.1400
3.	PROTEIN	PROTEIN (TOTAL)		0.1900	0.1600	0.1450	0.0400	0.1300	0.1200	0.0100	0.1600	0.1550
4.	SALT	SALT		0.0000	0.0000	0.0000	0.0000	0.0000	0.0000	0.0000	0.0000	0.0000
5.	USDA-AW	USDA ADDED WATER	C	-0.0450	0.0500	0.1200	-0.0100	0.0450	0.0900	0.0000	-0.0300	0.0800
6.	BIND	CALCULATED BIND	C	4.5600	1.4400	2.3200	0.1200	2.7300	2.1600	0.0100	1.2000	0.9300
7.	COLOR	CALCULATED COLOR	C	3.8000	4.6400	2.6100	0.0400	2.4700	1.2000	0.0100	2.5600	5.1150
8.	COLLAGEN	CALC COLLAGEN	C	0.0437	0.1152	0.1015	0.0240	0.0325	0.0360	0.0040	0.1120	0.0418
9.	M+P+F	MOISTURE+PROTEIN+FAT	C	0.9950	0.9900	0.9950	0.9900	0.9950	0.9900	1.0000	0.9900	0.9950
10.	BIND-NX	UNIV-GA BIND INDEX		24.0000	9.0000	16.0000	3.0000	21.0000	18.0000	1.0000	7.5000	6.0000
11.	COLOR-NX	UNIV-GA COLOR INDEX		20.0000	29.0000	18.0000	1.0000	19.0000	10.0000	1.0000	16.0000	33.0000
12.	COLL-NX	COLLAGEN/PROT RATIO		0.2300	0.7200	0.7000	0.6000	0.2500	0.3000	0.4000	0.7000	0.2700

SEQ NO.	SHORT NAME	ATTRIBUTE LONG NAME	ATT. TYPE	P-JOWL	P-LIP	P-LIVER	P-PIC	P-NECK	P-PICHRT	P-SNOUT	P-SPLEEN	P-STOM
1.	MOISTURE	MOISTURE		0.3000	0.6200	0.7400	0.6000	0.6000	0.7000	0.5500	0.7000	0.0000
2.	FAT	FAT		0.6150	0.2200	0.0600	0.2400	0.2300	0.1100	0.3000	0.1450	0.0000
3.	PROTEIN	PROTEIN (TOTAL)		0.0750	0.1550	0.1900	0.1550	0.1600	0.1800	0.1450	0.1550	0.0000
4.	SALT	SALT		0.0000	0.0000	0.0000	0.0000	0.0000	0.0000	0.0000	0.0000	0.0000
5.	USDA-AW	USDA ADDED WATER	C	0.0000	0.0000	-0.0200	-0.0200	-0.0400	-0.0200	-0.0300	0.0800	0.0000
6.	BIND	CALCULATED BIND	C	0.3450	0.1550	0.3800	3.1000	3.0400	4.1400	0.3625	0.1085	0.0000
7.	COLOR	CALCULATED COLOR	C	0.1275	0.0775	9.3100	2.4800	2.5600	3.4200	0.0725	8.6800	0.0000
8.	COLLAGEN	CALC COLLAGEN	C	0.0322	0.1395	0.1805	0.0356	0.0400	0.0396	0.1160	0.1395	0.0000
9.	M+P+F	MOISTURE+PROTEIN+FAT	C	0.9900	0.9950	0.9900	0.9950	0.9900	0.9900	0.9950	1.0000	0.0000
10.	BIND-NX	UNIV-GA BIND INDEX		4.6000	1.0000	2.0000	20.0000	19.0000	23.0000	2.5000	0.7000	0.0000
11.	COLOR-NX	UNIV-GA COLOR INDEX		1.7000	0.5000	49.0000	16.0000	16.0000	19.0000	0.5000	56.0000	0.0000
12.	COLL-NX	COLLAGEN/PROT RATIO		0.4300	0.9000	0.9500	0.2300	0.2500	0.2200	0.8000	0.9000	0.0000

(Continued)

Table 20-24 (Continued)

SEQ NO.	SHORT NAME	ATTRIBUTE LONG NAME	ATT. TYPE	P-TNGTRM	P-TONGUE	P-50	P-80	P-95	C-MDB	T-MDB	V-BNLS	S-BEBR
1.	MOISTURE	MOISTURE		0.0000	0.0000	0.3200	0.5600	0.0000	0.6600	0.0000	0.0000	0.0500
2.	FAT	FAT		0.0000	0.0000	0.5900	0.2800	0.0000	0.2050	0.0000	0.0000	0.0000
3.	PROTEIN	PROTEIN (TOTAL)		0.0000	0.0000	0.0850	0.1500	0.0000	0.1250	0.0000	0.0000	0.0100
4.	SALT	SALT		0.0000	0.0000	0.0000	0.0000	0.0000	0.0000	0.0000	0.0000	0.0000
5.	USDA-AW	USDA ADDED WATER	C	0.0000	0.0000	-0.0200	-0.0400	0.0000	0.1600	0.0000	0.0000	0.0100
6.	BIND	CALCULATED BIND	C	0.0000	0.0000	1.0200	2.8500	0.0000	1.8750	0.0000	0.0000	0.0000
7.	COLOR	CALCULATED COLOR	C	0.0000	0.0000	0.7650	2.5500	0.0000	0.6250	0.0000	0.0000	0.0000
8.	COLLAGEN	CALC COLLAGEN	C	0.0000	0.0000	0.0289	0.0360	0.0000	0.0437	0.0000	0.0000	0.0000
9.	M+P+F	MOISTURE+PROTEIN+FAT	C	0.0000	0.0000	0.9950	0.9900	0.0000	0.9900	0.0000	0.0000	0.0600
10.	BIND-NX	UNIV-GA BIND INDEX		0.0000	0.0000	12.0000	19.0000	0.0000	15.0000	0.0000	0.0000	0.0000
11.	COLOR-NX	UNIV-GA COLOR INDEX		0.0000	0.0000	9.0000	17.0000	0.0000	5.0000	0.0000	0.0000	0.0000
12.	COLL-NX	COLLAGEN/PROT RATIO		0.0000	0.0000	0.3400	0.2400	0.0000	0.3500	0.0000	0.0000	0.0000

SEQ NO.	SHORT NAME	ATTRIBUTE LONG NAME	ATT. TYPE	S-BOLO	S-FRANK	S-SALAMI	S-SAUS	I-CORN	I-CURE	I-SALT	X-CEREAL	X-ISP
1.	MOISTURE	MOISTURE		0.0000	0.0800	0.0000	0.0000	0.2600	0.0000	0.0000	0.0000	0.2900
2.	FAT	FAT		0.0000	0.0000	0.0000	0.0000	0.0000	0.0000	0.0000	0.0000	0.0000
3.	PROTEIN	PROTEIN (TOTAL)		0.0000	0.1500	0.0000	0.0000	0.0000	0.0000	0.0000	0.0000	0.7100
4.	SALT	SALT		0.0000	0.0000	0.0000	0.0000	0.0000	0.0000	1.0000	0.0000	0.0000
5.	USDA-AW	USDA ADDED WATER	C	0.0000	-0.5200	0.0000	0.0000	0.2600	0.0000	0.0000	0.0000	-2.5500
6.	BIND	CALCULATED BIND	C	0.0000	3.0000	0.0000	0.0000	0.0000	0.0000	0.0000	0.0000	31.9500
7.	COLOR	CALCULATED COLOR	C	0.0000	1.5000	0.0000	0.0000	0.0000	0.0000	0.0000	0.0000	0.0000
8.	COLLAGEN	CALC COLLAGEN	C	0.0000	0.0000	0.0000	0.0000	0.0000	0.0000	0.0000	0.0000	0.0000
9.	M+P+F	MOISTURE+PROTEIN+FAT	C	0.0000	0.2300	0.0000	0.0000	0.2600	0.0000	0.0000	0.0000	1.0000
10.	BIND-NX	UNIV-GA BIND INDEX		0.0000	20.0000	0.0000	0.0000	0.0000	0.0000	0.0000	0.0000	45.0000
11.	COLOR-NX	UNIV-GA COLOR INDEX		0.0000	10.0000	0.0000	0.0000	0.0000	0.0000	0.0000	0.0000	0.0000
12.	COLL-NX	COLLAGEN/PROT RATIO		0.0000	0.0000	0.0000	0.0000	0.0000	0.0000	0.0000	0.0000	0.0000

SEQ NO.	SHORT NAME	ATTRIBUTE LONG NAME	ATT. TYPE	X-NFDM	X-PREMUL	X-SPC	X-TVP	X-TSPC	WATER
1.	MOISTURE	MOISTURE		0.0000	0.5030	0.2000	0.0000	0.0000	1.0000
2.	FAT	FAT		0.0000	0.4190	0.0000	0.0000	0.0000	0.0000
3.	PROTEIN	PROTEIN (TOTAL)		0.0000	0.0620	0.7000	0.5000	0.7000	0.0000
4.	SALT	SALT		0.0000	0.0100	0.0000	0.0000	0.0000	0.0000
5.	USDA-AW	USDA ADDED WATER	C	0.0000	0.2550	-2.6000	-2.0000	-2.8000	1.0000
6.	BIND	CALCULATED BIND	C	0.0000	12.3380	28.0000	20.0000	28.0000	0.0000
7.	COLOR	CALCULATED COLOR	C	0.0000	0.0000	0.0000	0.0000	0.0000	0.0000
8.	COLLAGEN	CALC COLLAGEN	C	0.0000	0.0000	0.0000	0.0000	0.0000	0.0000
9.	M+P+F	MOISTURE+PROTEIN+FAT	C	0.0000	0.9840	0.9000	0.5000	0.7000	1.0000
10.	BIND-NX	UNIV-GA BIND INDEX		0.0000	199.0000	40.0000	40.0000	40.0000	0.0000
11.	COLOR-NX	UNIV-GA COLOR INDEX		0.0000	0.0000	0.0000	0.0000	0.0000	0.0000
12.	COLL-NX	COLLAGEN/PROT RATIO		0.0000	0.0000	0.0000	0.0000	0.0000	0.0000

Appearancewise, the frankfurters look similar, because each of them has been formulated to the same color index (190). However, textural differences occur between the four formulations due to the addition of by-products and higher collagen meats. Note the addition of beef hearts and pork lips to the BP frankfurters. With the addition of these meats, the product standards (specifications) had to be revised slightly. The bind index lower limit was reduced from 180 to 130 and the upper limit for collagen was increased from 3.75 to 4.20 to accommodate the lower-quality meats in the formulation (Tables 20-8, 20-13, and 20-21). The end result is that the two cheaper frankfurter formulations that contain by-products have a lower-bind index and a subsequently softer texture (slightly mushy).

The cost of each of the four types of frankfurters represents the *least cost* to produce, given the constraints placed on each type and the available raw materials.

PREBLENDING

Preblending is a term used in the sausage industry to describe the grinding and mixing of raw meat materials with salt and/or water and/or nitrite for later use in sausage formulations. Preblended meats are typically held in a cooler from 8 to 72 hours before the preblends are used in actual sausage production. During this time, processors use representative sampling techniques to carefully analyze the preblends for fat content. Some processors perform a variety of other laboratory tests on the preblends to insure consistent uniformity in quality of the finished product. By knowing the exact compositional makeup of each preblended batch, processors can combine lean-meat preblends and fat-meat preblends to consistently produce the desired fat content, bind, color, etc., from batch to batch. Precise control of compositional factors such as fat, moisture, and protein content is extremely important for maximizing profit and maintaining regulatory compliance with government regulations. Preblending operations are also well adapted to linear programming techniques, since there is time to collect the necessary data on raw material lots after preblending and before further processing.

Problem Areas in Sausage Production

Modern sausage production is plagued by several common problems. Preblending helps to eliminate these problems.

Fat and Water Binding

Insufficient time is allowed for salt-soluble protein extraction before fat meats and water are added, thus reducing the bind capability.

Raw Material Variations

Individual raw material analyses often vary by more than ± 5 percent in fat and moisture from lot to lot, causing similar swings in finished product analyses, particularly in small batches.

Production Scheduling

Since each finished product batch is prepared from materials on hand at a particular time, material flow becomes complex. It is difficult to have the right meats and equipment available at the right time.

Production Efficiency

Small batches and common use of equipment for different purposes impair efficient utilization of resources.

Extensive Product Line

Manufacture of many individual finished products, each with its own formula and special instructions, causes confusion and inefficiencies in the sausage kitchen.

Theory of Preblending

Modern preblending attempts to alleviate these production problems using some simple concepts and procedures. These include:

High-Brine Concentration

The high salt-to-lean content of meat in the preblend improves salt-soluble protein extraction.

Extended Period of Extraction

Storage for 8 to 72 hours allows time for optimum protein extraction and fat and water binding.

Large-Batch Size

Preblends are prepared in batch sizes equivalent to 2 to 10 finished-product batches, resulting in good mixing and economy of scale.

Lab Analysis

The preblend is sampled when dumped, and a lab analysis is performed for moisture, fat, and protein during the holding period.

Correct to Finished Product

Each finished product batch is reformulated to bring fat and water content back into specification. Correction is accomplished using standard lean and fat meats, which may themselves be preblended or thoroughly characterized. Spices and other ingredients are then added to make individual finished products.

How Preblending Works

Preblending and correction systems have evolved as a means of helping the sausage maker in several problem areas of production and quality control.

Fat and Water Binding

Preblending and storage of lean meats with salt improves the bind characteristics by up to 50 percent.

Raw Material Variations

The chemical analysis and actual weights of raw materials typically vary by several percent from batch to batch. When preblended in large batches, these fluctuations tend to average out and are compensated for by the actual analysis taken during the holding phase and the subsequent correction at the final blend.

Problem Meats

High-collagen meats, such as cheek and head meat, heart meat, shanks, etc., are more fully emulsified during the preblending step and present less of a problem in the finished product. Double blending helps guarantee adequate mixing.

Production Variations

Weighing errors, variations in blending times, and other production variations tend to average out in the larger batches common to preblending and can be corrected for after-chemical analysis.

Production Scheduling

Preblending formulas account for 80 to 90 percent of the meats used and are typically held constant over a week of production. Production schedules can be made in advance and do not suffer as much from material movement problems that characterize reformulation on a batch-by-batch basis. Variations in the scheduled use of materials are limited to the correction step, where only a few meats are used. Once preblended, storage life is increased to several days, and the preblend can be stored in standard quantities.

Production Efficiency

The preblend is usually prepared in large batch sizes and then poured into tubs for use in several finished product batches. Economy of scale reduces labor and equipment costs, reduces lab analysis workload, and minimizes material movement. Preblending and correction can be carried out in different shifts. More automation through interconnected scales, grinders, flakers, mixers, and blenders can be used.

Extensive Product Line

Preblends can be made in large batches for use in a variety of similar products, reducing the number of manufacturing steps. Blends to a given label, such as "beef," or "loaf base," are commonly prepared, with the individual finished product requirements handled by different spices and constraints in the correction chop or blend.

Preblending Techniques

Preblending can be carried out by utilizing any of several different strategies to decrease final production variations. These strategies may be based upon:

- Raw material.
- Lean or fat content.
- Species, *e.g.*, beef vs. pork.
- Label, such as "pork & beef," "pork, beef, water," "beef," etc.
- Individual finished product.

The method used depends on the importance of various factors, including equipment costs, available storage, handling requirements, production efficiency, and simplicity.

Preblending of raw material is usually associated with a company that prepares a narrow range of products from a limited number of meats in large quantities. Each meat ingredient is segregated and preblended by lot, each with its own analysis, thereby controlling variations and allowing optimization of the final formulation of finished products.

Generally, the industry has opted for preblending by label or finished product as the most effective use of equipment, labor, and storage resources and as the method giving the maximum production efficiency and simplicity. The large-batch sizes and fixed instructions of this method give good control of material and production variations.

Formulation Procedures for Preblending

A typical sequence of events in the purchasing/production cycle is:

Tuesday: Have sales meeting to determine next week's production requirements.

Wednesday: Get availabilities and prices. Standard products, cut-and-kill, and boning are estimated and accounted for.

Thursday: Procurement department starts purchasing bill-of-materials resulting from the multi-product formulation.

Friday: Raw materials are delivered and quality control checks made on raw materials (*i.e.,* fat content, condition).

Next week: Preblends are made by formula. Correct to finished product on a batch-by-batch basis. Adjust preblend formulas for abnormal events.

Formulating the Preblend

Each preblend will have its own formulation specifications. When preblending to a finished product, these specifications may be as simple as including only the fat and USDA added water at the finished product percentage limits. Materials offered to the preblend would include all meats offered to the finished product, salt, cure, and water.

The general principle in formulating a preblend is to arrive at a recipe which will bring the preblend-correction steps back into accord with the overall least cost solution, if the preblend is at its target analysis, and minimizing the cost to correct when the preblend is off its target analysis. The Least Cost Formulator® system distributed by Least Cost Formulations, Ltd.,[10] uses the following procedure to accomplish this purpose:

- Set up preblend product specifications as outlined.
- Identify meats offered to finished product which are suitable for use as corrector meats. Exclude from these cheek, head, and shank meat and offal.
- Inform the formulation system which meats must be used in total in the preblend, which can be used as correctors, and which ingredients must not be used in the preblend.
- Set up a basic corrector specification. This would include added ingredients and product requirements on fat, added water, etc. Modify this basic specification at formulation to include the actual preblend and corrector meats.

Once the preblend and corrector specifications have been set up, routine steps would include:

- Perform a special preblend least cost formulation based on the finished product solution from the multi-product run. The computer will prepare the preblend at least cost and choose the corrector meats to be added at the final blend.

---

[10]Least Cost Formulations, Ltd., 827 Hialeah Drive, Virginia Beach, VA 23464.

- If the preblend specifications are being adjusted, verifying the corrector will be feasible if the preblend is on or off target analysis in the usual range (typically up or down 2.5 to 3 percent in fat).
- When the actual preblend analysis is obtained from the lab, run the corrector formulation, using the cost and actual analysis of the preblend and offering the corrector meats available.

Making a Preblend

The following is a typical procedure for preparing a preblend:

- Grind meats (coarse grind fat meat to $5/16$ to $5/8$ inch, fine grind lean meat to $1/8$ to $1/4$ inch, and flake frozen meats).
- Sample ground raw materials for chemical analysis (vendor control).
- Mix in blender to 80 to 85 percent finished weight of final product. Include all problem meats (cheek, head, shank, etc.), 75 to 80 percent of other meats, all salt (as finely divided and pure as possible), one half or all of the cure, and sufficient water to work meat (approximately 10 percent of total preblend weight). The materials should be blended adequately (7 to 10 minutes) with a resulting pH of 6 or higher.
- Sample preblend for laboratory analysis. Perform fat, moisture, and protein tests to as high an accuracy as possible. Make sure the sample is representative.
- Dump into tubs. Sampling could be done on a grab-sample basis during this step. The same preblend may be used for various products.
- Store 8 to 72 hours in cooler. Protein extraction increases 15 percent the first day, 10 percent the second, and 5 percent the third. Longer storage can make up for slight underblending. Shorter storage gives better water bind; longer storage gives better fat bind. The best policy is to plan on holding the preblend 12 to 16 hours, with the shelf life of the preblend used to handle contingencies and Monday start-up needs.

Correcting to Finished Product

- Run corrector formulation based on preblend laboratory analysis.
- Mix in chopper or blender: the preblend (fixed at 80 to 85 percent of finished weight), lean and fat corrector meats, and remaining cure. Work in water, spices, and other ingredients and rework (use less than 5 percent in quality products).
- Chop or run through emulsifier.
- Stuff.

The corrector meats will vary in usage, depending on the actual chemical analysis of the preblend. Therefore, deviations from projected usages will occur, and the meats chosen should include consideration for increased or decreased needs. The corrector meats should themselves be preblended or well mixed and characterized by good sampling and lab analysis so that finished product fluctuations are minimized.

Example of Preblend-Corrector Formulation

Suppose the multi-product formulation of LCF all-meat franks (AM-FRANK) for next week's production is as given in Table 20-25.

The preblend (P-AM-FRK) is designed to have the same fat content (29 percent) and added water (9.5 percent) as AM-FRANK, with all the salt and half the cure, but with no spices or corn syrup. All meats except pork or beef head, cheek, shank, or diaphragm meat can be used as correctors. The least cost preblend formulation is shown in Table 20-26. Comparing these two tables (20-25 and 20-26) shows the preblend has left out the following meats as correctors:

P-Belly	pork belly strips	70.5 lbs.
P-Ham	pork ham trimmings	79.9 lbs.

Table 20-27 shows the corrector formulation with the preblend at its target analysis. Note the corrector puts back exactly the same meats left out by the preblend in the same quantities to arrive at the overall least cost finished product formulation. No better can be done!

When the preblend is off-target in its analysis, the corrector will not arrive at the same formula as the original finished-product formulation. However, a properly designed preblend-corrector pair will be insensitive to relatively wide variations in preblend analysis, typically causing cost variations off overall least cost by less than 0.5 percent per pound. Table 20-28 shows the effects on the corrector meats and cost of various cases of the actual preblend analysis relative to target.

Production preblend-corrector procedures, coupled with multi-product formulation for purchasing, give superior control of raw material flows and result in superior finished-product uniformity and increased-production efficiency.

When preblending is used in production, special procedures are necessary in the Least Cost Formulator® system to generate preblend and correction reports and to maintain optimum costs in the face of analysis variations.

Percent given as off-target is amount by which fat and moisture are deviant from calculated values, in opposite directions.

☞ NOTE: For + or − fat deviations of preblend from target of 2.75 ✐ percent, the cost to correct varies only by + or − 0.5¢ per pound from the original targeted optimum formulation. That is, the preblend-corrector

Table 20-25. Product Formulation Material Usage Report for AM-FRANK

CALC. #16 BASED ON COST #1

		Calculated Cost ($ Per Lb.)		
		Cost #1	Cost #2	Cost #3
Block Weight:		0.6784	0.6885	0.0000
Gross Weight:		0.5110	0.5186	0.0000
Finished Weight:		0.5615	0.5699	0.0000

Seq. No.	Material Name Short	Long	Usage	Cost #1	Extended Cost #1	Percentage of BW	Percentage of GW	Percentage of FW	Limits on Use Lo-limit	Limits on Use Hi-limit	Type
1	B-Cheek	Bf cheek meat	70.0	0.9600	67.20	7.69	5.79	6.36	.0000	Big	% GW
2	P-Belly	Pk belly strips	295.6	0.4100	121.21	32.47	24.46	26.88	.0000	Big	% GW
3	P-Ham	Pk fresh ham trim	152.3	0.7000	106.58	16.72	12.60	13.84	.0000	Big	% GW
4	P-Head	Pk head meat	78.3	0.6600	51.65	8.60	6.47	7.11	.0000	Big	% GW
5	P-Pic	Pk bnls picnics	314.3	0.8100	254.55	34.52	26.00	28.57	.0000	Big	% GW
		Block weight:	910.4 lb.		601.19	100.00	75.32	82.76			
6	S-Frank	Frank spice	20.0	0.3400	6.80	2.20	1.65	1.82	20.0000	20.0000	Lb.
7	I-Salt	Salt	21.0	0.0550	1.15	2.31	1.74	1.91	.0000	3.0000	% GW
8	I-Cure	Dry cure (w/salt)	4.0	0.2500	1.00	0.44	0.33	0.36	4.0000	4.0000	Lb.
9	I-Corn	Corn syrup	25.0	0.3000	7.50	2.75	2.07	2.27	25.0000	25.0000	Lb.
10	Water	Water and ice	228.4	0.0000	0.00	25.09	18.89	20.76	.0000	Big	% GW
		Gross weight:	1208.8 lb.		617.64	132.77	100.00	109.89			
		Shrinkage loss:	108.8			11.95	9.00	9.89			
		Finished weight:	1100.0 lb.		617.64	120.83	91.00	100.00			

Seq. No.	Requirement Description	Value	Limits on Range Low-limit	Limits on Range Hi-limit	Type	Penalty Cost	Activity Coeff.	Requirement Range Low	Requirement Range High
1	Moisture	54.0569	-Big	Big	$ FW	0.0000	0.0000	0.0000	0.0000
2	Fat	29.0000	26.0000	29.0000	$ FW	0.0081	0.0000	28.3423	30.6148
3	Protein	11.1392	-Big	Big	$ FW	0.0000	0.0000	0.0000	0.0000
4	Salt	2.2500	2.2500	2.2500	$ FW	0.0082	0.0000	1.3492	3.6376
5	Bind	170.0001	170.0000	Big	$ FW	0.0003	0.0000	164.9930	179.6605
6	Color	190.0000	190.0000	Big	$ FW	0.0004	0.0000	175.9622	229.2399
7	Collagen	3.7500	-Big	3.7500	$ FW	0.0168	0.0000	3.5693	4.1047
8	USDA-AW	9.5000	-Big	9.5000	$ FW	0.0087	0.0000	8.5992	11.7409
9	Group P>W	50.6308	.0000	Big	$ GW	0.0000	0.0000	48.2365	56.6190
10	Group W>B	13.1028	.0000	Big	$ GW	0.0000	0.0000	2.6617	14.9610
11	Group B	5.7909	3.0000	Big	$ GW	0.0000	0.0000		6.5429
12	Group Z	0.0000	-Big	25.0000	$ BW	0.0000	0.0000	2046.5442	-2046.5442
13	Group K	16.2849	-Big	Big	$ BW	0.0010	0.0000	0.0000	0.0000

Note: "Penalty-cost" is the amount in $ per lb. by which the formulation cost could be reduced if the requirement limit was changed by 1.0 as given. "Value" is given in the same units as the limits.

Table 20-26. Product Formulation Material Usage Report for P-AM-FRK

Product names—Short: P-AM-FRK Long: AM-FRANK PREBLEND Type: Min
Last updated: 1/20/83 Last calculated: 1/20/83

CALC. #2 BASED ON COST #1

— Calculated Cost ($ Per Lb.) —

	Cost #1	Cost #2	Cost #3
	0.6816	0.6926	0.0000
	0.5887	0.5982	0.0000
	0.5887	0.5982	0.0000

Seq. No.	Short	Long		Usage	Cost #1	Extended Cost #1	BW	GW	FW	Lo-limit	Hi-limit	Type
1	B-Cheek	Bf cheek meat		70.0	0.9600	67.20	9.21	7.95	7.95	.0000	Big	% GW
2	P-Belly	Pk belly strips		225.1	0.4100	92.30	29.62	25.58	25.58	.0000	Big	% GW
3	P-Ham	Pk fresh ham trim		72.4	0.7000	50.66	9.52	8.22	8.22	.0000	Big	% GW
4	P-Head	Pk head meat		78.3	0.6600	51.65	10.30	8.89	8.89	.0000	Big	% GW
5	P-Pic	Pk bnls picnics		314.3	0.8100	254.55	41.35	35.71	35.71	.0000	Big	% GW
		Block weight:	760.0 lb.			516.36	100.00	86.37	86.37			
6	I-salt	Salt		21.0	0.0550	1.15	2.76	2.39	2.39	.0000	Big	% GW
7	I-cure	Dry cure (w/salt)		2.0	0.2500	0.50	0.26	0.23	0.23	.0000	Big	% GW
8	Water	Water and ice		97.0	0.0000	0.00	12.76	11.02	11.02	.0000	Big	% GW
		Finished weight:	880.0 lb.			518.02	115.79	100.00	100.00			

Block Weight:
Gross Weight:
Finished Weight:

Percentage of — GW — FW
Limits on Use — Hi-limit

LAST CALC. #2 USED COST #1

Seq. No.	Requirement Description	Value	Low-limit	Hi-limit	Type	Penalty Cost	Activity Coeff.	Low	High
1	Moisture	56.2281	-Big	Big	% FW	0.0000	0.0000	0.0000	0.0000
2	Fat	29.0000	27.0000	29.0000	% FW	0.0138	0.0000	26.9214	30.8834
3	Protein	11.6820	-Big	Big	% FW	0.0096	0.0000	0.0000	0.0000
4	Salt	2.5994	-Big	Big	% FW	0.0003	0.0000	0.3409	3.5739
5	Bind	176.5755	-Big	Big	% FW	0.0005	0.0000	159.9910	176.4293
6	Color	208.0656	-Big	Big	% FW	0.0175	0.0000	172.7367	222.4250
7	Collagen	4.1401	-Big	Big	% FW	0.0117	0.0000	3.4480	4.0006
8	USDA-AW	9.5000	8.0000	9.5000	% FW	0.0000	0.0000	5.7476	12.9002
9	Group P	78.4116	-Big	Big	% GW	0.0000	0.0000	32.1007	51.5115
10	Group W	11.0203	-Big	Big	% GW	0.0000	0.0000	0.0000	15.1009
11	Group B	7.9545	-Big	Big	% GW	0.0000	0.0000	5.2755	20.7028

Limits on Range — Hi-limit
Requirement Range — Low / High

Note: "Penalty-cost" is the amount in $ per lb. by which the formulation cost could be reduced if the requirement limit was changed by 1.0 as given. "Value" is given in the same units as the limits.

Table 20-27. Fast Formulation for AM-FRANK with Preblend at Target

Product: C-AM-FRK AM-FRANK Corrector   Using Cost #1 Qty. #1
Cost on FW: $0.56153 Scaled to 1100.00 FW   Shrink Rate: 9.0000 % GW
Based on Original Costs: $0.09057   0.09129   0.00000   Calc #1

Seq. No.	Material Name	Description	Usage	Limits on Use Low-limit	Hi-limit	Type	Cost	Marginal Cost	Rounding Unit	Quantity Available	Fixed (Y/N)
1.	S-Frank	Frank spice	20.0	20	20	Lb.	0.34000	−1.4333	0.000	Big	N
2.	I-Cure	Dry cure (w/salt)	2.0	2.00000	2.00000	Lb.	0.25000	−0.9166	0.000	Big	N
3.	I-Corn	Corn syrup	25.0	25	25	Lb.	0.30000	−0.5633	0.000	Big	N
4.	Water	Water and ice	131.4	.00000	Big	% GW	0.00000	0.0000	0.000	Big	N
5.	P-Belly	Pk belly strips	70.5	.00000	Big	% GW	0.41000	0.0000	1.000	1550.0	N
6.	P-Ham	Pk fresh ham trim	79.9	.00000	Big	% GW	0.70000	0.0000	1.000	3800.0	N
7.	P-AM-FRK	AM-FRANK preblend	880.0	880	880	Lb.	0.58870	−0.0668	0.000	Big	N

Gross weight (GW): 1208.8
Shrinkage weight: 108.8 $ 0.51099

Finished weight (FW): 1100.0 $ 0.56153

21 Batches can be made before going out-of-stock on any item

Seq. No.	Requirement Description	Value	Limits Imposed Low-limit	Hi-limit	Type	Penalty Cost
1.	Moisture	54.0584	-Big	Big	% FW	0.0000
2.	Fat	29.0000	26.0000	29.0000	% FW	−0.0138
3.	Protein	11.1376	-Big	Big	% FW	0.0000
4.	USDA-AW	9.5000	-Big	9.5000	% FW	−0.0117

Table 20-28. Preblend and Corrections of LCF All-Meat Frank Formula

Preblend Fat Off-Target by	Cost of Corrected Product/Cwt.	Corrector Meats and Water Used		
		Water	P-Belly	P-Ham
	($)	. . . .	*(pounds)*	. . . .
On-target	56.16	131.4	70.5	27.9
2.5% Fat	56.58	150.2	8.8	122.7
2.5% Lean	55.73	112.6	132.2	28.4
2.75% Fat	56.63	152.0	2.5	127.2
2.75% Lean	55.69	110.6	138.3	32.9

procedures are extremely insensitive to variations in the preblend analysis. This indicates proper design of preblend-corrector formulation pairs.

A MODERN SAUSAGE KITCHEN

The mechanization necessary for efficient sausage production is difficult for the layperson to visualize, especially the majority of consumers who enjoy tasty hot dogs but are prone to complain if they seem to cost too much. Most of these consumers probably never take time to think about how the product became transformed from the living matter it once was into the delicious human food product they enjoy.

Americans do love hot dogs—to the tune of 12 billion in a recent year, or nearly 54 per person per year. Another way of looking at that consumption is on a per second basis; 380 hot dogs are being eaten every second of every day all year in the United States.

Figures 20-9 through 20-15 illustrate equipment and processes employed in modern sausage kitchens.

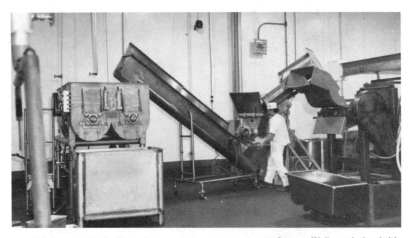

20-9. Portable incline screw conveyor moves meats from a Weiler grinder (with add-on sides) to a Griffith blender. (Courtesy *The National Provisioner*)

Fig. 20-10. When smooth emulsion products are made, the meats are moved to this high-wall batching screw conveyor mounted to a Toledo scale. This has ample space above the conveyor discs to accumulate a batch before unloading into the silent cutter (colloid mill). (Courtesy *The National Provisioner*)

Fig. 20-11. A 750-liter Seydelmann vacuum chopper. Large choppers such as the one above can make fine- or coarse-cut emulsion products. During chopping, the lid is closed by hydraulic arms, and a vacuum is drawn. Chopping under vacuum is reported to improve color stability, shelf life, and product uniformity. (Photo courtesy Wimmer's, Inc., Westpoint, Nebraska)

Fig. 20-12. Two Famco linkers link delicate natural sheep casings to a uniform size. A Vemag continuous vacuum stuffer can be seen in the background. (Photo courtesy Wimmer's, Inc., Westpoint, Nebraska)

Fig. 20-13. At this linking station, natural casing products such as mettwurst are separated into link sizes of approximately the same weight with the aid of a Pratco clipper. (Courtesy *The National Provisioner*)

Fig. 20-14. Frankfurters in cellulose casings have been stuffed and linked with this Townsend machine. (Courtesy *The National Provisioner*)

Fig. 20-15. Six-station trolley cages are moved into position for cooking and smoking in fully automatic Alkar cook cabinets. (Courtesy *The National Provisioner*)

Chapter 21

Meat as a Food

THE WORD *meat,* when used as a general food term, has a rather broad implication. In this text its meaning is limited to the edible parts of the carcass of animals and their organs and glands.

STRUCTURE OF MEAT

Meat is composed of three distinct muscle types: skeletal, smooth, and cardiac. Identifying characteristics of each muscle type are listed in Table 21-1. Skeletal muscles, the most important of the three types from an economic standpoint, are directly or indirectly attached to bone. These muscles facilitate movement and/or give support to the body. Smooth muscles are commonly referred to as visceral muscles and are found throughout the digestive and reproductive tracts of the animal. Smooth muscle tissues are also present throughout the blood vessels, capillaries, and arteries of the circulatory system. Cardiac muscle, as the name would imply, is exclusively heart muscle.

Table 21-1. Muscle Types and Their Characteristics

	Skeletal	Smooth	Cardiac
Method of control	Voluntary	Involuntary	Involuntary
Banding pattern	Striated	Non-striated	Striated
Nuclei/cell	Multinucleated	Single nucleus	Single nucleus

The edible organs and glands are designated as variety meats to contrast them with the skeletal meats of the carcass. They consist of the heart, tongue, liver, pancreas (sweetbreads), thymus (veal sweetbreads), kidney, spleen, brain, and the walls of the stomach (tripe) (see Chapter 10).

Skeletal muscles are a very complex contractile system made up of cylindrical, multinucleated muscle fibers (cells) of varying lengths surrounded by a layer of connective tissue known as the *endomysium.* Bundles of these muscle fibers are enclosed in a sheath of connective tissue known as the *perimysium,* while the entire muscle is surrounded by a denser connective tissue sheath called the *epimysium.*

Figures 21-1, 21-2, and 21-3 represent three-dimensional schematic illustrations of skeletal muscle demonstrating the structural complexity of what we call meat. The order of decreasing size of the functional parts of the muscle are as follows: muscle, muscle bundle, muscle fiber (or cell), myofibril, myofilament. The nature of this textbook is not targeted to the level of understanding that these figures depict; however, these figures effectively demonstrate the basic known structure of muscle tissue.

Many of the features seen in Figures 21-1, 21-2, and 21-3 can be seen in the scanning electron micrographs of muscle tissue shown in Figure 21-4. When muscle tissue is viewed under the transmission electron microscope, a definite banding pattern is observed (Figure 21-5). This pattern (striations) is caused by the structural alignment of the myofilaments. Thick filaments (composed primarily of myosin) and thin filaments (composed primarily of actin) create periods of dark and light patterns. These myofilaments overlap somewhat and slide together to enable the muscle to contract. A uniform repeating unit is also readily

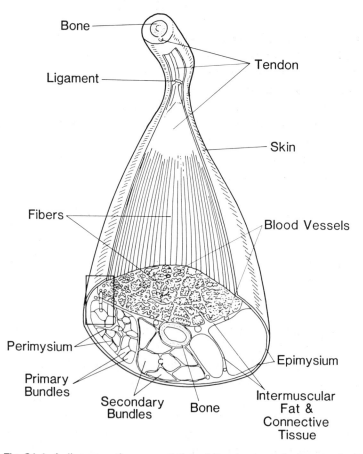

Fig. 21-1. A diagrammatic representation of the structure of a leg muscle. Redrawn with permission from illustration provided by Dr. Richard Rowe, C.S.I.R.O., Brisbane, Australia. Redrawn by Jerry Graslie, South Dakota State University.

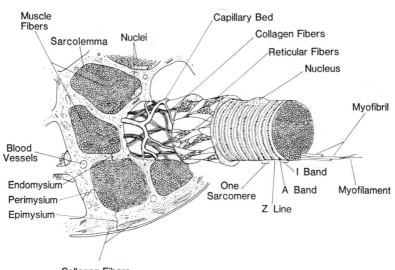

Fig. 21-2. An enlarged portion from the rectangular inset on Figure 21-1, showing a muscle fiber and associated structures which would be seen with an ordinary light microscope. Redrawn with permission from illustration provided by Dr. Richard Rowe, C.S.I.R.O., Brisbane, Australia. Redrawn by Jerry Graslie, South Dakota State University.

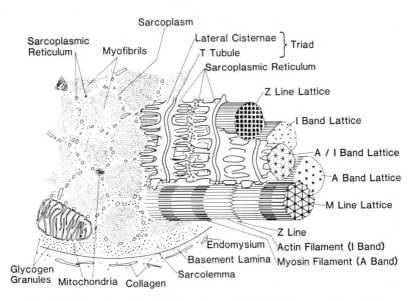

Fig. 21-3. An enlarged portion from the myofibril-myofilament region on Figure 21-2, showing the structure of a muscle fiber as seen with the aid of an electron microscope. Redrawn with permission from illustration provided by Dr. Richard Rowe, C.S.I.R.O., Brisbane, Australia. Redrawn by Jerry Graslie, South Dakota State University.

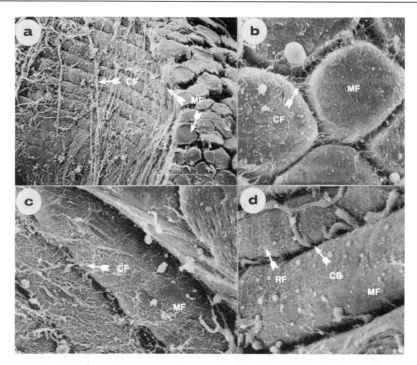

Fig. 21-4. Scanning electron micrographs of bovine skeletal muscle. (a) Muscle fibers (MF) run from the upper left to the lower right. Dense collagen fibers (CF) of the perimysium can be seen surrounding this muscle bundle. Magnification = 175×. (b) Cross-sectional view of muscle fibers (MF). Collagen fibers (CF) of the endomysium are evident between the muscle fibers. Magnification = 1000×. (c) Transverse view of muscle fibers (MF) running from upper left to lower right. Collagen fibers (CF) of the endomysium surround each muscle fiber. Magnification = 1000×. (d) Transverse view of muscle fibers (MF) showing a capillary bed (CB) surrounding a fiber. Fuzzy-appearing reticular fibers (RF) surround the surface of each muscle fiber. Magnification = 1230×. (Micrographs by Kevin W. Jones and Barbara Schrag, South Dakota State University)

evident, as observed in Figure 21-5. The repeating unit of myofibrils is known as a sarcomere. A sarcomere will typically measure 2.5 to 2.6 µm in length or .0001 inch. Since muscle fibers will often run the full length of a muscle, millions of sarcomeres are present in just one muscle fiber and will act in unison to cause that fiber to contract. The number of muscle fibers within a muscle will vary according to the size and function of the muscle. Typically, muscles will contain several thousand to over a million muscle fibers. In the simple flexing of your arm muscle, billions of sarcomeres are involved in various states of contraction. A sarcomere is separated from the next sarcomere by what is known as the Z line or Z disc. This is a region where thin filaments from one sarcomere are anchored to the thin filaments of the next sarcomere in order that the muscle will act in coordinated unison.

It becomes increasingly evident from this brief histological description that the smaller and more numerous the muscles, the greater the amount of connec-

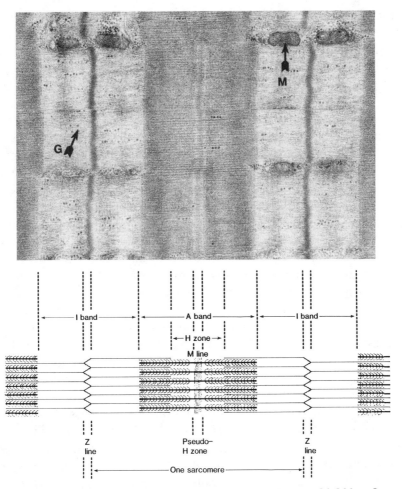

Fig. 21-5. Microstructure of skeletal muscle. Magnification: 30,300×. Sarcomere length: 2.5 microns. M—Mitochondria. G—Glycogen granules. (Micrograph courtesy James A. Burbach, Ph.D., Department of Laboratory Medicine, University of South Dakota. Reference: *The Meat We Eat*, 1985.)

tive tissue. Since connective tissues are far less tender than the cell contents, it follows that their presence in large quantities characterizes the less-tender cuts of meat.

BIOCHEMISTRY AND POST-MORTEM PHYSIOLOGY OF MEAT

In order for the meat science student to comprehend *why* and *how* certain post-mortem conditions (*e.g.*, cold shortening, PSE meat, DFD meat, muscle pH decline, rigor mortis, etc.) affect meat quality, a basic understanding of some biochemical events is necessary. Therefore, a brief review of muscle biochemistry is in order.

Muscle is a highly complex and specialized tissue, and its functioning is a classical example of conversion of chemical to mechanical energy in living systems. It stands to reason that there must be a continuous supply of energy within the muscle, as well as a means of storing reserves to be converted to energy. Think of your car, its engine, and the fuel as an analogy. Without fuel, the car will obviously go nowhere. The fuel itself supplies no energy until the spark plug initiates a chemical reaction (combustion). The chemical energy can then be transformed into useful mechanical energy through piston action on the crankshaft, then to the transmission, and finally, to the wheels which turn to make the car go. A muscle operates under similar circumstances, whether it be a human or animal muscle. The fuel is stored in the muscle in the form of glycogen granules (see Figure 21-5). Glycogen granules are composed of thousands of glucose molecules (6 carbon sugar) linked end to end. The mitochondria of the cells (Figure 21-5) can be compared to the car's engine. This is where the energy transfer takes place. The stimulus (the car's spark plug) that drives the reaction(s) that convert chemical energy to mechanical energy is a compound known as adenosine triphosphate (ATP). The myofilaments would correspond to the car's pistons where the ATP acts with the aid of calcium (Ca^{++}) to create what is known as the *power stroke* which is the initiation of muscle contraction. The power stroke occurs in unison throughout the entire muscle, which may be compared to the car's crankshaft where the energy is transformed into a usable state. Your brain may be considered the car's driver, since it controls which muscles are to be contracted (or relaxed) for locomotion purposes, just as the car's driver decides when to make the car go or stop by using the transmission shifter, accelerator, or brake. The transmission shifter can be likened to the nervous system of an animal whereby nervous impulses or signals tell the muscle what to do, just as the shifter tells the transmission what to do. The accelerator may be compared to the hormone adrenalin (also called epinephrine), which is secreted from the brain and signals the muscle to increase the glycolytic breakdown of glycogen to produce more energy. Similarly, the car's accelerator tells the carburetor to allow more fuel into the combustion chamber to produce more energy. The car's brake may be compared to the hormone norepinephrine, also secreted from the brain, which signals the muscle to slow down the glycolytic breakdown of glycogen so that less energy is produced. Of course, the muscles need a skeleton for support, just as a car needs a frame and axles. Ultimately, the muscle fuel (glycogen) is converted to carbon dioxide (CO_2) and water (H_2O) in the process, liberating a considerable amount of energy. The same is true in the car analogy where gasoline is ultimately converted into CO_2 and H_2O and in the process liberating mechanical energy known to most people as horsepower. This is an extremely oversimplified synopsis of what biochemists call *glycolysis,* the metabolic catabolism (breakdown) of glycogen.

Not only is glycolysis an almost universal central pathway of glucose catabolism in animals but also it occurs in plants and most microorganisms. In actuality, two pathways of glycolysis occur with many intermediate steps and reactions. These pathways are known as aerobic glycolysis (the presence of oxygen) and anaerobic glycolysis (the absence of oxygen). Figure 21-6 represents a simplified version of the possible pathways in the catabolism of glycogen.

In the live animal, the aerobic pathway of glycolysis predominates, which is much more efficient in terms of energy production than the anaerobic pathway (see Figure 21-6). The energy produced in the form of ATP is utilized for muscle contraction (and relaxation) used in movement, metabolic reactions, and transport of materials across membranes. Energy (kcal/mol) is liberated from ATP as illustrated by the following equation:

$$ATP + H_2O \longrightarrow ADP + Pi - 7.3 \text{ kcal/mol}$$

ADP refers to adenosine diphosphate and Pi refers to inorganic phosphate.

 NOTE: One kilocalorie (kcal) is the amount of energy required to raise the temperature of 1000 g of H_2O 1° C.

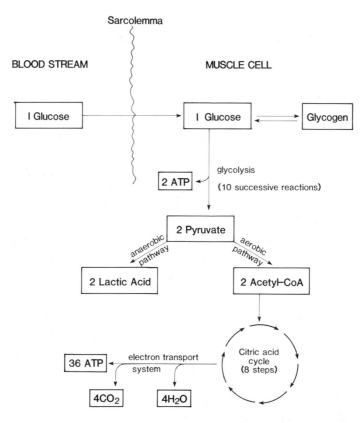

Fig. 21-6. Simplified version of the catabolic fate of muscle glycogen. One glucose molecule following the aerobic pathway can generate 38 molecules of ATP (36 plus 2 from glycolysis). However, 1 glucose molecule following the anaerobic pathway will yield only 2 ATP molecules and lactic acid as the end product instead of CO_2 and H_2O.

When an animal is slaughtered, life processes gradually cease, causing dramatic physiological changes to occur in meat. These changes can ultimately affect the eating quality of meat. In normal situations when an animal is slaughtered, the muscle undergoes a gradual decline in pH (from about 7.0 to 5.5). This decline is caused by the depletion of the animal's glycogen reserves held within the muscle and their conversion to lactic acid, the end product of anaerobic glycolysis. When an animal is bled, oxygen is no longer available to the muscle cells, and anaerobic glycolysis takes over. Lactic acid builds up, and the pH drops. What happens to the ATP generated during anaerobic glycolysis? It is depleted faster than it can be regenerated and is responsible for rigor mortis (stiffening of the muscles). Earlier we stated that ATP was the chemical energy involved in muscle contraction and relaxation. As ATP supplies are depleted, the actin and myosin filaments (see Figure 21-5) lock together at specific binding sites, preventing further contraction or relaxation of the muscle due to antagonistic forces. The rate of ATP depletion closely coincides with the drop in muscle pH.

POST-MORTEM QUALITY PROBLEMS

Dark, Firm, and Dry (DFD) Meat

The final pH that is attained is known as the *ultimate pH* of the meat and has an important influence on the meat's color, texture, water-holding ability, and tenderness. If an animal undergoes vigorous stress or exercise of the muscles prior to slaughter, the glycogen content within the muscles will be reduced substantially. A higher ultimate pH will occur in this animal after slaughter, and the meat will be *dark, firm,* and *dry.* This is a somewhat common event in beef, and those carcasses are termed *dark cutters.* This condition is also found in pork carcasses occasionally. It is thought that the dark color of high pH meat is due to its greater water-holding capacity, which causes muscle fibers to be swollen. The swollen state of the fibers causes more incidental light to be absorbed by the meat surface, hence a darker color. Dark cutters are severely discounted by the packing industry, due to poor consumer appeal of this meat; therefore, it is important to minimize stress and rough handling prior to slaughter. Table 21-2 describes certain parameters related to DFD and PSE conditions.

Pale, Soft, and Exudative (PSE) Meat

The rate of pH decline is an important factor affecting meat quality as well as the ultimate pH. A too rapid post-mortem (after death) drop in muscle pH is associated with the *pale, soft,* and *exudative* conditions that are somewhat common in pork. PSE meat is characterized by a soft, mushy texture, a poor water-holding capacity, and a pale color. These conditions are caused by a more rapid decline to a lower ultimate pH. The looser muscle structure associated with a

Table 21-2. Post-mortem Quality Defects and Physical Parameters

	At Death				24 Hours After Death			
Condition	Muscle Glycogen	Muscle Color	Lactic Acid	Muscle pH	Muscle Glycogen	Muscle Color	Lactic Acid	Muscle pH
Normal	1.0%	Purple	Low	7.0	0.1%	Bright red	Moderate	5.6
DFD pigs	0.3%	Purple	Low	7.0	0.1%	Dark, firm, dry	Low	6.5
Dark cutters	0.5% to 0.2%	Dark purple	Low	7.0	0.1%	Shady, dark or black	Low	6.3 to 6.7
PSE pigs	0.6%	Dark red	Medium	6.4	0.1%	Pale, soft, exudative	High	5.1

lower water-holding capacity is responsible for a greater reflectance of incidental light and hence a pale color. PSE conditions are also known to be highly stress related. Market hogs having the hereditary stress condition known as porcine stress syndrome (PSS) produce carcasses with a high incidence of PSE meat (see Chapter 3).

Fiery Fat and Splotched Lean in Beef

Short-term, violent excitement in cattle (usually from running and/or fighting) causes blood to fill capillaries in subcutaneous fat and muscle. If an animal is killed immediately, the muscle is fiery, pink, or bloodshot. If it is held and slaughtered one or two days later, the fat and muscle will usually be normal.

Cold Shortening

Cold shortening is a problem associated with too rapid chilling of the carcass and results in a dramatic decrease in the tenderness of the muscles. This problem affects beef and lamb carcasses more than pork carcasses, probably because a thinner external fat cover is generally present in beef and lamb. In addition, bovine and ovine muscles have a higher percentage of red muscle fibers (slow twitch) as opposed to white muscle fibers (fast twitch). Red muscle fibers may be more prone to cold shortening than white fibers, due to the different metabolic nature of the two fiber types.

Meat that has undergone cold shortening will often have a fivefold increase in toughness. Under rapid chilling conditions, such as in a 20° to 25° F cooler, muscle will shorten (or contract) to a much greater degree than when the cooler temperature is higher (32° to 36° F). Under normal circumstances, sarcomeres will consistently average 2.4 to 2.6 μm in length (see Figure 21-5); however, highly contracted muscle sarcomeres which have undergone cold shortening will typically measure 1.6 to 1.8 μm or less in length. These shortened muscle fibers will have a correspondingly larger diameter than muscle fibers in the relaxed state. The decrease in tenderness of such muscles is thought to be a function of two factors: the larger muscle fiber diameter and the increased content of endomysial collagen per cross-sectional area, due to the bunching of the endomysium as shortening occurs. Cold shortening is not a major problem in the U.S. meat industry.

Two methods are known to reduce or eliminate cold shortening problems. These methods are known as *delayed chilling* and *electrical stimulation,* both of which are discussed in the pages to follow.

Thaw Rigor

Thaw rigor is a condition somewhat similar to cold shortening, only it is much more severe. Thaw rigor is a phenomenon which occurs when muscle is frozen pre-rigor, while it still contains high levels of ATP. Upon thawing of such muscle, a very severe and rapid contraction of the muscle occurs, rendering it unpalatable. The muscle becomes *supercontracted,* making it very tough and almost inedible. This occurs because the freezing process holds the muscle rigid in an unnatural relaxed state while ATP is still present. The ATP becomes inactive at these frozen temperatures. The sarcoplasmic reticulum (an organelle surrounding myofibrils, see Figure 21-3) becomes damaged during the freezing process and allows uncontrolled leaching of Ca^{++} ions into the sarcoplasm. With high levels of Ca^{++} and ATP present within the myofibril at the same time, rapid and severe muscle contraction occurs. At the present time, thaw rigor is not a major concern among meat processors, because very little meat is frozen in the pre-rigor state. However, as new ideas and technology such as hot boning are developed and implemented by the meat industry, it could become a problem in the future.

Heat Ring

Heat ring is a problem associated with beef carcasses which have a relatively thin layer of external fat. The outer portion of the exposed rib-eye muscle (after ribbing) appears darker in color, coarser in texture, and often sinks, or drops, down (see Figure 21-7). Heat ring is caused by differential chilling rates of the rib-eye muscle. The outer portion chills faster and thus undergoes a slower glycolytic rate, pH decline, and rigor onset than the inner portion of the muscle, which is better insulated. Electrical stimulation of the carcass dramatically reduces the incidence of heat ring due to the accelerated rate of pH decline and subsequent increased speed of rigor onset (see following discussion on electrical stimulation).

POST-MORTEM TECHNOLOGY

Our improved understanding of post-mortem biophysical changes in muscle has led to the development of new technology. This technology has capitalized on natural physical events to improve the quality of meat, the efficiency of

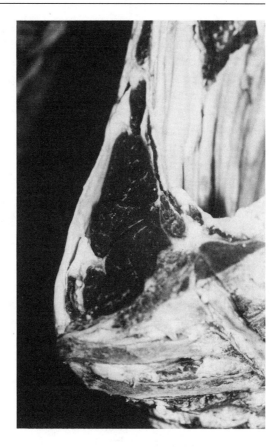

Fig. 21-7. An example of heat ring in a ribbed beef carcass. Note the sunken appearance of the outer portion of the rib-eye muscle.

slaughtering and processing, and thus the ultimate slim-profit margin. Electrical stimulation, delayed chilling, and hot boning are all relatively new developments in the meat industry.

Electrical Stimulation

Electrical stimulation refers to the passage of an electrical current through a carcass and is presently a widely accepted process in the beef packing industry. Electrical stimulation of beef carcasses has been shown to improve carcass quality grade, tenderness, lean color, texture, and firmness; to reduce heat ring problems; to assist in a more thorough bleeding of the carcass; and to make the hide-pulling process easier (see discussion in Chapter 5).

The stimulation process is known to accelerate anaerobic glycolysis, thereby increasing the rate of pH decline and accelerating the rigor mortis process. This accelerated glycolytic activity is thought to be due to the rupturing of the sarcoplasmic reticulin (SR) during the violent contractions associated with stimulation. The SR is a highly specialized membrane that surrounds each myofibril and acts as a calcium (Ca^{++}) pump. Normally, the SR functions to pump Ca^{++}

out of the sarcoplasm (the bathing fluid of the myofibril). Ca^{++} is required in addition to ATP for the myosin filaments to attach to the actin filaments and achieve muscle contraction. The SR sequesters this Ca^{++}, causing the myosin and actin filaments to detach, thus allowing relaxation of the sarcomere to occur. However, when the SR is ruptured, Ca^{++} cannot be sequestered, and the myosin filament cross bridges cannot be detached. This causes a more rapid depletion of ATP stores and a more rapid pH decline (due to the conversion of glycogen to lactic acid). These factors (pH decline and ATP depletion) both cause an accelerated rate of rigor onset to occur.

For the beef packer, perhaps the greatest advantage of stimulation is an economic consideration. Electrical stimulation has been demonstrated to improve the lean color and enhance the definition of the marbling naturally present within the *longissimus* muscle. The end result is that packers will often see an increase (up to 5 percent) in the number of carcasses that grade USDA Choice or higher.

The mechanisms leading to improved tenderness are not completely understood; however, it is thought that the tenderness advantages may be caused by two factors. The first factor is related to the release of lysozomal enzymes (due to ruptured membranes) which degrade Z lines and certain other structural components, thereby increasing the tenderness. The second factor is thought to be the physical tearing and breaking of some of the myofibrils due to severe muscle contractions during the stimulation process (see Chapter 5, Figure 5-7). Tenderness advantages do not appear to be consistent between species as illustrated by an experiment conducted by Stiffler et al. (1982).[1] These scientists reported that sensory panelists found stimulated meat to be more tender for beef and goat (26 and 32 percent improvement respectively), while lower improvements were noted for lamb and pork (12 and 3 percent respectively). In addition, some muscles of the carcass show little or no effect on tenderness due to stimulation, while others show moderate improvements.

The effects of electrical stimulation on flavor and juiciness characteristics are also inconsistent. Some studies have shown an increase in flavor scores with stimulation, others show no change. Juiciness scores generally show no effect due to stimulation, or they decrease slightly.

Other assumed advantages for using electrical stimulation, such as reduced chilling time and energy, easier skinning, and improved shelf life, have not been fully substantiated. The increased loss of blood associated with the stimulation is viewed as both an advantage and a disadvantage, depending on the individual. Some feel that this extra loss of blood would occur anyway and that it is better to lose it on the kill floor than in the chill cooler. Others suggest that this loss of blood represents additional shrinkage to the packer. More research is needed in these areas.

---

[1]Stiffler et al. 1982. *Electrical Stimulation: Purpose, Application and Results.* Bulletin B-1375, Texas Agric. Ext. Serv., College Station.

Delayed Chilling and High-Temperature Conditioning

Delayed chilling refers to the holding of carcasses for two to four hours following slaughter at room temperature (60° to 70° F) before transfer to the chill cooler. High-temperature conditioning usually indicates the holding of the carcasses for two to four hours at a temperature above 70° F but less than 95° F, following slaughter. The latter process is seldom practiced. The purpose of these practices is to accelerate the glycolytic process and thus the evolution of rigor. Metabolic activity has been shown to occur at an accelerated rate under these temperature conditions. However, both processes may shorten the product's shelf life if adequate sanitation and USDA approved bacteriostatic agents are not applied to the carcass (*e.g.*, a dilute chlorine solution is sometimes sprayed onto carcasses). Delayed chilling is not a common practice with large-volume packers, because the additional time requirements are not as conducive to high-chain speed operations. The advantages of delayed chilling can be accomplished more efficiently by electrical stimulation. Perhaps the greatest potential use of the delayed-chilling concept is in conjunction with hot boning (boning before the carcass has been chilled).

Hot Boning

With the exception of the "whole-hog" sausage industry, hot boning has not been widely accepted in this country, even though tremendous savings in energy have been demonstrated. The reasons for lack of acceptance in the beef packing industry are primarily related to grading and marketing problems of hot-boned meat. The grading standards (see Chapter 11) have no provisions to deal with such a system. If and when this technology is adopted as standard practice by the meat industry, it will likely happen in the pork industry first, followed by cow slaughter plants, since a large portion of these commodities (pork and cow beef) are generally further processed prior to consumption, and grades are thus of lesser importance.

COMPOSITION OF MEAT

Lean muscle of all species consists of approximately 20 percent protein, 72 percent water, 7 percent fat, and 1 percent ash. These proportions change as the animal or bird is fattened, resulting in a reduction in the percentage of protein and water and a proportionate increase in fat.

Meat Proteins

Muscle tissue contains many different proteins serving many different functions. Meat proteins can be grouped into three general classifications: (1) myofibrillar, (2) sarcoplasmic, and (3) stromal proteins. Each of these protein classes varies as to the functional properties that it contributes to meat.

Myofibrillar Proteins

The myofibrillar proteins consist primarily of myosin, actin, tropomyosin, m-protein, alpha-actinin, beta-actinin, c-protein, and troponin T, I, and C, as well as other minor proteins associated with the myofibril, but which are present in very small quantities. Myosin and actin are the two most important myofibrillar proteins, since they are the most abundant in this class and are directly responsible for muscle contraction and relaxation. These two proteins combine in a ratio of 1:3 (actin:myosin) to form a more complex protein known as actomyosin in post-rigor muscle. Myofibrillar proteins are also known as salt-soluble proteins, due to their ability to be solubilized in salt solutions.

Sarcoplasmic Proteins

The sarcoplasmic proteins consist of myoglobin, hemoglobin, cytochrome proteins, and a wide variety of endogenous enzymes. These proteins are also known as water-soluble proteins. Myoglobin is the most important protein in this group from a functional standpoint by contributing the red color to meat. Myoglobin serves the function of transporting oxygen to the mitochondria from the bloodstream and returning carbon dioxide to the bloodstream where hemoglobin returns these elements to the lungs for exchange. Myoglobin is present in the sarcoplasm, or bathing fluid, of the muscle cell. Hemoglobin is the protein found in red blood cells or erythrocytes. It carries oxygen or carbon dioxide to or from the cells to the lungs. In meat, approximately 30 to 50 percent of the blood will remain in the capillary network surrounding the muscle fiber, thus becoming part of the meat.

Stromal Proteins

The stromal proteins, or connective tissue proteins, consist primarily of collagen and elastin. Collagen is the single most abundant protein found in mammalian species, being present in bone, skin, tendons, cartilage, and muscle. It is a unique and specialized protein serving a variety of functions. For example, the cornea of the eye is almost pure collagen. The primary functions of collagen are to provide strength and support and to act as an impervious membrane (as in skin). In meat, collagen is the major factor influencing the tenderness of the muscle. Collagen is not broken down easily except with moist-heat cookery methods. Elastin is found in arterial walls and provides elasticity to those tissues. Elastin is sometimes referred to as "yellow" connective tissue due to its color. The *ligamentum nuchae* or backstrap is almost pure elastin. Elastin is not degraded by moist heat cookery methods as collagen is. Fortunately, muscle tissue contains very little elastin.

Another group of substances that are related to proteins (but are not true proteins) are the nitrogenous substances and nucleopeptides. These water-soluble components have little nutritive value in themselves but are physical and chemical substances that excite the flow of gastric juices. Along with fat, they provide a great deal of the aroma and flavor of meat. Because of this stimulating effect, broths are served as the first course to prepare the stomach for the heavier food to follow. Examples of this group of substances are creatine, creatinine, and the purines. More of these components are present in older animals, and they are particularly abundant in the more active muscles that make up the less-tender cuts. They impart to game animals that so-called "gamey flavor."

Amino Acids

Amino acids are the basic building blocks of all proteins. Proteins are hydrolyzed by the digestive juices into amino acids, a form in which they can be readily absorbed into the bloodstream. To date there are 28 to 30 recognized amino acids (more or less depending on source). Of these, 20 are commonly found in most proteins and 10 are considered essential to human life. Table 21-3 lists the known amino acids and identifies which are essential for humans. Three of the less common non-essential amino acids are present in meat, specifically in connective tissue. These are hydroxyproline, hydroxylysine, and cystine.

Table 21-3. Amino Acids

Essential	Non-essential	Less Common Non-essential
Arginine (Arg)	Alanine (Ala)	Cystine (Cys)$_2$
Histidine (His)	Asparagine (Asn)	Hydroxyproline (Hyp)
Isoleucine (Ile)	Aspartic acid (Asp)	Hydroxylysine (Hyl)
Leucine (Leu)	Cysteine (Cys)	Citrulline
Lysine (Lys)	Glutamine (Gln)	B-alanine
Methionine (Met)	Glutamic acid (Glu)	Aminobutyric acid
Phenylalanine (Phe)	Glycine (Gly)	Diaminopimelic acid
Threonine (Thr)	Proline (Pro)	Dihydroxyphenylalanine
Tryptophan (Trp)	Serine (Ser)	Ornithine
Valine (Val)	Tyrosine (Tyr)	Taurine

It is noteworthy that all of the essential amino acids have been found prevalent in muscle, heart, liver, and kidney tissue in quantities considered to be very adequate, such that meat protein has a *high biological value*.

The ability of the human body to resist disease is dependent upon its ability to produce antibodies—substances which attack specific foreign bodies. Using new analytical methods, scientists discovered that the antibody molecule is actually a molecule of globulin (a class of proteins). Since the blood globulin, as well as the albumins, is built up from the amino acids in food, it suggests to the scientist that the same conditions must apply for the proper synthesis of antibody globulin. Work completed at this writing has given proof that supplying sufficient amino acids in the diet to maintain the protein reserves of the body is an important factor in acquiring immunity to a disease.

The protein content of raw meat varies from around 16 to 20 percent in beef, veal, lamb, and pork, with some ducks and geese as low as 11 percent and chickens and turkeys as high as 25 percent.

Fats

Considerable energy in the average diet is supplied by animal fats which are highly digestible. Fats also supply needed fatty acids (*e.g.,* linoleic acid), carry the fat-soluble vitamins (A, D, E, K), and provide protection and insulation for the human body. Aside from these values, fat plays a role in adding palatability to the lean in meat because of the flavor and aroma it generates.

The chemical difference between saturated and unsaturated fats lies in the existence of double bonds between the carbon atoms in the fatty acids. When the carbon valence is satisfied with hydrogen, the fat is saturated, whereas if it lacks this hydrogen, it is an unsaturated fat. Highly unsaturated fats are soft and oily and may lower the grade of pork carcasses. Since iodine will unite with the carbon at the double bond, its addition to fat will denote the degree of unsaturation, the extent of which is designated as the iodine number. Another method of determining unsaturation is to run polarized light through a prepared piece of fat and read its refractive ability. More recent methods use gas-liquid chromatography or high-pressure liquid chromatography (HPLC) to separate and quantitate the fatty acids.

Fats may be hardened by a process in which the missing hydrogen atom(s) are combined with the unsaturated carbon(s) by the use of a catalyst (sodium methoxide) which saturates the carbon double bond. This is known as the hydrogenation process and is now in general use for the hardening of vegetable oils and lard.

The melting points of fat vary with the class of the domestic animal and the kind of feed it received.

The following shows the range of melting points from the different classes of animals:

Pork

      Back fat .... 86°-104° F
      Leaf fat .... 110°-118° F

Beef

      External fat .... 89°-110° F
      Kidney fat .... 104°-122° F

Lamb

    External fat .... 90°-115° F

    Kidney fat .... 110°-124° F

Poultry

    Abdominal fat .... 80°-110° F

Internal fats are more saturated and have higher melting points than external fats. External fats are more unsaturated than internal fats in order to maintain the liquid state on the outer layers of the live animal where body temperature is lower.

Nutrient Composition

Tables 21-4 and 21-5 give nutrient values of the edible portion of meat and poultry products. A publication by Murphy, Watt, and Rizek[2] described concisely the present state of nutrient tables and how they can be best utilized. It is recommended for the student interested in the complete detail of food composition. Other references are: Watt and Merrill, *Composition of Foods—Raw, Processed, Prepared,* Agricultural Handbook No. 8, USDA, 1963, with revised sections in Poultry, 1979; Sausages, 1980; and Pork, 1983; and Adams, *Nutritive Value of American Foods in Common Units,* Agricultural Handbook No. 456, USDA, 1975. As indicated in previous chapters, the composition of our meat animals, and thus their carcasses and the meat we eat, has changed and continues to change toward a leaner, higher protein product. When utilizing a nutrient table, be sure that it is not outdated.

Meats have been criticized for a lack of unsaturated fats and for their high content of saturated fats; however, saturated fats aid in the keeping quality of meats because they are less subject to oxidation. Nevertheless, meats do contain the unsaturated fats in substantial quantity as shown in Table 21-6.

The table shows that roughly 33 percent of the fat in beef and lamb is unsaturated, whereas pork and poultry contain about 50 percent unsaturated fats. Only trace amounts of linoleic acid are found in beef and lamb, whereas the fats in pork and poultry contain 25 to 33 percent linoleic acid. As can be noted from these data, most of the meats show fats to contribute between 35 and 50 percent of the total calories in the product, whereas the extremes show 19 percent for chicken breasts and 62 percent for regular hamburger. Cholesterol contents are generally proportional to fat; however, there are exceptions, for example, lamb and turkey legs.

---

[2]"Tables of Food Composition: Availability, Uses and Limitations." *Food Technology.* 27:40-51, 1973.

Table 21-4. Nutritive Values of the Edible Part of Meat (Typical Serving)[1]

(Dashes in the columns for nutrients show that no suitable value could be found, although there is reason to believe that a measurable amount of the nutrient may be present.)

	(oz.)	(g)	Water (%)	Food Energy (Cal.)	Protein	Fat (total)	Satu-rated	Unsaturated Oleic	Unsaturated Linoleic	Choles-terol	Calcium	Iron	Phos-phorus	Potas-sium	Sodium (mg)	Vita-min A	Thia-mine	Ribo-flavin	Niacin	Ascor-bic Acid
Bacon, 3 slices cooked	0.7	19	13	110	6	9	3	5	1.0	16	2	0.3	64	92	303	—	.13	.05	1.4	6*
Beef[2] (all cooked)																				
Chuck, lean & fat	3.0	85	42	350	21	29	12	14	1.1	88	16	2.3	116	199	53	—	.06	.18	2.0	—
Chuck, lean	2.1	60	52	180	18	11	5	5	0.5	64	11	2.0	100	169	42	—	.05	.16	1.7	—
Bottom rd., lean & fat	3.0	85	51	235	25	14	6	7	0.5	82	4	2.4	172	245	44	—	.06	.18	2.5	—
Bottom rd., lean	2.7	76	55	170	24	7	3	3	0.3	73	4	2.3	166	235	40	—	.06	.17	2.4	—
Ground beef, 10% fat	3.0	85	55	230	23	15	6	7	0.6	80	10	2.4	196	474	41	—	.08	.20	5.1	—
Ground beef, 21% fat	2.9	82	51	245	20	17	7	8	0.7	77	9	2.1	159	369	39	—	.07	.17	4.4	—
Rib roast, lean & fat	3.0	85	45	330	18	28	12	14	1.0	70	7	2.0	148	245	50	—	.08	.14	3.1	—
Rib roast, lean	2.1	60	56	160	15	10	4	5	0.4	46	5	1.7	127	216	40	—	.07	.13	2.9	—
Eye, round, lean & fat	3.0	85	58	205	22	12	5	6	0.5	60	4	1.4	184	309	49	—	.08	.14	2.7	—
Eye, round, lean	2.6	75	63	135	21	5	2	2	0.2	50	4	1.3	176	298	45	—	.08	.13	2.6	—
Steak, sirloin, lean & fat	3.0	85	52	250	23	17	7	8	0.6	74	9	2.2	208	311	50	—	.10	.21	3.4	—
Steak, sirloin, lean	2.5	71	59	155	21	7	3	3	0.3	60	7	2.1	197	295	45	—	.09	.20	3.2	—
Corned (med. fat)	3.0	85	59	185	22	10	4	5	0.4	80	17	3.7	90	51	802	—	.02	.20	2.9	—
Dried, chipped	2.5	71	48	145	24	4	2	2	0.2	46	14	2.3	287	142	3053	—	.05	.23	2.7	—
Heart, lean	3.0	85	61	160	27	5	2	1	1.0	233	5	5.0	154	197	88	26	.21	1.04	7.6	1
Liver	3.0	85	56	185	22	9	3	4	1.3	372	9	7.5	405	323	156	45390	.22	3.56	14.0	23*
Chicken[3] (cooked)																				
Broiler, flesh, fried	3.0	85	58	184	26	8	2	3	1.8	79	14	1.1	172	216	76	49	.07	.17	8.2	—
Broiler, breast & skin	3.5	98	62	197	30	8	2	3	1.7	84	14	1.0	206	235	68	91	.06	.12	12.5	—
Broiler, breast, flesh only	3.0	86	65	122	27	3	1	1	0.6	73	11	0.9	169	189	54	15	.05	.08	10.1	—
Broiler, drumstick & skin	1.9	57	65	116	14	6	2	2	1.3	48	6	0.7	91	119	47	52	.04	.11	3.1	—
Broiler, drumstick, flesh only	1.5	42	62	82	12	3	1	1	0.8	40	5	0.6	78	105	40	26	.03	.10	2.6	—
Stewing hen, boneless	3.0	85	56	199	26	10	3	3	2.4	71	11	1.2	174	171	66	94	.04	.23	5.4	—
Duck,[3] flesh & skin, roasted	3.1	87	52	293	17	25	8	11	3.2	72	10	2.3	135	177	52	183	.15	.23	4.2	—
Goose,[3] flesh & skin, roasted	3.4	94	52	287	24	21	6	10	2.4	86	13	2.7	254	309	66	66	.07	.30	3.9	—
Lamb[2] (cooked)																				
Arm chop, lean & fat	2.2	63	44	220	20	15	7	6	0.9	77	16	1.5	132	195	46	—	.04	.16	4.4	—
Arm chop, lean	1.7	48	49	135	17	7	3	3	0.4	59	12	1.3	111	162	36	—	.03	.13	3.0	—
Loin chop, lean & fat	2.8	80	54	235	22	16	7	6	1.0	78	16	1.4	162	272	62	—	.09	.21	5.5	—
Loin chop, lean	2.3	64	61	140	19	6	3	2	0.4	60	12	1.3	145	241	54	—	.08	.18	4.4	—

(Continued)

Table 21-4 (Continued)

Food	(oz.)	(g)	Water (%)	Food Energy (Cal.)	Protein	Fat (total)	Satu-rated	Unsaturated Oleic	Linoleic	Choles-terol	Calcium	Iron	Phos-phorus	Potas-sium	Sodium	Vita-min A	Thia-mine	Ribo-flavin	Niacin	Ascor-bic Acid
							(g)								(mg)					
Lamb² (cooked) (continued)																				
Leg, lean & fat	3.0	85	59	205	22	13	6	5	0.8	78	8	1.7	162	273	57	—	.09	.24	5.5	—
Leg, lean	2.6	73	62	140	20	6	2	2	0.4	65	6	1.5	15	247	50	—	.08	.20	4.6	—
Rib, lean & fat	3.0	85	47	315	18	26	12	11	1.5	77	19	1.4	139	224	60	—	.08	.18	5.5	—
Rib, lean	2.0	57	60	208	15	7	3	3	0.5	50	12	1.0	111	179	46	—	.05	.13	3.5	—
Pork² (cooked)																				
Ham, fresh, lean & fat	3.0	85	53	250	21	18	6	8	2.0	79	5	0.9	210	280	50	10	.54	.27	3.9	—
Ham, fresh, lean	2.5	72	60	160	20	8	3	4	1.0	68	5	0.8	202	269	46	10	.50	.25	3.6	.28
Loin chop, lean & fat	3.1	87	50	275	24	19	7	9	2.2	84	3	0.7	184	311	61	10	.87	.24	4.3	.27
Loin chop, lean	2.5	72	57	165	23	8	3	3	0.9	71	4	0.7	176	302	56	10	.83	.22	4.0	.28
Rib roast, lean & fat	3.0	85	51	270	21	20	7	9	2.3	69	9	0.8	190	313	37	10	.50	.24	4.2	.26
Rib roast, lean	2.5	71	57	175	20	10	3	4	1.2	56	8	0.7	182	300	33	10	.45	.22	3.8	.28
Picnic, lean & fat	3.0	85	47	295	23	22	8	10	2.4	93	6	1.4	162	286	75	10	.46	.26	4.4	.26
Picnic, lean	2.4	67	54	165	22	8	3	4	1.0	76	5	1.3	151	271	68	—	.40	.24	4.0	.26
Cured ham, lean & fat	3.0	85	58	205	18	14	5	7	1.5	53	6	0.7	182	243	1009	—	.51	.19	3.8	—
Cured ham, lean	2.4	68	66	105	17	4	1	2	0.4	37	5	0.6	154	215	902	—	.46	.17	3.4	—
Canned ham	3.0	85	67	140	18	7	2	4	0.8	35	6	0.9	188	298	908	—	.82	.21	4.3	—
Sausage																				
Bologna (beef & pork)	2 slices	57	54	180	7	16	6	8	1.4	31	7	0.9	52	103	581	—	.10	.08	1.5	12*
Braunschweiger (pork)	2 slices	57	48	205	8	18	6	9	2.1	89	5	5.3	96	113	652	8010	.14	.87	4.8	6*
Frankfurter (beef & pork)	1 frank	45	54	145	5	13	5	6	1.2	23	5	0.5	39	75	504	—	.09	.05	1.2	12*
Brown & serve (beef & pork)	1 link	13	45	50	2	5	2	2	0.5	9	1	0.1	14	25	105	—	.05	.02	0.4	—
Salami (beef & pork)	2 slices	57	60	145	8	11	5	5	1.2	37	7	1.5	66	113	607	—	.14	.21	2.0	7*
Dry salami (pork & beef)	2 slices	20	35	85	5	7	2	3	0.6	16	2	0.3	28	76	372	—	.12	.06	1.0	5*
Pork	1 link	13	45	50	3	4	1	2	0.5	11	4	0.2	24	47	168	—	.10	.03	0.6	0.2*
Vienna	1 sausage	16	60	45	2	4	2	2	0.3	8	2	0.1	8	16	152	—	.01	.02	0.3	—
Turkey³ (cooked)																				
Flesh & skin	2.9	80	62	166	23	8	2	3	2.0	66	21	1.4	162	224	55	—	.04	.13	4.1	—
Light, with skin	4.0	112	63	220	32	9	3	3	2.3	85	24	1.6	233	320	70	—	.06	.15	7.0	—
Dark, with skin	2.5	71	60	157	20	8	3	3	2.2	64	23	1.6	139	195	54	—	.04	.17	2.5	—
Veal (cooked)																				
Cutlet	3.0	85	55	230	23	14	6	6	1.0	86	10	0.7	211	259	57	—	.11	.26	6.6	—
Rib roast	3.0	85	60	185	23	9	4	4	0.6	86	9	0.8	196	258	56	—	.06	.21	4.6	—

*Ascorbate added in processing.

¹Except where indicated, data by calculations from USDA Home and Garden Bulletin No. 72 (revised 1984).

²For lean and fat, outer layer of fat on the cut was removed to within approximately ½ inch of the lean. Deposits of fat within the cut were not removed. For lean, all separable fat was removed.

³Data derived from Agric. Handbook No. 8-5, *Composition of Foods, Poultry Products.* 1979. USDA.

Table 21-5. Nutritive Content of the Edible Part of Meat[1]

(Per 100 Grams)

	Protein	Fat				Choles-terol	Calories
		Saturated	Unsaturated		Total		
			Mono	Poly			
	.... (g)					(mg)	
Bacon, 3 slices cooked	30.5	17.4	23.7	5.8	49.2	85	576
Beef[2] (all cooked)							
Chuck, lean & fat	24.9	13.9	16.2	1.2	33.6	103	409
Chuck, lean	30.7	7.7	8.5	0.8	18.6	106	299
Bottom rd., lean & fat	29.5	6.8	7.8	0.6	16.4	96	274
Bottom rd., lean	31.8	4.0	4.5	0.4	9.8	96	224
Ground beef, 10% fat	27.2	7.1	7.9	0.7	17.3	94	272
Ground beef, 21% fat	24.2	8.4	9.4	0.8	20.3	94	290
Rib roast, lean & fat	20.9	13.7	16.1	1.2	33.2	82	389
Rib roast, lean	25.3	7.1	7.9	0.7	17.3	76	264
Eye, round, lean & fat	26.2	6.0	7.0	0.5	14.5	70	243
Eye, round, lean	28.3	2.8	3.1	0.3	6.9	66	183
Steak, sirloin, lean & fat	26.8	8.4	9.7	0.8	20.2	87	297
Steak, sirloin, lean	30.1	4.1	4.6	0.4	10.0	85	219
Corned (med. fat)	25.3	5.0	5.8	0.4	12.0	94	216
Dried, chipped	34.3	2.6	2.9	0.3	6.3	65	203
Heart, lean	31.3	1.8	1.5	1.1	5.7	274	188
Liver	26.4	2.9	4.2	1.5	10.6	438	229
Chicken,[3] (cooked)							
Broiler, flesh, fried	30.6	2.5	3.4	2.2	9.1	94	219
Broiler, breast & skin	29.8	2.2	3.0	1.7	7.8	84	197
Broiler, breast, flesh only	31.0	1.0	1.2	0.8	3.6	85	165
Broiler, drumstick & skin	25.3	2.9	4.1	2.4	10.6	83	204
Broiler, drumstick, flesh only	28.6	2.1	2.9	2.0	8.1	94	195
Stewing hen, boneless	30.4	3.1	4.1	2.8	11.9	83	237
Duck,[3] flesh & skin, roasted	19.0	9.7	12.9	3.7	28.4	84	337
Goose,[3] flesh & skin, roasted	25.2	6.9	10.3	2.5	21.9	91	305
Lamb[2] (cooked)							
Arm chop, lean & fat	31.0	10.9	9.6	1.4	24.0	122	348
Arm chop, lean	36.0	6.0	5.5	0.9	14.0	122	279
Loin chop, lean & fat	27.0	9.1	8.0	1.2	20.0	97	294
Loin chop, lean	30.0	4.1	3.8	0.6	10.0	94	215
Leg, lean & fat	26.0	6.6	5.8	0.9	15.0	92	244
Leg, lean	28.0	3.3	3.0	0.5	8.0	89	191
Rib, lean & fat	21.0	14.2	12.5	1.8	31.0	90	368
Rib, lean	26.0	5.7	5.2	0.8	13.0	88	232
Pork[2] (cooked)							
Ham, fresh, lean & fat	25.0	7.5	9.5	2.3	20.7	93	294
Ham, fresh, lean	28.3	3.8	5.0	1.3	11.0	94	220
Loin chop, lean & fat	27.4	8.0	10.2	2.5	22.1	97	316
Loin chop, lean	32.0	3.6	4.7	1.3	10.5	98	231

(Continued)

Table 21-5 (Continued)

	Protein	Fat				Choles-terol	Calories
		Saturated	Unsaturated				
			Mono	Poly	Total		
	. . . . (g)					(mg)	
Pork[2] (cooked) (continued)							
Rib roast, lean & fat	24.7	8.5	10.8	2.7	23.6	81	318
Rib roast, lean	28.2	4.8	6.2	1.7	13.8	79	245
Picnic, lean & fat	26.8	9.3	11.7	2.9	25.5	109	345
Picnic, lean	32.3	4.2	5.5	1.5	12.2	114	248
Cured ham, lean & fat	21.6	6.0	7.9	1.8	16.8	62	243
Cured ham, lean	25.1	1.8	2.5	0.6	5.5	55	157
Canned ham	20.9	2.8	4.1	0.9	8.4	41	167
Sausage							
Bologna (beef & pork)	11.7	10.7	13.4	2.4	28.3	55	316
Braunschweiger (pork)	13.5	10.9	14.9	3.7	32.1	156	359
Frankfurter (beef & pork)	11.3	10.8	13.7	2.7	29.2	50	320
Brown & serve (beef & pork)	13.8	13.0	17.2	3.9	36.3	71	396
Salami (beef & pork)	13.9	8.1	9.2	2.0	20.1	65	250
Dry salami (pork & beef)	22.9	12.2	17.1	3.2	34.3	79	418
Pork	19.7	10.9	13.9	3.8	31.1	83	369
Vienna	10.3	9.3	12.6	1.7	25.2	52	279
Turkey[3] (cooked)							
Flesh & skin	28.1	2.8	3.2	2.5	9.7	82	208
Light, with skin	28.6	2.3	2.8	2.0	8.3	76	197
Dark, with skin	27.5	3.5	3.7	3.1	11.5	89	221
Veal							
Cutlet	27.1	4.8	4.8	0.7	11.1	101	216
Rib roast	27.2	7.1	7.1	1.2	16.9	101	269

[1]Except where indicated, data derived by calculations from USDA Home & Garden Bul. No. 72 (revised 1984).

[2]For lean and fat, outer layer of fat on the cut was removed to within approximately ½ inch of lean. Deposits of fat within the cut were not removed. For lean, all separable fat was removed.

[3]Data derived from Agric. Handbook No. 8-5, *Composition of Foods, Poultry Products*. 1979. USDA.

Carbohydrates

The liver is the carbohydrate reservoir of the animal body, containing about one half of all the carbohydrates found in the body. They are stored as glycogen and represent 2 to 8 percent of the liver weight. The remaining half is distributed in the muscles, with a limited amount in the bloodstream and other tissues.

The exact changes that take place in energy metabolism by the conversion of glycogen to glucose to lactic acid are somewhat complex and are controlled and mediated by enzymes and hormones. Glycogen changes to lactic acid and the process is reversible in the live animal but not in the dressed meat (see earlier discussion in this chapter). Because of this, the lactic acid content of a carcass increases during initial stages of aging or ripening.

Table 21-6. Nutrient Content of Cooked Meats[1]—85-Gram Portion

Meat	Protein	Fat		Cholesterol	Calories	
		Total	Unsaturated		Total	From Fat
		(g)		*(mg)*		
Beef						
Hamburger, lean	23	10	4	—	185	90
Hamburger, regular	21	17	8	80	245	153
Round, lean	26	3	1	77	137	27
Sirloin, lean	27	6	2	—	173	54
Lamb						
Chop, lean	24	7	2	85	161	63
Leg, lean	24	6	2	85	156	54
Shoulder, lean	23	8	2	85	173	72
Pork						
Chop, lean	27	12	7	75	230	108
Roast, lean	25	12	6	75	219	108
Chicken						
Flesh, only	26	8	3	74	184	72
Breast, flesh only	27	3	2	67	142	27
Drumstick, flesh only	24	6	3	77	164	54
Goose						
Flesh & skin	22	19	10	—	259	171
Turkey						
Breast	24	6	3	65	159	54
Legs	24	8	5	85	176	72

[1]As derived from data in Table 21-3 and Feeley et al. 1972. *Cholesterol Content of Foods.* J. American Diet. Assn. 61:134-149.

Water and Minerals

Fat tissue is low in moisture; therefore, the higher the finish, the lower the total water content of a carcass or cut. Mature fat beef may contain as little as 45 percent moisture, while veal may run as high as 72 percent. An important point to remember is that the leaner the meat demanded by the consumer, the more water purchased at meat prices, rather than fat.

Muscle itself is a poor source of calcium, since the calcium content of the body is centered in the bone. It seems logical that the use of mechanical boning machines which recover boneless meat product with calcium levels up to 0.75 percent calcium would be permitted by FSIS. Muscle meats, and more particularly glandular meats, are exceptionally rich in iron and phosphorus. Iron is essential in the formation of red corpuscles, a lack of which causes anemia, while phosphorus is an essential constituent of body cell tissue and is necessary for the assimilation of calcium from other sources. Liver is an excellent source of easily assimilated iron and is prescribed in the diet of anemia sufferers. Muscle contains less than half as much iron as liver.

It has been found that animals also require molybdenum, nickel, selenium, chromium, copper, fluorine, manganese, zinc, cobalt, magnesium, and iodine for normal functioning; therefore, these elements are also in animal tissues. Only a trace of manganese is found in muscle, with liver showing considerably more. Muscle contains a small amount of aluminum; liver contains a slightly larger amount.

The presence of zinc in the pancreas led to further research which has shown that it is a necessary mineral in the diet. Liver was found to be about four times as rich in this element as muscle. Copper, which has been shown to increase iron utilization, is found in small quantities in muscle and in slightly larger quantities in liver.

All of the other trace minerals are found in animal tissues, usually in amounts significant to human dietary needs.

Vitamins

It was not until the latter part of the nineteenth century that scientists discovered that dietary factors other than the proteins, carbohydrates, fats, and minerals were vital for health maintenance. In 1912, Casimir Funk, a Polish biochemist, coined the word *vitamine* to cover this group of dietary essentials, because he wished to designate a particular one which he believed at the time to be an amine. Since that time many new factors or vitamins have been discovered, most of which have been isolated, identified, and chemically synthesized.

Vitamin A (Anti-xerophthalmia)

Vitamin A or retinol is an alcohol of high molecular weight which is soluble in oils and fats but nearly insoluble in water. It is somewhat stable to heat, acids, and alkalis but is destroyed by light and by mild oxidation. It occurs in animal tissues chiefly in the form of fatty acid esters. Alpha, beta, and gamma carotene ($C_{40}H_{56}$) and cryptoxanthin, the yellow coloring matter in corn, green feeds, vegetables, and fruits, are called precursors or "provitamin A," and the animal body is able to convert them into vitamin A. Beta carotene is most prevalent and should yield two molecules of vitamin A, but frequently is poorly converted. This vitamin received early recognition as a required nutrient.

Vitamin A is considered an essential factor in keeping the epithelial tissues and the mucous membranes of the respiratory and genitourinary tracts and the cornea and conjunctiva of the eye in healthy condition and in preventing night blindness.

Vitamin A promotes growth, aids in the resistance to infection, tones the nervous system, and is essential for successful reproduction. Sheep and calf livers are particularly rich in vitamin A, followed by beef, lamb, hog and pig livers, kidney, and chicken liver in the order named. Cod liver oil, butter, cheese, eggs, and fish roe are also excellent sources, followed by beef fat, cream, ice cream, and whole milk. Apricots, broccoli, carrots, kale, spinach, pumpkins, yellow squash, sweet potatoes, and turnip greens, along with certain fruits, are sources of carotenoid pigments which can be transformed into vitamin A in the animal body.

It is estimated that ¼ pound of calf liver or ½ pound of beef liver will supply the daily requirement of this vitamin.

Vitamin B₁—Thiamine (Antineuritic)

Vitamin B_1 is a thiazol-pyrimidine compound called thiamine and is soluble in water but insoluble in oils and fats. It exists in pyrophosphate ester form in animal tissue and is an important coenzyme which plays a role in carbohydrate metabolism. Vitamin B_1 or thiamine was isolated in 1926, was chemically synthesized in 1936, and is sold as the salt thiamine hydrochloride.

A deficiency of this vitamin is the cause of a defect in nerve metabolism known as beriberi (polyneuritis in animals). Symptoms include loss in weight, loss of appetite, slowing of the heart beat, impaired intestinal functioning, impaired reproductive functioning, and failure of lactation. Thiamine promotes growth, stimulates appetite, aids digestion and assimilation, and is essential for normal functioning of nerve tissue.

Pork is an excellent source of thiamine. One center-cut pork chop contains 118 milligrams of thiamine, more than the entire daily requirement for women or children. From 12 to 50 percent of this vitamin may be lost during the cooking process, and some of the thiamine is extracted from the meat by the water in which the meat is cooked. In this case, the meat juice, or broth, contains the dissolved thiamine and should not be discarded. Fried meat shows a smaller loss of the vitamin. Yeast, bran, cereal grains, and legume seeds are rich vegetable sources. Liver, meat, bacon, fish, eggs, milk, and oysters are also considered good sources of thiamine, as are fruits and vegetables.

One ½-pound serving of round steak or two hamburger patties will furnish 30 milligrams of vitamin B_1.

Riboflavin (Vitamin B₂, Formerly G)

Riboflavin, a yellowish-green, fluorescent, water-soluble pigment, a compound of flavin and the pentose sugar ribose, is another of the growth promoting factors of the B complex. It was isolated in 1933 and chemically synthesized in 1935. Riboflavin is stable to heat, mineral acids, and oxidizing agents but is rather sensitive to light. A deficiency of this factor causes stunted growth, premature aging, unwholesomeness of the skin, and a general lowering of the muscle tone of the body. Riboflavin is now known to take part in a number of enzyme systems in the animal body, all of which play important roles in tissue oxidation. It has been found to be a valuable agent in addition to nicotinic acid and thiamine in the treatment of certain cases of pellagra.

Veal and beef liver, followed by beef kidney, lamb liver, pork liver, and pork kidney, are rich sources of riboflavin. Beef heart, milk, oysters, eggs, sardines, yeast, whey (dried), crabs, legumes, prunes, and strawberries are also excellent sources. Ham, bacon, chicken, fish, lamb, beef, cereals, and certain fruits and vegetables are good sources. Very little loss of the vitamin occurs in cooking. It is estimated that the meat in the ordinary diet furnishes about 20 percent of the necessary daily riboflavin requirement.

Nicotinic Acid (Niacin)

Niacin is a simple compound occurring as a white powder or in needle crystal form and is soluble in water and alcohol. Its biological importance was discovered in 1937, and it is made synthetically. It is also called the anti-black tongue factor, since it is a cure for black tongue in dogs. In the form of nicotinamide, niacin also plays an important part in oxidative enzyme systems in body tissues. A deficiency of nicotinic acid over an extended period will cause pellagra, dermatitis, and glossitis and contribute to a form of insanity in humans.

Nicotinic acid is heat stable and is found abundantly in pork, beef, veal, and lamb liver. Pork and beef kidney rank next, followed by pork and beef heart, pork meat, veal, chicken, beef, and lamb. Salmon, wheat germ, whey (dried), and yeast are also excellent sources. Other sources are buttermilk, eggs, haddock, milk, kale, peas, potatoes, tomatoes, and turnip greens; however, niacin from plant sources is less available.

Liver (¼ pound) or veal, pork, or beef (½ pound) per day is reported to furnish the daily human nicotinic acid requirement.

Vitamin B_6 (Pyridoxine)

Pyridoxine has been called the rat acrodynia factor, having been found to be essential for the maintenance of a healthy skin in rats, for amino acid metabolism, and for the utilization of unsaturated fatty acids. It was isolated in the crystalline state in 1938 and is available commercially. It is water-soluble, stable to heat, acids, and alkali but is destroyed by light and ultraviolet irradiation. Pyridoxine ($C_8H_{11}NO_3$) has a base structure similar to nicotinic acid, but with additions, that is, 2-methyl, 3-hydroxy, 4, 5-dihydroxymethyl pyridine.

Pyridoxine deficiency symptoms in the human have been restricted primarily to infants and young children subsisting on certain prepared dietary formulas. It has been established that pyridoxine derivatives function in certain enzyme systems which have to do with the transfer of amino (NH_2) groups in metabolism of nitrogen compounds. Lean meat and kidney are reported to be slightly more potent sources of vitamin B_6 than liver, with heart and brains furnishing lesser amounts. Egg yolk, wheat germ, and yeast are also excellent sources. Fish, milk, legumes, and wheat are good sources. It is estimated that meat furnishes a large share of the daily requirement of this vitamin.

Choline

Choline is usually included as a member of the B complex. It is β-hydroxy-ethyl–trimethyl ammonium hydroxide and functions in normal fat metabolism. It is one of several substances which help to prevent perosis and fatty livers. Liver, pancreas, egg yolk, and meat are considered rich sources of choline.

Folic Acid (Pteroylglutamic Acid)

Names formerly used for folic acid include vitamin M, vitamin Bc, Factor R, and L. casei factor. It has been found to be essential for the development of red and white blood cells. Pure folic acid crystals are used medicinally for the treatment of macrocytic anemias. It was synthesized in 1945, is only moderately heat stable, and is present in green leaves. Pteroylglutamic acid is the name of the basic folic acid molecule. At one end of the molecule is the double ring or pterin group, a previously known yellow compound. The central grouping is a para-aminobenzoic acid ring formerly thought to be protective against graying hair. The third component in the folic acid molecule is at least one unit of glutamic acid, a normal constituent of most proteins. Liver, kidney, beef, veal, yeast, green and leafy vegetables, wheat, and soybeans are good sources of folic acid.

Vitamin B_{12} (Anti-pernicious Anemia Factor)

The animal protein factor (APF) is largely represented by vitamin B_{12}, or cobalamin, a vitamin which contains an inorganic base, namely, cobalt. Variety meats contribute the most vitamin B_{12}. Muscle meats and fish are good sources, with eggs contributing a lesser amount. Gastric juice in the stomach provides the intrinsic factor for improving its absorption. A lack of vitamin B_{12} (extrinsic factor) or a lack of intrinsic factor leads to pernicious anemia. Folic acid may affect the storage of the vitamin. Both vitamin B_{12} and folic acid contribute to the prevention and treatment of human macrocytic anemia (enlarged red blood cells), a condition found in pernicious anemia, sprue, and the anemia that can occur during pregnancy.

Pantothenic Acid

Pantothenic acid, a B vitamin, plays an important role in the metabolism of fats and other fat compounds such as cholesterol. It is a combination of a derivative of butyric acid and β-amino propionic acid and is available as a synthetic preparation in the form of dextrorotatory calcium pantothenate. Experiments with rats, dogs, and pigs indicate that requirements of this vitamin are 5 to 10 times those of thiamine or riboflavin. Clinical evidence indicates that humans on diets low in pantothenic acid show anemias, sore tongues, and foot irritations.

Kidney, liver, and beef heart were found to be potent sources, followed by beef spleen, beef pancreas, and beef tongue, with the major cuts having 1/10 the potency of liver or kidney. Pantothenic acid is destroyed by prolonged dry heat.

Vitamin C (Antiscorbutic Factor)

Vitamin C is a product of hexose sugar metabolism (except in humans) and is also known as ascorbic acid. It is water-soluble and has been isolated and chemically synthesized. Animals normally synthesize adequate amounts but humans do not. A deficiency of vitamin C causes scurvy, and, under stress conditions, its requirements may greatly increase. Sprouting plants are rich in this vitamin, as are citrus fruits and some vegetables. Meats are only a fair source of vitamin C. The greatest concentration in animal tissue exists in the adrenals, corpus lutea, and the thymus. Open-kettle cooking or wilting destroys considerable vitamin C. Ascorbic acid is used to hasten the development of the characteristic cured-meat color and also retards or prevents nitrosamine formation in cured meats (see Chapter 19).

Vitamin D (Antirachitic Factor)

Vitamin D is formed in plants following irradiation of ergosterol to form calciferol or vitamin D_2. Other forms of this sterol group are now well known, such as vitamins D_3, D_4, and D_5. Vitamin D_3 is formed in human and animal bodies when the skin containing 7-dehydrocholesterol is exposed to direct sunlight or ultraviolet light. The resulting cholecalciferol is insoluble in water and soluble in oils and fats. It is further modified for specific functions to 25-hydroxy and 1,25-dihydroxy derivatives by controlled feedback mechanisms in the liver and kidney.

The functions of vitamin D metabolites are to regulate calcium and phosphorus metabolism and they are, therefore, essential to normal bone growth and tooth development. They help prevent rickets in infants and children and softening of the bones in adults and are involved in growth and reproduction. Rich sources of vitamin D are cod liver oil, fish, egg yolk, irradiated foods, and fortified milk. Pork liver and beef liver are considered good sources; calf liver and other meat products contain only fair amounts.

Vitamin K (Antihemorrhagic Factor)

Vitamin K was discovered in 1935 and can be made synthetically. Only two forms, designated as vitamin K_1 from plants and vitamin K_2 from bacteria, have been isolated from natural sources. Vitamin K is fat-soluble and heat- and light-stable. It is essential for the production of prothrombin, a blood coagulant, and aids in the prevention of hemorrhage as in newborn infants and in the prevention of obstructive jaundice. A form of vitamin K that can be injected intramuscularly prior to surgical operations has been developed.

The K vitamins were first isolated from alfalfa leaf meal and from putrefying fish, the former being the principal practical source. Hog liver is also a rich source. Cabbage, carrot greens, spinach, soybean oil, tomatoes, hempseed, cauliflower, rice, bran, kale, and egg yolk are considered good sources. The principal commercial sources are a synthetic product known as menadione and its sodium bisulfite, water-soluble derivative.

Chapter 22

Preparing and Serving Meats

METHODS OF COOKING MEAT

THE ULTIMATE satiety or enjoyment encountered in eating meat is largely dependent on how it is cooked. The very highest-quality steaks can be rendered unpalatable by overcooking or by other improper cooking methods. On the other hand, the lowest-quality cuts of meat can be turned into an exquisite dining experience when proper cookery guidelines are correctly followed.

A national committee of investigators has simplified and standardized the cooking of meat into two fundamental methods.

- *Dry heat*, in which the meat is surrounded by dry air in the oven or under the broiler, a method that is adaptable to the preparation of the more-tender cuts of meat.
- *Moist heat*, in which the meat is surrounded by hot liquid or steam, a method suitable to the preparation of the less-tender cuts of meat.

Wide variability exists in the preferred degree of doneness by consumers—especially for beef and somewhat for lamb. This variation in consumer preferences created a need for standardization of end-point cooking temperatures with well defined degrees of doneness. In 1979, the National Live Stock and Meat Board published a color guide illustrating six degrees-of-doneness standards (see color section near the center of the book). These degrees of doneness define a maximum internal temperature to which each degree should be cooked and are termed *very rare, rare, medium rare, medium, well done,* and *very well done.* The following descriptions are also helpful in categorizing each degree of doneness for broiled steaks.

Very rare: 130° F; red color throughout except for narrow, reddish pink layer under meat surface.
Rare: 140° F; red in center third; reddish pink to outer surface.
Medium rare: 150° F; reddish pink in center third; pink to light brown to outer surface.
Medium: 160° F; light pink in center; light brown to outer surface.
Well done: 170° F; light brown in center; darker brown to outer surface.

Very well done: 180° F; darker brown throughout; dry texture; charred
surface.

Dry Heat Method

Broiling

Broiling is employed with the more-tender steaks and chops and cured pork.
It consists of a direct exposure of the meat to heat, either from above or from
below, as with outdoor charcoal broiling, or from an oven broiler. The seasoning
may be applied before or after broiling, the latter being preferable. (See Table
22-1 for broiling times and temperatures.)

Table 22-1. Timetable for Broiling[1, 2]

		Approximate Total Cooking Time	
Cut and Thickness	Weight	Rare	Medium
		. . . . *(min.)*	
Beef			
Chuck steak (high			
quality)—1 in.	1½ to 2½ lb.	24	30
1½ in.	2 to 4 lb.	40	45
Rib steak—1 in.	1 to 1½ lb.	15	20
1½ in.	1½ to 2 lb.	25	30
2 in.	2 to 2½ lb.	35	45
Rib-eye steak—1 in.	8 to 10 oz.	15	20
1½ in.	12 to 14 oz.	25	30
2 in.	16 to 20 oz.	35	45
Top loin steak—1 in.	1 to 1½ lb.	15	20
1½ in.	1½ to 2 lb.	25	30
2 in.	2 to 2½ lb.	35	45
Sirloin steak—1 in.	1½ to 3 lb.	20	25
1½ in.	2¼ to 4 lb.	30	35
2 in.	3 to 5 lb.	40	45
Porterhouse steak—			
1 in.	1¼ to 2 lb.	20	25
1½ in.	2 to 3 lb.	30	35
2 in.	2½ to 3½ lb.	40	45
Filet mignon—1 in.	4 to 6 oz.	15	20
1½ in.	6 to 8 oz.	18	22
Ground beef patties			
1 in. thick by 3 in.	4 oz.	15	25
Pork (smoked)			
Ham slice—tendered			
½ in.	¾ to 1 lb.		10-12
1 in.	1½ to 2 lb.		16-20
Loin chops—		Generally	
¾ to 1 in.		cooked	15-20
Canadian bacon		well done	
¼-in. slices		(170° F)	6-8
½-in. slices			8-10
Bacon			4-5

(Continued)

Table 22-1 (Continued)

Cut and Thickness	Weight	Approximate Total Cooking Time	
		Rare	Medium
		.... *(min.)*	
Pork (fresh)		Generally	
Rib or loin chops	¾ to 1 in.	cooked	20-25
Shoulder steaks	½ to ¾ in.	well done	20-22
		(170° F)	
Lamb			
Shoulder chops—			
1 in.	5 to 8 oz.	Lamb chops	12
1½ in.	8 to 10 oz.	are not	18
2 in.	10 to 16 oz.	usually	22
Rib chops—1 in.	3 to 5 oz.	served rare	12
1½ in.	4 to 7 oz.		18
2 in.	6 to 10 oz.		22
Loin chops—1 in.	4 to 7 oz.		12
1½ in.	6 to 10 oz.		18
2 in.	8 to 14 oz.		22
Ground lamb patties			
1 in. thick by 3 in.	4 oz.		18

[1]This timetable is based on broiling at a moderate temperature (350° F). Rare steaks are broiled to an internal temperature of 140° F; medium to 160° F; well done to 170° F. Lamb chops are broiled to an internal temperature of 150° F to 165° F, depending on personal preference. Ham is cooked to 160° F. The time for broiling bacon is influenced by personal preference as to crispness.

[2]National Live Stock and Meat Board.

Panbroiling and Panfrying

Panbroiling and panfrying are suitable for the same cuts used for broiling, but heat reaches the cuts indirectly. The meat is placed in a heavy fry pan and is browned on both sides. After browning, the temperature is lowered and the cuts may be turned occasionally until done. The fat is poured off as it accumulates in panbroiling, and the fat remains in the pan in panfrying. Panfrying and panbroiling require about half the cooking time as broiling.

Roasting

Roasting is adapted to the preparation of the more-tender cuts such as beef ribs, beef sirloin, top round, sirloin tip, veal leg, veal rump, veal loin, veal shoulder, pork loin, pork shoulder, leg of lamb, sirloin lamb roll, loin lamb roll, rolled shoulder of lamb, and fresh or cured pork. It is accomplished by placing the cut (preferably not less than 2½ inches thick) in an open roasting pan with the fat side up so that it will be self-basting. No water is added; neither is a lid used to cover the roast. Smoked pork, fresh beef, veal, and lamb are roasted at an oven temperature of 300° to 325° F, whereas fresh pork is roasted at an oven temperature of 325° to 350° F. (See Table 22-2 for roasting times and temperatures.)

Table 22-2. Timetable for Roasting[1]

Cut	Approx- imate Weight	Oven Temperature Constant	Interior Temperature When Removed from Oven	Approx- imate Cooking Time
	(lb.)	. . . . *(°F)*		*(min. per lb.)*
Beef				
Standing rib[2]	6 to 8	300-325	140 (rare)	23 to 25
			160 (medium)	27 to 30
			170 (well)	32 to 35
	4 to 6	300-325	140 (rare)	26 to 32
			160 (medium)	34 to 38
			170 (well)	40 to 42
Rolled rib	5 to 7	300-325	140 (rare)	32
			160 (medium)	38
			170 (well)	48
Delmonico (rib eye)	4 to 6	350	140 (rare)	18 to 20
			160 (medium)	20 to 22
			170 (well)	22 to 24
Tenderloin, whole	4 to 6	425	140 (rare)	45 to 60 (total)
Tenderloin, half	2 to 3	425	140 (rare)	45 to 50 (total)
Rolled rump (high quality)	4 to 6	300-325	150-170	25 to 30
Sirloin tip	3½ to 4	300-325	140-170	35 to 40
(high quality)	4 to 6	300-325	140-170	30 to 35
Veal				
Leg	5 to 8	300-325	170	25 to 35
Loin	4 to 6	300-325	170	30 to 35
Rib (rack)	3 to 5	300-325	170	35 to 40
Rolled shoulder	4 to 6	300-325	170	40 to 45
Pork (fresh)				
Loin				
Center	3 to 5	325-350	170	30 to 35
Half	5 to 7	325-350	170	35 to 40
Blade loin or sirloin	3 to 4	325-350	170	40 to 45
Rolled	3 to 5	325-350	170	35 to 45
Picnic shoulder	5 to 8	325-350	170	30 to 35
Rolled	3 to 5	325-350	170	35 to 40
Cushion style	3 to 5	325-350	170	30 to 35
Boston butt	4 to 6	325-350	170	40 to 45
Leg (fresh ham)				
Whole (bone-in)	12 to 16	325-350	170	22 to 26
Whole (rolled)	10 to 14	325-350	170	24 to 28
Half (bone-in)	5 to 8	325-350	170	35 to 40
Spareribs		325-350	well done	1½ to 2½ hrs. (total)
Pork (smoked)				
Ham (cook before eating)				
Whole	10 to 14	300-325	160	18 to 20
Half	5 to 7	300-325	160	22 to 25
Shank or butt portion	3 to 4	300-325 -	160	35 to 40
Ham (fully cooked)[3]				
Half	5 to 7	325	130	18 to 24

(Continued)

Table 22-2 (Continued)

Cut	Approximate Weight	Oven Temperature Constant	Interior Temperature When Removed from Oven	Approximate Cooking Time
	(lb.)	.... (°F)		(min. per lb.)
Pork (smoked) (continued)				
Picnic shoulder	5 to 8	300-325	170	35
Shoulder roll	2 to 3	300-325	170	35 to 40
Canadian bacon	2 to 4	325	160	35 to 40
Lamb				
Leg	5 to 8	300-325	175-180	30 to 35
Shoulder	4 to 6	300-325	175-180	30 to 35
Rolled	3 to 5	300-325	175-180	40 to 45
Cushion	3 to 5	300-325	175-180	30 to 35
Rib	1½ to 3	375	170-180	35 to 45

[1]National Live Stock and Meat Board.

[2]Ribs which measure 6 to 7 inches from chine bone to tip of rib.

[3]Allow approximately 15 minutes per pound for heating whole ham to serve hot.

The use of longer cooking times at lower temperatures in roasting has been found to cut down considerably on the shrinkage incident compared to higher oven temperatures. Basting is eliminated by placing the fat side up, or placing loose fat or bacon strips on the top of lean cuts. Searing does not assist materially in keeping the meat juices from escaping, but it gives the meat color and aroma.

Moist Heat Method

Braising

Water, meat or vegetable stock, or sour cream or milk may be used to furnish the moisture. Braising is employed on the less-tender cuts such as the blade and arm roast of beef or steak from the same cuts; the heel of the round of beef; round and flank steak of beef; the steaks of veal, such as round, sirloin, blade, and arm veal steak; veal loin and rib chops; pork chops (both loin and rib); blade and arm pork steak from the pork shoulder; breast of lamb; neck slices of lamb; and lamb trotters.

The meat to be braised is first seasoned, dredged with flour (if desired), and browned, and the necessary liquid is added. The kettle or cooking utensil is covered, and the cut is cooked either in the oven or on top of the range at a simmering temperature. This method is commonly called pot roasting. (See Table 22-3 for braising times.)

Cooking in Liquid

This method of preparing small or large pieces of meat is suitable for such

Table 22-3. Timetable for Braising[1]

Cut	Average Weight or Thickness	Approximate Total Cooking Time
Beef		
Pot roast		
Arm or blade	3 to 4 lb.	2½ to 3½ hrs.
Boneless	3 to 5 lb.	3 to 4 hrs.
Swiss steak	1½ to 2½ in.	2 to 3 hrs.
Cubes for kabobs	2-in. cubes	1½ to 2½ hrs.
Short ribs	Pieces (2 in. × 2 in. × 4 in.)	1½ to 2½ hrs.
Round steak	¾ in.	1 to 1½ hrs.
Stuffed steak	½ to ¾ in.	1½ hrs.
Pork		
Chops	¾ to 1½ in.	45 to 60 min.
Spareribs	2 to 3 lb.	1½ hrs.
Tenderloin		
Whole	¾ to 1 lb.	45 to 60 min.
Fillets	½ in.	30 min.
Shoulder steaks	¾ in.	45 to 60 min.
Lamb		
Breast—stuffed	2 to 3 lb.	1½ to 2 hrs.
Breast—rolled	1½ to 2 lb.	1½ to 2 hrs.
Riblets		1½ to 2½ hrs.
Neck slices	¾ in.	1 hr.
Shanks	¾ to 1 lb. each	1 to 1½ hrs.
Shoulder chops	¾ to 1 in.	45 to 60 min.
Veal		
Breast—stuffed	3 to 4 lb.	1½ to 2½ hrs.
Breast—rolled	2 to 3 lb.	1½ to 2½ hrs.
Veal riblets		2 to 3 hrs.
Veal birds	½ in. (×2 in. ×4 in.)	45 to 60 min.
Chops	½ to ¾ in.	45 to 60 min.
Steaks or cutlets	½ to ¾ in.	45 to 60 min.
Shoulder chops	½ to ¾ in.	45 to 60 min.
Shoulder cubes	1 to 2 in.	45 to 60 min.

[1]National Live Stock and Meat Board.

cuts as beef shank (soup bones), beef plate and brisket, veal shank and breast, lamb shank and breast, pork spareribs, and fresh or smoked pork shoulder (butts and picnics).

In the case of meat that is cut into small pieces for stews, the seasoning is added, and the pieces are browned (this is optional) in their own or added fat and then covered with hot water; in some cases tomato juice is added. The kettle is covered, and the meat is allowed to cook at a simmering temperature. If vegetables are to be added, they should be added just long enough before the meat is tender so that they will not be overdone. The liquid can be thickened so that it may be served separately or with the stew. (See Table 22-4 for cooking in liquid and Table 22-5 for variety meats cooking times.)

Table 22-4. Timetable for Cooking in Liquid[1]

Cut	Average Weight	Approximate Time per Pound	Approximate Total Cooking Time
	(lb.)	(min.)	(hrs.)
Smoked ham (old style and country cured)			
Large	12 to 16	20	
Small	10 to 12	25	
Half	5 to 8	30	
Smoked ham (tendered)			
Shank or butt half	5 to 8	20 to 25	
Smoked picnic shoulder	5 to 8	45	
Fresh or corned beef	4 to 6	40 to 50	
Beef for stew			2½ to 3½
Veal for stew			2 to 3
Lamb for stew			1½ to 2

[1]National Live Stock and Meat Board.

Microwave Cooking

A regular oven is heated by gas or electricity, and the heated air inside the oven cooks the food. In an electronic (microwave) oven, a magnetron (think of a vacuum tube) produces microwaves which are absorbed by the food, causing the molecules within the food to vibrate against each other. These microwaves are a low-level form of radiant energy, just as are radio waves, visible light, and infrared heat; they all have long wave lengths, so their radiant energy is nonionizing, meaning that it has no cumulative harmful effect on humans. The friction that is created causes heat penetration within the food itself, thus cooking it. Microwaves are reflected by metal (the oven walls), transmitted through glass, paper, pottery, and plastic (the materials the food is to be cooked in) and absorbed by the food. This explains why only the food gets hot, leaving oven walls and pan cool.

The primary advantage of microwave cooking is speed; cooking time is usually cut in half. This includes the actual cooking time plus the "standing time" that most foods require for heat equalization after cooking—conventionally cooked food continues to cook for a while after it is removed from the oven.

When meat is prepared using microwave cookery methods, the recommended practice is to place the cut (steak, chop, or roast) in a plastic cooking bag and cook at a low to medium power setting. If the microwave oven is not equipped with a rotating turnstyle device, the cooking dish should be rotated 180° halfway through the cooking process. Use of a non-melting cooking bag is especially important in microwave cookery of pork. The bag creates a steamy cooking environment which helps insure greater temperature uniformity of the meat cut being cooked. Without the cooking bag, "hot" and "cold" spots may occur in the chop or roast. It has been demonstrated that viable *trichinae* can

Table 22-5. Timetable for Cooking Variety Meats[1]

Kind	Broiled	Braised[2]	Cooked in Liquid
	(min.)		
Liver			
Beef			
3- to 4-pound piece		2 to 2½ hrs.	
Sliced		20 to 25 min.	
Veal (calf), sliced	8-10		
Pork			
Whole (3 to 3½ pounds)		1½ to 2 hrs.	
Sliced		20 to 25 min.	
Lamb, sliced	8-10		
Kidney			
Beef		1½ to 2 hrs.	1 to 1½ hrs.
Veal (calf)	10-12	1 to 1½ hrs.	¾ to 1 hr.
Pork	10-12	1 to 1½ hrs.	¾ to 1 hr.
Lamb	10-12	¾ to 1 hr.	¾ to 1 hr.
Heart			
Beef			
Whole		3 to 4 hrs.	3 to 4 hrs.
Sliced		1½ to 2 hrs.	
Veal (calf)			
Whole		2½ to 3 hrs.	2⅓ to 3 hrs.
Pork		2½ to 3 hrs.	2½ to 3 hrs.
Lamb		2½ to 3 hrs.	2½ to 3 hrs.
Tongue			
Beef			3 to 4 hrs.
Veal (calf)			2 to 3 hrs.
Pork } usually sold			
Lamb } ready-to-serve			
Tripe			
Beef	10-15[3]		1 to 1½ hrs.
Sweetbreads	10-15[3]	20 to 25 min.	15 to 20 min.
Brains	10-15[3]	20 to 25 min.	15 to 20 min.

[1]National Live Stock and Meat Board.
[2]On top of range or in a 300° F to 325° F oven.
[3]Time required after precooking in water.

survive in these "cold" pockets when pork cuts register an internal temperature of 160° F. Research at Iowa State University[1] has shown that no viable *trichinae* survive when pork is microwave cooked to an internal temperature of 160° F or higher using a cooking bag (not sealed) and a low to medium power setting.

Because there is no heat in the oven itself, meat and other foods which require the hot air of a conventional oven for browning and crisping may not be as satisfactory—unless the speed factor is more important than a crisp, brown exte-

---

[1]"Microwave Cooking of Pork." 1984. Proceedings of the Meat Industry Research Conference. National Live Stock and Meat Board.

rior. One alternative is to cook the meat in a microwave oven until it is halfway done, then use a regular oven or broiler for quick browning and crisping.

A microwave oven can be used to thaw frozen foods quickly, and later they can be cooked in a regular range. The microwave oven is especially good for large, slow-thawing roasts and poultry. Precooked frozen foods can be quickly thawed or reheated; this method has been used successfully for years in many restaurants.

Microwave cookery is not yet the panacea for the home or the institutional cook, but it does offer tremendous advantages. The energy efficiency of microwave ovens is an additional advantage that is often overlooked by the homemaker. Microwave ovens will typically use 40 to 50 percent less energy to cook a given quantity of food than a conventional oven. More research is needed and is planned on the application of microwave cookery of meats.

Convection Ovens

A convection oven is similar to a conventional electric oven or range except that the hot air within is circulated with a blower or fan. This effectively reduces cooking time, due to a more rapid transfer of heat. Convection ovens therefore offer the advantages of faster cooking times and require less energy to operate when compared to a regular oven. Unlike in a microwave, meat will brown uniformly just as it does in a regular oven. Convection ovens have become more popular in recent years for these reasons.

COOKING LOSSES

It is common knowledge that the major loss in weight in cooked meat is due to the loss in moisture evaporated by the heat. This change in weight alters the percentage of protein, fat, and ash of the cooked meat as compared to the fresh meat. Another weight loss is that of the melted fat. This will be affected in large part by the degree of doneness.

The work of Leverton and Odell of Oklahoma in 1956-1958 showed that evaporation loss during cooking varied from 1.5 to 54.5 percent with an average range between 15 and 35 percent. The cooked lean meat with all separable fat removed contained from 5 to 10 percent fat and up to 35 percent protein.

The ratio of fat to lean is more important to people of moderate means than to those with higher incomes. It is easy to say, "You must have fat to have quality in meat" and "You don't have to eat the fat,"—but it is rather difficult to compromise on fat with a lean wallet.

REMINDERS

The guess work in determining the doneness of meat is eliminated by the use of a meat thermometer, a very much appreciated kitchen accessory. It is particularly useful in determining the doneness of roasts. Experiments conducted in commercial and college laboratories on the time required to prepare various meats are based on the use of the meat thermometer and automatically regulated oven and broiler temperatures. This does not mean that expert cooks with years of cooking experience have any need for these gadgets, but beginners must have specific answers and directions. Some of these are as follows:

- Bacon is more desirable when panbroiled below its smoke point (290° to 300° F) in its own grease until limp, but not quite crisp and a light golden brown in color. Draining off the grease as the bacon fries may cause scorching or burning; however, bacon broiled on a rack about 4 inches below the flame retains more of its original thiamine (vitamin B_1).
- A temperature of 300° to 325° F is best for deep-fat frying.
- Meat should not be boiled. A simmering temperature of 185° to 205° F should be used. This includes soup making.
- A constant oven temperature of 300° F for roasts of beef, veal, and lamb and 325° to 350° F for pork produces more tender and palatable roasts than higher temperatures do.
- High oven temperatures affect the cost of the meat by increasing fuel consumption and shrinkage in meat poundage. It's very disheartening to open an oven and find the roast charred and about half its original size.
- The shape being the same, the larger cut will require a longer total cooking period but fewer minutes per pound.
- Meat cooked without undue shrinkage is juicier and more highly flavored. Burned meat results in damaged proteins.
- Searing meat does not hold in the juices, but it is done by many cooks to brown the outside of a roast and develop aroma.
- The water-soluble B vitamins leak into the drippings. Low-temperature cooking will result in more of these vitamins remaining in the meat. Gravy should never be discarded. It is the valuable by-product in meat cookery.
- Where controlled temperatures are possible, a constant broiler temperature of 350° to 400° F gives the best results. Without controls, a "high" setting or continuously "on" will have to be regulated by opening the oven door and/or adjusting the height of the rack below the heat source.
- Boneless or rolled cuts require from 5 to 10 minutes more time per pound to cook than do bone-in cuts.
- Roasts with long bones require less time than do thick, chunky cuts.
- Retention of the B vitamins in properly roasted meat averages: thiamine—70 percent, riboflavin and niacin—90 percent.
- Meat that is cooked without undue shrinkage has higher nutritional value.
- The tenderness of steaks generally decreases with the decrease in carcass grade (Prime, Choice, Good, Standard, Comercial, Utility).

- Cooking time can be decreased by the use of metal skewers and by unventilated ovens.
- Small roasts under 2 pounds and less than 2 inches thick are not economical; they tax the patience of the retailer. Many fine dishes can be made from leftovers. And how about those handy cold cuts of roast beef, pork, lamb, and veal for sandwiches or midnight snacks?
- Expensive cuts such as steak for grinding will increase the cost of hamburger. The cheapest cut, mixed with sufficient beef suet or fat back, will be as delicious and even more nutritious. This applies to personal marketing and not to telephone-order buying.
- For a delicious meat loaf, a good combination is ¼ pound of fresh pork for each pound of beef to be ground together. Pork shoulder and boneless beef shank, neck, or plate are good buys.
- Meat should not be removed from its package upon delivery; it should be placed immediately under refrigeration. Aged meats left to lie in a warm kitchen for half a day may develop an odor that is offensive, indicating microbial growth and a possible cause for food infection or intoxication (see Chapter 18).
- A good kitchen scale makes an efficient short-weight detective.
- Lard should be placed in a well-sealed container and stored under refrigeration.
- A dark color in meat does not necessarily mean that it is spoiled or of poor quality (see Chapter 21).
- Monosodium glutamate, the vegetable protein derivative, accentuates the natural flavor of food and has been considered as essential as table salt by the Chinese and Japanese.

HANDLING FROZEN MEAT IN THE HOME AND INSTITUTION

Below zero temperatures (the lower the better) are best for holding frozen meat.

Frozen meat that has been thawed under refrigeration need not be used immediately as is commonly recommended, because repeated tests have shown that such meat will keep as long as fresh meat properly refrigerated.

Refreezing meat does not materially affect its quality. Tests were made in which beef, properly wrapped in a good grade of locker paper, film, or aluminum foil, was thawed in the unopened package until it became warm, was refrozen, and later was rethawed and held in that condition at 38° F in a household refrigerator for an additional week, and was still in excellent condition when cooked. This does not mean that you should become careless, but it also suggests that you need not become panicky about using all the meat in a package that has been thawed if it is more than is needed for that meal. Rewrap it, re-freeze it, and use it at another time.

Every time frozen meat is thawed it will lose some of the meat juices. If the position of the thawed meat package is reversed (turned over) when replaced in the zero compartment for refreezing, these juices will be reabsorbed to a large extent.

Zero temperatures materially depress the proteolytic enzyme action that breaks down connective tissue during the aging process.

The so-called freezer burn on meats is caused by a considerable dehydration (moisture loss) of the meats or parts thereof, due to a poor grade of wrapping pa-per, improper wrapping, or holes in the paper. In badly dehydrated meat, water added in the cooking to replace that which was lost will help, but the meat has lost considerable flavor and tends to be tough and stringy.

Remember that it is not the lean meat but the fat that changes in flavor in zero storage. The oxygen in air will combine with unsaturated fats and break them down into free fatty acids and aldehydes, giving them a stale, rancid flavor. This flavor is, in part, absorbed by the lean. That is the reason for using a good grade of wrapping paper, one that is moisture-vapor proof, and employing the drugstore method of wrapping or vacuum-packaging to exclude the air.

It is rather foolish to go to all this trouble, using expensive paper and tape and taking valuable time to do a good job, and then fling the package into a bas-ket and rip it.

When roasting an unthawed cut, allow additional time equal to one third to one half the recommended time for unfrozen cuts (see Table 22-2). If a meat thermometer is used, it should be inserted after the meat is partially cooked and the frost is out of the center.

If thawing meat before cooking is preferred, refer to Table 22-6 for a timeta-ble for defrosting frozen meat. Thawing in a refrigerater is the recommended practice.

In a study of cooking times, yields, and temperatures of frozen roasts con-ducted at the University of Illinois[2] and cosponsored by the National Associa-tion of Meat Purveyors and the National Live Stock and Meat Board, the final conclusion was that "Roasts cooked from the frozen state yield as much as

Table 22-6. Timetable for Defrosting Frozen Meat[1]

Meat	In Refrigerator	Room Temperature[2]
Large roast	4 to 7 hrs. per lb.	2 to 3 hrs. per lb.
Small roast	3 to 5 hrs. per lb.	1 to 2 hrs. per lb.
1-inch steak	12 to 14 hrs.	2 to 4 hrs.

[1]National Live Stock and Meat Board.
[2]Generally not recommended; better to plan ahead.

---

[2]*Roasting Frozen Meat*, National Live Stock and Meat Board.

roasts partially or completely thawed prior to cooking.'' It was further observed, after cooking 860 roasts weighing almost 4 tons, that roasting from a frozen state requires between 1.3 and 1.45 times as long to cook as from a chilled state.

COOKING POULTRY

Preparation

Frozen birds should be thawed to remove the giblets and internal fat. Sprinkle the inside of the bird with salt. Place a stuffing in the body cavity of the bird (avoid packing), and draw the ends of the drumsticks down against the opening, using a cord that laps over the back of the tail and the ends of the legs. Ducks, pheasants, and guinea fowl will require stitching to close the opening, but the legs of all fowl must be tied close to the body to avoid overcooking and drying. The loose skin over the crop may be filled with the stuffing and the edge fastened with a cord.

Fold the wing tips back on the wings. Rub the breast and legs with butter or margarine and sprinkle with salt, preferably celery salt. Dust lightly with flour, if desired. Ducks and geese need no added fat.

Roasting

The procedure will vary with the age of the bird. Old birds should be steamed or braised for 1½ to 2 hours. This is done by placing the bird breast-up on the rack in the roaster. Cover the bottom of the roaster with hot water, place lid on roaster and braise in an oven temperature of 250° to 275° F. It may be necessary to add water several times during the braising process. Some prefer to braise the stuffed bird, and others add the stuffing after the braising period. After this steaming period, the lid and water are removed, and the oven temperature is adjusted to 325° F for the remainder of the roasting.

Young birds or those having a flexible tip on the rear end of the breast bone are placed on the roasting rack, with no lid and no water added. The position of the bird in the conventional method of roasting is with the breast up. The more recent practice is to place the bird on its side or squarely on its breast using a V-shaped rack for support. Basting at 45-minute intervals with pan drippings may be desirable or an aluminum foil tent (loosely covered) may be used to minimize drying. The tent should be removed near the end of the roasting time to facilitate browning and allow undesirable volatiles to escape. Ducks and geese are self-basting, and the skin should be pricked with a fork during the roasting process to allow some of the fat to drain. Time guide lines for roasting young birds are given in Table 22-7.

A bird cooked according to a time schedule may or may not be done because of various conditions such as age, weight, oven temperature, air circulation, etc. Some of the indications of doneness are a slightly shrunken flesh be-

Table 22-7. Roasting Timetable for Young Birds

Kind	Weight	Oven Temperature	Approximate Total Cooking Time
	(lb.)	*(°F)*	*(hrs.)*
Chicken	4 to 5	350	1½ to 2
Duck	5 to 6	350	2 to 2½
Goose	10 to 12	325	3 to 4
Guinea	2 to 2½	350	1½
Turkey	6 to 9	325[1]	2½ to 3[1]
	10 to 13	325	3 to 4
	14 to 17	325	4 to 5
	18 to 25	325	6 to 8

[1]Temperature recommended by the National Turkey Federation. Usually, 20 minutes per pound is required.

neath the skin, a flexibility of the leg joint, and the absence of any pink juice when the flesh of the thigh is pricked with a fork or a skewer. Large birds with thick thighs may have the drumsticks released from the cord that binds them to the body when the roasting period is about three-fourths completed. This permits the heat to circulate more readily around the thick, meaty thighs but can cause the drumsticks to dry out.

The giblets are simmered to tenderness before they are added to the gravy. The neck may be cooked with the giblets. The liver needs only about 15 minutes of cooking and should be added during the last period. Allow 1 to 1½ hours for chicken giblets and 2 to 3 hours for turkey giblets. If the giblets are to be incorporated with the stuffing, they should be cooked the previous day.

Stuffing

The ingredients that furnish the bulk to a stuffing are starchy in nature and consist of bread crumbs, boiled rice, or mashed potatoes. To get added richness of flavor, melted butter or some melted fat taken from the body of the bird is added. The seasoning vegetables consist of celery, parsley, and onion. The spices or herbs that are in favor consist of thyme, sweet marjoram, pepper, and sage. Other ingredients that add variety to a stuffing are oysters, nuts, mushrooms, dried apricots or prunes, sausage, raisins, diced salt pork fried crisp, and sliced apples.

The dry stuffing is made of medium dry crumbs without milk or water added. The moist stuffing is made with crumbs with milk or water added or with a base of boiled rice or potatoes.

A 4- to 5-pound bird will require about 4 cups of crumbs, and a 14- to 15-pound turkey will require from 10 to 12 cups of crumbs. An ordinary 1-pound loaf of bread (two to four days old) will make approximately 4 to 5 cups of crumbs. Use 1 cup less of boiled rice than bread crumbs, because the rice will swell.

Oyster Stuffing

(12-lb. turkey)

1½ pt. oysters
¾ cup butter or other fat
⅛ cup chopped parsley
1 tbs. chopped onion

8-10 cups bread crumbs
½ tsp. savory seasoning
1 to 2 tsp. celery salt

Heat the oysters for several minutes, then drain. Cook the parsley and onion for several minutes in the melted fat, and add it and the drained oysters to the bread crumbs.

Sausage Bread Stuffing

(12-lb. turkey)

1 lb. sausage
2 eggs
1 cup milk
7 cups bread crumbs

2 tbs. diced onion
1 tsp. salt
4 tbs. chopped parsley
1 cup diced celery

Panfry the sausage until brown, and drain off the fat. Beat the eggs slightly, and add hot milk. Pour the egg mixture over the remaining ingredients.

Savory Stuffing

(12-lb. turkey)

¾ cup butter or other fat
1 pt. chopped celery
½ cup chopped parsley
1 small onion chopped

8-10 cups bread crumbs
1-2 tsp. savory seasoning
1-2 tsp. salt
pepper to taste

Cook the celery, parsley, and onion in the melted fat for several minutes. Add to the bread crumbs and dry seasoning and mix. Add nuts, if desired. If chestnuts are used, boil them in water for 15 minutes, and remove the shell and brown skin while they are still hot.

Broiling

Young, plump birds split into halves or quarters are the only ones suitable for this purpose. The distance of the broiler rack from the flame or heating element will vary with different ovens, but a temperature of 375° to 400° F is most desirable. This makes it necessary to have the broiler rack 3 inches from the heat in some ovens and from 4 to 7 inches away in others. The speed at which the bird browns will govern the distance to use. A 2-pound broiler should cook in 35 to 45 minutes.

Coat the bird with melted fat, season with salt and pepper, and sprinkle with flour if desired. If a barbecue sauce is to be used, it should be applied during the last half of the cooking period to avoid excessive browning before the meat is fully cooked. Start with the skin side away from the broiler heat and turn several times as it browns. A good practice is to partly cook the bird in a 350° F oven and then broil. Small 3- to 5-pound turkeys, squab, guinea, and ducklings are broiled in the same manner.

Frying

Panfrying is widely practiced. Various methods are used and all have enthusiastic supporters. One method consists of steaming the disjointed bird in a 300° F oven until practically tender and then dipping each joint in a beaten egg and rolling it in bread, cracker crumbs, or corn meal. Place in a thick skillet that contains melted butter or ½ inch of melted fat and brown quickly. Salt and pepper are added to the steamed bird.

Another method is to bread each joint, put the thickest pieces in the pan first, and have sufficient fat to come up around each piece. Cover the pan to avoid spattering, and cook at moderate or medium heat until brown, turning each piece as it browns. This requires from 20 to 25 minutes for chicken. It may be finished in a moderate oven (325° F).

Turkey Steak

The method of cutting the steak governs its preparation. Steaks, made by cutting crossways to the boneless breast, are dipped in egg; seasoned with celery salt, pepper, and a pinch of rubbed parsley; and rolled in crumbs. Brown the steak in a skillet with ¼ to ½ inch of fat, and then steam in a roaster until tender (30 to 45 minutes).

If the boneless steaks have been tenderized by running them through a steak machine, they can be seasoned and broiled, or breaded, seasoned, and fried in deep fat. A short steaming period will make them more tender, but it is not necessary. The secret in flavoring turkey steak lies in the use of celery salt and finely ground parsley.

Stewing

Stewing is employed because it produces a more tender and flavorful product. The flavor is due in large part to the type of bird used for stewing. The old bird is high in flavor. The same thing is true of all meat animals. Flavor intensity is known to increase with the age of the animal or fowl. What is more delicious than a fat hen that has been stewed until she can no longer hold her beautiful form but simply disintegrates when speared with a fork! Or that tough old rooster whose morning crow announced the dawn and disrupted many a laggard's sleep—dismember him with strong hands and stony heart, place him in a kettle half filled with slighly salted water, cover and simmer and simmer. The old codger may need more water but stint him not—give him more. Subjected to 4 or 5 hours or maybe less of this treatment, served with a flour-thickened gravy, and his gastronomic appeal has such shocking import that it may bring remorse. Cooked any other way and he would be damned. And what a soup he makes—even his feet! He has uses no end. Chicken fricassee, chicken gumbo, chicken noodle soup, chicken consommé, cream of chicken soup, jellied chicken, chicken sandwich, chicken salad, chicken chop suey, chicken timbales, chicken risotto, chicken mousse, chicken soufflé, chicken croquettes, chicken à la king, chicken loaf, creamed chicken, curried chicken, and add your own.

CARVING POULTRY

The conventional method of carving pursued by most people is done with the bird on its back. The more recent method is to carve the bird as it rests on its side. With the bird on its back, the first step is to turn the platter with the legs of the bird pointing toward you. Grasp the end of the leg with the fingers of your left hand and cut between the leg and the body (Figure 22-1). Pull the point of the knife through the joint, and sever the skin between the leg and back. Lift the leg to a second plate, if the platter space is limited, and separate the drumstick from the second joint or thigh. The dark meat is sliced from the second joint and also from the drumstick if it is too large for a single serving. Remove the wing by cutting around the area where it appears to join the body, and force it toward the back.

If the bird is on its side, remove the wing between the first and second joint, leaving the second joint attached to the bird. Remove the drumstick, leaving the thigh attached to the body. This ends the most difficult part of the carving operation. Many a tragedy has occurred in disjointing a bird.

Slicing the breast meat by slicing down and away from you is the most comfortable method if you stand to carve a bird resting on its back. Most experts

Fig. 22-1. Carving poultry (breast up). Removing the leg. (Photo by Peter Killian; courtesy USDA)

recommend placing the fork squarely across the breastbone toward the end of the keel (Figure 22-2). This places the left hand, which is steadying the bird with the fork, in a position that does not interfere with the right hand.

When you are in a sitting position, however, the breast of the bird must be next to you; in which case your right hand is working under your left arm. To avoid this unnatural position, point the front of the bird toward you so you can place the fork into the opposite breast several inches below the keel, and slice as you did in the standing position. The slices of white breast meat can be arranged opposite the cuts of dark meat on a separate plate. When sufficient servings have been made or when one side of the bird has been carved, shove the platter away from you, and have the plates put in its place. Put a spoonful of stuffing on the plate and a portion of white and dark meat on top of or beside it. Some prefer to serve as they remove a slice of breast meat but this slows up the serving, since it means extra handling of tools between each operation.

The illustrations on carving a bird on its side show that this method has the advantage of making the breast easier to carve and eliminates handling the second wing joint and the thigh as separate pieces (Figures 22-3 through 22-10).

Fig. 22-2. Carving poultry (breast up). Slicing the breast meat. (Photo by Peter Killian; courtesy USDA)

Fig. 22-3. When carving a bird resting on its side, remove the wing portion first. Grasp the wing tip firmly between thumb and fingers, lift up, and sever between the first and second joints. Drop the wing tip and first-joint portion to the side platter. Leave the second joint attached to the bird.

Fig. 22-4. Remove the drumstick. Grasp the end of the drumstick and lift it up and away from the body, disjointing it at the thigh; then transfer the drumstick to the side platter for slicing the meat. Leave the thigh attached to the bird.

Fig. 22-5. Slice the drumstick meat. Hold the drumstick upright, cut down parallel with the bone, and turn the leg to get uniform slices.

Fig. 22-6. Slice the thigh meat. Anchoring the fork where it is most convenient to steady the bird, cut slices parallel to the body until the bone is reached, and transfer the slices to the side platter. Run the point of the knife around the thigh bone, lift up with the fork, and use either fork or fingers to place the bone on the side platter. Then slice the remaining thigh meat.

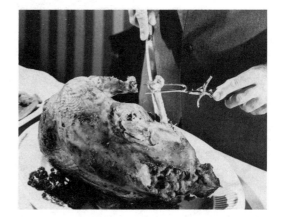

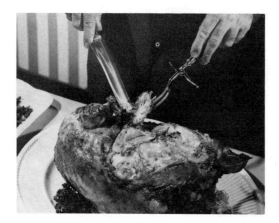

Fig. 22-7. Remove the "oyster," the choice dark meat above the thigh and adjoining backbone. Use the point of the knife to lift it out of its spoon-shaped cradle.

Fig. 22-8. Cut short breast slices until the wing socket is exposed. Sever the second joint of the wing, and transfer it to the side platter. Slice the meat in the same manner as you did the drumstick meat.

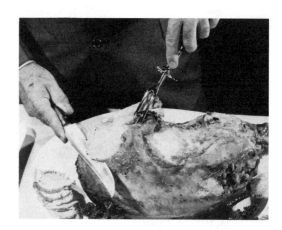

Fig. 22-9. Continue slicing the breast meat. Steady the bird with a fork. Cut thin slices of breast meat until enough slices have been provided or until the breast bone has been reached.

Fig. 22-10. Remove the dressing. Slit the thin tissues of the thigh region with the tip of the knife, and make an opening large enough for a serving spoon to enter. The dressing in the breast end may be served by laying the skin back onto the platter with the dressing uppermost.

The breast of duck or goose is too shallow to be carved in the same manner as turkey, chicken, or guinea. Instead, cut long, thin slices with the grain and parallel to the ridge or keel bone and then cut them into portions across the grain if they are too large. Another method consists of lifting the entire breast from the keel, loosening it with the point of the knife, and placing it on a separate plate. Portions for serving are made by cutting across the grain of the meat.

Use a napkin, not the tongue, to wipe your fingers. Remember, the children are watching!

BARBECUING

Barbecuing is a popular means of preparing meat for those small backyard gatherings as well as very large community picnics or "barbecues." Following are some helpful suggestions on how to prepare meat for large groups of people.

Open Fire

The cut (generally ham or beef round) is attached to a metal rod that is mechanically rotated close to a layer of burning charcoal which glows through the grates. Steaks, chops, and kabobs are grilled on the grate.

Indirect Heat

The cut is coated with a ½-inch layer of dough and placed in an oven (400° F) to roast. This is the least wasteful method since there is no charred meat, and the product is very tasty and juicy.

Although entire hindquarters of beef are barbecued by the open-fire method, a more tasty product will be secured if prepared according to a method prescribed by the American Hereford Cattle Breeders' Association. It is known as the *trench method*.

The Open-Trench (Pit) Method

Barbecuing meat by this method has become one of the most popular means of preparing meat for large rural gatherings, particularly of the livestock interests. It gained considerable political stature when J. A. Walton was governor of Oklahoma. He used it to feed over 100,000 people at his inauguration.

To prepare the barbecued beef, pork, lamb, buffalo, deer, antelope, duck, goose, chicken, rabbit, squirrel, and opossum served on this occasion required 1 mile of trenches.

The cooking principle involved in this method of barbecuing is a combination of dry heat roasting and steaming. The steam is formed from the moisture in the meat and held in the sealed pit.

Building the Trench

Use a heavy soil containing plenty of clay; sandy soil will require a brick lining to prevent a cave-in. Make the trench 3½ feet deep and 3 to 3½ feet wide. The length will depend upon the number of people to be served. A liberal serving is considered to be ½ pound (on a fresh meat basis) per person, or 50 pounds for 100 people. To barbecue 100 pounds of meat requires 3 feet of pit length; 200 pounds—5 feet; 400 pounds—10 feet; 600 pounds—15 feet; 800 pounds—20 feet; etc.

Provide covers for the pit. These may consist of pieces of corrugated sheet metal or rough boards. In case the latter are used, they should be covered with tarpaulins to keep the dirt from sifting through, since the final seal will be made by using about a foot of dirt over the top. If steam leaks occur, they are plugged with more dirt.

Making the Bed of Hot Coals

Dry oak or hickory wood, measuring from 4 to 5 inches in diameter and cut in 2- to 3-foot lengths, is best in producing the 15- to 18-inch bed of hot coals. Apple wood is satisfactory, but the soft and resinous woods are not. Any chunks of wood that are not burned to coals should be removed from the pit or moved to one end by the use of a long rod with a hooked end. Allow four or five hours for producing the bed of coals. It requires twice the volume of the pit in wood to make the desired bed of coals, or 1 cord (1 ton) per 7 feet of pit length. Inefficient cooking, to say the least!

Coating with Sand

The hot coals must have an overcoating of dry sand or fine gravel to a depth of 1 inch. If the sand or gravel is moist, place sheet iron over part of the pit, spread the sand on it, and stir occasionally to dry while the wood is burning. Wet sand will produce too much smoke.

The Meat

Any of the better grades of meat, poultry, or game are suitable for barbecuing, although beef is the most popular. The boneless cuts are a decided advantage for speed in carving, which is necessary when serving large groups. It is important to have each cut as nearly the same thickness as possible in order that all the cuts will cook uniformly.

The meat must be liberally seasoned with salt and pepper before being partially wrapped in aluminum foil. Use .0015 gauge foil with a lengthwise drugstore lock wrap, but leave the ends partly open to form a tube that will not scoop up sand. Place the tube-style wrapped meats on the hot sand by using a three-tined fork, the tines of which are bent into a right-angled curve to hold the roast. Place the creased fold part of the aluminum foil down on the fork and roll it off into position on the hot sand with the crease up. When all the meat is in position, place the cover over the pit and seal it with 8 to 12 inches of dirt. Be certain that the framework or sheet iron covering is strongly reinforced to prevent the top from falling into the pit.

Allow 12 hours for barbecuing and 4 to 5 hours for building the bed of coals. Do *not* open the pit until shortly before serving is to begin.

Barbecue Sauce

Regardless of what the authors may think of sauces or condiments (other than salt and pepper, they destroy the flavor of naturally flavorful meat), the majority rules and sauce it must be.

A sauce (pickle marinade) recommended by authorities with experience in its use (it makes any slow-cooked roast mouth watering) is made as follows:

Barbecue Sauce

Ingredient	*oz./lb. meat*	*g/lb. meat*	*g/kg meat*
Catsup	1.80	50	110
Worcestershire sauce	0.45	13	29
Prepared mustard	0.15	4	9
Prepared barbecue sauce			
(on sale in stores)	0.80	22	48
Salt	0.20	6	13

1 lb. = 454 g; 1 kg = 2.2 lb.

Mix together with approximately 20 percent of ingredient volume of water, which is used to rinse out ingredient bottles.

One gallon of sauce should be sufficient for 40 pounds of meat which is almost enough for 100 hungry people. Pump the roasts to 110 percent of starting weight, and cover with the remainder. Hold at 32° to 40° F for two or three days, then cook in a smokehouse, in a pit, or on a rotisserie. Warm the cover sauce, and serve it over the sliced meat or beside it.

Barbecue Menu for 100 People

The late Professor J. W. Cole of the University of Tennessee reported the following needs:

> Meat—50 pounds (boned and rolled)
> Buns—200 (sliced almost through and buttered)
> Potatoes—6 pounds potato chips, or 30 pounds scalloped potatoes, or
>   100 pounds baked potatoes
> Beans—30 pounds, baked
> Salad—30 pounds potato salad, with pickles, eggs, etc., or
>   20 pounds cabbage salad, with dressing, or
>   15 to 20 pounds lettuce salad, with dressing
> Pickles—1 gallon
> Coffee—7 to 8 gallons (2 pounds of regular grind in a cloth bag, placed
>   in a 10-gallon cream can with water and boiled for three to five
>   minutes)
> Dessert—100 cups of ice cream, cup cakes, or fruit in season

Serving

Have separate tables for those doing the carving. The serving tables, each 3 feet by 10 feet, set on trestles, should be covered with clean wrapping paper, and the paper plates, paper napkins, and plastic forks and spoons placed at the head ends. Follow this by a systematic arrangement, such as buns, meat, potatoes, salad, relishes, dessert, and beverage. The serving may be run as self-service, or attendants may fill each plate completely (except for relish and beverage) before handing it to the guest.

Homemade Barbecue Grills

Use regular-size concrete building blocks for the walls, placing them end to end from two to three tiers high. The width of space between the lateral walls varies from 3 to 5 feet. The ends may be open or closed. Pressed charcoal briquettes are lodged in piles on the ground or gravel base and lighted with lighter fluid. Start the fire one to two hours before serving time. When the briquettes show gray areas (15 to 20 minutes), the piles can be leveled (Figure 22-11).

Metal reinforced grills are made with a frame of 1-inch pipe, with a grill surface 3 feet wide and 4 feet long, made of #9 gauge wire with 2-inch mesh. All sections of the grill have long handles. The cuts of steaks, chops, half-chickens, burgers, kabobs, etc., are placed on the grill, and when it is time to turn the cuts (indicated by the appearance of moisture droplets on the upper surface of the patties, chops, or steaks) a second grill is placed over the meat. This permits two persons, one on either side of the grill, to turn and baste the cuts (Figure 22-12). Hot butter containing some additional salt and some pepper is a good basting for practically all meats. It can be brushed on or sprayed on hot. In the case of chicken, the sauce consists of ½ pint of water, 1 pint of vinegar, ½ pound of butter, and 1 ounce of salt (sufficient for 10 chicken halves). If you insist on spicing up your pork chops while they're on the grill, try this tasty pork barbecue spice: 1 pound salt, 3 ounces garlic salt, 3 ounces onion salt, 2 ounces chili powder, 4 ounces paprika. Sprinkle a moderate amount on each side of the chops as they are grilling. Many pork producer groups sell this or a similar spice mixture.

Fig. 22-11. Grill walls made of concrete blocks.

Fig. 22-12. The double flip-over grill.

A slightly different grill, with the added advantage of portability, obviously lacking in a concrete block grill, is shown in Figure 22-13. This trailer is made of heavy gauge steel and has held up well during several years of use and travel around Illinois.

A pig roast is a popular and economical means of feeding large groups of people. Roasting pigs in the weight range of 100 to 150 pounds are ideal, due to

Fig. 22-13. A portable grill. (Courtesy University of Illinois Hoof and Horn Club)

their greater trimness at lighter weights. Figure 1 pound of carcass weight per person in calculating the meat needs of the group to be fed. An innovative portable arrangement for roasting whole hogs, beef rounds, or any large cut on a rotisserie is shown in Figure 22-14. The whole hog should be cooked to 165° F, measured at the innermost portion of the shoulder. When cooking on a rotisserie, it is essential that the whole hog be securely fastened. This is accomplished by looping #9 gauge wire around the loin and backbone (Figure 22-15). Chicken wire is then tightly wrapped around the carcass (Figure 22-16). It will require approximately 60 to 80 pounds of charcoal and 10 to 12 hours to cook a 200-pound roaster using these methods.

Fig. 22-14. Portable hog roaster with rotisserie.

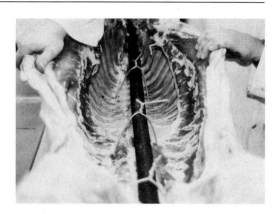

Fig. 22-15. Securing roasting pig to rotisserie bar.

Fig. 22-16. Wrapping new chicken wire around roaster.

CARVING RED MEAT

Carving should not be an objectionable task; it should be a proud accomplishment. Demonstrating carving dexterity will invariably provoke the commendations of guests, which is certainly not objectionable. There is a definite technique or way of carving different cuts that is best explained by drawings (see Figures 22-18 through 22-25).[3] In addition to knowing how to carve, remember that:

- The carving knife must be sharp (it must not be sharpened at the table).
- Whenever possible, carving should be done across the grain of the meat.
- The carving platter should be of ample size (it is embarrassing to serve cuts from the tablecloth or the lap).
- The purpose of the fork is to hold the cut, *not* to dull the knife.
- Small, loose, striated pieces will make a fool of any instrument except a pair of molars or a meat grinder and should be ignored.

---

[3]All prints on carving are courtesy of the National Live Stock and Meat Board.

If a piece of meat will not hold together, or if it desires to run a race around the platter, you should not attract the attention of the guests by condemning the meat, or the cook, or your own shortcomings as a carver, but should continue to work swiftly and quietly.

In order to do a commendable job, the carver must have elbow room and plenty of platter space. If it is more convenient to stand while carving, you should do so. Standing is particulary convenient for those sporting oversize waistlines. Figure 22-17 shows an arrangement of the carver's place with one tantalizing tumbler placed in the danger zone. Your salad, sherbet, water, coffee, or whatever food is before each guest should not be placed before you until you have finished your task. If you are also required to serve the vegetables, the dishes containing them should be conveniently grouped to your right or left in some sensible pattern that has practicality for its theme rather than artistic effect.

Fig. 22-17. Arrangement of plates and utensils for carving. The tumbler should be moved out of the danger zone prior to carving.

Serving can be hastened and more carving space made available if someone else will serve the vegetables and divert the guests' attention from your carving by injecting them with a conversational hypo. (This will not "take" on those who came to learn how or how not to carve but will earn your gratitude.)

A carver should always appear at ease, and this is not possible when the hand holding the fork is crossed over the hand doing the carving. In the case of pot roasts from the chuck, carving is simplified by cutting out solid chunks and turning them in position to make possible the carving of neat slices across the grain. It is not bad form to serve small slices, but it is rather embarrassing to serve large, straggly pieces with trailers.

Porterhouse steak is a cut that requires a little thought in carving, because the tenderloin muscle is large enough for only a single serving. A good way to handle this situation is to remove the T-shaped bone and then carve across the tenderloin and loin muscle, giving a piece of each as a serving.

Carving a ham or leg of lamb is simplified by first slicing the meat from the front of the ham or leg (just above the stifle joint and anterior to the *femur*). This permits the cut to be placed on the flat carved surface with the back or meaty part of the cut on top. Now the carving should begin just above the hock and the slicing continued without the slices being removed until they are sufficient for completing the service. Then, a wedge-shaped piece next to the first slice at the hock should be removed. This makes room for flattening the knife so that all the slices can be cut and lifted from the bone with one final carving motion. For further carving, the roast can be placed on its side and sliced as before.

A familiarity with anatomy is of course of inestimable value in efficient carving. By efficient carving is meant the greatest number of neat slices. The homemaker who takes pride in neatness and gastronomic appeal will not be satisfied with the arguments that meat is meat; it all goes to the same place; it has to be cut and mangled anyway; and the small brown pieces are the best.

A rolled rib roast is placed on the platter with the larger cut surface down.

- Use the standard carving set or the slicer and carver's helper.
- With the guard up, push the fork firmly into the roast on the left side an inch or two from the top.
- Slice across the grain toward the fork from the far right side (first illustration). Uniform slices of ⅛- to ⅜-inch thick make desirable servings.
- As each slice is carved, lift it to the side of the platter or to another hot serving platter (second illustration).
- Remove each cord only as it is approached in making slices. Sever it with the tip of the blade, loosen it with the fork, and place it to one side.

When a standing rib roast is purchased, the meat retailer will, on request, remove the short ribs and separate the backbone from the ribs. The backbone can then be removed in the kitchen after roasting. This makes the carving much easier, as only the rib bones remain.

The roast is placed on the platter with the small cut surface up and the rib side to your left (Figure 22-19).

- Either the standard carving set or the

Fig. 22-18. Carving a rolled rib roast of beef.

roast meat slicer and carver's helper can be used on this roast.

- With the guard up, insert the fork firmly between the two top ribs. From the far outside edge, slice across the grain toward the ribs (first illustration). Make the slices ⅛- to ⅜-inch thick.

- Release each slice by cutting closely along the rib with the knife tip (second illustration).

- After each cut, lift the slice on the blade of the knife to the side of the platter (third illustration). If the platter is not large enough, have another hot platter nearby to receive the slices.

- Make sufficient slices to serve all guests before transferring the servings to individual plates.

A lamb crown roast is made from the rack, or rib section, of the lamb. A pork crown is made from the rib sections of two or more loins of pork. Either cut is carved in a method similar to that of the lamb crown roast (Figure 22-20).

- Use a standard carving set.

- Move to the side of the platter any garnish in the center which may interfere with carving. Dressing can be cut and served along with the slices.

- Steady the roast by placing the fork firmly between the ribs.

- Cut down between the ribs, allowing one rib to each slice (first illustration).

- Lift the slice on the knife blade, using the fork to steady it (second illustration).

The leg of lamb should be placed before you so that the shank bone is to your right, and the thick, meaty section, or cushion, is on the far side of the platter. Different roasts will not always have the same surface upper-

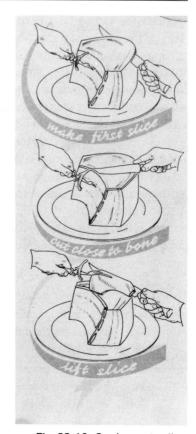

Fig. 22-19. Carving a standing rib roast of beef.

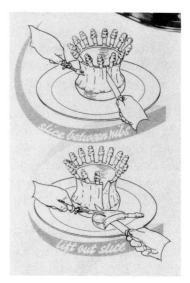

Fig. 22-20. Carving a crown roast of lamb.

most because of the difference in right and left legs; however, this does not affect the method of carving. The illustrations show a right leg of lamb resting on the large, smooth side (Figure 22-21).

- A standard carving set is a convenient size for this roast.
- Insert the fork firmly in the large end of the leg, and carve two or three lengthwise slices from the near thin side (first illustration).
- Turn the roast so that it rests on the surface just cut. The shank bone now points up from the platter.
- Insert the fork in the left side of the roast. Starting at the shank end, slice down to the leg bone. Parallel slices may be made until the aitch bone is reached (second illustration). A desirable thickness is ¼ to ⅜ inch.
- With the fork still in place, run the knife along the leg bone, and release all the slices.

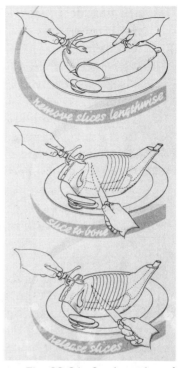

Fig. 22-21. Carving a leg of lamb.

It is much easier to carve a pork loin roast if the backbone is separated from the ribs. This is done at the market by sawing across the ribs close to the backbone. The backbone becomes loosened during roasting (note in the first illustration that it has fallen away from the ribs [Figure 22-22]).

- The standard carving set is preferred for carving the pork loin, although a smaller size may be used.
- Before the roast is brought to the table, remove the backbone by cutting between it and the rib ends (second illustration).
- The roast is placed on the platter so that the rib side faces you. This makes it easy to follow the rib bones, which are the guides for slicing. Make sure of

Fig. 22-22. Carving a loin roast of pork.

the slant of the ribs before you carve, as all the ribs are not perpendicular to the platter.

- Insert the fork firmly in the top of the roast. Cut closely against both sides of each rib. You alternately make one slice with a bone and one without. Roast pork is more tempting when sliced fairly thin. In a small loin, each slice may contain a rib; if the loin is large it is possible to cut two boneless slices between ribs.
- Two slices for each person is the usual serving.

The ham is placed on the platter with the fat or decorated side up. The shank end should always be to your right. The thin side of the ham, from which the first slices are made, will be nearest or farthest from you, depending on whether the ham is from a right or a left side of pork. The illustration shows a left ham with the first slices cut nearest you. The diagram shows the bone structure and direction of the slices (Figure 22-23).

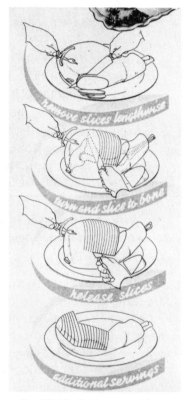

Fig. 22-23. Carving a roast ham.

- Use a standard carving set or the slicer and carver's helper on the baked ham.
- Insert the fork, and cut several slices parallel to the length of the ham on the nearest side (first illustration).
- Turn the ham so that it rests on the surface just cut. Hold the ham firmly with the fork and cut a small wedge from the shank end (second illustration). By removing this wedge, you will find the succeeding slices are easier to cut and to release from the bone.
- Keep the fork in place to steady the ham and cut thin slices down to the leg bone (second illustration).
- Release slices by cutting along bone at right angles to slices (third illustration).

• For more servings, turn the ham back to its original position, and slice at right angles to the bone (fourth illustration).

Contrary to most carving rules, a steak is carved with the grain. A steak need not be cut across the grain, because the meat fibers are tender and already relatively short (Figure 22-24).

• Use the steak set with a knife-blade of 6 to 7 inches.
• Holding the steak with the fork inserted at the left, cut closely around the bone (first illustration). Then lift the bone to the side of the platter where it will not interefere with the carving.
• With the fork in position, cut across the full width of the steak (second illustration). Make wedge-shaped portions, widest at the far side. Each serving will be a piece of the tenderloin and a piece of the large muscle.
• Serve the flank end last, if additional servings are needed (third illustration).

Fig. 22-24. Carving a porterhouse steak.

In order to protect the cutting edge of the knife, as well as the platter, use a small cutting board which is almost a necessity when you are carving a steak.

The blade pot roast contains at least part of one rib and a portion of the blade bone. The long cooking process softens the tissues attached to the bones; therefore the bones can be slipped out easily before the roast is placed on the table (Figure 22-25).

• Either the steak set or the standard carving set may be used for carving the pot roast.
• Hold the pot roast firmly with the fork inserted at the left and separate a section by running the knife between two muscles, then close to the bone, if the

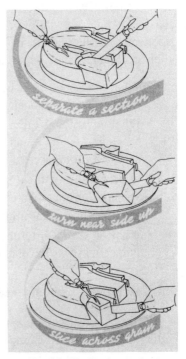

Fig. 22-25. Carving a chuck or blade roast.

bone has not been removed (first illustration).

- Turn the section just separated so that
  the grain of the meat is parallel with
  the platter (second illustration). This
  enables you to cut the slices across the
  grain of the meat.
- Holding the piece with the fork, cut
  slices ¼- to ⅜-inch thick (third illustration).
- Separate the remaining sections of the
  roast (note the direction of the meat
  fibers, and carve across the grain).
- Two or three slices, depending on size,
  are served to each person.

SOME INSTITUTIONAL MANAGEMENT PROBLEMS

The operation of dining commons in many types of institutions presents
many problems in dealing with grades of meat, the form in which they should be
purchased, their cutting, and their preparation. Where economy, because of
budget limitations, is the paramount issue, quantity and quality of meat purchases will naturally suffer. This does not mean that hash from cutter and canner
stock must be the main dish. It is a challenge to the person in charge to make the
kitchen a laboratory in which to discover:

- What grades of meat best suit special needs from the standpoint of complete utilization, consumer satisfaction, and cost per serving.
- What advantages there may be in buying meat in the carcass as compared
  to wholesale and boneless cuts, fresh as compared to frozen, and from the
  packer as compared to the jobber.
- What cuts furnish satisfactory roasts, steaks, etc., at the lowest cost per
  serving (for example, chucks as compared to rounds for roasts).
- What method of cutting is best adapted to the utilization of all the meat.
- What new and different ways there may be in preparing and serving the
  same cuts so as to relieve the monotony so often prevalent.

The Cutting and Cooking Test

The most businesslike approach to the solution of these problems is to make
cutting tests on carcasses and cuts of different grade and price levels, and thereby determine actual costs of servings. For example, if boneless beef is required
for roasts and good grade chucks are quoted at $0.79, good grade rounds at

$1.09, and 6- to 8-pound chuck rolls at $1.25, which is the most satisfactory and economical buy for that purpose? Will a $0.79 chuck, when boned, furnish chuck rolls under $1.25 per pound, and can chuck or round be cut to better advantage? Also, what will servings cost and how do the two compare in palatability and ease in serving? The answers are found by cutting and preparing the meat for the oven and then dividing the usable weight into the total cost. The cooked product when served will give further information as to shrinkage and actual number of servings secured per pound of fresh meat. (Consult Tables 22-8 and 22-9 for data on servings and cost per serving.) For information on how to go about answering these questions, the reader is referred to Chapter 12.

What Grade of Meat Should I Purchase?

The answer depends upon whether you are serving paying or nonpaying clients. Choice and Good grades should be used for the former, but tasty meals can be prepared more economically from Standard, Commerical, and Utility grades of meat. A good policy is to use the better grades for chops, steaks, and roasts and have some of the Utility grade on hand to incorporate with the more wasty cuts, thereby enabling the use of most of the excess fat. Whether hindquarters or forequarters should be used for this fat-saving purpose will depend upon the difference in price between the two, hindquarters being preferred by many. Hindquarters will usually average about 2 percent less bone than the forequarters.

Table 22-8. Servings per Pound to Expect from a Specific Cut of Meat[1, 2]

Beef			
Cut	Serving	Cut	Serving
Steaks		Pot Roasts	
Chuck (arm or blade)	2	Arm (chuck)	2
Club	2	Blade (chuck)	2
"Cubed"	4	Chuck, boneless	2½
Filet mignon	3	Cross rib	2½
Flank	3	Other Cuts	
Porterhouse	2	Brisket	3
Rib	2	Cubes	4
Rib eye (Delmonico)	3	Loaf	4
Round	3	Patties	4
Sirloin	2½	Short ribs	2
T-bone	2	Variety Meats	
Top loin	3	Brains	5
Roasts		Heart	5
Rib, standing	2	Kidney	5
Rib eye (Delmonico)	3	Liver	4
Rump, rolled	3	Sweetbreads	5
Sirloin tip	3	Tongue	5

(Continued)

Table 22-8 (Continued)

Pork

Cut	Serving	Cut	Serving
Chops and Steaks		Picnic shoulder (bone-	
Blade chops or steaks	3	in) fresh or smoked	2
Boneless chops	4	Sirloin	2
Fresh ham (leg) steaks	4	Smoked shoulder roll (butt)	3
Loin chops	4	Other Cuts	
Rib chops	4	Back ribs	1½
Smoked (rib or loin) chops	4	Bacon (regular), sliced	6
Smoked ham (center slice)		Canadian-style bacon	5
steaks	5	Country-style back ribs	1½
Roasts		Cubes (fresh or smoked)	4
Ham (leg), fresh, bone-in	3	Hocks (fresh or smoked)	1½
Ham (leg), fresh, boneless	3½	Pork sausage	4
Ham, smoked, bone-in	3½	Spareribs	1½
Ham, smoked, boneless	5	Tenderloin (whole)	4
Ham, smoked, canned	5	Tenderloin (fillets)	4
Boston shoulder (rolled),		Variety Meats	
boneless	3	Brains	5
Loin blade	2	Heart	5
Loin (rolled), boneless	3½	Kidney	5
Loin, center	2½	Liver	4
Loin smoked	3		

Lamb

Cut	Serving	Cut	Serving
Chops and Steaks		Shoulder (boneless)	3
Leg chops (steaks)	4	Other Cuts	
Loin chops	3	Breast	2
Rib chops	3	Breast (riblets)	2
Shoulder chops	3	Cubes	4
Sirloin chops	3	Shanks	2
Roasts		Variety Meats	
Leg (bone-in)	3	Heart	5
Leg (boneless)	4	Kidney	5
Shoulder (bone-in)	2½		

[1]National Live Stock and Meat Board.

[2]The servings per pound are only a guide to the average amount to buy to provide 3 to 3½ ounces of cooked lean meat. The cooking method and cooking temperature, the degree of doneness, the difference in the size of bone in the bone-in cuts, and the amount of fat trim are some of the factors that vary and will affect the yield of cooked lean meat.

Table 22-9. Cost for a Serving of Meat at Various Price Levels[1]

Cost per Pound	Approximate Cost per Serving							
	1½ Servings per Pound	2 Servings per Pound	2½ Servings per Pound	3 Servings per Pound	3½ Servings per Pound	4 Servings per Pound	5 Servings per Pound	6 Servings per Pound
	dollars							
.79	.53	.40	.32	.26	.23	.20	.16	.13
.89	.59	.45	.36	.30	.25	.22	.18	.15
.99	.66	.50	.40	.33	.28	.25	.20	.17
1.09	.73	.55	.44	.36	.31	.27	.22	.18
1.19	.79	.60	.48	.40	.34	.30	.24	.20
1.29	.86	.65	.52	.43	.37	.32	.26	.22
1.39	.93	.70	.56	.46	.40	.35	.28	.23
1.49	.99	.75	.60	.50	.43	.37	.30	.25
1.59	1.06	.80	.64	.53	.45	.40	.32	.27
1.69	1.13	.85	.68	.56	.48	.42	.34	.28
1.79	1.19	.90	.72	.60	.51	.45	.36	.30
1.89	1.26	.95	.76	.63	.54	.47	.38	.32
1.99	1.33	1.00	.80	.66	.57	.50	.40	.33
2.09	1.39	1.05	.84	.70	.60	.52	.42	.35
2.19	1.46	1.10	.88	.73	.63	.55	.44	.37
2.29	1.53	1.15	.92	.76	.65	.57	.46	.38
2.39	1.59	1.20	.96	.80	.68	.60	.48	.40
2.49	1.66	1.25	1.00	.83	.71	.62	.50	.42
2.59	1.73	1.30	1.04	.86	.74	.65	.52	.43
2.69	1.79	1.35	1.08	.90	.77	.67	.54	.45
2.79	1.86	1.40	1.12	.93	.80	.69	.56	.46
2.89	1.92	1.45	1.16	.97	.83	.72	.58	.48
2.99	1.99	1.50	1.20	1.00	.86	.74	.60	.50
3.09	2.05	1.55	1.24	1.03	.88	.77	.62	.51

[1]National Live Stock and Meat Board.

The tables on the retail cut yields of the beef carcass (Chapter 14) and of lamb (Chapter 15) should be helpful in estimating the yield of edible meat from different grades of carcasses and cuts.

In the case of institutions having nonpaying clients or wards, it may be necessary, because of limited appropriations, to use the Standard, Commercial, and Utility grades of meat.

Chapter 23

Meat Judging and Evaluation[1]

INTERCOLLEGIATE meat judging contests were inaugurated and sponsored by the National Live Stock and Meat Board in the fall of 1926 when the first contest of its kind was held in connection with the International Livestock Exposition, Chicago, Illinois. This contest has since been moved to Madison, Wisconsin, and more recently to Dakota City, Nebraska. The Meat Board now sponsors similar contests at Emporia, Kansas; Dallas, Texas; Timberville, Virginia; and Denver, Colorado. These contests, in many instances, are concurrent with the livestock expositions held at those places.

The success of intercollegiate livestock judging contests, and the fact that more agricultural colleges were teaching meat courses, prompted R. C. Pollock, then general manager of the board, to get intercollegiate meat judging on its way. Ten teams competed that first year as compared to 26 teams in 1960 and 21 teams in 1984. It has accomplished what it set out to do, which is to give college students who meet the eligibility rules set up by a rules committee the opportunity to put to a test the meat knowledge they acquired in their respective institutions and to gain a wider knowledge of the meat industry. As was to be expected, it did more than that. Student met student, and these students in turn met leaders in the industry. The meat industry (packers) became interested in these young students with the result that many college-trained individuals are now holding responsible positions in the industry.

Tables 23-1 through 23-6 list the participating schools and their ranking since each contest was started.

AK-SAR-BEN MEAT ANIMAL EVALUATION CONTEST

A contest designed to allow students to compete in an overall program was begun in 1964. The program includes meat judging, breeding animal judging, and

---

[1]The authors are greatly appreciative of Dr. H. Dwight Loveday, University of Tennessee, and Roger Johnson, South Dakota State University, for their critiques of Figures 23-1 through 23-8 and the associated terminology. The authors are also appreciative of Dr. Dell Allen, Kansas State University, for his original work from which these figures were modified in the preparation of this chapter.

Table 23-1. International—Rank of

Institution	26	27	28	29	30	31	32	33	34	35	36	37	38	39	40	41	46	47	48	49	50	51	52	53
Auburn	–	–	–	–	–	–	–	–	–	–	–	–	–	–	–	–	–	–	–	–	–	–	–	–
Brigham Young	–	–	–	–	–	–	–	–	–	–	–	–	–	–	–	–	–	–	–	–	–	–	–	–
Calif. Poly.	–	–	–	–	–	–	–	–	–	–	–	–	–	–	–	–	–	–	–	–	–	–	–	–
Clemson	–	–	–	–	–	–	–	–	–	–	–	–	–	–	–	–	–	–	–	–	–	–	–	–
Colorado	7	–	–	–	–	–	–	–	–	–	–	–	–	–	–	–	–	–	–	–	–	–	–	–
Connecticut	–	–	–	–	–	–	–	–	–	–	–	–	–	–	–	–	–	13	–	17	–	–	–	21
Cornell	–	–	–	–	–	–	–	–	–	–	–	–	–	–	–	–	–	–	–	–	–	–	–	9
Florida	–	–	–	–	–	–	–	–	–	–	–	–	–	–	–	–	–	–	–	–	–	–	–	–
Georgia	–	–	–	–	–	–	–	–	–	–	–	–	–	–	–	–	–	–	–	–	–	–	–	–
Idaho	–	–	–	–	–	–	–	–	–	–	–	–	–	–	–	–	–	–	–	–	–	–	–	–
Illinois	10	6	4	8	7	5	–	–	–	–	–	5	7	11	2*	11	–	11	8	5	9	7	7	12
Illinois State	–	–	–	–	–	–	–	–	–	–	12	12	14	15	–	–	–	–	–	–	–	–	–	–
Iowa	2	1	5	2	8	2	3	2	7	3	10	3	2	14	10	5	2*	3	9	2	2	1	4	3
Kansas	–	3	2	4	6	1	4	6	1	2	9	10	5	2	5*	10	4	12	4	10	5	11	5	15
Kentucky	–	–	–	–	–	–	–	–	–	–	–	–	–	–	–	–	–	–	–	19	12	9	2	6
Maryland	–	–	–	–	–	–	–	–	–	–	–	–	–	–	–	–	–	–	–	–	–	–	–	22
Massachusetts	–	–	–	–	–	–	8	5	9	9	7	9	11	13	–	8	–	10	11	18	4	17*	19	20
Michigan	–	–	–	–	–	–	–	–	–	–	–	–	–	–	–	–	11	7	6	9	19	13	8	8
Minnesota	9	–	–	–	–	–	–	8	6	10	11	7	6	6*	13	7	10	16	12	15	20	12	13	10*
Mississippi	–	–	–	–	–	–	–	–	–	–	–	–	–	–	–	–	–	–	–	–	–	–	–	–
Missouri	4	4	7	5	1	9	–	–	–	5	1	–	12	3	4	6	6	9	13	11	8	3*	6	7
Nebraska	1	9	1	3	2	4	2	3	2	1	4	1	1	8	1	2	–	8	17	4	13	10	11*	17
New Mexico	–	–	–	–	–	–	–	–	–	–	–	–	–	–	–	–	–	–	–	–	–	–	–	–
North Carolina	–	–	–	–	–	–	–	–	–	–	–	–	–	–	–	–	2*	6	18	21	11	15	16	18
North Dakota	–	12	–	–	–	–	–	–	–	–	–	–	–	–	–	–	–	–	–	–	–	–	–	19
Ohio	–	–	3	6	4	–	1	–	8	11	8	6	3	10	9	13	5	1	2	13	6	16	3	2
Oklahoma	3	5	8*	–	–	–	–	9	–	8	3	8	4	5	2*	1	1	2	1	1	7	2	9	1
Ontario	–	–	–	–	5	8	7	1	4	6	2	4	10	12	–	–	–	–	3	16	17	3*	18	23
Panhandle	–	–	–	–	–	–	–	–	–	–	–	–	–	–	–	–	–	–	–	–	–	–	–	–
Pennsylvania	6	7	8*	1	9	3	9	7	5	4	5	11	8	4	11	12	8	15	15	20	15	19	20	13
Purdue	–	–	–	–	–	–	–	–	–	–	–	–	–	–	–	–	–	–	–	–	–	–	–	–
South Dakota	5	2	6	7	–	7	5	4	3	7	6	2	13	9	7	4	7	4	10	14	16	5	10	10*
Tennessee	–	–	–	–	–	–	–	–	–	–	–	–	–	–	–	–	–	–	16	6	10	17*	11*	14
Texas A & M	–	–	–	–	–	–	–	–	–	–	–	–	–	6*	8	3	–	–	5	12	14	14	14	16
Texas Tech	–	–	–	–	–	–	–	–	–	–	–	–	–	–	–	–	–	–	–	7	18	20	15	5
VPI & SU	–	–	–	–	–	–	–	–	–	–	–	–	–	–	–	–	–	–	–	–	–	–	–	–
Washington	–	–	–	–	–	–	–	–	–	–	–	–	–	–	–	–	–	–	–	–	–	–	–	–
West Virginia	–	10	–	9	3	6	6	–	–	–	–	–	–	–	12	14	12	14	14	8	1	8	17	24
Wisconsin–Madison	–	11	–	–	–	–	–	–	–	–	–	–	9	1	5*	9	9	5	7	3	3	6	1	4
Wisconsin–River Falls	–	–	–	–	–	–	–	–	–	–	–	–	–	–	–	–	–	–	–	–	–	–	–	–
Wyoming	8	8	–	–	–	–	–	–	–	–	–	–	–	–	–	–	–	–	–	–	–	–	–	–

*Tie.

Meat Judging Teams, 1926-1984

54	55	56	57	58	59	60	61	62	63	64	65	66	67	68	69	70	71	72	73	74	75	76	77	78	79	80	81	82	83	84
-	-	-	-	-	-	-	-	23	-	-	-	-	-	-	-	-	-	-	-	-	-	-	-	15*	11	14	18	15	23*	-	11
-	-	-	-	-	-	-	-	23	-	-	-	18	-	-	-	-	-	-	-	-	-	-	-	-	-	-	-	-	-	-
-	-	-	-	-	-	-	-	24	-	-	-	-	-	-	-	-	-	-	-	-	-	-	-	-	-	-	-	-	-	-
-	-	-	-	11	25	24*	13	21	20	20	-	-	-	-	-	-	-	-	-	-	-	-	-	-	-	-	-	-	-	-
-	-	-	-	-	-	-	-	-	-	-	-	-	-	-	-	-	-	-	-	-	-	13	9	9	6	3	1	9	7	4
10	18	22	24	25	6	21	19	22	19	19	19	16	22	20	20	22	17	15	17	19	20	21	8	11	19	-	18	22	21	-
16	21	6	18	19	22	26	-	10	-	16	-	-	-	-	-	-	-	-	-	-	-	-	-	-	-	-	-	-	-	-
-	23	9	17	23	17*	13	20	19	-	-	10*	-	7*	13	2	19	-	18	10	20	21	18	19	16*	15	-	23	24	22	-
-	-	-	-	-	-	-	-	-	-	-	13	-	-	-	-	-	-	-	-	-	-	-	-	-	-	-	-	-	-	15
13*	9	10	6	12	16	20	3	4	6	3	9	14	15	7	10	2	8	11	2	7	1	2	7	1	4	6	3	1	8	3
2	1	1	5	3	8	1	6	1	10	-	6	11	2	10	9	10	13	14	4	13	7	11	15*	3	12	8	8	5	2	6
9	7	11	2	16*	5	6	10	9	16	9	5	2	4	2	3	3	5	1	5	4	5	3	1	2	11	4	9	6	17	10
4	17	12	20	18	24	5	9	6	15	-	7*	15	10	21	19	17	-	19	20	14	19	17	-	24	21	10	6	20	16	18
-	-	-	-	-	-	-	-	-	-	-	-	-	-	-	-	-	-	-	-	-	-	-	-	-	-	-	-	-	-	-
21	8	24	23	21	23	22	23	-	-	-	-	-	-	-	-	-	-	-	-	-	-	-	-	-	-	-	-	-	-	-
7	19	8	11	6	12	19	11	12	7	5	2	9*	7*	3	11	9	15	6	7	9	8	7	13	4	2	2	5	8	10	9
11	11	7	13	16*	19	15	21	-	-	12	17	17	16	-	17	18	-	20	21	18	17*	19	-	-	-	19	20	21	-	19
12	12	17	4	5	11	17	6	3	-	11	11*	8	12	8	14	16	3	9	12	17	16	9	17	8	9	14	11	12	18	8
15	14	15	21	15	17*	18	16	8	18	15	14	9*	1	11	4*	8	4	7*	9	1	2	5	14	10	7	7	2	4	9	14
-	-	-	-	-	-	-	-	-	-	-	-	-	6	9	-	-	-	-	-	-	-	-	-	3	-	-	-	-	-	-
20	20	13	9	20	7	9	-	-	-	-	-	-	-	-	-	-	-	-	-	-	-	-	-	-	-	-	-	-	-	-
17	13	14	14	24	3	14	15	11	9	13	12	7*	13	17	13	21	16	12	15	16	14	21	4	13	13	17	16	10	12	13
13*	3	16	10	8	4	3	4	10	4	14	16	19	18	5	18	11	10	7*	11	6	15	14	11	18	18	12	19	14*	19	16
1	6	3	3	2	1	6	1	5	1	1	1	5	3	1	1	5	12	4	14	3	6	1	5	7	10	5	12	3	6	5
22	22	20	-	22	26	-	-	-	-	-	-	-	-	-	-	-	18	-	-	-	-	-	-	-	-	-	-	-	-	-
-	-	-	-	-	-	-	-	-	-	-	-	-	-	-	-	-	18	-	-	-	-	-	-	-	-	-	-	-	-	-
18	10	4	15	10	10	16	17	18	17	17	10*	4	20	18	8	13	14	17	13	12	9	10*	20	21	16	16	17	16	14	20
-	-	-	-	-	-	-	-	-	-	-	-	-	-	15	15	20	9	-	22	9	4	4	3	15	5	11	7	14*	13	-
19	15	21	8	9	9	11	14	13	12*	10	7*	6	5	6	4*	4	7	2	3	11	12	6	6	-	-	9	13	11	5	1
5	5	18	12	13	14	4	2	20	11	6	3	10	21	15	12	15	6	5	6	8	10	8	12	6	8	-	10	17	11	-
6	4	5	7	1	2	12	5	7	12*	2	4	1	11	4	7	1	1	3	1	2	3	10*	2	5	1	1	14	2	1	2
8	16	23	19	-	-	24*	18	25	-	21	-	-	19	19	16	7	-	-	19	22	-	-	-	22	-	-	4	7	4	7
-	-	19	16	7	13	23	22	15	14	18	15	3	14	12	6	6	2	16	8	15	11	12	10	23	20	13	24	19	15	17
-	-	-	-	-	-	-	-	8	8	11*	13	9	14	-	12	-	-	-	-	-	-	-	-	-	-	-	-	-	-	-
-	-	-	22	-	15	7	12	17	2	7	18	12	17	-	-	-	-	-	-	-	-	-	-	-	-	-	-	-	-	-
3	2	2	1	4	20	2	7	2	3	4	8	7*	-	-	-	-	-	10	16	21	17*	15	22	19	17	-	-	-	-	-
-	-	-	-	-	-	-	-	-	-	-	-	-	-	-	-	-	-	-	-	-	-	-	15*	11	14	18	15	23*	-	-
-	-	-	-	-	-	-	-	-	-	-	-	-	-	-	-	-	-	-	-	-	-	-	18	14	-	15	21	13	3	12

Table 23-2. American Royal—Rank of

Institution	27	28	29	30	31	32	33	34	35	36	37	38	39	40	41	46	47	48	49	50	51	52	53
Auburn	–	–	–	–	–	–	–	–	–	–	–	–	–	–	–	–	–	–	–	–	–	–	–
Colorado	–	–	–	–	–	–	–	–	–	–	–	–	–	–	–	–	–	–	–	–	–	–	–
Florida	–	–	–	–	–	–	–	–	–	–	–	–	–	–	–	–	–	–	–	–	–	–	–
Georgia	–	–	–	–	–	–	–	–	–	–	–	–	–	–	–	–	–	–	–	–	–	–	–
Idaho	–	–	–	–	–	–	–	–	–	–	–	–	–	–	–	–	–	–	–	–	–	–	–
Illinois	1	2	1	5	3	–	–	–	–	–	8	8	2	6	9	–	7	14	4	5*	3	7	13
Illinois State	–	–	–	–	–	–	–	–	–	–	–	11	–	–	–	–	–	–	–	–	–	–	–
Iowa	3	3	3	1	2	2	1	2	6	4	5	10	8	10	6	5	10	8	6	1	7	8	6
Kansas	4	4	2	2	1	1	2	3	4	5	3	4	6	9	7	2	8	4	5	4	6	13	15
Kentucky	–	–	–	–	–	–	–	–	–	–	–	–	–	–	–	–	–	–	–	14	–	14	3
Louisiana	–	–	–	–	–	–	–	–	–	–	–	–	–	–	–	–	–	–	–	–	–	–	–
Michigan	–	–	–	–	–	–	–	–	–	–	–	–	–	–	–	–	9	2*	10	13	–	–	–
Minnesota	–	–	–	–	–	–	4	1	3	3	4	9	7	7	2	–	–	–	–	–	–	11	11
Mississippi	–	–	–	–	–	–	–	–	–	–	–	–	–	–	–	–	–	–	–	–	–	–	–
Missouri	2	1	5	3	4	3	3	5	1	2	6	6	1	8	8	6	5	11	14	9	8	4	2
Nebraska	5	–	4	4	–	–	–	–	–	1	1	1	4	3	3	–	6	10	11	10	11	10	8
New Mexico	–	–	–	–	–	–	–	–	–	–	–	–	–	–	–	–	–	–	–	–	–	–	–
North Dakota	–	–	–	–	–	–	–	–	–	–	–	–	–	–	–	–	–	–	–	–	–	–	–
Ohio	–	–	–	–	–	–	–	–	–	–	–	–	–	–	–	–	4	5	3	8	10	5	5
Oklahoma	6	6	–	–	–	–	5	4	5	6	2	5	3	1	1	1	3	2*	1	5*	1	1	4
Panhandle	–	–	–	–	–	–	–	–	–	–	–	–	–	11	–	–	–	–	–	–	–	–	–
Pennsylvania	–	–	6	–	5	4	–	–	2	7	–	7	9	2	5	7	–	13	12	11	–	15	14
Purdue	–	–	–	–	–	–	–	–	–	–	–	–	–	–	–	–	–	–	–	–	–	–	–
South Dakota	8	5	–	–	–	–	–	–	–	–	7	2	5	5	4	3	2	6	7	2	5	9	12
Tennessee	–	–	–	–	–	–	–	–	–	–	–	–	–	–	–	–	–	7	9	–	–	6	7
Texas A & M	–	–	–	–	–	–	–	–	–	–	–	3	–	–	–	–	–	9	3	7	4	3	9
Texas Tech	–	–	–	–	–	–	–	–	–	–	–	–	–	–	–	–	–	12	13	12	9	12	10
VPI & SU	–	–	–	–	–	–	–	–	–	–	–	–	–	–	–	–	–	–	–	–	–	–	–
Washington	–	–	–	–	–	–	–	–	–	–	–	–	–	–	–	–	–	–	–	–	–	–	–
West Virginia	–	–	–	–	–	–	–	–	–	–	–	–	–	–	–	–	–	–	–	–	–	–	–
Wisconsin	–	–	–	–	–	–	–	–	–	–	–	–	4	–	4	1	1	2	3	2	2	1
Wyoming	7	–	–	–	–	–	–	–	–	–	–	–	–	–	–	–	–	–	–	–	–	–	–

*Tie.

Meat Judging Teams, 1927-1984

54	55	56	57	58	59	60	61	62	63	64	65	66	67	68	69	70	71	72	73	74	75	76	77	78	79	80	81	82	83	84
–	–	–	–	–	–	–	–	–	–	–	–	–	–	–	–	–	–	–	–	–	–	19	–	–	13	–	–	–	–	–
–	–	–	–	–	–	–	–	–	–	–	–	–	–	–	–	–	–	–	–	–	–	3	4	5	3	3	5	6	1	11
–	–	–	–	–	–	–	–	–	–	–	14	–	8	18	10	17	–	–	–	–	–	–	–	–	–	20	–	–	–	–
–	–	–	–	–	–	–	–	–	–	–	–	–	–	–	–	–	–	–	–	–	–	–	–	–	–	–	–	–	9	9
–	–	–	–	–	–	–	–	–	–	–	–	–	12	–	–	–	–	–	–	–	–	–	–	–	–	–	–	–	–	–
13	14	15	10	10	15	14	9*	2	4	7	11	12*	12	9	3	5	8*	8	4	7	1	5	2	3	1	2	10	1	3	1
5	6	4	9	5	5	4	5	5	8	–	12	2	10*	14	15	9	10	1	5	13	8	7	12	9	8*	4	15	7	5	4
9	8	9	5	3	1	1	4	4	2	2	10	3	2	1	2	1	2	2	2	2	2	2	6	6	4	6*	16	3	12	10
8	10	10	8	7	12	12	1*	7	10	–	8	8	13	20	21	12	–	12	14	14	14	–	13	20	20	12	7*	16	11	14
–	–	–	13	9	10	–	16	12	3	3	1	5	1	2	4	7	5	11	8	6	11	13	15	4	2	9	2	2	7	8
6	13	7	17	13	17	7	17	13	11	–	–	–	–	–	–	–	–	–	–	–	13	15	–	19	22	19	18	19	19	18
–	–	–	–	16	11	16	15	–	–	–	9	18	10*	15	20	–	–	–	–	–	–	–	23	15	16	17	–	–	–	–
2	7	12	7	8	7	11	6	6	–	8	13	4	9	6	16	–	7	10	15	12	12	16	9	15	17	8	7*	12	14	2
11	15	13	14	6	13	10	14	15	13	–	17	17	7	8	1	10	6	4	9	4	4	6	14	13	8*	6*	3	4	10	16
–	–	–	–	–	–	–	–	–	–	–	–	7	3	7	–	–	–	–	7	–	–	9*	8	11	7	–	–	–	–	–
–	12	11	12	18	16	15	11	9	7	–	15	14	14	10	18	11	15	7	13	10	17	11	7	14	12	13*	12	10	17	12
10	5	3	6	12	8	9	12	14	9	10	6	10	17	4	19	13	11	9	10	16	7	17	19	18	14	17	13	15	15	15
4	2	6	4	1	3	2	3	1	1	5	2	1	4	3	7	8	4	3	6	3	5	4	5	12	6	15	14	11	8	3
15	11	2	11	17	14	17	13	11	12	–	16	15	16	16	17	16	14	14	16	5	9	14	17	22	16	11	21	17	18	13
–	–	–	–	–	–	–	–	–	–	–	–	–	–	13	15	13	–	18	9	10	12	3	7	10	10	1	13	6	–
14	16	16	16	14	9	6	8	8	14	9	3	11	6	12	11	2	3	6	3	8	18	9*	10	10	19	5	6	5	13	6
1	1	8	1	11	6	8	9*	–	–	6	4	16	–	13	14	14	12	13	11	11	6	8	11	1	5	13*	9	14	–	–
7	4	5	2	4	2	5	7	10	6	4	5	6	5	11	5	4	1	5	1	1	3	1	1	2	11	1	4	18	4	7
12	9	14	15	15	18	13	–	16	–	–	–	–	15	21	8	3	–	–	12	17	16	–	–	8	–	–	11	9	2	5
–	–	–	–	–	–	–	–	–	–	–	–	–	–	5	9	6	8*	15	17	15	15	18	18	21	18	18	19	–	–	–
–	–	–	–	–	–	–	–	–	–	–	–	–	–	6	–	–	–	–	–	–	–	–	–	–	–	–	–	–	–	–
–	–	–	–	–	–	–	–	–	–	–	–	–	12	–	–	–	–	–	–	–	–	–	–	–	–	–	–	–	–	–
3	3	1	3	2	4	3	1*	3	5	1	7	9	–	–	–	–	–	–	–	–	–	–	–	20	17	–	–	–	–	–
–	–	–	–	–	–	–	–	–	–	–	–	12*	–	17	–	–	–	–	–	–	–	–	16	16	21	21	20	8	16	17

Table 23-3. Eastern—Rank of Meat Judging Teams, 1950-1984

Institution	50	51	52	53	54	55	56	57	58	59	60	61	62	63	64	65	66	67	68	69	70	71	72	73	74	75	76	77	78	79	80	81	82	83	84
Auburn	—	—	—	—	—	—	—	—	—	—	—	—	—	—	—	—	—	—	—			—	—	—	—	—	9	7	10	6	6	9*	4	5	6
Clemson	—	—	—	—	—	—	—	—	—	—	—	—	—	—	—	—	—	—	10			10	11	10	12	10	12	12	8	13	13	13	13	14	—
Connecticut	—	—	—	5	7	5	7	12	12	9	6	11	12	10	10	11	6	—	9			—	—	—	—	—	—	—	—	—	—	—	—	—	—
Cornell	—	—	—	7	1	6	8	3	2	10	5	11	8	9	6	—	—	—	—			—	9	11	11	12	5	—	—	—	—	—	—	—	—
Florida	—	—	—	—	—	—	—	11	5	7	11	5	1	—	8	9	—	2	8			—	9	11	11	12	5	3	11	9	9	11	12	13	—
Georgia	—	—	—	—	—	—	—	—	—	—	—	—	—	—	9	—	—	—	—			—	1	7	2	2	1	6	5	—	1	—	10	7	7
Illinois	—	—	—	—	—	—	—	—	—	—	—	—	4	1	9	6	—	1	2			1	1	—	2	2	1	5	5	1	1	—	2	3	5
Kentucky	6	8	6	4	—	—	—	7	—	—	13	9	—	—	—	7	—	1	—			—	4	8	10	8	11	—	—	12	—	—	6	11	6	12
Maryland	5	8	6	4	8	9	10	—	13	13	14	—	—	—	—	—	—	—	—			—	—	—	—	—	—	—	—	—	—	6	—	—	—
Massachusetts	7	4	—	8	9	9	9	—	7	11	8	—	—	—	—	—	—	—	—			—	—	—	—	—	—	—	—	—	—	—	—	—	—
Michigan	—	—	7	3	3	4	4	6	3	3	2	8	2	2	4	2	1	4	1			3	2	3	8	3	3	4	6	4	4	7	1	2	1
Mississippi	—	—	—	—	—	—	—	—	—	—	—	—	—	—	—	—	—	—	5			9	7	9	6	9	7	—	—	10	10	9*	5	—	8
Missouri	—	—	—	—	—	—	—	—	—	—	—	—	—	—	—	—	—	—	—			7	5	12	—	—	—	—	—	—	—	—	—	—	4
Nebraska	—	—	5	9	4	7	3	—	8	8	7	—	—	—	—	—	—	—	—			—	—	—	—	1	—	—	—	—	—	—	—	—	—
North Carolina	2	3	5	9	4	7	3	1	8	8	7	—	—	—	—	—	—	—	—			—	—	—	—	—	—	—	—	—	—	—	—	—	—
Ohio	1	2	2	2	5	3	5	5	10	2	1	7	5	8	1	3	—	5	7			6	3	4	3	7	10	8*	13	5	5	8	6	11	9
Pennsylvania	3	7	3	6	6	2	2	2	4	4	4	4	6	6	7	1	4	7	4			8	8	2	4	5	4	11	7	11	11	12	7	8	10
Purdue	—	—	—	—	—	—	—	—	—	—	—	—	—	—	—	—	—	—	—			4	—	—	5	4	2	2	4	3	3	4	9	9	—
Rutgers	—	—	—	—	—	—	—	9	6	—	—	—	—	—	—	—	—	—	—			—	—	—	—	—	—	—	—	—	—	—	—	—	—
South Dakota	—	—	—	—	—	—	—	—	—	—	—	—	—	—	—	—	—	—	—			—	—	—	—	—	—	—	—	—	—	3	—	4	2
Tennessee	6	5	—	—	—	8	—	—	—	—	—	1	10	7	5	4	5	6	3			5	6	6	7	6	6	10	3	7	7	5	3	12	—
Texas A & M	—	—	—	—	—	—	—	—	—	—	—	—	—	—	—	—	—	—	—			—	—	1	—	1	—	1	1	2	2	—	—	—	—
Texas Tech	—	—	—	—	—	—	—	—	—	—	—	—	—	—	—	—	—	—	—			—	—	—	—	—	—	—	—	—	—	—	—	1	3
VPI & SU	—	—	—	—	10	10	6	4	11	6	9	10	9	5	3	10	2	3	6			2	10	5	9	11	8	8*	9	12	12	14	8	10	11
West Virginia	4	1	4	—	—	—	6	10	—	12	12	6	7	4	2	8	3	6	—			—	—	—	—	—	—	—	—	—	—	—	—	—	—
Wisconsin	—	—	1	1	2	1	1	3	1	1	3	2	3	3	—	5	—	—	—			—	—	—	—	—	—	—	—	—	—	—	—	—	—

*Tie.

Note — Columns 69 and 70: NO CONTEST.

Table 23-4. Pacific International—Rank of Meat Judging Teams, 1960-1975[1]

Institution	60	61	62	63	64	65	66	67	68	69	70	71	72	73	74	75
Brigham Young	5	4	1	—	4	5	2	4	6	7	3	2	5	5	—	—
Calif. Poly. (Pomona)	—	—	—	—	—	7	—	1	2	3	—	—	3	—	—	—
Calif. Poly. (San Luis Obispo)	2	—	5	4	7	6	—	—	—	—	—	—	—	—	—	—
Chico State	—	—	—	—	8	8	5	5	—	—	—	—	—	—	2	2
Colorado	—	—	—	—	—	—	—	—	—	—	—	—	—	—	8	—
Fresno State	3	2	2	3	9	—	—	—	—	—	—	3	4	7	6	—
Idaho	7	5	3	1	2	1	3	2	4	4	—	4	—	—	7	—
Montana	4	6	—	—	6	4	4	—	3	6	—	4	6	4	—	—
New Mexico	—	—	—	—	—	—	—	—	—	2	—	—	—	2	3	—
Oregon	6	3	—	—	3	3	—	—	3	—	—	—	7	1	—	—
Texas A & M	—	1	—	—	—	—	1	3	—	—	—	—	—	—	1	1
Washington	1	1	4	2	1	2	1	—	1	1	2	5	2	6	5	—
Wyoming	—	—	—	—	5	—	—	—	—	5	1	1	1	3	4	3

[1]Pacific International has been replaced by the National Western.

Table 23-5. National Western—Rank of Meat Judging Teams, 1976-1984

Institution	76	77	78	79	80	81	82	83	84
Brigham Young	11	11	–	13	–	–	–	–	–
Calif. State (Fresno)	12	–	–	–	–	–	–	–	–
Colorado	4	1	1	1	1	3	5	8	9
Kansas	3	3	6	6*	6	4	2	6	2
Missouri	–	–	10	8	–	–	–	–	–
Montana	–	–	13	14	–	–	–	–	–
Nebraska	6	7	3	4*	4*	2	3	–	7
New Mexico	5	6	4	4*	2	8	9	9	11
North Dakota	7	8	11	10	8	7	8	10	10
Oklahoma	2	4	7	6*	4*	5	1	1	4
Panhandle	–	–	–	11	–	–	–	–	–
South Dakota	–	5	–	–	7	–	–	7	3
Texas A & M	1	2	2	2	3	1	10	2	8
Texas Tech	10	–	9	12	–	6	6	3	1
West Texas State	9	10	8	3	10	–	–	–	–
Wyoming	8	9	5	9	9	9	4	4	6
Tarleton State	–	–	12	–	–	–	7	5	5

*Tie.

market animal evaluation. The following explanation of this contest was written by R. G. Kauffman, University of Wisconsin, one of the persons who was involved in this contest from the start and who has remained instrumental in the operation of the contest. (Kauffman has been trying unsuccessfully to remain anonymous, but he really deserves credit for the continued success of this contest.)

The coordinated approach to meat animal evaluation was initiated to assist and encourage students of animal science to be more aware of the relationships and limitations that exist when evaluating breeding and market animals, and to help them more fully appreciate the importance of carcass excellence as it related to production, as well as meat processing, merchandising and consumption, This program was specifically designed to stimulate college teaching and to motivate students to seek a more complete understanding of meat animal evaluation—from conception to consumption.

The basic idea took root April 21, 1955, in Chicago, Illinois, when the National Live Stock and Meat Board, through encouragement by concerned educators, sponsored the first of several clinics to provide students an opportunity to evaluate market livestock, before and after slaughter. From this beginning, there has been a continued growth of interest, support and participation.

The idea for the Ak-Sar-Ben MAEC developed when educators and businessmen of the livestock and meat industry designed an exercise that would emphasize all aspects of meat animal evaluation. It was decided that breeding livestock, market livestock and carcasses should be included and that such a program be educational, stimulating and competitive. They organized the first one through the cooperation of the Rath Packing Company, Waterloo, Iowa, in 1964. Forty students representing six universities competed through the cooperation of Farmbest, Inc., and Iowa Beef Processors, Denison, Iowa. In 1968, the contest was moved to Omaha, Nebraska, where 117 students from 11 universities competed under the sponsorship of the Knights of Ak-Sar-Ben, Safeway Stores, Inc., Union Stock Yards Co., and Wilson & Co.

In 1984, students representing 18 universities competed in three divisions of the contest to compare their knowledge. Table 23-7 lists the winning schools in the AK-SAR-BEN contest since its inception.

THE SETUP

The rules and regulations governing intercollegiate meat judging contests are undergoing constant change. They are briefly as follows:

- Any college or university having adequate instruction in meats is eligible to enter a team composed of four members, men or women, either or both. The eligibility of each team member is determined by his or her institution. Blank forms, placing, and reason cards are provided by the management. The contestant must supply the pencil or pen, preferably the latter.
- Each year the winning team is awarded the custody of a perpetual trophy which must be won three times for permanent possession. A plaque is awarded as the permanent property of the winning team, and place ribbons are awarded to each of the 10 high teams.
- An appropriate emblem is awarded annually to the highest ranking individual in total points and reasons score and to the highest ranking individual in each of the major classes—beef, pork, and lamb; beef carcass grading; and combined beef and IMPS (Institutional Meat Purchase Specifications) classes (see page 406).
- There are ribbons for the 10 highest contestants in total points and ribbons for the 5 highest contestants in each division.

Conduct of Contestants

The contestants must abide by the regulations and may be disqualified by the group leader for any violation upon warning and continued violation. The rules forbid talking among contestants; the use of gimmicks such as grade-guide cards, photographs, or measuring rulers; the handling of beef and pork cuts; or touching the rib eye in the carcass-grading class. This has been made necessary

Table 23-6. Southwestern Exposition—Rank of

Institution	38	39	40	41	42	48	49	50	51	52	53	54	55	56	57	58	59	60
Abilene Christian	–	–	–	–	–	–	–	–	–	–	–	–	–	9	–	–	14	–
Arizona	–	–	–	–	–	–	–	–	–	–	–	–	–	–	–	–	–	13
Auburn	–	–	–	–	–	–	–	–	–	–	–	–	–	–	–	–	–	–
Calif. Poly.	–	–	–	–	–	–	–	–	–	–	–	–	–	–	–	–	13	10
Clemson	–	–	–	–	–	–	–	–	–	–	–	–	–	–	–	–	–	–
Colorado State	–	–	–	–	–	–	–	–	–	–	–	–	–	–	–	–	–	–
East Texas State	–	–	–	–	–	–	–	–	–	–	–	–	–	–	–	–	–	–
Florida	–	–	–	–	–	–	–	–	–	–	–	–	–	–	–	–	–	–
Iowa	–	–	–	–	–	–	–	–	3	–	2	4	5	5	2	4	9	5
Kansas State	–	–	–	–	–	–	–	4	–	–	4	–	6	–	3	3	1	3
Kentucky	–	–	–	–	–	–	–	–	–	–	–	5	–	3	5	8	7	8
Louisiana	3	3	3	–	–	–	–	–	–	–	–	–	–	–	–	–	–	–
Mississippi	–	–	–	–	–	–	–	–	–	–	–	–	–	–	–	–	11	12
Missouri	–	–	–	–	–	–	–	–	–	–	–	6	–	–	–	–	–	–
Nebraska	–	–	–	1	1	1	4	–	–	–	7	7	–	–	–	10*	4	7
New Mexico	–	–	–	–	–	–	–	–	–	–	–	–	–	–	–	10*	12	14
North Dakota	–	–	–	–	–	–	–	–	–	–	–	–	–	–	–	–	–	–
Oklahoma State	2	2	1	2	3	2	1	1	1	1	1	1	2	1	6	1	3	2
Panhandle	–	–	4	4	–	–	–	–	–	–	–	–	–	–	–	–	–	–
Purdue	–	–	–	–	–	–	–	–	–	–	–	–	–	–	–	–	–	–
South Dakota	–	–	–	–	–	–	–	–	–	–	8	9	8	8	9	5	6	9
Tarleton State	–	–	–	–	–	–	–	–	–	–	–	–	–	–	–	–	–	–
Tennessee	–	–	–	–	–	–	–	–	–	–	–	3	3	4	7	9	8	6
Texas A & M	1	1	2	3	2	4	5	2	4	4	6	8	4	6	4	2	2	4
Texas Tech	4	4	5	5	4	3	2	5	5	2	5	10	7	7	8	7	10	11
VPI & SU	–	–	–	–	–	–	–	–	–	–	–	–	–	–	–	–	–	–
West Texas State	–	–	–	–	–	–	–	–	–	–	–	–	–	–	–	–	–	–
Wisconsin	–	–	–	–	–	–	3	3	2	3	3	2	1	2	1	6	5	1
Wyoming	–	–	–	–	–	–	–	–	–	–	–	–	–	–	–	–	–	–

*Tie.

Meat Judging Teams, 1938-1984

61	62	63	64	65	66	67	68	69	70	71	72	73	74	75	76	77	78	79	80	81	82	83	84
–	–	–	–	–	–	–	–	–	–	–	–	–	–	–	–	–	–	–	–	–	–	–	–
–	–	8	5	–	–	–	–	11	–	–	–	–	–	–	–	–	–	–	–	–	–	–	–
–	–	–	–	–	–	–	–	–	–	–	–	–	–	–	–	–	–	–	–	–	–	11	11
–	–	–	–	–	–	–	6	–	–	–	–	–	–	–	–	–	–	–	–	–	–	–	–
–	–	–	–	–	–	–	–	–	–	12	–	–	–	–	–	–	–	–	–	–	–	–	–
–	–	–	8	–	6	–	–	–	–	–	–	–	7	–	3	6	1	2	1	1	–	2	2
–	–	–	–	–	–	–	–	–	–	–	–	–	–	10	–	–	–	–	5	–	–	–	–
–	–	–	–	–	–	7	8	6	–	–	7	8	8	–	–	–	–	–	–	–	–	–	–
2	–	7	–	–	–	–	–	–	–	–	–	–	–	–	–	–	–	–	–	–	–	–	–
1	3	2	3	3	5	1	1	2	1	3	2	3	2	5	7	3*	8	7	–	–	2	7	12
10*	10	–	–	6	–	5	–	8	–	–	9	–	–	9	–	–	–	–	–	–	–	–	–
–	–	–	–	–	7	8	–	–	–	–	–	–	–	–	–	–	–	–	–	–	–	–	–
10*	–	–	7	7	9	11	–	12	–	11	10	10	10	11	12	–	–	–	–	–	–	–	–
–	–	–	–	–	–	–	–	–	–	–	8	–	–	–	–	–	–	–	6	5	–	–	9
–	–	–	–	–	–	–	3	3	8	4	3	5	4	4	4	8	3	6	4	–	6	3	10
8	9	10	–	8	3	4	5	4	7	6	5	2	6	3	5	5	4	4	–	–	–	10	13
9	8	–	–	–	–	–	–	–	–	–	–	–	–	–	–	–	–	–	–	–	–	–	–
5	1	1	1	1	1	2	2	1	2	2	6	4	3	2	2	1	5	5	3	4	5	1	4
–	–	–	–	–	–	–	10	10	10	–	–	–	–	–	–	–	–	10	–	–	–	–	–
–	–	–	–	–	–	–	–	–	–	–	11	–	–	–	–	–	–	–	–	–	–	–	–
4	6	5	–	5	2	6	9	–	3	7	4	6	5	7	6	3*	7	9	–	–	–	6	7
–	–	–	–	–	–	–	–	–	–	–	–	–	–	–	9	9	9	–	–	–	4	9	5
3	5	4	4	4	–	–	–	–	11	9	–	7	–	6	10	7	–	–	–	–	–	–	6
7	2	6	2	2	4	3	7	7	4	1	1	1	1	1	1	2	2	1	2	2	3	5	3
–	7	9	–	–	–	10	11	–	9	8	–	9	9	8	11	–	10	8	–	3	1	8	1
–	–	–	–	–	–	–	4	5	5	5	–	–	–	–	–	–	–	–	–	–	–	–	–
–	–	–	–	–	–	–	–	–	–	–	–	–	–	–	8	10	6	3	7	–	–	–	–
6	4	3	6	–	–	–	–	–	–	–	–	–	–	–	–	–	–	–	–	–	–	–	–
–	–	–	9	–	8	9	10	9	6	–	–	–	–	–	–	–	–	–	–	6	–	4	8

Table 23-7. Past Winners, AK-SAR-BEN Meat Animal Evaluation Contest, 1964-1984

Year	No. Schools	Market Animal Division	Breeding Animal Division	Carcass Division	Overall
1964	6	Illinois	Illinois	Illinois	ILLINOIS
1965	7	Iowa State	Illinois	Michigan State	ILLINOIS
1966	8	So. Dakota State	Kansas State	So. Dakota State	SO. DAKOTA STATE
1967	11	Iowa State	Illinois	So. Dakota State	ILLINOIS
1968	11	Wisconsin (Madison)	Iowa State	Michigan State	MICHIGAN STATE
1969	14	Michigan State	Kansas State	Kansas State	MICHIGAN STATE
1970	14	Minnesota	Illinois	So. Dakota State	(SO. DAKOTA STATE – WISCONSIN) Tie
1971	19	Iowa State	Illinois	Kansas State	ILLINOIS
1972	20	Texas A & M	So. Dakota State	So. Dakota State	ILLINOIS
1973	17	Wisconsin (Madison)	Illinois	Illinois	IOWA STATE
1974	21	So. Dakota State	Oklahoma State	Kansas State	SO. DAKOTA STATE
1975	19	Nebraska	Purdue	Illinois	MINNESOTA
1976	18	So. Dakota State	Missouri	So. Dakota State	SO. DAKOTA STATE
1977	21	Idaho	Illinois	Illinois	ILLINOIS
1978	16	So. Dakota State	Iowa State	Illinois	ILLINOIS
1979	20	Michigan State	So. Dakota State	Colorado State	ILLINOIS
1980	19	So. Dakota State	Michigan State	Purdue	PURDUE
1981	18	So. Dakota State	Illinois	Nebraska	ILLINOIS
1982	21	Nebraska	Illinois	Illinois	SO. DAKOTA STATE
1983	22	So. Dakota State	Illinois	Iowa State	IOWA STATE
1984	18	Iowa State	Purdue	Illinois	ILLINOIS

by the large number of contestants working over the limited classes. Use the eye, not the hand.

Table 23-8 lists the classes that will confront each contestant and which are to be placed on their merits, ignoring bruises or faulty trim or workmanship. Each class to be judged consists of four specimens.

Table 23-8. Classes for a Standard Contest

Class	Placing	Reasons
Beef carcasses	50	50
Beef carcasses	50	–
Beef cuts	50	50
Beef cuts	50	–
Pork carcasses	50	50
Pork carcasses	50	–
Fresh skinned hams	50	50
Lamb carcasses	50	50
Lamb carcasses	50	–
Beef carcass Quality grading (15)	150	
Score per carcass		
Correct 10 points		
⅓ grade off 8 points		
⅔ grade off 5 points		
Full grade off 0 points		
Beef carcass Yield grading (15)	150	
Score per carcass		
Correct 10 points		
Off 0.1 YG 9 points		
Off 0.2 YG 8 points		
Off 0.3 YG 7 points		
Off 0.4 YG 6 points		
Off 0.5 YG 5 points		
Off 0.6 YG 4 points		
Off 0.7 YG 3 points		
Off 0.8 YG 2 points		
Off 0.9 YG 1 point		
Off 1.0 YG 0 points		
Total	$\overline{750}$	$\overline{250}$
IMPS classes (2)[1]	100	
Grand total		1,100 points

[1]IMPS classes were added to contests starting in 1983.

Time

Fifteen minutes is allowed for note taking and placing each of the five reason classes. Contestants will stand back from all classes except grading classes for four minutes, to observe general appearance. A two-minute warning is given before placing cards are collected. Ten minutes is allowed for judging each of the four nonreason classes. Contestants are given a two-minute standback and a one-minute warning in nonreason classes.

MEAT JUDGING

For the serious student of meat judging, the *Meat Evaluation Handbook,* a 70-page manual containing more than 190 pictures in color, produced by the National Live Stock and Meat Board and edited by members of the American Meat Science Association, is highly recommended.

Judging carcasses differs from grading in several respects; the carcasses must be rated or placed in the order in which the one surpasses the other on the basis of muscling, trimness, and quality. Since the four specimens may fall in the same grade, or with an assured spread of not more than one full grade, a more critical examination and evaluation of such factors as muscling and trimness as they affect yield must be emphasized. Quality factors to be considered include marbling, maturity, lean color, and texture.

Meat Judging in Three Easy Steps

1. Three *big* things are important in *meat.*
   a. Trimness
   b. Muscling (conformation)
   c. Quality → Palatability
2. What tells you
   a. Trimness?
      (1) *Trimness* over rib eye, lower rib, flank, brisket, round, chuck.
      (2) If you see *blue,* you are seeing muscle in the round, not a thick covering of fat. *Penalize outside* fat and *intermuscular* fat (between the muscles), *i.e., seam* fat.
   b. Muscling?
      (1) Rib-eye area.
      (2) *Bulge and plumpness* in round (ham) and chuck (shoulder) and through loin and rib; generally *width, thickness,* and *depth.*
   c. Quality?
      (1) Color—*beef:* bright cherry red
                    pork: bright grayish pink
                    lamb: bright reddish pink
      (2) Firmness of lean and fat
      (3) Smooth texture
      (4) Marbling—fine specks or strands *within* the muscle
      (5) Youth—cartilage between or on ends of bones
                    blood in bones
                    bright, light muscle color
3. Put 2a, 2b, and 2c together! The leanest carcass or cut with the most muscling and quality wins, and you place the others accordingly.

   More emphasis on *muscle, leanness (quantity):*
      chucks, rounds, hams, pork carcasses

More emphasis on *quality:*
  ribs, loins

  Beef carcasses and lamb carcasses are about equal—balance them off.

Good luck!

BEEF CARCASS JUDGING

This judging routine may apply to other classes as well. The first impression is more often right than wrong, so take a good look at the carcasses from a distance in the time allotted. Don't stand there flexing your leg muscles for a jump toward the carcasses at the sound of the whistle to see how discourteous you can be to the opposition. The reason the first general impression is likely to be correct lies in the fact that it takes into consideration the outstanding points of excellence or inferiority as far as overall muscling and trimness are concerned.

This is one routine that might work. Make note of your snap placing on muscling and trimness. When the horn blows and the hunt is on, take a look at the rib-eye muscle of each carcass, and on the card below the snap placing put the numerical position of the way the carcasses rate in size, color, marbling, and firmness of the rib eye. Step back while the hounds bay and study the card, which may look something like this:

no difference				maturity
1	3	2	4	muscling
3	1	4	2	trimness
1	3	2	4	size of rib eye
3	1	4	2	color
1	3	4	2	marbling
1	3	4	2 (W)	firmness
1	3	4	2	final

The (W) in this case means watery or soft. Remember, don't touch. Use the eyes—watery eye (not supposed to be a joke). Give the class a final inspection to check the placing, and note whether stage of maturity is sufficient to cause a switch to be made. If so, it is because the committee made a slip; the carcasses are supposed to be of practically the same age.

It will be evident from what has been said that looking for the indications of quality on the inside of the carcass was unnecessary. All the factors of quality are evident in the rib-eye muscle, with the exception of coarseness of bone. This saves time that can be used for note taking in the reason classes. See Figure 23-1 for points to consider in beef carcass judging.

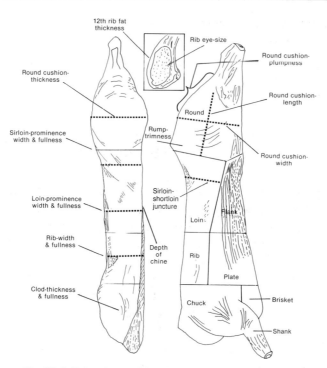

Fig. 23-1. Points for consideration in judging beef carcasses.

TRIMNESS

 1. Rib eye
 2. Round cushion
 3. Rump
 4. Sirloin
 5. Loin edge
 6. Plate
 7. Rib
 8. Chuck
 9. Kidney, heart, and pelvic fat
10. Cod or udder
11. Brisket

MUSCLING

 1. Round cushion—length, width, thickness, plumpness
 2. Sirloin—width, fullness, prominence
 3. Loin—fullness, width, depth of chine
 4. Rib eye—size, shape
 5. Rib—fullness, width
 6. Chuck—thickness, fullness of clod

QUALITY

 1. Marbling—amount, fineness, distribution
 2. Color—youthfulness, brightness of lean
 3. Firmness of lean
 4. Texture of lean—fineness
 5. External fat—whiteness, firmness
 6. Bone—youthfulness

PORK CARCASS JUDGING

Trimness

The reason that the authors consider trimness first and muscling closely second in judging all classes of carcasses, with the exception of quality grading beef carcasses, is that the most obvious thing to the eye should come first. In pork, trimness definitely is the first thing considered, because it is the main factor used in grading. Therefore, determine from the backfat thickness, the grade into which the carcass falls. To do this, one must be familiar with the backfat thickness that is associated with the muscling of a carcass designated for the particular grade. (See Chapter 11.)

It becomes a matter of judgment as to the merits of one carcass over another carcass of the same or a near grade.

- Look for an even distribution of backfat. Many hogs have a tendency to lay the fat on heavily over the shoulder. On ribbed carcasses, consider fat depth over the eye at the tenth rib.
- A certain minimum thickness of belly, particularly in the region of the ham pocket, is necessary for a uniform bacon slab; however, it is fat that makes the belly thicker, thus only a minimum thickness of approximately 1½ inches is necessary. Any additional is waste fat.
- Look for a trim, neat jowl.
- Look for a firm, white fat that is not greasy to the touch. Extremely soft pork must be discounted rather severely.

The final decision between two specimens may hinge on the general distribution of the external finish. An overfat pork carcass is wider through the belly than at either the ham or the shoulder, due to excess backfat and belly fat. In such a carcass the ham and shoulder blend smoothly into the loin and belly from both ends.

Muscling

Note the form or shape of the carcass in respect to its length, width, and thickness through the ham, loin, and shoulder. All of these characteristics give balance and uniformity to a carcass and are reflected in the yield of lean and fat cuts.

The length of the carcass is not too important, provided it is in balance and conforms to standard measurements for the grade. It so happens that most of the carcasses that conform to the top-grade specifications are between 29 and 33 inches long (from the first rib to the forward end of the aitch bone), and weigh between 140 and 185 pounds.

Discount long ham hocks, tapering or banjo hams, flat hams, and carcasses that are too flat (those which lack loin development).

Quality

A carcass of superior quality is one that is bright in appearance, as evidenced by the bright pink color of the flesh on the inside of the belly, and in the ham face and in the loin eye (when exposed). Dark muscle is associated with such things as maturity, intact male sex condition, excessive drying, and bacterial action, all of which are undesirable.

Look for the indications of marbling on the inside of the carcass. Feathering between the ribs is the best indicator of marbling in unribbed pork carcasses. If the carcass is ribbed, look at the eye itself.

Most pork has lower levels of marbling due to the young age at which hogs are marketed. Excessive backfat thickness is no guarantee that the lean is marbled. Actually, the only sure method of determining marbling in a pork carcass is to see a cross section of the loin-eye muscle. Since the size (area) of this muscle and the thickness of the fat covering it are very important in predicting carcass composition, pork carcasses are often ribbed for judging purposes. They must be ribbed for certified litter testing in the development of breeding stock for meat-type hogs. See Figure 23-2 for points to consider in pork carcass judging.

LAMB CARCASS JUDGING

Trimness

The amount of finish necessary to make a Choice quality lamb is very small. Consumers generally are not very tolerant of fats, particularly lamb and mutton fat. The great difficulty with highly finished lamb or mutton carcasses is that the deposition of intermuscular masses of fat outstrips intercellular marbling. This is particularly true of the shoulder.

Sufficient white, brittle fat to cover the back with ⅛ to ¼ inch of fat with a lighter covering over the leg and shoulder is adequate finish for a quality lamb carcass. A papery (no fat under the fell) back on a carcass shows lack of finish. A fiery color to the fat is slightly objectionable. When the carcass has been ribbed at the 12th rib, the fat cover over the rib eye and the lower rib gives an excellent indication of the overall trimness of the carcass.

Muscling

Thickness and meatiness in lamb carcasses are important, because the economic value of cuts from small carcasses is dependent in a large measure on the ratio of lean to bone. In Choice carcasses, the separable fat should not run over 30 percent (25 to 30 percent), and the lean should amount to 50 or 55 percent of the carcass, as against 15 to 20 percent bone.

The carcass should be thick and uniformly wide. Carcasses that are slightly rangy are not objectionable, provided they are uniformly wide and thick. Neat,

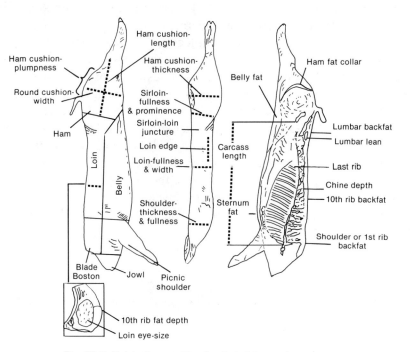

Fig. 23-2. Points for consideration in judging pork carcasses.

TRIMNESS

1. Backfat—first rib, last rib, last lumbar
2. Sirloin-loin juncture
3. Ham collar
4. Loin edge
5. Shoulder—clear plate
6. Belly
7. Sternum
8. Jowl
9. Tenth rib fat depth (ribbed carcasses only)

MUSCLING

1. Length of side
2. Ham cushion—length, width, thickness, plumpness
3. Sirloin—fullness, prominence
4. Lumbar lean—size
5. Loin—fullness, width
6. Chine—depth
7. Shoulder—thickness, fullness
8. Loin eye—size (ribbed carcasses only)

QUALITY

1. Feathering—amount
2. Color—grayish pink
3. Belly—firmness, thickness
4. Loin eye—color, texture, firmness, marbling (ribbed carcasses only)

smooth shoulders, well fleshed over the blades and covered with a thin layer of white fat, are preferred to narrow shoulders.

The legs should be plump. Tapering legs are not characteristic of a valuable lamb carcass. Only a light fat covering over the outside of the leg can be expected in even the most highly finished carcasses. A slight crease over the backbone is indicative of a well fleshed back, but a prominent backbone indicates a shallow muscling and a small rib eye. Flat lamb shoulders with prominent blades at the top of the shoulders are not characteristic of excellent conformation. The loin and rib rack should be broad, thick, full, and well turned in the rib to give the carcass a neat, trim appearance. When the carcass has been ribbed, the area of the rib eye itself gives a good indication of the overall muscling in the carcass.

Pot-bellied carcasses are objectionable because they increase the amount of cheap flank and breast meat. The necks should be short and thick, rather than long and thin.

Quality

The break joint must show four well defined red ridges, indicating youth, as does redness in the ribs.

If the carcass is ribbed, which is very desirable, marbling, a major determinant of quality, can be readily and accurately observed. Feathering between the ribs in the chest cavity and fat streaks in the flank are indices of marbling; however, feathering has been eliminated as a quality grading factor. Firmness is associated with finish, and thin carcasses are naturally soft because the hard fat is absent. The fat should be firm, white, and waxy and be evenly distributed over the entire carcass; however, if the fat is oily and soft, the carcass is lacking in firmness, even if it is well finished. The flank should be firm and dry, and the inside of the flank should show a few fat streaks and have a bright reddish pink color to the flesh.

It is the opinion of the authors that lambs should be ribbed. This practice has been followed for years in the commercial carcass contests and is the only way to do a complete, accurate job of carcass evaluation and judging. It would make for more accurate student appraisal of meat quantity and quality in lamb carcasses, which is still too much guesswork as it now stands without ribbing. Contest lambs for FFA and 4-H should definitely be ribbed. See Figure 23-3 for points to consider in lamb carcass judging.

JUDGING WHOLESALE (PRIMAL) BEEF CUTS

Judging Beef Chucks (Square-Cut)

This primal cut is utilized primarily for pot roasts, Swiss and braised steaks, boiling beef, stew beef, dried beef, and ground meat. The blade and arm ends of the chuck present a considerable cut area for judging trimness and muscling dif-

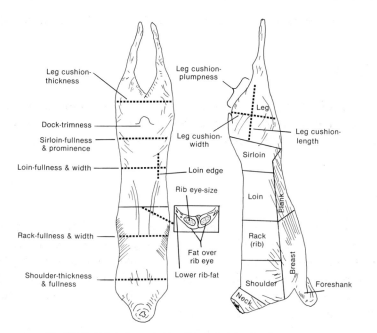

Fig. 23-3. Points for consideration in judging lamb carcasses.

TRIMNESS

1. Rib eye (ribbed carcasses only)
2. Lower rib (ribbed carcasses only)
3. Leg cushion
4. Sirloin
5. Dock
6. Loin
7. Loin edge
8. Rack
9. Shoulder—top and lower portions
10. Kidney, heart, and pelvic fat
11. Cod or udder
12. Flank
13. Breast

MUSCLING

1. Leg cushion—length, width, thickness, plumpness
2. Sirloin—fullness, prominence
3. Loin—fullness, width
4. Rib eye—size, shape (ribbed carcasses only)
5. Rack—fullness, width
6. Shoulder—thickness, fullness

QUALITY

1. Flank streaking—primary and secondary flanks, amount, fineness
2. Color—reddish pink, youthfulness, brightness of lean
3. Firmness
4. Feathering—amount, fineness (eliminated in 1983 as a quality-grading factor)
5. Marbling—amount, fineness, distribution (ribbed carcasses only)
6. Texture—fineness (ribbed carcasses only)

ferences. Quality should have less emphasis than muscling or trimness in beef chucks.

Trimness

Discount soft, oily fats. A tendency toward an undesirable heavy fat deposit over the clod muscle in the center of the shoulder is evident in highly finished chucks. Large fat deposits between muscles (intermuscular fat) must be discounted.

Muscling

Uniformity of depth is important. The arm end should be rounded and heavily muscled and should not fall away too rapidly into the cross-rib region (arm end of the fourth and fifth chuck ribs, formerly called the English cut). The blade end should be deep and give the appearance of plumpness as opposed to flatness. A very good indication of meatiness and plumpness is the prominence of the shoulder joint.

The neck should be short and blend in with the rest of the chuck. Long, flat necks are objectionable.

Quality

The first thing to do is to inspect the blade bone to see that it is still white and cartilaginous at the fifth rib. Then look at the chine bones to see if the ends (buttons) are still soft and white and if the bone itself is red and porous. It is always a good procedure in judging a wholesale cut to first determine the age of the animal, because a hard bone, regardless of the excellence of muscling and trimness or the superiority of the marbling, will degrade a cut into the Commercial grade. (Under present intercollegiate rules, handling is not permitted, but it will not be necessary to do so because cuts from only young animals are used.)

The color of the lean should be bright cherry red. Degrade the darker colors of red. Marbling should be slight to modest on the blade end. The same degree of marbling will not be in evidence on the arm end, since the muscles on the cut surface were attached to the much exercised shank muscles. The surface of the meat should present a smooth, velvety appearance and should be firm and not watery. See Figure 23-4 for points to consider in judging beef chucks.

Judging Beef Rounds

The beef round is the source of the very popular round steaks and roasts; popular because they contain so little bone and fat. More emphasis should be placed on cutability than quality in beef rounds; however, quality is slightly more important in rounds than in chucks.

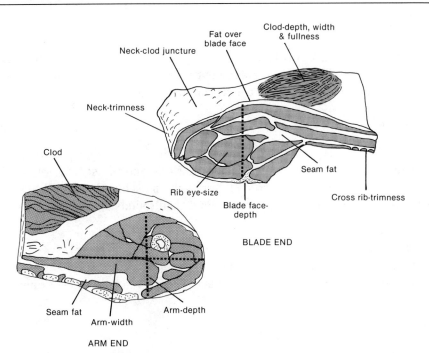

Fig. 23-4. Points for consideration in judging beef chucks.

TRIMNESS

1. Clod
2. Neck-clod juncture
3. Neck
4. Arm end
5. Blade end
6. Throat region
7. English cut or cross rib
8. Seam fat in arm end
9. Seam fat in blade end

MUSCLING

1. Clod—width, depth, fullness
2. Neck-clod juncture—depth, fullness
3. Neck—fullness, meatiness
4. Blade end—depth, size, meatiness
5. Arm end—depth, width, size
6. Rib eye—size

QUALITY

1. Blade end:
   a. Color—youthfulness, brightness of lean
   b. Marbling—amount, fineness, distribution
   c. Firmness of lean
   d. Texture of lean—fineness
2. Arm end:
   Same as blade end
3. External fat—whiteness, firmness, uniformity

Trimness

The external fat covering is generally rather sparsely distributed over the round but tends to be heavy or patchy over the rump in the more highly finished beef. Select for smoothness and trimness in this area.

Muscling

Muscling is very important, since it determines the poundage of round steak that can be cut from the area between the rump and the stifle joint. An ideal round is plump, wide, and deep and carries the muscling down well toward the hock (full at the heel). Flat, tapering, and dished rounds are heavily discounted.

Quality

Firmness, marbling, and acceptable color are quality considerations for desirable rounds. See Figure 23-5 for points to consider in judging beef rounds.

Judging Beef Ribs

Beef ribs constitute the highest priced cut in the forequarter. They are suitable for dry-heat roasting, and the steaks cut from the top grades are becoming increasingly popular. Acceptable quality is very important in beef ribs and should have equal or greater emphasis than muscling or trimness.

Muscling and Trimness

The external finish should be firm and dry and evenly distributed over the entire cut. Excessive external fat covering should be discounted. Excessive intermuscular fat in the blade end or in the lower rib on the loin end should be discounted.

The rib-eye muscle should be proportionally large, in relation to the size of the cut, and oval in shape. A kidney-shaped eye muscle is undesirable. A large rib-eye muscle will make a meaty-appearing rib. The blade end of the rib should be deep and meaty, not flat. The contour of the external part of the rib down to the short-rib section should be gradual, not dipped or dished. A combination of these desirable features of conformation results in a deep, well balanced cut that will yield a high proportion of the desirable rib eye *(longissimus* muscle).

Quality

Inspect the bone to determine age, particularly the presence or absence of the buttons on the ends of the feather bones. Hard bone, flat ribs, a yellow cast to the fat, and a dark color to the lean spell *cow ribs*.

Determine the degree of marbling, the firmness, the color, and the texture. See Figure 23-6 for points to consider in judging beef ribs.

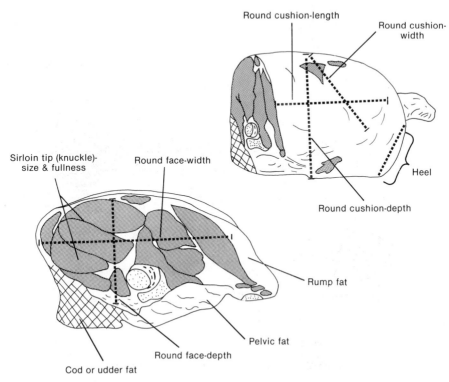

Fig. 23-5. Points for consideration in judging beef rounds.

TRIMNESS

1. Round face
   a. Rump
   b. Sirloin tip or knuckle
2. Cushion
3. Heel
4. Cod or udder fat
5. Seam fat
6. Pelvic fat

MUSCLING

1. Cushion—length, width, depth, plumpness
2. Round face—depth, width, size, meatiness
   a. Rump—size
   b. Sirloin tip or knuckle—size, fullness
3. Heel—plumpness, meatiness

QUALITY

1. Color—youthfulness, brightness of lean
2. Marbling—amount, fineness, distribution
3. Firmness of lean
4. Texture of lean—fineness
5. External fat—whiteness, firmness, uniformity

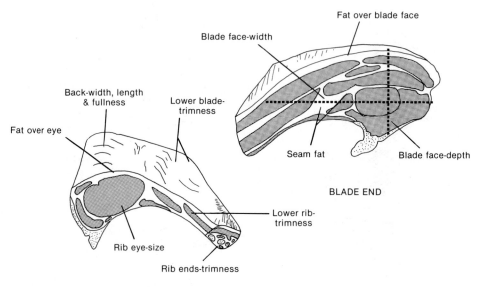

Fig. 23-6. Points for consideration in judging beef ribs.

TRIMNESS

1. Rib eye
2. Lower rib
3. Rib ends or short ribs
4. Lower blade
5. Back
6. Blade end
7. Seam fat in blade

MUSCLING

1. Rib eye—size, shape
2. Back—width, length, fullness
3. Blade—width, depth, size, meatiness

QUALITY

1. Rib eye (loin end):
    a. Marbling—amount, fineness, distribution
    b. Color—youthfulness, brightness of lean
    c. Firmness of lean
    d. Texture of lean—fineness
2. Blade end:
    Same as loin end
3. External fat—whiteness, firmness, uniformity

Judging Beef Loins

Quality is very important in this wholesale cut because from it are secured the most tender and most expensive steaks in the entire beef carcass. Quality should have more emphasis than cutability; however, the latter should not be ignored. The wholesale loin represents 17 percent of the carcass weight.

Trimness

Highly finished cattle have heavy external fat deposits on this wholesale cut and particularly over the edge of the shortloin. A dip or depression in front of the hip indicates trimness. Select for smoothness and a minimum amount of patchiness. In the trimmed loin (flank off, kidney and suet out), the remainder of the kidney fat should be hard and brittle. The external fat should have similar qualities. A lack of external and internal kidney and pelvic (channel) fat is associated with excellent cutability. See Figure 23-7 for points to consider in judging beef loins.

Muscling

The thicker and heavier the muscling on a loin, the greater the yield of steak. Look for a full, rounded, meaty sirloin end blending well into the shortloin. The shortloin should show fullness with a large, oval eye muscle on the rib end. Degrade flat sirloin ends, prominent hips, and depressed and shallow shortloins.

Quality

Marbling is highly desirable in any cut of meat, but it is doubly so in the loin cuts. When the choice is to be made between a fine-textured or a coarse-textured marbling, if the total amount appears to be the same, give preference to the former. Texture is also very important, as it affects the tenderness of the steak. Color is probably more important in steak than in any other cut, since consumers see such a large area of exposed meat and are more apt to register a gripe if they do not like the color, since it is a high-priced item.

The meat should be firm. Firmness increases and moisture decreases as marbling increases. The reason, as stated elsewhere in this book but which can stand repetition, is that the moisture content of fat is 8 percent and of lean 60 to 70 percent. Another reason for desiring firmness in any wholesale cut, and particularly those that are cut into steaks, is that it is easier to cut a firm steak of even thickness without having it flop over the knife or ooze away from the knife edge.

Do not fail to inspect the chine bone to determine if it can qualify for the top grades for age, or whether a white, hard fused bone in the *sacral* region will degrade it into the Commercial grade.

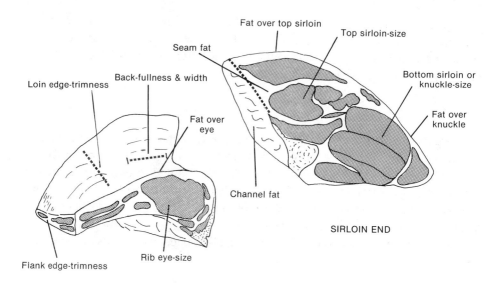

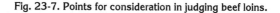

Fig. 23-7. Points for consideration in judging beef loins.

TRIMNESS

1. Eye
2. Loin edge
3. Flank edge
4. Back or shortloin
5. Sirloin end
   a. Top sirloin
   b. Bottom sirloin
6. Seam fat in sirloin end
7. Channel fat (pelvic fat)

MUSCLING

1. Loin eye—size, shape
2. Back or shortloin—fullness, width
3. Sirloin end—depth, width, size, meatiness
   a. Top sirloin—size
   b. Bottom sirloin (knuckle)—size

QUALITY

1. Loin eye (rib end):
   a. Marbling—amount, fineness, distribution
   b. Color—youthfulness, brightness of lean
   c. Firmness of lean
   d. Texture of lean—fineness
2. Sirloin end:
   Same as rib end
3. External fat—whiteness, firmness, uniformity

JUDGING HAMS

Trimness

Commercial hams are trimmed into skinned hams, thus removing the major part of the fat from about two thirds of the surface area of the ham; however, variation in trimming could result in excess fat remaining under the butt (rump) face, which should be discounted. Any mention of trimness should also include reference to the amount of fat over the heel and collar of the ham. Discount for a heavy layer of heel and collar fat. Also discount prominent amounts of inter-muscular fat showing in the butt (rump) face.

Muscling

Basically, a ham may be considered in three parts: butt (rump), center cut, and shank. Individually, each part represents one third of the weight of the ham. Pricewise, the center cut is the profit item, the butt is a 10 to 15 percent over-cost item, and the hock is a loss item.

This price picture sets up the pattern of what a ham should have as far as conformation is concerned. A short, slim hock with a moderate bulge at the heel means less weight in this loss item. A long (from aitch bone to 1 inch above stifle joint) center cut that is deep (the distance through the ham from inside to out-side) and has a good proportionate depth will make for more profit. A ham butt that is full fleshed rather than pointed will throw more weight into this cut and more nearly cover the loss in the shank. The overall appearance of a ham of the desired conformation features meatiness, plumpness, depth, and general trim-ness, with as much weight as possible represented in the more valuable center section.

Quality

The quality of the lean is an important factor to consider in the judging of fresh hams, because it relates to the final quality and smokehouse yield of the cured cut. Quality in fresh hams is reflected in the firmness and color of the lean and fat. Discount soft, oily, and off-color fat.

The texture of the lean is very similar for hogs in the same age bracket. The grain and color of the lean change with the increased age of the animal. A desir-able color for pork is considered to be bright pink but is more often a grayish pink tinged with red, and in many cases, the muscles next to the *ilium* bone *(ili-opsoas, gluteus profundus, gluteus accessorius)* may be dark, giving what is termed a *two-toned condition*. Other things being equal, the brighter, more even-ly colored lean is given preference. See Figure 23-8 for points to consider in judging fresh hams. Hams that are very pale in color and soft and watery in tex-ture should be severely discounted (*e.g.,* see PSE pork, pages 714-715).

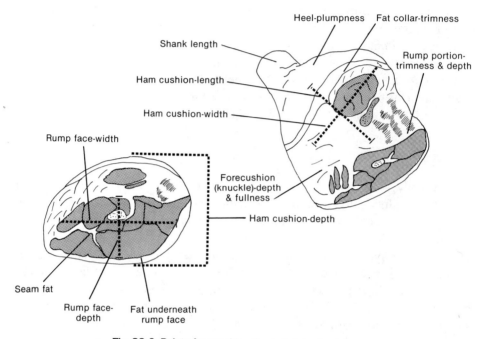

Fig. 23-8. Points for consideration in judging fresh hams.

TRIMNESS

1. Butt (rump) face—underneath
2. Knuckle or forecushion
3. Collar
4. Cushion
5. Rump portion
6. Heel
7. Seam fat in butt (rump) face

MUSCLING

1. Cushion—length, width, depth, plumpness
2. Butt (rump) face—depth, width, size
3. Knuckle or forecushion—depth, fullness
4. Heel—plumpness, meatiness

QUALITY

1. Firmness of butt (rump) face
2. Color of butt (rump) face—grayish pink, brightness of lean, uniformity
3. Marbling—amount, fineness, distribution
4. Texture of butt (rump) face—fineness

SPECIFICATION CLASSES

In an effort to improve the usefulness of meat judging as an educational activity, the intercollegiate coaches' association elected to include Institutional Meat Purchase Specifications (IMPS) classes in the 1982 contests on a trial basis. The idea was very well received by the beef-fabrication industry as a more meaningful training experience of college graduates entering the meat field. At the 1982 Reciprocal Meats Conference, a committee was established to finalize rules and guidelines for IMPS classes in the 1983 and subsequent contests. The following rules were agreed upon by the committee.

RULES AND GUIDELINES

- No rulers or copies of the specifications will be provided or used during the contest.
- There will be a total of 10 cuts per contest. The cuts will be divided into two classes of five each, with a time of 10 minutes allowed per class.
- Score cards will have an accept category and a reject category. The reject category will be divided into reject by cut, trim, and grade.
- Each cut will be worth 10 points for a total of 100 points per individual for the contest. (A total of 400 team points per contest.)
- Each cut will be labeled with the correct IMPS number and name.
- Each contestant should assume that unexposed areas of a cut are acceptable.
- On cuts which are fabricated to be used as a reject item, the defect should be obvious in order to eliminate any questions as to the reason for rejecting the item.
- Factors that would be cause for reject because of conditions such as ragged edges, bone dust, etc., will not be used in the contest.

REASONS

As a contestant, your ability to tell by oral or written word the reasons why one carcass or cut is superior to another depends upon your training and experience, your knowledge of meat terminology, and your method of presenting the reasons.

Written reasons are unlike oral reasons in that the individuals who write them have certain information before them that does not require repetition. The card upon which reasons are written is divided into four equal parts, headed respectively by First, Second, Third, and Fourth. If No. 2 carcass or cut is placed first by a contestant, it is suggested that the carcass placed under it be indicated as 2/1, meaning No. 2 over No. 1. This makes it easier for the person who is reading and grading the reasons.

As long as you stick to the facts and present them clearly so the judge can follow the reasoning without glancing back to see how they were placed, you can be considered to be in good form. The amount of knowledge you possess will show up in your use of meat terms and your applications of facts to economics. No two people express themselves alike.

Fifteen minutes is allotted to writing reasons, so don't waste time but be sure to write legibly. An intermission of three minutes will be given between writing reasons on each class to allow you to review your notes on the next class.

It is no crime to find equal qualities in both carcasses or some of each in all four carcasses and admit it. The crime lies in improvising, making a false statement, or referring to the wrong carcass. Tell the truth, as you see it, in a simple manner that can be easily followed by the committee member who must grade the reasons.

A common fault of contestants is that of giving what is termed *sterotyped* reasons. This is generally true of those who have not had sufficient training and who have acquired a vocabulary of meat terms that are more or less meaningless to them. As a result, they know how to say or describe certain things, and these are repeated verbatim for each carcass or cut.

Some are taught a *descriptive* method that simply describes the merits or faults without direct comparison. This method may be rather effective if some judges are insistent that contestants see certain points that these judges had in mind when the official placing was made. By following such a method, a contestant is less likely to miss these key points, but it weakens the effectiveness of a set of reasons because it sidesteps argument.

The most convincing and effective reasons are those presented through *comparisons* in which a contestant presents the superior qualities of one carcass or cut over another.

Even though it is important to have the class placed correctly, you can obtain a high reasons score on an incorrect placing if you have correctly analyzed the class and emphasized these points in your reasons in an eye-catching manner.

In most meats contests, the difference between first and fifth place is usually found in the reasons scores which makes reasons writing a very important aspect of meat judging. To write reasons exactly as you saw the class without any excess trivia requires much time, work, and dedication in practicing.

Reasons have to be graded rapidly so that you must present your thoughts in a well organized, clear, concise manner that is easily readable and understood.

It is important for you to develop your own style of reasons writing with these points in mind. Opening and closing statements must be strong, since these are the first and last things a judge reads. Don't worry too much about different sets of your reasons looking quite similar, since every set of reasons is corrected by a different judge; however, you must work on varying your comparisons within a class. Do not present all comparisons in the same manner or continually repeat the same terms. The point is to work on developing your own individual style, yet leave some room for variation within classes.

The most important part of writing reasons is accuracy. This means to double-check your placing to make sure you are writing about the same cut or carcass you are thinking about. Don't shoot in the dark or tell fibs just to have something to fill space. These practices will cut your score greatly.

Stress the more important factors first, and make these factors the *primary* basis for placing one over another. Omit unimportant details if time and space don't permit; however, in a *close* placing, small details may have their place in helping make the decision and should then be included.

Cover all the points under each category (trimness, muscling, and quality) systematically. In order to be easily understood, it is not a good policy to mix trimness, muscling, and quality statements in your reasons.

Judging usually involves placing one over another because the first had a greater balance or more desirable traits than the second; however, some classes of cuts and carcasses require more emphasis on cutability (trimness and muscling), due to the way they are to be utilized.

Following is an expansion of the reasoning for division of classes according to ultimate meat use which was mentioned earlier in this chapter. You should place slightly more emphasis on the cutability traits, depending on their ultimate use. This does not mean that you should totally exclude the other factors.

Quality Classes	*Equal*	*Cutability Classes*
Beef ribs	Beef carcasses	Beef rounds
Beef loins	Lamb carcasses	Beef chucks
		Hams
		Pork carcasses

General Rules for Writing Reasons

1. Use the past tense throughout reasons.
2. Use a variety of connecting words.
3. Use a variety of verbs.
4. Use a variety of words denoting *degrees of difference.*
5. In 2, 3, and 4 above, remember the meaning of the words you use. Don't plug in words indiscriminately.
6. As long as you are expending all this effort to write a full page of words, use a little more effort to write something worthwhile and meaningful.
7. Make opening and closing comparisons especially strong, because the opening comparison will influence the judge's thinking through the reasons, and the final comparison will be made immediately before the judge records the reasons score.

Penmanship and Grammar

Writing reasons will give you training in neatness and good penmanship. Through the corrections made, spelling and good sentence structure will also be a part of the training. As oral reasons give you training in speaking, written reasons give you training in writing.

Grants

In most cases, *grants* are necessary in your thorough analysis of a class. Grants are admissions that an exhibit that has been placed below another has its merits, also. In an extremely close placing, the grants may take as much consideration in your reasons as the individual placed above it; however, in most cases, the grants are brief and cover only a few points.

Do not neglect grants if they are important in the comparison.

Terminology

Before written reasons can even be attempted, a knowledge of the terminology used for carcasses and cuts is essential. If you don't know the accepted carcass nomenclature, how can you accurately convey an idea to anyone else?

General and specific terms or statements may be used singly or several may be used in sequence; however, if you make a broad general statement, never forget to ask yourself where, how, and why to describe this statement more specifically.

Quality

Specific	*General*
1. A higher degree	1. More youthful
2. A greater amount of	2. Higher quality
3. More extensive	3. Firmer
4. More evenly dispersed	
5. More evenly distributed	
6. Redder rib bones	
7. Lighter, brighter cherry red beef color	
Lighter, brighter grayish pink pork color	
Lighter, brighter reddish pink lamb color	
8. Smoother, finer textured	
9. More finely dispersed marbling	
10. More evenly dispersed feathering	

11. Softer, more pearly white buttons
    (beef)
12. Whiter, flakier fat
13. More uniformly covered
14. Firmer, thicker flank
    Firmer, thicker side

Last Place

1. Underfinished
2. Lacked quality
3. Wasty

4. Dark colored
5. Soft, oily
6. Hard-boned

Muscling

Specific

1. Bulging
2. Thicker
3. Deeper
4. Fuller
5. Plumper
6. Wider
7. Shorter
8. Longer
9. Shorter shanked
10. Shorter necked
11. Larger
12. Deeper chined

General

1. More symmetrical
2. Meatier
3. Heavier muscled
4. Thicker fleshed

Last Place

1. Ill-shaped
2. Small
3. Angular
4. Lacked meatiness

5. Long shanked, thin, tapering, round
6. Long, thin fleshed ham (leg)
7. Poorly balanced

Useful Terms

Comparative Verbs

1. As shown, showed
2. Displayed
3. Possessed
4. Exhibited
5. Indicated
6. Demonstrated
7. Lacked

Grants

1. Realize
2. Although
3. Grant
4. However
5. Recognize
6. Admit

Degrees of Comparison

1. Unsurpassed
2. Somewhat
3. Much
4. Greater
5. Distinctly
6. Larger
7. Extremely
8. Superior
9. Excessive
10. Higher degree
11. Little
12. Slightly
13. Smaller
14. Limited
15. Lesser
16. Lower

Connective

1. Furthermore
2. In addition
3. Also
4. Carrying into
5. Along with
6. Resulting in
7. As evidenced by
8. Blending into
9. Characterized by
10. For being
11. Contributing to
12. Coupled with
13. Shown by

Commonly Misspelled Words

1. Blade
2. Bulging
3. Carrying
4. Chine
5. Conformation
6. Desirably
7. Exudative
8. Feathering
9. Heel
10. Length
11. Loin
12. Meatier
13. Muscling
14. Quality
15. Quantity
16. Symmetrical
17. Thoracic
18. Trimmer
19. Value

Notes

A good share of successful written reasons can be attributed to accurate and thorough note taking. Since reasons must be written in a limited time, the notes taken during judging need to be organized and clear. They should contain enough phrase terms to be easily incorporated into the reasons and should remind you of the class you are writing on.

Take notes *systematically* in a logical order; use abbreviations and short sentences to save time. Divide your note cards into sections for trimness, muscling, and quality of each pair; however, wait until you see the class before making these divisions, because you want obvious characteristic listed first. By doing this, you can later write your reasons in order directly from your note cards. Underline important facts and double underline to show more emphasis in your notes.

SAMPLE REASONS

The following reasons are copies from actual contests, and the scores indicated were awarded in intercollegiate competition as recent as the 1984 International. Keep in mind that the scores listed represent one official's opinion of the accuracy and quality of each respective set of reasons.

SAMPLE

Score 48

REPORT OF REASONS

Heavy Beef Carcasses

Placings: 1st 4   2nd 3   3rd 1   4th 2

FIRST 4/3—In this class of heifer carcasses, I placed 4 over 3 as 4 was a higher quality carcass, as revealed by a higher concentration of marbling in the rib eye. Furthermore, 4 had a wider, slightly thicker cushioned round and a slightly larger rib eye. Also, 4 had less fat over the round collar and chuck, along with less flank and brisket fat. I admit 3 had less fat over the rump and slightly less over the loin edge, at the eye, and over the lower rib, coupled with much less udder fat.

SECOND 3/1—3 placed over 1 as 3 was a heavier muscled, trimmer carcass that would yield a higher percentage of boneless trimmed retail cuts. 3 featured a wider, thicker, plumper cushioned round; a larger rib eye; and a thicker clodded chuck. Then too, 3 exhibited definitely less fat at the rib eye and over the lower rib and chuck. I concede 1 revealed a superior degree of marbling in the rib eye. Also, 1 had less fat over the round and round collar, coupled with less udder and brisket fat.

THIRD 1/2—1 easily placed over 2 as 1 was an especially trimmer carcass that would yield a higher percentage of the trimmed most valuable primals. 1 exemplified less fat over the round, round collar, rump, sirloin-shortloin region and was especially trimmer at the eye and lower rib. Furthermore, 1 was trimmer over the chuck and had less udder, plate, brisket, and heart fat. In addition, 1 had a longer cushioned round. I grant 2 had a thicker cushioned round, a slightly larger eye, and a thicker forequarter, partially due to fat.

FOURTH 2—I realized 2 was a high quality carcass as manifested by a high degree of marbling in a firm, fine textured rib eye; however, I placed 2 last as it was the fattest carcass in the class. 2 displayed the most fat over the round, loin edge, at the eye and over the chuck, coupled with the most udder and brisket fat. Accordingly, 2 would yield the lowest percentage of boneless trimmed retail cuts.

REPORT OF REASONS

Beef Ribs

Placings: 1st 4   2nd 3   3rd 2   4th 1

FIRST 4/3—I placed 4 easily over 3 as 4 was a trimmer, heavier muscled, higher cutability rib. 4 was trimmer at the rib eye, lower rib, back, lower blade, and rib ends. Furthermore, 4 possessed a larger rib eye and a deeper chined, longer, wider back. In addition, 4 had more marbling in the rib eye and a firmer, finer textured blade end. I admit 3 revealed a finer distribution of marbling in a more youthful colored, finer textured rib eye.

SECOND 3/2—3 placed over 2 as 3 was a trimmer, heavier muscled rib that would yield a higher percentage of trimmed roasts and steaks. 3 was leaner at the rib eye, back, lower blade, rib ends, and blade end, combined with less seam fat in the blade end. Furthermore, 3 presented a deeper, wider, meatier blade end that contained a larger eye. Moreover, 3 had a higher degree of marbling in both ends and a more youthful colored rib eye.

THIRD 2/1—2 placed over 1 as 2 was a higher quality rib that would have a higher merchandising value. 2 possessed a greater amount of marbling in both ends, more evenly distributed marbling in the rib eye, and a finer textured blade end. Furthermore, 2 had a wider, meatier blade end. Also, 2 was trimmer over the rib ends. I grant 1 was trimmer at the rib eye. Then too, 1 had a deeper blade end.

FOURTH 1—I realize 1 was trim at the rib eye; however, I placed 1 last as it had the least amount of marbling in the rib eye and a coarse textured blade end. Furthermore, 1 had too much fat over the back and lower blade, with too much seam fat in the blade end. Moreover, 1 had a small eye in the blade end. Therefore, 1 would yield the lowest percentage of boneless trimmed retail cuts.

REPORT OF REASONS

Pork Carcasses

Placings: 1st 3   2nd 2   3rd 4   4th 1

FIRST 3/2—I placed 3 over 2 as 3 was a heavier muscled carcass as indicated by a thicker, plumper cushioned ham and a fuller, meatier sirloin. Furthermore, 3 had less backfat at the last lumbar, last rib and first rib, coupled with less sternum fat. I admit 2 was trimmer over the ham collar and in the belly region. Also, 2 had a thicker shoulder partially due to fat. Moreover, 2 revealed more rib feathering.

SECOND 2/4—2 placed over 4 as 2 combined trimness and muscling to a greater degree. 2 possessed less backfat at the last rib combined with less fat over the loin edge and in the flank, belly, and sternum regions and had less jowl. Also, 2 had a brighter colored flank. I concede 4 presented a wider cushioned ham and a thicker, meatier shoulder. Moreover, 4 had less backfat at the first rib, along with less fat over the ham collar. In addition, 4 revealed more marbling in the lumbar lean.

THIRD 4/1—In placing 4 over 1, 4 was a trimmer carcass as evidenced by less backfat from ham to shoulder, less fat in the flank, belly, and sternum regions and had less leaf fat. Additionally, 4 had a thicker cushioned ham and a thicker, meatier shoulder. I grant 1 exhibited a greater amount of exposed lumbar lean. Also, 1 revealed a higher concentration of rib feathering.

FOURTH 1—I placed 1 last as it had the most backfat over the lumbar, center loin, and clear plate regions and had too much fat in the flank, belly, and sternum regions. Therefore, 1 would yield the lowest lean to fat ratio.

S A M P L E

Score   44

REPORT OF REASONS

Fresh Hams

Placings: 1st 2   2nd 4   3rd 1   4th 3

FIRST 2/4—I placed 2 over 4 as 2 was an especially heavier muscled ham as exemplified by a greater amount of exposed lean in the butt face, a much meatier knuckle, a deeper, more bulging center section, and a plumper, meatier heel. Also, 2 had slightly less intermuscular fat in the butt face. Moreover, 2 revealed a more grayish-pink colored rump face. I concede 4 was a higher quality ham as shown by a higher degree of marbling in a slightly firmer butt face, along with less muscle separation in the rump face. In addition, 4 was slightly trimmer over and along the butt face and had a longer, wider cushion.

SECOND 4/1—4 placed over 1 as 4 combined muscling and quality to a higher degree. 4 manifested an especially deeper butt face; a meatier forecushion; a definitely plumper, deeper center section; and a fuller, meatier heel. Moreover, 4 revealed more evenly dispersed marbling in a firmer butt face with less muscle separation. I acknowledge 1 was leaner under the rump face and was slightly trimmer over the ham collar. Then too, 1 had a wider butt face and a wider, longer cushion. Also, 1 had a brighter colored rump face.

THIRD 1/3—In placing 1 over 3, 1 was a leaner ham as indicated by especially less fat under and along the rump face, over the forecushion, ham collar, and heel, along with slightly less intermuscular fat in the rump face. Furthermore, 1 exhibited a meatier, fuller knuckle and a longer, much wider center section. Then too, 1 had slightly more marbling in the butt face. I admit 3 presented a slightly deeper rump face and a deeper cushion and fuller heel, both partially due to fat. Then too, 3 displayed a more grayish-pink colored butt face.

FOURTH 3—I realized 3 revealed a grayish-pink colored rump face; however, I placed 3 last as it was the fattest ham in the class. 3 had the most fat under and along the butt face, over the knuckle, ham collar, and heel. Also, 3 had a short cushion and a light muscled heel. Accordingly, 3 would yield the lowest lean to fat ratio in the class.

SAMPLE

Score <u>45</u>

REPORT OF REASONS

Light Lamb Carcasses

Placings: 1st <u>4</u>   2nd <u>2</u>   3rd <u>3</u>   4th <u>1</u>

FIRST <u>4/2</u>—I placed 4 over 2 as 4 was an especially heavier muscled carcass that would yield a higher percentage of trimmed leg and loin. 4 possessed a heavier muscled leg with more inside and outside flare, a definitely larger rib eye, and a thicker fleshed rack and shoulder. Moreover, 4 showed a more youthful colored rib eye and rib bones, along with a higher degree of primary and secondary flank streaking. Also, 4 had less kidney and pelvic fat. I admit 2 had less fat over the leg, dock, sirloin, loin, and loin edge, especially over the lower rib and shoulder, along with less crotch fat.

SECOND <u>2/3</u>—2 placed over 3 in a close decision as 2 was a trimmer carcass, as evidenced by slightly less fat at the rib eye, over the dock, loin, loin edge, and lower rib and rack. Furthermore, 2 had a slightly thicker cushioned leg. I concede 3 had a wider cushioned leg, a larger rib eye, and a meatier shoulder. Also, 3 had less internal fat. Then too, 3 had a higher degree of marbling in the eye and a greater amount of flank lacing.

THIRD <u>3/1</u>—3 easily placed over 1 as 3 combined trimness and muscling to a higher degree and would yield a higher percentage of boneless trimmed retail cuts. 3 exhibited less fat over the sirloin and loin and slightly less at the rib eye, coupled with less flank, breast, kidney, and pelvic fat. In addition, 3 had a meatier sirloin, a definitely larger eye, and a thicker fleshed rack and shoulder. Also, 3 had a more youthful colored rib eye. I grant 1 showed redder, rounder rib bones and a brighter colored flank.

FOURTH <u>1</u>—I realized 1 had a high degree of marbling and youthful colored rib bones; however, I placed 1 last as it combined trimness and muscling to the lowest degree. 1 had the most fat over the loin and at the rib eye, in conjunction with the most breast fat. Also, 1 had the smallest rib eye in the class. Accordingly, 1 would yield the lowest percentage of boneless trimmed retail cuts.

HAM AND BACON SHOWS

The American Association of Meat Processors (AAMP),[2] at each annual convention, holds a National Cured Meats Show. Many states have a similar competition prior to the national show. Any meat processor who is a member of AAMP is eligible to enter a ham, bacon, or sausage product.

CLASSES OF ENTRIES

Fifteen classes of products were open for competition among AAMP mem-

---

[2]P.O. Box 269, Elizabethtown, PA 17022.

bers in 1984. These included seven classes of ham, two classes of bacon, three classes of sausage, one class of dried beef, and one class of smoked turkey. In addition, a new class, the American Meat Platter Competition, was introduced in 1983. The fifteen classes are:

I—Country Hams, Smoked, Lightweight (under 16 pounds)
II—Country Hams, Smoked, Heavyweight (16 pounds and over)
III—Country Hams, Unsmoked, Lightweight (under 16 pounds)
IV—Country Hams, Unsmoked, Heavyweight (16 pounds and over)
V—Mild Cured, Commercial Style Hams, Lightweight (under 16 pounds)
VI—Mild Cured, Commercial Style Hams, Heavyweight (16 pounds and over)
VII—Specialty Commercial Hams
VIII—Country Bacon
IX—Commercial Bacon
X—Summer Sausage, Uncooked
XI—Summer Sausage, Cooked
XII—Cooked Ring Bologna
XIII—Cured and/or Smoked Dried Beef
XIV—Smoked Turkey
XV—American Meat Platter Competition

STANDARDS OF EVALUATION

Since judges may be changed from year to year and there may be considerable difference of opinion as to the characteristics of the ideal or average product, entries are judged according to the following standards (a total of 1,000 points is possible for each class):

Country Hams, Smoked (Classes I and II)

Hams must be whole and must be neither cut nor sliced. To be eligible for this class, hams must have been dry cured by the dry cover cure (not pumped or brine-cured) method, held for a minimum of 90 days, smoked, and considered suitable for safe storage without refrigeration.

Country Hams, Unsmoked (Classes III and IV)

Hams must be whole and must be neither cut nor sliced. To be eligible for this class, hams must have been dry cured by the dry cover cure (not pumped or brine-cured) method, held for a minimum of 90 days, unsmoked, and considered suitable for safe storage without refrigeration.

Classes I, II, III, and IV are evaluated on the following standards:

General Appearance: 150 Points

Eye appeal, conformation, trim, lean-to-fat ratio, uniform smoke coloring, and cutability or yield shall be the main points considered. Scoring for general appearance will be done in two phases, allowing up to 75 points for eye appeal, conformation, and trim before cutting and 75 points for cutability and yield after cutting. Hams will be cut 1 inch posterior to and parallel with the H-bone.

Aroma: 150 Points

Off or foreign odors should downgrade the ham much more than should a lack of aroma, since much of the aroma of country ham develops with cooking. A definitely good, smooth aroma should upgrade the ham.

Texture: 150 Points

A country ham should be firm but neither too hard nor too soft. High-moisture content or too coarse a grain in the ham is undesirable.

Inside Color: 150 Points

In country hams, the gray color of the dry-salt ham or the sugar-cured ham without the use of nitrate or nitrite is not objectionable and should not be down-graded if the color is uniform and otherwise acceptable. The red color of the ham is very attractive; however, if it is too red it is an indication of an excessive amount of nitrate or nitrite and should be taken into consideration.

Flavor: 400 Points

The ham should have the very definite tang and mellow flavor associated with a true country ham. The presence of off-flavors such as rancidity, bacterial spoilage, etc., will severely downgrade the product. A full half-slice from the cushion side of the ham will be fried and two taste samples taken, one from the lower muscle and one from the top muscle, unless the ham is sour or has an off-flavor in the first sample, making it too inferior to be considered in the top four places. When a full half-slice is fried, the amount of moisture cooked out of the ham is very obvious, and, if it is excessive, the ham should be downgraded accordingly, since this is very objectionable in a premium-priced true country ham.

Mild Cured, Commercial-Style Ham (Classes V and VI)

Hams must be entered whole with bone in and must be neither cut nor sliced. To be eligible for this class, hams should be cured mildly and may or may not be tenderized. This class is considered to consist principally of pumped hams which may or may not require refrigeration.

General Appearance: 200 Points

Hams will be graded on eye appeal, conformation, trim, and cutability or yield. Scoring for general appearance will be done in two phases, allowing up to 100 points for eye appeal, conformation, and trim before cutting and 100 points for cutability and yield after cutting. Hams will be cut 1 inch posterior to and parallel with the H-bone.

Aroma: 150 Points

Off, foreign, sour, or sharp odors should downgrade the ham; the good, mellow aroma that pleases the judge should upgrade it.

Texture of Cut Surface: 100 Points

Excessive moisture is objectionable. The ham should not be too coarse-grained. There should be no excess of fat marbled in the ham.

Inside Color: 150 Points

The color should be uniform and appealing. It should be neither too light nor too red, and bruises should score quite heavily against color. If a ham should be noticeably cooler than the rest of the hams, giving it an advantage in color and firmness, this should be considered and graded accordingly.

Flavor: 400 Points

A full half-slice of ham from the cushion side of the ham will be fried and two taste samples taken, one from the lower muscle and one from the top muscle, unless the ham is sour or has an off-flavor in the first sample, making it too inferior to be considered in the top four places. The ham should be neither too salty nor too bland and should have the good, mellow flavor expected in a good commercial ham. The presence of off-flavors such as rancidity, bacterial spoilage, etc., will severely downgrade the product.

Specialty Commercial Hams (Class VII)

Hams must be whole boneless hams and must be neither cut nor sliced. To be eligible for this class, hams may be shaped, formed, or pressed but *must consist of the meat from only one ham* and may not be chunked, flaked, or reconstituted; shall be cured mildly; may or may not be tenderized; may or may not require refrigeration. Boiled hams and cuts other than pork hams are specifically excluded.

☞   NOTE: The regulations for this class of hams are under consideration   ✎
for change. Check with AAMP for the latest rules.

General Appearance: 200 Points

Hams will be graded on eye appeal, conformation, trim, and cutability or yield. Scoring for general appearance will be done in two phases, allowing up to 100 points for eye appeal, conformation, and trim before cutting and 100 points for cutability and yield after cutting. Hams will be cut only once, and the angle of the cut will be at the option of the exhibitor.

Aroma: 150 Points

Off, foreign, sour, or sharp odors should downgrade the ham; the good mellow aroma that pleases the judge should upgrade it.

Texture of Cut Surface: 100 Points

Excessive moisture is objectionable. The ham should not be too coarse-grained. There should be no excess of fat marbled in the ham.

Inside Color: 150 Points

The color should be uniform and appealing. It should be neither too light nor too red, and bruises should score quite heavily against color. If a ham should be noticeably cooler than the rest of the hams, giving it an advantage in color and firmness, this should be considered and graded accordingly.

Flavor: 400 Points

A full half-slice of ham will be fried and two taste samples taken, each from a different part of the slice. The ham should be neither too salty nor too bland but should have the good, mellow flavor expected in a good ham. The presence of off-flavors such as rancidity, bacterial spoilage, etc., will severely downgrade the product.

Country Bacon (Class VIII)

Bacon must be whole and must be neither cut nor sliced. To be eligible for this class, bacon will be of standard approved trim with skin on or off, be cured heavily enough by either the dry or the brine-soaking (immersion) method to insure good keeping quality without refrigeration, and be smoked sufficiently to give a rather dry surface and firm feel.

General Appearance: 150 Points

Manner in which the belly was trimmed, eye appeal, conformation, quality of workmanship, and a deep outside color should be the main points considered. Judges will severely penalize a belly showing excessive trimming, poor workmanship in trimming, robbing of lean areas, and lack of uniformity of outside color.

Fat to Lean (After Cut): 150 Points

Lean bacon is very desirable and should be scored accordingly, unless it is so lean that it would indicate a poor-quality hog.

Aroma: 100 Points

Off or sour odors should downgrade the bacon; a good, smooth, ample aroma should upgrade it.

Texture: 100 Points

Country bacon should be firm but not too hard and should be relatively dry to the touch. Temperature of the bacon should be considered in scoring texture, since chilled bacon is more firm.

Inside Color: 100 Points

Lean should be light red, fat should be white, color should be uniform, and bruises should downgrade the bacon.

Flavor: 400 Points

Bacon should have a full, rich flavor that indicates the full cure and smoke of country bacon. The presence of off-flavors such as rancidity, bacterial spoilage, etc., will severely downgrade the product.

Commercial Bacon (Class IX)

Bacon must be whole and must be neither cut nor sliced. To be eligible for this class, bacon will be of standard approved trim with skin on or off and be cured by any accepted method. If the number of entries is high, the class may be divided into heavy and light divisions.

General Appearance: 150 Points

Amount of trimming, workmanship, eye appeal, conformation, and outside color should be the main points considered. Bacon may be skinned or un-skinned, but skinned bellies should exhibit careful workmanship in skinning. Judges will severely penalize a belly showing excessive trimming, poor work-manship in trimming, and robbing of lean areas.

Fat to Lean (After Cut): 150 Points

Lean bacon is very desirable and should be scored accordingly, unless it is so lean that it would indicate a poor-quality hog.

Aroma: 100 Points

Off or sour odors should downgrade the bacon; a good, smooth aroma should upgrade it.

Texture: 100 Points

Pumped bacon may be downgraded in texture if it appears to have excess moisture. Temperature of the bacon should be considered in scoring texture, since cooler bacon is more firm.

Inside Color: 100 Points

Lean should be light red, fat should be white, color should be uniform, and bruises should downgrade the bacon.

Flavor: 400 Points

Bacon should have a pleasing rich flavor, neither too salty nor too bland. The presence of off-flavors such as rancidity, bacterial spoilage, etc., will se-verely downgrade the product.

Summer Sausage, Uncooked (Class X)

Sausages must be whole and must be neither cut nor sliced. To be eligible for this class, summer sausage will be uncooked and must have been subjected to one of the methods of eliminating trichina specified in Section 318.10 of the

Federal Meat Inspection Regulations, copies of which are available upon request from AAMP (see Chapter 3, page 51).

☞ NOTE: Each Class X entry shall be accompanied by an ingredients ✍ statement listing ingredients in descending order of predominance and a copy of the processing schedule for the product or, for products containing pork, a certificate indicating that the product was prepared with pork that was certified trichina-free.

Summer Sausage, Cooked (Class XI)

Sausages must be whole and must be neither cut nor sliced. To be eligible for this class, summer sausage will have been cooked to a temperature of not less than 137° F and must have been subjected to one of the methods of eliminating trichina specified in Section 318.10 of the *Federal Meat Inspection Regulations,* copies of which are available upon request from AAMP (see Chapter 3, page 51).

☞ NOTE: Each Class XI entry shall be accompanied by an ingredients ✍ statement listing ingredients in descending order of predominance and a copy of the processing schedule for the product or, for products containing pork, a certificate indicating that the product was prepared with pork that was certified trichina-free.

Cooked Ring Bologna (Class XII)

To be eligible for this class, bologna must be whole and must be neither cut nor sliced. It must consist of one ring with a uniform casing diameter of *not less than* 38 millimeters (approximately 1.52 inches) nor more than 46 millimeters (approximately 1.84 inches). The product may be beef and/or pork and must have been heated to a minimum of 137° F.

☞ NOTE: Each Class XII entry shall be accompanied by a copy of the ✍ processing schedule (including times and temperatures) or, for products containing pork, a certificate indicating that the product was prepared with pork that was certified trichina-free in accordance with Section 318.10 of the *Federal Meat Inspection Regulations,* copies of which are available upon request from AAMP (see Chapter 3, page 51).

Classes X, XI, and XII are evaluated on the following standards:

External Appearance: 250 Points

- Uniformity of color: 75 points
- Uniformity of shape (diameter, etc.): 75 points
- Lack of defects (*e.g.,* air pockets, wrinkles, fat caps, etc.): 100 points

Internal Appearance: 350 Points

- Uniformity of color: 75 points
- Uniformity of texture: 50 points
- Lean-to-fat ratio: 50 points
- Lack of obvious defects: 100 points
- Aroma: 75 points

Eatability: 400 Points

- Taste: 300 points
- Texture/mouth feel: 100 points

Flavors such as rancidity, bacterial spoilage, etc., and the presence of strong aftertastes will severely downgrade the product.

Cured and/or Smoked Dried Beef (Class XIII)

Dried beef must be whole and must be neither cut nor sliced. To be eligible for this class, beef must be from beef rounds, cured and heat processed with or without smoke so that the product may be eaten without further cooking but may require refrigerated storage. The products entered must be made from single muscle pieces from the beef round or the clod and must not be a sectional or flaked and formed product.

General Appearance: 200 Points

For eye appeal, surface should be a bright mahogany red, free from hardening; workmanship should show care with no second cuts or dried tag ends. Meat should be firm and of appropriate texture.

Aroma: 100 Points

Product should have a desirable cured aroma. If smoked, the smoke should not be overpowering. Off odors will be considered highly objectionable.

Texture: 100 Points

Interior texture should be firm, similar to a dry-cured ham with no moisture showing on cut surface. Meat should be fine in texture. Product should hold together well with thin slicing.

APPEARANCE: 320 POINTS

- Overall impression: 80 points
- Decoration, the effective use of color: 80 points
- Visual impact: 80 points
- Design balance: 80 points

WORKMANSHIP: 320 POINTS

- Effective use of materials: 80 points
- Creativity, uniqueness: 80 points
- Arrangement and workmanship: 80 points
- Durability of display: 80 points

MERCHANDISABILITY: 320 POINTS

- Variety of products: 80 points
- Durability of products: 80 points
- Selection of complementing foods: 80 points
- Serving ease: 80 points

SUITABILITY: 40 POINTS

- For number of persons to be served: 40 points

NOTE: If the entry is deemed suitable for the indicated number to be served, all 40 points will be assigned; if it is unsuitable, no points will be assigned under this heading.

MEAT CONTESTS FOR ALL INTERESTED STUDENTS

Competition is undoubtedly one of the greatest instruments for arousing interest among young people. It has proved so effective among the youth of the nation that it has actually become a so-called final examination for many courses offered in schools and colleges.

This method of fostering interest in the youth of the United States has been applied by the Future Farmers of America and the 4-H Club Congress to various phases of their work. Contests of national importance that have to do with meat are the Meat Judging Contest and the Meat Identification Contest held for Future Farmers of America and for the national 4-H organization at the American Royal Livestock Show at Kansas City, Missouri. The national 4-H organization also sponsors a contest at the National Western in Denver, Colorado.

Inside Color: 200 Points

Inside color should be a uniform dark red, free from heat rings. Uncured spots will be a decided fault. Product should be free from two-toning. Fat should be white.

Flavor: 400 Points

Flavor should be appropriate for the type of product. Flavor should be mildly salty. The presence of off-flavors such as rancidity, bacterial spoilage, etc., will severely downgrade the product. Product will be sampled as thin-sliced without further cooking. In all cases, cured-meat flavor should predominate, not the flavor of smoke or salt alone. Fat, if present, should be free from rancidity. Bland flavors or off-flavors will be severely discriminated against.

Smoked Turkey (Class XIV)

To be eligible for this class, turkey must be brine cured and smoked, must be fully cooked (minimum final internal temperature of product during processing, 155° F), must be whole, with bone in. If the number of entries is large, the class may be divided into heavy and light divisions.

General Appearance: 200 Points

Turkeys will be graded on eye appeal, conformation, plumpness of breast, straightness of keel bone, smoked color, and yield. Skin should be intact, should not be too dry, and should have a uniform smoked color. Bruises and pinfeathers will downgrade the product.

Aroma: 150 Points

Off, foreign, sour, or sharp odors will downgrade the turkey; the good, mellow aroma that pleases the judge will upgrade it.

Texture of Cut Surface: 100 Points (White Meat: 50 Points; Dark Meat: 50 Points)

While excessive moisture is objectionable, the turkey meat should not be too dry. Excessive dryness in the turkey and a basted turkey will be cause to downgrade the product. The turkey should not be too coarse-grained.

Inside Color: 150 Points (White Meat:
100 Points; Dark Meat: 50 Points)

The color of the white meat should be a uniform light pink, and bruises will score quite heavily against color. The color of dark meat should be a uniform light red color, and bruises will score quite heavily against color. Any uncured areas will downgrade the product.

Palatability Characteristics: 400 Points (White
Meat: 200 Points; Dark Meat: 200 Points)

A slice will be taken from one half of the breast near the keel bone in line with the point of the keel bone. One slice of dark meat will be taken from the midpoint of the length of the thigh bone. Cold slices will be tasted. The turkey should be neither too salty nor too bland; should have the good, mellow flavor expected in a good smoked turkey, with no unpleasant aftertaste; and should be tender. The presence of off-flavors such as rancidity, bacterial spoilage, etc., will severely downgrade the product.

American Meat Platter Competition (Class XV)

The purpose of the platter competition is to encourage and develop the production and marketing of an expanding variety of meat products by demonstrating the convenience, versatility, variety, elegance, workmanship, creativity, and cost effectiveness that can be achieved through meat platter merchandising.

Rules

Each entry will consist of a platter on which is arranged ready-to-eat cuts of meat in accordance with the following:

- Platters will be of a disposable material suitable for the display of meat and other food products. Show management will not be responsible for returning the platters to entrants.

- Fruits, vegetables, cheeses, and other foods may be incorporated into the meat platter displays, either as part of the major food items being displayed or as decorations and garnishes; however, meat items should be predominant.

- No raw or uncooked meat may be included in any display.

- A designation will be provided with each entry indicating how many people the platter is designed to serve.

- To retain freshness, platters may be covered with transparent plastic domes that do not touch the product and that can be easily removed by simply being lifted off; however, platters may not be covered with plastic films or wraps.

- Meats that are unprofessionally cut will be downgraded.

- Garnishes and decorations should harmonize with the entry in quantity, taste, color, and presentation.

- To enhance the keeping quality of entries, thin, spiced aspic (meat stock jelly) may be used; overcooking should be avoided.

- Rare meats must be sufficiently cooked so that, when covered with aspic, no blood or juice seeps out. Oozing meat, vegetable, or fruit juices will result in downscoring.

- Unprofessionally cut garnishes of vegetables or fruit will detract from the score.

- The use of butter, oleomargarine, mayonnaise and/or salad dressing is permitted; however, appropriate concern for food safety should be exercised.

- Limited use of non-edible items (such as foil-covered fiberglass) to achieve raised platforms, etc., is permissible.

- Overloading of entries should be avoided.

- The size of the entry should correspond to the number of persons for whom it is designed.

- If sauces are used, the containers in which they are placed should be more than two-thirds full so that they will not spill in the event the platter are moved in the course of the judging.

- All entries should remain intact during display and awards. It is understood that the entries will not be returned to the entrants but may be able for consumption by conventioneers at the close of the display pe

- All pork products entered in the competition must have been subjec one of the methods of eliminating trichina specified in Section 31 the *Federal Meat Inspection Regulations,* copies of which are a from AAMP on request (see Chapter 3, page 51).

- The entrant signifies acceptance of all of the above rules by regis the competition.

Scoring

Each entry will be judged for appearance, workmanship, merch and suitability for number of persons to be served. Evaluation wil by a numerical score assigned for each of several characteristics ferent headings with the maximum number of points (1,000) as fol

The FFA Meat Judging Contest requires the contestant to judge five classes of carcasses or wholesale cuts instead of the nine classes used in intercollegiate competition. These judging classes consist of one class each of the following: four beef carcasses, four wholesale cuts of beef, four pork carcasses, four wholesale cuts of pork, and four lamb carcasses. The contestant also quality and yield grades 10 beef carcasses.

There are three contestants to a team, and each contestant is allowed 10 minutes for making the placings and filling in the official placing card.

The FFA Meat Identification Contest is for the purpose of determining the contestant's knowledge of the various cuts of meat and the edible by-products of meat-producing animals.

Each contestant is given a meat identification card upon which are listed code numbers for species, primal (wholesale) cuts, retail cuts, and edible meat by-products. A group of any 25 retail cuts constitute this class, the cuts being numbered from 1 to 25 consecutively. The contestant must write the number of the species, the primal cut, and the retail cut opposite the label number indicated for the displayed cut.

Each contestant is allowed 20 minutes for identification. Three contestants compose a team, and the same team or a different team may compete in each contest.

One point is given for each cut that is correctly identified as to species; two additional points are awarded for each cut correctly identified as to the primal cut from which it came; and three additional points are awarded for each retail trade name that is correctly identified. There are 150 possible points in the meat identification section of the national contest. An example of the meat identification score card is illustrated in Figure 23-9.

COACHING A TEAM FOR MEAT IDENTIFICATION

Charts, showing the location of the wholesale and retail cuts of a carcass, are of value in memorizing the location of the cuts in the carcass. The study of photographs alone for identification of cuts is of doubtful value. Whether purchaser or student, it is absolutely necessary that one comes in direct contact with the product through actual purchasing, by learning the cuts through meat displays, or by doing the actual cutting. It is rather useless for vocational schools to enter meat judging or meat identification teams in national contests with no coaching other than from photographs.

Suggested Coaching Methods

One of the most effective ways of securing the necessary training for students is to place them as helpers for certain hours during the week, with competent and interested meat retailers, with the understanding that the students be

MEAT IDENTIFICATION CARD

Class Name_____

Contestant Name_____

Class Number_____

Contestant Number_____

Select Species, Primal Cut, and Retail Name from the listings below and fill in the column blanks beside the cut number. The score column is for tabulators use only.

EXAMPLE: Beef round rump roast bnls

Species—1 point

1—Beef   2—Pork   3—Lamb

Primal Cuts—2 points

1. Brisket	6. Leg	12. Shoulder
2. Chuck	7. Loin	13. Side (Belly)
3. Flank	8. Plate	14. Jowl
4. Foreshank and	9. Rib	15. Variety Meats
Breast	10. Round	16. Various**
5. Ham	11. Shank	

Retail Trade Names—3 points

Roasts/Pot Roasts

1. American Style	25. Rump (Bnls)	**Chops**
2. Arm	26. Rump Portion/Half	54. Arm
3. Arm-Picnic	27. Shank Portion/Half	55. Blade
4. Back Ribs	28. Short Ribs	56. Loin
5. Blade	29. Spare Ribs	57. Rib
6. Blade-Boston	30. Shoulder (Bnls)	58. Sirloin
7. Breast	31. Shoulder-Roll	
8. Brisket (Whole)	32. Sirloin	**Variety Meats**
9. Center Rib	33. Sirloin Half	59. Brains
10. Cross Cuts	34. Square Cut (Whole)	60. Heart
11. Cross Rib	35. Tenderloin (Whole)	61. Kidney
12. Foreshank	36. Tip	62. Liver
13. Fresh Side	37. Tip-cap off	63. Sweet Bread
14. Fresh Leg (Ham)		(Thymus)
15. Frenched Style	**Steaks**	64. Tongue
16. Heel of Round	38. Arm	
17. Loin	39. Blade	**Processed Cuts**
18. Mock Tender	40. Bottom Round	**Non-Smoked**
19. Picnic (Whole)	41. Center Slice	65. Beef for Stew
20. Plate-Rolled (Bnls)*	42. Eye	66. Fresh Ground Sausage
21. Rib	43. Flank	67. Corned Brisket (Bnls)
22. Riblets	44. Porterhouse	68. Ground Beef
23. Roast-Large End	45. Round	69. Ground Lamb Patties
24. Rump	46. Sirloin	
	47. Sirloin (Bnls)	**Smoked (or Cured) Pork**
	48. Smallend	70. Rump Half
	49. T-Bone	71. Shank Half
	50. Tenderloin	72. Center Slice
	51. Top Loin (Bnls)	73. Loin Chops
	52. Top Round	74. Canadian Style Bacon
	53. Top Sirloin (Bnls)	75. Picnic (Whole)
		76. Shoulder Roll
		77. Slab Bacon
		78. Sliced Bacon
		79. Jowl

Cut No.	Spe-cies	Pri-mal	Retail Name	Score
00	1	10	25	
1				
2				
3				
4				
5				
6				
7				
8				
9				
10				
11				
12				
13				
14				
15				
16				
17				
18				
19				
20				
21				
22				
23				
24				
25				

*(Bnls) = Boneless   ** = From Various Primal Cuts

Fig. 23-9. From national FFA contests, 1982 to 1984, national FFA organization, Instruction in Vocational Agriculture, U.S. Dept. of Education, Washington, D.C.

given every opportunity to learn the names of the wholesale and retail cuts and
to help make them whenever convenient. The students work with or without
pay, according to agreement. Several weeks or a month of such training for students will teach them more than is possible by any other method.

Another method that is more convenient, but not as fruitful, is one in which
arrangements are made with one or more meat retailers to give the students
practice in identifying available cuts one evening a week.

Still another method that can be employed with some degree of satisfaction
is one in which the coach makes frequent shopping tours with one or two members of the class to engage in identifying showcase display cuts. Under this
method, it is of course a prerequisite that the instructor must know the meats.
More than two students in a group is unwieldy and may crowd the otherwise already busy shop.

Under any system of coaching, a list of the cuts as published by the group
that is responsible for the contest should be used so there will be no time lost in
identifying cuts other than those indicated on the chart.

Where there are more trained contestants than the number required to constitute a team, an elimination contest can be arranged by the meat department of
the state agricultural college.

Identification Features

The first task that confronts a contestant is that of determining whether the
cut is beef, pork, veal, or lamb. These four kinds of meat are recognized by:

Color of the Lean

- Beef varies from bright to dark red.
- Lamb is light reddish pink. Mutton is brick red.
- Pork is grayish pink to grayish red.
- Veal is pinkish brown. The older the veal, the more it borders on reddish
  brown.

Size of the Cut

- Beef cuts are large in size.
- Lamb cuts are small in size.
- Pork and veal cuts run similar in size.

Type of Fat

- Beef has a white or cream-white (yellow in the lower grades), firm, and
  rather dry fat.

- Lamb has a chalk-white, brittle, rather dense fat, usually covered with the "fell," a colorless connective tissue membrane.
- Pork has a characteristic white, greasy fat.
- Veal is readily recognized by the absence of fat.

Having identified the cut as to kind, the next task is to identify the cut, both as to name and as to the wholesale cut from which it is derived. Refer to the color section in the center of this text for retail cut identification assistance. This requires a familiarity with anatomy to determine location by the shape of the bone and the shape and contour of the muscles. The contestant must remember that the difference between a roast and a steak of the same name is one of thickness. Steaks are generally from ½ to 1½ inches thick, whereas roasts are over 2 inches thick.

Chapters 12, 13, 14, 15, and 16 provide the basis for gaining a foundation in the knowledge of cut identification and should be read before one attempts a great deal of direct contact. Variety meat pictures appear in Figure 10-1.

Index

C

Inside Color: 200 Points

Inside color should be a uniform dark red, free from heat rings. Uncured spots will be a decided fault. Product should be free from two-toning. Fat should be white.

Flavor: 400 Points

Flavor should be appropriate for the type of product. Flavor should be mildly salty. The presence of off-flavors such as rancidity, bacterial spoilage, etc., will severely downgrade the product. Product will be sampled as thin-sliced without further cooking. In all cases, cured-meat flavor should predominate, not the flavor of smoke or salt alone. Fat, if present, should be free from rancidity. Bland flavors or off-flavors will be severely discriminated against.

Smoked Turkey (Class XIV)

To be eligible for this class, turkey must be brine cured and smoked, must be fully cooked (minimum final internal temperature of product during processing, 155° F), must be whole, with bone in. If the number of entries is large, the class may be divided into heavy and light divisions.

General Appearance: 200 Points

Turkeys will be graded on eye appeal, conformation, plumpness of breast, straightness of keel bone, smoked color, and yield. Skin should be intact, should not be too dry, and should have a uniform smoked color. Bruises and pinfeathers will downgrade the product.

Aroma: 150 Points

Off, foreign, sour, or sharp odors will downgrade the turkey; the good, mellow aroma that pleases the judge will upgrade it.

Texture of Cut Surface: 100 Points (White Meat: 50 Points; Dark Meat: 50 Points)

While excessive moisture is objectionable, the turkey meat should not be too dry. Excessive dryness in the turkey and a basted turkey will be cause to downgrade the product. The turkey should not be too coarse-grained.

Inside Color: 150 Points (White Meat:
100 Points; Dark Meat: 50 Points)

The color of the white meat should be a uniform light pink, and bruises will score quite heavily against color. The color of dark meat should be a uniform light red color, and bruises will score quite heavily against color. Any uncured areas will downgrade the product.

Palatability Characteristics: 400 Points (White
Meat: 200 Points; Dark Meat: 200 Points)

A slice will be taken from one half of the breast near the keel bone in line with the point of the keel bone. One slice of dark meat will be taken from the midpoint of the length of the thigh bone. Cold slices will be tasted. The turkey should be neither too salty nor too bland; should have the good, mellow flavor expected in a good smoked turkey, with no unpleasant aftertaste; and should be tender. The presence of off-flavors such as rancidity, bacterial spoilage, etc., will severely downgrade the product.

American Meat Platter Competition (Class XV)

The purpose of the platter competition is to encourage and develop the production and marketing of an expanding variety of meat products by demonstrating the convenience, versatility, variety, elegance, workmanship, creativity, and cost effectiveness that can be achieved through meat platter merchandising.

Rules

Each entry will consist of a platter on which is arranged ready-to-eat cuts of meat in accordance with the following:

- Platters will be of a disposable material suitable for the display of meat and other food products. Show management will not be responsible for returning the platters to entrants.
- Fruits, vegetables, cheeses, and other foods may be incorporated into the meat platter displays, either as part of the major food items being displayed or as decorations and garnishes; however, meat items should be predominant.
- No raw or uncooked meat may be included in any display.
- A designation will be provided with each entry indicating how many people the platter is designed to serve.

- To retain freshness, platters may be covered with transparent plastic domes that do not touch the product and that can be easily removed by simply being lifted off; however, platters may not be covered with plastic films or wraps.
- Meats that are unprofessionally cut will be downgraded.
- Garnishes and decorations should harmonize with the entry in quantity, taste, color, and presentation.
- To enhance the keeping quality of entries, thin, spiced aspic (meat stock jelly) may be used; overcooking should be avoided.
- Rare meats must be sufficiently cooked so that, when covered with aspic, no blood or juice seeps out. Oozing meat, vegetable, or fruit juices will result in downscoring.
- Unprofessionally cut garnishes of vegetables or fruit will detract from the score.
- The use of butter, oleomargarine, mayonnaise and/or salad dressing is permitted; however, appropriate concern for food safety should be exercised.
- Limited use of non-edible items (such as foil-covered fiberglass) to achieve raised platforms, etc., is permissible.
- Overloading of entries should be avoided.
- The size of the entry should correspond to the number of persons for whom it is designed.
- If sauces are used, the containers in which they are placed should be no more than two-thirds full so that they will not spill in the event the platters are moved in the course of the judging.
- All entries should remain intact during display and awards. It is understood that the entries will not be returned to the entrants but may be available for consumption by conventioneers at the close of the display period.
- All pork products entered in the competition must have been subjected to one of the methods of eliminating trichina specified in Section 318.10 of the *Federal Meat Inspection Regulations,* copies of which are available from AAMP on request (see Chapter 3, page 51).
- The entrant signifies acceptance of all of the above rules by registering for the competition.

Scoring

Each entry will be judged for appearance, workmanship, merchandisability, and suitability for number of persons to be served. Evaluation will be indicated by a numerical score assigned for each of several characteristics under the different headings with the maximum number of points (1,000) as follows:

APPEARANCE: 320 POINTS

- Overall impression: 80 points
- Decoration, the effective use of color: 80 points
- Visual impact: 80 points
- Design balance: 80 points

WORKMANSHIP: 320 POINTS

- Effective use of materials: 80 points
- Creativity, uniqueness: 80 points
- Arrangement and workmanship: 80 points
- Durability of display: 80 points

MERCHANDISABILITY: 320 POINTS

- Variety of products: 80 points
- Durability of products: 80 points
- Selection of complementing foods: 80 points
- Serving ease: 80 points

SUITABILITY: 40 POINTS

- For number of persons to be served: 40 points

☞   NOTE: If the entry is deemed suitable for the indicated number to be   ☜
served, all 40 points will be assigned; if it is unsuitable, no points will be
assigned under this heading.

MEAT CONTESTS FOR ALL INTERESTED STUDENTS

Competition is undoubtedly one of the greatest instruments for arousing interest among young people. It has proved so effective among the youth of the
nation that it has actually become a so-called final examination for many courses
offered in schools and colleges.

This method of fostering interest in the youth of the United States has been
applied by the Future Farmers of America and the 4-H Club Congress to various
phases of their work. Contests of national importance that have to do with meat
are the Meat Judging Contest and the Meat Identification Contest held for Future Farmers of America and for the national 4-H organization at the American
Royal Livestock Show at Kansas City, Missouri. The national 4-H organization
also sponsors a contest at the National Western in Denver, Colorado.